Sustainable Green Catalytic Processes

Scrivener Publishing
100 Cummings Center, Suite 541J
Beverly, MA 01915-6106

Publishers at Scrivener
Martin Scrivener (martin@scrivenerpublishing.com)
Phillip Carmical (pcarmical@scrivenerpublishing.com)

Sustainable Green Catalytic Processes

Edited by
Mousumi Sen
Department of Applied Chemistry, Amity University, Uttar Pradesh, Noida, India

Scrivener
Publishing

WILEY

This edition first published 2024 by John Wiley & Sons, Inc., 111 River Street, Hoboken, NJ 07030, USA and Scrivener Publishing LLC, 100 Cummings Center, Suite 541J, Beverly, MA 01915, USA

For more information about Scrivener publications please visit www.scrivenerpublishing.com.

Wiley Global Headquarters
111 River Street, Hoboken, NJ 07030, USA

For details of our global editorial offices, customer services, and more information about Wiley products visit us at www.wiley.com.

Library of Congress Cataloging-in-Publication Data

ISBN 978-1-394-21255-2

Cover image: *Wikimedia Commons*
Cover design by Russell Richardson

Set in size of 11pt and Minion Pro by Manila Typesetting Company, Makati, Philippines

Printed in the USA

10 9 8 7 6 5 4 3 2 1

Contents

Preface

This book offers a concise description of the application of catalysts for sustainable transformation of the environment in various essential aspects of science and technology. Sustainable green catalytic processes are emerging as the means to empower the existing protocols with a greener, sustainable, and environmentally benign versions that hold enormous potential in industry and academia. Growing international concerns about environmental pollution and global warming have directed the attention of scientists worldwide toward approaches for developing sustainable protocols, and the need for employing greener and more sustainable catalytic approaches that are environmentally greener and more eco-friendly than present ones. Green and sustainable catalysts possess higher selectivity and activity, efficient recovery from the reaction medium, recyclability, and cost-effectiveness, and are prepared using environmentally benign preparation techniques.

The most potent instrument in organic synthesis and the cornerstone of green chemistry is catalysis, which has broadened the possibilities for sustainable, organic transformations in future. The challenges associated with the application of synthetic transformations include the hazardous and cost inefficiency of reaction conditions. Green catalytic conditions, mostly dealing with utilization of safer driven chemical reactions, relax the processes in terms of ambiance and environment. Moreover, the application of photo- and biocatalytic reactions in organic synthesis has received a lot of hype in recent years, with the light-mediated access to bioactive and medicinal compounds along with natural products attracting the focus of the scientific world.

The application in drug derivatization and degradation has explicitly boosted the potency of photo- and biocatalytic processes. The concept of nano photocatalyst for drug-free therapeutics has added a new dimension to the process. The catalyst has played a vital role, from the improvement of reaction conditions to enhanced selectivity of products with a decrease in the creation of byproducts. However, all the catalytic approaches are not always greener. Some employ toxic metals that are not environmentally friendly and are difficult to recycle, which ultimately limits their application on an industrial scale.

The major challenge now is to frame a greener design and development process to create catalysts, and then to use those catalysts in chemical processes that are sustainable. Several catalysts are used in this, including, but not limited to, metal-catalysis, organometallic catalysis, electrocatalysis, photocatalysis, nanocatalysis, biocatalysis, homogeneous and heterogeneous catalysis, and others. Thus, this book highlights the developments made toward designing new catalysts and presents the advancements in the field of chemical synthesis using greener catalytic routes with far-reaching applications.

Chapter 1 focuses on the creation of ecologically friendly chemical processes that depend heavily on green and sustainable catalytic reactions. As the need to lessen the environmental impact of industrial activities becomes more widely acknowledged, catalysis offers a viable solution by permitting effective and selective transformations with less waste creation. The idea of an "atom economy" and the use of renewable feedstock help to further advance sustainability and lessen reliance on fossil fuels. These developments in green catalysis lay the groundwork for the creation of more effective chemical processes that are also cleaner, promoting a more sustainable future.

The second chapter describes various organic transformations that take place in water. Although significant advancements occur every day in the field of green chemistry, there is still much to discover. To attain a sustainable future, there is an urgent requirement to focus on the development of diverse, eco-friendly protocols, especially those which can be carried out in H_2O. Enzyme catalysis needs to be explored extensively as well.

Chapter 3 is about the Reversible Addition-Fragmentation Chain Transfer (RAFT) polymerization, which operates through the reversible deactivation of radicals using special transfer agents. Photocatalysts in atom transfer radical polymerization (ATRP) have been found to mitigate downsides, such as biocidal UV light, low oxygen tolerance, and applicability to limited monomers.

The RAFT technique has been integrated with photooxidation/reduction catalytic reaction, where the modification has benefited in terms of low energy consumption, mild reaction condition, controllability, and high tolerance towards oxidation. Also note that while the inclusion of organic or inorganic photo redox catalysts to enhance the oxygen-tolerance of RAFT polymerization has been widely accepted, a metal-free approach has been additionally developed which is considered "greener". Further, the sustainability of the RAFT polymerization technique has been improved by expanding its applicability to design bio-renewable plastics.

Chapter 4 evaluates new approaches like C-H bond activation, photoredox catalysis, borrowed hydrogen catalysis, selective (stereo- and regio-) syntheses, as well as other novel, inventive general protocols for the synthesis and functionalization of heterocyclic compounds that have greatly aided in drug discovery and pharmaceutical research. Therefore, this chapter provides an overview of several cutting-edge heterocyclic synthesis techniques currently used in the field of drug development and medicinal chemistry.

The study of asymmetric organo-catalysis is currently quite popular. Numerous appealing characteristics exist in it, including new forms of activation, gentle reaction conditions, and processes free of transition metals. These organo-catalysts more effectively uphold the sustainability and green chemistry principles. Asymmetric organocatalytic synthesis of enantioenriched profens has made amazing progress, although it has yet to reach its full potential. Thus, the fifth chapter discusses the future advancements in the synthesis of profens and how their derivatives are expected to significantly aid further research into innovative organo-catalyst systems that require reduced catalytic loadings.

Chapter 6 discusses the advancements made in recent years for the greener synthesis of chemical compounds using nanocatalysts. The chapter focuses on three major ideas that are considered important in the green synthesis of chemical compounds: (i) synthesis of green nanocatalysts; (ii) greener reaction conditions; and (iii) the use of green nanocatalyst along with greener reaction conditions.

The seventh chapter frames the recent advances in heterogeneous photocatalysis and its applications. It delves into the fundamental principles that underpin photocatalysis, encompassing a comprehensive examination of photocatalyst categories, the mechanisms governing photocatalysis, and the factors that influence its efficiency. Subsequently, this section explores the diverse applications of heterogeneous photocatalysis, spanning pollutant degradation, CO2 reduction, water splitting, and organic synthesis. The mechanisms and challenges associated with each application are discussed, along with recent advances in enhancing the photocatalytic performance of catalysts.

The chapter also presents the future directions and challenges in heterogeneous photocatalysis, which include the development of more efficient and stable catalysts, exploring new photocatalytic reactions, optimizing reaction conditions, and highlighting the importance of heterogeneous photocatalysis in addressing global environmental and energy challenges.

Briefly put, compared to traditional chemical processes, biocatalytic processes are more cost- and environmentally friendly, as well as sustainable. By applying the ideas of biochemistry, chemical science may solve the energy crisis, make industrial production infinitely more efficient, and provide mankind with wings.

Chapter 8 emphasizes the applications and improved properties of the nanostructures obtained using green synthesis, while Chapter 9 explores the mechanisms underlying the total synthesis and functionalization of natural products with light-driven reactions. This section explores the important phtoredox catalysis process, which is used extensively for the total synthesis of natural products of their functionalization. Because, the utility of natural products has been expanding their application in several fields like human and veterinary medicine, agriculture, the food and cosmetic industries, etc. The widespread application of visible-light-mediated photocatalysis is evolving as a new chemical tool for various C-C bond formation reactions.

Upcoming directions could include the increased use of these photocatalytic approaches due to its atom economical, green, and efficient characteristics. The introduction of "Green Metrics" has played a pivotal role, underscoring their critical contribution to advancing sustainable development. The tenth chapter introduces numerous metrics from across different fields, many have seen limited use for specific, one-time purposes.

Chapter 11 discusses the bio-based economy as the best approach to converting waste biomass, similar to how agricultural forestry by-products and food supply chain waste convert into liquid fuels, common chemicals, and biopolymers. Carbon-based nanocomposites are used as bio-catalysis, which are environmentally friendly.

Every technology's success is determined by its accessibility of use and economic feasibility. This is the first evaluation of its sort to shed light on the energy and exergy evaluation of catalytic pyrolysis. As an alternative, exploration towards biocatalysts has begun because of their benefits, including milder reaction conditions, recyclability, selectivity, and biodegradability. Today, these changes could help bio-based chemistry by making it easier to develop helpful new ideas.

The goal of the twelfth chapter is to provide a brief overview of the various aspects of computational approaches, which provides a robust framework embedded with green chemistry. The development of extensive databases of chemical and environmental data can help to increase the precision of computational models and make it easier to create pharmaceutical products that are more environmentally friendly.

Chapter 13 highlights various green synthesis techniques and the principles of green chemistry that validate its power and potential to be highly beneficial to humans and the environment. The discipline of green chemistry has demonstrated that the innovation of thorough experts worldwide is producing an interesting and new chemistry that commands the drive of this transformation in theory and practice.

The fourteenth chapter presents various photocatalysts with high activity and functionality, better physicochemical stability, and superior biocompatibility, as well as potential applications in the bio-medical field. These compounds can also be used to manage biomedical waste. Various standard physicochemical techniques have been introduced to create nano-photocatalysts, but there may be impediments, such as the use of unsafe chemicals, arduous methods, expensive instruments, and time-consuming reactions. Furthermore, while researchers have explored the use of photocatalysts in drug delivery, their use in gene transport has not yet been well studied and requires further inquiry.

Chapter 15 concerns the sustainable development of a modern society that largely depends on the improvement of biorefinery platforms and green renewable fuels that reduce environmental pollution, as well as meet the global energy demand. This section explains how the chemistry behind the biofuel generation process will help design robust strategies for large-scale production of green fuels derived from biomass and waste for industrial applications.

The sixteenth chapter aims to detoxify industrial wastewater and reuse it through techniques like photo-, bio-, and nanocatalysts. The catalytic method is likely to be effective for treating drinking water and domestic and commercial wastes. Overall, this section projects a path of rapid development that provides a broad overview of recent applications of nanotechnologies and biotechnologies in wastewater treatment.

Chapter 17 presents a concise overview of several catalytic properties exhibited by hypervalent iodine (III/V). This section collates the reactions documented in the field of asymmetric hypervalent iodine chemistry, with a particular emphasis on the synthesis of chiral iodine (III and V) reagents or chiral iodoarenes, and explores the reactivity and proposed mechanisms of these reactions.

The eighteenth chapter sheds light on the significance of high-turnover palladium catalysts in the acceleration of C-H activation for sustainable green catalysis. The integration of efficient catalysts like palladium not only enables the synthesis of complex molecules but also reduces the environmental impact of chemical processes.

The last chapter broadly overviews the recent developments and novel applications of thin film fabrication techniques used in dye-sensitized solar cells (DSSCs) fabrication and discusses their advantages and challenges.

In conclusion, I am very grateful for all the hard work and effort made by the many contributors to this book. I thank all the authors for sharing their insightful research and information. I am particularly thankful to Aarushi Sen for her unending encouragement and support throughout the making of this book. Finally, I am most grateful to Martin Scrivener of Scrivener Publishing for his help and for making this book possible. I thank him for his patience and consistent support throughout the journey.

Dr. Mousumi Sen
July 2024

1

Green and Sustainable Catalytic Reaction Processes Including New Reaction Medium-Enriched Atom Utilization

Amit and Mousumi Sen

Department of Chemistry, Amity Institute of Applied Sciences, Amity University, Uttar Pradesh, India

Abstract

The creation of ecologically friendly chemical processes depends heavily on green and sustainable catalytic reactions. As the need to lessen the environmental impact of industrial activities is becoming more widely acknowledged, catalysis offers a viable solution by permitting effective and selective transformations with less waste creation. With a special emphasis on the creation of novel reaction media and the enrichment of atom use, this abstract focuses on the essential elements of environmentally friendly and sustainable catalytic reactions. Since using a greener reaction medium can considerably increase the overall sustainability and efficiency of catalytic processes, it has attracted a lot of interest. These media, which have special features that enable increased selectivity, simpler separation, and less environmental impact, include ionic liquids, supercritical fluids, and water. To further reduce the usage of risky and volatile organic solvents and the total carbon footprint of the reaction, efforts have been made to design and develop alternative solvents or solvent-free systems. Ionic liquids have demonstrated significant potential as green reaction media due to their low vapor pressure, excellent thermal stability, and recyclable nature. They provide increased catalytic activity and selectivity while reducing the atmospheric release of volatile organic compounds (VOCs). Water, a plentiful and renewable solvent, has gained popularity as a desirable reaction medium for many catalytic processes. Its application decreases waste production and removes the need for hazardous and combustible organic solvents. The maximization of atom use is a fundamental component of environmentally friendly and sustainable catalytic reactions. Low atom efficiency in traditional catalytic processes frequently results in significant waste production. Recent developments, however, have concentrated on creating catalytic systems that enhance atom usage while minimizing waste generation. This can be done by creating catalysts with high selectivity, using cascade reactions to change several reactants at once, and coming up with methods to value-add waste streams or by-products. The overall effectiveness and sustainability of the reaction can be significantly increased by using catalysts that support high atom economy. Incorporating renewable feedstocks, such as molecules generated from biomass, also improves atom use and lessens reliance on fossil fuels. The idea of an "atom economy" and the use of renewable feedstocks help to further advance sustainability and lessen reliance on fossil fuels. These developments in green catalysis lay the groundwork for the creation of more effective chemical processes that are also cleaner, promoting a more sustainable future.

Email: mousumi1976@gmail.com

Mousumi Sen (ed.) Sustainable Green Catalytic Processes, (1–28) © 2024 Scrivener Publishing LLC

Keywords: Sustainable, catalytic, environmental, renewable, atom economy, selectivity

1.1 Introduction

A viable strategy for creating new chemicals and materials that adhere to environmental norms is using green and sustainable catalytic reactions. Three key components of an environmentally friendly and sustainable catalytic reaction process are the use of sustainable and renewable raw materials, the creation of effective catalytic systems, and the application of atom-efficient reactions. Additionally, generating long-lasting and effective catalytic reactions might greatly benefit from the creation of novel reaction media [1]. The process of green and sustainable catalytic reactions includes atom-efficient reactions as a key component. With little waste and great atom utilization, these processes utilize all or almost all of the atoms in the starting materials. With this method, there is less of a need for energy-intensive separation and purification procedures, which means less energy use and less negative environmental effects. Another crucial component of environmentally friendly and sustainable catalytic reactions is the utilization of renewable raw materials. Sustainable chemical and material production can be accomplished by using renewable raw materials like biomass and trash [2]. For catalytic reactions to be sustainable, efficient catalytic systems must also be developed. Effective catalytic systems can speed up reactions, use less catalyst, and increase product yields. To achieve sustainable and effective catalytic reactions, new reaction media such as ionic liquids (ILs), supercritical fluids, and deep eutectic solvents can also be crucial. These media can decrease waste production, increase energy efficiency, and enhance reaction selectivity [3]. The most prevalent solvent on Earth, water, has become a popular medium for environmentally friendly catalytic processes. Due to its low cost, non-toxicity, and inflammability, it demonstrates great green credentials. Different catalysts' hydrophobicity and hydrophilicity can be adjusted to support specific aqueous medium reactions. Additionally, improved selectivity and reduced energy demands are frequently produced via water-mediated catalysis, thus boosting the sustainability of the process. Another prominent green reaction medium that has attracted interest recently is supercritical carbon dioxide (scCO_2). Carbon dioxide transforms into a supercritical fluid with gas-like diffusivity and liquid-like density under particular pressure and temperature conditions. scCO_2 is an eco-friendly substitute for conventional organic solvents, since it is nontoxic, nonflammable, and simple to recover after the reaction [4]. The family of solvents known as ionic liquids is made up only of ions. They are suitable for a variety of processes due to their low vapor pressure, good thermal stability, and excellent solubilizing qualities. By choosing various cations and anions, ILs can be customized to improve the solvation of reactants and catalysts, enabling cleaner and more effective catalytic processes. Due to their low cost, low toxicity, and ease of production, deep eutectic solvents (DESs) have become recognized as viable green reaction media. DESs are created by combining two or more components, often a hydrogen bond acceptor (such as urea) and a hydrogen bond donor (such as choline chloride). They are extremely adaptable as reaction media due to the specific interactions that result in the production of a eutectic mixture with properties different from those of its separate parts [5]. Numerous environmentally friendly catalytic reactions have been created as a result of the fusion of green catalysts and cutting-edge reaction media. Each of these illustrates the promise of green catalysis in diverse areas of organic synthesis, medicines, and materials research. They range from C-C and C-X bond formation, reduction and oxidation reactions, to multicomponent processes. In conclusion, resource conservation and environmental sustainability are two major worldwide concerns that need to be addressed. Green and sustainable catalytic reactions, with an emphasis on enhanced atom utilization in innovative reaction media, constitute a groundbreaking solution. Researchers and businesses alike may plow the way toward a greener and more affluent future where chemistry acts as a catalyst for change by leveraging the power of eco-friendly catalysts and cutting-edge solvents. The advancements and applications of particular green catalytic reactions will be covered in more detail in the sections that follow, with a focus on their revolutionary potential for creating a sustainable future [6].

1.2 Background

1.2.1 Traditional Catalytic Reactions

Traditional catalytic reactions are chemical processes that need a catalyst to speed up the process and/or modify the reaction's selectivity. Catalytic reactions are more effective and economical ways to create compounds, since the catalyst is not used up in the reaction and may be reused [7]. The idea of catalysis has been around for centuries, ever since the first scientists noticed how particular compounds affected chemical reactions. However, important advancements in our understanding of catalysis did not occur until the late 18th and early 19th centuries. The groundwork for the creation of catalytic theory was built by scientists like Humphry Davy and Johann Wolfgang Döbereiner, who conducted groundbreaking research [8]. A typical illustration of a conventional catalytic reaction is hydrogenation. In this procedure, a metal catalyst made of platinum, palladium, or nickel is used to catalyze the reaction between hydrogen gas (H_2) and unsaturated hydrocarbons such alkenes or alkynes. The catalyst makes it easier for the hydrocarbon's double or triple bonds to break, allowing hydrogen atoms to combine with the carbon atoms to form saturated hydrocarbons. In order to improve the texture and stability of liquid vegetable oils, hydrogenation is frequently employed in the food industry to turn them into solid fats like margarine [9]. The traditional catalytic reaction known as oxidation, in which a material loses electrons or joins with oxygen, is another significant one. Transition metal catalysts like iron or manganese are frequently used to promote this reaction. For instance, the industrial manufacturing of nitric acid, a critical chemical used to make fertilizers, explosives, and other compounds, requires the oxidation of ammonia to produce nitric oxide (NO) and then nitrogen dioxide (NO_2).

The Haber–Bosch process, which creates ammonia from nitrogen and hydrogen, is an illustration of a conventional catalytic reaction. Iron is used in this process as a catalyst, allowing the reaction to take place at lower temperatures and pressures than would otherwise be necessary [10].

The transformation of complicated hydrocarbons into useful products is a major function of catalysis in the petroleum industry. One noteworthy example is the catalytic cracking process, which converts big hydrocarbons from crude oil into smaller, more valuable hydrocarbons like petrol and diesel. In this procedure, zeolite catalysts are frequently used because of their great selectivity and stability. Traditional catalytic reactions are important for the industry, but they are also essential for environmental conservation. Vehicle catalytic converters transform dangerous exhaust gases like nitrogen oxides (NOx) and carbon monoxide (CO) into less dangerous substances like nitrogen, carbon dioxide, and water vapor using precious metals like platinum, palladium, and rhodium. Research into new catalyst materials and novel reaction routes is advancing quickly in the field of catalysis. In order to lower energy usage, waste production, and environmental effect, scientists are working to create more efficient and selective catalysts. Interdisciplinary research projects combining chemistry, physics, materials science, and engineering are responsible for this ongoing advancement. The efficient production of necessary chemicals, fuels, and minerals has been made possible by conventional catalytic reactions, which have altered industries and contemporary civilization. These processes have wide-ranging effects on numerous industries, ranging from catalytic cracking to ammonia synthesis and hydrogenation to oxidation. The future of humanity will be more sustainable and prosperous—thanks to the continued research and development in catalysis [11].

1.2.2 Significance of Green and Sustainable Catalytic Reactions

Due to growing worries about how human activity affects the environment, green and sustainable catalytic reactions are very important in today's society [12]. The following are some of the main implications of environmentally friendly and sustainable catalytic reactions (Figure 1.1):

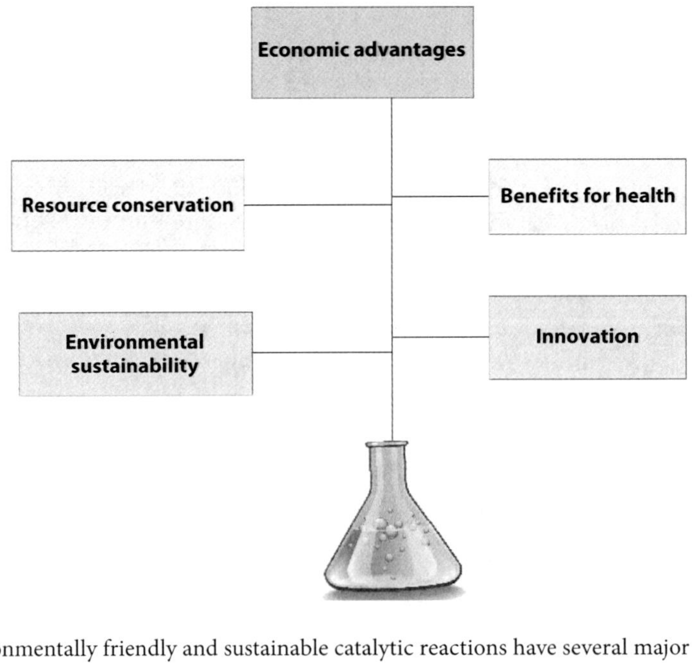

Figure 1.1 Environmentally friendly and sustainable catalytic reactions have several major ramifications.

1. **Environmental Sustainability:** By using fewer harmful chemicals, producing less waste, and using less energy, green and sustainable catalytic reactions help to lessen their negative effects on the environment.
2. **Resource Conservation:** By utilizing raw materials and renewable resources, these catalytic reactions lessen reliance on nonrenewable resources [13].
3. **Economic Advantages:** By lowering costs related to waste disposal and energy consumption, environmentally friendly and sustainable catalytic reactions can offer economic advantages. In the process of producing sustainable materials, they might also open up new economic prospects.
4. **Benefits for Health:** By decreasing exposure to dangerous chemicals, these catalytic reactions help to lower the health risks connected to their use.
5. **Innovation:** The development of new catalysts, reaction mechanisms, and process designs are encouraged by green and sustainable catalytic reactions, which can result in more effective and environmentally friendly chemical production [14].

1.2.3 New Reaction Medium-Enriched Atom Utilization

In the area of environmentally friendly and sustainable catalytic reactions, substantial research is being done on the use of new reaction medium-enriched atoms. The effectiveness and viability of catalytic reactions can be significantly impacted by the use of innovative reaction media [15].

Deep eutectic solvents, supercritical fluids, and ionic liquids are a few examples of novel reaction media (Figure 1.2) that have recently been studied and compared (Table 1.1). These materials offer special qualities that can increase atom utilization, boost selectivity, and cut down on waste production [16].

1.2.4 Overview of Research Problem

Using enriched atoms and new reaction media, the research problem on green and sustainable catalytic reactions focuses on the environmental and sustainability issues raised by conventional catalytic processes [20]. The goal is to create and improve catalytic reactions that are more energy-, environmentally, and commercially feasible [21].

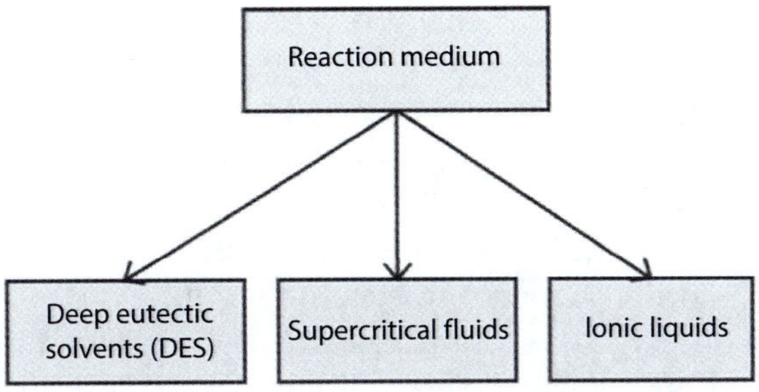

Figure 1.2 Novel reaction media include deep eutectic solvents, supercritical fluids, and ionic liquids.

Table 1.1 Comparison of deep eutectic solvents (DESs), supercritical fluids, and ionic liquids based on a few common criteria.

Reaction medium	State at standard conditions	Melting point	Applications	Reference
Deep eutectic solvents (DESs)	Liquid condition under ordinary pressure and room temperature	Possess reduced melting points	Utilized in electrochemistry, the extraction of metals from biomass, and pharmaceutical manufacturing	[17]
Supercritical fluids	Depending on the particular fluid and its critical temperature, a gaseous or liquid condition at standard pressure and temperature	Have no definite melting point	Used for medicinal purposes, essential oil extraction, and decaffeination	[18]
Ionic liquids	Liquid condition under ordinary pressure and room temperature	Melting points lower than 100°C	Used as lubricants, separation agents, electrolytes in batteries, and in electrochemistry and catalysis	[19]

Key elements of the research issue are as follows:

1. **Environmental Impact:** Conventional catalytic processes frequently use solvents and reagents that are toxic or bad for the environment, which results in the production of hazardous waste and contamination. The goal of the research is to find and create greener reaction media with reduced or no negative environmental effects.
2. **Energy Efficiency:** Finding methods to lower energy usage during the reaction processes is a key component of sustainable catalysis. This could involve researching renewable energy sources to power catalytic reactions or creating catalysts that can operate at lower temperatures [22].
3. **Atom Economy:** A key component of green catalysis is the use of enriched atoms. A reaction system's overall efficiency is increased by making the most of its atoms in order to reduce waste production.
4. **Catalyst Design:** The goal of the research is to develop new, effective, selective, and reusable catalysts. This will lessen the need for excessive catalyst dosages and prevent the usage of potentially dangerous substances.

5. **Renewable Feedstocks:** Investigating the use of renewable feedstocks as the initial components in catalytic reactions promotes sustainability and lessens reliance on finite resources.
6. **Green Chemistry Principles:** The study issue is in line with green chemistry's guiding principles, which emphasize the value of waste minimization, ecologically friendly methods, and the use of safer chemicals [23].

Interdisciplinary approaches incorporating knowledge from the fields of engineering, chemistry, and environmental sciences are necessary to meet these difficulties. By encouraging the use of green and sustainable catalytic processes in a variety of applications, the aim is to help the chemical industry become more environmentally friendly and sustainable [24].

1.3 Literature Review

1.3.1 Sustainable Catalysis

The creation and application of catalytic procedures that are socially, economically, and environmentally sound are referred to as sustainable catalysis. Utilizing catalysts that are nontoxic, regenerative, and effective at transforming raw materials into finished goods with high selectivity and yield constitutes sustainable catalysis [25]. With sustainable catalysis, the production of waste and the usage of dangerous chemicals are all reduced or eliminated. It also involves the utilization of renewable raw resources, such as biomass, and the creation of new catalytic processes that have a high atom economy, meaning that they employ a high percentage of the starting material to make the desired output [26]. The use of heterogeneous catalysts, which are easily removed from the reaction mixture and reused numerous times, is one of the important methods in sustainable catalysis. This reduces waste formation and boosts cost-effectiveness. High selectivity and yield can also be achieved using heterogeneous catalysts that are tuned to particular processes [27].

Overall, research on sustainable catalysis is crucial because it attempts to advance the creation of chemical processes that are more economically and environmentally sound. The production of chemicals and materials could be revolutionized through sustainable catalysis, creating a more sustainable future [28].

1.3.2 Green Catalytic Processes

Green catalytic reactions are those that are developed and put into use in a sustainable and environmentally beneficial manner. Green catalytic processes strive for excellent selectivity and yield while minimizing or eliminating the use of potentially harmful compounds, creating less waste, and consuming less energy [29].

There are a number of methods for creating green catalytic processes, such as the following:

1. **Developing New Catalysts:** The development of novel catalysts that are more efficient and selective can lead to the generation of less waste and reduce

energy usage. Natural minerals and other renewable resources can be used to make catalysts, which can help keep the process sustainable [30].

2. **Utilizing Renewable Raw Materials:** Utilizing sustainable raw materials that are renewable, such as biomass or garbage, might lessen the process's negative effects on the environment [31].
3. **Using Nontoxic Solvents:** Using nontoxic solvents, including water or green solvents, can lessen the process's negative effects on the environment and encourage sustainability.
4. **Employing Heterogeneous Catalysis:** Less waste is produced and expenses are reduced when heterogeneous catalysis, which uses a solid catalyst, is easily removed from the reaction mixture [32].
5. **Developing Biocatalysts:** Enzymes are an example of a biocatalyst that may catalyze reactions under benign conditions, eliminating the need for harsh chemicals and fostering sustainability [33].

1.3.3 Atom Utilization in Catalysis

The efficient use of atoms in a catalytic reaction to increase the yield of the desired product and reduce the creation of undesirable by-products is known as atom utilization in catalysis. Atom economics, which refers to the proportion of atoms from the beginning material that end up in the finished product, is closely related to the idea of atom utilization [34]. Effective atom utilization in catalysis is crucial for a number of reasons, including lowering the amount of waste produced during the reaction, utilizing fewer raw materials, and lessening the process's negative effects on the environment. As a result of using atoms more effectively, less expensive reagents or catalysts are required, which can also result in cost savings [35].

Utilizing highly selective catalysts, which encourage the desired reaction pathway and minimize the creation of undesirable by-products, is one technique to increase atom utilization in catalysis. This can be accomplished by selecting the catalyst carefully, enhancing the circumstances of the reaction, and managing the stoichiometry of the reactants [36]. Utilizing catalysts made to function at low catalyst loadings is another technique to increase atom utilization during catalysis. High-activity catalysts that run at low loadings can improve reaction efficiency and lower waste production [37].

1.3.4 Reaction Medium for Green Catalysis

In green catalysis, the choice of reaction medium is crucial, since it has a big impact on the process's overall effectiveness and sustainability. By utilizing safe and sustainable reaction media, green catalysis aims to reduce its negative effects on the environment and encourage environmentally friendly practices. This essay will examine some of the most popular reaction media used in green catalysis and how they help the chemical industry become more environmentally friendly and sustainable. Green catalysis is the use of catalysts and reaction conditions that limit the use of potentially dangerous solvents, limit the use of energy, and limit the production of trash. Therefore, for green catalysis, the choice of the reaction media is crucial [38].

Water, ethanol, isopropanol, and supercritical carbon dioxide are some of the most often used green solvents in green catalysis. Since it is so plentiful and safe for the environment, water is frequently utilized as a reaction medium in green catalysis. Water is an ideal solvent for high-temperature reactions because it is a polar solvent that can dissolve a wide variety of organic and inorganic compounds and has a high boiling point [39].

The reaction medium supercritical carbon dioxide is another one that shows promise for green catalysis. When carbon dioxide is exposed to particular temperature and pressure conditions, it transforms into the dense phase known as $scCO_2$. It is a flexible solvent for numerous reactions, since it combines the characteristics of both gases and liquids. $scCO_2$ is abundantly available from natural sources, nontoxic, and inflammable, making it safe for the environment. Its distinctive solvation capabilities make it possible to selectively extract some substances and can improve the catalytic activity and selectivity of some catalysts. Additionally, the separation of $scCO_2$ from reaction by-products is simple, allowing catalyst renewal and streamlining the purification procedure. As a result, $scCO_2$ has drawn attention as a sustainable and environmentally friendly substitute for conventional organic solvents. Ionic liquids have also shown promise as reaction media for environmentally friendly catalysis. Ionic liquids are molten salts made up of cations and anions; they are excellent for a variety of catalytic reactions due to their adjustable properties. They do not emit harmful fumes and are not combustible or volatile, and they frequently have very little vapor pressure. Ionic liquids have a wider range of applications, since they may dissolve a variety of substrates, including polar and nonpolar molecules. They can also stabilize catalysts and metal nanoparticles, making it simple to recover and recycle catalysts. Ionic liquids are used in green catalysis; however, their use may lead to questions regarding their biodegradability and potential toxicity. As a result, efforts are being undertaken to design and use more ecologically friendly ionic liquids. In some circumstances, the greenest method of preparing reaction media is solvent-free or solid-state catalysis. By removing the need for any liquid medium, solvent-free catalysis minimizes waste production and lowers the energy required for solvent recovery. On the other side, solid-state catalysis includes performing reactions on solid catalyst surfaces, enabling increased selectivity and recyclability. Both strategies boost atom utilization efficiency (AUE) while minimizing their negative effects on the environment, making them appealing choices for green catalysis [40–42]. In green catalysis, ethanol and isopropanol are frequently employed as environmentally friendly solvents. Both solvents are great choices for environmentally friendly reactions because they are both readily biodegradable and relatively nontoxic.

Overall, the type of reaction being performed, the characteristics of the catalyst, and the desired result all influence the choice of the reaction medium. But for green catalysis, water, ethanol, isopropanol, and supercritical carbon dioxide are all fantastic options [43].

1.4 Environmental Impact of Catalytic Reactions

Various industrial processes and chemical transformations depend heavily on catalytic reactions, but they can also have a big influence on the environment. Although catalysis can be more environmentally friendly than non-catalytic alternatives, it is important to be mindful of any possible environmental problems brought on by catalytic reactions [44]. Key environmental effects include the following:

1. **Energy Requirement:** Energy input is frequently needed for catalytic reactions, particularly in processes that involve high temperatures and pressures. The energy source employed for these reactions may emit carbon, which would increase greenhouse gas emissions and the effects of climate change.
2. **Production and Disposal of Catalysts:** The creation of catalysts can be an energy- and resource-intensive process. The creation of the catalyst may require mining, refining, and processing of raw materials, which might result in habitat loss, water pollution, and energy consumption, depending on the catalyst's composition. In addition, if the used catalysts contain dangerous or poisonous components, their disposal may be problematic [45].
3. **Production of Waste:** Catalytic reactions can result in the production of waste products such as by-products, unreacted starting materials, and used catalysts. These wastes have the potential to pollute the environment and contaminate the air, water, and soil if improperly managed.
4. **Toxicity and Pollution:** Some catalytic reactions may use hazardous compounds as catalysts or produce toxic by-products, which, if released into the environment, can be harmful to human health and ecosystems. Pollutants can have a negative impact on aquatic life, wildlife, and human populations by leaking into water bodies or entering the atmosphere.
5. **Greenhouse Gas Emissions:** Some catalytic processes can result in the release of greenhouse gases, which fuels the warming and changing of the climate. For instance, the creation and usage of specific catalysts as well as the energy needed to power catalytic reactions could emit greenhouse gases like carbon dioxide [46].
6. **Water Use:** For cooling and processing, some catalytic reactions need a lot of water. Utilizing too much water can put a burden on the local water supply, especially in areas where there is a water shortage or a drought.
7. **Land Use and Habitat Destruction:** The extraction of raw materials for the manufacture of catalysts and the construction of facilities for catalytic processes can result in land use changes and habitat destruction, which can have an impact on biodiversity and ecosystem balance.
8. **Risk of Accidents:** Some catalytic reactions could include dangerous substances or high pressure, raising the possibility of mishaps or leaks that might endanger people on the job, nearby communities, and the environment [47].

Researchers, businesses, and politicians are aggressively exploring sustainable catalysis methods to reduce these negative environmental effects. Green catalytic processes strive to use less energy, use less hazardous catalysts, increase atom efficiency, and provide environmentally friendly reaction media. Catalyst recycling and reuse can also cut down on resource use and waste production. The environmental impact of catalytic processes can be considerably decreased by using more environmentally friendly procedures and technology, which will help create a cleaner and more sustainable future [48].

1.4.1 Impact of Traditional Catalytic Reactions

The creation of materials, chemicals, and fuels, as well as other facets of contemporary civilization, has been significantly influenced by conventional catalytic processes. These responses have made it possible to produce goods in huge quantities effectively and affordably, which has helped the world economy expand [49]. Catalytic reactions have been essential in the manufacture of fuels, polymers, and other chemical compounds in the petrochemical industry. The subject of refining crude oil represents one of the most significant contributions. Heavy hydrocarbons have been transformed into lighter, more valued products like petrol and diesel using catalytic breaking and reforming techniques. This helps countries that are largely dependent on oil exports expand economically while also meeting the need for transport fuels on a worldwide scale [50]. Additionally, conventional catalytic processes have had a significant impact on the pharmaceutical sector. Catalysis is used to selectively and effectively produce a number of necessary pharmaceuticals and therapies. Palladium and platinum are two transition metal catalysts that have become essential instruments in the synthesis of pharmaceuticals because they allow for the precise assembly of complicated molecular structures. As a result, the creation of new medications has been accelerated, improving the choices for treating many disorders. It is impossible to ignore how classical catalytic processes affect the environment. The attempts to reduce waste and manage pollution have benefited greatly from these reactions. Catalytic converter application in automobile exhaust systems is one such instance. Before being released into the environment, these devices transform dangerous pollutants like nitrogen oxides and carbon monoxide into less dangerous ones. As a result, catalytic converters have considerably improved air quality and lessened the negative effects of vehicle emissions on the environment [51]. Additionally, catalytic processes have been used to reduce greenhouse gas emissions. Through catalytic processes like methanation and hydrogenation, carbon dioxide, a primary cause of global warming, can be changed into valuable chemicals. These processes allow carbon dioxide to be converted into useful compounds or even synthetic fuels, potentially providing a solution to the problem of carbon dioxide usage and climate change. Traditional catalytic reactions have effects that go beyond those that are favorable to the environment and industry. They have also affected the development of energy-saving procedures, resulting in more affordable and resource-efficient industrial techniques. Catalysts enable processes to run at lower temperatures and pressures, saving energy consumption and overall operational costs. This is accomplished by lowering the activation energy needed for chemical reactions. This has improved many sectors' economic viability and competitiveness. It is important to recognize, though, that conventional catalytic processes also have their limitations. Some catalysts could be expensive or demand scarce and limited resources, which might prevent their mainstream adoption. Enzymatic catalysis and heterogeneous catalysis are two examples of alternate catalytic materials and methods that have been investigated, since it is difficult to create effective catalysts for some reactions [52].

In conclusion, conventional catalytic reactions have had a substantial influence on a variety of facets of daily life as well as industrial processes. Catalysis has been a key factor in many developments, from the creation of necessary chemicals and fuels to the creation of lifesaving medicines [53]. Additionally, the advantages for the environment, such as pollution reduction and a decrease in greenhouse gas emissions, underscore the critical part catalytic reactions play in sustainable development. Catalysis is likely to stay at the forefront of innovation as research advances, paving the way for a more wealthy and sustainable future. For instance, poisonous solvents like benzene, toluene, and xylene, which can be hazardous to both the environment and human health, are used in the production of various chemicals and products. In addition, the usage of high temperatures and pressures can contribute to the creation of pollutants, including greenhouse gases, which cause climate change and air pollution [54].

1.4.2 Environmental Benefits of Green Catalytic Reactions

Compared to conventional catalytic reactions, green catalytic reactions have a number of advantages for the environment, such as the following:

1. **Reduced Usage of Dangerous Solvents:** Green catalytic reactions frequently make use of solvents that are more plentiful, easily biodegradable, and less toxic. As a result, there is less chance of environmental contamination and dangerous chemical exposure for people.
2. **Lower Energy Consumption:** Green catalytic reactions frequently take place at lower temperatures and pressures, which cuts down on energy use and the emissions of greenhouse gases that go along with it [55].
3. **Minimized Waste Generation:** Green catalytic processes frequently generate less waste and less by-products, which lessens their negative effects on the environment [56].
4. **Improved Selectivity and Efficiency:** When using green catalysts, the desired reaction can be achieved with less energy and resources expended [57].
5. **Reduced Reliance on Rare or Precious Metals:** The environmental impact of mining and refining rare or valuable metals is minimized through the use of green catalysts, which frequently utilize renewable or earth-available minerals.

Overall, green catalytic reactions outperform conventional catalytic reactions in terms of environmental advantages, making them a promising strategy for environmentally friendly chemical synthesis and production [58].

1.4.3 Role of Reaction Medium in Reducing Environmental Impact

In order to minimize the negative effects of chemical reactions on the environment, the reaction medium must be carefully chosen. The solvent or mixture of solvents in which the reaction occurs is known as the reaction medium, and it can have a substantial impact on the reaction's efficacy, selectivity, and environmental impact [59]. Green reaction media are often chosen for their high efficacy, low toxicity, and minimal influence on the environment. Water, ethanol, isopropanol, supercritical carbon dioxide, and ionic liquids are a few examples of green reaction media [60].

Since it is so plentiful and safe for the environment, water is frequently employed as a reaction medium in green chemistry. Water is an ideal solvent for high-temperature reactions because it is a polar solvent that can dissolve a wide variety of organic and inorganic compounds and has a high boiling point [61]. Other often utilized green solvents that are easily biodegradable and nontoxic include ethanol and isopropanol. They are frequently employed in processes that call for nonpolar or barely polar solvents [62].

1.5 Experimental Section

1.5.1 Materials and Methodology

Such reactions should be designed with an eye toward minimizing waste, conserving energy, and employing catalysts that are safe for the environment.

1.5.1.1 Materials

1. **Catalysts:** In environmentally friendly and sustainable catalytic reactions, the choice of catalyst is crucial. The catalyst should ideally be accessible, affordable, and nontoxic. Due to their high activity and selectivity, transition metals including nickel, palladium, and copper are frequently utilized in catalytic processes [63].
2. **Solvents:** Using solvents in catalytic reactions can have a substantial negative influence on the environment. Traditional solvents, which could be dangerous or have a high volatility, can be replaced by green solvents including water, ionic liquids, and supercritical carbon dioxide [64].
3. **Substrates:** The environmental sustainability of a catalytic process can also be influenced by the substrate choice. To lessen dependency on fossil fuels and cut greenhouse gas emissions, substrates made of renewable feedstocks, such as carbon dioxide or biomass, can be employed [65].

1.5.1.2 Methodology

1. **Reaction Conditions:** To reduce waste and energy usage, the reaction conditions should be optimized. Energy requirements can be decreased by conducting reactions at low temperatures or in ambient conditions, and waste can be minimized by maximizing the stoichiometry of the process [66].
2. **Recycling:** The catalyst should be recyclable to cut down on waste production and reaction costs. To reduce the negative effects on the environment, catalyst recovery techniques such as filtration or centrifugation should be chosen [67].
3. **Monitoring the Process:** Monitoring the process can help to improve the conditions for reactions and cut down on waste. Real-time monitoring of the reaction progress using *in situ* monitoring techniques, such as spectroscopy or chromatography, can help to pinpoint the ideal conditions for the reaction and minimize the quantity of waste produced.

Overall, waste reduction, reduced energy consumption, and the use of environmentally friendly materials should be given priority in the materials and methodology utilized in green and sustainable catalytic processes. Thus, it is conceivable to create catalytic processes that are both effective and environmentally responsible [68].

1.5.2 Preparation of the Reaction Medium

The reaction medium should be chosen with the intention of minimizing waste, cutting down on energy use, and employing eco-friendly products. Here are some recommendations for creating a sustainable and green reaction medium:

1. **Make Use of Nontoxic or Renewable Solvents:** Making a green and sustainable reaction medium requires careful consideration of the solvent used. Traditional solvents, such as dichloromethane or chloroform, might pose concerns to the environment and human health because of their toxicity or high volatility. Alternatives to conventional solvents include water, ionic liquids, and supercritical carbon dioxide. These solvents can lessen environmental effects and increase safety because they are nontoxic, are nonflammable, and have low volatility [69].
2. **Reduce Solvent Usage:** An important source of environmental impact might come from the amount of solvent utilized in a reaction. Reduce the amount of solvent utilized and the amount of trash produced by using solvent-free or solvent-minimized conditions. In some circumstances, the reaction can be carried out in a solid form, which completely does away with the requirement for a solvent [70].
3. **Use Renewable Feedstocks:** To lessen dependency on fossil fuels and cut greenhouse gas emissions, renewable feedstocks can be employed as substrates in catalytic processes, such as carbon dioxide or biomass. In order to increase the reaction's sustainability, these feedstocks can also be utilized as reaction media [71].
4. **Optimize Reaction Conditions:** To minimize waste and save energy, reaction conditions like temperature and pressure should be optimized. Reaction energy can be reduced by lowering reaction pressure or temperature, and waste can be minimized by maximizing the stoichiometry of the reaction [72].
5. **Catalyst Recovery:** The catalyst employed in the reaction needs to be recyclable in order to cut down on waste production and reaction costs. To reduce the negative effects on the environment, catalyst recovery techniques such as filtration or centrifugation should be chosen [73].

1.5.3 Reaction Conditions

The design and optimization of such reactions heavily depend on the reaction conditions. Green and sustainable catalytic processes can involve a variety of reaction conditions, including the following:

1. **Temperature and Pressure:** Optimizing reaction conditions will reduce the amount of energy needed to initiate the reaction. Lower pressures and temperatures can increase a reaction's effectiveness while using less energy [74].
2. **Solvent:** Less harmful or even recyclable solvents are frequently used in green and sustainable catalytic reactions. Supercritical fluids, ionic liquids, and water are a few examples of solvents that are safe for the environment and can be employed in catalytic reactions [75].
3. **Catalyst Loading:** To reduce waste and increase efficiency, the amount of catalyst employed in a reaction should be optimized. Low catalyst loadings may cause slower reaction rates, whereas high catalyst loadings may result in more waste being produced [76].
4. **Catalyst Recovery and Reuse:** To lessen the quantity of waste produced, catalysts should be readily recoverable and recyclable. Using catalysts that are immobilized on a support or employing a scavenger to remove the catalyst after the reaction can accomplish this [77].
5. **Reaction Selectivity:** It is important to design the reaction conditions to favor the desired outcome while reducing the development of unwanted by-products. Controlling reaction parameters like temperature, pressure, and reactant ratios can accomplish this [78].

Overall, in order to maximize efficiency and reduce waste generation, green and sustainable catalytic reactions call for careful consideration of the reaction conditions.

1.5.4 Analytical Techniques

Monitoring the development of green and sustainable catalytic reactions, improving the circumstances of the reaction, and describing the catalysts and products all need the use of analytical techniques. In green and sustainable catalytic processes, a few typical analytical methods are as follows:

1. **Nuclear Magnetic Resonance (NMR):** NMR spectroscopy is an effective method for characterizing catalysts and end products as well as keeping track of the status of reactions. In addition to their reaction kinetics and processes, it can reveal details on the structure and makeup of molecules [79].
2. **Fourier-Transform Infrared Spectroscopy (FTIR):** Catalysts, reactants, and products all include functional groups that can be recognized and described using FTIR spectroscopy. Additionally, it can be used to monitor the status of reactions and improve their conditions [80].
3. **Gas Chromatography (GC):** For separating and quantifying the constituents of a reaction mixture, gas chromatography is a potent analytical method. It can be used to measure any by-product or contaminant as well as keep track of how reactants are changing into products [81].
4. **High-Performance Liquid Chromatography (HPLC):** Another separation method that is frequently applied in environmentally friendly and sustainable

catalytic reactions is HPLC. It can be used to monitor reaction progress and optimize reaction conditions. It is especially helpful for isolating and measuring tiny molecules, such as catalysts and products [82].

5. **X-Ray Diffraction (XRD):** The crystal structure of catalysts and products can be identified using XRD. It can reveal details regarding the makeup, purity, crystallinity, and reaction kinetics and processes of many materials [83].
6. **Transmission Electron Microscopy (TEM):** The form, size, and distribution of nanoparticles and other catalysts are visualized and described using TEM. It can reveal details about the stability and lifetime of a material as well as its catalytic activity and selectivity [84].

1.6 Results and Discussion

1.6.1 Catalyst Characterization

Designing, optimizing, and comprehending green and sustainable catalytic reactions all depend heavily on catalyst characterization. In these processes, some typical methods for catalyst characterization include the following:

1. **Surface Area and Pore Size Analysis:** Techniques like nitrogen adsorption/desorption isotherms and mercury intrusion porosimetry can be used to calculate the surface area and pore size distribution of a catalyst. Using this knowledge, the catalyst production can be improved, and its performance in catalytic reactions can be better understood [85].
2. **Elemental Analysis:** X-ray fluorescence (XRF) and inductively coupled plasma (ICP) spectroscopy are two examples of elemental analysis methods that can be used to determine the composition and purity of a catalyst. The catalyst synthesis can be improved using this knowledge, and its reproducibility can be guaranteed [86].
3. **Scanning Electron Microscopy (SEM) and Transmission Electron Microscopy:** Powerful imaging methods like SEM and TEM can provide details about the shape, size, and distribution of catalysts. Additionally, they can be utilized to examine the structure of supported catalysts and the impact of various reaction conditions on the catalyst's structure [87].
4. **Thermogravimetric Analysis (TGA):** TGA is used to investigate the catalysts' thermal stability. The reaction conditions can be improved using this knowledge, and the impact of various reaction parameters on the stability of the catalyst can be examined [88].

1.6.2 Catalytic Activity and Efficiency

The term "catalytic activity" describes a catalyst's capacity to speed up a chemical reaction by reducing the amount of energy needed for the reaction to begin, all while remaining unharmed or irreversibly changed. By offering a different reaction pathway with a lower activation energy, a catalyst facilitates the easier conversion of reactant molecules into products [89]. The effectiveness of a catalyst in accelerating a chemical reaction is known as catalytic efficiency. It is typically described as the comparison between the rates of the catalyzed and uncatalyzed reactions under identical conditions. The effectiveness of the catalyst in accelerating the process increases with catalytic efficiency. Although they are similar ideas, catalytic activity and catalytic efficiency are not the same thing. If a catalyst must be used in vast quantities to produce a meaningful rate increase, it may have high activity but low efficiency. On the other hand, a catalyst may have low activity but high efficiency if it just needs a tiny quantity to significantly increase the pace of reaction. Therefore, while assessing a catalyst's performance, it is crucial to take both activity and efficiency into account [90]. In several environmentally friendly processes, supported metal nanoparticles have proven to be extremely effective catalysts. Since nanoparticles have a large surface area, there are more active sites available, which increases catalytic activity. These catalysts can also be immobilized on solid supports, increasing efficiency and sustainability by enabling catalyst separation, recycling, and reuse [91]. Furthermore, the development of green catalysts might make use of naturally occurring and renewable resources, eliminating the need for rare and expensive metals. Another fascinating area of research in green catalysis is the creation of bio-inspired catalysts, which replicate the enzymatic reactions seen in nature. These catalysts frequently display excellent selectivity and activity, providing a more efficient and long-lasting method of carrying out chemical reactions. Atom utilization efficiency and catalytic efficiency are strongly connected. Maximizing the utilization of atoms in the reactants and reducing waste generation are crucial objectives in green catalysis. Based on a catalyst's capacity to produce desired products with high conversion and yield while consuming little raw material, its efficiency can be assessed [92]. Tandem and cascade catalysis are two methods for enhancing catalytic efficiency. These methods combine various chemical processes in a single catalytic system, enabling the progressive transformation of raw materials into finished goods without the need for intermediate isolation and purification stages. Tandem and cascade catalyses boost atom utilization effectiveness and overall reaction sustainability by combining various catalytic processes. Another element affecting the catalytic effectiveness of green processes is recyclable materials. Multiple reaction cycles are possible with reusable catalysts, which lower the demand for new catalysts and further reduce resource use. In addition, recycling and catalyst recovery limit trash generation, making the process more environmentally friendly. Designing more effective catalysts has gotten significantly easier because of computational technologies like density functional theory (DFT). DFT calculations are helpful in the rational design of catalysts with improved activity and selectivity because they can forecast reaction processes and pinpoint important intermediates. By streamlining the catalyst development process and minimizing trial-and-error experimentation, this method helps to make catalysis more environmentally friendly and long-lasting [93].

To sum up, catalytic activity and efficiency are key factors in environmentally friendly and long-lasting catalytic reactions. An effective reaction medium and selective catalysts work together to maximize atom utilization efficiency while minimizing environmental effects [94]. Researchers can create a more environmentally friendly and economically viable chemical industry by concentrating on the invention of green catalysts and improving reaction conditions [95].

1.6.3 Atom Utilization Efficiency

Atom utilization efficiency is a crucial statistic for assessing how long catalytic reactions can last. In comparison to the total number of reactant atoms employed in the reaction, it describes the proportion of reactant atoms that are incorporated into the desired product. High atom utilization efficiency is preferred in environmentally friendly and resource-efficient catalytic reactions because it reduces waste and maximizes resource consumption. AUE can be increased by using catalysts that can only selectively activate the desired functional groups of a molecule. Using cascade reactions—multiple reactions carried out in a single pot—allows for the effective conversion of starting materials into final products without the need for preliminary isolation and purification processes. This method is one way to increase AUE. Additionally, using abundant and renewable feedstocks can enhance AUE and make the reaction more sustainable [96].

The goal of green and sustainable catalysis is to create effective chemical processes that do not harm the environment. Traditional chemical processes can result in significant waste and demand enormous amounts of raw materials, which cause resource depletion and environmental damage. As opposed to this, sustainable catalytic reactions encourage the development of catalysts that can speed up atom-efficient conversions and produce high yields with little waste [97]. The selection of the catalyst is one of the major variables affecting the effectiveness of atom utilization in green catalytic reactions. In order to maximize the targeted reaction pathway while minimizing unintended side reactions that result in waste creation, a catalyst must be selective. The catalyst should also be reusable to allow for numerous reaction cycles, increasing atom efficiency and lowering the total environmental impact [98]. Due to their large surface area and adaptable reactivity, supported metal nanoparticles have become effective catalysts in numerous environmentally friendly reactions. These catalysts may be easily separated from one another and recovered for reuse by being immobilized on solid supports. These catalysts can considerably increase atom utilization efficiency, lowering the requirement for extra reactants and reducing the production of undesirable by-products. The selection of solvents is a crucial factor in green catalytic processes [99]. Traditional organic solvents frequently pose concerns to human health and the environment. The efficiency of atom utilization can be improved by switching to more environmentally friendly solvents like water or supercritical carbon dioxide. The quantity of trash produced during the reaction process can be reduced by using these solvents because they are more benign. Other methods to increase atom efficiency in catalytic reactions include catalyst immobilization on solid supports and the use of flow systems. Continuous operation improves atom utilization efficiency in flow systems by facilitating the recycling of catalysts and reactants as well as better control over reaction conditions. Additionally, the rational design of catalysts has benefited from the use of computational methods like density functional theory. DFT calculations can forecast the reaction's mechanism and selectivity, allowing for the creation of catalysts that are more effective and have greater atom utilization rates [100].

In conclusion, atom utilization efficiency is a crucial variable in environmentally friendly and long-lasting catalytic reactions. Chemists can develop more effective and environmentally friendly catalytic processes by selecting selective recyclable catalysts and eco-friendly solvents, using tandem and cascade catalyses, and utilizing computational techniques [101].

Enhancing atom utilization efficiency helps create a greener, more sustainable chemical industry while also reducing waste and conserving resources. We may anticipate even more groundbreaking solutions to tackle environmental issues and build a more sustainable future as green catalysis research develops [102].

1.6.4 Reaction Mechanism

The series of discrete stages by which reactant molecules are changed into product molecules is known as a reaction's mechanism [103]. The reaction mechanism is significant in green and sustainable catalytic reactions because it can affect the effectiveness and sustainability of the reaction [104].

Using mechanisms involving commonly available nontoxic chemicals and catalysts is one method for constructing sustainable catalytic reactions [105]. For instance, substituting water for organic solvents as a solvent can increase a reaction's sustainability by minimizing waste and the use of potentially dangerous substances [106]. Utilizing processes that include moderate reaction conditions, including low temperatures and pressures, can lower the reaction's energy requirements and lessen the production of waste products. Further reducing the reaction's environmental impact is employing catalysts that are simple to recover and reuse [107, 108].

Understanding the reaction mechanism and the catalyst's function in supporting the intended reaction pathway is essential to creating sustainable catalytic reactions (Table 1.2). Characterizing the reactive intermediates and transition states involved in the reaction as well as optimizing the reaction conditions to favor the chosen pathway while minimizing undesirable side reactions can be done in this manner [109, 110].

1.6.5 Comparison with Traditional Catalytic Reactions

Catalytic reactions that are environmentally friendly and sustainable are made to be as effective, selective, and sustainable as possible. They offer a number of benefits over conventional catalytic processes, including the following:

1. **Reduced Environmental Impact:** Green and sustainable catalytic processes have a lower environmental impact because they use fewer toxic or dangerous materials, produce less waste, and use less energy [111].
2. **Increased Selectivity:** By making sustainable catalytic processes more selective, undesirable waste and by-products can be prevented from forming.
3. **Lower Cost:** Because green and sustainable catalytic processes utilize less resources and less energy than conventional catalytic reactions, they may be more cost-effective [112, 113].
4. **Use of Renewable Resources:** Utilizing biomass as a feedstock for sustainable catalytic reactions can help minimize the need for fossil fuels [114].
5. **Increased Atom Utilization Efficiency:** Higher atom utilization efficiency in sustainable catalytic processes indicates that more reactant atoms may be used to create the required products, resulting in less waste and more resource efficiency [115].

Table 1.2 Comparison of traditional catalytic processes with green, sustainable reactions using similar parameters [118–126].

Parameters	Green and sustainable catalytic reactions	Traditional catalytic reactions
Environmental impact	Utilizing sustainable resources and safe catalysts will help to reduce any environmental harm. Eco-friendly solvents are given priority, and toxic by-products are decreased.	Frequently produce more trash and employ hazardous or poisonous catalysts, both of which have a negative impact on the environment
Energy efficiency	Generally place an emphasis on energy efficiency and strive to run under calm conditions with minimal energy input	Might involve harder conditions and demand more energy
Catalysts	Use more environmentally friendly catalysts that can be recycled, such as enzymes, supported metals, or organocatalysts	May rely on pricey rare metal catalysts that are less eco-friendly and may only be available in limited quantities
Renewable feedstocks	Frequently use renewable feedstocks, such as CO_2, natural products, or chemicals generated from biomass	May be dependent on nonrenewable resources made from fossil fuels
Toxicity	Prioritize using nontoxic or low-toxicity materials to minimize dangers to the environment and human health	May include more harmful substances, raising the risk of pollution
Reaction conditions	Prefer less extreme reaction conditions, which minimize the requirement for high pressures and temperatures	To reach the needed conversion rates, tougher circumstances could be necessary.
Waste minimization	Make an effort to reduce waste production through catalyst recycling and selective reactions	Due to challenging catalyst recovery and less selective reactions, could produce more waste.

Traditional catalytic processes, on the other hand, could call for higher temperatures, pressures, or more dangerous or poisonous chemicals or solvents. If not appropriately managed, they may produce more trash and have a greater negative influence on the environment [116, 117].

Overall, green and sustainable catalytic reactions provide an economical and efficient alternative to conventional catalytic reactions that is also more sustainable and kind to the environment [127].

1.7 Summary and Outlook

1.7.1 Summary

Green and sustainable catalytic reactions attempt to promote chemical transformations with minimal negative effects on the environment while maximizing efficiency, representing a paradigm change in the science of chemistry. The creation of eco-friendly catalysts, favorable reaction environments, and novel reaction media forms the basis of this strategy.

By enabling selective and effective reactions, green catalysts, such as metal-based, organocatalysts, and biocatalysts help to reduce the creation of toxic by-products. These catalysts encourage atom utilization, a cornerstone of green chemistry, which raises yields and lowers waste production. Green catalysis encourages resource conservation and energy efficiency by maximizing the amount of atoms integrated into the target product and minimizing atom loss in side reactions.

The use of innovative reaction medium is one of the fundamental components of green catalytic reactions. Greener substitutes for conventional organic solvents are used, including water, supercritical carbon dioxide, ionic liquids, and deep eutectic solvents. The nontoxicity, high stability, and biodegradability of these novel media are special benefits that further increase the sustainability of the catalytic processes.

1.7.2 Outlook

Green and sustainable catalytic reactions have a bright future ahead of them—thanks to continuous research and technological advancements. The significance of green chemistry and sustainable practices will only rise as environmental concerns do, making these catalytic reactions even more pertinent. The development of new and enhanced catalysts will continue, with an emphasis on maximizing their selectivity, recyclable nature, and overall environmental impact. Furthermore, the sustainability of catalytic processes will be further improved by the incorporation of renewable resources and biocatalysis.

Greener reaction medium design and implementation initiatives will continue, with an emphasis on creating more effective and financially viable substitutes. Due to its abundance and eco-friendliness, using water as a reaction medium in particular has a lot of potential. Additionally, the development of cutting-edge technologies like computational modeling and machine learning will be extremely important in hastening the discovery and improvement of green catalytic reactions. These technologies allow for a greater comprehension of reaction mechanisms, catalyst design, and reaction conditions, resulting in processes that are more effective and long-lasting. In order to comply with more stringent environmental restrictions and consumer demands for eco-friendly products, industries will increasingly use green and sustainable catalytic concepts. This trend is not just seen in academia. Greener practices can reduce the environmental impact of many industries, from energy generation and materials synthesis to pharmaceuticals and fine chemicals.

Overall, green and sustainable catalytic reactions have a promising future, and academics, businesses, and governments working together can hasten the transition to a more environmentally friendly and long-lasting chemical landscape. We can work toward a greener and more prosperous future for future generations by utilizing enriched atom utilization and green catalysis in innovative reaction media.

References

1. Wu, Y., Wang, Q., Liu, Z., Yu, S., Xu, J., Ma, D., Palladium-catalyzed C-C bond formation reaction in ionic liquids. *Green Chem.*, 19, 1, 80–83, 2017.
2. Zhao, H., Baker, G.A., Holmes, S., Song, Z., Continued exploration and development of deep eutectic solvents including their relevance to that of ionic liquids. *ACS Sustain. Chem. Eng.*, 1, 12, 1667–1681, 2012.
3. Zhao, D., Liao, L., Liu, D., Carbon dioxide as a green solvent for chemical reactions. *Green Chem.*, 17, 3, 1460–1478, 2015.
4. Wang, T., Li, S., Xu, H., Yan, N., Supported platinum catalysts for the hydrogenation of carbon dioxide to formic acid. *Green Chem.*, 21, 17, 4652–4661, 2019.
5. Anastas, P.T. and Warner, J.C., *Green chemistry: Theory and practice*, Oxford University Press, New York, 1998.
6. Sheldon, R.A., Green solvents for sustainable organic synthesis: State of the art. *Green Chem.*, 9, 12, 1273–1283, 2007.
7. Polshettiwar, V. and Varma, R.S., Green chemistry by nano-catalysis. *Green Chem.*, 12, 5, 743–754, 2010.
8. Li, C.J., Organic reactions in aqueous media with a focus on carbon-carbon bond formations: A decade update. *Chem. Rev.*, 105, 8, 3095–3165, 2005.
9. Hutchings, G.J., Nano-catalysis in the green chemistry era. *Green Chem.*, 10, 9, 890–906, 2008.
10. Yamaguchi, K., Mizuno, N., Kamata, K., Recent advances in environmentally benign catalytic oxidations of alcohols. *Green Chem.*, 9, 12, 1328–1337, 2007.
11. Palkovits, R., Antonetti, C., Kantarelis, E., Catalytic processes in ionic liquids—The challenge of leaching. *ChemSusChem*, 4, 11, 1584–1596, 2011.
12. Wang, W., Lu, Z., Xu, L., Tang, B., Aqueous phase catalysis using metal nanoparticles. *Chem. Soc. Rev.*, 40, 7, 3722–3745, 2011.
13. Dupont, J., Catalytic and sustainable oxidation of alcohols by supported gold nanoparticles. *Chem. Sci.*, 3, 4, 916–920, 2002.
14. Török, B. and Kopf, J., Catalysis in supercritical carbon dioxide. *Chem. Rev.*, 112, 5, 2662–2714, 2012.
15. Sheldon, R.A., Green and sustainable manufacture of chemicals from biomass: State of the art. *Green Chem.*, 14, 10, 2890–2892, 2012.
16. Gu, Y., Li, C., Xu, J., Yan, N., Solid base catalysis for biodiesel production: A green and sustainable approach. *Green Chem.*, 15, 8, 2095–2114, 2013.
17. Polshettiwar, V., Baruwati, B., Varma, R.S., Nanoparticle-supported and magnetically recoverable nickel catalyst: A robust and sustainable hydrogenation protocol. *Green Chem.*, 11, 8, 1270–1274, 2009.
18. Clark, J.H. and Macquarrie, D.J., *Handbook of green chemistry and technology*, John Wiley & Sons, Paris, France, 2002.
19. Li, Y., Han, B., Liu, Z., Yang, Y., Zeolite-catalyzed biomass conversion processes: From fundamental developments to commercialization. *Chem. Rev.*, 112, 9, 5750–5783, 2012.
20. Zhao, X. and He, L., New developments in heterogeneous catalysis for bio-oil upgrading. *Green Chem.*, 14, 6, 1584–1612, 2012.
21. Pletcher, D. and Walsh, F.C., *Industrial electrochemistry*, 2nd ed., Springer, Springer Netherlands, 2010.
22. Vilela, S.M.F., Silvestre, A.J.D., Gandini, A., Mondello, L.P., Tranchida, P.Q., *The power of analytical separation techniques in green chemistry*, CRC Press, Boca Raton, Florida, 2010.
23. Martín-Aranda, R.M. and Ovín-Ania, M. Á., The importance of the solvent: Some examples of green solvents for syntheses and their effects. *Curr. Org. Chem.*, 15, 8, 1302–1315, 2011.

24. Riisager, A. and Fehrmann, R., Sustainable chemical processing using supported liquid phase catalysts. *Chem. Soc. Rev.*, 38, 3, 662–675, 2009.
25. Anastas, P.T. and Warner, J.C., *Green Chemistry: Theory and Practice*, 1st ed., Oxford University Press, Walton Street, Oxford, 2000.
26. Poliakoff, M., Fitzpatrick, J.M., Farren, T.R., Anastas, P.T., Green Chemistry: Science and Politics of Change. *Science*, 297, 5582, 807–810, 2002.
28. Sheldon, R.A., Green and sustainable solvents in chemical processes. *Chem. Rev.*, 114, 2, 801–818, 2014.
29. Palkovits, R. and Antonetti, C., Green Catalytic Oxidations in Ionic Liquids. *ACS Catal.*, 1, 7, 718–724, 2011.
30. Polshettiwar, V. and Varma, R.S., Greener and expeditious synthesis of quinazolinones in glycerol. *Green Chem.*, 11, 9, 1277–1280, 2009.
31. Zhang, Y., Yuan, Y., Tang, C., Li, Z., Enriched Atom Utilization in Catalytic Reactions: Selective Ring-Opening of Epoxides with Carbon Dioxide. *ChemSusChem*, 9, 12, 1385–1390, 2016.
32. Bruijnincx, P.C.A. and Weckhuysen, B.M., Shaping the future of biomass-derived fuels and chemicals by means of heterogeneous catalysis. *ACS Catal.*, 3, 10, 2456–2473, 2013.
33. Black, D.A. and Arndtsen, B.A., Copper-catalyzed coupling of imines, acid chlorides, and alkynes: A multicomponent route to propargylamides. *Org. Lett.*, 6, 1107–1110, 2004.
34. Sakaguchi, S., Kubo, T., Ishii, Y., A three-component coupling reaction of aldehydes, amines, and alkynes. *Angew. Chem. Int. Ed.*, 40, 2534–2536, 2001.
35. Fischer, C. and Carreira, E.M., Direct addition of TMS-acetylene to aldimines catalyzed by a simple, commercially available Ir(I) complex. *Org. Lett.*, 3, 4319–4321, 2001.
36. Li, C.J., Wei, C., Highly efficient Grignard-type imine additions via C–H activation in water and under solvent-free conditions. *Chem. Communs.* 3, 268–269, 2002.
37. Wei, C. and Li, C.J., Direct addition of tms-acetylene to aldimines catalyzed by a simple, commercially available Ir(I) Complex. *Proc. Natl. Acad. Sci. U.S.A.*, 101, 5749–5754, 2004.
38. Zeng, T., Yang, L., Hudson, R. *et al.*, An intramolecular relay catalysis strategy for Knoevenagel condensation and 1,3-dipolar cycloaddition domino reactions. *Org. Lett.*, 13, 442–445, 2011.
39. Wei, C. and Li, C.J., A highly efficient three-component coupling of aldehyde, alkyne, and amines via C–H activation catalyzed by gold in water. *J. Am. Chem. Soc.*, 125, 9584–9585, 2003.
40. Skouta, R. and Li, C.J., Gold-catalyzed reactions of C–H bonds. *Tetrahedron*, 22, 4917–4938, 2008.
41. Kung, K.K.Y., Li, G.L., Zou, L. *et al.*, Gold-mediated bifunctional modification of oligosaccharides via a three-component coupling reaction. *Org. Biomol. Chem.*, 10, 925–930, 2012.
42. Wei, C., Li, Z., Li, C.J., The first silver-catalyzed three-component coupling of aldehyde, alkyne, and amine. *Org. Lett.*, 5, 4473–4475, 2003.
43. Li, Z., Wei, C., Chen, L. *et al.*, Efficient three-component coupling catalysed by mesoporous copper–aluminum based nanocomposites. *Tetrahedron Lett.*, 45, 2443–2446, 2004.
44. Zeng, T., Chen, W.W., Cirtiu, C.M. *et al.*, Fe_3O_4 nanoparticles: a robust and magnetically recoverable catalyst for three-component coupling of aldehyde, alkyne and amine. *Green Chem.*, 12, 570–573, 2010.
45. Shore, G., Yoo, W.J., Li, C.J. *et al.*, Perspectives on green synthesis and catalysis. *Chem. Eur. J.*, 16, 126–133, 2010.
46. Zhao, L. and Li, C.J., Functionalizing Glycine Derivatives by Direct C–C Bond Formation. *Angew. Chem. Int. Ed.*, 47, 7075–7078, 2008.
47. Zhao, L., Basle, O., Li, C.J., Cu-catalyzed cross-dehydrogenative coupling: a versatile strategy for C-C bond formations via the oxidative activation of sp(3) C-H bonds. *Proc. Natl. Acad. Sci. U.S.A.*, 106, 4106–4111, 2009.

48. Bonfield, E.R. and Li, C.J., Perspectives on green synthesis and catalysis. *Adv. Synth. Catal.*, 350, 370–374, 2008.
49. Knopfel, T.F. and Carreira, E.M., The first conjugate addition reaction of terminal alkynes catalytic in copper: conjugate addition of alkynes in water. *J. Am. Chem. Soc.*, 125, 6054–6055, 2003.
50. Chen, L. and Li, C.J., 2-Iodoxybenzenesulfamides: new pseudo benziodoxole-based hypervalent iodine reagents, *Tetrahedron Lett.*, 45, 2771–2774, 2004.
51. Chen, L. and Li, C.J., The first palladium-catalyzed 1,4-addition of terminal alkynes to conjugated enones. *Chem. Commun.*, 20, 2362–2364, 2004.
52. Wang, H., Hearne, Z., Knauber, T. *et al.*, Perspectives on green synthesis and catalysis. *Tetrahedron*, 71, 5866–5870, 2015.
53. Zhou, L., Chen, L., Skouta, R. *et al.*, Application of a novel thioesterification reaction to the synthesis of chemokine CCL27 by the modified thioester method. *Org. Biomol. Chem.*, 6, 2969–2977, 2008.
54. Jia, C., Kitamura, T., Fujiwara, Y., Catalytic functionalization of arenes and alkanes via C-H bond activation. *Acc. Chem. Res.*, 34, 633–639, 2001.
55. Huang, C.Y., Kang, H., Li, J. *et al.*, En route to intermolecular cross-dehydrogenative coupling reactions. *J. Org. Chem.*, 84, 12705–12721, 2019.
56. Murahashi, S.I., Komiya, N., Terai, H., Ruthenium-catalyzed oxidative cyanation of tertiary amines with hydrogen peroxide and sodium cyanide. *Angew. Chem. Int. Ed.*, 44, 6931–6933, 2005.
57. Li, Z. and Li, C.J., CuBr-Catalyzed Efficient Alkynylation of sp^3 C-H Bonds Adjacent to a Nitrogen Atom. *J. Am. Chem. Soc.*, 126, 11810–11811, 2004.
58. Li, C.J. and Li, Z., Green chemistry: The development of cross-dehydrogenative coupling (CDC) for chemical synthesis. *Pure Appl. Chem.*, 78, 935–945, 2006.
59. Li, Z. and Li, C.J., Electrolytic Polymerization of the Ethylenedioxy-Substituted Terthiophene–Fullerene Dyad. *Org. Lett.*, 6, 4997–4999, 2004.
60. Li, Z., MacLeod, P.D., Li, C.J., Cross-dehydrogenative coupling reaction using copper oxide impregnated on magnetite in deep eutectic solvents. *Tetrahedron: Asymmetry*, 17, 590–597, 2006.
61. Twyman, R.M., *Principles of Proteomics*, BIOS Scientific Publishers, Milton Park near Abingdon, Oxfordshire, South of Oxford, England, 2004.
62. Li, Z. and Li, C.J., CuBr-catalyzed direct indolation of tetrahydroisoquinolines via cross-dehydrogenative coupling between sp3 C-H and sp2 C-H bonds. *J. Am. Chem. Soc.*, 127, 6968–6969, 2005.
63. Li, Z., Bohle, D.S., Li, C.J., Cu-catalyzed cross-dehydrogenative coupling: A versatile strategy for C–C bond formations via the oxidative activation of sp^3 C–H bonds. *Proc. Natl. Acad. Sci. U.S.A.*, 103, 8928–8933, 2006.
64. Deng, G., Zhao, L., Li, C.J., Ruthenium-catalyzed oxidative cross-coupling of chelating arenes and cycloalkanes. *Angew. Chem. Int. Ed.*, 47, 6278–6282, 2008.
65. Deng, G. and Li, C.J., Sc(OTf)3-Catalyzed Direct Alkylation of Quinolines and Pyridines with Alkanes. *Org. Lett.*, 11, 1171–1174, 2009.
66. Basle, O. and Li, C.J., Copper catalyzed oxidative alkylation of sp^3 C–H bond adjacent to a nitrogen atom using molecular oxygen in water. *Green Chem.*, 9, 1047–1050, 2007.
67. Zeng, T., Song, G., Moores, A. *et al.*, Bare Magnetic Nanoparticles: Sustainable Synthesis and Applications in Catalytic Organic Transformations. *Synlett*, 2010, 2002, 2010.
68. Sasamoto, N., Dubs, C., Hamashima, Y. *et al.*, Preparation of dicationic palladium catalysts for asymmetric catalytic reactions. *J. Am. Chem. Soc.*, 128, 14010–14011, 2006.

69. Zhang, Y. and Li, C.J., Remodeling of Fumagillol: Discovery of an oxygen-directed oxidative mannich reaction, *J. Am. Chem. Soc.*, 128, 4242–4243, 2006.
70. Basle, O., Borduas, N., Dubois, P. *et al.*, Toward rational understandings of α-C-H functionalization: Energetic studies of representative tertiary amines. *Chem. Eur. J.*, 16, 8162–8166, 2010.
71. Condie, A.G., Gonzalez-Gomez, J.C., Stephenson, C.R.J., Visible-light photoredox catalysis: Aza-henry reactions via C–H functionalization. *J. Am. Chem. Soc.*, 132, 1464–1465, 2010.
72. Trost, B.M. and Crawley, M.L., Asymmetric transition-metal-catalyzed allylic alkylations: applications in total synthesis. *Chem. Rev.*, 103, 2921–2944, 2003.
73. Trost, B.M., Strege, P.E., Weber, L. *et al.*, Allylic Alkylation: Preparation of π-Allylpalladium Complexes from Olefins. *J. Am. Chem. Soc.*, 100, 3407–3415, 1978.
74. Li, Z. and Li, C.J., Cu(II)-Catalyzed Functionalizations of Aryl C–H Bonds Using O_2 as an Oxidant. *J. Am. Chem. Soc.*, 128, 56–57, 2006.
75. Li, Z., Cao, L., Li, C.J., Enantioselective oxidative cross-dehydrogenative coupling of tertiary amines to aldehydes. *Angew. Chem. Int. Ed.*, 46, 6505–6507, 2007.
76. Walling, C., Intermediates in the reactions of fenton type reagents. *Acc. Chem. Res.*, 31, 155–157, 1998.
77. Zhang, Y. and Li, C.J., Highly efficient direct alkylation of activated methylene by cycloalkanes. *Eur. J. Org Chem.*, 2007, 4654–4657, 2007.
78. Clark, J.H., , Chemistry goes green. *Nat. Chem.*, 1, 12–13, 2009.
79. Werz, D.B., *Carbohydrate Chemistry: State of the Art and Challenges for Drug Development*, pp. 195–213, World Scientific, London, 2015.
80. Li, C.J. and Chen, L., Organic chemistry in water. *Chem. Soc. Rev.*, 35, 68–82, 2006.
81. Huang, B., Yao, X., Li, C.J., Perspectives on green synthesis and catalysis. *Adv. Synth. Catal.*, 348, 1528–1532, 2006.
82. Zhang, Q., Cheng, M., Hu, X. *et al.*, Copper-catalyzed tandem A3 -coupling–isomerization–hydrolysis reactions of aldehydes and terminal alkynes leading to chalcones, *J. Am. Chem. Soc.*, 132, 7256–7257, 2010.
83. Dai, X.J. and Li, C.J., , En route to a practical primary alcohol deoxygenation, *J. Am. Chem. Soc.*, 138, 16, 5433–5440, 2016.
84. Chen, Z., Zeng, H., Girard, S.A. *et al.*, Formal direct cross-coupling of phenols with amines, *Angew. Chem. Int. Ed.*, 54, 14487–14491, 2015.
85. Chen, Z., Zeng, H., Gong, H. *et al.*, Palladium-catalyzed reductive coupling of phenols with anilines and amines: Efficient conversion of phenolic lignin model monomers and analogues to cyclohexylamines, *Chem. Sci.*, 6, 4174–4178, 2015.
86. Zeng, H., Cao, D., Qiu, Z. *et al.*, Palladium-catalyzed formal cross-coupling of diaryl ethers with amines: Slicing the 4-O-5 linkage in lignin models, *Angew. Chem. Int. Ed.*, 57, 3752–3757, 2018.
87. Qiu, Z., Lv, L., Li, J. *et al.*, Direct conversion of phenols into primary anilines with hydrazine catalyzed by palladium, *Chem. Sci.*, 10, 4775–4781, 2019.
88. Guin, J., Frohlich, R., Studer, A., Thiol-catalyzed stereoselective transfer hydroamination of olefins with N-aminated dihydropyridines, *Angew. Chem. Int. Ed.*, 47, 779 € –782, 2008.
89. K€ark€as, M.D., Nitrene Photochemistry of Manganese N-Haloamides. *ACS Catal.*, 7, 4999–5022, 2017.
90. Dominguez-Huerta, A., Perepichka, I., Li, C.J., Direct conversion of phenols into primary anilines with hydrazine catalyzed by palladium. *Chem. Commun.*, 1, 45, 2018.
91. Liu, M., Wang, Y., Kong, X. *et al.*, Chirotechnology, Industrial Synthesis of Optically Active Compounds, Marcel Dekker, New York. *iScience*, 17, 208–216, 2019.
92. Badr, O., Ancestral sequence reconstruction enhances gene mining efforts for industrial ene reductases by expanding enzyme panels with thermostable catalysts. *Appl. Energy*, 41, 95, 1992.

93. Dutta, K., Chaudhari, V., Li, C.J. *et al.*, Droplet microfluidics and directed evolution of enzymes: An intertwined journey. *Appl. Catal. Gen.*, 595, 117430, 2020.
94. Chaudhari, V., Dutta, K., Li, C.J. *et al.*, Rapid Screening of Diverse Biotransformations for Enzyme Evolution. JACS *Mol. Catal.*, 482, 110606, 2020.
95. Li, L., Fan, S., Mu, X. *et al.*, The Coming of Age of de Novo Protein Design. *J. Am. Chem. Soc.*, 136, 7793–7796, 2014.
96. Dutta, K., Li, L., Gupta, P. *et al.*, Roche's approach to green metrics in drug development. *Catal. Commun.*, 106, 16–19, 2018.
97. Omae, I., Green metrics in pharmaceutical development. *Curr. Res. Green Sustain. Chem.*, 33, 100564, 2022.
98. Darensbourg, D.J. and Holtcamp, M.W., Ultrasound promoted one- pot synthesis of 2-amino-4, 8-dihydropyrano [3, 2-b] pyran-3-carbonitrile scaffolds in aqueous media: A complementary 'green chemistry 'tool to organic synthesis. *Ultrason. Sonochem.*, 20, 1, 401–407, 2013.
99. Eghbali, N. and Li, C.J., Contributions of organic electrosynthesis to green chemistry. *C.R. Chim.*, 14, 7-8, 745–765, 2011.
100. Yoo, W.J. and Li, C.J., Solvent-free microwave organic synthesis as an efficient procedure for green chemistry. *C.R. Chim.*, 7, 2, 103–112, 2004.
101. Keith, J.K., Timothy, K.S., Stephen, L.M., Jeremy, J.Y., Vincent, T.M., Vishal, B.S., Dac-Trung, N., Lars, J.J., Dušica, V., Stephan, C.S., Jayme, H., Karlie, R.S., Ajay, P., Cristian, G.B., Jeremy, S.E., Ewy, A.M., Tudor, I.O., Pharos 2023: an integrated resource for the understudied human proteome. *Nucleic Acids Res.*, 51, D1, D1405–D1416, 2023.
102. Kanehisa, M., Furumichi, M., Sato, Y., Kawashima, M., Ishiguro-Watanabe, M., KEGG for taxonomy-based analysis of pathways and genomes. *Nucleic Acids Res.*, 51, D587–D592, 2023.
103. Caspi, R., Altman, T., Billington, R., Dreher, K., Foerster, H., Fulcher, C.A., Holland, T.A., Keseler, I.M., Kothari, A., Kubo, A., Krummenacker, M., Latendresse, M., Mueller, L.A., Ong, Q., Paley, S., Subhraveti, P., Weaver, D.S., Weerasinghe, D., Zhang, P., Karp, P.D., The MetaCyc Database of metabolic pathways and enzymes and the BioCyc collection of Pathway/Genome Databases. *Nucleic Acids Res.*, 42, 1, D459–D471, 2014.
104. Colacot, T.J., New Trends in Cross-Coupling: Theory and Applications, Royal Society of Chemistry, Cambridge, 2014.
105. Dijkstra, H.P.,. van Klink, G.P.M., van Koten, G., The use of ultra- and nanofiltration techniques in homogeneous catalyst recycling. *Acc. Chem. Res.*, 35, 798–810, 2002.
106. Iakovou, E., Karagiannidis, A., Vlachos, D., Toka, A., Malamakis, A., Waste biomass-to-energy supply chain management: A critical synthesis. *Waste Manage.*, 30, 1860–1870, 2010, https://doi.org/10.1016/j.wasman.2010.02.030.
107. Sheldon, R.A. and van Pelt, S., Enzyme immobilisation in biocatalysis: Why, what and how. *Chem. Soc. Rev.*, 42, 6223–6235, 2013, https://doi.org/10.1039/c3cs60075k.
108. Bornscheuer, U.T. and Kazlauskas, R.J., Catalytic promiscuity in biocatalysis: Using old enzymes to form new bonds and follow new pathways. *Angew. Chem. Int. Ed.*, 43, 6032–6040, 2004, https://doi.org/10.1002/anie.200460416.
109. Lin, Y., Ren, J., Qu, X., Catalytically active nanomaterials: A promising candidate for artificial enzymes, *Acc. Chem. Res.*, 47, 1097–1105, 2014, https://doi.org/10.1021/ar400250z.
110. Sun, H., Zhou, Y., Ren, J., Qu, X., Carbon Nanozymes: Enzymatic Properties, Catalytic Mechanism, and Applications. *Angew. Chem. Int. Ed.*, 57, 9224–9237, 2018, https://doi.org/10.1002/ anie.201712469.
111. Kita, Y. and Dohi, T., Pioneering Metal-Free Oxidative Coupling Strategy of Aromatic Compounds Using Hypervalent Iodine Reagents. *Chem. Rec.*, 15, 886–906, 2015.

112. Kurti, L., Herczegh, P., Visy, J., Simonyi, M., Antus, S., Pelter, A., New Insights Into the Mechanism of Phenolic Oxidation with Phenyliodonium(III) Reagents. *J. Chem. Soc., Perkin Trans.*, 1, 1, 379–380, 1999.
113. Dohi, T., Maruyama, A., Takenaga, N., Senami, K., Minamitsuji, Y., Fujioka, H., Caemmerer, S.B., Kita, Y.A., Chiral Hypervalent Iodine(III) Reagent for Enantioselective Dearomatization of Phenols. *Angew. Chem., Int. Ed.*, 47, 3787–3790, 2008.
114. Koposov, A.Y., Boyarskikh, V.V., Zhdankin, V.V., Amino Acid Derived Iodobenzene Dicarboxylates: Reagents for Oxidative Conversion of Alkenes to Amino Acid Esters. *Org. Lett.*, 6, 3613–3615, 2004.
115. Boye, A.C., Meyer, D., Ingison, C.K., French, A.N., Wirth, T., Novel Lactonization with Phenonium Ion Participation Induced by Hypervalent Iodine Reagents. *Org. Lett.*, 5, 2157–2159, 2003.
116. Hartwig, J.F., Discovery and Understanding of Transition-Metal-Catalyzed Aromatic Substitution Reactions. *Synlett,* 9, 1283–1294, 2006.
117. S.H. Yang, C.S. Li, C.H. Cheng, Halide exchange reactions between aryl halides and alkali halides catalyzed by nickel metal. *J. Org. Chem.*, 52, 691–694, 1987.
118. T.D. Sheppard, Metal-catalysed halogen exchange reactions of aryl halides. *Org. Biomol. Chem.* 7, 1043–1052, 2009.
119. Kalyani, D., Dick, A.R., Anani, W.Q., *et al.*, A simple catalytic method for the regioselective halogenation of arenes. *Org. Lett.*, 8, 2523–2526, 2006.
120. T.S. Mei, R. Giri, N. Maugel, *et al.*, Squaramide-catalyzed enantioselective Michael addition of diphenyl phosphite to nitroalkenes. *Angew. Chem. Int. Ed.*, 47, 5215–5219, 2008.
121. A.S. Dudnik, N. Chernyak, C. Huang, *et al.*, A General Strategy Toward Aromatic 1,2-Ambiphilic Synthons: Palladium-Catalyzed ortho-Halogenation of PyDipSi-Arenes. *Angew. Chem. Int. Ed.*, 49, 8729–8732, 2010.
122. Newman, S.G., Howell, J.K., Nicolaus, N. *et al.*, Nickel-Catalyzed Enantioselective Carbamoyl Iodination: A Surrogate for Carbamoyl Iodides. *J. Am. Chem. Soc.*, 133, 14916–14919, 2011.
123. Klapars, A., Buchwald, S.L., Copper-Catalyzed Halogen Exchange in Aryl Halides: An Aromatic Finkelstein Reaction. *J. Am. Chem. Soc.*, 124, 14844–14845, 2002.
124. Li, L., Liu, W., Zeng, H., *et al.*, Photo-induced Metal-Catalyst-Free Aromatic Finkelstein Reaction. *J. Am. Chem. Soc.*, 137, 8328–8331, 2015.
125. Nagib, D.A., MacMillan, D.W.C., Trifluoromethylation of arenes and heteroarenes by means of photoredox catalysis. *Nature* 480, 7376, 224–228 , 2011.
126. Ji, Y., Brueckl, T., Baxter, R.D., *et al.*, Innate C-H trifluoromethylation of heterocycles. *Proc. Natl. Acad. Sci. U.S.A.*, 108, 2011, 14411–14415, 2011.
127. Ohkubo, K., Fujimoto, A., Fukuzumi, S., Visible-Light-Induced Oxygenation of Benzene by the Triplet Excited State of 2,3-Dichloro-5,6-dicyano-p-benzoquinone. *J. Am. Chem. Soc.*, 135, 5368–5371, 2013.

2

Green Catalysis for Chemical Transformation: Need for the Sustainable Development

Dripta De Joarder[1], Rajarshi Sarkar[1] and Dilip K. Maiti[2]*

[1]*Department of Chemistry, School of Advanced Sciences, VIT-AP University, Andhra Pradesh, India*
[2]*Department of Chemistry, University of Calcutta, Kolkata, India*

Abstract

The conceptual foundation of Green Chemistry was first formulated by Paul Anastas and John Warner. They put forward the famous Twelve Principles of Green Chemistry that elucidate various concepts such as energy efficiency, atom economy, renewable feedstock, safer chemicals, waste reduction, and monitoring of pollution at real time. These postulates shifted the conscience of the scientific community in a fundamental way, as it became impossible to remain oblivious to the harmful effects of various chemicals on human health and their roles in environmental pollution and degradation. Not only the scientific community, various governments and industries around the world especially in developed nations started searching for green alternatives for common chemical transformations. One of the key approaches in Green Chemistry is catalysis that helps in the reduction of usage of stoichiometric reagents and minimization of toxic wastes generated in the chemical processes. Although humans have been using catalysts for thousands of years, however, the requirement for efficient catalysts with greater turnover number and turnover frequency has been necessitated by the rapid socioeconomic growth around the world. Catalytic transformations are intrinsically "Green" and can be made even more eco-friendly by carrying them out in aqueous phase or in solid phase and thereby completely removing solvents. Industry and academia have taken a keen interest in environmentally benign chemistry due to the fact that different rare and precious metals such as Au, Rh, and Ru can be used as catalysts in modest quantities to carry out numerous essential yet complicated reactions. Not only are organic solvents a source of major environmental pollution but they pose a serious threat to human health and safety, as they are toxic, flammable, and explosive, as well as carcinogenic, mutagenic, and teratogenic. However, chemists in the industry and academia are compelled to use them, as most of the organic building blocks and starting materials are insoluble in water. Therefore, organic catalysis that can be carried out in H_2O has been receiving constant attention from the scientific community, as it minimizes and/or reduces the usage of harmful organic solvents. Research and development of catalytic processes in H_2O are growing extensively, as it not only addresses sustainability apprehensions but also the distinct selectivities that various characteristics of water bring to a particular transformation. Additionally, other than being inexpensive, green, and eco-friendly, water also facilitates convenient product separation and easy catalyst recycling. This book chapter will focus on the aspects of catalysis in water medium.

**Corresponding author*: dkmchem@caluniv.ac.in

Mousumi Sen (ed.) Sustainable Green Catalytic Processes, (29–58) © 2024 Scrivener Publishing LLC

Keywords: Green Chemistry, catalysis, metal catalyzed reactions, reactions in water medium, organic transformations, heterocycle synthesis

2.1 Introduction

Green Chemistry has established a framework for chemists and chemical engineers to contribute to the vast scope of global sustainability.[1] Even though critical advancements have been made in the field to address various obstacles faced by the synthetic community, one of the key challenges we face continues to be decreasing our dependence on organic solvents that are highly toxic and environmentally hazardous.[2,3] Water has been the alternative option because of its abundance, cost, and availability.[4] Moreover, all reactions in nature are always carried out in aqueous media. But a literature survey revealed that there are only a few reports that use H_2O as the reaction medium understandably due to solubility, reagent compatibility, and other issues.[5] This chapter aims to sum up recently published miscellaneous organic methodologies that are specifically executed in water. Another emerging area of research that is gaining popularity is biocatalysis, where enzymes, enzyme mimics, or cells are used as catalysts.[6–9] Instances of biocatalysis will be elaborated as well.

Selective carbon-carbon bond cleavage has remained one of the most difficult transformations to accomplish owing to its competition with C-H activation. Usually, such transformations require extremely harsh conditions.[10, 11] Zhou *et al.*[12] reported an Fe-catalyzed coupling reaction (Scheme 2.1) between benzothiazole (**1**) and oxime ester (**2**) in aqueous media involving C–C bond dissociation in the oxime ester moiety by means of a single-electron transfer. The iron triflate-catalyzed reaction was conducted in water without the requirement of any inert gas environment that provides a mild synthetic protocol to synthesize alkyl nitrile-substituted benzothiazole derivatives. They applied this strategy to obtain various important building blocks including 4-heterocyclic-3-arylbutanoic acid. This protocol was successful in producing an excellent yield of 87% in a gram-scale experiment, which further enhances its usability.

Unsymmetrical 1,3-diynes are usually prepared by Cadiot–Chodkiewicz coupling of terminal alkynes of either 1-haloalkynes or 1,1-dibromoalkenes catalyzed by Pd/Cu-,[13] Ni/Cu-,[14] Cu-,[15] and Et_2Zn[16] systems. These transition metal-catalyzed processes display a very high degree of selectivity; however, they require various harmful organic solvents such as MeCN, DMF, and NMP to be carried out. Therefore, the development of an eco-friendly and benign protocol to carry out the Cadiot–Chodkiewicz coupling remains a challenging task. In this context, Liao *et al.*[17] reported an iron-catalyzed Cadiot–Chodkiewicz reaction

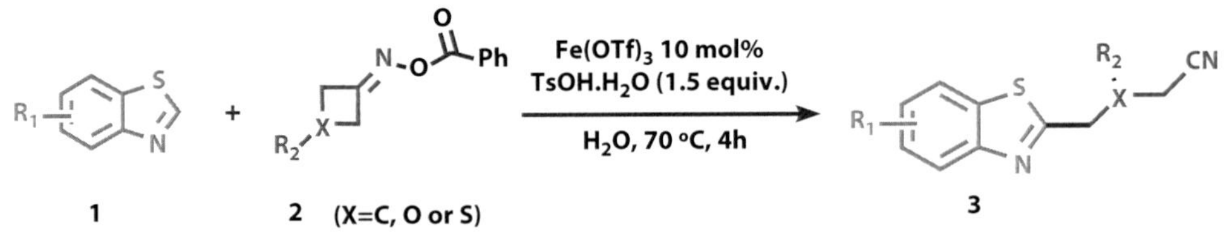

Scheme 2.1 Construction of alkyl nitrile-substituted benzothiazole derivatives.

protocol by using a water-soluble cationic ligand, L (**7**) in aqueous media (Scheme 2.2). They used readily available $FeCl_3{\cdot}6H_2O$ to couple terminal alkynes (**5**) with 1-bromoalkynes (**4**) in aqueous media in the presence of air and Zn as the reducing agent to produce the desired compound (**8**). The reaction was also utilized to synthesize various unsymmetrical 1,3,5-hexatriynes (**9**) by coupling of (**6**) with (**4**) and in a one-pot reaction.

Spiroindolenine derivatives are important biologically active molecules that are prevalent in a variety of biologically useful scaffolds.[18] Due to their importance, numerous methods have been employed to synthesize them.[19] In majority of the procedures published, tedious workup and the use of organic solvents as the reaction media are frequently reported. Consequently, a green protocol to produce spiroindolenine is a long-standing requirement to construct such structural motifs. In 2021, a team of scientists led by Jiang *et al.*[20] delineated a novel approach (Scheme 2.3) for co-catalyzed integration of (**10**) with (**11**) in H_2O to assemble (**12**). This protocol is particularly advantageous due to excellent yields, high atom efficiency, gentle reaction condition, wide substrate scope, and wider functional group tolerance. Various water-soluble cobalt salts such as cobalt oxalate (CoC_2O_4), $CoCl_2$, and $Co(OAc)_2{\cdot}4H_2O$ were successfully employed to carry out the conversion, with $Co(OAc)_2{\cdot}4H_2O$ producing the highest yield.

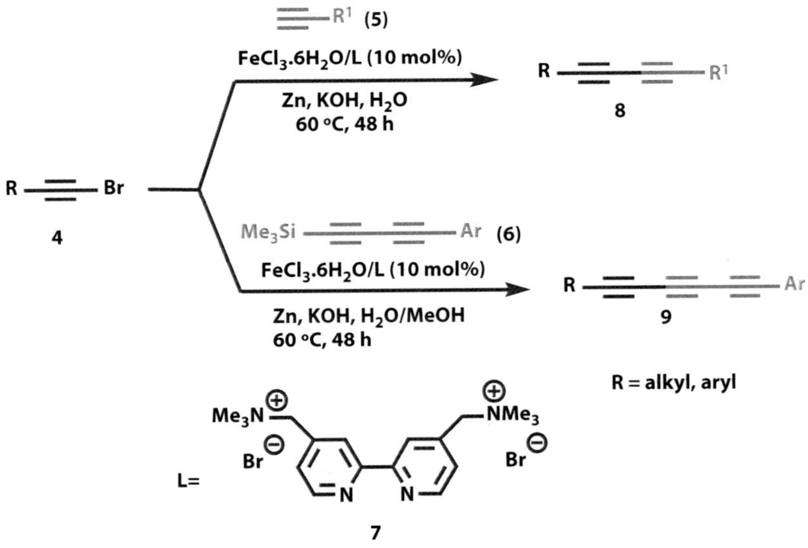

Scheme 2.2 Iron-catalyzed Cadiot–Chodkiewicz protocol.

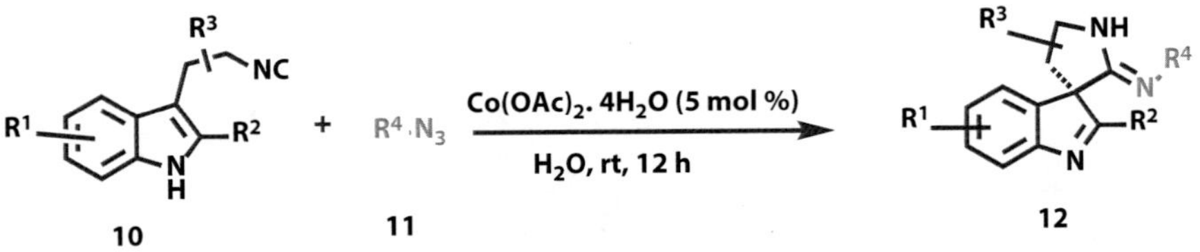

Scheme 2.3 Co-catalyzed coupling toward spiroindolenine derivatives.

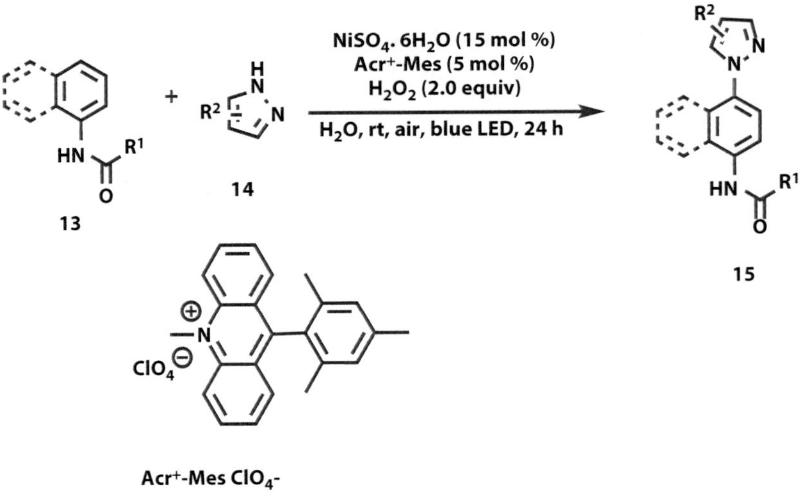

Scheme 2.4 Ni(II)-mediated cross-coupling reaction.

Pyrazole derivatives have been of great interest recently because of their frequent occurrences in bioactive molecules.[21] Therefore, it is crucial to formulate simple yet efficient protocols to synthesize functionalized pyrazoles. As depicted in Scheme 2.4, in 2018, Yu *et al.*[22] reported a nickel-(II)-catalyzed regioselective eco-friendly approach for the C–N cross-coupling between (**13**) and (**14**) to obtain C-N cross-coupling product (**15**) by using a photoredox process in aqueous media. A variety of photocatalysts were utilized such as eosin Y, Acr+-MesClO4 –, methylene blue, and fluorescein in combination with various nickel(II) salts. After the screening process, picolinamide was finalized as the directing group. However, it was discerned that pyrazole moieties with EWGs (–NO_2 and –CN) were unable to undergo the C–N cross-coupling.

Regarding aminals, the *N,N* analogs of acetals and serve diverse roles in synthetic chemistry, the dearth of gentle and systematic strategies for a wide construction of aminals severely limits their application in organic synthesis.[23] Wang *et al.*[24] reported a benign protocol for the synthesis of aminals in water medium (Scheme 2.5). They reacted furfural (**16**) with morpholine (**17**) to obtain aminal. After screening of an array of metal salts, Cu(OTf)2 was chosen to be the catalyst, as it produced (**18**) in quantitative analytical yield after 2 minutes. The reaction was equally susceptible to electron-donating and electron-withdrawing groups.

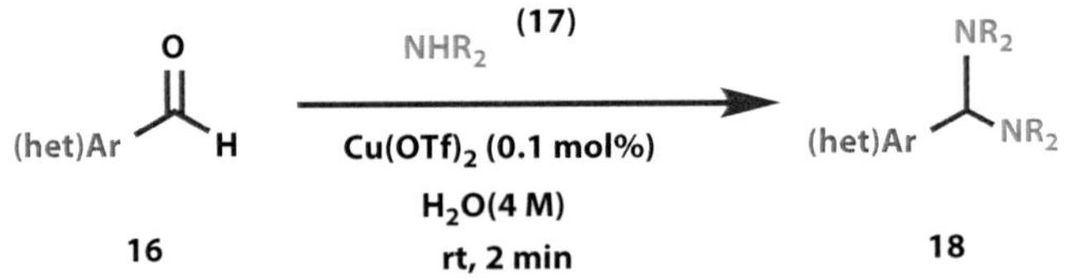

Scheme 2.5 Copper-catalyzed preparation of aminals.

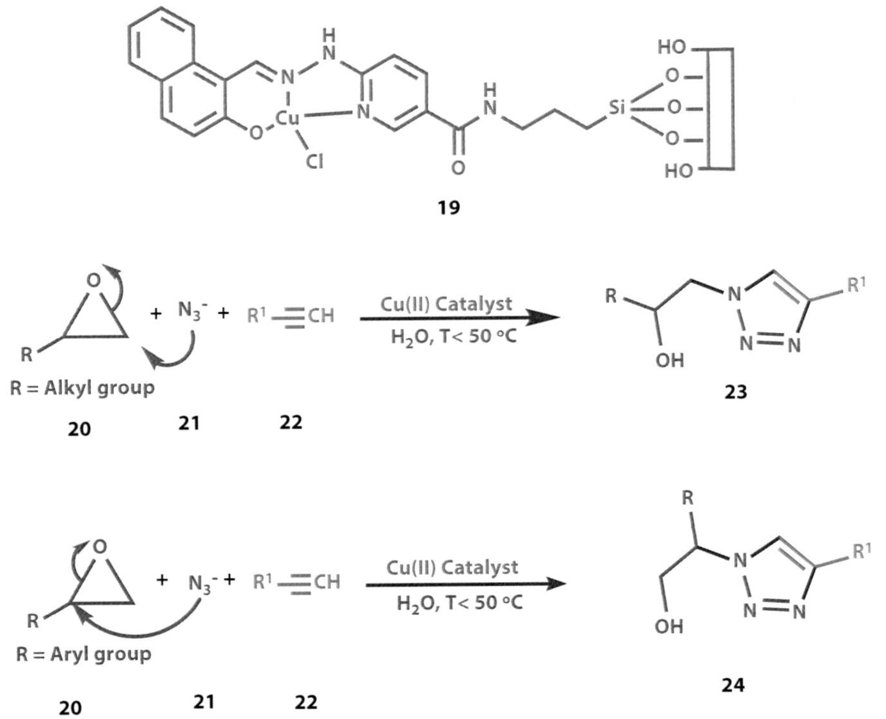

Scheme 2.6 Copper(II)-catalyzed click reaction.

In recent years, pharmaceutical and medicinal chemistry has paid a lot of attention to the eco-friendly construction of N-based heterocyclic molecules. In this regard, 1,2,3-triazoles[24] play a significant role in medicinal chemistry by exhibiting an array of biological and pharmacological activities.[25-26] Additionally, this class of heterocycles has drawn a lot of interest in materials science because of its distinct chemical and structural characteristics. They have found wide applicability in commercial sectors.[27] Consequently, the creation of effective, straightforward, unique, and environmentally friendly processes for the manufacture of 1,2,3-triazole derivatives is desirable in synthetic organic chemistry. Emami *et al.*[28] have reported the synthesis of a copper(II) complex that was advantageously supported on silica gel by the amidification reaction to obtain a new catalyst (**19**). Thus, the obtained Cu catalyst was a stable, inexpensive, easily producible, and highly active catalyst. It was employed to synthesize β-hydroxy-1,2,3-triazoles from a 3-ingredient azide–epoxide–alkyne cycloaddition reaction in H_2O medium (Scheme 2.6).

Castillo *et al.*[29] reported a novel monophosphine Cu(I) compound attached with bis(pyrazolyl)methane (**29**). They used it as a catalyst in the 3-component click reaction (Scheme 2.7) to create (**28**) with up to 93% yield from the components taking part in the process, (**25**), (**26**), and (**27**). The catalyst operates completely efficiently under ultrasonic conditions, is highly stable, and is compatible with oxygen and water. The scope of reactant studies revealed that in all cases, (**28**) was acquired in high transformations (up to 99%) and good yields (up to 93%). The purification process was simple, and a wide variety of functional groups were tolerated and therefore could be used as platforms to obtain various other synthetically crucial moieties.

Scheme 2.7 Cu(I)-catalyzed tricomponent click reaction.

Polycyclic π-extended heteroarenes are crucial components in many physiologically active molecules, natural products, and organic moieties,[30] and extended 3-oxindoles (3-indolinones) are specifically important due to their usage in pharmaceutical and dye industries.[31] Therefore, C-H activation of different nitrogen-containing heteroarenes is crucial in this regard.[32] Wang *et al.*[33] have reported a protocol (Scheme 2.8) of CH alkylation of naphthylamine derivatives in the *ortho*-position with diazo moieties catalyzed by ruthenium(II)-complex $[RuCl_2(p\text{-cym})]_2$ in water medium to produce extended 2,2-disubstituted 3-oxindoles. A key aspect of this protocol is that it is the earliest instance of C-H alkylation via carbenoid insertion in aqueous media. This technique is tolerated by a multitude of alkyl groups (such as 1°-, 2°-, 3°-) and aromatic compounds. This reaction is particularly sustainable due to the usage of readily available naphthylamines. Moreover, thus obtained compounds (**32**) showed promising antitumor activities.

Flavans are a crucial subclass of flavonoid compounds containing a saturated 2-arylchroman core. They exhibit various pharmacological properties, such as cytoprotective, antiviral, and anticancer activities.[34] Hence, the development of convenient and systematic protocols for the enantioselective creation of flavan moieties is highly alluring. In this regard, the metal-mediated asymmetric transfer hydrogenation (ATH) and asymmetric hydrogenation (AH) protocols are recognized as two competent and ingenious tools for the enantioselective fabrication of flavonoids.[35]

Demidoff *et al.*[36] have reported a Ru(II)-catalyzed C-C/C-O reduction of chalcones in one-pot using water mediated by ATH (Scheme 2.9). The usage of the oxo-substituted-Ru(II) precatalyst along with the hydrogen source HCO_2Na facilitated the reduction of a variety of enone groups with EWGs and EDGs at different locations of both aromatic rings producing 1,3-diarylpropan-1-ols in excellent yield with high and enantioselectivities. Will's 3-C-tethered-Ru(II) complex (R,R)-F was initially chosen because of its greater robustness. The oxo-tethered-Ru(II) precatalyst (R,R)-H of Ikariya was tested under

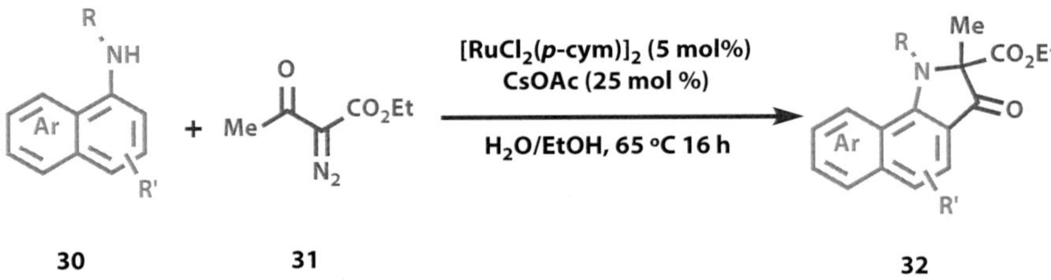

Scheme 2.8 Ru(II)-catalyzed C-H alkylation via carbenoid insertion.

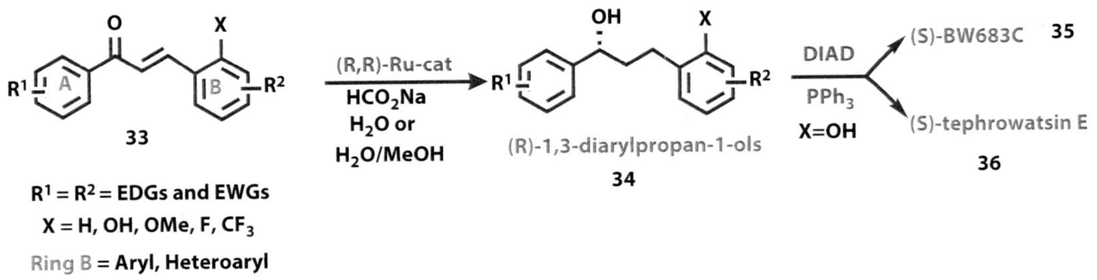

Scheme 2.9 Ru(II)-catalyzed C-C/C-O reduction of chalcones via ATH in water.

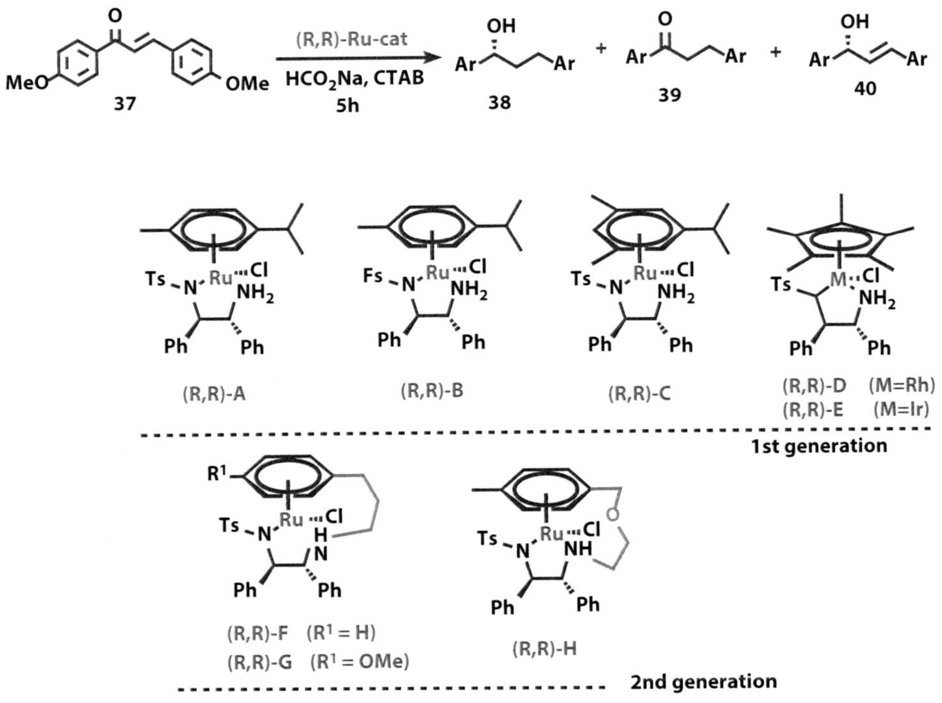

Scheme 2.10 Noyori–Ikariya-type catalysts.

identical circumstances; alcohol (R)-38 was produced with the highest enantiomeric purity (Scheme 2.10).

Aromatic moieties are key components to produce advanced chemicals required for various industrial processes.[37] At present, aromatic compounds are majorly produced using fossil fuels. With limited fossil resources and increased global awareness about this environmental issue, it is critical to develop sustainable methods for producing aromatic compounds. Lignin is nature's greatest aromatic heteropolymer.[38] Owing to this, lignin is contemplated to be one of the inexhaustible sources of aromatics that can possibly reduce reliance on natural oils.[39] Nevertheless, due to the chemical structure of lignin, selective depolymerization remains quite a task.[40] Therefore, a lignin depolymerization protocol without the presence of any external oxidant/reductant under mild condition and in water medium is desirable from both industrial and green chemistry aspect.

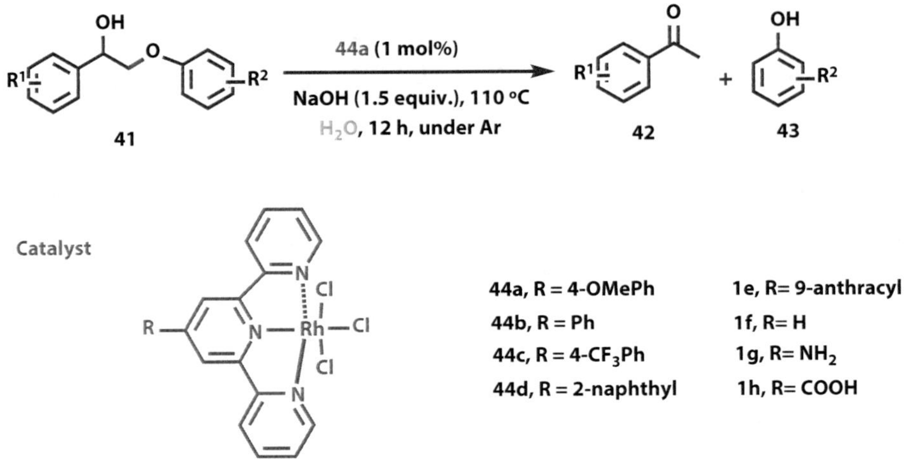

Scheme 2.11 Rh(III)-mediated depolymerization of lignin.

Liu *et al.*[41] have synthesized structurally modular Rh-terpyridine complexes that are capable of catalyzing redox-neutral cleavage of the C–O bonds of β-O-4 lignin model substrates under mild conditions in aqueous media (Scheme 2.11). The flexibility of the catalyst was then investigated using a number of lignin model compounds. Under optimal conditions, the -OMe group at *o*-, *m*-, and *p*-positions on both aromatic moieties are tolerated, which are typically found in natural lignin. The reaction protocol using the Rh-terpyridine catalytic system was able to depolymerize actual lignin and raw poplar wood powder.

Due to their abundance in prominent medications and bioactive chemicals, as well as their versatility of functional availability, 1,2-amino alcohols are a very crucial structural motif.[42] Over the last few decades, these features have drawn attention to synthesize diverse vicinal amino alcohols.[43] Shi *et al.*[44] reported a green protocol of catalytic alkene aminohydroxylation in water (Scheme 2.12). This chemical cascade reaction is catalyzed by rhodium complex and involves a one-pot aziridination/ring opening process. Importantly, water simultaneously serves as a reaction medium, a diluent of AcOH produced *in situ*, a reactant, and a Brønsted acid catalyst, resulting in high regioselectivity and chemoselectivity along with strong reactivity. Furthermore, the benign reaction conditions and possibility of late-stage functionalization highlight its potential to obtain more complex compounds.

Alkenes are a constituent of many of the commonly used organic compounds in the chemical industry, as well as a variety of bioactive substances. Despite the fact that the preparation of E-alkenes is widely documented, the methodologies to create Z-alkenes have

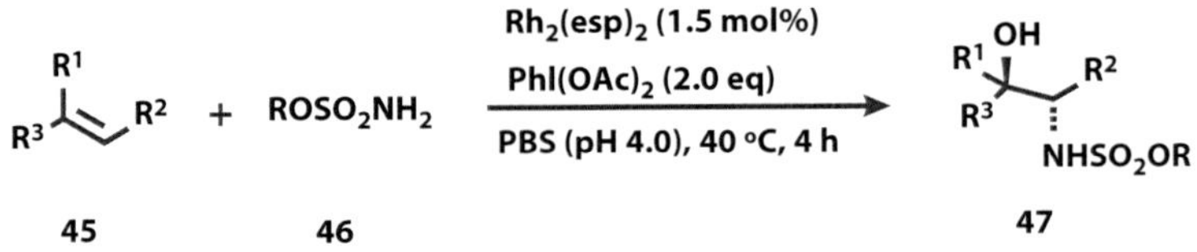

Scheme 2.12 Rh-catalyzed alkene aminohydroxylation.

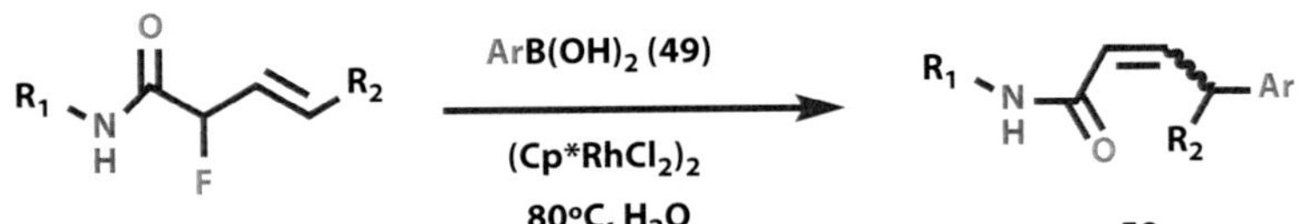

Scheme 2.13 Rh(III)-catalyzed fabrication of α,β-unsaturated amides.

received less attention due to their low thermodynamic stability. As a result, the manufacture of Z-olefins is a difficult and crucial task for the scientific community.[45]

Dr. Bouzbouz developed a highly efficient environmentally responsible methodology (Scheme 2.13) for the Rh(III)-mediated γ-selective and stereospecific reaction utilizing (**48**) and (**49**), resulting in (**50**) with high Z selectivity and a stereogenic center via the formation of a Csp^2-Csp^3 bond.[46]

At first, α-fluoroenamide was tested in water as an allylating ingredient for (**49**) in the presence of several catalysts. A control reaction with phenylboronic acid as a nucleophile in the absence of the catalyst indicated the crucial role of the catalyst. Using the complex, they were able to generate a moiety having a high Z/E ratio in favor of the Z compound.

In order to generalize those conditions, a series of substituted N-benzylamides (**48**) were tested in para, meta, and ortho positions with arylboronic acid, mono- or disubstituted arylboronic acids bearing both electron-withdrawing and electron-donating groups (Cl, Br, F, CF_3, $COCH_3$, $NHCOCH_3$, Me). The intended products (**50**) were produced with strong Z/E selectivity (4.4/1 to 10.5/1) and good yield (75%). However, when other boronic acids with ortho substituents were tested, no reaction was seen likely due to steric effects. Further study revealed that the reaction is responsive to the electronic/steric needs of both the participating scaffolds. The intended products (**50**) were obtained in high yields and with favorable Z/E ratios. A variety of phenylboronic acids were used to study a series of primary α-fluoroenamide substrates (**48**), yielding the corresponding products (**50**) in moderate to good yields (up to 81%) and excellent Z/E ratios (15.5/1 to 43/1).

To illustrate the practicality and operational simplicity of the developed system, fluorinated alkene arylation technique, the group performed large-scale operations. This large-scale method yielded high yields, good Z/E ratios, and 99% ee in racemic and stereospecific variants (Scheme 2.14).

Many natural or bioactive substances contain isoindolin-1-ones.[47] Consequently, the synthesis of this family of scaffolds has piqued the interest of many researchers, and several synthetic methods have been developed. Known techniques suffer from harsh circumstances, inaccessible substrates, or waste generation. In 2018, Wu *et al.*[48] communicated a study (Scheme 2.15) wherein they showed that isoindolin-1-ones could be created using a Rh(III)-catalyzed reaction in H_2O via an efficient C–H functionalization/allene formation/cyclization route.

Initially at 55°C, (**55**) and (**56**) were reacted in an array of solvents with [Cp*RhC12]2 as a catalyst and NaOAc as a base. Studies revealed that reactions in the presence of HOAc produced larger yields of (**58**). They chose water as the solvent for further research because of its low cost and environmental friendliness.

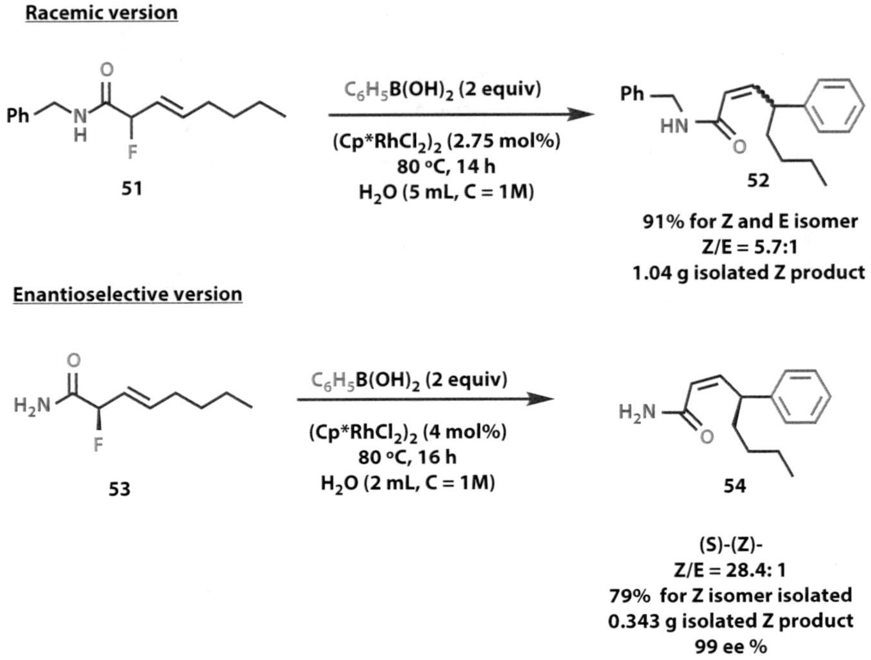

Scheme 2.14 Allylic arylations in gram scale.

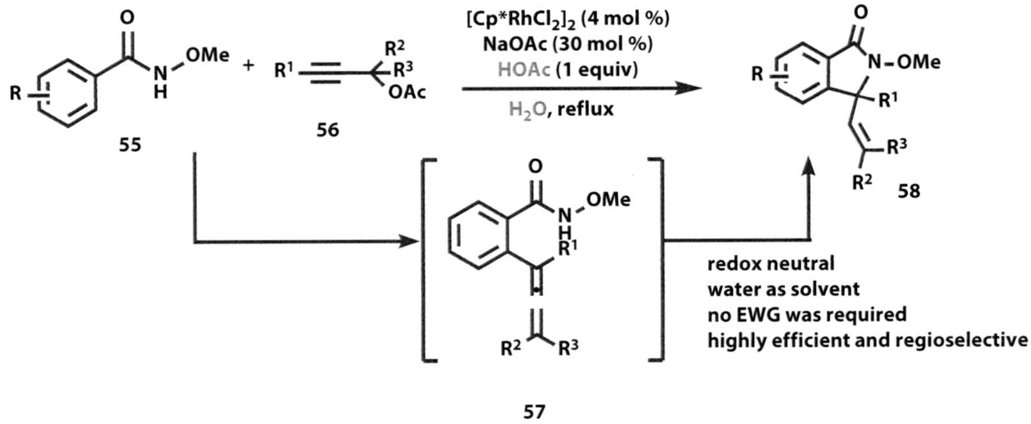

Scheme 2.15 Rh-mediated eco-friendly construction of isoindolin-1-ones.

After obtaining the optimal reaction conditions, the universality of the protocol was investigated. EWGs such as Br, CO_2Me, and CF_3 and EDGs such as OMe, ^{t}Bu, and CH_3 in the aryl unit in the amides all survived the reaction condition, yielding good yields of the corresponding (**58**).

Additionally, various N-substituted benzamides were investigated employing optimum reaction conditions, and N-OMe amide remains the best: N-OBn benzamide was also shown to furnish the target molecule but with a somewhat lower yield, whereas $NOC(CH_3)_3$, N-Pr, and N-OPiv benzamide could not.

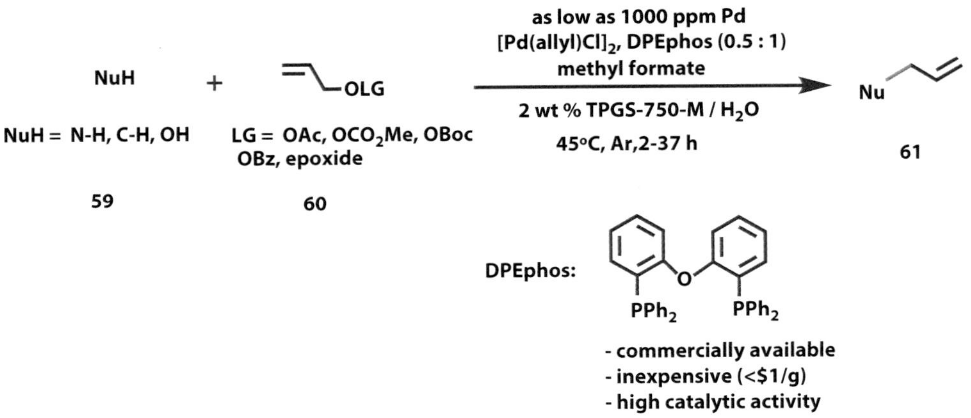

Scheme 2.16 Palladium-catalyzed Tsuji–Trost reaction.

For decades, "Tsuji Trost" reactions[49] have been extensively explored in synthetic chemistry. Owing to their capacity to insert synthetically beneficial allyl group into diverse functional scaffolds employing mild circumstances, as well as the possibility for intramolecular cyclizations, they are important as a tool for complex scaffold construction and catalysis. A maximum number of protocols created for this reaction are carried out in common organic reaction media and use Pd loadings in the 15 mol% range, which is plainly unsustainable.

Lee *et al.*[50] delineated a green procedure (Scheme 2.16) for carrying out this reaction in recyclable H_2O medium and relied on ppm levels (typically 1,000 ppm or 0.1 mol%) of a Pd catalyst available in the market.

Several nitrogen-based nucleophiles were tested, with yields ranging from 70% to 96%. Also, a wide range of carbon nucleophiles, including 1,3-dicarbonyl-bearing molecules such α-nitroketones and malonates, took part easily in the reaction. Allylic O-Bz and cinnamyl-O-Bz derivatives were eventually shown to have significantly higher yields without dimerization.

In order to illustrate the method's potential for application, multigram reactions at 1,000 ppm Pd in H_2O at 45°C were carried out after precomplexing the catalyst in a minimum of toluene. This trial furnished the targeted product in 84% yield.

In 2023, Zhou *et al.*[51] published an article (Scheme 2.17) reporting that $PdCl_2$(Ln@β-CD), a cheap, easily available, and extremely stable catalyst complex, can endure a diverse range of starting materials and produce good yields from Suzuki coupling in the

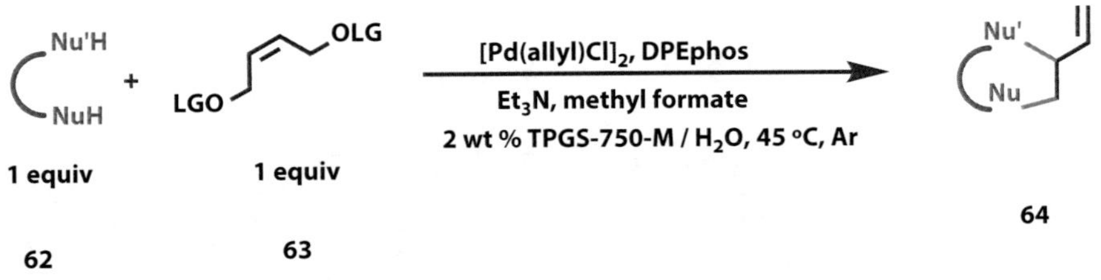

Scheme 2.17 Intramolecular cyclization reactions.

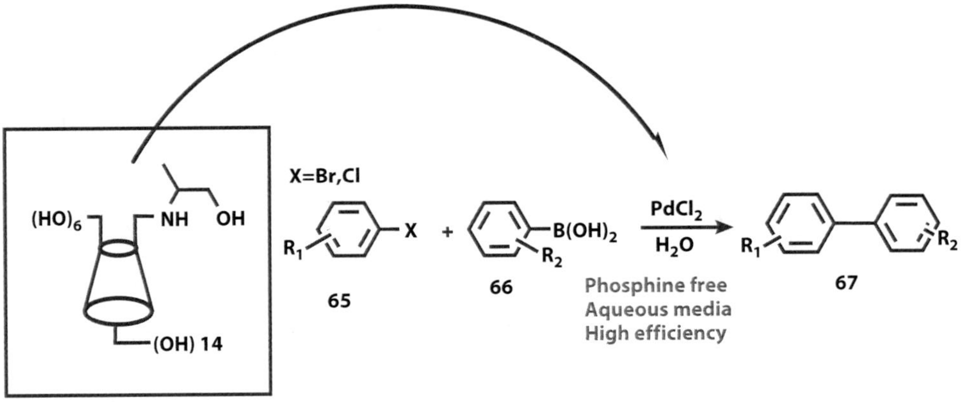

Scheme 2.18 $PdCl_2$(Ln@β-CD)-mediated Suzuki reaction in H_2O.

presence of H_2O and air. The synthesis, crystal structure, and characterization of mono-6-(L-aminopropanol)-deoxy-β-cyclodextrin (Ln@β-CD) were illustrated in the paper.

In synthetic chemistry, "cross-coupling" has been accepted as a favored approach for creating C-C bonds.[52] Cross-coupling reactions catalyzed by palladium are known as highly adaptable protocols for C-C bond production.[53] It has been shown that using Pd-PX_3 species[54] is an outstanding method for Suzuki couplings with very low Pd requirements. As phosphine ligands are costly, poisonous, and typically hydrophobic and air-sensitive, cross-coupling processes under phosphine-free circumstances remain a significant barrier for researchers.[55]

Despite the global popularity of Suzuki coupling reactions for carbon skeleton synthesis, there are enhancements that might be made to make this process even more effective.

The complex's catalytic activity at 0.01 mol% loading was not adequate for the coupling reaction of (**65**) with (**66**), and only 71% of the target compound (**67**) was produced when *p*-nitrochlorobenzene was used as the substrate (Scheme 2.18). After screening various bases, it was discovered that K_3PO_4 and Li_2CO_3 performed best when employing the $PdCl_2$(Ln@-CD) complex (0.01 mol%) in water for 4 hours. The optimal reaction temperature was 90°C.

The transition metal-mediated allylic substitution reactions[49] of a wide variety of nucleophiles via a π-allyl metal intermediate are of major significance because of diverse applications and practicality as an easy and systematic tool for the formation of C-C, C-N, and C-O bonds,[56] which are important in organic chemistry.

Benzathiazole moieties are abundant in a variety of bioactive molecules and contain diverse biological and pharmacological activities.[57] It is of the utmost importance to create

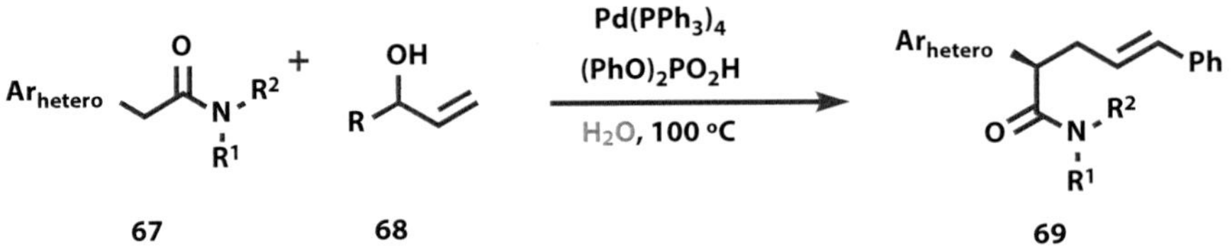

Scheme 2.19 Palladium-catalyzed allylic substitution reaction.

green approaches for the production of molecularly diverse benzothiazoles. In 2019, Pan *et al.*[58] have disclosed (Scheme 2.19) an allylic substitution reaction of (**67**) with (**68**) that is catalyzed by palladium and Brønsted acid and takes place in H_2O. The reactions take place in a highly selective way with a broad substrate scope, and both linear and branched allylic alcohols were tolerated in the reaction.[58]

Model studies revealed that the reaction of ethyl 2-(benzo[d]thiazol-2-yl)acetamide and 1-phenylprop-2-en-1-ol in the presence of 4 mol% $Pd(PPh_3)_4$ as catalyst and 8 mol% $(PhO)_2PO_2H$ as additive in H_2O at 100°C furnished the best yield. An array of allylic alcohols were tested, and all reactions went well with minimal or no undesired products. Allylic alcohols substituted with arene-bearing groups of different electron demand were all tolerated well in this transformation, furnishing good to excellent yields (57%–99%).

Interestingly, halogen atoms at the *o*-, *m*-, or *p*- positions of the Ph ring survived the reaction and produced the required alkylated target molecules in high yields (91%–98%). Furthermore, starting materials with thienyl and fused-ring aryl groups produced the target molecules in 82% and 97% yields, respectively. Although the aliphatic aldehyde-derived branched allylic alcohol may be used in the process, the allylated product was produced in lower yields.

It was pointed out that the reaction could be performed with ease on a gram scale (with 5.0 mmol of benzothiazolylacetamide 1a) and produced a high yield (88%). Furthermore, the resultant compounds could be transformed in a variety of ways as delineated in Scheme 2.20.

The essential scaffold of Evans-type chiral auxiliaries, which have found enormously broad uses in chiral strategies, is oxyazolidinone.[59] Meanwhile, it can be found in a variety of physiologically active chemicals, including the antibiotic linezolid. There are various other metal catalyzed reaction that are already available in the literature. However, most of them suffer from flaws such as the use of toxic organic solvents, exorbitant and exotic catalysts, and harsh reaction conditions. Moreover, due to their decreased nucleophilicity, only a limited number of protocols are found to be suitable for primary propargylic amines.[60] Alkyl-substituted internal triple bonds remain a problematic substrate class for the alkyne moiety. Finally, no procedure has been documented that uses H_2O medium as the solvent.[61] As a result, developing a synthetic process that is broad in substrate scope, economical in

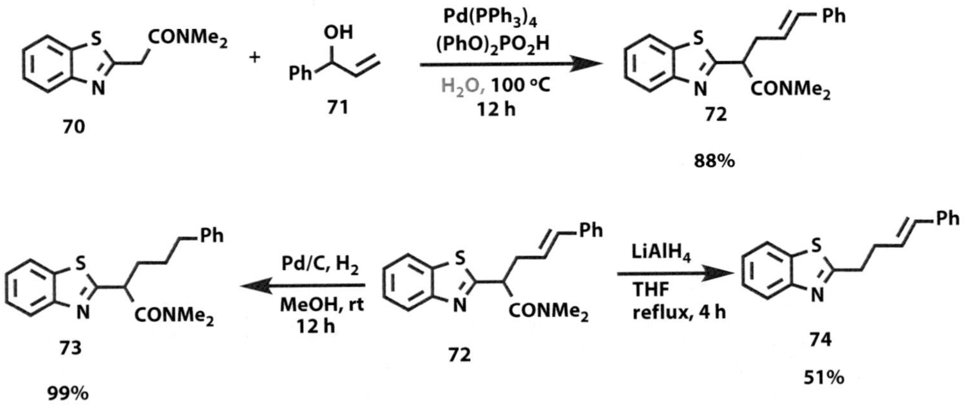

Scheme 2.20 Gram-scale reaction.

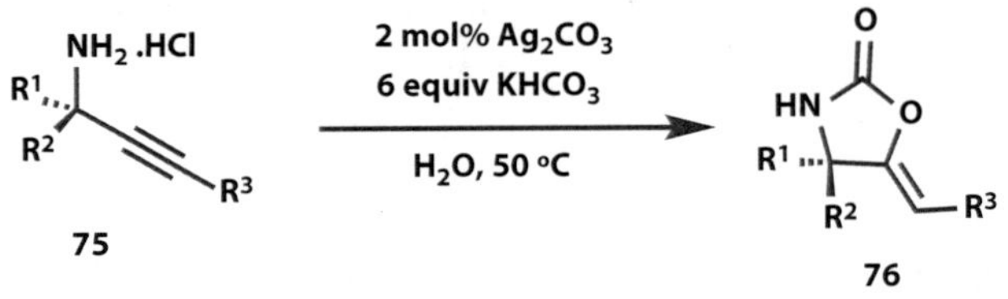

Scheme 2.21 Silver (I)-mediated carboxylative cyclization.

reagent and catalyst costs, and meets the standards of eco-friendly chemistry remains a challenge. To address some of these issues, Qin *et al.*[62] proposed a green synthesis of (**76**), which is carried out in degassed H_2O.

In view of its water solubility and simplicity of handling, (**75**) was selected as the reference substrate for optimization. When 10 equivalents of $NaHCO_3$ were utilized, the reaction progressed smoothly to 72% conversion and 86% yield under 70°C with 10 mol% Ag(I) as the catalyst. The olefin in product (**76**) was identified to be of Z configuration by NMR. The reaction is well tolerated by various functional groups (Scheme 2.21).

N-alkylated amines are useful reactive intermediates with a wide range of uses in the commercial sector.[63] Although a number of transition metal-catalyzed protocols[64] have been fashioned, these strategies typically require reaction at high temperature and in organic solvents and the use of a base, which limits their applicability. A group of scientists led by Dr. Zhuofeng Ke has created a family of bifunctional Cp*Ir complexes (**79a-e**) as represented in Scheme 2.22.[65] These compounds are effective base-free catalysts for the N-alkylation of anilines or sulfonamides with alcohols in H_2O medium and provide a wide range of (hetero)aromatic amines or *p*-methylbenzenesulfonamide in good-to-excellent isolation yields.

For their initial studies, they chose benzyl alcohol and aniline as the model substrate. The control studies in aqueous environment at 110°C to furnish N-benzylaniline revealed that the N-alkylation process did not progress satisfactorily without a catalyst or when a simple $[Cp^*IrCl_2]_2$ catalyst was used.

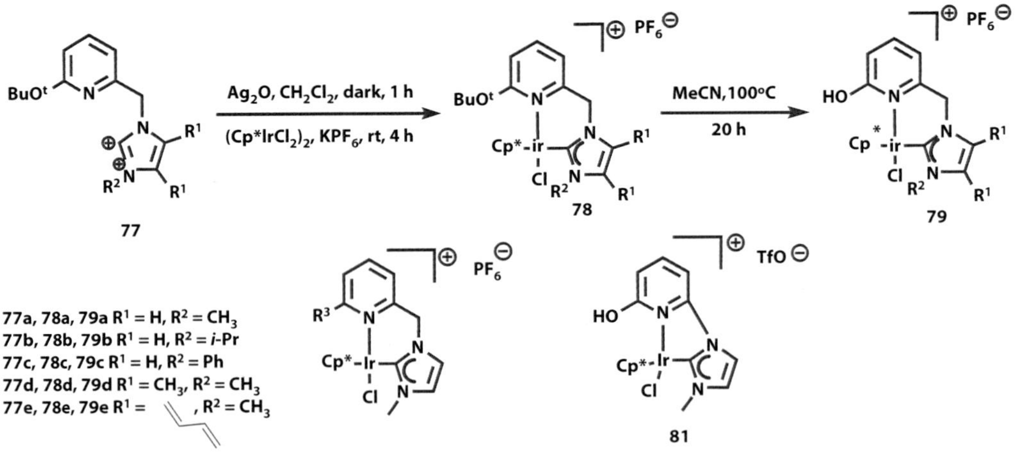

Scheme 2.22 Ir (III) complexes.

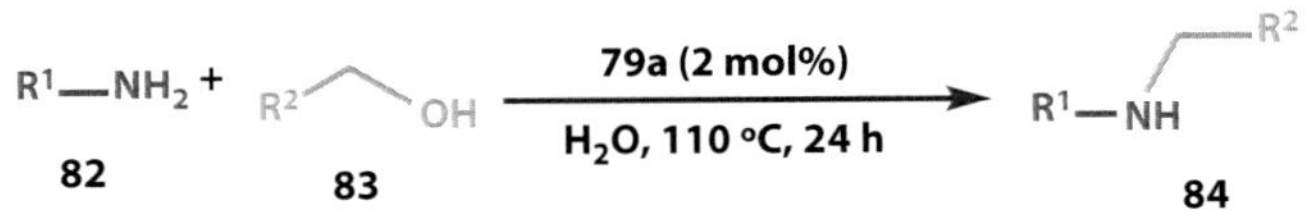

Scheme 2.23 Fabrication of di-alkyl amines.

To elaborate the possible usefulness of this catalytic system, they expanded the scope of the methodology to a larger variety of amines and alcohols. N-alkylation of (**82**) with either electron-donating/withdrawing substituents at the *p*- position with benzylic alcohol led to the target di-alkyl amine compounds with 70%–90% yield. However, when heteroaryl or aliphatic alcohols were utilized, the protocol did not generate satisfactory results, and the related products (**84**) were separated in 25%–70% yields (Scheme 2.23).

The N-alkylation of (**85**) with alcohols was then studied to broaden the generality of the catalytic system (Scheme 2.24). The intended products **86** were obtained in 43%–92% yields by reacting **85** with **83** possessing an electron-donating/withdrawing group.

An important reaction that produces highly valuable spiroindole alkaloids is the Gold(I)-catalyzed1 spirocyclization of indoles that are tethered to alkynes.[66]

Sabat *et al.*[67] demonstrated (Scheme 2.25) dearomative cyclizations of N-propargyl tryptamine and tryptophan derivatives to the corresponding spirocyclic scaffolds in aqueous

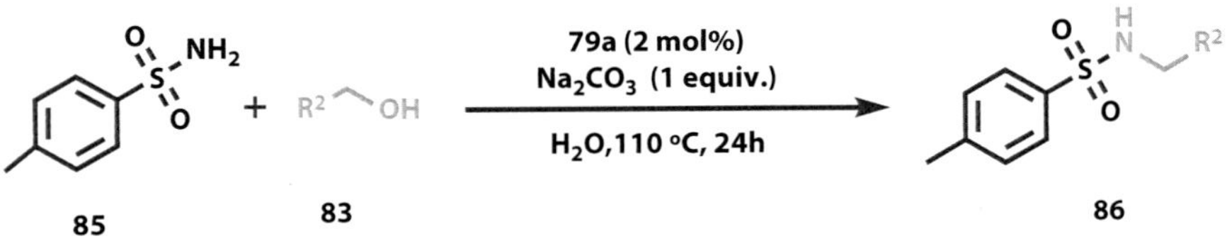

Scheme 2.24 N-alkylation of sulfonamides.

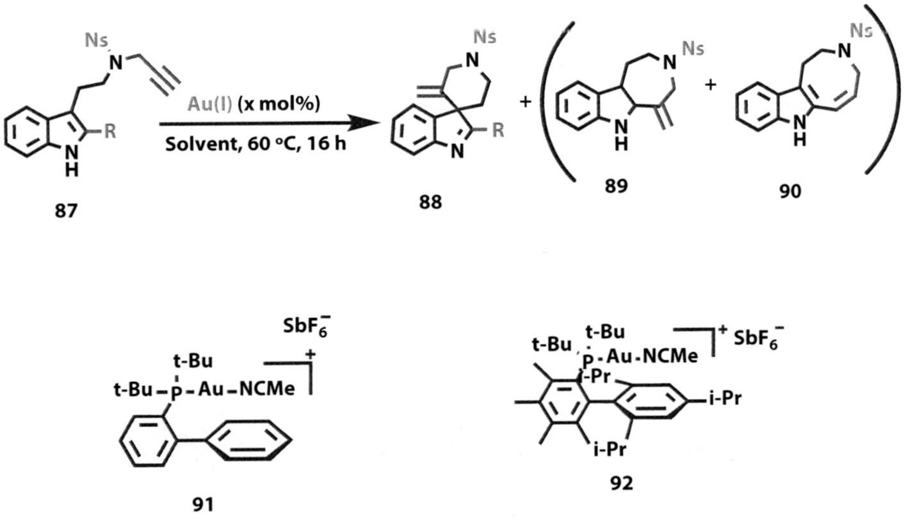

Scheme 2.25 Preparation of spirocyclic derivatives in H_2O.

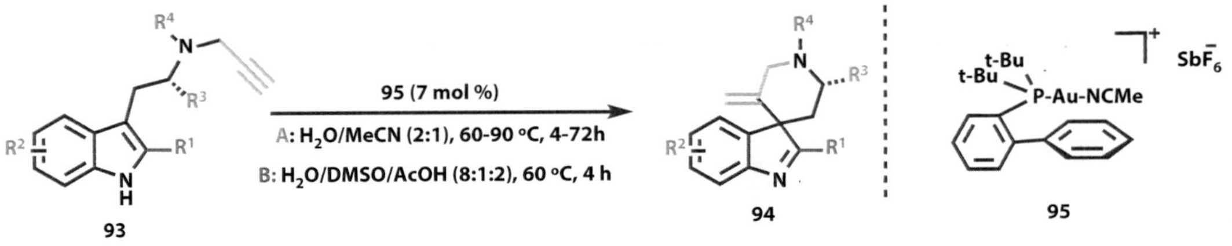

Scheme 2.26 Au-catalyzed spirocyclizations of tryptamine and tryptophan derivatives.

environments under the influence of gold. In particular, the group has shown that the particular technique and isolation of the C2-unsubstituted spiroindolenines are made possible by the use of water medium.[67]

They primarily focused on the complicated task of cyclizing C2-unsubstituted tryptamine **87** in aqueous phase under gold-catalyzed conditions. It was noticed that the reaction produced three separate products, notably the target (**88**) along with (**89**) and (**90**). Although the reaction generated multiple products, the major product was found to be (**88**) when the procedure was carried out in micellar conditions at 60°C. An array of C2-substituted tryptamines (**93**) was then used as a substrate in the cyclizations, using 7 mol% of **95** to warrant full transformations with high yields within a very short time (Scheme 2.26).

Indoles and their derivatives can be found in an extensive range of bioactive naturally occurring moieties.[68] Owing to their good pharmacological effects, (**98**) and (**99**) have gotten a lot of attention. Not unexpectedly, many synthetic techniques for these sulfur-containing indole derivatives exist.[69, 70]

Majority of these procedures necessitate the use of equimolar amount of oxidants and unreal thiolating agents, which produce stoichiometric side products. In this scenario, Zhu *et al.*[71] reported an environmentally friendly protocol for producing highly functionalized indole-based sulfides. Under aqueous micellar conditions, Zhu *et al.*[71] discovered temperature-controlled region-selective benzylic and C3-thiolation of (**96**).

In the beginning, (**96**) and *p*-thiocresol were utilized as benchmark substrates, and the attainability of the thiolation protocol in water was investigated using 5 mol% *p*-dodecylbenzenesulfonic acid (DBSA) as Brønsted acidic surfactant-combined catalyst (Scheme 2.27).

To their amazement, they were able to obtain benzylic thiolation product and C3-thiolation product in 62% and 20% yields, respectively. Then, in water at room temperature, they tested a variety of acid catalysts, among which indium trifluoromethanesulfonate [$In(OTf)_3$] outperformed the others, yielding the required benzylic thiolation product in 89% yield with only a trace quantity of C3-thiolation product.

Next, they investigated the regioselective benzylic sulfenylation of 2-indolylmethanols using the optimal thiolation conditions. It was observed that thiophenols with both

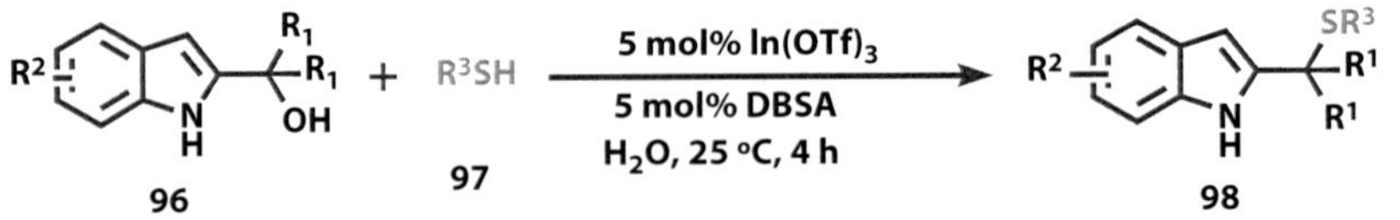

Scheme 2.27 Regioselective $In(OTf)_3$-mediated benzylic sulfenylation of **98**.

electron-donating groups (Me, ^{t}Bu, and MeO) and electron-withdrawing groups (F, Cl, Br, and NHCOMe) were well tolerated. The extent of C3-sulfenylation of 2-indolylmethanols was then investigated using improved circumstances.

The practical applicability of the developed methodology was established by carrying out large-scale reactions of the benzylic and C3-thiolation of (**96**) with (**97**) (Scheme 2.28). After the reaction, the required products were precipitated out and were easily recovered by decanting the water layer without using an organic solvent. The reaction medium containing DBSA and $In(OTf)_3$ can be used five times without much variance in the product yield.

Saini *et al.*[72] have reported (Scheme 2.29) an efficient, straightforward, and eco-friendly protocol for therapeutic core **104**, **105** by using bio-organic catalyst taurine in water medium. The said synthetic methodology has many advantages such as shorter reaction times, excellent yield, high atom economy, convenient isolation of product, usage of water as solvent, and mild reaction condition involving nontoxic, inexpensive, metal-free bioorganic catalyst taurine. Taurine or 2-aminoethane sulfonic acid is a water-soluble organic motif found in various animals and humans in copious amount. Taurine can attain the zwitterionic shape in aqueous solution, which has been attributed to its catalytic activity and, in recent times, it has emerged as one of the sought-after green catalysts for various organic cascade transformations.[73] During the screening studies for the synthesis of 105 moieties, authors performed a 3-component one-pot reaction among (**100**), (**101**), and (**102**), and the reaction was carried out in various solvents and in various concentrations of taurine. It was observed that 28 mol% (0.035 g) of taurine in refluxing H_2O produced the best yield. After optimization studies, different aliphatic and aromatic aldehydes with

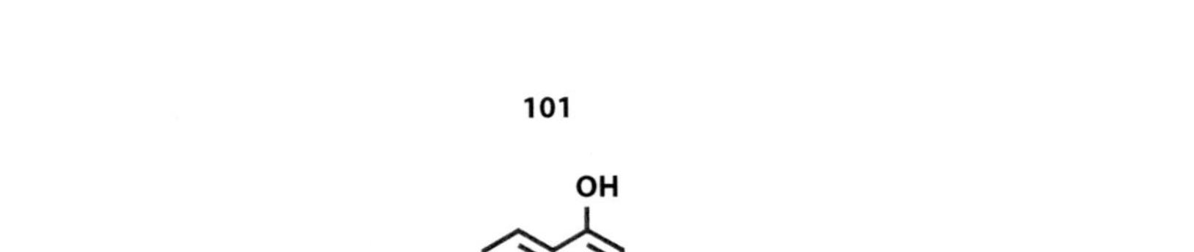

Scheme 2.28 Regioselective $In(OTf)_3$-mediated C3-sulfenylation of **96**.

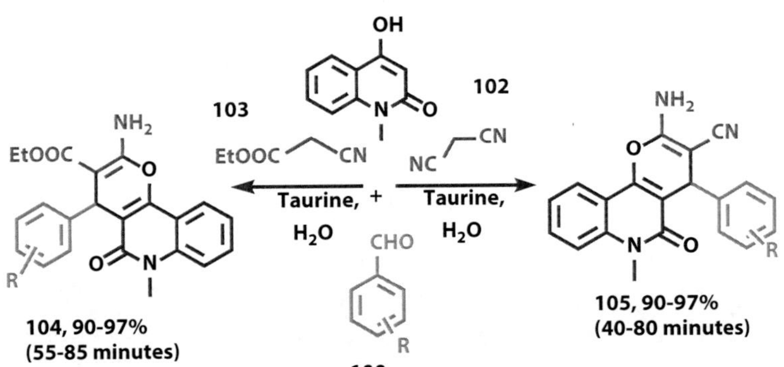

Scheme 2.29 Manufacture of pyrano[3,2-c]quinolone scaffolds.

electron-donating/withdrawing groups were utilized to obtain derivatives of desired scaffolds in excellent yields.

2.2 Conclusion

In summary, this book chapter has illustrated various articles published in the last 5 years describing various organic transformations that are taking place in H_2O medium. One reported biocatalysis has been elaborated as well. Although significant advancements are being made every day in the field of "Green Chemistry," there is still a huge scope for the scientific community. To attain a sustainable future, thus, there is an urgent requirement to focus on the development of diverse eco-friendly protocols especially those that can be carried out in H_2O. Enzyme catalysis needs to be explored extensively as well.

References

1. Ganesh, K.N., Zhang, D., Miller, S.J., Rossen, K., Chirik, P.J., Kozlowski, M.C., Zimmerman, J.B., Brooks, B.W., Savage, P.E., Allen, D.T., Voutchkova-Kostal, A.M., Green Chemistry: A Framework for a Sustainable Future. *Org. Process Res. Dev.*, 25, 7, 1455–1459, 2021.
2. Anastas, P. and Eghbalia, N., Green Chemistry: Principles and Practice. *Chem. Soc. Rev., 39*, 301–312, 2010.
3. Clark, J.H., Green chemistry: challenges and opportunities. *Green Chem., 1*, 1–8, 1999.
4. Cortes-Clerget, M., Yu, J., Kincaid, J.R.A., Walde, P., Gallou, F., Lipshutz, B.H., Water as the reaction medium in organic chemistry: from our worst enemy to our best friend. *Chem. Sci., 12*, 4237–4266, 2021.
5. Kitanosono, T., Koichiro Masuda, K., Xu, P., Kobayashi, S., Catalytic Organic Reactions in Water toward Sustainable Society. *Chem. Rev.*, 118, 2, 679–746, 2018.
6. Sheldon, R.A. and Woodley, J.M., Role of Biocatalysis in Sustainable Chemistry. *Chem. Rev.*, 118, 2, 801–838, 2018.
7. Yi, D., Bayer, T., Badenhorst, C.P.S., Wu, S., Doerr, M., Höhne, M., Bornscheuer, U.T., Recent trends in biocatalysis. *Chem. Soc. Rev., 50*, 8003–8049, 2021.
8. Hanefeld, U., Hollmann, F., Paul, C.E., Biocatalysis making waves in organic chemistry. *Chem. Soc. Rev.*, 51, 594–627, 2022.
9. Wu, S., Snajdrova, R.S., Moore, J.C., Baldenius, K., Bornscheuer, U.T., Biocatalysis: Enzymatic Synthesis for Industrial Applications. *Angew. Chem., Int. Ed.*, 60, 88–119, 2021.
10. Morcillo, S.P., Radical-Promoted C–C Bond Cleavage: A Deconstructive Approach for Selective Functionalization. *Angew. Chem., Int. Ed., 58*, 14044–14054, 2019.
11. Yu, X.-Y., Chen, J.-R., Xiao, W.-J., Visible Light-Driven Radical-Mediated C–C Bond Cleavage/Functionalization in Organic Synthesis. *Chem. Rev., 121*, 506–561, 2021.
12. Zhou, W., Long, Y., Xiang, H., Xu, B., Zhou, X., Iron-Catalyzed C–C Bond Cleavage of Oximes for Direct Coupling of Benzothiazole in Water. *J. Org. Chem., 88*, 4875–4879, 2023.
13. (a) Wityak, J. and Chan, J.B., Synthesis of 1,3-Diynes Using Palladium-Copper Catalysis. *Synth. Commun., 21*, 977–979, 1991. (b) Shi, W., Luo, Y., Luo, X., Chao, L., Zhang, H., Wang, J., Lei, A., Investigation of an Efficient Palladium-Catalyzed C(sp)-C(sp) Cross-Coupling Reaction Using Phosphine-Olefin Ligand: Application and Mechanistic Aspects. *J. Am. Chem. Soc., 130*, 14713–14720, 2008. (c) Weng, Y., Cheng, B., He, C., Lei, A., Rational Design of a Palladium-Catalyzed Csp–Csp Cross-Coupling Reaction Inspired by Kinetic Studies. *Angew. Chem., Int. Ed.*,

51, 9547–9551, 2012. (d) Xu, Z., Ai, J., Cai, M., An Efficient Heterogeneous Palladium(0)-Catalysed Cross-Coupling Between 1-Bromoalkynes and Terminal Alkynes Leading to Unsymmetrical 1,3-Diynes. *J. Chem. Res.*, 42, 133–137, 2018. (e) Nishihara, Y., Ikegashira, K., Hirabayashi, K., Ando, J.-I., Mori, A., Hiyama, T., Coupling Reactions of Alkynylsilanes Mediated by a Cu(I) Salt: Novel Syntheses of Conjugate Diynes and Disubstituted Ethynes. *J. Org. Chem.*, *65*, 1780–1787, 2000.

14. Mukherjee, N., Kundu, D., Ranu, B.C.A., Co-operative Ni–Cu System for Csp–Csp and Csp–Csp2 Cross-Coupling Providing a Direct Access to Unsymmetrical 1,3-Diynes and Enzynes. *Chem. Commun.*, *50*, 15784–15787, 2014.
15. (a) Chinta, B.S. and Baire, B., A Systematic Study on the Cadiot–Chodkiewicz Cross Coupling Reaction for the Selective and Efficient Synthesis of Hetero-diynes. *RSC Adv.*, *6*, 54449–54455, 2016. (b) Marino, J.P. and Nguyen, H.N., Bulky Trialkylsilyl Acetylenes in the Cadiot-Chodkiewicz Cross-Coupling Reaction. *J. Org. Chem.*, 67, 6841–6844, 2002. (c) Knutson, P.C., Fredericks, H.E., Ferreira, E.M., Synthesis of 1,3-Diynes via Cadiot–Chodkiewicz Coupling of Volatile, *in Situ* Generated Bromoalkynes. *Org. Lett.*, 20, 6845–6849, 2018. (d) Asha, S., Anjana, S., Thomas, A.M., Thomas, M.E., Rohit, K.R., Krishnan, K.K., Ujwaldev, S.M., Anilkumar, G., A Convenient Route to 1,3-Diynes Using Ligand-Free Cadiot–Chodkiewicz Coupling Reaction at Room Temperature under Aerobic Conditions. *Synth. Commun.*, 49, 256–265, 2019. (e) Wang, S., Yu, L., Li, P., Meng, L., Wang, L., Copper(I) Iodide Catalyzed Cross-Coupling Reaction of Terminal Alkynes with 1-Bromoalkynes: A Simple Synthesis of Unsymmetrical Buta-1,3-diynes. *Synthesis*, 2011, 1541–1546, 2011. (f) Jiang, H.-F. and Wang, A.-Z., Copper-Catalyzed Cross- Coupling Reactions of Bromoalkynols with Terminal Alkynes in Supercritical Carbon Dioxide. *Synthesis*, 2007, 1649–1654, 2007.
16. Krishnan, K.K., Ujwaldev, S.M., Thankachan, A.P., Harry, N.A., Anilkumar, G.A., Novel Zinc-Catalyzed Cadiot-Chodkiewicz Cross-Coupling Reaction of Terminal Alkynes with 1-Bromoalkynes in Ethanol Solvent. *Mol. Catal.*, *440*, 140–147, 2017.
17. Liao, Y.-A., Peng, W.-S., Liu, L.-J., Ye, T. –Y., Fu, J.-H., Chan, Y.-T., Tsai, F.-Y., Iron-Catalyzed Cadiot–Chodkiewicz Coupling with High Selectivity in Water under Air. *J. Org. Chem.*, 87, 13698–13707, 2022.
18. (a) Sasaki, Y., Kato, D., Boger, D.L., Asymmetric Total Synthesis of Vindorosine, Vindoline, and Key Vinblastine Analogues. *J. Am. Chem. Soc.*, 132, 13533, 2010. (b) Mori, M., Nakanishi, M., Kajishima, D., Sato, Y., A Novel and General Synthetic Pathway to Strychnos Indole Alkaloids: Total Syntheses of (−)-Tubifoline, (−)-Dehydrotubifoline, and (−)-Strychnine Using Palladium-Catalyzed Asymmetric Allylic Substitution. *J. Am. Chem. Soc.*, 125, 9801, 2003. (c) Dounay, A.B., Humphreys, P.G., Overman, L.E., Wrobleski, Total Synthesis of the Strychnos Alkaloid (+)-Minfiensine: Tandem Enantioselective Intramolecular Heck–Iminium Ion Cyclization A.D., *J. Am. Chem. Soc.*, 130, 5368, 2008. (d) He, W.-G., Hu, J.-D., Wang, P.-Y., Chen, L., Ji, K., Yang, S.-Y., Li, Y., Xie, Z.-L., Xie, W.-Q., Highly Enantioselective Tandem Michael Addition of Tryptamine-Derived Oxindoles to Alkynones: Concise Synthesis of Strychnos Alkaloids. *Angew. Chem., Int. Ed.*, 57, 3806, 2018. (e) Moreno, J., Picazo, E., Morrill, L.A., Smith, J.M., Garg, N.K., Enantioselective Total Syntheses of Akuammiline Alkaloids (+)-Strictamine, (−)-2(S)-Cathafoline, and (−)-Aspidophylline A. *J. Am. Chem. Soc.*, 138, 1162, 2016. (f) Wang, C., Zhang, S.-J., Wang, Y., Huang, S.-H., Hong, R., Total synthesis of strictamine: a tutorial for novel and efficient synthesis. *Org. Chem. Front.*, 5, 447, 2018.
19. (a) Xu, Z.-R., Wang, Q., Zhu, J.-P., Metamorphosis of cycloalkenes for the divergent total synthesis of polycyclic indole alkaloids. *Chem. Soc. Rev.*, 47, 7882, 2018. (b) Saya, J.M., Ruijter, E., Orru, R.V.A., Total Synthesis of Aspidosperma and Strychnos Alkaloids through Indole Dearomatization. *Chem. Eur. J.*, 25, 8916, 2019. (c) Bonjoch, J. and Sole, D., Synthesis of Strychnine. *Chem. Rev.*, 100, 3455, 2000. (d) Kizil, M., Patro, B., Callaghan, O., Murphy, J.A.,

Hursthouse, M.B., Hibbs, D., Tandem Radical Cyclizations on Iodoaryl Azides: Synthesis of the Core Tetracycle of Aspidosperma Alkaloids. *J. Org. Chem.*, 64, 7856, 1999. (e) Zhou, S.-Z., Bommezijn, S., Murphy, J. A., Formal Total Synthesis of (±)-Vindoline by Tandem Radical Cyclization. *Org. Lett.*, 4, 443, 2002. (f) Mori, M., Nakanishi, M., Kajiishima, D., Sato, Y., A Novel and General Synthetic Pathway to Strychnos Indole Alkaloids: Total Syntheses of (−)-Tubifoline, (−)-Dehydrotubifoline, and (−)-Strychnine Using Palladium-Catalyzed Asymmetric Allylic Substitution. *J. Am. Chem. Soc.*, 125, 9801, 2003. (g) Boonsombat, J., Zhang, H., Chughtai, M.J., Hartung, J., Padwa, A., A General Synthetic Entry to the Pentacyclic Strychnos Alkaloid Family, Using a [4 + 2]-Cycloaddition/Rearrangement Cascade Sequence. *J. Org. Chem.*, 73, 3539, 2008. (h) Mizoguchi, H., Oikawa, H., Oguri, H., *Nat. Chem.*, 6, 57, 2014. (i) Wang, Y.-S., Lin, J.-S., Wang, X.-Y., Wang, G.-H., Zhang, X.-H., Yao, B., Zhao, Y.-D., Yu, P.-F., Lin, B., Liu, Y.-X., Cheng, M.-S., Brønsted Acid-Catalyzed Tandem Cyclizations of Tryptamine-Ynamides Yielding 1H-Pyrrolo[2,3-d]carbazole Derivatives. *Chem. Eur. J.*, 24, 4026, 2018. (j) Du, K., Yang, H., Guo, P., Feng, L., Xu, G.-Q., Zhou, Q.-H., Chung, L.W., Tang, W.-J., Efficient syntheses of (−)-crinine and (−)-aspidospermidine, and the formal synthesis of (−)-minfiensine by enantioselective intramolecular dearomative cyclization. *Chem. Sci.*, 8, 6247, 2017. (k) Feng, L.-W., Ren, H., Xiong, H., Wang, P., Wang, L.-J., Tang, Y., Reaction of Donor-Acceptor Cyclobutanes with Indoles: A General Protocol for the Formal Total Synthesis of (±)-Strychnine and the Total Synthesis of (±)-Akuammicine. *Angew. Chem., Int. Ed.*, 56, 3055, 2017. (l) Pereira, J., Barlier, M., Guillou, C., Formal Total Syntheses of Aspidosperma Alkaloids via a Novel and General Synthetic Pathway Based on an Intramolecular Heck Cyclization. *Org. Lett.*, 9, 3101, 2007. (m) Kuehne, M.E. and Xu, F., Total synthesis of strychnan and aspidospermatan alkaloids. 3. The total synthesis of (.+-.)-strychnine. *J. Org. Chem.*, 58, 7490, 1993. (n) Guo, B.-B., Ge, L., Huang, G.-X., Zhao, L., Chen, J., Cao, W.-G., Wu, X.-Y., Expedient Introduction of β-Methoxyacrylate Unit onto 3-Substituted Indoles and Application of the Resulting Indole-Dienes in Organocascade toward Indolino-Polycyclics. *Adv. Synth. Catal.*, 357, 4013, 2015. (o) Chen, W., Yang, X.-D., Tan, W.-Y., Zhang, X.-Y., Liao, X.-L., Zhang, H.-B., Total Synthesis of (−)-Vindorosine. *Angew. Chem., Int. Ed.*, 56, 12327, 2017. (p) Jones, S. B., Simmons, B., Mastracchio, A., MacMillan, D.W.C., Collective synthesis of natural products by means of organocascade catalysis. *Nature*, 475, 183, 2011. (q) Laforteza, B. N., Pickworth, M., MacMillan, D.W.C., Enantioselective Total Synthesis of (−)-Minovincine in Nine Chemical Steps: An Approach to Ketone Activation in Cascade Catalysis. *Angew. Chem., Int. Ed.*, 52, 11269, 2013. (r) Zhu, J., Cheng, Y.-J., Kuang, X.-K., Wang, L.-J., Zheng, Z.-B., Tang, Y., Highly Efficient Formal [2+2+2] Strategy for the Rapid Construction of Polycyclic Spiroindolines: A Concise Synthesis of 11-Demethoxy-16-epi-myrtoidine. *Angew. Chem., Int. Ed.*, 55, 9224, 2016.

20. Jiang, S., Wen-Bin Cao, W.-B., Li, H.-Y., Xiao-Ping Xu, X.-P., Ji, S.-J., Convenient synthesis of spiroindolenines from tryptamine-derived isocyanides and organic azides by cobalt catalysis in pure water. *Green Chem.*, *23*, 2619–2623, 2021.
21. (a) Taban, I.M., Elshihawy, H.E.A.E., Torun, B., Zucchini, B., Williamson, C.J., Altuwairigi, D., Ngu, A.S.T., McLean, K.J., Levy, C.W., Sood, S., Marino, L.B., Munro, A.W., Carvalho, L.P.S., Simons, C., Novel Aryl Substituted Pyrazoles as Small Molecule Inhibitors of Cytochrome P450 CYP121A1: Synthesis and Antimycobacterial Evaluation. *J. Med. Chem.*, 60, 10257, 2017. (b) Wang, Y., Zhi, Y., Jin, Q., Lu, S., Lin, G., Yuan, H., Yang, T., Wang, Z., Yao, C., Ling, J., Guo, H., Li, T., Jin, J., Li, B., Zhang, L., Chen, Y., Lu, T., Discovery of 4-((7H-Pyrrolo[2,3-d]pyrimidin-4-yl)amino)-N-(4-((4-methylpiperazin-1-yl)methyl)phenyl)-1H-pyrazole-3-carboxamide (FN-1501), an FLT3- and CDK-Kinase Inhibitor with Potentially High Efficiency against Acute Myelocytic Leukemia. *J. Med. Chem.*, 61, 1499, 2018. (c) Brown, A.W., Fisher, M., Tozer, G.M., Kanthou, C., Harrity, J. P. A., Sydnone Cycloaddition Route to Pyrazole-Based Analogs of Combretastatin A4. *J. Med. Chem.*, 59, 9473, 2016.

22. You, G., Wang, K., Wang, X., Wang, G., Sun, J., Duan, G., Xia, C., Visible-Light-Mediated Nickel(II)-Catalyzed C–N Cross-Coupling in Water: Green and Regioselective Access for the Synthesis of Pyrazole-Containing Compounds. *Org. Lett.*, *20*, 4005–4009, 2018.
23. (a) Xu, Y., Shi, D., Wang, X., Yu, S., Pu, L., Development of Aldehyde-Based Fluorescent Probes for Highly Selective Recognition of 1,3-Diaminopropane. *Eur. J. Org. Chem.*, 2017, 4990–4994, 2017. (b) Hu, L., Yu, S., Wang, Y., Yu, X., Pu, L., Enhanced Fluorescence of 3,3′-Diformyl BINOL by Functional Secondary Amines. *Org. Lett.*, 19, 3779–3782, 2017.
24. (a) Wang, Y.-C., Xie, Y.-Y., Qu, H.-E., Wang, H.-S., Pan, Y.-M., Huang, F.-P., Ce(OTf)3-Catalyzed [3 + 2] Cycloaddition of Azides with Nitroolefins: Regioselective Synthesis of 1,5-Disubstituted 1,2,3-Triazoles. *J. Org. Chem.*, 79, 4463–4469, 2014. (b) Zhu, A., Xing, X., Wang, S., Yuan, D., Zhu, G., Geng, M., Guo, Y., Zhang, G., Li, L., Multi-component syntheses of diverse 5-fluoroalkyl-1,2,3-triazoles facilitated by air oxidation and copper catalysis. *Green Chem.*, 21, 3407–3412, 2019. (c) Huang, K., Sheng, G., Lu, P., Wang, Y., From 1-Sulfonyl-4-aryl-1,2,3-triazoles to 1-Allenyl-5-aryl-1,2,3-triazoles. *J. Org. Chem.*, 82, 5294–5300, 2017.
25. (a) Tian, Y., Liu, Z., Liu, J., Huang, B., Kang, D., Zhang, H., Clercq, E.D., Mans, D.D., Pannecouque, C., Lee, K.H., Chen, C.H., Zhan, P., Liu, X., Targeting the entrance channel of NNIBP: Discovery of diarylnicotinamide 1,4-disubstituted 1,2,3-triazoles as novel HIV-1 NNRTIs with high potency against wild-type and E138K mutant virus. *Eur. J. Med. Chem.*, 151, 339–350, 2018. (b) Mohammed, I., Kummetha, I.R., Singh, G., Sharova, N., Lichinchi, G., Dang, J., Stevenson, M., Rana, T.M., 1,2,3-Triazoles as Amide Bioisosteres: Discovery of a New Class of Potent HIV-1 Vif Antagonists. *J. Med. Chem.*, 59, 7677–7682, 2016.
26. Buckle, D.R., Outred, D.J., Rockell, C.J.M., Smith, H., Spicer, B.A., Studies on v-triazoles. 7. Antiallergic 9-oxo-1H, 9H-benzopyrano [2, 3-d]-v-triazoles. *J. Med. Chem.*, 26, 251–254, 1983.
27. (a) Duan, T., Fan, K., Fu, Y., Zhong, C., Chen, X., Peng, T., Qin, J., Triphenylamine-based organic dyes containing a 1,2,3-triazole bridge for dye-sensitized solar cells via a 'Click' reaction. *Dyes Pigm.*, 94, 28–33, 2012. (b) Debia, N.P., Saraiva, M.T., Martins, B.S., Beal, R., Goncalves, P.F.B., Rodembusch, F.S., Alves, D., Ludtke, D.S., Synthesis of Amino Acid-Derived 1,2,3-Triazoles: Development of a Nontrivial Fluorescent Sensor in Solution for the Enantioselective Sensing of a Carbohydrate and Bovine Serum Albumin Interaction. *J. Org. Chem.*, 83, 1348–1357, 2018. (c). Ghosh, D., Rhodes, S., Hawkins, K., Winder, D., Atkinson, A., Ming, W., Padgett, C., Orvis, J., Aiken, K., Landge, S., A simple and effective 1,2,3-triazole based "turn-on" fluorescence sensor for the detection of anions. *New J. Chem.*, 39, 295–303, 2015. (d) Meisner, Q.J., Accardo, J.V., Hu, G., Clark, R.J., Jiang, D.-e.; Zhu, L. Fluorescence of Hydroxyphenyl-Substituted "Click" Triazoles. *J. Phys. Chem. A*, 122, 2956–2973, 2018. (e) Sinopoli, A., Black, F.A., Wood, C.J., Gibson, E.A., Elliott, P., II, Investigation of a new bis(carboxylate) triazole-based anchoring ligand for dye solar cell chromophore complexes. *Dalton Trans.*, 46, 1520–1530, 2017.
28. Emami, M., Bikas, R., Noshiranzadeh, N., Kozakiewicz, A., Lis, T., Cu(II)-Hydrazide Coordination Compound Supported on Silica Gel as an Efficient and Recyclable Heterogeneous Catalyst for Green Click Synthesis of β-Hydroxy-1,2,3-triazoles in Water. *ACS Omega*, *5*, 13344–13357, 2020.
29. Castillo, J.-C., Bravo, N.-F., Tamayo, L.-V., Mestizo, P.-D., Hurtado, J., Macías, M., Portilla, J., Water-Compatible Synthesis of 1,2,3-Triazoles under Ultrasonic Conditions by a Cu(I) Complex-Mediated Click Reaction. *ACS Omega*, *5*, 30148–30159, 2020.
30. (a) Ito, H., Segawa, Y., Murakami, K., Itami, K., Polycyclic Arene Synthesis by Annulative π-Extension. *J. Am. Chem. Soc.*, 141, 3– 10, 2019. (b) Miao, Q., *Polycyclic Arenes and Heteroarenes: Synthesis, Properties, and Applications*, Wiley, 2015. (c) Mamane, V., Hannen, P., Furstner, A., Synthesis of Phenanthrenes and Polycyclic Heteroarenes by Transition-metal Catalyzed Cycloisomerization Reactions. *Chem. - Eur. J.*, 10, 4556–4575, 2004. (d) Wang, X.G., Zhang, J., Chen, D., Wang, B., Yang, X.F., Ma, Y.M., Szostak, M., Rh(III)-Catalyzed C–H Amidation of

2-Arylindoles with Dioxazolones: A Route to Indolo[1,2-c]quinazolines. *Org. Lett.*, 21, 7038–7043, 2019. (e) Zhang, J., Zhuang, Y.Y., Ma, Y.M., Yang, X.F., Szostak, M., Palladium-Catalyzed Synthesis of Benzothiophenes via Cross-Dehydrogenative Coupling of 4-Arylthiocoumarins and Pyrones. *Adv. Synth. Catal.*, 361, 5709–5714, 2019.

31. (a) Boger, D.L. and Machiya, K., Total Synthesis of (+)-Duocarmycin SA. *J. Am. Chem. Soc.*, 114, 10056–10058, 1992. (b) Yamada, K., Kurokawa, T., Tokuyama, H., Fukuyama, T., Total Synthesis of the Duocarmycins. *J. Am. Chem. Soc.*, 125, 6630–6631, 2003. (c) MacMillan, K.S. and Boger, D.L., Fundamental Relationships between Structure, Reactivity, and Biological Activity for the Duocarmycins and CC-1065. *J. Med. Chem.*, 52, 5771–5780, 2009. (d) Wu, P.L., Hsu, Y.L., Jao, C.W., Indole Alkaloids from Cephalanceropsis gracilis. *J. Nat. Prod.*, 69, 1467–1470, 2006. (e) Lee, J.H., So, J.H., Jeon, J.H., Hoi, E.B., Lee, Y.R., Chang, Y.T., Kim, C.H., Bae, M.A., Ahn, J.H., Synthesis of a New Fluorescent Small Molecule Probe and its Use for *in Vivo* Lipid Imaging. *Chem. Commun.*, 47, 7500–7502, 2011. (f) Ferreira, E.S.B., Hulme, A.N., McNab, H., Quye, A., The Natural Constituents of Historical Textile Dyes. *Chem. Soc. Rev.*, 33, 329–336, 2004. (g) Randell, N.M., Radford, C.L., Yang, J., Quinn, J., Hou, D., Li Y.; Kelly, T. L. Effect of Acceptor Unit Length and Planarity on the Optoelectronic Properties of Isoindigo-Thiophene Donor- Acceptor Polymers. *Chem. Mater.*, 30, 4864–4873, 2018. (h) Ganguly, A., Zhu, J., Kelly, T.L., Effect of Cross-Conjugation on Derivatives of Benzoisoindigo, an Isoindigo Analogue with an Extended π-System. *J. Phys. Chem. C*, 121, 9110–9119, 2017. For reviews on 3-oxindoles (3- indolinones), see: (i) Joule, J.A. and Mills, K., *Heterocyclic Chemistry*, Wiley-Blackwell, 2010. (j) Sundberg, R.J., *The Chemistry of Indoles*, Academic Press, 1970, For further representative methods for oxindole synthesis, see: (k) Hennessy, E.J. and Buchwald, S.L., Synthesis of Substituted Oxindoles from α-Chloroacetanilides *via* Palladium- Catalyzed C–H Functionalization. *J. Am. Chem. Soc.*, 125, 12084–12085, 2003. (l) Klein, J.E.M.N., Perry, A., Pugh, D.S., Taylor, R.J.K., First C-H Activation Route to Oxindoles using Copper Catalysis. *Org. Lett.*, 12, 3446–3449, 2010. (m) Wei, W.T., Zhou, M.B., Fan, J.H., Liu, W., Song, R.J., Liu, Y., Hu, M., Xie, P., Li, J.H., Synthesis of Oxindoles by Iron-Catalyzed Oxidative 1,2-Alkylarylation of Activated Alkenes with an Aryl C(sp2)-H Bond and a C(sp3)-H Bond Adjacent to a Heteroatom. *Angew. Chem., Int. Ed.*, 52, 3638–3641, 2013. (n) Li, Y., Wang, K., Ping, Y., Wang, Y., Kong, W., Nickel-Catalyzed Domino Heck Cyclization/Suzuki Coupling for the Synthesis of 3,3-Disubstituted Oxindoles. *Org. Lett.*, 20, 921–924, 2018. (o) Cao, Z.Y., Zhou, F., Zhou, J., Development of Synthetic Methodologies via Catalytic Enantioselective Synthesis of 3, 3-Disubstituted Oxindoles. *Acc. Chem. Res.*, 51, 1443–1454, 2018, and references cited therein.

32. For selected reviews on C–H activation, see: (a) Yu, J.Q., *Science of Synthesis: Catalytic Transformations via C–H Activation*, Thieme, 2015. (b) Davies, H.M.L. and Morton, D., Recent Advances in C–H Functionalization. *J. Org. Chem.*, 81, 343–350, 2016. (c) Rossi, R., Bellina, F., Lessi, M., Manzini, C., Cross-coupling of Heteroarenes by C-H Functionalization: Recent Progress towards Direct Arylation and Heteroarylation Reactions involving Heteroarenes Containing One Heteroatom. *Adv. Synth. Catal.*, 356, 17–117, 2014. (d) Rouquet, G. and Chatani, N., Catalytic Functionalization of C(sp2)–H and C(sp3)–H Bonds by Using Bidentate Directing Groups. *Angew. Chem., Int. Ed.*, 52, 11726–11743, 2013. (e) Yeung, C.S. and Dong, V.M., Catalytic Dehydrogenative Cross-coupling: forming Carbon–Carbon Bonds by Oxidizing Two Carbon– Hydrogen Bonds. *Chem. Rev.*, 111, 1215–1292, 2011. (f) Lyons, T. and Sanford, M., Palladium- catalyzed Ligand directed C–H Functionalization Reactions. *Chem. Rev.*, 110, 1147–1169, 2010. (g) Ackermann, L., Vicente, R., Kapdi, A.R., Transitionmetal-catalyzed Direct Arylation of (Hetero)Arenes by C-H bond Cleavage. *Angew. Chem., Int. Ed.*, 48, 9792–9826, 2009. (h) Gandeepan, P., Muller, T., Zell, D., Cera, G., Warratz, S., Ackermann, L., 3d Transition Metals for C–H Activation. *Chem. Rev.*, 119, 2192–2452, 2019. (i) Sambiagio, C., Schonbauer, D., Blieck, R., Dao-Huy, T., Pototschnig, G., Schaaf, P., Wiesinger, T., Farooq Zia,

M., Wencel- Delord, J., Besset, T., Maes, B.U.W., Schnurch, M.A., Comprehensive Overview of Directing Groups Applied in Metal catalysed C–H Functionalisation Chemistry. *Chem. Soc. Rev.*, *47*, 6603–6743, 2018.

33. Wang, X., Zhang, J., He, Y., Chen, D., Wang, C., Yang, F., Wang, W., Ma, Y., Szostak, M.Ruthenium(II)-Catalyzed Ortho-C–H Alkylation of Naphthylamines with Diazo Compounds for Synthesis of 2,2-Disubstituted π-Extended 3-Oxindoles in Water. *Org. Lett.*, *22*, 5187–5192, 2020.
34. (a) Gangopadhyay, A., Natural Flavans and (Iso)Flavanones with Anticancer Activity: A Review. *Curr. Org. Chem.*, 25, 1028–1046, 2021. (b) Han, S., Li, L., Song, S., Daphne Giraldii Nitsche (Thymelaeaceae): Phytochemistry, Pharmacology and Medicinal Uses. *Phytochemistry*, 171, 12231, 2020. (c) Cazarolli, L., Zanatta, L., Alberton, E., Bonorino Figueiredo, M.S., Folador, P., Damazio, R., Pizzolatti, M., Barreto Silva, F.R., Flavonoids: Prospective Drug Candidates. *Mini Rev. Med. Chem.*, 8, 1429–1440, 2008. (d) Saini, K.S. and Ghosal, S., Naturally Occurring Flavans Unsubstituted in the Heterocyclic Ring. *Phytochemistry*, 23, 2415–2421, 1984.
35. Caleffi, G.S., Demidoff, F.C., Nájera, C., Costa, P.R.R., Asymmetric Hydrogenation and Transfer Hydrogenation in the Enantioselective Synthesis of Flavonoids. *Org. Chem. Front.*, 9, 1165–1194, 2022.
36. Demidoff, F.C., Caleffi, G.S., Figueiredo, M., Costa, P.R.R., Ru(II)-Catalyzed Asymmetric Transfer Hydrogenation of Chalcones in Water: Application to the Enantioselective Synthesis of Flavans BW683C and Tephrowatsin E. *J. Org. Chem.*, *87*, 14208–14222, 2022.
37. (a) Dai, T., Li, C., Li, L., Zhao, Z.K., Zhang, B., Cong, Y., Wang, A., Selective Production of Renewable para-Xylene by Tungsten Carbide Catalyzed Atom-Economic Cascade Reactions. *Angew. Chem., Int. Ed.*, 57, 1808–1812, 2018. (b) Maneffa, A., Priecel, P., Lopez-Sanchez, J.A., Biomass-Derived Renewable Aromatics: Selective Routes and Outlook for p-Xylene Commercialisation. *ChemSusChem*, 9, 2736–2748, 2016. (c) Settle, A.E., Berstis, L., Rorrer, N.A., Roman-Leshkov, Y., Beckham, G.T., Richards, R.M., Vardon, D.R., Heterogeneous Diels–Alder catalysis for biomass-derived aromatic compounds. *Green Chem.*, 19, 3468–3492, 2017. (d) Xu, Y., Liu, D., Liu, X., Conversion of syngas toward aromatics over hybrid Fe-based Fischer-Tropsch catalysts and HZSM-5 zeolites. *Appl. Catal.*, A, 552, 168–183, 2018.
38. (a) Zakzeski, J., Bruijnincx, P.C.A., Jongerius, A.L., Weckhuysen, B.M., The Catalytic Valorization of Lignin for the Production of Renewable Chemicals. *Chem. Rev.*, 110, 3552–3599, 2010. (b) Li, C., Zhao, X., Wang, A., Huber, G.W., Zhang, T., Catalytic Transformation of Lignin for the Production of Chemicals and Fuels. *Chem. Rev.*, 115, 11559–11624, 2015.
39. (a) Tuck, C.O., Perez, E., Horvath, I.T., Sheldon, R.A., Poliakoff, M., Valorization of Biomass: Deriving More Value from Waste Science, 337, 695–699, 2012. (b) Huber, G.W., Iborra, S., Corma, A., Synthesis of Transportation Fuels from Biomass: Chemistry, Catalysts, and Engineering. *Chem. Rev.*, 106, 4044–4098, 2006
40. (a) Zhang, Z.R., Song, J.L., Han, B.X., Catalytic Transformation of Lignocellulose into Chemicals and Fuel Products in Ionic Liquids. *Chem. Rev.*, 10, 6834–6880, 2017. (b) Guo, H., Miles-Barrett, D.M., Neal, A.R., Zhang, T., Li, C., Westwood, N.J., Unravelling the enigma of ligninOX: can the oxidation of lignin be controlled? *Chem. Sci.*, 9, 702–711, 2018. (c) Melero, J.A., Iglesias, J., Garcia, A., Biomass as renewable feedstock in standard refinery units. Feasibility, opportunities and challenges Energy. *Environ. Sci.*, 5, 7393–7420, 2012.
41. Liu, Y., Li, C., Miao, W., Tang, W., Xue, D., Xiao, J., Zhang, T., Wang, C., Rhodium-terpyridine catalyzed redox-neutral depolymerization of lignin in water. *Green Chem.*, *22*, 33–38, 2020.
42. (a) Spiegel, S., Milstien, S., Sphingosine-1-phosphate: an enigmatic signalling lipid. *Nat. Rev. Mol. Cell Biol.*, 4, 397–4067, 2003. (b) Gante, J., Peptidomimetics—Tailored Enzyme Inhibitors. *Angew. Chem., Int. Ed.*, 33, 1699–1720, 1994. (c) Blattmann, H. and Mulhaupt, R., Multifunctional β-amino alcohols as bio-based amine curing agents for the isocyanate- and phosgene-free synthesis of 100% bio-based polyhydroxyurethane thermosets. *Green Chem.*,

18, 2406–2415, 2016. (d) Heravi, M.M., Lashaki, T.B., Fattahi, B., Zadsirjan, V., Application of asymmetric Sharpless aminohydroxylation in total synthesis of natural products and some synthetic complex bio-active molecules. *RSC Adv.*, 8, 6634–6659, 2018.
43. (a) Bergmeier, S.C., The Synthesis of Vicinal Amino Alcohols. *Tetrahedron*, 56, 2561–2576, 2000. (b) Karjalainen, O.K. and Koskinen, A.M.P., Diastereoselective synthesis of vicinal amino alcohols. *Org. Biomol. Chem.*, 10, 4311–4326, 2012. (c) Gupta, P. and Mahajan, N., Biocatalytic approaches towards the stereoselective synthesis of vicinal amino alcohols. *New J. Chem.*, 42, 12296–12327, 2018.
44. Shi, Y., Wang, Y., Lu, X., Zhang, Y., Wu, Y., Zhong, F., Rhodium-catalyzed aminohydroxylation of unactivated alkenes in aqueous media for the benign synthesis of 1, 2-amino alcohols. *Green Chem.*, *21*, 780–784, 2019.
45. For review of strategies to access Z-alkenes, see: (a) Siau, W.-Y., Zhang, Y., Zhao, Y., Stereoselective synthesis of Z-alkenes. *Top. Curr. Chem.*, 327, 33, 2012. (b) Oger, C., Balas, L., Durand, T., Galano, J.-M., Are alkyne reductions chemo-, regio-, and stereoselective enough to provide pure (Z)-olefins in polyfunctionalized bioactive molecules? *Chem. Rev.*, 113, 1313, 2013.
46. Casalta, C. and Bouzbouz, S., Rhodium(III) Catalyzed Regioselective and Stereospecific Allylic Arylation in Water by β-Fluorine Elimination of the Allylic Fluoride: Toward the Synthesis of Z-Alkenyl-Unsaturated Amides. *Org. Lett.*, 22, 2359–2364, 2020.
47. (a) Stierle, A., Hershenhorn, J., Strobel, G., Zinniol-related phytotoxins from Alternaria cichorii. *Phytochemistry*, 32, 1145, 1993. (b) Li, E., Jiang, L., Guo, L., Zhang, H., Che, Y., Pestalachlorides A–C, antifungal metabolites from the plant endophytic fungus Pestalotiopsis adusta. *Bioorg. Med. Chem.*, 16, 7894, 2008. (c) Li, J.C., Li, G.P., Yang, J.H., Dai, Y., Duan, Y.X., Zhang, J.S., A New Alkaloids from Corydalis impatiens Asian. *J. Chem.*, 24, 2815, 2012. (d) Almeida, C., Hemberger, Y., Schmitt, S.M., Bouhired, S., Natesan, L., Kehraus, S., Dimas, K., Gutschow, M., Bringmann, G., Konig, G.M., Marilines A–C: Novel Phthalimidines from the Sponge-Derived Fungus Stachylidium sp. *Chem. - Eur. J.*, 18, 8827, 2012.
48. Wu, S., Wu, X., Fu, C., Ma, S., Rhodium(III)-Catalyzed C–H Functionalization in Water for Isoindolin-1-one Synthesis. *Org. Lett.*, 20, 2831–2834, 2018.
49. (a) Tsuji, J., Takahashi, H., Morikawa, M., Organic Syntheses by Means of Noble Metal Compounds XVII. Reaction of π-Allylpalladium Chloride with Nucleophiles. *Tetrahedron Lett.*, 6, 4387–4388, 1965. (b) Trost, B.M. and Fullerton, T.J., New Synthetic Reactions. Allylic Alkylation. *J. Am. Chem. Soc.*, 95, 292–294, 1973.
50. Lee, N.R., Moghadam, F.A., Braga, F.C., Lippincott, D.J., Zhu, B., Gallou, F., Lipshutz, B.H., Sustainable Palladium-Catalyzed Tsuji–Trost Reactions Enabled by Aqueous Micellar Catalysis. *Org. Lett.*, 22, 4949–4954, 2020.
51. Zhou, X., Guo, X., Jian, F., Wei, G., Highly Efficient Method for Suzuki Reactions in Aqueous Media. *ACS Omega*, 3, 4418–4422, 2018.
52. Peng, C., Wang, Y., Wang, J., Palladium-Catalyzed Cross- Coupling of α-Diazocarbonyl Compounds with Arylboronic Acids. *J. Am. Chem. Soc.*, 130, 1566–1567, 2008.
53. Chatterjee, A. and Ward, T.R., Recent Advances in the Palladium Catalyzed Suzuki–Miyaura Cross-Coupling Reaction in Water. *Catal. Lett.*, 146, 820–840, 2016.
54. (a) Jansa, J., Reznícek, T., Dostál, L., Ruzicková, Z., Bures, F., Jambor, R., Amidophosphine-stabilized palladium complexes catalyse Suzuki-Miyaura cross-couplings in aqueous media: Efficient catalysts for Suzuki-Miyaura reaction in aqueous media. *Appl. Organomet. Chem.*, 30, 1036–1042, 2016. (b) Ge, J., Jiang, J., Yuan, C., Zhang, C., Liu, M., Palladium nanoparticles stabilized by phosphine ligand for aqueous phase room temperature Suzuki-Miyaura coupling. *Tetrahedron Lett.*, 58, 1142–1145, 2017. (c) Chen, T., Guan, J.T., Zhang, Z.Y., Chen, J.J., Liu, F.,

Liu, X., Alkylimidazole-Based Phosphines as Efficient Ligands for Palladium-Catalyzed Suzuki Reactions. *J. Heterocycl. Chem.*, 55, 551–555, 2018.

55. (a) Kazi, S.A., Campi, E.M., Hearn, M.T.W., A convenient and efficient one pot synthesis of unsymmetrically substituted p-terphenyls via a phosphine-free double Suzuki cross-coupling protocol using 1,4-dibromo-2-nitrobenzene as the substrate. *Tetrahedron*, 74, 1731–1741, 2018. (b) López-Saucedo, F., Flores-Rojas, G.G., González-Sebastián, L., Reyes-Martínez, R., German-Acacio, J.M., Avila-Sorrosa, A., Hernández-Ortega, S., Morales-Morales, D., Palladium complexes bearing pyridylthioether ligands. Synthesis and application as efficient phosphine-free catalysts in Suzuki-Miyaura couplings. *Inorg. Chim. Acta*, 473, 83–93, 2018. (c) Jerome, P., Sathishkumar, P.N., Bhuvanesh, N.S.P., Karvembu, R., Towards phosphine-free Pd(II) pincer complexes for catalyzing Suzuki-Miyaura cross-coupling reaction in aqueous medium. *J. Organomet. Chem.*, 845, 115–124, 2017.
56. (a) Trost, B.M., Pd Asymmetric Allylic Alkylation (AAA). A Powerful Synthetic Tool. *Chem. Pharm. Bull.*, 50, 1–14, 2002. (b) Horton, D.A., Bourne, G.T., Smythe, M.L., The Combinatorial Synthesis of Bicyclic Privileged Structures or Privileged Substructures. *Chem. Rev.*, 103, 893–930, 2003. (c) Takeuchi, R. and Kezuka, S., Iridium-Catalyzed Formation of Carbon-Carbon and Carbon-Heteroatom Bonds. *Synthesis*, 39, 3349–3366, 2006. (d) Harutyunyan, S.R., den Hartog, T., Geurts, K., Minnaard, A.J., Feringa, B.L., Catalytic Asymmetric Conjugate Addition and Allylic Alkylation with Grignard Reagents. *Chem. Rev.*, 108, 2824–2852, 2008. (e) Alexakis, A., Backvall, J.E., Krause, N., Pamies, O., Dieguez, M., Enantioselective Copper-Catalyzed Conjugate Addition and Allylic Substitution Reactions. *Chem. Rev.*, 108, 2796–2823, 2008. (f) Lu, Z. and Ma, S., Metal-Catalyzed Enantioselective Allylation in Asymmetric Synthesis. *Angew. Chem. Int. Ed.*, 47, 258–297, 2008. (g) Hartwig, J.F. and Stanley, L.M., Mechanistically Driven Development of Iridium Catalysts for Asymmetric Allylic Substitution. *Acc. Chem. Res.*, 43, 1461–1475, 2010. (h) Hartwig, J.F. and Pouy, M.J., Iridium-Catalyzed Allylic Substitution. *Top. Organomet. Chem.*, 34, 169–208, 2011. (i) Weaver, J.D., Recio, A., Grenning, A.J., Tunge, J.A., Transition Metal-Catalyzed Decarboxylative Allylation and Benzylation Reactions. *Chem. Rev.*, 111, 1846–1913, 2011. (j) Liu, W.-B., Xia, J.-B., You, S.-L., Iridium-Catalyzed Asymmetric Allylic Substitution Reactions. *Top. Organomet. Chem.*, 38, 169–208, 2011. (k) Chen, D.-F., Han, Z.-Y., Zhou, X.-L., Gong, L.-Z., Asymmetric Organocatalysis Combined with Metal Catalysis: Concept, Proof of Concept, and Beyond. *Acc. Chem. Res.*, 47, 2365–2377, 2014. (l) Kim, S.W., Zhang, W., Krische, M.J., Catalytic Enantioselective Carbonyl Allylation and Propargylation via Alcohol-Mediated Hydrogen Transfer: Merging the Chemistry of Grignard and Sabatier. *Acc. Chem. Res.*, 50, 2371–2380, 2017. (m) Fu, J.-K., Huo, X.-H., Li, B.-W., Zhang, W.-B., Cooperative Bimetallic Catalysis in Asymmetric Allylic Substitution. *Org. Biomol. Chem.*, 15, 9747–9759, 2017. (n) Cheng, Q., Tu, H.-F., Zheng, C., Qu, J.-P., Helmchen, G., You, S.-L., Iridium-Catalyzed Asymmetric Allylic Substitution Reactions. *Chem. Rev.*, 119, 1855–1969, 2019.
57. (a) Rouf, A. and Tanyeli, C., Bioactive Thiazole and Benzothiazole Derivatives. *Eur. J. Med. Chem.*, 97, 911–927, 2015. (b) Wu, X.-J., Kassie, F., Mersch-Sundermann, V., Induction of Apoptosis in Tumor Cells by Naturally Occurring Sulfur-Containing Compounds. . *Mutat. Res.-Rev. Mutat.*, 589, 81–102, 2005. (c) Facchinettil, V., Reis, R.D.R., Gomes, C.R.B., Vasconcelos, T. R. A., Chemistry and Biological Activities of 1,3-Benzothiazoles. *Mini-Rev. Org. Chem.*, 9, 44–53, 2012. (d) Hutchinson, I., Jennings, S.A., Vishnuvajjala, B.R., Westwell, A.D., Stevens, M.F.G., Antitumor Benzothiazoles. 16.1 Synthesis and Pharmaceutical Properties of Antitumor 2-(4-Aminophenyl)benzothiazole Amino Acid Prodrugs. *J. Med. Chem.*, 45, 744–747, 2002. (e) Hargrave, K.D. and Hess, F.K., Oliver, J. T. N-(4-Substituted-thiazolyl)oxamic Acid Derivatives, New Series of Potent, Orally Active Antiallergy Agents. *J. Med. Chem.*, 26, 1158–1163, 1983. (f) Gronowitz, S. and Gronowitz, S. (Eds.), Part 3, (Chapter 2), in: *The Chemistry of Heterocyclic Compounds: Thiophene and its Derivatives*, vol. 44, Wiley, New

York, NY, USA, 1991. (g) Patt, W.C., Hamilton, H.W., Taylor, M.D., Ryan, M.J., Taylor, D.G.J., Connolly, C.J.C., Doherty, A.M., Klutchko, S.R., Sircar, I., Steinbaugh, B.A., Structure-Activity Relationships of a Series of 2-Amino-4-thiazole-Containing Renin Inhibitors. *J. Med. Chem.*, 35, 2562–2572, 1992. (h) Jaen, J.C., Wise, L.D., Caprathe, B.W., Tecle, H., Bergmeier, S., Humblet, C.C., Heffner, T.G., Meltzner, L.T., Pugsley, T.A., 4-(1,2,5,6-Tetrahydro-1-alkyl-3-pyridinyl)-2-thiazolamines: a Novel Class of Compounds with Central Dopamine Agonist Properties. *J. Med. Chem.*, 33, 311–317, 1990. (i) Sharma, R.N., Xavier, F.P., Vasu, K.K., Chaturvedi, S.C., Pancholi, S.S., Synthesis of 4-Benzyl-1,3-thiazole Derivatives as Potential Anti-Inflammatory Agents: An Analogue-Based Drug Design Approach. *J. Enzyme Inhib. Med. Chem.*, *24*, 890–897, 2009. (j) Kempf, D.J., Sham, H.L., Marsh, K.C., Flentge, C.A., Betebenner, D., Green, B.E., McDonald, E., Vasavanonda, S., Saldivar, A., Wideburg, N.E., Kati, W.M., Ruiz, L., Zhao, C., Fino, L.M., Patterson, J., Molla, A., Plattner, J.J., Norbeck, D.W., Discovery of Ritonavir, a Potent Inhibitor of HIV Protease with High Oral Bioavailability and Clinical Efficacy. *J. Med. Chem.*, *41*, 602–617, 1998.

58. Pan, S., Wu, B., Hu, J., Xu, R., Jiang, M., Zeng, X., Zhong, G., Palladium-Catalyzed Allylic Substitution Reaction of Benzothiazolylacetamide with Allylic Alcohols in Water. *J. Org. Chem.*, 84, 10111–10119, 2019.
59. (a) Evans, D.A., Bartroli, J., Shih, T.L., Enantioselective aldol condensations. 2. Erythro-selective chiral aldol condensations via boron enolates. *J. Am. Chem. Soc.*, 103, 2127, 1981. (b) Heravi, M.M. and Zadsirjan, V., Oxazolidinones as chiral auxiliaries in asymmetric aldol reactions applied to total synthesis. *Tetrahedron. Asymmetry*, 24, 1149–1188, 2013.
60. (a) Yoshida, S., Fukui, K., Kikuchi, S., Yamada, T., Silver-catalyzed preparation of oxazolidinones from carbon dioxide and propargylic amines. *Chem. Lett.*, 38, 786, 2009. For a Ag/DBU dual catalyst protocol, see: (b) Yoshida, M., Mizuguchi, T., Shishido, K., Synthesis of Oxazolidinones by Efficient Fixation of Atmospheric CO2 with Propargylic Amines by using a Silver/1,8-Diazabicyclo[5.4.0]undec-7-ene (DBU) Dual-Catalyst System. *Chem.-Eur. J.*, 18, 15578, 2012. (c) Brunel, P., Monot, J., E.Kefalidis, C., Maron, L., Martin-Vaca, B., Bourissou, D., Valorization of CO2: Preparation of 2-Oxazolidinones by Metal–Ligand Cooperative Catalysis with SCS Indenediide Pd Complexes. *ACS Catal.*, 7, 2652, 2017.
61. Li, C.-J. and Chan, T.-H., *Comprehensive Organic Reactions in Aqueous Media*, 2nd Ed., Wiley-Interscience, 2007.
62. Qin, J.-F., Wang, B., Lin, G.-Q., Silver(I)-catalysed carboxylative cyclisation of primary propargylic amines in neat water using potassium bicarbonate as the carboxyl source: environment-friendly synthesis of Z-5-alkylidene-1,3-oxazolidin-2-ones. *Green Chem.*, *21*, 4656–4661, 2019.
63. (a) Vitaku, E., Smith, D.T., Njardarson, J.T., *J. Med. Chem.*, 57, 10257–10274, 2014. (b) Abdel-Magid, A.F. and Mehrman, S.J., *Org. Process Res. Dev.*, 10, 971–1031, 2006. (c) Seayad, A., Ahmed, M., Klein, H., Jackstell, R., Gross, T., Beller, M., *Science*, 297, 1676–1678, 2002. (d) Kobayashi, S. and Ishitani, H., *Chem. Rev.*, 99, 1069–1094, 1999.
64. (a) Celaje, J.J.A., Zhang, X., Zhang, F., Kam, L., Herron, J.R., Williams, T.J., A Base and Solvent-Free Ruthenium-Catalyzed Alkylation of Amines. *ACS Catal.*, 7, 1136–1142, 2017. (b) Marichev, K.O. and Takacs, J.M., Ruthenium-Catalyzed Amination of Secondary Alcohols Using Borrowing Hydrogen Methodology. *ACS Catal.*, 6, 2205–2210, 2016. (c) Xie, X. and Huynh, H.V., Tunable Dehydrogenative Amidation versus Amination Using a Single Ruthenium-NHC Catalyst. *ACS Catal.*, 5, 4143–4151, 2015. (d) Enyong, A.B. and Moasser, B., Ruthenium-Catalyzed N-Alkylation of Amines with Alcohols under Mild Conditions Using the Borrowing Hydrogen Methodology. *J. Org. Chem.*, 79, 7553–7563, 2014. (e) Fernandez, F.E., Puerta, M.C., Valerga, P., Ruthenium(II) Picolyl-NHC Complexes: Synthesis, Characterization, and Catalytic Activity in Amine N-alkylation and Transfer Hydrogenation Reactions. *Organometallics*, 31,

6868–6879, 2012. (f) Hamid, M.H.S.A., Allen, C.L., Lamb, G.W., Maxwell, A.C., Maytum, H.C., Watson, A.J.A., Williams, J.M.J., Ruthenium-Catalyzed N-Alkylation of Amines and Sulfonamides Using Borrowing Hydrogen Methodology. *J. Am. Chem. Soc.*, 131, 1766–1774, 2009. (g) Liu, P., Liang, R., Lu, L., Yu, Z., Li, F., Use of a Cyclometalated Iridium(III) Complex Containing a N∧C∧N-Coordinating Terdentate Ligand as a Catalyst for the α-Alkylation of Ketones and N-Alkylation of Amines with Alcohols. *J. Org. Chem.*, 82, 1943–1950, 2017. (h) Li, J.-Q. and Andersson, P.G., Room temperature and solvent-free iridium-catalyzed selective alkylation of anilines with alcohols. *Chem. Commun.*, 49, 6131–6133, 2013. (i) Bartoszewicz, A., Marcos, R., Sahoo, S., Inge, A.K., Zou, X., Martin-Matute, B., A highly active bifunctional iridium complex with an alcohol/alkoxide-tethered N-heterocyclic carbene for alkylation of amines with alcohols. *Chem. – Eur. J.*, 18, 14510–14519, 2012. (j) Michlik, S. and Kempe, R., New Iridium Catalysts for the Efficient Alkylation of Anilines by Alcohols under Mild Conditions. *Chem. – Eur. J.*, 16, 13193–13198, 2010. (k) Zhang, G., Yin, Z., Zheng, S., Cobalt-Catalyzed N-Alkylation of Amines with Alcohols Guoqi Zhang, Zhiwei Yin and Shengping Zheng. *Org. Lett.*, 18, 300–303, 2016. (l) Mastalir, M., Tomsu, G., Pittenauer, E., Allmaier, G., Kirchner, K., Co(II) PCP Pincer Complexes as Catalysts for the Alkylation of Aromatic Amines with Primary Alcohols. *Org. Lett.*, 18, 3462–3465, 2016. (m) Rosler, S., Ertl, M., Irrgang, T., Kempe, R., Cobalt-Catalyzed Alkylation of Aromatic Amines by Alcohols. *Angew. Chem., Int. Ed.*, 54, 15046–15050, 2015. (n) Yan, T., Feringa, B.L., Barta, K., Benzylamines via Iron-Catalyzed Direct Amination of Benzyl Alcohols. *Catal., A.C.S.*, 6, 381–388, 2016. (b) Pan, H.-J., Ng, T.W., Zhao, Y., Iron-catalyzed amination of alcohols assisted by Lewis acid. *Chem. Commun.*, 51, 11907–11910, 2015. (o) Yan, T., Feringa, B.L., Barta, K., Iron catalysed direct alkylation of amines with alcohols. *Nat. Commun.*, 5, 5602, 2014. (p) Bala, M., Verma, P.K., Sharma, U., Kumar, N., Singh, B., Iron phthalocyanine as an efficient and versatile catalyst for N-alkylation of heterocyclic amines with alcohols: one-pot synthesis of 2-substituted benzimidazoles, benzothiazoles and benzoxazoles. *Green Chem.*, 15, 1687–1693, 2013. (q) Elangovan, S., Neumann, J., Sortais, J.-B., Junge, K., Darcel, C., Beller, M., Efficient and selective N-alkylation of amines with alcohols catalysed by manganese pincer complexes. *Nat. Commun.*, 7, 12641, 2016. (r) Bruneau-Voisine, A., Wang, D., Dorcet, V., Roisnel, T., Darcel, C., Sortais, J.-B., Mono-N-methylation of anilines with methanol catalyzed by a manganese pincer-complex. *J. Catal.*, 347, 57–62, 2017.

65. Huang, M., Li, Y., Liu, J., Xiao-Bing Lan, X.-B., Liu, Y., Zhao, C., Ke, Z., A bifunctional strategy for N-heterocyclic carbene-stabilized iridium complex-catalyzed N-alkylation of amines with alcohols in aqueous media. *Green Chem.*, 21, 219–224, 2019.
66. (a) Pape, A.R., Kaliappan, K.P., Kundig, E.P., Transition-Metal-Mediated Dearomatization Reactions. *Chem. Rev.*, 100, 2917–2940, 2000. (b) Zhuo, C.-X., Zhang, W., You, S.-L., Catalytic Asymmetric Dearomatization Reactions. *Angew. Chem., Int. Ed.*, 51, 12662–12686, 2012. (c) Roche, S.P., Youte Tendoung, J.-J., Treguier, B., Advances in dearomatization strategies of indoles. *Tetrahedron*, 71, 3549– 3591, 2015. (d) James, M. J., O'Brien, P., Taylor, R. J. K., Unsworth, W.P., Synthesis of Spirocyclic Indolenines. *Chem. - Eur. J.*, 22, 2856–2881, 2016. (e) Bariwal, J., Voskressensky, L.G., Van der Eycken, E.V., Recent advances in spirocyclization of indole derivatives. *Chem. Soc. Rev.*, 47, 3831–3848, 2018. (f) Lovering, F., Bikker, J., Humblet, C. J., Escape from Flatland: Increasing Saturation as an Approach to Improving Clinical Success. *J. Med. Chem.*, 52, 6752–6756, 2009. (g) Lovering, F., Escape from Flatland 2: complexity and promiscuity. *Med. Chem.Commun.*, 4, 515–519, 2013. (h) Bandini, M., Gold-catalyzed decorations of arenes and heteroarenes with C–C multiple bonds. *Chem. Soc. Rev.*, 40, 1358–1367, 2011. (i) Ohno, H., Gold-Catalyzed Cascade Reactions of Alkynes for Construction of Polycyclic Compounds Isr. *J. Chem.*, 53, 869–882, 2013. (j) Dorel, R. and Echavarren, A.M., Gold(I)-Catalyzed Activation of Alkynes for the Construction of Molecular Complexity. *Chem. Rev.*, 115, 9028–9072, 2015. (k) James, M.J., Cuthbertson, J.D., O'Brien, P., Taylor, R.J.K.,

Unsworth, W.P., Silver(I)- or copper(II)-mediated dearomatization of aromatic ynones: direct access to spirocyclic scaffolds. *Angew. Chem., Int. Ed.*, 54, 7640–7643, 2015. (l) Zhuo, C.-X., Zhou, Y., Cheng, Q., Huang, L., You, S.-L., Enantioselective Construction of Spiroindolines with Three Contiguous Stereogenic Centers and Chiral Tryptamine Derivatives via Reactive Spiroindolenine Intermediates. *Angew. Chem., Int. Ed.*, 54, 14146–14149, 2015. (m) Wang, Y., Zheng, C., You, S.-L., *Angew. Chem., Int. Ed.*, 56, 15093–15097, 2017. (n) Fedoseev, P., Coppola, G., Ojeda, G.M., Van der Eycken, E.V., Synthesis of spiroindolenines by intramolecular ipso-iodocyclization of indol ynones. *Chem. Commun.*, 54, 3625–3628, 2018. (o) Grugel, C.P. and Breit, B., Rhodium-Catalyzed Diastereo- and Enantioselective Tandem Spirocyclization/Reduction of 3-Allenylindoles: Access to Functionalized Vinylic Spiroindolines. *Org. Lett.*, 21, 9672–9676, 2019.

67. Sabat, N., Soualmia, F., Retailleau, P., Benjdia, A., Berteau, O., Guinchard, X., Gold-Catalyzed Spirocyclization Reactions of N-Propargyl Tryptamines and Tryptophans in Aqueous Media. *Org. Lett.*, 22, 4344–4349, 2020.
68. (a) Regina, G.L., Bai, R., Rensen, W.M., Cesare, E.D., Coluccia, A., Piscitelli, F., Famiglini, V., Reggio, A., Nalli, M., Pelliccia, S., Pozzo, E.D., Costa, B., Granata, I., Porta, A., Maresca, B., Soriani, A., Iannitto, M.L., Santoni, A., Li, J., Cona, M.M., Chen, F., Ni, Y., Brancale, A., Dondio, G., Vultaggio, S., Varasi, M., Mercurio, C., Martini, C., Hamel, E., Lavia, P., Novellino, E., Silvestri, R., Toward Highly Potent Cancer Agents by Modulating the C-2 Group of the Arylthioindole Class of Tubulin Polymerization Inhibitors. *J. Med. Chem.*, 56, 123, 2013. (b) Nuth, M., Guan, H., Zhukovskaya, N., Saw, Y.L., Ricciardi, R.P., Design of Potent Poxvirus Inhibitors of the Heterodimeric Processivity Factor Required for Viral Replication. *J. Med. Chem.*, 56, 3235, 2013. (c) Guan, Q., Han, C., Zuo, D., Zhai, M., Li, Z., Zhang, Q., Zhai, Y., Jiang, X., Bao, K., Wu, Y., Zhang, W., Synthesis and evaluation of benzimidazole carbamates bearing indole moieties for antiproliferative and antitubulin activities. *Eur. J. Med. Chem.*, 87, 306, 2014. (d) Wen, Z., Xu, J., Wang, Z., Qi, H., Xu, Q., Bai, Z., Zhang, Q., Bao, K., Wu, Y., Zhang, W., 3-(3, 4, 5-Trimethoxyphenylselenyl)-1H-indoles and their selenoxides as combretastatin A-4 analogs: Microwave-assisted synthesis and biological evaluation. *Eur.J. Med. Chem.*, 90, 184, 2015. and synthesis of antiviral drug Arbidol analogues with significantly improved affinity to influenza hemagglutinin. *Bioorg. Med. Chem. Lett.*, 27, 3744, 2017. (f) Fink, S.L., Vojtech, L., Wagoner, J., Slivinski, N.S.J., Jackson, K.J., Wang, R., Khadka, S., Luthra, P., Basler, C.F., Polyak, S. J., The Antiviral Drug Arbidol Inhibits Zika Virus. *Sci. Rep.*, 8, 8989, 2018. (g) Famiglini, V., Regina, G.L., Coluccia, A., Pelliccia, S., Brancale, A., Maga, G., Crespan, E., Badia, R., Riveira-Munoz, E., Este, J.A., Ferretti, R., Cirilli, R., Zamperini, C., Botta, M., Schols, D., Limongelli, V., Agostino, B., Novellino, E., Silvestri, R., Indolylarylsulfones Carrying a Heterocyclic Tail as Very Potent and Broad Spectrum HIV-1 Non-nucleoside Reverse Transcriptase Inhibitors. *J. Med. Chem.*, 57, 9945, 2014. (h) Regina, G.L., Coluccia, A., Brancale, A., Piscitelli, F., Gatti, V., Maga, G., Samuele, A., Pannecouque, C., Schols, D., Balzarini, J., Novellino, E., Silvestri, R., Indolylarylsulfones as HIV-1 Non-Nucleoside Reverse Transcriptase Inhibitors: New Cyclic Substituents at Indole-2-carboxamide. *J. Med. Chem.*, 54, 1587, 2011. (i) Ragno, R., Coluccia, A., Regina, G.L., Martino, G.D., Piscitelli, F., Lavecchia, A., Novellino, E., Bergamini, A., Ciaprini, C., Sinistro, A., Maga, G., Crespan, E., Artico, M., Silvestri, R., Design, molecular modeling, synthesis, and anti-HIV-1 activity of new indolyl aryl sulfones. Novel derivatives of the indole-2-carboxamide. *J. Med. Chem.*, 49, 3172, 2006. (j) Martino, G.D., Regina, G.L., Coluccia, A., Edler, M.C., Barbera, M.C., Brancale, A., Wilcox, E., Hamel, E., Artico, M., Silvestri, R., Arylthioindoles, Potent Inhibitors of Tubulin Polymerization. *J. Med. Chem.*, 47, 6120, 2004. (k) Silvestri, R., Martino, G.D., Regina, G.L., Artico, M., Massa, S., Vargiu, L., Mura, M., Loi, A.G., Marceddu, T., Colla, P.L., Novel indolyl aryl sulfones active against HIV-1 carrying NNRTI resistance mutations: synthesis and SAR studies. *J. Med. Chem.*, 46, 2482,

2003. (l) Unangst, P.C., Connor, D.T., Stabler, S.R., Weikert, R.J., Carethers, M.E., Kennedy, J.A., Thueson, D.O., Chestnut, J.C., Adolphson, R.L., Conroy, M.C., Novel indolecarboxamido-tetrazoles as potential antiallergy agents. *J. Med. Chem.*, 32, 1360, 1989. (m) Potin, D., Parnet, V., Teulon, J.-M., Camborde, F., Caussade, F., Meignen, J., Provost, D., Cloarec, A., The design and synthesis of thrombin inhibitors: analogues of MD805 containing non-polar surrogates for arginine at the P1 position. *Bioorg. Med. Chem. Lett.*, 10, 805, 2000. (n) Li, C., Lee, D., Graf, T.N., Phifer, S.S., Nakanishi, Y., Burgess, J.P., Riswan, S., Setyowati, F.M., Saribi, A.M., Soejarto, D.D., Farnsworth, N.R., Falkinham, J.O., Kroll, D.J., Kinghorn, A.D., Wani, M.C., Oberlies, N.H., A hexacyclic ent-trachylobane diterpenoid possessing an oxetane ring from Mitrephora glabra. *Org. Lett.*, 7, 5709, 2005. (o) Hutchinson, J.H., Riendeau, D., Brideau, C., Chan, C., Delorme, D., Denis, D., Falgueyret, J.-P., Fortin, R., Guay, J., Hamel, P., Jones, T.R., Macdonald, D., McFarlane, C.S., Piechuta, H., Scheigetz, J., Tagari, P., Therien, M., Girard, Y., Substituted thiopyrano[2,3,4-c,d]indoles as potent, selective, and orally active inhibitors of 5-lipoxygenase. Synthesis and biological evaluation of L-691,816. *J. Med. Chem.*, 36, 2771, 1993.

69. (a) Vasquez-Cespedes, S., Ferry, A., Candish, L., Glorius, F., Heterogeneously Catalyzed Direct C–H Thiolation of Heteroarenes. *Angew. Chem. Int. Ed.*, 54, 5772, 2015. (b) Wu, Z., Li, Y.-C., Z. Ding, W.-, Zhu, T., Liu, S.-Z., Ren, X., Zou, L.-H., Copper-Catalyzed Regioselective Sulfenylation of Indoles with Arylsulfonyl Chlorides Asian. *J. Org.Chem.*, 5, 625, 2016. (c) Guo, Y.-J., Tang, R.-Y., Li, J.-H., Zhong, P., Zhang, X.-G., Palladium-Catalyzed Annulation of 2-(1-Alkynyl)benzenamines with Disulfides: Synthesis of 3-Sulfenylindoles. *Adv. Synth. Catal.*, 351, 2615, 2009. (d) Li, Z., Hong, L., Liu, R., Shen, J., Zhou, X., Copper-catalyzed chalcogeno-amination of 2-alkynylanilines with dichalcogenides for one-step synthesis of 3-sulfenylindoles and 3-selenylindoles. *Tetrahedron Lett.*, 52, 1343, 2011. (e) Liu, Y., Zhang, Y., Hu, C., Wan, J.-P., Wen, C., Synthesis of 3-sulfenylated indoles by a simple NaOH promoted sulfenylation reaction. *RSC Adv.*, 4, 35528, 2014. (f) Sang, P., Chen, Z., Zou, J., Zhang, Y., K2CO3 promoted direct sulfenylation of indoles: a facile approach towards 3-sulfenylindoles. *Green Chem.*, 15, 2096, 2013. (g) Zhou, H., Zhu, D., Xing, Y., Huang, H., Sulfur-Assisted and 1,8-Diazabicyclo[5.4.0]undec-7-ene (DBU)-Catalyzed Cyclization of 2-Alkynylanilines for the Metal-Free Synthesis of Indole Derivatives. *Adv. Synth. Catal.*, 352, 2127, 2010. (h) Chen, M., Luo, Y., Zhang, C., Guo, L., Wang, Q., Wu, Y., Graphene oxide mediated thiolation of indoles in water: a green and sustainable approach to synthesize 3-sulfenylindoles. *Org. Chem. Front.*, 6, 116, 2019. (i) Hostier, T., Ferey, V., Ricci, G., Pardo, D.G., Cossy, *J., Chem. Commun.*, 51, 13898, 2015. (j) Ge, W. and Wei, Y., Iodine-catalyzed oxidative system for 3-sulfenylation of indoles with disulfides using DMSO as oxidant under ambient conditions in dimethyl carbonate. *Green Chem.*, 14, 2066, 2012. (k) Liu, X., Cui, H., Yang, D., Dai, S., Zhang, G., Wei, W., Wang, H., Iodine-catalyzed Direct Thiolation of Indoles with Thiols Leading to 3-Thioindoles Using Air as the Oxidant. *Catal. Lett.*, 146, 1743, 2016. (l) Katrun, P., Hongthong, S., Hlekhlai, S., Pohmakotr, M., Reutrakul, V., Soorukram, D., Jaipetchb, T., Kuhakarn, C., Iodine–PPh3-mediated C3-sulfenylation of indoles with sodium sulfonates. *RSC Adv.*, 4, 18933, 2014. (m) Kumaraswamy, G., Rajua, R., Narayanarao, V., Metal- and base-free syntheses of aryl/alkylthioindoles by the iodine-induced reductive coupling of aryl/alkyl sulfonyl chlorides with indoles. *RSC Adv.*, 5, 22718, 2015. (n) Guo, T. and Wei, X.-N., Ammonium iodide-mediated sulfenylation of 4-hydroxycoumarins or 4-hydroxyquinolinones with a sulfonyl chloride as a sulfur source. *Synlett*, 28, 2499, 2017. (o) Azeredo, J.B., Godoi, M., Martins, G.M., Silveira, C.C., Braga, A.L., A Solvent- and Metal-Free Synthesis of 3-Chacogenyl-indoles Employing DMSO/I2 as an Eco-friendly Catalytic Oxidation System. *J. Org. Chem.*, 79, 4125, 2014. (p) Choudhury, P., Roy, B., Basu, B., Sustainable and Site-Selective C–H Sulfenylation of Aromatic Compounds with Thiol using Catalytic Graphene Oxide and NaI Asian. *J. Org. Chem.*, 6, 1569, 2017. (q) Zhang, H., Bao, X., Song, Y., Qu, J.,

Wang, B., Iodine-catalysed versatile sulfenylation of indoles with thiophenols: controllable synthesis of mono- and bis-arylthioindoles. *Tetrahedron*, 71, 8885, 2015.

70. (a) Cai, Y., Li, J., Chen, W., Xie, M., Liu, X., Lin, L., Feng, X., Catalytic Asymmetric Sulfenylation of Unprotected 3-Substituted Oxindoles. *Org. Lett.*, 14, 2726, 2012. (b) Nalbandian, C.J., Miller, E.M., Toenjes, S.T., Gustafson, J.L., A conjugate Lewis base-Brønsted acid catalyst for the sulfenylation of nitrogen containing heterocycles under mild conditions. *Chem. Commun.*, 53, 1494, 11, 2017. (c) Hostier, T., Ferey, V., Ricci, G., Pardo, D.G., Cossy, J., TFA-promoted direct C–H sulfenylation at the C2 position of non-protected indoles. *Chem. Commun.*, 51, 13898, 2015. (d) Wu, Q., Zhao, D., Qin, X., Lan, J., You, J., Synthesis of di(hetero)aryl sulfides by directly using arylsulfonyl chlorides as a sulfur source. *Chem. Commun.*, 47, 9188, 2011. (e) Chen, M., Huang, Z.-T., Zheng, Q.-Y., Visible light-induced 3-sulfenylation of N-methylindoles with arylsulfonyl chlorides. *Chem. Commun.*, 48, 11686, 2012. (f) Shi, Q., Li, P., Zhang, Y., Wang, L., Visible light-induced tandem oxidative cyclization of 2-alkynylanilines with disulfides (diselenides) to 3-sulfenyl- and 3-selenylindoles under transition metal-free and photocatalyst-free conditions. *Org.Chem. Front.*, 4, 1322, 2017. (g) Ravi, C., Joshi, A., Adimurthy, S., C3 Sulfenylation of N-Heteroarenes in Water under Catalyst-Free Conditions. *Eur. J. Org. Chem.*, 25, 3646, 2017. (h) Yang, Y., Zhang, S., Tang, L., Hu, Y., Zha, Z., Wang, Z., Catalyst-free thiolation of indoles with sulfonyl hydrazides for the synthesis of 3-sulfenylindoles in water. *Green Chem.*, 18, 2609, 2016. (i) Yang, F.-L. and Tian, S.-K., Iodine-Catalyzed Regioselective Sulfenylation of Indoles with Sulfonyl Hydrazides. *Angew. Chem. Int. Ed.*, 52, 4929, 2013.

71. Zhu, S., Zhang, Y., Luo, J., Wang, F., Cao, X., Huang, S., Temperature-Controlled Regioselective Thiolation of 2-Indolylmethanols under Aqueous Micellar Conditions. *Green Chem.*, *22*, 657–661, 2020.

72. Saini, K., Raigar, A.K., Manju, Jangid, D.K., Mathur, J., Dhadda, S., Guleria, A., Tandem Protocol for the Synthesis of Pyrano[3,2-c]quinolone Derivatives Using Taurine as a Green Bio-Organic Catalyst in Aqueous Medium. *J. Org. Chem.*, *87*, 13734–13743, 2022.

73. (a) Agarwal, S., Kalal, P., Sethiya, A., Soni, J., Taurine: A Water Friendly Organocatalyst in Organic Reactions. *Mini-Rev. Org. Chem.*, 19, 617–628, 2022. (b) Biglari, M., Shirini, F., Mahmoodi, N.O., Zabihzadeh, M., Langarudi, M.S.N., Alipour Khoshdel, M., Taurine/Choline Chloride Deep Eutectic Solvent as a Novel Eco-Compatible Catalyst to Facilitate the Multi-Component Synthesis of Pyrano[2,3-d]Pyrimidinone (Thione), Hexahydroquinoline, and Biscoumarin Derivatives. *Polycyclic Aromat. Compd.*, 42, 1452–1473, 2022. (c) Chate, A.V., Rudrawar, P.P., Bondle, G.M., Sangeshetti, J.N., 2-Aminoethanesulfonic acid: An efficient organocatalyst for green synthesis of spirooxindole dihydroquinazolinones and novel 1,2-(dihydroquinazolin-3(4H)isonicotinamides in water. *Synth. Commun.*, 50, 226–242, 2020. (d) Jadhav, C. K., Nipate, A. S., Chate, A. V., & Gill, C. H. An Organocatalytic Newer Synthetic Approach toward the Access of Dihydropyrido[2,3-d] Pyrimidine in Water: A Perfect Synergy for Eco-compatible Organic Synthesis. Polycyclic Aromatic Compounds, 42, 10, 7368–7384., 2022. (e) Khan, M.M., Saigal, B., Shareef, S., Khan, S., Sahoo, S.C., One-pot practical method for synthesis of functionalized 4H-chromen-5-one derivatives under catalyst and solvent-free conditions. *Synth. Commun.*, 48, 2683–2694, 2018. (f) Verma, K., Tailor, Y.K., Khandelwal, S., Rushell, E., Agarwal, M., Kumar, M., Efficient and environmentally sustainable domino protocol for the synthesis of diversified spiroheterocycles with privileged heterocyclic substructures using bio-organic catalyst in aqueous medium. *Mol. Diversity*, 24, 1355–1365, 2020.

3

Green Avenues in Controlled Radical Polymerization for Precision Synthesis of Macromolecules

Pratibha Sharma[1]* and Amit Kumar[2]

[1]*Department of Chemistry, Amity Institute of Applied Sciences, Amity University, Noida, Uttar Pradesh, India*

[2]*Department of Materials Science and Engineering, Indian Institute of Technology Delhi, Hauz Khas, Delhi, India*

Abstract

Polymers have become the most widely used material in several forms such as adhesives, protective coatings, foams, packaging materials, electronics, medical equipment and are mainly produced via free-radical polymerization (FRP). It is essential to note that a specific application mandates a unique polymer architecture, which may be attained by using controlled radical polymerization (CRP) strategies, that allows precision synthesis of macromolecules and provides control over the chain microstructure. In addition, CRP strategies offer operability in mild reaction conditions as well as adaptability to a broad range of chemical environments. Furthermore, CRP strategies provide control over the molecular weight, molecular weight distribution, molecular architecture (regular network, star, combs, cycle, and brushes), and molecular composition (alternate pattern, grafting, and block copolymer). CRP processes operate through a reversible deactivation radical propagation (RDRP) mechanism, which involves an equilibrium between the active and dormant species. It is interesting to note that CRP processes have not only overcome the challenges from the industrial viewpoint but also additionally taken into account the principles of green chemistry. In view of the same, cutting-edge innovations in CRP processes with the intense focus on green efficient routes for the precision synthesis of polymeric architectures have been highlighted in this chapter.

Keywords: Controlled, deactivation, dispersity, polymerization, radical, reversible

3.1 Introduction

The term "macromolecules" was first coined by Nobel Laureate Hermann Staudinger, the founder of polymer chemistry, in his published report of 1920 [1]. Since then, researchers and industrialists have made enormous efforts in developing interesting macromolecular architectures to meet the demands of sectors such as strong constructions, packaging, communication, health, and energy. The ubiquity of polymers in the lives of humans can be

**Corresponding author*: pratibhasharma.venky@gmail.com

Mousumi Sen (ed.) Sustainable Green Catalytic Processes, (59–78) © 2024 Scrivener Publishing LLC

stipulated from their huge demand, and subsequently from their annual production, which has been found to be approximately 500 million metric tons (MTs) per year. The production of plastics has witnessed an increase of 35% since the past decade and is expected to display a continuous rise up to 700 million metric tons in 2030, which can be considered as 80 kg of polymers per human being [2]. Further, the applicability of a particular polymer is decided by its macromolecular architecture, which includes chemical composition, topology, and functionalities. Furthermore, the characteristics of polymer such as processibility or mechanical performance are governed by its molecular weight and molecular weight distribution (often referred to as dispersity), which are controlled by the synthetic methodology adopted in their production. Enormous amount of work has been performed to modify synthetic strategies adopted in the production of polymers, with an aim to achieve control over the chain microstructures and to deliver advanced polymers with a narrow molecular weight distribution.

Majority of the polymer production is performed via free-radical polymerization technique, which is associated with numerous side reactions and, therefore, delivers polymer with variable characteristics. Interestingly, "living" polymerization technique emerged as a significant approach to deliver polymers with narrow molecular weight distributions, as these reactions do not involve termination of radicals and transfer reactions. However, early reports revealed that the ionic nature of the living polymerization technique renders certain drawbacks to this strategy, which include sensitivity to moisture, carbon dioxide, and acidic/basic impurities. In addition, performing copolymerization reactions via ionic mechanism is challenging, due to differences in the reactivities of the starting monomers, which limits the applicability of this technique. It is to be noted that "livingness" in radical polymerizations can be alternatively achieved by eliminating the chain termination reactions, which seems difficult initially, due to the fast kinetics of bimolecular termination of radicals. In view of the same, control in the radical polymerization is conferred by establishing a dynamic equilibrium in propagating radicals and dormant species (as presented in Figure 3.1), where the latter species has potential to undergo reactivation to form active radicals [3]. Mechanistically, the radicals are allowed to undergo rapid exchange with special transfer agents via degenerative transfer [4] and is termed as reversible deactivation radical polymerization (RDRP). It is to be noted that

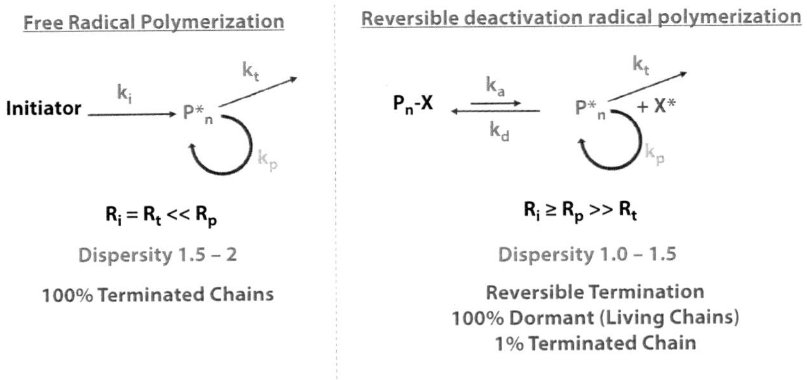

Figure 3.1 General reaction scheme showing the mechanism of conventional free-radical polymerization and reversible deactivation radical polymerization (RDRP) (reproduced with permission from ref. [5]).

radical polymerization methodologies that are classified as "controlled radical polymerization" (CRP) techniques are highly celebrated, as they have potential to deliver polymers with very low dispersity (1–1.5).

Over the years, RDRP has received great attention and resulted in the development of a variety of RDRP techniques for designing a multitude of macromolecular architectures. Among the variety of RDRP techniques, the most popular and successful ones are atom transfer radical polymerization (ATRP), reversible addition fragmentation chain transfer (RAFT) polymerization, and nitroxide-mediated polymerization (NMP). It is worth mentioning that the applicability of RDRP techniques has acquired the readiness level of large-scale production of polymers, [6] which was made possible by the continuous devotion of polymer chemists and chemical engineers towards optimizing the processes and overcoming a variety of obstacles, such as limited commercially available RDRP agents, long duration of reactions, sensitivity to oxygen, high amounts of remaining catalysts, and the complicated purification procedures. The efforts made in the advancement of the RDRP techniques from the view point of polymer chemistry and engineering have been excellently reviewed in detail by Zhou *et al.* [7] Apart from the precision synthesis of polymers and the attribute of scalability, RDRP techniques offer flexibility to adapt sustainability in the methodologies. Semsarilar and Perrier [8] have reported that RDRP techniques have addressed all the 12 principles of green chemistry. Firstly, RDRP techniques have satisfied the criteria of atom economy by providing a control over the architecture of the polymer, as well as limiting the production of side products, which is considered as a waste of raw materials. Further, the process allows the use of safer solvents such as aqueous solvents, as well as the polymerization can be performed in either homogeneous environment or dispersion conditions. Furthermore, RDRP techniques have proven to be energy-efficient, as some of the techniques allow polymerization at room temperature. Moreover, RDRP techniques allow the polymerization of a variety of functional monomers and do not require additional protection/deprotection steps. Additionally, catalysts that are employed in RDRP techniques can be recycled after polymerization. Furthermore, RDRP allows the production of polymeric chains with low molecular weights, which can be conveniently eliminated using bioactive media. Most importantly, the chemistry utilized in RDRP does not mandate the use of rigorous reaction conditions. Green advancements made towards improving the sustainability of a variety of RDRP techniques have been reviewed in detail, [8–10] and the recent reports on adopting either environment-friendly or energy-efficient synthetic protocols for polymerization have been discussed in this chapter.

3.2 Green Advances in Atom Transfer Radical Polymerization Technique

First reported in 1995, [11–13] ATRP has observed extensive upgradation and is well known for engineering functional polymers, such as block copolymers, [14] as well as has been reported for grafting of interesting complex structures onto the surface of nanoparticles to impart smart characteristics [15]. Generally, ATRP involves a reversible reaction of metal complex (metal is in low oxidation state) with an alkyl halide, which then results in the formation of the radical species and the metal complex additionally coordinated with

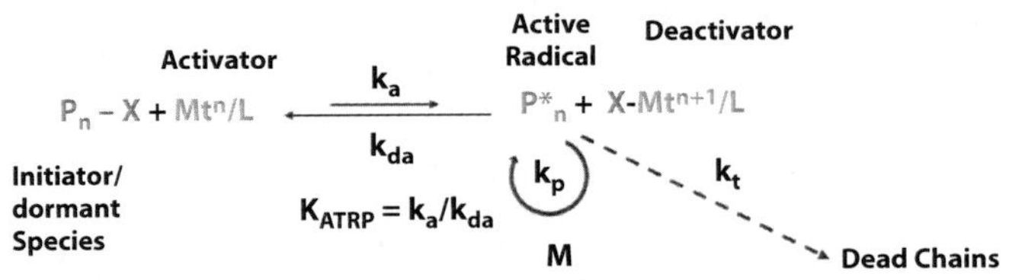

Figure 3.2 General reaction scheme representing atom transfer radical polymerization (ATRP) (reproduced with permission from ref. [7]).

the halide ligand (metal is in higher oxidation state) (Figure 3.2). It is to be noted that metal complexes based on Ti, [16] Mo, [17] Re, [18] Ru, [19] Fe, [20] Rh, [21] Ni, [22] Pd, [23] Co, [24] and Cu [25] have been successfully employed to perform polymerization of a variety of monomers via ATRP. The "livingness" in ATRP is proffered by activation of the alkyl halide adduct of the unsaturated monomer for its propagation with the incoming monomer and is subsequently followed by its deactivation to regenerate the dormant species. Interestingly, ATRP equilibrium can be established from any of the sides, either by combining alkyl halide with a metal complex in lower oxidation state or by combining radical source with a metal complex in higher oxidation state, and the latter is referred to as reverse ATRP, which offers to employ a stable catalyst, thereby making the process convenient [26–28].

ATRP has been considered as one of the preferred processes for the production of speciality polymers that can be utilized as coatings and surfactants, as well as to develop materials for biomedical applications. However, present-day polymer industry emphasizes on greener synthetic approaches, and the same is reflected in the enormous reports being published in this area. In this direction, Tsarevsky and Matyjaszewski [9] listed a few issues associated with ATRP, which should be overcome in order to enhance the sustainability associated with this polymerization technique. It is to be noted that ATRP is a metal-based catalyst-mediated polymerization and thus utilizes a large amount of metal-based catalysts. In view of the same, the removal of the catalyst is considered of primary importance particularly in the polymer that has been designed for electronic or biomedical application; otherwise, the end product is contaminated with a significant amount of metal complexes. In this context, Campalani *et al.* recently reported the use of nitrogen-doped carbon nanoparticles as metal-free photosensitizer, in addition to the Cu(II) complex, which allowed the use of small quantities (1% mol based on the monomer) of metal catalyst for the preparation of polymer from [2-(methacryloyloxy)ethyl] trimethylammonium chloride. It was reported that on irradiation, the carbon dots undergo photoexcitation to their excited state, where the latter can transfer electrons to aid the reduction of the Cu(II) to Cu(I) complex, which is responsible to initiate the radical polymerization of alkyl bromide [Figure 3.3(a)]. It is to be noted that using light as stimulus for initiating any reaction is an abundant and noninvasive environment-friendly technique [29]. Further, the possibility of performing ATRP in water was evaluated to contribute in the sustainable development of polymerization, and the water/methanol (1:1) mixture was found to be an effective reaction medium. The photosensitizer was hydrothermally derived by heating aqueous solutions of citric acid and diethylenetriamine. It is interesting to note that the polymerization was activated under irradiance of UV as well as visible light, which offers inexpensive and affordable methodology for the production of polymethacrylates.

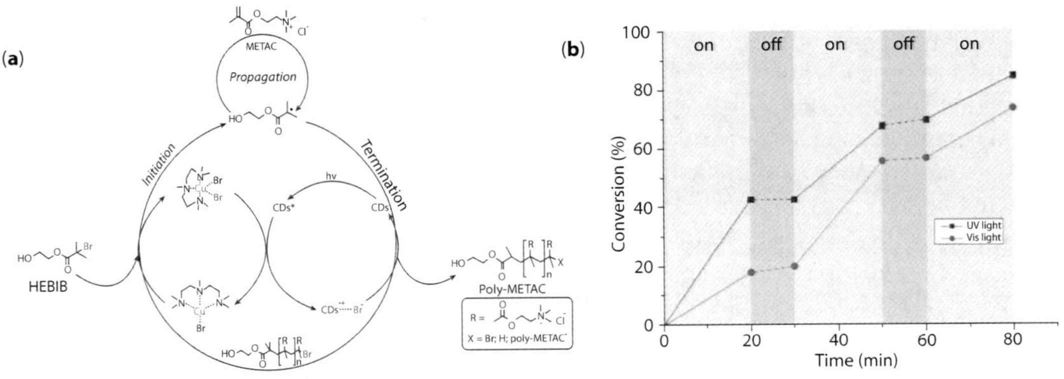

Figure 3.3 (a) Mechanism proposed for polymerization of 2-(methacryloyloxy) ethyl] trimethylammonium chloride) via photo-induced ATRP using carbon dots as photosensitizer and Cu(II) complex. (b) Monomer conversion (%) determined with respect to time using UV (black plot) as well as visible light (red plot) irradiation. Experiments were performed in on–off conditions, and yellow regions refer to changes in monomer conversion during light on while blue regions to light off. (*Reproduced with permission from* [30].)

Furthermore, monomer conversion of 89% with 1.4 dispersity was determined in 1 hour of UV irradiation, while the visible light resulted in 70% monomer conversion with 1.5 dispersity [Figure 3.3(b)] [30].

It is essential to note that the presence of oxygen in ATRP interferes with the initiation as well as propagation of radicals by generating peroxy radicals [31]. In addition, the presence of oxygen can lead to oxidation of the active form of catalyst [Cu(I)], thereby converting it into inactive form [Cu(II)], resulting in inhibition of the polymerization [32]. In view of the same, polymerization via ATRP technique requires deoxygenation reaction condition, which is usually accomplished via purging of inert gas or multiple freeze-pump thaw cycles [11]. However, this becomes difficult for polymerization of biohybrids, as deoxygenation of the chemical environment can lead to degradation of biomolecules such as proteins and nucleic acids [33, 34]. In view of the same, the ATRP technique has been modified with reducing agents [35–38], or has been integrated with electro-, [39] photo-, [40] or mechanical techniques [41] to regenerate the Cu(I)/L activator. Among these methods, integration of ATRP with photocatalytic technique (photo-induced ATRP) has been highly appreciated due to mild reaction conditions, as well as the integration offers the ability to temporally and spatially control polymerization [42–44]. Conventionally, photo-induced ATRP is performed in the presence of an amine-based ligand, which assists the reduction of Cu(II)L to Cu(I) L, where the latter species further reacts with P-X, forming carbon radical and X-Cu(II)/L [45]. Furthermore, employing an amine-based excess ligand has been additionally reported to improve the oxygen tolerance of this approach [46, 47]. It is to be noted that the reaction that is stimulated using lights conventionally employs biocidal UV light for polymerization, which imposes difficulties in designing biohybrid architectures, and is applicable to certain functionalities [48]. Interestingly, combining catalytic activity of copper complexes with the photocatalysts in ATRP has been found to overcome the abovementioned downsides such as the need of biocidal UV light, low oxygen tolerance, and applicability to limited monomers [49–53]. Furthermore, another important desire for "greener" ATRP is performing the polymerization in water, as it is considered as one of the most important aspects of green chemistry, as well as the aqueous reaction medium is ideal for deriving biohybrid polymers. However, it has been observed that equilibrium constants of ATRP are usually on the higher side in water, which can increase the concentration of radicals, and consequently, the number of dead chains will increase [54]. Furthermore, the presence of water can cause significant dissociation of the X-Cu(II)/L to the free halide ion and Cu(II)/L, where the latter will not be able to serve as a deactivator for the propagating radical [55, 56]. In addition, water can cause damage to P-X by hydrolyzing it to P-OH, where the latter will not be able to propagate further [57–59]. Due to these reasons, limited reports are available where photo-induced ATRP of hydrophilic acrylates has been attempted in water medium [60–62]. Recently, Kapil *et al.* [63] reported the polymerization of water-soluble acrylates via dual catalyzed ATRP technique using biorelevant conditions. In this report, green light (λ–520 nm) was employed to induce ATRP of oligo(ethylene oxide) methyl ether methacrylate, while ppm levels of Eosin Y, Cu(II)/Me_6TREN (where Me_6TREN is tris[2-(dimethylamino) ethyl]amine) were utilized as photo- and copper catalyst, respectively. Interestingly, without performing deoxygenation, high monomer conversions were observed within 40 minutes, as well as molecular weight distribution was found to be narrow with PDI in the range of 1.17–1.23. The scope of the technique was expanded to polymerization of 2-hydroxyethyl acrylate (HEA), 2-(methylsulfinyl)ethyl acrylate (MSEA), and zwitterionic carboxy betaine acrylate (CBA) monomers, where monomer conversions were found in the range of 54%–78% with PDI in the range of 1.16–1.22. Further, protein–polymer hybrids were additionally synthesized by conjugating α-chymotrypsin macroinitiators having 7 and 12 ATRP initiator sites with poly(ethylene oxide) methyl ether acrylate at 15°C–18°C and without rigorous deoxygenation conditions.

ATRP can be additionally made an environmentally benign process by using an iron-based catalyst, which was first reported in 1997 [64, 65]. Furthermore, iron-mediated ATRP processes have been found to be associated with advantages such as easy handling, appreciable tolerance toward oxygen, and better biocompatibility due to nontoxic nature of iron catalysts [66]. Typically, the iron-mediated ATRP technique involves the regeneration of Fe(II) complexes from a catalyst having iron in III oxidation state and is aided by a reducing agent. Furthermore, the choice of ligand is very crucial in homogenizing the iron catalyst in the reaction media, as well as is responsible in adjusting the redox potential of the metal center, which controls the dynamics of atom transfer in the ATRP. In this direction, phosphorus-based compounds are usually employed as ligands [67]; however, they have been found to be toxic, as well as expensive, and therefore, ligands based on organic acids, such as pyromellitic acid, iminodiacetic acid, succinic acid, isophthalic acid, have been reported as alternatives [68–71]. Interestingly, iron-mediated ATRP polymerization can be performed in the absence of ligands, where the need of the ligand is sufficed by the presence of polar solvents, such as N-methyl-2-pyrrolidone, dimethylformamide, acetonitrile, polyethylene glycol, and dimethyl sulfoxide [72]. Dadashi-Silab *et al.* [73] and Schroeder *et al.* [74] reported that iron-based anionic and cationic species are generated via disproportionation or displacement of ligand by molecules of polar solvents, which then governs the ATRP equilibrium. It is to be noted that certain polar solvents such as dimethylformamide are considered highly toxic, as well as impose difficulty in their removal. In this direction, bio-based green solvents such as γ-valerolactone have been reported for the iron-catalyzed AGET ATRP [where AGET is activators (re)generated by electron transfer] of methyl methacrylate (MMA) monomer [75]. Recently, Song *et al.* [76] have reported the iron-mediated AGET ATRP of MMA by using ethyl lactate as a green solvent and $FeBr_3$ as the catalyst without using any additional ligand. The combination of ethyl lactate with Fe species was able to display monomer conversion of 61.1% with 1.62 dispersity. It was further reported that using ethyl lactate undoubtedly enhanced the sustainability of the polymerization; however, using a high dosage of catalyst and poor controllability of polymerization in the later stages still need to be addressed.

3.3 Green Advances in Reversible Addition Fragmentation Chain Transfer Polymerization Technique

The RAFT polymerization technique has been found to be versatile and can be applied to a huge library of monomers, such as styrene derivative, acrylates, acrylamides, methacrylates, methacrylamides, and vinyl esters. Further, RAFT polymerization has been found to be tolerant to a wide variety of reaction conditions and, therefore, does not require any special free-radical polymerization setups. The degenerative process in RAFT polymerization proceeds by using an organic chain transfer agent referred to as RAFT agent, which comprises a thiocarbonylthio functionality. Figure 3.4 shows the general structure of RAFT agents, where **R** group is involved in initiating the growth of polymeric chains, while the **Z** group is responsible for activating the thiocarbonyl functionality toward radical addition, as well as in the stabilization of the resultant adduct radical. The mechanism generally accepted for RAFT polymerization has been presented in Figure 3.4, where the first step

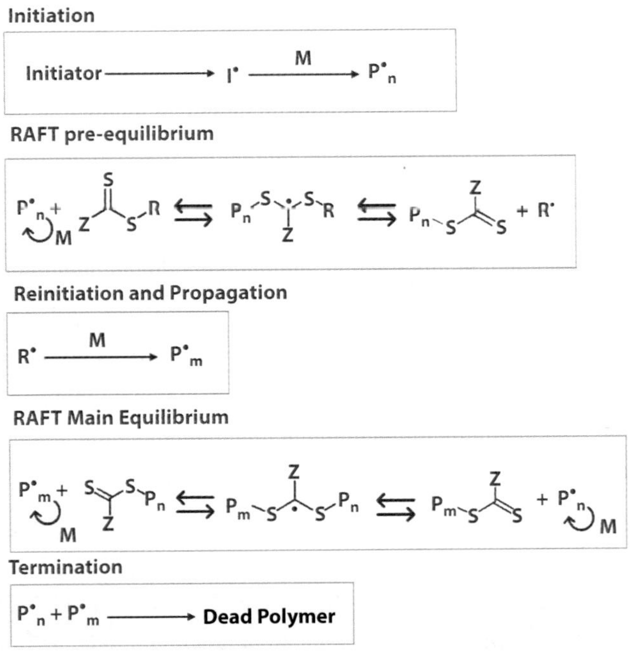

Figure 3.4 Representation of general structures of RAFT agents and steps involved in RAFT polymerization. (Reproduced from (ref [77]) with permission from the Royal Society of Chemistry).

is initiation step, which involves the generation of radical from initiator and oligomeric radicals are additionally produced. This is followed by the reversible reaction of radical addition of oligomeric radicals on the highly reactive C=S bond of RAFT agent. Further, the product undergoes fragmentation where the oligomeric RAFT agent and a reinitiating R radical are generated. Furthermore, reinitiation involves growing of polymer chains by the addition of monomers, which are then exchanged rapidly with the species capped with a thiocarbonylthio functionality. It is this step that limits the termination of polymer chains; however, disproportionation mechanism can still operate to terminate the growth of polymer chains. It is to be noted that the structures of **R** and **Z** groups in RAFT agents are highly crucial in deciding the success of RAFT polymerization. It is preferred that the **R** group of RAFT agent should be a better leaving group, relative to the propagating radical, due to its involvement in the pre-equilibrium stage of the polymerization.

Semsarilar and Perrier [8] have excellently reviewed the sustainable growth of RAFT polymerization and have summarized the reports on performing RAFT polymerization in environment-friendly conditions. In a section of the article, it was suggested that sustainability of any reaction originates from the possibility of performing it in bulk conditions, where it is expected that all starting materials will be converted in the product. This ideology will certainly obviate the need for purification steps, however, uncontrollable exothermic reaction conditions encountered due to gel effect (Trommsdorff–Norrish effect) are usually considered as a downside with bulk polymerization. Interestingly, employing RAFT agents in radical polymerization has been observed to overcome the downside associated with the gel effect, as well as offers aid in increasing the monomer conversion [78–82].

As already mentioned, the presence of oxygen in the chemical environment of the radical polymerization becomes challenging due to the quenching of radical species generated in the initiation and propagation steps. In view of the same, Shanmugam *et al.* [83] and Xu *et al.* [84] reported the modified polymerization technology by integrating RAFT polymerization technique with the photooxidation/reduction catalytic reaction (referred to as PET-RAFT), which is activated in the presence of visible/infrared light. Following this, several investigations were reported additionally by varying catalysts in PET-RAFT polymerization technique such as quantum dots, [85] organic dyes, [86] metal oxides, [87] metalloporphyrins, [88] $Ru(bpy)_3Cl_2$, [89] and bacteriochlorophyll a [90]. Furthermore, the integration of these two techniques benefited in terms of low energy consumption, mild reaction condition, controllability, and high tolerance toward oxidation. Further, the photo-redox catalyst has potential to reduce the molecular oxygen to superoxide, which is inert, and hence renders operability without the need of deoxygenation of the reaction medium. PET-RAFT has additionally achieved selectivity in the photo-activation reactions, which can be utilized in preparing polymers with sequential control. Shanmugam *et al.* explored zinc porphyrins (zinc tetraphenylporphine) as photo-catalyst, where the catalyst was able to selectively activate PET-RAFT polymerization of styrene, which was ascribed to a preferential interaction between them, and the same can be extended to the area of the molecular recognition [88]. Recently, Sun *et al.* [91] fabricated a biosensor platform based on the "trinity" molecular recognition mode by adopting a Zn(TCPP) PET-RAFT polymerization signal amplification strategy for detecting miRNA-21 with a detection limit of 4.48 aM.

It is to be noted that inclusion of organic or inorganic photo-redox catalysts to enhance the oxygen tolerance of RAFT polymerization has been widely accepted; however, developing a simple method where the deoxygenation protocol is metal-free will be considered more "greener." In this direction, Huang *et al.* [92] reported a two-component system comprising of 2-(dimethylamino)ethyl methacrylate (DMAEMA) and 4-cyanopentanoic acid dithiobenzoate (CPADB), where the former acted as a monomer and an electron donor, while the latter acted as an iniferter agent (initiator-transfer-terminator) for polymerization control, as well as an intrinsic photo-redox catalyst for deoxygenation.

Another aspect of enhancing the sustainability of the polymerization technique is to widen its applicability in designing biorenewable plastics. In this direction, Hatton [93] has reviewed numerous reports where RAFT polymerization technique has been employed to develop novel plastics from different biomass resources. Additionally, Versteeg *et al.* [94] demonstrated the use of RAFT polymerization technique for polymerization of α-methylene-butyrolactone, which is a biomass-derived monomer having five-membered γ-butyrolactone ring, and obtained macromolecular architectures with molecular weights in the range of 10,000–20,000 g mol^{-1}, having dispersity less than 1.5. Further, it was reported that the polymerization of α-methylene-butyrolactone was performed in supercritical carbon dioxide due to its low toxicity and nonflammable characteristics. In addition, supercritical CO_2 is associated with high diffusivity, which can be ascribed to its unique characteristics such as its viscosity is similar to that of gas and density is of a liquid. Furthermore, the conversion of the monomer was found to be 85% in supercritical CO_2, while polymerization in DMF resulted in a monomer conversion of approximately 65% [94]. Additionally, Liu *et al.* [95] recently reported the use of RAFT polymerization technique to develop amphiphilic lignin copolymers, for their use as stabilizing solid particles in pickering emulsions, which are associated with low energy consumption, as the requirement of agitation and sonication is removed. In this report, RAFT chain transfer agents with photo-initiating properties were developed first by modifying the hydroxyl groups of lignin with carbon disulfide and brominating agents. This was followed by the evaluation of the developed lignin-based chain transfer agents for photo-initiating RAFT polymerization of acrylamide and then were employed to derive amphiphilic copolymers using a variety of monomers under UV light irradiance. The prepared copolymers were finally utilized to prepare nanoparticles using a nanoprecipitation technique, which then were employed to form oil-in-water pickering emulsions [95].

3.4 Green Advances in Nitroxide-Mediated Polymerization Technique

Among RDRP techniques, nitroxide-mediated polymerization exhibits high scale-up potential, as the process does not necessarily use transition metal-based catalysts or complex thiol-based transfer agents [96, 97]. Consequently, NMP does not need additional purification steps for removing the metal-based catalysts, which makes this technique favorable for designing macromolecular architectures for sensitive biological or electronic applications. Nitroxide-mediated polymerization is based on the reversible activation–deactivation of a stable free radical, which facilitates controlled radical

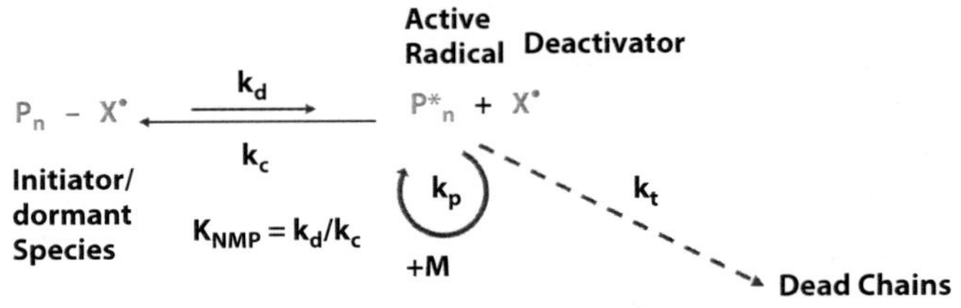

Figure 3.5 General reaction scheme of nitroxide-meditated polymerization technique (reproduced with permission from ref. [7]).

polymerization of monomers, and the general reaction scheme is presented as Figure 3.5. Initially, NMP was considered effective in the polymerization of styrene derivatives only; however, emerging designs of initiators such as N-tert-butyl-N-[1-diethylphosphono-(2,2-dimethylpropyl)] (referred to as SG1), 2,2,5-trimethyl-4-phenyl-3-azahexane-3-oxy (referred to as TIPNO), and 3-(((2-cyanopropan-2-yl)oxy)-(cyclohexyl)amino)-2,2-dimethyl-3-phenylpropanenitrile (referred to as Dispolreg 007) have expanded its applicability to acrylates, [98] methacrylates, [99, 100] dienes, [101] vinyl acetate, [102] acrylamides [103]. In view of the versatility displayed by NMP, this technique has been employed to design advanced materials for lithium batteries, organic electronics, photolithography, drug delivery systems, and antimicrobial surfaces [104].

Further, mediating the synthesis of polymers by nitroxide radicals is considered as a highly "greener" choice, as these species display low toxicity. Durand *et al.* [105] determined the cytotoxicity of nitroxides and reported that six-membered TEMPO derivatives are relatively less toxic than five-membered cyclic nitroxide radicals at millimolar concentration. It is to be noted that nitroxides have been reported to exhibit mutagenic behavior at millimolar levels [106]; however, these levels of concentrations are difficult to achieve by the polymers prepared via nitroxide-meditated polymerization in a biological system [107]. Furthermore, NMP requires a reaction temperature in the range of 70°C–130°C for formation of radical species, which suggests the energy intensiveness of this technique. In this regard, making attempts toward developing new NMP initiators in order to reduce the reaction temperature, and subsequently, implementing the "green chemistry" principle of low energy requirement, cannot be considered as an effective advancement, as the energy to maintain a reaction temperature contributes as a very small fraction of production costs. In 2021, Zaremski and Melik-Nubarov [10] made the first attempt to review the green advancements reported for the NMP technique and accentuated the use of photopolymerization, aqueous media, supercritical carbon dioxide, and bio-sourced monomers for polymerization.

As mentioned above for other RDRP techniques, utilizing monomers derived from natural resources have been considered as one of the important vectors to direct the green development of the polymerization techniques. In this direction, Luk and Maríc reported the mini-emulsion polymerization of farnesene via the NMP technique, which is a bio-sourced monomer, using succinimidyl-modified BlocBuilder (NHS-BB) and D7 initiators. It was reported that the work is associated with three sustainable aspects, which include the use of a bio-sourced monomer, the use of aqueous medium for polymerization, and performing the polymerization at atmospheric pressure by considering the lower volatility of farnesene monomer. Additionally, poly(far) was subjected to chain extension with isobornyl methacrylate (iBOMA) using D7 initiator to form a fully bio-sourced di-block copolymer. Furthermore, farnesene and iBOMA have been reported to be bio-based alternatives of butadiene and styrene monomer, respectively, and thus, the process can be adopted to design a highly useful material like poly(styrene-block-butadiene-block-styrene) (SBS) [108]. Moreover, the technique has been additionally adopted to prepare bio-sourced copolymers from isobornyl methacrylate and a mixture of methacrylic esters with an average alkyl side chain length of 13 units (C13MA)) using the Dispolreg 007 initiator, where the monomer conversion was found to be 94% and dispersity was found to be in the range of 1.5–1.73 [100].

Recently, the use of aqueous medium for NMP polymerization was demonstrated by Lages *et al.*, where nitroxide-meditated seeded emulsion copolymerization of dibenzo[c,e] oxepane-5-thione (DOT) with n-butyl acrylate (n-BA) as well as with styrene, using a two-step emulsion process, was reported. It was mentioned that n-BA was first subjected to polymerization in water using BlocBuilder alkoxyamine as initiator and DOWFAX 8390 as the surfactant, which resulted in a stable suspension of low-molecular weight living seeds. This was followed by the chain extension using the mixture of DOT and n-BA resulting in a stable latex, which retained their colloidal stability up to a month [109].

Furthermore, light-activated polymerization has been regarded as one of the interesting approaches to reduce the energy requirement. In addition, the modification allows the prevention of side reactions such as disproportionation reactions between the initiator and propagating radicals, which has direct impact on the molecular weight distribution of the polymers. In view of the same, macroscale 3D printing systems have adapted photo-RDRP techniques with a generic aim to enable the extension, customization, and reconfiguration of 3D objects [110]. Recently, Belqat *et al.* [111] reported the use of nitroxide-meditated photopolymerization of acrylates by employing benzophenone-based alkoxyamine initiator (labeled as PA1) to generate 4D objects via 3D direct laser writing technique. Furthermore, it was mentioned that 4D feature in objects is rendered by the "livingness" associated with the 3D objects, which allows their customization [111].

3.5 Conclusions and Future Perspective

Performing a radical polymerization to its completion stage as well as delivering a product with high-end group fidelity used to be considered impossible. Interestingly, CRP techniques have made this possible to achieve, as well as provide significant tolerance towards oxygen. In addition, CRP techniques have the potential to deliver polymers with very low dispersity relative to the polymers obtained from conventional free-radical polymerization. Among the variety of CRP techniques developed so far, the most successful ones are atom transfer radical polymerization, reversible addition fragmentation chain transfer polymerization, nitroxide-mediated polymerization, which operates through reversible-deactivation of radicals using special transfer agents. Several reports revealed that ATRP can be successfully performed in environmentally friendly reaction media, including water, and supercritical carbon dioxide. In addition, reports where the catalytic activity of copper complexes has been combined with the photocatalysts in ATRP, have reported to overcome the downsides such as biocidal UV light, low oxygen tolerance, and applicability to limited monomers. Furthermore, iron-mediated ATRP processes have also emerged as a greener approach due to easy handling, appreciable tolerance toward oxygen, and better biocompatibility due to nontoxic nature of iron catalysts. RAFT polymerization is another CRP technique where the degenerative process proceeds by using an organic chain transfer agent referred to as RAFT agents that comprises thiocarbonylthio functionality. RAFT polymerization technique has been integrated with photooxidation/reduction catalytic reaction, where the modification has benefited in terms of low energy consumption, mild reaction condition, controllability, and high tolerance toward oxidation. It is to be noted that inclusion of organic or inorganic photo-redox catalysts to enhance the oxygen tolerance of RAFT polymerization has been widely accepted; however, a metal-free approach has been additionally developed that is considered "greener." Furthermore, the sustainability of RAFT polymerization technique has been improved by expanding its applicability to design biorenewable plastics. Reports on NMP polymerization technique for the preparation of adhesives, coatings, dispersers, and compatibilizers are available. Further, the applicability of nitroxide-mediated photopolymerization has expanded to a variety of monomers, and therefore, the technique has been utilized to design advanced materials for lithium batteries, organic electronics, photolithography, drug delivery systems, and antimicrobial surfaces. Unfortunately, NMP has received late attention to improve its sustainability, as per the principles of green chemistry. In view of the same, limited reports are available related to the utilization of renewable starting materials in NMP techniques, and therefore, more attention from polymer chemists is required.

References

1. Staudinger, H., Über Polymerisation. *Ber. Dtsch. Chem. Ges. A B Ser.*, 53, 1073–1085, 1920, https://doi.org/10.1002/cber.19200530627.
2. Britt, P.F., Coates, G.W., Winey, K.I., Byers, J., Chen, E., Coughlin, B., Ellison, C., Garcia, J., Goldman, A., Guzman, J., Hartwig, J., Helms, B., Huber, G., Jenks, C., Martin, J., McCann, M., Miller, S., O'Neill, H., Sadow, A., Scott, S., Sita, L., Vlachos, D., Waymouth, R., Report of the Basic Energy Sciences Roundtable on Chemical Upcycling of Polymers, United States, 2019, https://doi. org/10.2172/1616517.
3. Matyjaszewski, K., Introduction to living polymeriz. Living and/or controlled polymerization. *J. Phys. Org. Chem.*, 8, 197–207, 1995, https://doi.org/10.1002/poc.610080403.
4. Matyjaszewski, K., Gaynor, S., Wang, J.-S., Controlled Radical Polymerizations: The Use of Alkyl Iodides in Degenerative Transfer. *Macromolecules*, 28, 2093–2095, 1995, https://doi.org/10.1021/ma00110a050.
5. Dworakowska, S., Lorandi, F., Gorczyński, A., Matyjaszewski, K., Toward Green Atom Transfer Radical Polymerization: Current Status and Future Challenges. *Adv. Sci.*, 9, 2106076, 2022, https://doi.org/10.1002/advs.202106076.
6. Matyjaszewski, K. and Spanswick, J., Controlled/living radical polymerization. *Mater. Today*, 8, 26–33, 2005, https://doi.org/10.1016/S1369-7021(05)00745-5.
7. Zhou, Y.-N., Li, J.-J., Wang, T.-T., Wu, Y.-Y., Luo, Z.-H., Precision polymer synthesis by controlled radical polymerization: Fusing the progress from polymer chemistry and reaction engineering. *Prog. Polym. Sci.*, 130, 101555, 2022, https://doi.org/10.1016/j.progpolymsci.2022.101555.
8. Semsarilar, M. and Perrier, S., 'Green' reversible addition-fragmentation chain-transfer (RAFT) polymerization. *Nat. Chem.*, 2, 811–820, 2010, https://doi.org/10.1038/nchem.853.

9. Tsarevsky, N.V. and Matyjaszewski, K., "Green" Atom Transfer Radical Polymerization: From Process Design to Preparation of Well-Defined Environmentally Friendly Polymeric Materials. *Chem. Rev.*, 107, 2270–2299, 2007, https://doi.org/10.1021/cr050947p.
10. Zaremski, M.Y. and Melik-Nubarov, N.S., Reversible Deactivation Radical Polymerization Mediated by Nitroxides and Green Chemistry. *Polym. Sci. Ser. C*, 63, 126–143, 2021, https:// doi. org/10.1134/S1811238221020120.
11. Wang, J.-S. and Matyjaszewski, K., Controlled/"living" radical polymerization. atom transfer radical polymerization in the presence of transition-metal complexes. *J. Am. Chem. Soc.*, 117, 5614–5615, 1995, https://doi.org/10.1021/ja00125a035.
12. Kato, M., Kamigaito, M., Sawamoto, M., Higashimura, T., Polymerization of Methyl Methacrylate with the Carbon Tetrachloride/Dichlorotris-(triphenylphosphine)ruthenium(II)/ Methylaluminum Bis(2,6-di-tert-butylphenoxide) Initiating System: Possibility of Living Radical Polymerizati. *Macromolecules*, 28, 1721–1723, 1995, https://doi.org/10.1021/ ma00109a056.
13. Percec, V. and Barboiu, B., "Living" Radical Polymerization of Styrene Initiated by Arenesulfonyl Chlorides and CuI(bpy)nCl. *Macromolecules*, 28, 7970–7972, 1995, https://doi.org/10.1021/ ma00127a057.
14. Mühlebach, A., Gaynor, S.G., Matyjaszewski, K., Synthesis of Amphiphilic Block Copolymers by Atom Transfer Radical Polymerization (ATRP). *Macromolecules*, 31, 6046–6052, 1998, https://doi.org/10.1021/ma9804747.
15. Liu, T., Jia, S., Kowalewski, T., Matyjaszewski, K., Casado-Portilla, R., Belmont, J., Water-Dispersible Carbon Black Nanocomposites Prepared by Surface-Initiated Atom Transfer Radical Polymerization in Protic Media. *Macromolecules*, 39, 548–556, 2006, https://doi. org/10.1021/ma051659y.
16. Kabachii, Y.A., Kochev, S.Y., Bronstein, L.M., Blagodatskikh, I.B., Valetsky, P.M., Atom Transfer Radical Polymerization with Ti(III) Halides and Alkoxides. *Polym. Bull.*, 50, 271–278, 2003, https://doi.org/10.1007/s00289-003-0157-9.
17. Stoffelbach, F., Claverie, J., Poli, R., Atom transfer radical polymerisation of styrene controlled by phosphine-containing coordination compounds of Mo(III)/Mo(IV). *C. R. Chim.*, 5, 37–42, 2002, https://doi.org/10.1016/S1631-0748(02)01306-1.
18. Kotani, Y., Kamigaito, M., Sawamoto, M., Living Radical Polymerization of Para-Substituted Styrenes and Synthesis of Styrene-Based Copolymers with Rhenium and Iron Complex Catalysts. *Macromolecules*, 33, 6746–6751, 2000, https://doi.org/10.1021/ma000624p.
19. Haas, M., Solari, E., Nguyen, Q.T., Gautier, S., Scopelliti, R., Severin, K., A Bimetallic Ruthenium Complex as a Catalyst Precursor for the Atom Transfer Radical Polymerization of Methacrylates at Ambient Temperature. *Adv. Synth. Catal.*, 348, 439–442, 2006, https://doi. org/10.1002/adsc.200505318.
20. Gibson, V.C., O'Reilly, R.K., Wass, D.F., White, A.J.P., Williams, D.J., Polymerization of Methyl Methacrylate Using Four-Coordinate (α-Diimine)iron Catalysts: Atom Transfer Radical Polymerization vs Catalytic Chain Transfer. *Macromolecules*, 36, 2591–2593, 2003, https://doi. org/10.1021/ma034046z.
21. Opstala, T., Zedníkb, J., Sedláčekb, J., Svobodab, J., Vohlídalb, J., Verpoorta, F., Atom Transfer Radical Polymerization of Styrene and Methyl Methacrylate Induced by RhI(cycloocta-1,5-diene) Complexes. *Collect. Czechoslov. Chem. Commun.*, 67, 1858–1871, 2002, https://doi. org/10.1135/cccc20021858.
22. Duquesne, E., Habimana, J., Degée, P., Dubois, P., Nickel-Catalyzed Supported ATRP of Methyl Methacrylate Using Cross-Linked Polystyrene Triphenylphosphine as Ligand. *Macromolecules*, 38, 9999–10006, 2005, https://doi.org/10.1021/ma051413t.

23. Lecomte, P., Drapier, I., Dubois, P., Teyssié, P., Jérôme, R., Controlled Radical Polymerization of Methyl Methacrylate in the Presence of Palladium Acetate, Triphenylphosphine, and Carbon Tetrachloride. *Macromolecules*, 30, 7631–7633, 1997, https://doi.org/10.1021/ma970890b.
24. Wang, B., Zhuang, Y., Luo, X., Xu, S., Zhou, X., Controlled/"Living" Radical Polymerization of MMA Catalyzed by Cobaltocene. *Macromolecules*, 36, 9684–9686, 2003, https://doi.org/10.1021/ma035334y.
25. Teodorescu, M. and Matyjaszewski, K., Controlled polymerization of (meth)acrylamides by atom transfer radical polymerization. *Macromol. Rapid Commun.*, 21, 190–194, 2000, https://doi. org/10.1002/(SICI)1521-3927(200003)21:4<190::AID-MARC190>3.0.CO;2-S.
26. Xia, J. and Matyjaszewski, K., Homogeneous Reverse Atom Transfer Radical Polymerization of Styrene Initiated by Peroxides. *Macromolecules*, 32, 5199–5202, 1999, https://doi.org/10.1021/ma990036f.
27. Xia, J. and Matyjaszewski, K., Controlled/"Living" Radical Polymerization. Homogeneous Reverse Atom Transfer Radical Polymerization Using AIBN as the Initiator. *Macromolecules*, 30, 7692–7696, 1997, https://doi.org/10.1021/ma9710085.
28. Wang, J.-S. and Matyjaszewski, K., "Living"/Controlled Radical Polymerization. Transition-Metal- Catalyzed Atom Transfer Radical Polymerization in the Presence of a Conventional Radical Initiator. *Macromolecules*, 28, 7572–7573, 1995.
29. Yagci, Y., Jockusch, S., Turro, N.J., Photoinitiated polymerization: advances, challenges, and opportunities. *Macromolecules*, 43, 6245–6260, 2010, https://doi.org/10.1021/ma1007545.
30. Campalani, C., Bragato, N., Morandini, A., Selva, M., Fiorani, G., Perosa, A., Carbon dots as green photocatalysts for atom transfer radical polymerization of methacrylates. *Catal. Today*, 418, 114039, 2023, https://doi.org/10.1016/j.cattod.2023.114039.
31. Bhanu, V.A. and Kishore, K., Role of oxygen in polymerization reactions. *Chem. Rev.*, 91, 99–117, 1991, https://doi.org/10.1021/cr00002a001.
32. Matyjaszewski, K., Coca, S., Gaynor, S.G., Wei, M., Woodworth, B.E., Controlled Radical Polymerization in the Presence of Oxygen. *Macromolecules*, 31, 5967–5969, 1998, https://doi.org/10.1021/ma9808528.
33. Baker, S.L., Kaupbayeva, B., Lathwal, S., Das, S.R., Russell, A.J., Matyjaszewski, K., Atom Transfer Radical Polymerization for Biorelated Hybrid Materials. *Biomacromolecules*, 20, 4271–4298, 2019, https://doi.org/10.1021/acs.biomac.9b01271.
34. Zhang, L., Zhang, Y., Cheng, J., Wang, L., Wang, X., Zhang, M., Gao, Y., Hu, J., Zhang, X., Lü, J., Li, G., Tai, R., Fang, H., Inert Gas Deactivates Protein Activity by Aggregation. *Sci. Rep.*, 7, 10176, 2017, https://doi.org/10.1038/s41598-017-10678-3.
35. Matyjaszewski, K., Dong, H., Jakubowski, W., Pietrasik, J., Kusumo, A., Grafting from Surfaces for "Everyone": ARGET ATRP in the Presence of Air. *Langmuir*, 23, 4528–4531, 2007, https://doi. org/10.1021/la063402e.
36. Flejszar, M., Ślusarczyk, K., Chmielarz, P., Smenda, J., Wolski, K., Wytrwal-Sarna, M., Oszajca, M., SI-ATRP on the lab bench: A facile recipe for oxygen-tolerant PDMAEMA brushes synthesis using microliter volumes of reagents. *Polym.*, 257, 125268, 2022, https:// doi.org/10.1016/j.polymer.2022.125268.
37. Sato, T., Dunderdale, G.J., Hozumi, A., Simple and Scalable Protocol for Producing Hydrophobic Polymer Brushes Beyond Wafer-Scale Dimensions toward Real-Life Applications. *ACS Appl. Polym. Mater.*, 3, 1395–1405, 2021, https://doi.org/10.1021/acsapm.0c01244.
38. Kang, H., Jeong, W., Hong, D., Antifouling Surface Coating Using Droplet-Based SI-ARGET ATRP of Carboxybetaine under Open-Air Conditions. *Langmuir*, 35, 7744–7750, 2019, https://doi.org/10.1021/acs.langmuir.9b00822.
39. De Bon, F., Fonseca, R.G., Lorandi, F., Serra, A.C., Isse, A.A., Matyjaszewski, K., Coelho, J.F.J., The scale-up of electrochemically mediated atom transfer radical polymerization without deoxy-genation. *Chem. Eng. J.*, 445, 136690, 2022, https://doi.org/10.1016/j.cej.2022.136690.

40. Aydogan, C., Yilmaz, G.S., Shegiwal, A., Haddleton, D.M., Yusuf, Y., Photo-induced Controlled/ Living Polymerizations. *Angew. Chemie Int. Ed.*, 61, e202117377, 2022, https://doi.org/10.1002/ anie.202117377.
41. Pan, X., Fantin, M., Yuana, F., Matyjaszewski, K., Externally controlled atom transfer radical polymerization. *Chem. Soc. Rev.*, 47, 5457–5490, 2018, https://doi. org/10.1039/C8CS00259B.
42. Chen, M., Zhong, M., Johnson, J.A., Light-Controlled Radical Polymerization: Mechanisms, Methods, and Applications. *Chem. Rev.*, 116, 10167–10211, 2016, https://doi.org/10.1021/acs. chemrev.5b00671.
43. Pan, X., Tasdelen, M.A., Laun, J., Junkers, T., Yagci, Y., Matyjaszewski, K., Photomediated controlled radical polymerization. *Prog. Polym. Sci.*, 62, 73–125, 2016, https://doi.org/10.1016/j. progpolymsci.2016.06.005.
44. Dadashi-Silab, S., Doran, S., Yagci, Y., Photoinduced Electron Transfer Reactions for Macromolecular Syntheses. *Chem. Rev.*, 116, 10212–10275, 2016, https://doi.org/10.1021/acs. chemrev.5b00586.
45. Ribelli, T.G., Konkolewicz, D., Bernhard, S., Matyjaszewski, K., How are Radicals (Re)Generated in Photochemical ATRP? *J. Am. Chem. Soc.*, 136, 13303–13312, 2014, https://doi.org/10.1021/ ja506379s.
46. Yang, Q., Lalevée, J., Poly, J., Development of a Robust Photocatalyzed ATRP Mechanism Exhibiting Good Tolerance to Oxygen and Inhibitors. *Macromolecules*, 49, 7653–7666, 2016, https://doi.org/10.1021/acs.macromol.6b01808.
47. Rolland, M., Whitfield, R., Messmer, D., Parkatzidis, K., Truong, N.P., Anastasaki, A., Effect of Polymerization Components on Oxygen-Tolerant Photo-ATR. *ACS Macro Lett.*, 8, 1546–1551, 2019, https://doi.org/10.1021/acsmacrolett.9b00855.
48. Neves-Petersen, M.T., Petersen, S., Gajula, G.P., UV Light Effects on Proteins: From Photochemistry to Nanomedicine, in: *Mol. Photochem. - Var. Asp*, pp. 125–158, 2011, https:// doi.org/10.5772/37947.
49. Szczepaniak, G., Jeong, J., Kapil, K., Dadashi-Silab, S., Yerneni, S.S., Ratajczyk, P., Lathwal, S., Schild, D.J., Das, S.R., Matyjaszewski, K., Open-air green-light-driven ATRP enabled by dual photoredox/copper catalysis. *Chem. Sci.*, 13, 11540–11550, 2022, https://doi.org/10.1039/ D2SC04210J.
50. Sun, M., Szczepaniak, G., Dadashi-Silab, S., Lin, T.-C., Kowalewski, T., Matyjaszewski, K., Cu-Catalyzed Atom Transfer Radical Polymerization: The Effect of Cocatalysts. *Macromol. Chem. Phys.*, 224, 2200347, 2023, https://doi.org/10.1002/macp.202200347.
51. Jazani, A.M., Schild, D.J., Sobieski, J., Hu, X., Matyjaszewski, K., Visible Light-ATRP Driven by Tris(2-Pyridylmethyl)Amine (TPMA) Impurities in the Open Air. *Macromol. Chem. Phys.*, 44, 2200855, 2023, https://doi.org/10.1002/marc.202200855.
52. Dadashi-Silab, S., Kim, K., Lorandi, F., Szczepaniak, G., Kramer, S., Peteanu, L., Matyjaszewski, K., Red-Light-Induced, Copper-Catalyzed Atom Transfer Radical Polymerization. *ACS Macro Lett.*, 11, 376–381, 2022, https://doi.org/10.1021/acsmacrolett.2c00080.
53. Luo, X., Wan, J., Meckbach, N., Strehmel, V., Li, S., Chen, Z., Strehmel, B., A Porphyrin-Based Organic Network Comprising Sustainable Carbon Dots for Photopolymerization. *Angew. Chemie - Int. Ed.*, 61, e202208180, 2022, https://doi.org/10.1002/anie.202208180.
54. Fantin, M., Isse, A.A., Matyjaszewski, K., Armando, G., ATRP in Water: Kinetic Analysis of Active and Super-Active Catalysts for Enhanced Polymerization Control. *Macromolecules*, 50, 2696–2705, 2017, https://doi.org/10.1021/acs.macromol.7b00246.
55. Fantin, M., Isse, A.A., Gennaro, A., Matyjaszewski, K., Understanding the Fundamentals of Aqueous ATRP and Defining Conditions for Better Control. *Macromolecules*, 48, 6862–6875, 2015, https://doi.org/10.1021/acs.macromol.5b01454.

56. Averick, S., Simakova, A., Park, S., Konkolewicz, D., Magenau, A.J.D., Mehl, R.A., Matyjaszewski, K., ATRP under Biologically Relevant Conditions: Grafting from a Protein. *ACS Macro Lett.*, 1, 6–10, 2012, https://doi.org/10.1021/mz200020c.
57. Coca, S., Jasieczek, C.B., Beers, K.L., Matyjaszewski, K., Polymerization of acrylates by atom transfer radical polymerization. Homopolymerization of 2-hydroxyethyl acrylate. *J. Polym. Sci. Part A Polym. Chem.*, 36, 1417–1424, 1998, https://doi.org/10.1002/(SICI)1099-0518(19980715)36:9<1417::AID-POLA9>3.0.CO;2-P.
58. Oh, J.K., Min, K., Matyjaszewski, K., Preparation of Poly(oligo(ethylene glycol) monomethyl ether methacrylate) by Homogeneous Aqueous AGET ATRP. *Macromolecules*, 39, 3161–3167, 2006, https://doi.org/10.1021/ma060258v.
59. Jones, G.R., Anastasaki, A., Whitfield, R., Engelis, N., Liarou, E., Haddleton, D.M., Copper-Mediated Reversible Deactivation Radical Polymerization in Aqueous Media. *Angew. Chemie - Int. Ed.*, 57, 10468–10482, 2018, https://doi.org/10.1002/anie.201802091.
60. De Bon, F., Barbosa, A.B., Fonseca, R.G., Fantin, M., Serra, A.C., Coelho, J.F.J., Large volume and oxygen tolerant photoinduced aqueous atom transfer radical polymerization. *Chem. Eng. J.*, 451, 138777, 2023, https://doi.org/10.1016/j.cej.2022.138777.
61. Bian, C., Zhou, Y.-N., Guo, J.-K., Luo, Z.-H., Aqueous Metal-Free Atom Transfer Radical Polymerization: Experiments and Model-Based Approach for Mechanistic Understanding. *Macromolecules*, 51, 2367–2376, 2018, https://doi.org/10.1021/acs.macromol.8b00348.
62. Bian, C., Zhou, Y.-N., Guo, J.-K., Luo, Z.-H., Photoinduced Fe-mediated atom transfer radical polymerization in aqueous media. *Polym. Chem.*, 8, 7360–7368, 2017, https://doi.org/10.1039/C7PY01762F.
63. Kapil, K., Jazani, A.M., Szczepaniak, G., Murata, H., Olszewski, M., Matyjaszewski, K., Fully Oxygen-Tolerant Visible-Light-Induced ATRP of Acrylates in Water: Toward Synthesis of Protein-Polymer Hybrids. *Macromolecules*, 56, 2017–2026, 2023, https://doi.org/10.1021/acs.macromol.2c02537.
64. Ando, T., Kamigaito, M., Sawamoto, M., Iron(II) Chloride Complex for Living Radical Polymerization of Methyl Methacrylate. *Macromolecules*, 30, 4507–4510, 1997, https://doi.org/10.1021/ma961478j.
65. Matyjaszewski, K., Wei, M., Xia, J., McDermott, N.E., Controlled/"Living" Radical Polymerization of Styrene and Methyl Methacrylate Catalyzed by Iron Complexes. *Macromolecules*, 30, 8161–8164, 1997, https://doi.org/10.1021/ma971010w.
66. Szczepaniak, G., Fu, L., Jafari, H., Kapil, K., Matyjaszewski, K., Making ATRP More Practical: Oxygen Tolerance. *Acc. Chem. Res.*, 54, 1779–1790, 2021, https://doi.org/10.1021/acs.accounts.1c00032.
67. Wang, Y., Kwak, Y., Matyjaszewski, K., Enhanced Activity of ATRP Fe Catalysts with Phosphines Containing Electron Donating Groups. *Macromolecules*, 45, 5911–5915, 2012, https://doi.org/10.1021/ma3010795.
68. Zhang, L., Cheng, Z., Shi, S., Li, Q., Zhu, X., AGET ATRP of methyl methacrylate catalyzed by FeCl3/iminodiacetic acid in the presence of air. *Polym.*, 49, 3054–3059, 2008, https:// doi.org/10.1016/j.polymer.2008.04.057.
69. Wang, G., Zhu, X., Cheng, Z., Zhu, J., Reverse atom transfer radical polymerization of methyl methacrylate with FeCl3/pyromellitic acid. *Eur. Polym. J.*, 39, 2161–2165, 2003, https://doi.org/10.1016/S0014-3057(03)00137-X.
70. Hou, C., Qu, R., Liu, J., Ying, L., Wang, C., High-molecular-weight polyacrylonitrile by atom transfer radical polymerization. *J. Appl. Polym. Sci.*, 100, 3372–3376, 2006, https://doi.org/10.1002/ app.23727.
71. Chen, H., Yang, L., Liang, Y., Hao, Z., Lu, Z., ARGET ATRP of acrylonitrile catalyzed by FeCl3/isophthalic acid in the presence of air. *J. Polym. Sci. Part A Polym. Chem.*, 47, 3202–3207, 2009, https://doi.org/10.1002/pola.23406.

72. Yang, D., He, D., Liao, Y., Xue, Z., Zhou, X., Xie, X., Iron-mediated AGET ATRP of methyl methacrylate in the presence of polar solvents as ligands. *J. Polym. Sci. Part A Polym. Chem.*, 52, 1020–1027, 2014, https://doi.org/10.1002/pola.27083.
73. Dadashi-Silab, S., Kim, K., Lorandi, F., Schild, D.J., Fantina, M., Matyjaszewski, K., Effect of halogen and solvent on iron-catalyzed atom transfer radical polymerization. *Polym. Chem.*, 13, 1059–1066, 2022, https://doi.org/DOI https://doi.org/10.1039/D1PY01601F.
74. Schroeder, H., Buback, J., Demeshko, S., Matyjaszewski, K., Meyer, F., Buback, M., Speciation Analysis in Iron-Mediated ATRP Studied via FT-Near-IR and Mössbauer Spectroscopy. *Macromolecules*, 48, 1981–1990, 2015, https://doi.org/10.1021/acs.macromol.5b00023.
75. Shen, X.R., Xia, D., Xiang, Y., Gao, J., γ-valerolactone (GVL) as a bio-based green solvent and ligand for iron-mediated AGET ATRP. *E-Polymers*, 19, 323–329, 2019, https://doi.org/10.1515/epoly-2019-0033.
76. Song, T., Xiang, Y., Gao, J., Shen, X., Iron-Mediated AGET ATRP of Methyl Methacrylate in Green Polar Ethyl Lactate Solvent without any External Ligand. *Polym. Sci. Ser. B.* 65, 103–110, 2023, https://doi.org/10.1134/S1560090423700756.
77. Lee, Y. Boyer, C. and Kwon, M. S., Photocontrolled RAFT polymerization: past, present, and future, *Chem. Soc. Rev.*, 52, 3035–3097, 2023, https://doi.org/10.1039/D1CS00069A.
78. Buback, M., Meiser, W., Vana, P., Mechanism of CPDB-Mediated RAFT Polymerization of Methyl Methacrylate: Influence of Pressure and RAFT Agent Concentration. *Aust. J. Chem.*, 62, 1484–1487, 2009, https://doi.org/10.1071/CH09219.
79. Heidenreich, A.J. and Puskas, J.E., Synthesis of arborescent (Dendritic) polystyrenes via controlled inimer-type reversible addition-fragmentation chain transfer polymerization. *J. Polym. Sci. Part A Polym. Chem.*, 46, 7621–7627, 2008, https://doi.org/10.1002/pola.23062.
80. Yu, B. and Lowe, A.B., Synthesis of di- and tri-tertiary amine containing methacrylic monomers and their (co)polymerization via RAFT. *J. Polym. Sci. Part A Polym. Chem.*, 47, 1877–1890, 2009, https://doi.org/10.1002/pola.23281.
81. Perrier, S.C.B.-K., Quinn, J.F., Vana, P., Davis, T.P., Origin of Inhibition Effects in the Reversible Addition Fragmentation Chain Transfer (RAFT) Polymerization of Methyl Acrylate. *Macromolecules*, 35, 8300–8306, 2002, https://doi.org/10.1021/ma0203445.
82. Wood, M.R., Duncalf, D.J., Findlay, P., Rannard, S.P., Perrier, S., Investigation of the Experimental Factors Affecting the Trithiocarbonate-Mediated RAFT Polymerization of Methyl Acrylate. *Aust. J. Chem.*, 60, 772–778, 2007, https://doi.org/10.1071/CH07171.
83. Shanmugam, S., Xu, J., Boyer, C., Photoinduced Electron Transfer–Reversible Addition–Fragmentation Chain Transfer (PET-RAFT) Polymerization of Vinyl Acetate and N-Vinylpyrrolidinone: Kinetic and Oxygen Tolerance Study. *Macromolecules*, 47, 4930–4942, 2014, https://doi.org/10.1021/ma500842u.
84. Xu, J., Shanmugam, S., Duonga, H.T., Boyer, C., Organo-photocatalysts for photoinduced electron transfer-reversible addition–fragmentation chain transfer (PET-RAFT) polymerization. *Polym. Chem.*, 6, 5615–5624, 2015, https://doi.org/10.1039/C4PY01317D.
85. McClelland, K.P., Clemons, T.D., Stupp, S.I., Weiss, E.A., Semiconductor Quantum Dots Are Efficient and Recyclable Photocatalysts for Aqueous PET-RAFT Polymerization. *ACS Macro Lett.*, 9, 7–13, 2020, https://doi.org/10.1021/acsmacrolett.9b00891.
86. Wu, C., Corrigan, N., Lim, C.-H., Jung, K., Zhu, J., Miyake, G., Xu, J., Boyer, C., Guiding the Design of Organic Photocatalyst for PET-RAFT Polymerization: Halogenated Xanthene Dyes. *Macromolecules*, 52, 236–248, 2019, https://doi.org/10.1021/acs.macromol.8b02517.
87. Hakobyan, K., Gegenhuber, T., McErlean, C.S.P., Müllner, M., Visible-Light-Driven MADIX Polymerisation via a Reusable, Low-Cost, and Non-Toxic Bismuth Oxide Photocatalyst. *Angew. Chemie - Int. Ed.*, 58, 1828–1832, 2019, https://doi.org/10.1002/anie.201811721.

88. Shanmugam, S., Xu, J., Boyer, C., Exploiting Metalloporphyrins for Selective Living Radical Polymerization Tunable over Visible Wavelengths. *J. Am. Chem. Soc.*, 137, 9174–9185, 2015, https://doi.org/10.1021/jacs.5b05274.
89. Xu, J., Jung, K., Boyer, C., Oxygen Tolerance Study of Photoinduced Electron Transfer–Reversible Addition–Fragmentation Chain Transfer (PET-RAFT) Polymerization Mediated by Ru(bpy)3Cl2. *Macromolecules*, 47, 4217–4229, 2014, https://doi.org/10.1021/ma500883y.
90. Shanmugam, S., Xu, J., Boyer, C., Light-Regulated Polymerization under Near-Infrared/Far-Red Irradiation Catalyzed by Bacteriochlorophyll a. *Angew. Chemie - Int. Ed.*, 128, 1048–1052, 2016, https://doi.org/10.1002/ange.201510037.
91. Sun, H., Liu, J., Kong, J., Zhang, J., Zhang, X., Ultrasensitive miRNA-21 Biosensor Based on Zn(TCPP) PET-RAFT Polymerization Signal Amplification and Multiple Logic Gate Molecular Recognition. *ACS Appl. Mater. Interfaces*, 15, 17716–17725, 2023, https://doi.org/10.1021/acsami.3c02428.
92. Huang, Y., Sun, Y., Weng, Y., Zhang, W., A Simple and Green Oxygen-Tolerant RAFT Polymerization without Additional Catalyst and Initiator. *ChemistrySelect*, 7, e202201583, 2022, https://doi. org/10.1002/slct.202201583.
93. Hatton, F.L., Recent advances in RAFT polymerization of monomers derived from renewable resources. *Polym. Chem.*, 11, 220–229, 2020, https://doi.org/10.1039/C9PY01128E.
94. Versteeg, F.G., Hegeman, N.C., Sebakhy, K.O., Picchioni, F., RAFT Polymerization of a Biorenewable/Sustainable Monomer via a Green Process. *Macromol. Rapid Commun.*, 43, 2200045, 2022, https://doi.org/10.1002/marc.202200045.
95. Liu, J., Shi, X., Ma, L., Zhang, D., Lai, C., Wang, C., Li, M., Ragauskas, A.J., Chu, F., Xu, Y., Facile design of renewable lignin copolymers by photoinitiated RAFT polymerization as Pickering emulsion stabilizers. *Green Chem.*, 25, 5428–5437, 2023, https://doi.org/10.1039/D3GC01500A.
96. Nicolas, J., Guillaneuf, Y., Lefay, C., Bertin, D., Gigmes, D., Charleux, B., Nitroxide-mediated polymerization. *Prog.Polym. Sci.*, 38, 63–235, 2013, https://doi.org/10.1016/j.progpolymsci.2012.06.002.
97. Grubbs, R.B., Nitroxide-Mediated Radical Polymerization: Limitations and Versatility. *Polym. Rev.*, 51, 104–137, 2011, https://doi.org/10.1080/15583724.2011.566405.
98. Vancoillie, G., Van Guyse, J.F.R., Voorhaar, L., Maji, S., Frank, D., Holder, E., Hoogenboom, R., Understanding the effect of monomer structure of oligoethylene glycol acrylate copolymers on their thermoresponsive behavior for the development of polymeric sensors. *Polym. Chem.*, 10, 5778–5789, 2019, https://doi.org/DOI https://doi.org/10.1039/C9PY01326A.
99. Schmidt, A.C., Turgut, H., Le, D., Beloqui, A., Delaittre, G., Making the best of it: nitroxide-mediated polymerization of methacrylates via the copolymerization approach with functional styrenics. *Polym. Chem.*, 11, 593–604, 2020, https://doi.org/10.1039/C9PY01458F.
100. Tajbakhsh, S., Hajiali, F., Marić, M., Nitroxide-Mediated Miniemulsion Polymerization of Bio-Based Methacrylates. *Ind. Eng. Chem. Res.*, 59, 8921–8936, 2020, https://doi.org/10.1021/acs.iecr.0c00840.
101. Métafiot, A., Pruvost, S., Gérard, J.-F., Defoort, B., Marić, M., Synthesis of isoprene-based triblock copolymers by nitroxide-mediated polymerization. *Eur. Polym. J.*, 134, 109798, 2020, https:// doi.org/10.1016/j.eurpolymj.2020.109798.
102. Banerjee, S., Domenichelli, I., Ameduri, B., Nitroxide-Mediated Alternating Copolymerization of Vinyl Acetate with tert-Butyl-2-trifluoromethacrylate Using a SG1-Based Alkoxyamine. *ACS Macro Lett.*, 5, 1232–1236, 2016, https://doi.org/10.1021/acsmacrolett.6b00707.
103. King, B. and Lessard, B.H., Controlled Synthesis and Degradation of Poly(N-(isobutoxymethyl) acrylamide) Homopolymers and Block Copolymers. *Macromol. React. Eng.*, 11, 1600073, 2017, https://doi.org/10.1002/mren.201600073.

104. Lamontagne, H.R. and Lessard, B.H., Nitroxide-Mediated Polymerization: A Versatile Tool for the Engineering of Next Generation Materials. *ACS Appl. Polym. Mater.*, 2, 5327–5344, 2020, https://doi.org/10.1021/acsapm.0c00888.
105. Durand, G., Choteau, F., Prosak, R.A., Rockenbauer, A., Villamena, F.A., Puccia, B., Synthesis, physical-chemical and biological properties of amphiphilic amino acid conjugates of nitroxides. *New J. Chem.*, 34, 1909–1920, 2010, https://doi.org/10.1039/C0NJ00024H.
106. Guo, X., Seo, J.-E., Bryce, S.M., Tan, J.A., Wu, Q., Dial, S.L., Moore, M.M., Mei, N., Comparative Genotoxicity of TEMPO and 3 of Its Derivatives in Mouse Lymphoma Cells. *Toxicol. Sci.*, 163, 214–225, 2018, https://doi.org/10.1093/toxsci/kfy022.
107. Delplace, V., Guégain, E., Harrisson, S., Gigmes, D., Guillaneuf, Y., Nicolas, J., A ring to rule them all: a cyclic ketene acetal comonomer controls the nitroxide-mediated polymerization of methacrylates and confers tunable degradability. *Chem. Commun.*, 51, 12847–12850, 2015, https:// doi.org/10.1039/C5CC04610F.
108. Luk, S.B. and Maríc, M., Polymerization of Biobased Farnesene in Miniemulsions by Nitroxide- Mediated Polymerization. *ACS Omega*, 6, 4939–4949, 2021, https://doi.org/10.1021/acsomega.0c05992.
109. Lages, M., Gil, N., Galanopoulo, P., Mougin, J., Lefay, C., Guillaneuf, Y., Lansalot, M., D'Agosto, F., Nicolas, J., Degradable Latexes by Nitroxide-Mediated Aqueous Seeded Emulsion copolymerization using a Thionolactone. *Macromolecules*, 56, 7973–7983, 2023, https://doi.org/10.1021/acs.macromol.3c01478.
110. Bagheri, A., Engel, K.E., Bainbridge, C.W.A., Xu, J., Boyer, C., Jin, J., 3D printing of polymeric materials based on photo-RAFT polymerization. *Polym. Chem.*, 11, 641–647, 2020, https://doi.org/10.1039/C9PY01419E.
111. Belqat, M., Wu, X., Morris, J., Mougin, K., Petithory, T., Pieuchot, L., Guillaneuf, Y., Gigmes, D., Clément, J.-L., Spangenberg, A., Customizable and Reconfigurable Surface Properties of Printed Micro-objects by 3D Direct Laser Writing via Nitroxide Mediated Photopolymerization. *Adv. Funct. Mater.*, 33, 2211971, 2023, https://doi.org/10.1002/adfm.202211971.

4

Catalytic Synthesis and Application of Heterocyclic and Heteroatom Compounds: Recent Advances

Nayeem Ahmed[1], Zeba N. Siddiqui[2*], Waqas A. Khan[3] and Hinna Hamid[3†]

[1]Department of Chemistry, University of Kashmir, Srinagar, India
[2]Department of Chemistry Aligarh Muslim University, Aligarh, India
[3]Department of Chemistry, Jamia Hamdard, Delhi, India

Abstract

Widely found in nature, heterocycles are necessary for life. The importance lies in the fact that pyrimidines and purines, which are found in the genetic material DNA, are heterocyclic in nature. Many heterocyclic substances, both man-made and natural, have pharmacological activity and are used in medicine. Many heterocyclic substances are used as herbicides, pesticides, fungicides, insecticides, and more. Pigments that produce chlorophyll and carry oxygen in hemoglobin are both heterocyclic substances. In view of their importance, large interest has been devoted toward their synthesis and many methods have been developed that are efficient and sustainable. Early stages of drug discovery necessitate rapid access to new chemical space for medicinal chemistry design hypotheses, such as new heterocyclic compounds or proper substitutions that increase the potency of a compound and lead us toward new drugs. The development of new synthetic methodologies and flexible protocols and the capability to introduce different functionalities by late-stage functionalization are frequently necessary during the medical trials in order to improve effectiveness, selectivity, metabolic stability, and other physicochemical properties.

Although well-recognized synthetic techniques are frequently used in drug discovery programs, the creation of novel pathways for the synthesis of heterocycles that permit a variety of bond formation techniques does have a considerable influence on the pharmaceutical sector. The bulk of the most popular reaction pathways in medicinal chemistry was revealed more than 20 years ago, with the most common synthetic approaches being amide bond formation, Suzuki–Miyaura coupling, and nucleophilic aromatic substitution. The predominance of these reaction pathways might be due to their clear substrate specificities, chemoselectivity, and tolerance of different functional groups, as well as the culture of medicinal chemistry, which places a premium on meeting deadlines and outsourcing reliable simple pathways for analog synthesis. Therefore, new pathways that are energy-efficient, sustainable, and save time are of great value and interest. New approaches like C-H bond activation, photoredox catalysis, borrowed hydrogen catalysis, MCRs, selective (stereo- and regio-) syntheses, and other novel inventive general protocols for the synthesis and functionalization of heterocyclic compounds have greatly aided in drug discovery and pharmaceutical research.

Therefore, this chapter aims to provide an overview of several cutting-edge heterocyclic synthesis techniques currently being used in the field of drug development and medicinal chemistry. The application of new methodologies for the formation of heterocyclic compounds, such as C-H bond

**Corresponding author*: Siddiqui_zeba@yahoo.co.in
†*Corresponding author*: hhamid@jamiahamdard.ac.in

Mousumi Sen (ed.) Sustainable Green Catalytic Processes, (79–96) © 2024 Scrivener Publishing LLC

activation, borrowing hydrogen catalysis, MCRs, and selective (stereo- and regio-) syntheses will be covered.

***Keywords*:** Heterocycles, medicines, bioactive compounds, drug discovery, catalysis

4.1 Introduction

Heterocyclic compounds have become indispensable in modern chemistry and pharmaceutical research due to their wide range of applications and significance in drug discovery programs. They are the most noticeable and distinct class of organic compounds, with more than 90% of medicines comprising of a heterocyclic core being developed after a deep understanding of the biological system [1]. These compounds have various uses in medicinal chemistry, including as dyes, disinfectants, corrosion inhibitors, and antioxidants and for the design and synthesis of copolymers [1]. Efficient approaches for designing pathways for newly discovered heterocycles and their moieties are crucial for accelerating the drug development process and expanding the drug-like chemical space [1, 2]. Heterocyclic compounds play a very important role in therapeutic chemistry and exhibit diverse biological activities, such as anticancer, anti-inflammatory, antifungal, antiallergic, antibacterial, anti-HIV, antiviral, and anticonvulsant properties [1]. The design of innovative synthetic pathways for functionalized heterocyclic compounds in collaboration between industry and academia has further shaped the pharmaceutical industry and drug discovery programs [2]. Recent advances in synthetic methodologies, such as C–H bond activation, photoredox catalysis, borrowing hydrogen catalysis, and MCRs, have greatly contributed to the design and functionalization of heterocyclic compounds and the delivery of projects in this field [2]. Furthermore, heterocycles are present in a significant majority of physiologically active chemical compounds and are important structural components of many anticancer drugs on the market today [1, 3]. Many pesticides, insecticides, and herbicides are based on heterocyclic compounds [4]. These compounds play a crucial role in protecting crops from pests and diseases, contributing to increased agricultural productivity. Heterocyclic compounds are also used as monomers or building blocks in the synthesis of polymers and materials [5]. They have unique properties that can lead to the development of novel materials with specific characteristics, such as flexibility, strength, and electrical conductivity.

Heterocyclic compounds are also prevalent in numerous natural products obtained from plants, animals, and microorganisms [6]. These natural products often possess bioactive properties and have been a rich source of lead compounds in drug discovery. Many flavoring agents and fragrances are derived from heterocyclic compounds. These compounds contribute to the taste and aroma of various foods, beverages, and perfumes [7].

The versatility and significance of heterocyclic compounds in these diverse applications make them an essential area of research and development in chemistry and related fields. Scientists continue to explore new heterocyclic structures and synthetic methods to discover novel compounds with improved properties and applications [8–10].

Drug research and development depend on having available a wide diversity of functionalized heterocyclic compounds. Even though there is a wealth of literature devoted toward the synthesis and functionalization of heterocyclic compounds, further research in this field is still urgently needed. Early stages of drug discovery necessitate quick access to novel heterocyclic compounds or desired substitutions, for instance, that can placate stringent physicochemical specifications, offer new directions for structure-based drug design, or permit availability of novel intellectual properties.

This chapter will therefore provide insight on several recent cutting-edge heterocyclic synthesis techniques currently being used in the field of drug development and medicinal chemistry. The modern protocols for the synthesis of heterocycles, such as C-H bond activation, enantioselective catalysis, MCRs, and photoredox chemistry, have been discussed.

4.1.1 Rhodium- and Silver-Catalyzed C-H Amidation

Metal-catalyzed C-H amidation techniques were used to create pleuromutilin, an oral antibiotic, (Scheme 4.1) C13 analogs [11]. Prior to the cyclization process, reduction of the vinyl side chain to an ethyl group was undertaken in order to prevent substantial side reactions and enable nitrogen introduction at C13 through nitrene generation and C-H bond insertion, although the yields were poor. However, the presence of the vinyl moiety allowed for C13 functionalization through intramolecular C-H amidation. According to the foundational work by Espino *et al.* [12], rhodium catalysis produced, at most, modest quantities of oxazolidinone; however, catalysis by Ag increased the C-H amidation yield to above 70% on a multigram scale [13]. Stereoselective insertion at C13 was seen in both instances.

4.1.2 C–H Bond Activation/[4 + 2] Annulation Catalyzed by Rh(III)

A potent technique for the synthesis of isoquinolones involves the C-H activation of aromatic carboxamides and aromatic N-pivaloyloxycarboxamides, followed by [4 + 2] annulation [14]. Early attempts to apply the approach to the annulation of nicotinamide derivatives, however, were hampered by slow rates of reaction and nonselective C-H bond activation at the C2 and C4 positions, which resulted in mixes of the regioisomeric compounds 1,6-naphthyridin-5(6H)-one and 2,7-naphthyridin-1(2H)-one (Scheme 4.2) [15]. Chemists at Amgen investigated nicotinamide N-oxide derivatives as substrates in the [4 + 2] annulation because they needed selective and efficient access to 1,6-naphthyridin-5(6H)-ones as drug scaffolds. They reasoned that the N-oxide would both help direct C-H activation to the 2-position and increase reaction rates because the heterocyclic ring is more electron-rich.

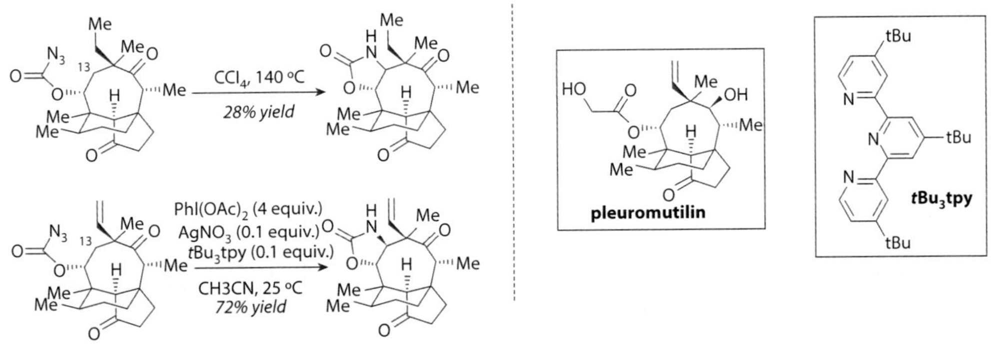

Scheme 4.1 Synthesis of pleuromutilin C13 analogs.

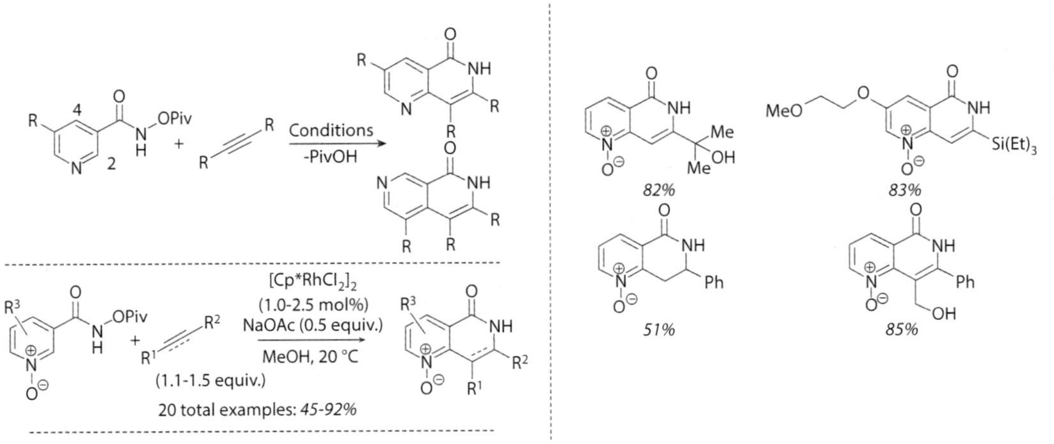

Scheme 4.2 Selective synthesis of 1,6-napthyridine-5-ones *via* [4 + 2] annulation of pyridine N-oxides catalyzed by rhodium.

Several examples show that this justification was accurate. Alkynes and alkenes were frequently used in effective reactions to produce compounds with excellent yields as a consequence of selective C2 activation. The results from earlier research with bulky and sp2-hybridized substituents favoring the site next to the lactam nitrogen were echoed in the regiochemistry of alkyne insertion. A wide range of substituents on the alkyne and alkene coupling components, as well as a variety of substituents on the nicotinamide, were tolerated in the reaction. The parent 1,6-naphthyri-din-5(6H)-ones were produced in high yields by reducing the N-oxide in moderate circumstances (Zn0, 30% aq. NH4Cl, THF, 20°C).

4.1.3 Enantioselective Copper-Catalyzed Hydroalkylation

Through a copper-catalyzed hydroalkylation, Wang *et al.* [16] created an enantioselective pathway for carbo- and heterocyclic compounds (Scheme 4.3). Numerous classes of compounds, such as appropriately modified cyclobutanes, cyclopentanes, indanes, and saturated

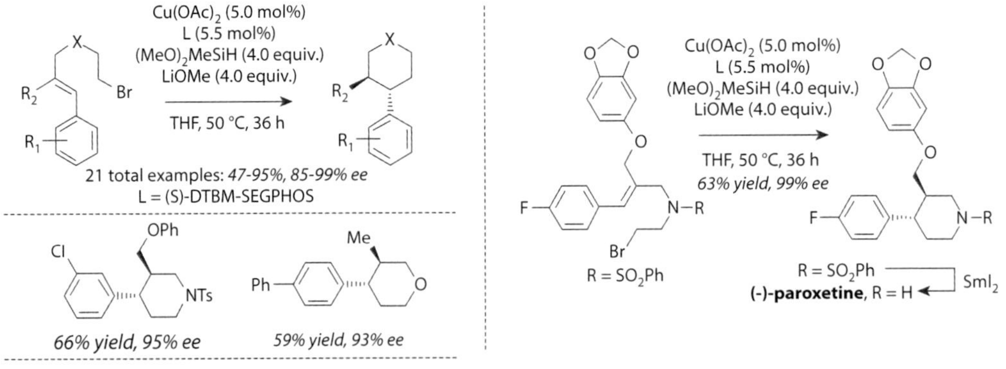

Scheme 4.3 Enantioselective Cu-catalysed hydroalkylation.

six-membered N- and O-containing heterocyclic compounds, were created starting with commonly accessible alkyl bromides and internal alkenes [16]. Esters, heterocycles, steric bulk, and even cis-substituted alkenes are all tolerated by the reaction. The transformation of precursor styrene with both electron-deficient and electron-rich functional groups has been demonstrated. Wang *et al.* [16] created the (-)-paroxetine that is a serotonin reuptake inhibitor as a proof of its usefulness. A single diastereomer containing SO_2PH group with good enantioselectivity was produced as a result of hydroalkylation of bromide substrate. The final target, (-)-paroxetine, was obtained in good yield when the sulfonyl group was later deprotected.

4.1.4 Multicomponent Copper-Catalyzed Synthesis of Quinoline Ring System

The quinoline motif has numerous biological and pharmacological activities, making it an important class of heterocyclic compounds for new drug development. Quinoline and its derivatives have several pharmacological activities, including antimalarial, anticancer, antibacterial, anthelmintic, antiviral, antifungal, anti-inflammatory, analgesic, cardiovascular, central nervous system, hypoglycemic, and miscellaneous activities. Due to these wide-ranging activities, numerous synthetic routes have been developed for the synthesis of quinoline and its derivatives. In particular, the present review highlights diverse methods for the synthesis of quinoline that can be used by synthetic chemists to achieve different synthetic modifications and access a new series of quinoline motifs for novel drug design [9]. These synthetic modifications are important because quinoline and its derivatives are versatile and have gained much attention in the area of drug development. However, the existing methods for the preparation of quinoline skeleton have limitations, and therefore, there is a need for more efficient methods to be developed. Quinoline nucleus is endowed with a variety of therapeutic activities, making it an important scaffold for the development of new drugs. Quinolines are being adopted by various research groups for their synthetic methodologies, and several established methods exist for the preparation of quinoline skeleton. In addition, new therapeutic agents have been developed using quinoline nucleus in recent years. With the occurrence of drug resistance in bacteria causing several diseases over the years, quinoline-derived drugs have seen new developments in medicine [9].

Recently, Qian *et al.* [17] from Amgen Inc. used azide-alkyne cycloaddition and Ullmann-type coupling to catalyze multicomponent Cu-catalyzed processes. They reported new poly-heterocyclic systems synthesized from commercially accessible reactants in a single pot utilizing a three-component cascade condensation (Scheme 4.4). The initial step of the suggested mechanism is a [3 + 2] cycloaddition reaction between an alkyne and an azide with the subsequent intermolecular amidation between a primary amide and an aryl bromide. A 2-quinolinone ring is created *via* an intramolecular aldol-dehydration (Camps cyclization). It was suggested that the Camps cyclization would be the last step, as it is made easier by the enhanced acidic nature of the CH_2 alpha to the carbonyl group after triazole synthesis. Diverse functional groups, such as esters, heteroaromatic groups, acetals, alcohols, basic amines, and halides are tolerated by the transition. One drawback is the use of secondary amides, which have low yields of product (30%) due to steric hindrance. Overall, the approach can produce two new rings, three new carbon-nitrogen bonds, and one new carbon-carbon bond in a single catalytic operation by combining a coupling reaction (Ullmann-type) with the click reaction of azide-acetylene.

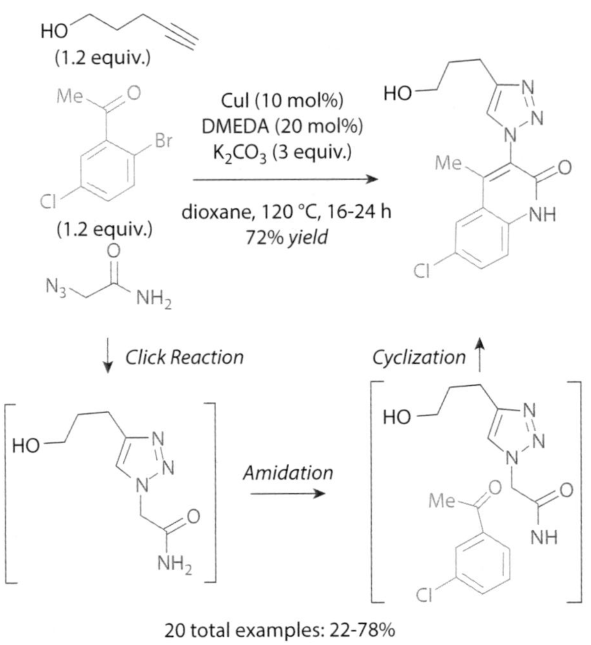

Scheme 4.4 Synthesis of heterocyclic compounds by the Cu-catalyzed multicomponent pathway.

4.1.5 Iridium-Catalyzed Enantioselective Synthesis of Vabicaserin

A 5-HT_{2C} receptor agonist called vabicaserin useful for the treatment of schizophrenia has successfully completed Phase 2 clinical trials (Scheme 4.5) [18]. Vabicaserin is structurally distinguished by a syn-fused cyclopentadiazepinoquinoline ring structure, which has proven difficult to synthesize. The intermediate quinolinium salt was produced *via* an oxidative multicomponent reaction, which Pfizer chemists characterized as being followed by an asymmetric hydrogenation [19] to produce the enantioselective synthesis of the vabicaserin derivative. In the first phase, iodine was used to create the quinolinium salt in an 89% yield. A phosphor-amidite-iridium system was created utilizing 2,6-di-tert-butylpyridine (2,6-DTBP). This system enabled decreasing of the catalyst amount to 0.5 mol%. It was discovered that the chloride anion increased enantioselectivity, possibly as a result of mixed halide catalysis [20]. After optimization, quinoline with reduced double bonds was acquired

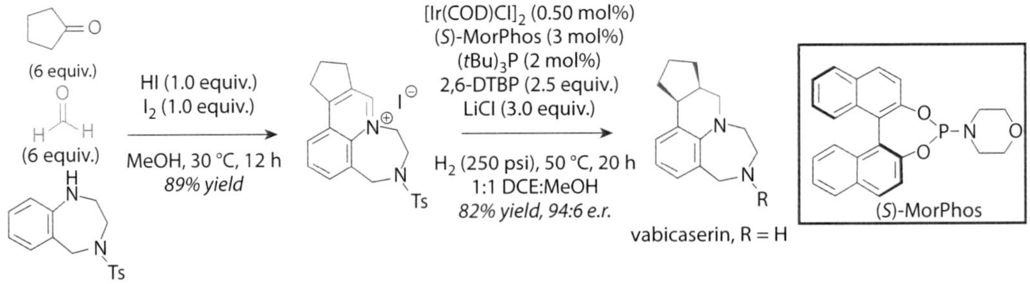

Scheme 4.5 Synthesis of vabicaserin, a potential drug for the treatment of schizophrenia.

in 82% yield and 92% ee. After employing HCl in acetic acid to remove the tosyl group, vabicaserin was subsequently produced in a 92% yield. The target was delivered in a final recrystallization at 99.9+% ee. Vabicaserin was produced overall *via* a four-step process using commercially accessible materials, an MCR, and catalytic asymmetric hydrogenation, with a 54% yield.

4.1.6 Catalytic C-H Bond Functionalization

The development of zinc sulfinates for quick and simple heterocycle diversification was reported by Baran's team and colleagues at Pfizer (Scheme 4.6). Zinc difluoromethanesulfinate reagent (DFMS) was used for the difluoromethylation of additional complicated heterocycles and was also described as an extension of the zinc sulfinate chemistry. It was also claimed that the nucleophilic radical zinc difluoromethylation reagent might be used as a litmus test to identify or forecast heteroaromatic medication candidates that are likely to undergo aldehyde oxidase metabolism [21–24].

Sulfinates consisting of small rings produced from different carboxylic acids are used in a Minisci process to functionalize heterocycles (Scheme 4.7) [25]. Using photolytic chemistry driven by visible light, the sulfinates were readily accessible from a variety of cyclic ethers and azetidine-based carboxylic acids. The process starts with the production of a Barton ester, which transforms into a sulfide when exposed to light. Several sulfinate reagents were produced by salt creation followed by $RuCl_3$-catalyzed oxidation, and they were employed as the radical source for the C-H bond functionalization.

4.1.7 C–H Bond Activation of Heterocycles by Photoredox Pathway

In order to create -aryl amines, MacMillan *et al.* developed a photoredox-catalyzed C-H arylation process for amines (Scheme 4.8) [26, 27]. Significant nitrogen-containing heterocyclic compounds that are common in best-selling medications are represented by this structural motif. As seen in Figure 4.1, the benzylic amine product is produced when cyclic and acyclic amines are exposed to benzonitriles when photocatalyst Ir(ppy)3 is present. In the catalytic cycle, the highly reducing *Ir(III)(ppy)3, which is itself produced by photoexcitation of the iridium photocatalyst Ir(III)(ppy)3, quickly reduces the 1,4-dicyanobenzene.

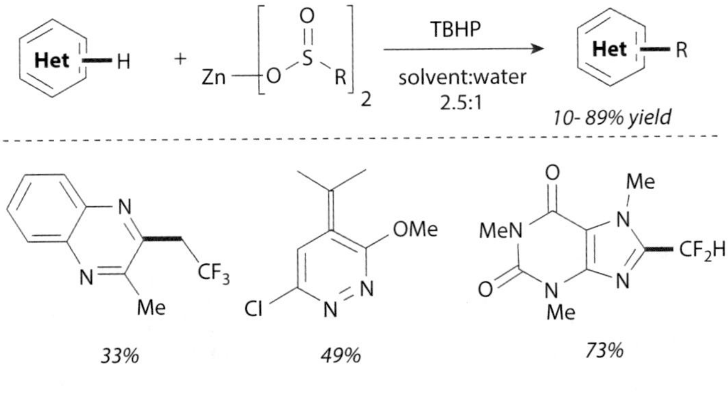

Scheme 4.6 Modification of heterocycles by C–H bond activation pathway using zinc sulfinates.

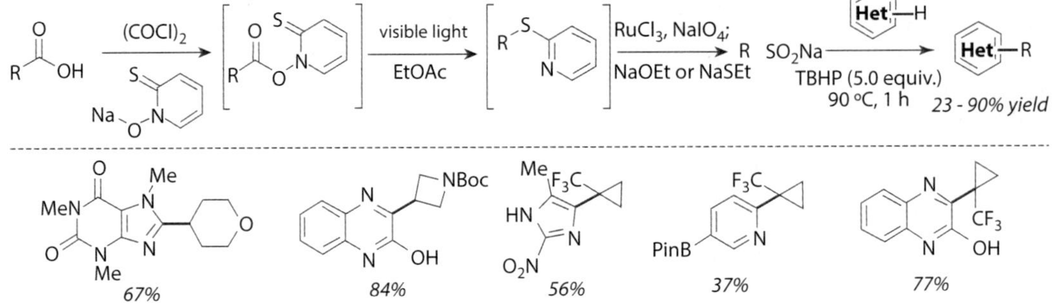

Scheme 4.7 Functionalization of heterocycles by sulfinate derived from carboxylic acids *via* C–H bond activation.

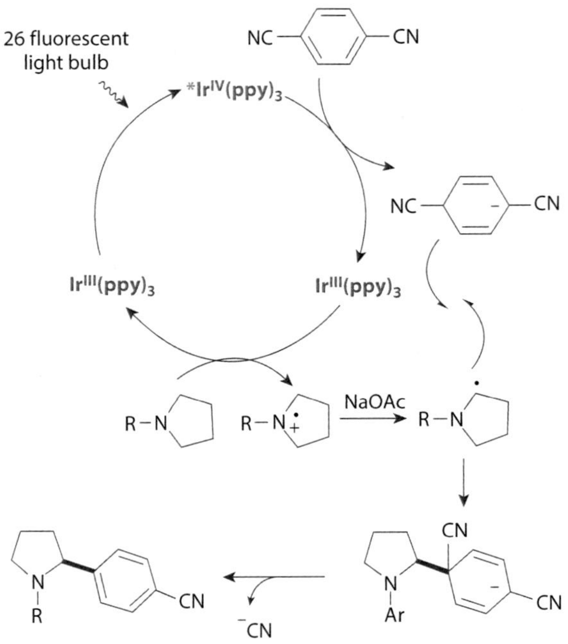

Figure 4.1 Schematic representation of the catalytic cycle showing photoredox α-arylation of amines (Reproduced with permission from Ref. [9]).

The single-electron oxidation of the oxidized Ir(IV)(ppy)3 species and subsequent deprotonation by acetate produce the -amino radical. A new C-C bond is subsequently produced by radical-radical coupling with the radical anion of 1,4-dicyanobenzene. The resulting -aryl amine is then produced by cyanide being expelled during aromatization. Different heterocyclic N-aryl amines and acyclic amines were effectively functionalized by using the optimized reaction conditions. Numerous five- or six-membered aromatic heterocycles with electron deficiencies were well-tolerated. Using the same process, linezolid, an antibiotic used to treat severe infections brought on by Gram-positive bacteria resistant to other antibiotics, was functionalized in the late stages. Later, the direct decarboxylative sp^2-sp^3 cross coupling of amino acids and aryl halides or cyanoarenes employed a similar strategy [28].

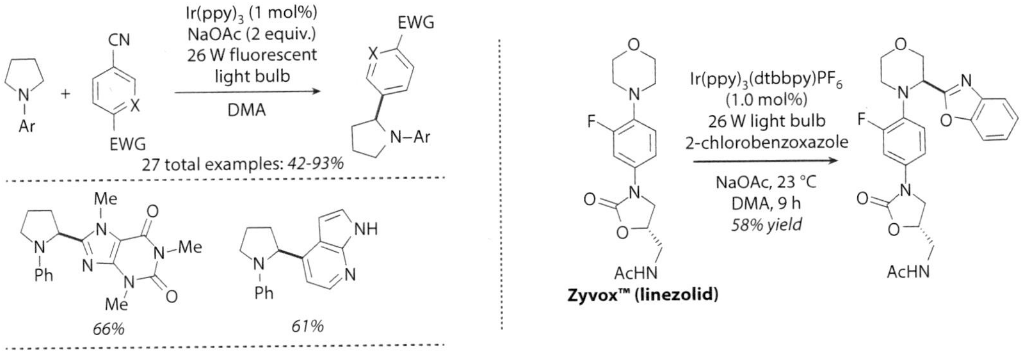

Scheme 4.8 Photoredox α-arylation of amines.

4.1.8 Cu-Catalyzed Synthesis of Indole-Fused Oxazinone-1,2,3-Triazoles

Triazole compounds have gained significant attention in modern synthetic chemistry and drug discovery due to their diverse biological activities and potential applications in the development of better therapeutic agents. The unique moiety of triazole compounds is responsible for their various biological activities, and further studies on their possible substitutions can lead to the synthesis of better agents with strong commitments in the future [29]. Triazole analogs have been reported to have potency for various pharmacological activities, including antimicrobial, antitumor, antiviral, anti-inflammatory, and analgesic activities, among others [29]. Triazole compounds favor hydrogen bonding and are stable for metabolic degradation, making them favorable in increasing solubility and binding biomolecular targets [30]. The resistance of the triazole ring to chemical hydrolysis and metabolic degradation makes it a better heterocyclic core for therapeutic agents. Additionally, triazole scaffolds can act as a spacer for covalent attachment of high-molecular weight biomacromolecules for exploring structure–function relationships. The potential applications of triazole compounds in drug discovery are vast, and their significance can be attributed to their broad range of biological activities and pharmacological properties.

An effective methodology for the synthesis of new heterocyclic compounds based on indole-fused oxazinone-1,2,3-triazole scaffolds is presented. This approach was inspired by the biological and pharmacological relevance of heterocyclic scaffolds [31]. It has been noted that 1,2,3-triazole and indole-fused oxazines utilize a multicomponent CuAAC-based process (copper-catalyzed azide-alkyne cycloaddition) when 18-crown-6 is present. Ketones were used in the iodotriazole synthesis. Following the creation of diiodoindole, the target product was produced in a modest amount of yield in a model reaction including sodium azide, phenylacetylene, and 10 mol% $CuSO_4.5H_2O$.

Different alkynes and sodium azide were utilized to investigate the reaction's potential, and the desired indole-fused oxazinone-1,2,3-triazole skeleton was produced in good to exceptional yields (Scheme 4.9). Alkyne reactants have been shown to easily integrate both electron-donating and electron-withdrawing groups, and several instances of indole-fused oxazinone-1,2,3-triazole skeletons have been created.

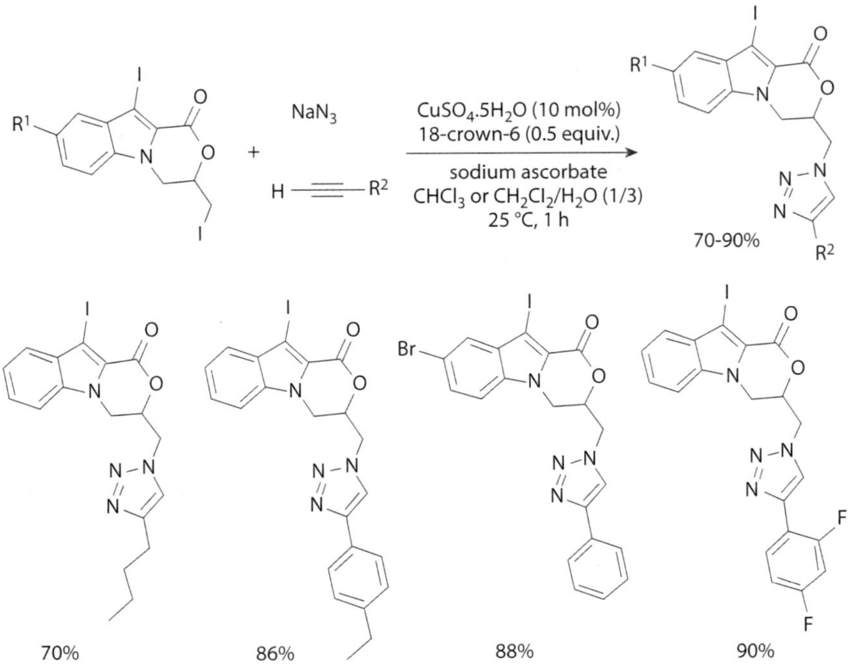

Scheme 4.9 CuAAC-based multicomponent reaction for the synthesis of indole-fused oxazinone-1,2,3-triazoles.

4.1.9 3-Isothiocyanato Oxindoles as Substrates for Catalytic Asymmetric Synthesis of Spiroindoles

Chiral spirooxindoles, which have a spiro-ring fused at the C3 position of the oxindole core, are among a family of special spirocyclic frameworks. With a wide spectrum of biological actions, including antitumor, antibacterial, anti-HIV, antimalarial, and antiviral, being antipyretic agents and sodium channel blockers, these privileged heterocyclic patterns have been abundantly identified in natural alkaloids and pharmaceutical compounds. Spirooxindoles are thought to be hard spirocyclic structures from a structural perspective, and they exhibit a strong affinity for three-dimensional proteins [32]. Another ongoing difficulty in organic synthesis is the creation of chiral spirocyclic frameworks. In this context, the effective synthesis of spirooxindoles, especially when done enantioselectively, has generated a great deal of interest in both the academic and industrial communities.

The 3-oxyindole isothiocyanate has been regarded as a universal structural component for the synthesis of spiroindoles ever since Yuan *et al.* [33] published their initial research in 2011. The 3-isothiocyanate-oxindole at the C3 position is very nucleophilic because of the strong electron-withdrawing characteristics of the NCS group and carbonyl group. NCS groups, on the other hand, show electrophilic characteristics. As a result, C=O, C=N, C=C, N=N, and CC are only a few of the electron-deficient unsaturated bonds that 3-oxoindole isothiocyanate may engage in [3+2] cyclization processes with (Scheme 4.10a). As of 2014, many organocatalytic or metal catalytic asymmetric processes involving

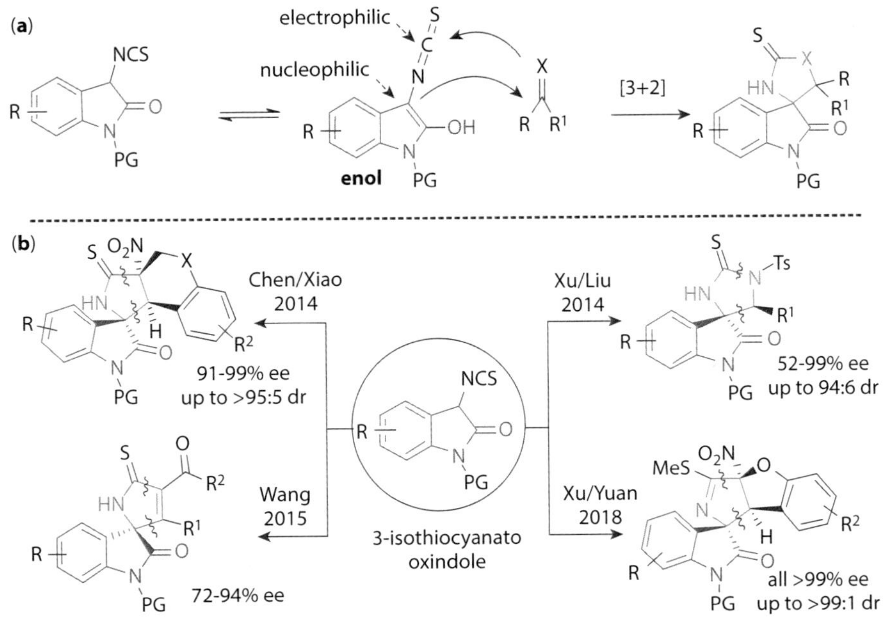

Scheme 4.10 Synthesis of chiral fused spirooxindoles from 3-isothiocyanato oxindoles.

3-isothiocyanate-indole have been established to produce chiral spiroindoles. For example, the Chen/Xiao group reported the asymmetric [3+2] cyclization of 3-nitro-2H-chromene catalyzed by Zn(OTf)2; the Xu/Liu group developed for imine synthesis a chiral amine catalyst system; the Wang group showed that alkynyl ketones can serve as suitable substrates; the Xu/Yuan group accomplished the enantioselective [3+2] cycloaddition of 2-nitrobenzofuran and 3-isothiocyanate-indole leading to dearomatization (Scheme 4.10b) [34].

4.1.10 Palladium-Catalyzed Synthesis of Chiral Fused Spirooxindoles

An essential intermediary for the cycloaddition process has been identified as Pd-containing 1,3-dipole, a π-allyl–Pd complex with an anion. In light of this, the synthesis of chiral spirooxindoles from 1,3-dipoles containing palladium has generated a lot of attention in the chemical world. In order to produce enantioenriched spirooxindoles, a highly stereoselective [3+2] cycloaddition of chiral palladium-containing N^1-1,3-dipoles with methyleneindolinones using palladium catalysis was reported. According to Scheme 4.11, prochiral substrates containing aryldiazonium salts underwent smooth carbonyl-directed Heck–Matsuda desymmetrization to produce spirooxindoles in good to exceptional yields with typically high diastereo- and enantioselectivities [35].

4.1.11 Co(III)-Catalyzed Synthesis of 2,3-Disubstituted *N*-Alkyl Indoles

Indoles have been found to play an essential role in human health and disease prevention. One significant area of research is the potential use of indole as a therapy for nonsteroidal

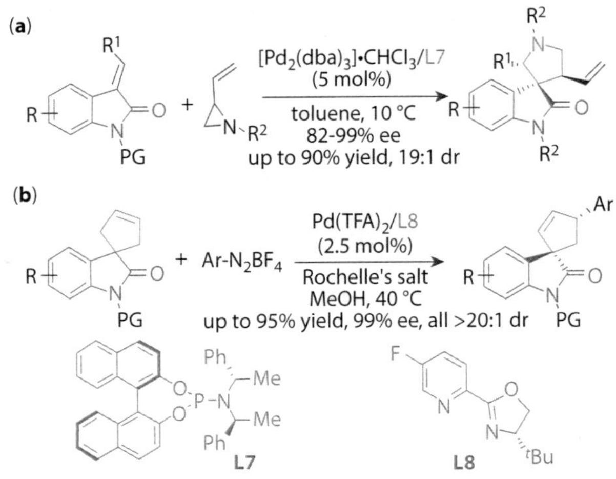

Scheme 4.11 Synthesis of spirooxindoles catalyzed by Pd.

anti-inflammatory drug (NSAID) enteropathy. Neutrophils are considered the key factor in the pathogenesis of NSAID enteropathy, and they cause inflammation and tissue damage by releasing various inflammatory mediators [36]. Indole has been shown to have a potential immunomodulatory effect on NSAID enteropathy. Despite indole's potential therapeutic benefits, it has a double-edged sword effect on human health and disease prevention. High concentrations of isatin (an indole derivative) can cause multiple system dysfunction and promote oxidative stress, inflammation, and other pathological changes. Furthermore, indole generated by liver metabolism has renal toxicity and cardiovascular toxicity at high concentrations. Nevertheless, indole can modulate aryl hydrocarbon receptor (AhR), which is crucial for brain development, immunity, metabolism, and cancer prevention. Indoles may also be helpful as therapeutic agents against pathogens such as EHEC. For instance, indole, ICA, and IAA downregulate the virulence of pathogenic *E. coli* strains and the production of Shiga toxins. Finally, the chemistry of indole derivatives can help in the design and synthesis of novel drugs useful in the mitigation of various disorders such as diabetes, cancer, tuberculosis, and malaria. Indole also helps in affixing drugs to the residues of the binding site of desired targets, and synthetic drug molecules contain an indole nucleus as part of their pharmacophore structure [37].

For the synthesis of 2,3-disubstituted N-alkyl indoles from N-nitroso anilines and alkynes (Scheme 4.12), Jiao *et al.* [38] created a Co(III)-based catalytic system. This procedure employed copper acetate as an additive and $[Cp^*Co(MeCN)_3](SbF_6)_2$ as a catalyst. Later, substrate scope tests utilizing 2,2,2-trifluoroethanol (TFE) as the solvent and aerobic conditions at 80°C were used to demonstrate the protocol's scope. When the p-position was either replaced with a halogen atom, an ERG/EWG, or an m-substituent, the selected nitrosoanilines were found to be well tolerated under the reaction circumstances. Additionally, several internal alkynes were investigated, and in some cases, a modest increase in temperature was needed. DMAD, alkyl-aryl alkynes, dialkyl arynes, (cyclopropylethynyl) benzene, etc. were also shown to be effective targets for the technique, but with lower yields. The ortho C-H bond is activated by the $Cp^*Co(OAc)_2$

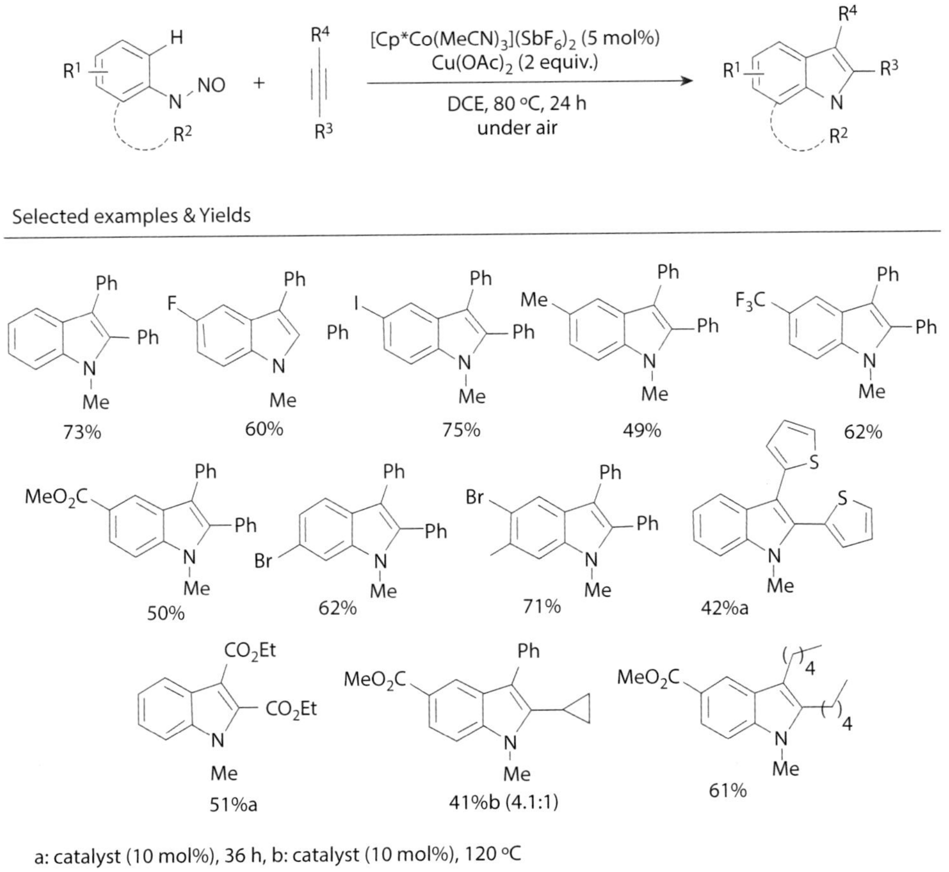

Scheme 4.12 $[Cp^*Co(MeCN)_3](SbF_6)_2$-catalyzed synthesis of indoles from N-nitrosoanilines.

species, with the directing effect of a NO group. The product is then released through a final acetate transfer step to the cobalt center after an alkyne insertion step and an acetate-mediated sp^2(C)-N bond formation step [38].

4.1.12 Co(III)-Catalyzed Synthesis of 2-Indolyl Propenols *via* C-H Bond Functionalization

By using a Co(III)-catalyzed C-H activation technique, Yu *et al.* [39] achieved an effective synthesis of 2-indolyl propenols from indoles and vinylethylene carbonates [39]. Using the ball milling method, the reaction performed effectively in orderly circumstances. By creating an indolyl-based acyclic Co(III)-intermediate to the Co-C bond, in which the vinyl group is introduced and then decarboxylated to generate the allylated indole, an *N*-pyrimidinyl group was able to introduce the directing effect. Different substituted vinylethylene carbonates were treated under the optimized circumstances (Scheme 4.13). Most of the time, E:Z selectivity 2.5:1 resulted in moderate to outstanding yields. However, it was somewhat unexpected to observe the stereoselectivity flipping in favor of the Z isomer

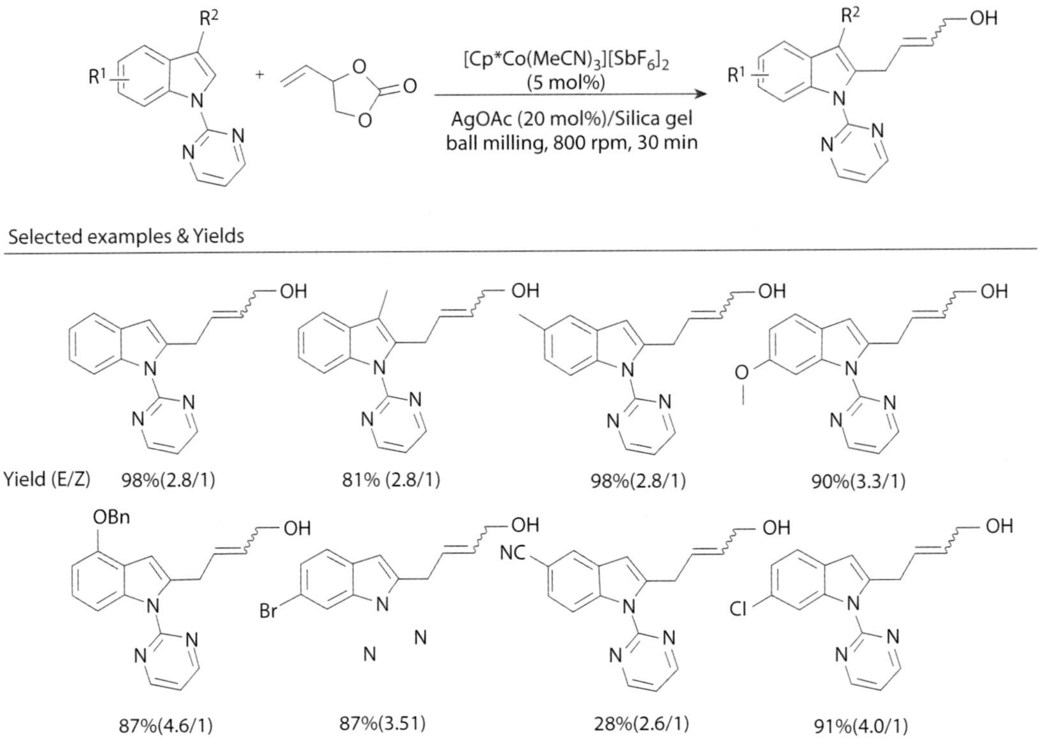

Scheme 4.13 Directing effects of pyrimidine-group toward the synthesis of 2-indolyl propenols *via* C–H bond functionalization.

throughout the process. The presence of a deactivating cyano-group reduced the yield, even though there was no discernible trend in the yields depending on the electronic nature of the substituents. In comparison to its other analogs, the yield was somewhat decreased when a methyl group was present in the third position, probably as a result of steric interactions [39].

4.2 Conclusion

The catalytic processes provide the means to achieve our desired products in a manageable time period and in good yields. The catalytic processes become more important when we talk from the perspective of green or sustainable chemistry. Therefore, the synthesis of heterocycles using different catalysts is going to be a top area of research encompassing fields like drug discovery and total synthesis.

The diversity of compounds that may be synthesized in large libraries, which traditionally has been dominated by a few dependable transformations, may be increased through robust heterocycle syntheses and functionalizations. This can significantly reduce design, synthesis, and screen cycle durations in preclinical research, increasing the efficiency of drug discovery. The drug development process will be accelerated by dependable synthetic pathways that extend from preclinical SAR production *via* bulk preparations of compounds of clinical interest. We believe that the fundamental developments in the field of heterocycles will continue to play a crucial role in providing fresh molecular ideas that will eventually lead to the creation of life-changing medications for patients and the development of newer catalytic methods to make already existing processes easy and sustainable.

References

1. Pitt, W.R., Parry, D.M., Perry, B.G., Groom, C.R., Heteroaromatic rings of future. *J. Med. Chem.*, 52, 2952, 2009.
2. (a) Brown, D.G. and BostrÖm, J., Analysis of past and present synthetic methodologies on medicinal chemistry: where have all the new reactions gone? *J. Med. Chem.*, 59, 4443, 2016. (b) Cooper, T.W.J., Campbell, I.B., Macdonald, S.J.F., factors determining the selection of organic reactions by medicinal chemists and the use of these reactions in arrays (small focused libraries). *Angew. Chem., Int. Ed.*, 49, 8082, 2010.
3. Walters, W.P., Green, J., Weiss, J.R., Murcko, M.A., What do medicinal chemists actually make? A 50-year retrospective. *J. Med. Chem.*, 54, 6405, 2011.
4. (a) Michaudel, Q., Ishihara, Y., Baran, P.S., Academia–industry symbiosis in organic chemistry. *Acc. Chem. Res.*, 48, 712, 2015. (b) Welch, C.J., Hawkins, J.M., Tom, J., Precompetitive collaboration on enabling technologies for the pharmaceutical industry. *Org. Process Res. Dev.*, 18, 481, 2014. (c) Fitzpatrick, D.E., Battilocchio, C., Ley, S.V., Enabling technologies for the future of chemical synthesis. *ACS Cent. Sci.*, 2, 131, 2016.
5. (a) Baumann, M., Baxendale, I.R., Ley, S.V., Nikbin, N., An overview of the key routes to the best selling 5-membered ring heterocyclic pharmaceuticals. *Beilstein J. Org. Chem.*, 7, 442, 2011. (b) Baumann, M. and Baxendale, I.R., An overview of the synthetic routes to the best selling drugs containing 6-membered heterocycles. *Beilstein J. Org. Chem.*, 9, 2265, 2013. (c) Ding, H.X., Leverett, C.A., Kyne Jr., R.E., Liu, K.K.-C., Fink, S.J., Flick, A.C., O'Donnell, C.J., Synthetic approaches to the 2013 new drugs. *Bioorg. Med. Chem.*, 23, 1895, 2015.
6. (a) Ding, H.X., Leverett, C.A., Kyne Jr., R.E., Liu, K.K.-C., Sakya, S.M., Flick, A.C., O'Donnell, C.J., Synthetic approaches to the 2012 new drugs. *Bioorg. Med. Chem.*, 22, 2005, 2014.(b) Ding, H.X., Liu, K.K.-C., Sakya, S.M., Flick, A.C., O'Donnell, C.J., Synthetic approaches to the 2011 new drugs. *Bioorg. Med. Chem.*, 21, 2795, 2013.
7. (a) Liu, K.K.-C., Sakya, S.M., O'Donnell, C.J., Flick, A.C., Ding, H.X., Synthetic approaches to the 2010 new drugs. *Bioorg. Med. Chem.*, 20, 1155, 2012. (b) Liu, K.K.-C., Sakya, S.M., O'Donnell, C.J., Flick, A.C., Li, J., Synthetic approaches to the 2009 new drug. *Bioorg. Med. Chem.*, 19, 1136, 2011.
8. Mei, T.-S., Kou, L., Ma, S., Engle, K.M., Yu, J.-Q., Heterocycle Formation via Palladium-Catalyzed C–H Functionalization. *Synthesis,* 44, 12, 1778-1791, 2012.
9. (a) Taylor, A.P., Robinson, R.P., Fobian, Y.M., Blakemore, D.C., Jones, L.H., Fadeyi, O., Modern advances in heterocyclic chemistry in drug discovery. *Org. Biomol. Chem.*, 14, 6611, 2016. (b) Rajesh, B.R.D. and Quinoline, Y., Heterocycles: Synthesis and Bioactivity, in: *Heterocycles - Synthesis and Biological Activities*, IntechOpen, 2020.
10. Kiser, E.J., Magano, J., Shine, R.J., Chen, M.H., Kilogram-Lab-Scale Oxindole Synthesis via Palladium-Catalyzed C–H Functionalization. *Org. Process Res. Dev.*, 16, 255, 2012.
11. Uccello, D.P., Miller, S.M., Dieterich, N.A., Stepan, A.F., Chung, S.W., Farley, K.A., Samas, B., Chen, J., Montgomery, J., II, The synthesis of C-13 functionalized pleuromutilins via C–H amidation and subsequent novel rearrangement product. *Tetrahedron Lett.*, 52, 4247, 2011.

12. Espino, C.G., Fiori, K.W., Kim, M., Du Bois, J., Expanding the scope of C–H amination through catalyst design. *J. Am. Chem. Soc.*, 126, 15378, 2004.
13. Cui, Y. and He, C., A Silver-catalyzed intramolecular amidation of saturated CH bonds. *Angew. Chem., Int. Ed.*, 43, 4210, 2004.
14. (a) Guimond, N., Gouliaras, C., Fagnou, K., Rhodium(III)-Catalyzed Isoquinolone Synthesis: The N–O Bond as a Handle for C–N Bond Formation and Catalyst Turnover. *J. Am. Chem. Soc.*, 132, 6908, 2010. (b) Hyster, T.K. and Rovis, T., Rhodium-Catalyzed Oxidative Cycloaddition of Benzamides and Alkynes via C–H/N–H Activation. *J. Am. Chem. Soc.*, 132, 10565, 2010. (c) Rakshit, S., Grohmann, C., Besset, T., Glorius, F., Rh(III)-Catalyzed Directed C–H Olefination Using an Oxidizing Directing Group: Mild, Efficient, and Versatile. *J. Am. Chem. Soc.*, 133, 2350, 2011.
15. (a) Wang, H. and Glorius, F., Mild rhodium(III)-catalyzed C-H activation and intermolecular annulation with allenes. *Angew. Chem., Int. Ed.*, 51, 7318, 2012. (b) Huckins, J.R., Bercot, E.A., Thiel, O.R., Hwang, T.-L., Bio, M.M., Rh(III)-Catalyzed C–H Activation and Double Directing Group Strategy for the Regioselective Synthesis of Naphthyridinones. *J. Am. Chem. Soc.*, 135, 14492, 2013.
16. (a) Wang, Y.-M., Bruno, N.C., Placeres, A.L., Zhu, S., Buchwald, S.L., Enantioselective synthesis of carbo- and heterocycles through a CuH-catalyzed hydroalkylation approach. *J. Am. Chem. Soc.*, 137, 10524, 2015. (b) Ascic, E. and Buchwald, S.L., Highly diastereo- and enantioselective CuH-catalyzed synthesis of 2,3-disubstituted indolines. *J. Am. Chem. Soc.*, 137, 4666, 2015.
17. (a) Qian, W., Wang, H., Allen, J., Copper-catalyzed domino cycloaddition/C-N coupling/cyclization/(C-H arylation): an efficient three-component synthesis of nitrogen polyheterocycles. *Angew. Chem., Int. Ed.*, 52, 10992, 2013. (b) Qian, W., Amegadzie, A., Winternheimer, D., Allen, J., One-pot synthesis of 3-Triazolyl-2-iminochromenes via a catalytic three component cascade reaction. *Org. Lett.*, 15, 2986, 2013.
18. (a) Dunlop, J., Watts, S.W., Barrett, J.E., Coupet, J., Harrison, B., Mazandarani, H., Nawoschik, S., Pangalos, M.N., Ramamoorthy, S., Schechter, L., Smith, D., Stack, G., Zhang, J., Zhang, G., Rosenzweig-Lipson, S., Characterization of vabicaserin (SCA-136), a selective 5-Hydroxytryptamine 2C receptor agonist. *J. Pharm. Exp. Tech.*, 337, 673, 2011. (b) Rosenzweig-Lipson, S., Dunlop, J., Marquis, K.L., *Drug News Perspect.*, 9, 565, 2005.
19. Dragan, V., McWilliams, J.C., Miller, R., Sutherland, K., Dillon, J.L., O'Brien, M.K., Asymmetric synthesis of vabicaserin via oxidative multicomponent annulation and asymmetric hydrogenation of a 3,4-substituted quinolinium salt. *Org. Lett.*, 15, 2942, 2013.
20. Wang, X.-S., Li, Q., Yao, C.-S., Tu, S.-J., An efficient method for the synthesis of Benzo[f]quinoline and Benzo[a]phenanthridine derivatives catalyzed by iodine by a three-component reaction of arenecarbaldehyde, Naphthalen-2-amine, and Cyclic Ketone. *Eur. J. Org. Chem.*, 3513-3518, 2008.
21. Fujiwara, Y., Dixon, J.A., O'Hara, F., Funder, E.D., Dixon, D.D., Rodriguez, R.A., Baxter, R.D., Herle, B., Sach, N., Collins, M.R., Ishihara, Y., Baran, P.S., Practical and innate carbon–hydrogen functionalization of heterocycles. *Nature*, 492, 95, 2012.
22. Coe, J.W., Brooks, P.R., Vetelino, M.G., Wirtz, M.C., Arnold, E.P., Huang, J., Sands, S.B., Davis, T., II, Lebel, L.A., Fox, C.B., Shrikhande, A., Heym, J.H., Schaeffer, E., Rollema, H., Lu, Y., Mansbach, S., Chambers, L.K., Rovetti, C.C., Schulz, D.W., Tingley, III, F.D., O'Neill, B.T., Varenicline: An $\alpha4\beta2$ nicotinic receptor partial agonist for smoking cessation. *J. Med. Chem.*, 48, 3474, 2005.
23. (a) Fujiwara, Y., Dixon, J.A., Rodriguez, R.A., Baxter, R.D., Dixon, D.D., Collins, M.R., Blackmond, D.G., Baran, P.S., A new reagent for direct difluoromethylation. *J. Am. Chem. Soc.*, 134, 1494, 2012. (b) O'Hara, F., Burns, A.C., Collins, M.R., Dalvie, D., Ornelas, M.A., Vaz,

A.D.N., Fujiwara, Y., Baran, P.S., A simple litmus test for aldehyde oxidase metabolism of heteroarenes. *J. Med. Chem.*, 57, 1616, 2014.

24. (a) Pryde, D.C., Dalvie, D., Hu, Q., Jones, P., Obach, R.S., Tran, T.D., Aldehyde oxidase: an enzyme of emerging importance in drug discovery. *J. Med. Chem.*, 53, 8441, 2010. (b) Garattini, E. and Terao, M., Increasing recognition of the importance of aldehyde oxidase in drug development and discovery. *Drug Metab. Rev.*, 43, 374, 2011. (c) Garattini, E. and Terao, M., The role of aldehyde oxidase in drug metabolism. *Expert Opin. Drug Metab. Toxicol.*, 8, 487, 2012. (d) Garattini, E. and Terao, M., Aldehyde oxidase and its importance in novel drug discovery: present and future challenges. *Expert Opin. Drug Discovery*, 8, 641, 2013. (e) St. Jean, D.J. and Fotsch, C., Mitigating heterocycle metabolism in drug discovery. *J. Med. Chem.*, 55, 6002, 2012. (f) Hutzler, J.M., Obach, R.S., Dalvie, D., Zientek, M.A., Strategies for a comprehensive understanding of metabolism by aldehyde oxidase. *Expert Opin. Drug Metab. Toxicol.*, 9, 153, 2013.
25. Gianatassio, R., Kawamura, S., Eprile, C.L., Foo, K., Ge, J., Burns, A.C., Collins, M.R., Baran, P.S., Simple Sulfinate Synthesis Enables C-H Trifluoromethylcyclopropanation. *Angew. Chem., Int. Ed.*, 53, 9851, 2014.
26. (a) Zeitler, K., Photoredox Catalysis with Visible Light *Angew. Chem., Int. Ed.*, 48, 9785, 2009. (b) Yoon, T.P., Ischay, M.A., Du, J., Visible light photocatalysis as a greener approach to photochemical synthesis. *Nat. Chem.*, 2, 527, 2010. (c) Narayanam, J.M.R. and Stephenson, C.R.J., Visible light photoredox catalysis: applications in organic synthesis. *Chem. Soc. Rev.*, 40, 102, 2011. (d) Tucker, J.W. and Stephenson, C.R.J., Shining light on photoredox catalysis: theory and synthetic applications. *J. Org. Chem.*, 77, 1617, 2012. (e) Xuan, J. and Xiao, W.-J., Visible-light photoredox catalysis. *Angew. Chem., Int. Ed.*, 51, 6828, 2012. (f) Angnes, R.A., Li, Z., Correia, C.R.D., Hammond, G.B., Recent synthetic additions to the visible light photoredox catalysis toolbox. *Org. Biomol. Chem.*, 13, 9152, 2015. (g) Xi, Y., Yi, H., Lei, A., Synthetic applications of photoredox catalysis with visible light. *Org. Biomol. Chem.*, 11, 2387, 2013.
27. McNally, A., Prier, C.K., MacMillan, D.W.C., Discovery of an α-Amino C–H Arylation Reaction Using the Strategy of Accelerated Serendipity. *Science*, 334, 1114, 2011.
28. (a) Zuo, Z., Ahneman, D.T., Chu, L., Terrett, J.A., Doyle, A.G., MacMillan, D.W.C., *Science*, 345, 619, 2014. (b) Zuo, Z. and MacMillan, D.W.C., Use of unprotected amino acids in metal-free tandem radical cyclization reactions: divergent synthesis of 6-alkyl/acyl phenanthridines†. *J. Am. Chem. Soc.*, 136, 5257, 2014.
29. (a) Boechat, N., Ferreira, V.F., Ferreira, S.B., Ferreira, M.L.G., da Silva, F.C., Bastos, M.M., Costa, M.S., LourenÇo, M.C.S., Pinto, A.C., Krettli, A.U., Aguiar, A.C., Teixeira, B.M., da Silva, N.V., Martins, P.R.C., Bezerra, F.A.F.M., Camilo, A.L.S., da Silva, G.P., Costa, C.C.P., Novel 1,2,3-Triazole Derivatives for Use against Mycobacterium tuberculosis H37Rv (ATCC 27294) Strain. *J. Med. Chem.*, 54, 5988, 2011. (b) Agalave, S.G., Maujan, R.S., Pore, V.S., Synthesis and antifungal activity of 1,5-disubstituted-1,2,3-triazole containing fluconazole analogues. *Chem. Asian J.*, 6, 2696, 2011. (c) Lima-Neto, R.G., Cavalcante, N.N.M., Srivastava, R.M., Mendonca, F.J.B., Wanderley, A.G., Neves, R.P., dos Anjos, J.V., Synthesis of 1,2,3-Triazole Derivatives and *in Vitro* Antifungal Evaluation on Candida Strains. *Molecules*, 17, 5882, 2012.
30. Senger, M.R., Gomes, L.C.A., Ferreira, S.B., Kaiser, C.R., Ferreira, V.F., Silva Jr., F.P., *Chem. Bio. Chem.*, 13, 1584, 2012. (b) Shanmugavelan, P., Nagarajan, S., Sathishkumar, M., Ponnuswamy, A., Yogeeswari, P., Sriram, D., Efficient synthesis and *in vitro* antitubercular activity of 1,2,3-triazoles as inhibitors of Mycobacterium tuberculosis, *Bioorg. Med. Chem. Lett.*, 21, 7273, 2011.
31. Mayooufi, A., Romdhani-Younes, M., Carcenac, Y., Thibonnet, J., Easy installation of 1,2,3-triazoles or iodo-1,2,3-triazoles onto indole-fused oxazinones via CuAAC-based MCR in the presence of 18-crown-6. *Synth. Commun.*, 49, 2168, 2019.
32. Han, W.-Y., Zhao, J.-Q., Zuo, J., Xu, X.-Y., Zhang, X.-M., Yuan, W.-C., Recent Advances of α-Isothiocyanato Compounds in the Catalytic Asymmetric Reaction. *Adv. Synth. Catal.*, 357, 3007, 2015.

33. Chen, W.-B., Wu, Z.-J., Hu, J., Cun, L.-F., Zhang, X.-M., Yuan, W.-C., Organocatalytic Direct Asymmetric Aldol Reactions of 3-Isothiocyanato Oxindoles to Ketones: Stereocontrolled Synthesis of Spirooxindoles Bearing Highly Congested Contiguous Tetrasubstituted Stereocenters. *Org. Lett.*, 13, 2472, 2011.
34. (a) Cai, H., Zhou, Y., Zhang, D., Xu, J., Liu, H., A Mannich/cyclization cascade process for the asymmetric synthesis of spirocyclic thioimidazolidineoxindoles. *Chem. Commun.*, 50, 14771, 2014. (b) Kato, S., Kanai, M., Matsunaga, S., Enantioselective synthesis of spirooxindoles via direct catalytic asymmetric aldol-type reaction of isothiocyanato oxindoles. *Heterocycles*, 88, 475, 2014. (c) Tan, F., Lu, L.-Q., Yang, Q.-Q., Guo, W., Bian, Q., Chen, J.-R., Xiao, W.-J., Enantioselective Cascade Michael Addition/Cyclization Reactions of 3-Nitro-2H-Chromenes with 3-Isothiocyanato Oxindoles: Efficient Synthesis of Functionalized Polycyclic Spirooxindoles. *Chem. - Eur. J.*, 20, 3415, 2014. (d) Wang, L., Yang, D., Li, D., Liu, X., Zhao, Q., Zhu, R., Zhang, B., Wang, R., Catalytic Asymmetric [3 + 2] Cyclization Reactions of 3-Isothiocyanato Oxindoles and Alkynyl Ketones Via an *in Situ* Generated Magnesium Catalyst. *Org. Lett.*, 17, 4260, 2015. (e) Zhao, J.-Q., Zhou, X.-J., Zhou, Y., Xu, X.-Y., M. Zhang, X.-, Yuan, W.-C., Diastereo- and Enantioselective Dearomative [3 + 2] Cycloaddition Reaction of 2-Nitrobenzofurans with 3-Isothiocyanato Oxindoles. *Org. Lett.*, 20, 909–912, 2018.
35. Li, T.-R., Cheng, B.-Y., Wang, Y.-N., Lu, L.-Q., Xiao, W.-J., Fan, S.-Q., Xiao, W.-J., Highly Stereoselective [3+2] Cycloadditions of Chiral Palladium-Containing N1-1,3-Dipoles: A Divergent Approach to Enantioenriched Spirooxindoles. *Chem. - Eur. J.*, 22, 6243, 2016.
36. For reviews on biologically important indole derivatives, see: (a) Lewis, S.E., Recent advances in the chemistry of macroline, sarpagine and ajmaline-related indole alkaloids. *Tetrahedron*, 62, 8655, 2006. (b) Higuchi, T. and Kawasaki, T., Simple indole alkaloids and those with a non-rearranged monoterpenoid unit. *Nat. Prod. Rep.*, 24, 84, 2007. (c) Bal, C., Baldeyrou, B., Moz, F., Lansiaux, A., Colson, P., Kraus-Berthier, L., Leonce, S., Pierre, A., Boussard, M.F., Rousseau, A., Wierzbicki, M., Bailly, C., Novel antitumor indenoindole derivatives targeting DNA and topoisomerase II. *Biochem. Pharmacol.*, 68, 1911, 2004.
37. (a) Boussard, M.F., Truche, S., Rousseau-Rojas, A., Briss, S., Descamps, S., Droual, M., Wierzbicki, M., Ferry, G., Audinot, V., Delagrange, P., Boutin, J.A., New ligands at the melatonin binding site *MT3*. *Eur. J. Med. Chem.*, 41, 306, 2006. DOI: https://doi.org/10.1016/j.ejmech.2005.12.002. (b) Shao, D., Zou, C., Luo, C., Tang, X., Li, Y., Synthesis and evaluation of tacrine–E2020 hybrids as acetylcholinesterase inhibitors for the treatment of Alzheimer's disease. *Bioorg. Med. Chem. Lett.*, 14, 4639, 2004.
38. Liang, Y. and Jiao, N., Cationic Cobalt(III) Catalyzed Indole Synthesis: The Regioselective Intermolecular Cyclization of N-Nitrosoanilines and Alkynes. *Angew. Chem., Int. Ed.*, 55, 4035, 2016.
39. Jiang, X., Chen, J., Zhu, W., Cheng, K., Liu, Y., Su, W.-K., Yu, C., Cobalt(III)-Catalyzed Fast and Solvent-Free C–H Allylation of Indoles Using Mechanochemistry. *J. Org. Chem.*, 82, 10665, 2017.

5

The Novel Trends in Asymmetric Catalysis: Green and Sustainable Catalysts

Surya Prakash Verma, Devashish Singh and Poonam Rajesh Prasad*

Department of Chemistry, Institute of Science, Banaras Hindu University, Varanasi, India

Abstract

The extraordinary growth of asymmetric synthesis, and in particular of asymmetric catalysis, over the past few decades is largely due to the persistent demand for enantiopure or enantioenriched chiral molecules in a variety of industrial areas, such as the pharmaceutical, agrochemical, cosmetic, food chemistry, and material science. Nonetheless, the three primary asymmetric catalysis methods metal, organo, and biocatalysis are actively developing and addressing brand-new problems today. Academic and industrial research has been giving green chemistry more attention as a result of the changing demands regarding environmental issues. The pharmaceutical and organic synthesis industries are continually working to develop enantioselective green procedures that use fewer steps of purification, fewer toxic chemicals, solvents, and less energy. The use of organocatalysts and nontoxic metal catalysts, reusable catalysts, multistep reactions, and continuous flow chemical processes are all heavily promoted in this context. Asymmetric photocatalysis and electrosynthesis were developed as a result of novel ligands and organocatalysts from biomass feedstocks, which were further encouraged by the requirement to utilize sustainable sources of energy and materials. Catalytic asymmetric synthesis is presently required for the enantioselective synthesis of pharmaceuticals, natural substances, and agrochemicals (primarily fungicides, herbicides, pheromones, and pesticides). The various catalytic asymmetric syntheses include enantioselective hydroformylation, enantioselective hydrogenation, asymmetric Sharpless epoxidation and dihydroxylation, isomerization, and organocatalytic asymmetric synthesis along with recent successful applications. Chiral agrochemicals are also a new generation of such compounds with specialized and unique modes of action in order to provide sustainable crop protection. Trends in the creation of more environmentally friendly and sustainable products include the application of photochemistry, alternative energy inputs (such as mechanical chemistry, ultrasound or microwave irradiation), and greener reaction media to the synthesis of organic and nanomaterial processes that have been condensed here over the last two decades. Organic synthesis processes that are catalyzed by basic water or reusable magnetic nanocatalysts include coupling reactions and the construction of heterocyclic molecules. The synthesis of nanoparticles benefits from biomimetic processes that include reducing and capping agents such as vitamins, sugars, plant polyphenols, and agricultural waste residues. Metal nanocatalysts (Pd, Au, Ag, Ni, Ru, Ce, etc.) immobilized on biodegradable materials like cellulose and chitosan or recyclable magnetic ferrites via ligands like dopamine or glutathione are being given particular attention. One of the most significant groups of nonsteroidal anti-inflammatory medications is the profane family (NSAIDs). There have been several different biocatalytic methods suggested for the manufacture of optically pure (S)-profens.

**Corresponding author*: poonamrp.chem@bhu.ac.in; poonam11a86@gmail.com

Mousumi Sen (ed.) Sustainable Green Catalytic Processes, (97–120) © 2024 Scrivener Publishing LLC

Keywords: Asymmetric catalysis, green and sustainable catalysts, enzymatic catalysis, green synthesis, asymmetric cross-couplings, asymmetric hydrogenation through transition metals, organocatalysis

5.1 Introduction

The global economy has been gradually expanding over the last 18 centuries, and the chemical sector has been steadily progressing. The huge loss of green resources and the pronounced growth in competition have resulted in the potential risk of the future unavailability of many renewable sources[1]. Due to the extensive use of chemical technology, the chemical industry has to deal with significant conflict on the harmful effects on environmental preservation and sustainable growth. In this context, a number of techniques have evolved that are helping to create fresh options for a sharp increase in the chemical transformation's effectiveness[2]. Enzymes are the most potential green catalysts to bring about the change in lowering hazardous waste and assisting the chemical industry get closer to a sustainable and environmentally friendly future among the different green approaches now accessible. Figure 5.1 demonstrates the links between the utilization of environmentally conscious chemistry and an enzymatic biocatalysis to deal with the difficulty of non-sustainability in a useful way. Green methods are typically used to create chemical compounds, reducing the necessity for extra purification procedures and the formation of transitional toxic waste via efficient and environmentally friendly methods that use heavily beneficial and selective enzymatic pathways as opposed to potentially harmful natural materials and processes[3]. In general, the term "green catalyst" mainly refers to the use, recovery, and reuse of enzymes as biocatalysts. By lowering the pricing, raising potential effectiveness, minimizing environmental impact, and encouraging for the creation of an all-encompassing sustainable technique, they are effectively emphasized for chemical transformation[4]. According to the 12 Principles of Green Chemistry, enzymatic biocatalysis is a possible tool for the production and processing of the products illustrated in Figure 5.2 below[5].

Figure 5.1 Modern industrial era: green chemistry and enzymatic biocatalysis.

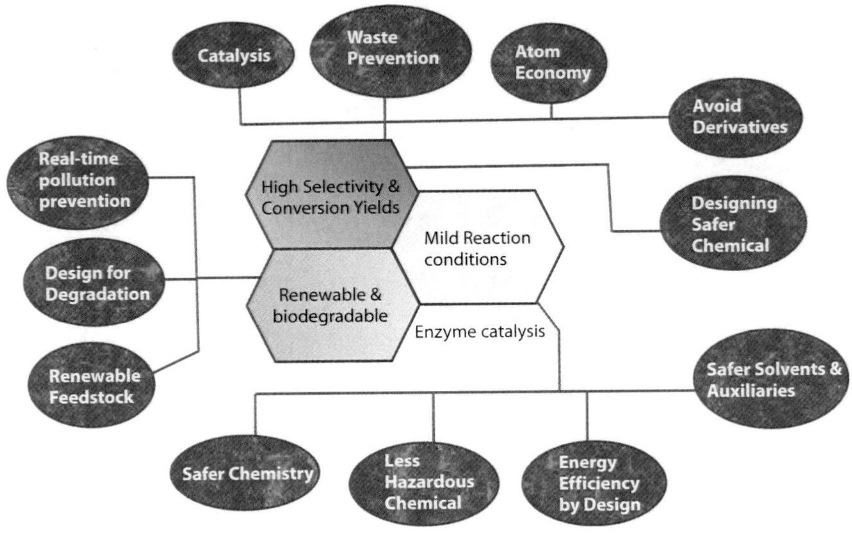

Figure 5.2 Enzymatic biocatalysis and the 12 principles of green chemistry are acquainted.

5.1.1 Correlation Between Enzymatic Catalysis and the Green Chemistry Principle

The importance of enzymes is directly related to how quickly they can accelerate chemical reactions when using biocatalysts, and it also depends on whether they can be produced from renewable sources like plants, animals, and microorganisms[6]. The importance of enzymatic capabilities is the reason why many industries, including chemical, food, textile, papers, biofuels, pharmaceuticals, agriculture, wastewater treatment, and healthcare, are gravitating toward biocatalysts (either enzymes or entire cells). The enzymatic method produces end product streams that are purer since it has excellent selectivity, activity, and specificity and requires mild reaction conditions and little energy. It also reduces problems with isomerization and rearrangement[7]. Additionally, by displaying chemo, region, and stereoselectivity, in addition to the reduction in by-product formation, biocatalysts can significantly reduce the functional group activation[8]. A variety of biopolymers is manufactured primarily on an industrial scale using biocatalysts (Figure 5.3)[9]. As a way for carrying out the chemical reaction as well as processing of the various commercially synthesized products, green techniques are used. Examples include the industrial production of semisynthetic antibiotics like penicillin and cephalosporin, the transformation of materials in the paper industry, and the utilization of the most valuable enzyme lipases due to their flexibility in various kinds of traditional chemical reactions[10, 11]. Additionally, greener instruments like the use of plant microbe-based fuel cell (PMFC) production for bioenergy play a big part in green chemistry by generating power with the aid of enzymes[12]. The potential for voltage generation by microbial strains like *R. leguminosarum, Azatobacter*, and others will increase economic value and take a greener approach to producing energy in the future[13, 14].

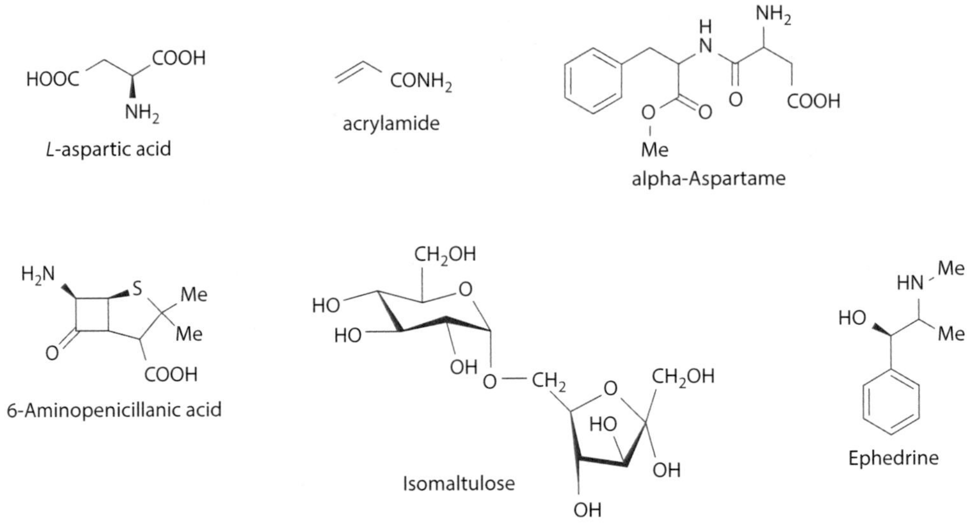

Figure 5.3 Compounds that have been created through substantial industrial application.

5.2 Role of Green Synthesis and Catalyst

With recent changes to their asymmetric synthesis, profens are kind of nonsteroidal anti-inflammatory medicines (NSAIDs) that have strong analgesic, antipyretic, and other pharmacological properties in addition to acting as an anti-inflammatory. For the synthesis of enantioenriched profens, a substantial number of catalytic asymmetric methods have been introduced. The development of possible enantioselective catalytic systems for the synthesis of profens and their related derivatives. We classified these transformations into three groups based on the type of reaction: organocatalytic asymmetric reactions, asymmetric cross-couplings, and asymmetric hydrogenations all include transition metals[15]. Due to its anti-inflammatory, analgesic, therapeutic, and antirheumatic activity, profens are applied to treat osteoarthritis, rheumatoid arthritis, antipyretic, and to lessen a blood clot[16–18]. Distinct side effects can result from various profens[19]. These may include gastrointestinal bleeding and ulcers[20], kidney disease[21, 22], and cardiac arrest[23, 24], and they depend of course on the formulation, dosage, and length of the drug's use. In addition to naproxen, which has long been accessible as a single enantiomer, ibuprofen and ketoprofen are currently available in enantioenriched form as well[25]. According to pharmacological mechanistic research, cyclooxygenase (COX) activity is inhibited to produce the anti-inflammatory action, which generates prostaglandins and thromboxane TXA2[26]. The (S)-enantiomer exhibits an anti-inflammatory effect, and additional study has demonstrated that the (R)-enantiomer can be racemized enzymatically. However, the magnitude of these flippings will depend on the medicine and the patient (at user-dependent rates)[27, 28]. The R-enantiomer of fenoprofen, for instance, undergoes racemization under physiological circumstances, but R-ketoprofen or R-flurbiprofen barely do[29, 30]. Early debate focused on the impact of the negative effects of the (R)-enantiomer and the varied pharmacokinetics of racemic mixtures[31]. These important investigations support the preference for enantiopure (S)-profen manufacturing, including ibuprofen, naproxen, and ketoprofen (Figure 5.4).

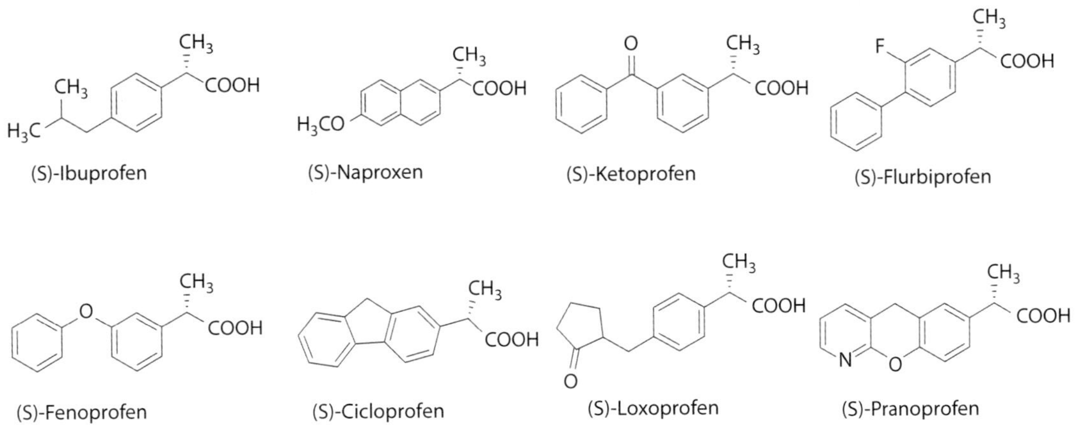

Figure 5.4 The common structure of (S)-profens.

Access to enantioenriched (S)-profens has been made possible through a number of methods. Enantioenriched precursors were used in the conventional method to produce the (S)-profen family, which resulted in high expenditures[32]. Another age-old method that is still in use makes use of racemization to recycle the less active (R)-enantiomer and traditional resolutions[33]. The third method uses asymmetric catalysis to produce (S)-profen directly from racemic or achiral substrates. Indubitably, the asymmetric catalysis strategy is the most effective tactic, despite the fact that it can be more difficult to produce. Even yet, a lot of work has gone into developing enantioenriched small-molecule catalysts for enantioselective synthesis[34].

Kourist *et al.*[35] have reviewed biocatalytic methods regarding the asymmetric preparation of profens. These biocatalysts are a result of remarkable advancements in protein engineering and enzyme discovery. In this review, they have created a metal and organo-catalyzed asymmetrical synthesis of (S)-proxen and/or its derived compounds. According to the form of reaction and the substances used, an evaluation will be firmed up. It is classified into three categories: asymmetric hydrogenation catalyzed by transition metals (TMs), asymmetrical cross-coupling TM-catalyzed, and organocatalytic synthesis. The benefits, applications, and restrictions of each approach are thoroughly covered in each section.

5.3 Asymmetric Hydrogenation Catalyzed by Transition Metals

One of the most effective techniques for producing enantioenriched profens was the asymmetric hydrogenation (AH) of conjugated carboxylic acids, which are catalyzed by TM. The type of TM/enantioenriched ligand combinations has a significant impact on the efficiency and enantioselectivity of asymmetrical TM catalysis. Other elements like solvent, substrate steric hindrance, and hydrogen pressure can also have an effect on enantioselectivity, sometimes rather dramatically. Mechanistic reaction pathways state that the pathways vary based on the metal and substrate is ambiguous. According to Scheme 5.1, the co-catalyzed asymmetric hydrogenation reaction was assumed to follow through the Cycle 1 pathway. It suggested that intermediate Co(0) oxidation state **(a)** undergoes ligand exchange and oxidative hydrogen addition to bind the unsaturated substrate in view of these and other

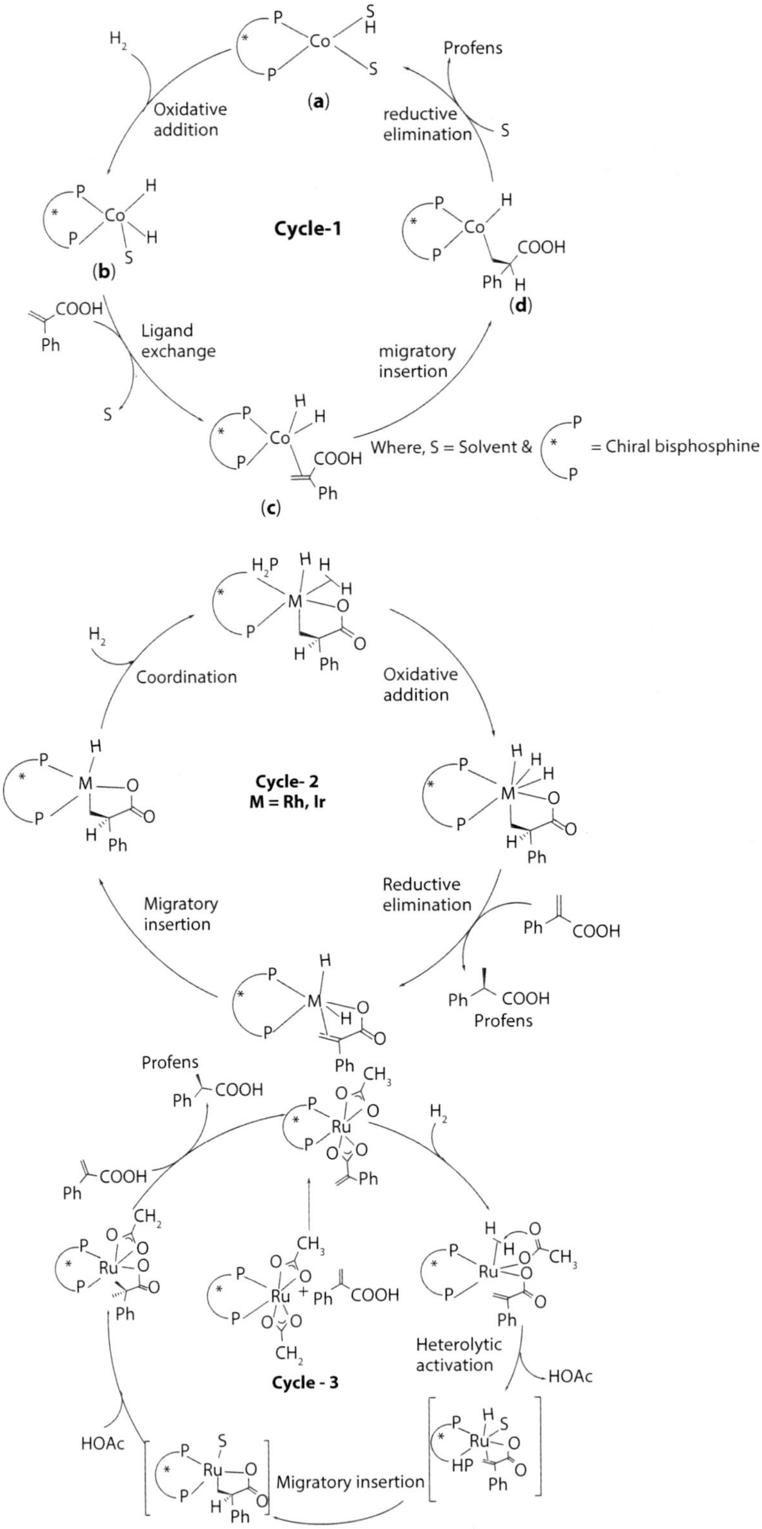

Scheme 5.1 Generic pathways of asymmetric hydrogenation processes mediated by transition metals.

experiments. To obtain the product and regenerate Co(0), migrating insertions are subsequently following that reductive elimination. Although the exact mechanism of the acid's chelation is unknown, interactions with esters result in decreased conversions. Studies show that Rh and Ir-catalyzed asymmetric hydrogenations generally adhere to Cycle 2 (Scheme 5.1), in contrast to cobalt catalysts[15]. As shown in Scheme 5.1 (Cycle 3), the production of dihydrogen adducts and carboxylate-assist hydrolysis from the bound H_2 are two potential reaction pathways. The substrate's C=C is probably chelated by the resultant monohydride complex, which is then followed by a transferring implantation. The enantioenriched saturated acid is made when the Ru-Carbon bond is protonated and exchanged via new substrate. It is important to note each of the carboxylate group of the reactant is crucial in these TM-catalyzed asymmetric hydrogenations (AHs), most likely because it binds to the catalyst. If the acids are swapped out for their equivalent esters, no hydrogenation takes place when using Ir, Rh, and Ru-based catalysts.

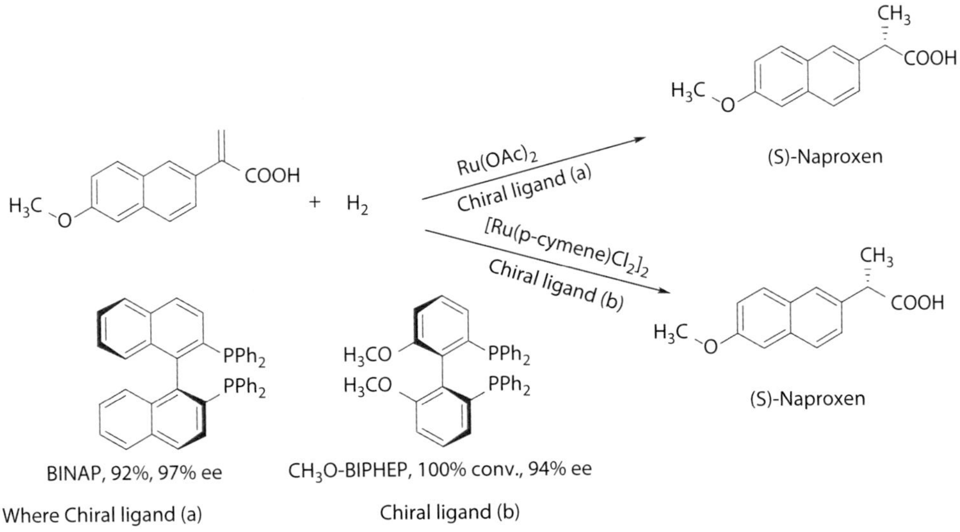

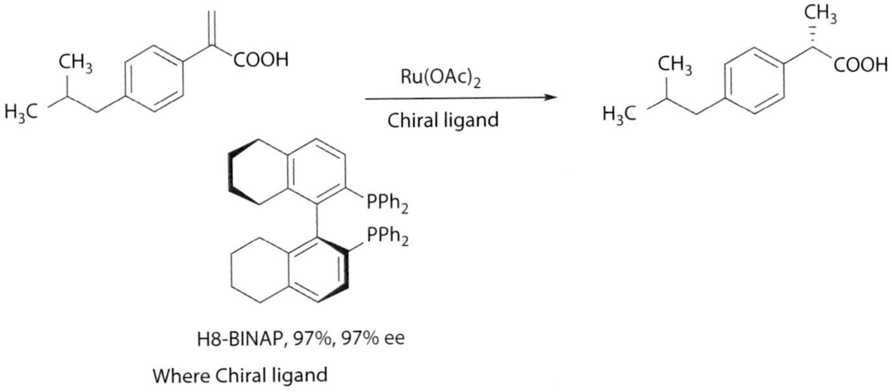

Example: Homogeneous catalysis from ruthenium-mediated asymmetric hydrogenations (AHs) for the preparation of enantioenriched profens.

5.3.1 The Asymmetric Hydrogenations are Caused by Cobalt

The extensive use of AH reactions, particularly in industrial processes, has spurred research into emerging cheaper earth-abundant TM catalysts to take the role of precious metals[36–38]. Chen and Lu[39] in 2018 reviewed the earth-rich first-row TMs, such as titanium and cobalt that were used in the catalytic AH of prochiral olefins. For example, using a cobalt catalyst, Schwarzwalder *et al.*[71] and Binder *et al.*[72] in 2020 reported the production of α,β-unsaturated carboxylic acids to form profens via asymmetric hydrogenations catalyzed by TM profens from α, β-unsaturated carboxylic acids (Scheme 5.2).

5.4 Asymmetric Cross-Couplings Catalyzed by TM

Growing development of enantioenriched compounds with tertiary and quaternary stereogenic centers has become possible with the use of asymmetric crossover involving two distinct coupling collaborators, catalyzed by TM[40]. There were only a few examples of enantioenriched profens directly synthesized among these processes. Because substrates containing carboxylic acids are acidic, coupling partners find it difficult to couple with them, which makes these reactions hard. TM-catalyzed (palladium, nickel, cobalt, iron, and copper) asymmetrical pairings have been used to produce enantioenriched profen derivative products, especially α-aryl amides α-aryl esters. These processes' by-products can be easily hydrolyzed into enantiopure profens without losing any enantiomeric excess. The monoarylation products of TM-catalyzed asymmetrical arylations of carbonyl-based substances present a challenge, since they comprise a readily deprotonated acidic α-hydrogen that racemized in basic circumstances.

5.4.1 Asymmetrical Cross-Couplings for Profen Derivative Products Under Pd Catalysis

Cross-coupling reactions with palladium as the catalyst have been widely applied in many different fields[41]. The coupling strategy may be superior to enantioselective hydrogenations, since it is more convergent and stereoselective cross-coupling processes create Carbon-Carbon bond. Hartwig *et al.* have given significant contribution to improve the

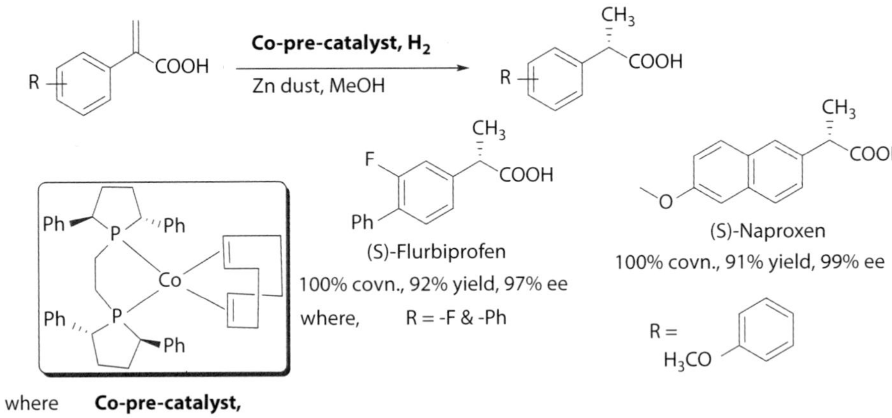

Scheme 5.2 Enantioenriched profens were created using Co-catalyzed AHs.

production of enantioenriched along with racemic compound profen derivative products by using Pd-catalyzed cross-coupling processes[42, 43]. They independently created a number of carboxylic acid esters that were α-arylations with aryl halides and pseudohalides using palladium as the catalyst. When the Evans auxiliary, which is widely known to regulate the stereochemistry of carbonyl derivatives, was present in trimethylsilyl enol ethers. The base-sensitive stereocenters did not epimerize when Zn(O-*t*Bu)$_2$ or ZnF_2 was present, according to Hartwig and coworkers (Scheme 5.3a). According to this discovery, silyl-enol ether arylation could be used to create enantioselective α-arylation procedures.

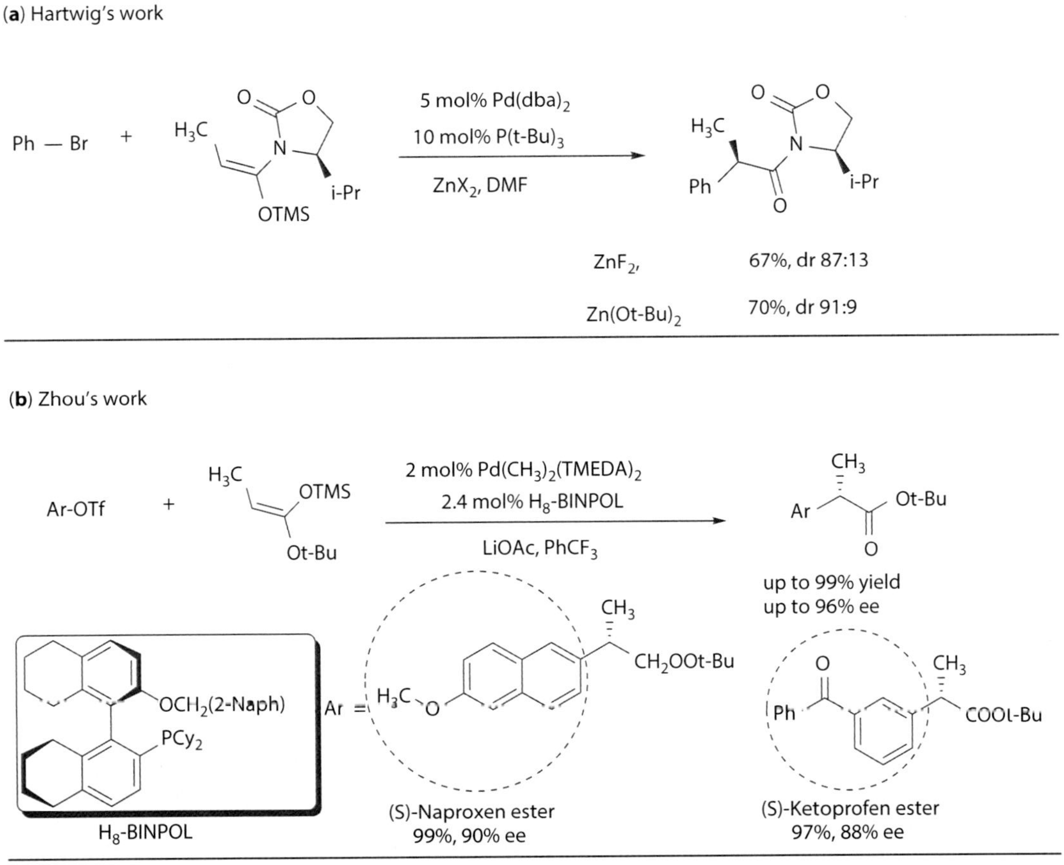

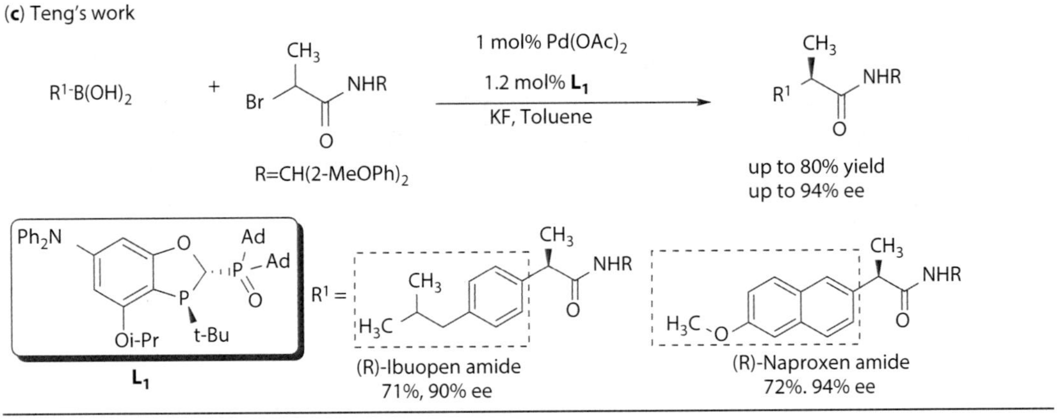

In 2011, Huang *et al.*[44] described to synthesize enantioenriched profen derivatives, in which TMS-silyl ketene acetals were asymmetrically arylated using aryl triflates (Scheme 5.3b). Due to the stereostatic resistance associated with silyl ketene acetals and the chiral monophosphine ligands generated from H8-BINOL, both have a significant effect on the product's enantiomeric excess. Because the weakly basic LiOAc utilised as an activator was insufficiently basic to deprotonate α-arylesters, no racemization was seen in this

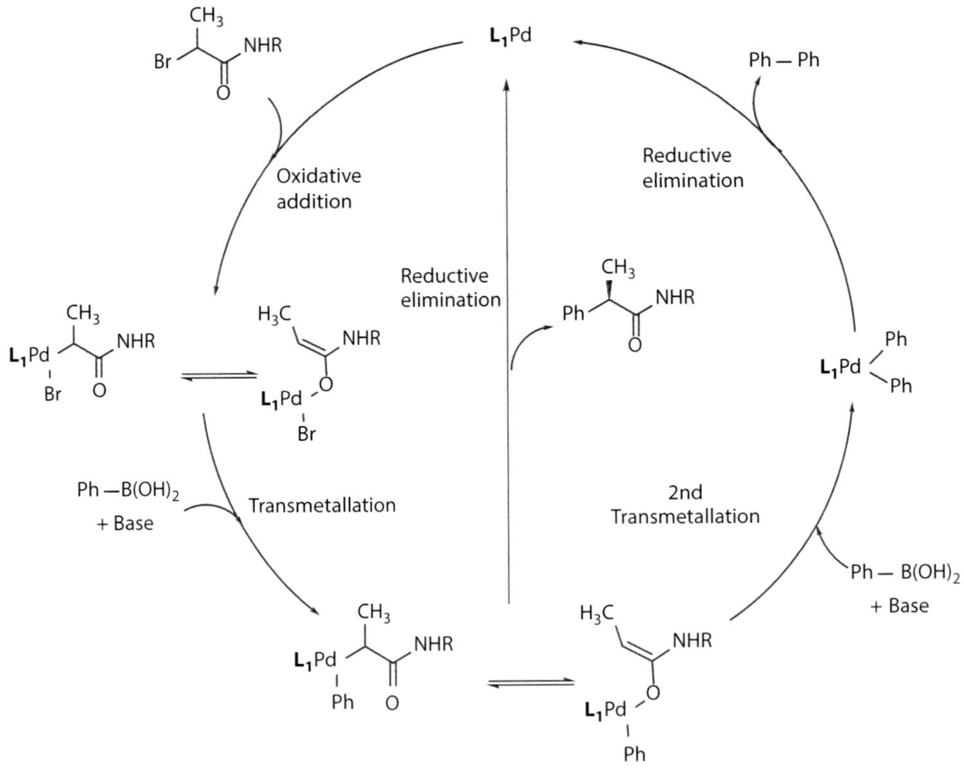

Scheme 5.3 Asymmetrical cross-coupling reactions for the synthesis of profens, catalyzed by palladium. The creation of the O-bound enol-at allows for the epimerization of the α-stereogenic center.

reaction. Li *et al.*[45] in 2019 described the preparation of enantioenriched profen derivatives through an enantioselective Suzuki–Miyaura cross-coupling among racemic α-bromo carboxamides and aryl-boronic acids. In the method they had generated P-chiral phosphorus ligands are available in a section 5.4.1 in Teng's work that are enantioenriched and have a dihydrobenzooxaphosphole backbone[46]. In this cross-coupling reaction, phosphorus ligand (**L**$_1$) was used that showed high efficacy (up to 80% yield) and enantioselectivity (up to 94% ee). The steric barrier of the N-aryl group also contributed to increasing levels in the ee. Furthermore, when the amide coupling products were hydrolyzed with 2 N HCl, enantioenriched profens such as ibuprofen (80%, 90% ee) and naproxen (85%, 93% ee) were formed.

5.4.2 Asymmetrical Cross-Couplings for Profen Derivative Products Under Fe Catalysis

Asymmetric catalysis has recently observed an increase in attention due to the construction of stereocenters using earth-abundant and affordable TM catalysts supported by enantioenriched ligands. Jin *et al.*[47] have described the first iron-catalyzed enantioselective cross-coupling reaction having aryl Grignard reagents and α-chloro-esters (Scheme 5.4). To positively affect the enantioselectivity, the enantioenriched phosphine ligand (BenzP) and the sterically demanding substrate OR group were both essential.

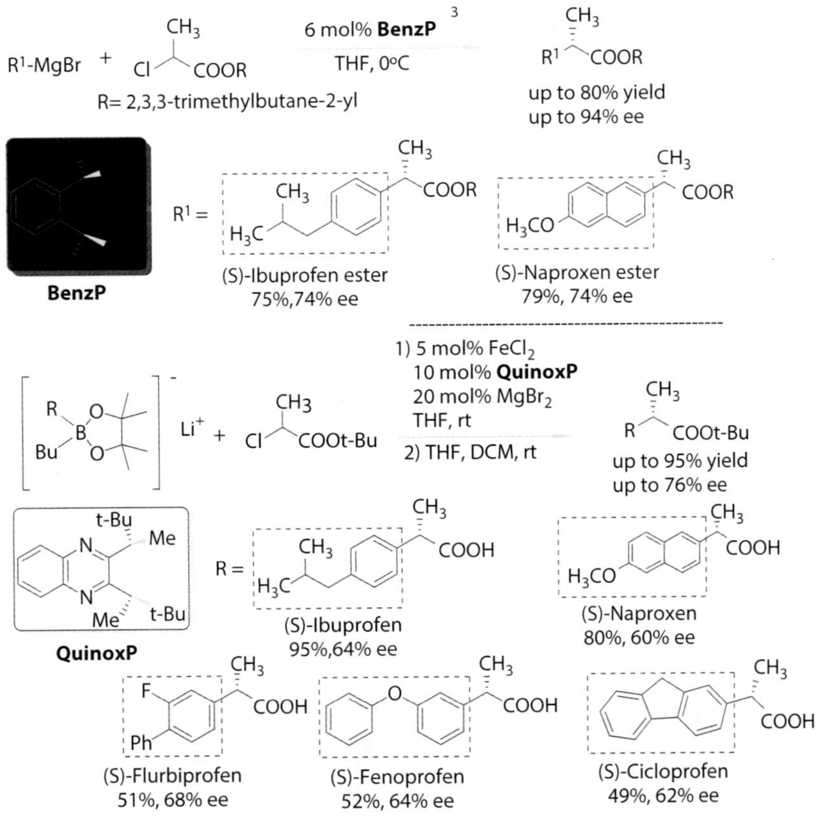

Although this asymmetric Kumada reaction with base metals indicated an entirely novel class of catalysts for the synthesis of enantioenriched profen derivatives. For instance, the (S)-ibuprofen ester and (S)-naproxen ester were produced in of 75% yield and 74% ee, respectively, and 79% yield and 74% ee, respectively. Nevertheless, this study provided motivation for other researchers working with base metal catalysts. This team published an additional example of iron-catalyzed enantioselective Suzuki–Miyaura coupling for the synthesis of optically active profens 4 years later. QuinoxP showed greater activity and enantioselectivity in this reaction when compared to BenzP, a chiral ligand in their prior research. Profens such as (S)-ibuprofen (95%, 64% ee), (S)-naproxen (80%, 60% ee), (S)-flurbiprofen (51%, 68% ee), (S)-fenoprofen (52%, 64% ee), and (S)-cicloprofen (49%, 62% ee) were created without losing optical purity after simple hydrolysis with TFA. Mechanistic studies indicated that a radical pathway was involved in these Fe-catalyzed asymmetric cross-couplings. The achiral α-carbonyl radical is produced after the oxidation of Fe to Fe(II) and the production of LFe-X, which is followed by halide abstraction from the α-bromo carbonyl molecule. The Fe-Ar bond is created via transmetallation from the 4-coordinate boron and the resultant Fe(II). The Fe(III) intermediate is produced by the oxidative capture of the α-carbonyl radical and can be used to complete the catalytic cycle through reductive elimination.

5.4.3 Asymmetrical Cross-Couplings for Profen Derivative Products Under Co-Catalysis

The first enantioselective cross-coupling reaction between aryl Grignard reagents and α-bromo-esters was reported by Mao *et al.*[48] in 2014 (Scheme 5.5). In the Kumada coupling, the benzyl-substituted bisoxazoline ligand (BiOX-1) showed outstanding enantioselectivity (up to 97%). Profen derivatives, such as (S)-fenoprofen ester and (S)-naproxen ester, were produced at respective yields of 87% with 93% ee and 57% with 90% ee. They also produced a different bisoxazoline ligand (BiOX-2) that showed comparable enantioselectivity and efficiency in this Kumada reaction. After lithium hydroxide induced hydrolysis in 87% yield without racemization, highly enantioenriched (S)-ibuprofen (64% yield, 98% ee) was produced. Enantiomeric purity increased from 82% to 98% when the mixture was recrystallized from n-hexane/dichloromethane (20:1). More recently, this team published the first enantioselective Negishi cross-coupling reaction involving aryl-zinc halides and α-bromo-esters that were catalyzed by cobalt[49]. Several α-arylalkanoic esters were produced using the enantioenriched bisoxazoline ligand (BiOX-3), with good enantioselectivities and yields (up to 97% ee and 98% yield), including the (S)-ibuprofen ester (90% yield with 93% ee) and the (S)-naproxen ester (83% yield with 89% ee). Although there have been several Co-catalyzed cross-coupling reactions, there is still a lack of mechanistic information. Radical clock research indicates that the mechanism of the cobalt-catalyzed asymmetric cross-coupling (Scheme 5.5) is similar to the iron-catalyzed process in Scheme 5.4, which involves the synthesis of radicals. It makes sense that Co(I) either experiences halide abstraction from the substrate or SET followed by halide binding. The more electrophilic Co(II) will be more appropriate for transmetallation[50]. The radical is oxidatively captured, increasing the Co oxidation state and preparing the procedure for reductive elimination.

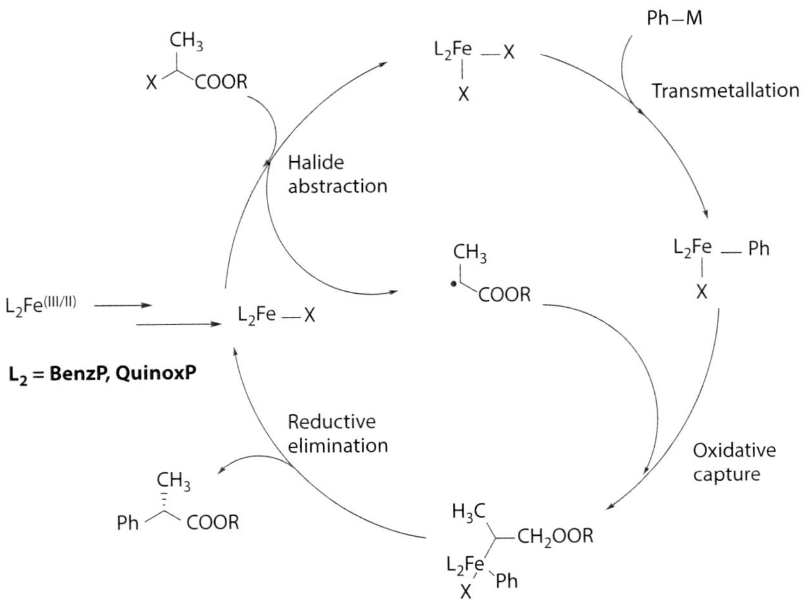

Scheme 5.4 Asymmetric cross-couplings using iron as the catalyst to form profens.

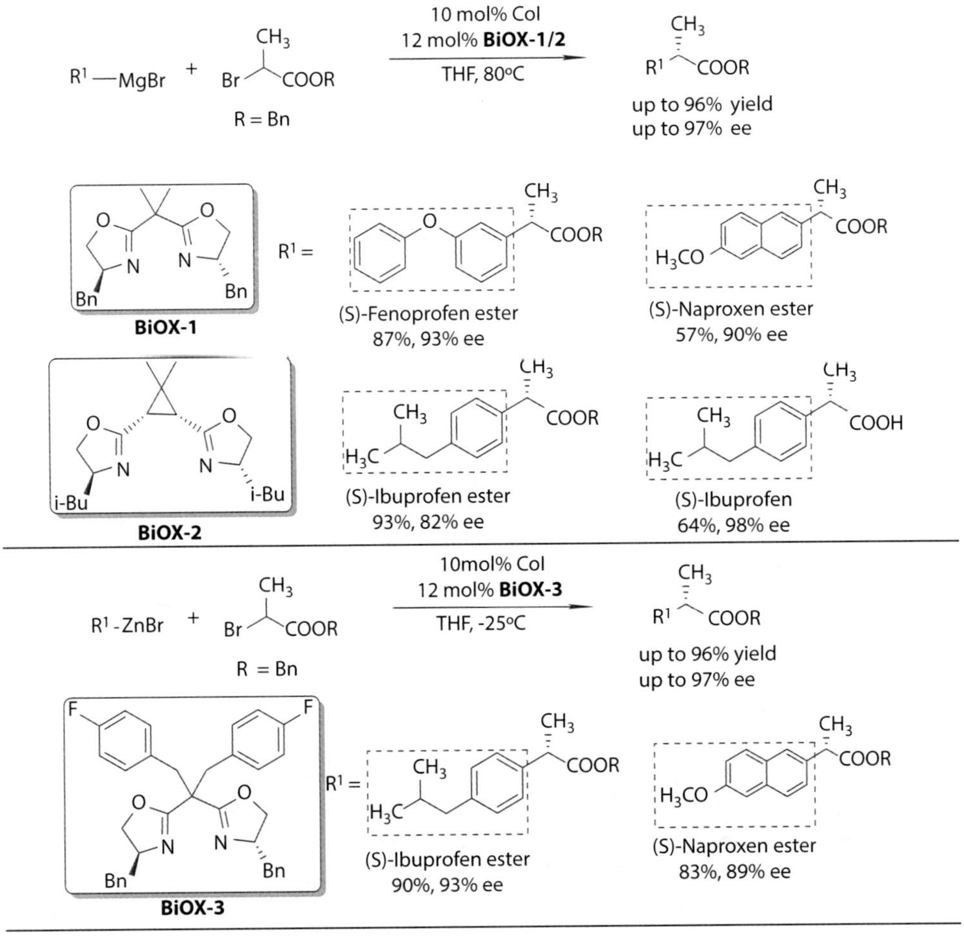

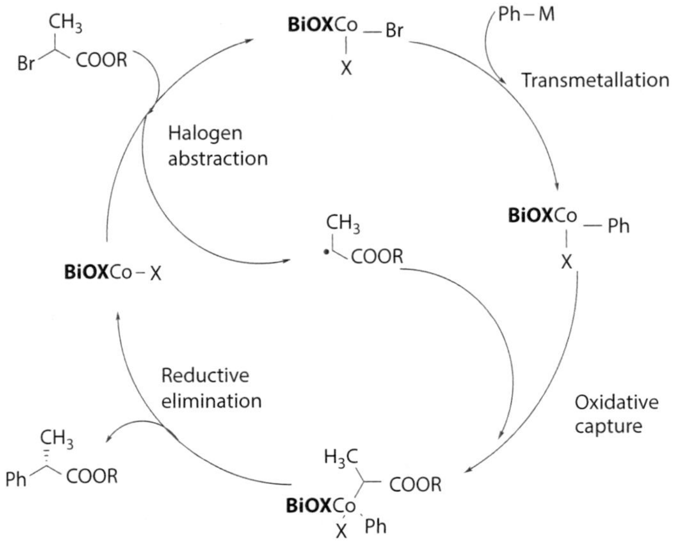

Scheme 5.5 Synthesis of enantioenriched profens using asymmetric cross-couplings catalyzed by cobalt.

5.4.4 Asymmetrical Cross-Couplings for Profen Derivative Products Under Ni Catalysis

Nickel catalysts have received a lot of attention recently due to their low toxicity, distinctive reactivity, and earth abundance[51]. Therefore, nickel salts have been investigated as catalysts in numerous cross-coupling and associated processes[52]. These reports on profen derivatives are the most pertinent ones. Lundgren *et al.*[73] and Wang *et al.*[74] identified an enantioselective α-arylation of α-substituted γ-butyrolactones with aryl chlorides in 2002 (Scheme 5.6a). In this chemical-based the pre-catalyst show in section 5.4.4 in (a) Buchwald's work $Ni(COD)_2$ (5 mol%) was combined with (S)-BINAP, and the resultant catalyst showed good efficacy (up to 99% yield) and enantioselectivity (up to 95% ee). The cost of this air-sensitive precursor is higher than NiX2-based precursors, which is a limitation to this catalytic system and many examples using $Ni(COD)_2$. The pro-nucleophile was deprotonated using the moderately strong base $NaN(SiMe_3)_2$, which limited the substrates to those that would proceed through the arylation process and produce a chiral quaternary center. Because the α-hydrogens are more acidic than the initial carbonyl compounds by approximately 5 pK units, it would be challenging to synthesize enantioenriched tertiary α-carbonyl stereocenters with this approach. An undesirable double arylation may also emerge from this process, which creates the achiral enolate. A number of nickel-catalyzed asymmetrical reactions involving cross-coupling among the aryl metal catalysts and racemic α-halo compounds were developed in order to circumvent the product deprotonation/racemization issue. Fu and his coworkers were principally responsible for creating this method[53–59]. They published the first time in 2008, there was an asymmetrical cross-coupling with organosilicon catalysts and -halo carbonyl derivatives, catalyzed by Ni. The best enantioenriched ligand was the diamine ligand (L2). It offered exceptional yields (up to 89%) and enantioselectivities (up to 99%) for the cross-coupled compounds. Two years later, they completed enantioselective Suzuki arylation of racemic α-chloro-amides with boron compounds using the identical ligands (L2).

Racemic α-halonitriles undergo their first stereo-convergent cross-coupling was created in 2012 explain in section 5.4.4 (b) Fu's work & (c) Reisman's work. The usage of the bisoxazoline ligand (BiOX-4) for this asymmetrical Negishi arylation confirmed successful effective application, yielding the desired products with up to 94% ee and 99% yield (Scheme 5.6b). This research laid the basis for the purpose to produce enantioenriched profens utilizing Fu's method, regardless if profen derivatives were not produced. Cross-electrophile coupling reactions are a significant development within transition metal-catalyzed coupling processes that had an impact on the synthesis of profens. These procedures combined two electrophiles with either an organic reductant or a reducing metal. Kadunce and Reisman[60] reported that heteroaryl iodides and α-chloro-nitriles are reductively coupled in an asymmetric method in 2015 using the cross-electrophile coupling method.

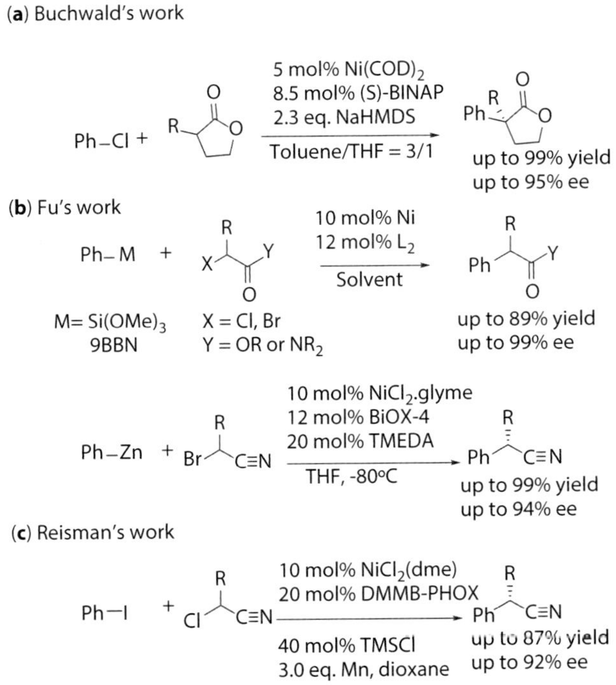

According to mechanistic research, the most likely pathway for the asymmetric reductive cross-coupling reaction to take place is a radical one because Ni oxidation states range from 0 to III. $\mathbf{L_2}$Ni(Ar)X is produced by the oxidative addition of the aryl halide to the enantioenriched $\mathbf{L_2}$Ni^0. In the meantime, an alkyl halide and a Ni(I) intermediate undergo SET to produce an α-carbonyl radical and alkyl halide was reduced by $\mathbf{L_2}$NiX, which was produced when $\mathbf{L_2}$Ni(Ar)X oxidatively captures this radical. $\mathbf{L_2}$Ni(Ar)X then produces via use of reductive elimination, this Ni(III) intermediate is created. The cycle is completed by using one of the reducing processes to convert the resultant Ni(II) to Ni(0) (Scheme 5.6).

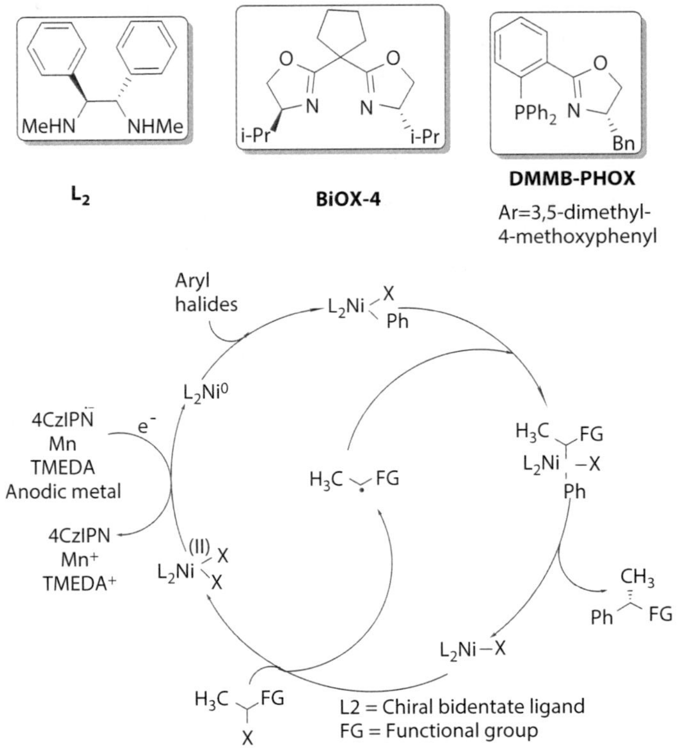

Scheme 5.6 Asymmetric cross-electrophile couplings with nickel catalysts directed toward enantioenriched profens.

5.4.5 Asymmetrical Cross-Couplings for Profen Derivative Products Under Cu-Catalysis

The quantity of copper on the Earth's crust made the use of copper catalysts appealing as well. In 2011, Bigot *et al.*[61] generated an asymmetric α-arylation of N-acyloxazolidinones and diaryliodonium salts that was copper (II) catalyzed. $Cu(OTf)_2$ and (R,R)-diphenyl-bisoxazoline were used to make the enantioenriched catalyst precursor known as **Precatalyst 1**, which had great efficiency (up to 99% yield) and enantioselectivities (up to 95% ee) (Scheme 5.7a). The operationally simple procedure offered a novel method for the preparation of enantioenriched profen derivatives, considering the mechanism of this reaction is unclear. In DCM at room temperature, derivatives of (R)-ibuprofen were produced with 99% yield and 93% efficiency. An α-arylation of N-acyloxazolidinones with diaryliodonium salts containing a copper-bisoxazoline catalyst was reported by Harvey *et al.*[62] almost simultaneously (Scheme 5.7b). The catalyst produced by using (S, S)-diphenyl-bisoxazoline (BiOX-5) presented remarkable efficiency (up to 96% yield) and enantioselectivities (up to 95% ee). This transformation in a mixed medium (DCM/toluene) at 0°C resulted in the creation of the (S)-naproxen derivative in 89% yield and 94% efficiency. A copper-catalyzed enantioselective conversion of benzylic C-H bonds into benzylic nitriles, precursors of profens, was described by Zhang *et al.*[63] in 2016 (Scheme 5.7c, left). (S,S)-dibenzyl bisoxazoline (BiOX-6) showed exceptional efficacy (up to 91% yield) and enantioselectivity (up to 98% ee) in this study. According to radical-probe tests, a radical relay pathway is most likely involved in this copper-catalyzed asymmetric reaction. An analogous case of copper/photoredox-catalyzed enantioselective decarboxylative cyanation of N-hydroxyphthalimide (NHP) esters was reported by Allen and MacMillan *et al.*[67] reported using a combination of organic and copper catalysts to achieve enantioselective α-arylation of aldehydes with diphenyliodonium triflates.. by Wang *et al.*[64]. 1 year later (Scheme 5.7c, right). The most commonly used structure was a chiral ligand-bearing bisoxazoline (BiOX) framework, similar to their earlier work. Among the ligands examined, BiOX-7 performed the reaction most efficiently in this case, producing the appropriate profen sources in outstanding quality ee (85%–96%) and moderate to excellent yields (59%–98%). By combining photo-redox catalysis and copper catalysis, several racemic compound aryl-propanoic acids were converted into the equivalent enantioenriched benzylic nitriles, including flurbiprofen, ketoprofen, pranoprofen, loxoprofen, carprofen, and zaltoprofen.

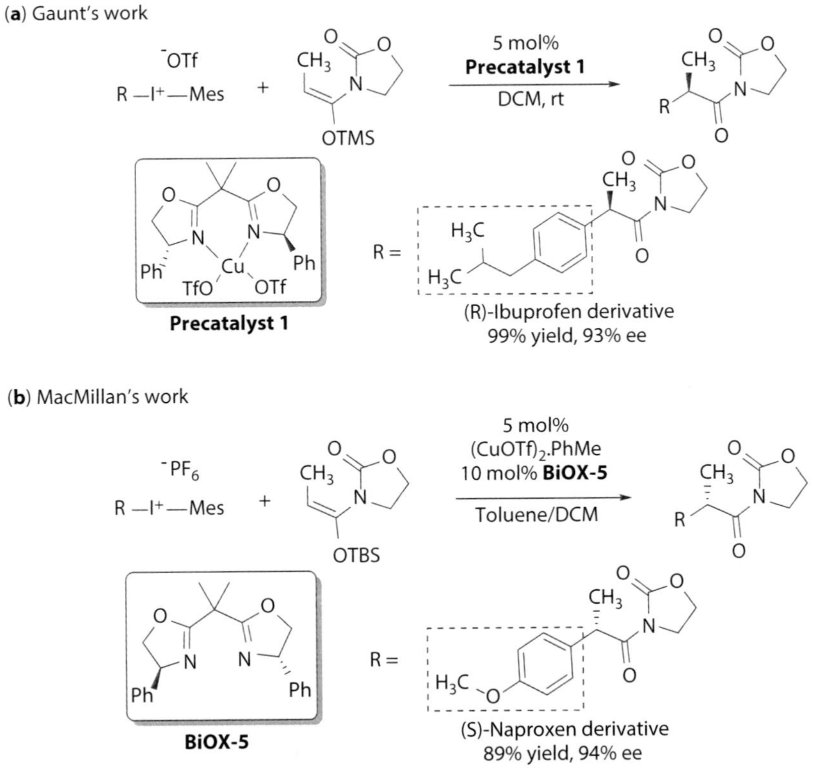

5.5 Approaches to Profens Through Organocatalysis

A potential but difficult strategy is the stereo-controlled production of C-C bonds using asymmetric organocatalysis. Earlier, transition metals and enzymatic catalysts have dominated the discussion of asymmetrical catalysis, attaining a new foundation due to the pioneering research done by List[65] and List *et al.*[66]. A different method for synthesizing profen derivatives has been made possible by the growth of catalysts based on tiny organic molecules for chemical processes, which were significantly grown during the past few years. Allen and MacMillan *et al.*[67] reported using a combination of organic and

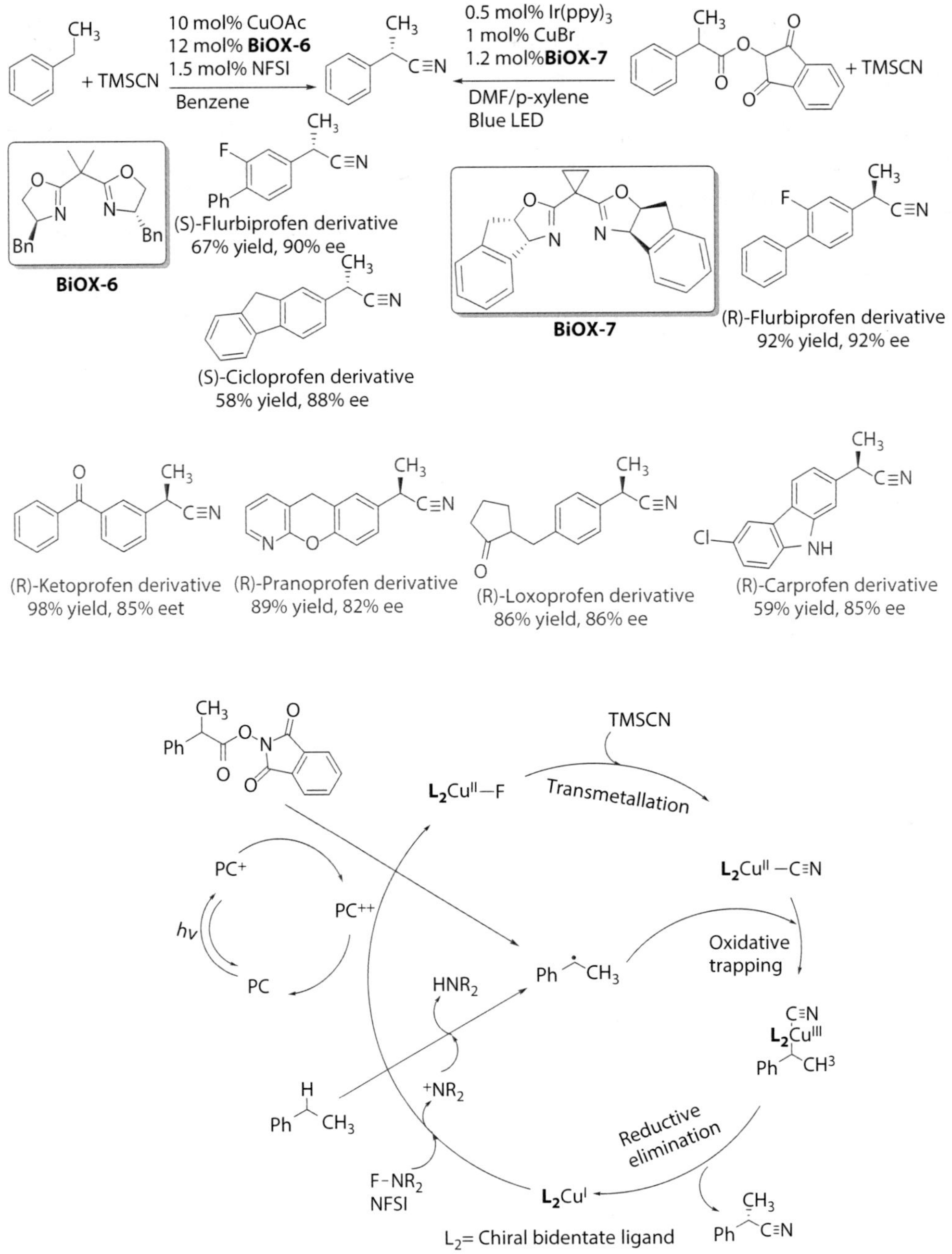

Scheme 5.7 Asymmetric cross-couplings with copper as the catalyst are used to create profens.

copper catalysts to achieve enantioselective α-arylation of aldehydes with diphenyliodonium triflates. (Scheme 5.8a). The covalent enamine intermediate is formed by the imidazolidinone catalyst (**OC-1**) and was essential for the development of a novel stereocenter and the enzymatic reaction of the carbonyl compounds. Later, aldehyde was converted to acid, (S)-ketoprofen was produced with an overall yield of 71% and an efficiency of 92%.

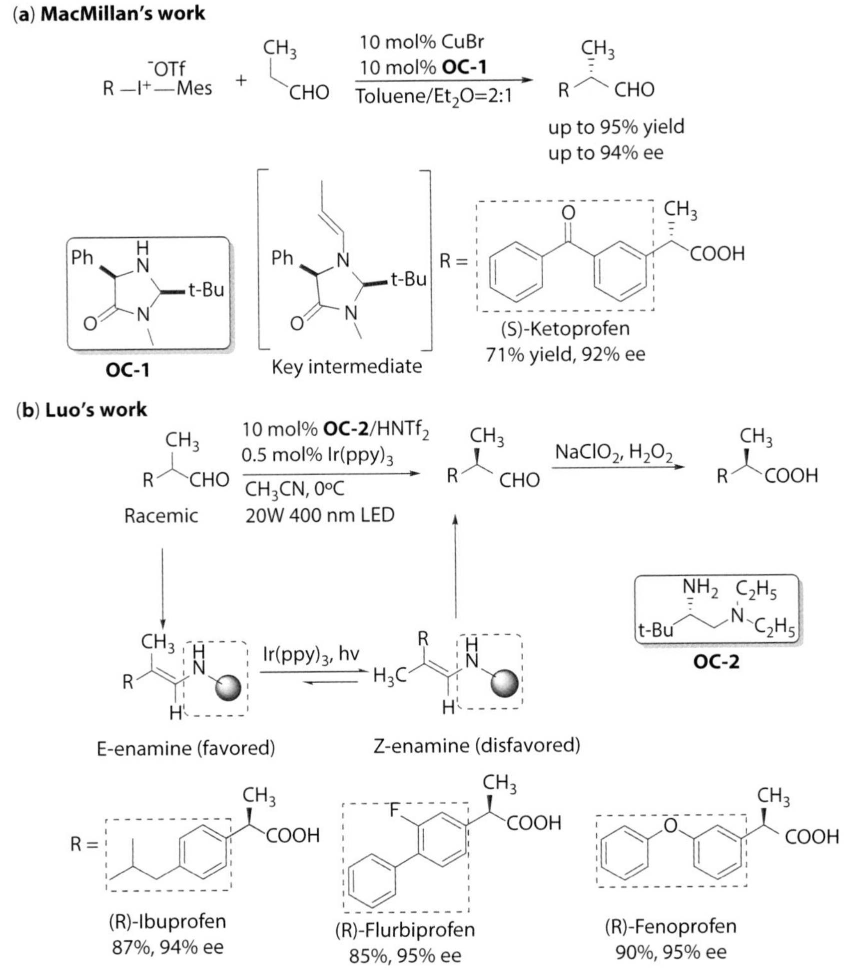

Scheme 5.8 Asymmetric organocatalytic synthesis of profens.

Huang *et al.*[68] effectively accomplished the deracemization of α-branched aldehydes by combining simple amino-catalysis (**OC-2**) with widely accessible photosensitizers (Scheme 5.8b). A direct method for creating enantiopure α-tertiary carbonyls, which were used as a direct predecessor to enantioenriched profens, was α-branched aldehydes catalytically deracemized. Mechanistic research proved that the two essential processes photochemical E/Z isomeric formation with the stereospecificity of enamine protonation by **OC-2** were involved during the deracemization of -branched aldehydes. It was possible to easily produce (R)-ibuprofen (87% yield, 94% ee), (R)-flurbiprofen (85% yield, 95% ee), and (R)-fenoprofen (90% yield, 95% ee) from their respective enantioenriched aldehydes.

5.6 Conclusions

With annual sales in the billions, the profen family of compounds is one of the most widely used pharmaceuticals worldwide[69]. Asymmetric catalysis has made significant advancements in the direction of their synthesis, spurred on by their amazing activity and application. However, there is still opportunity for advancement with regard to concerns such as catalyst cost, catalyst toxicity, ligand accessibility and stability, and green chemistry[70]. Using first-row transition metal catalysts has currently been developed in relation to the asymmetric TM-catalyzed procedures, and they are opening up new possibilities for the synthesis of highly enantioenriched profens. These metals are anticipated to be a feasible alternative to expensive transition metals [such as R (II), Ir(II), and Pd(II)] due to their rich abundance on the Earth surface and less toxicity. More convergent strategies, such as those using enantioselective C-C bond production, are also expected to continue to gain traction over enantioselective hydrogenations. The study of asymmetric organocatalysis is currently quite popular. Numerous appealing characteristics exist in it, including new forms of activation, gentle reaction conditions, and processes free of transition metals. These organocatalysts more effectively uphold the sustainability and green chemistry principles. Asymmetric organocatalytic synthesis of enantioenriched profens has made amazing progress, although it has not yet reached its full potential. Future advancements in the synthesis of profens and their derivatives are expected to be greatly aided by further research into innovative organocatalyst systems that require reduced catalytic loadings.

Acknowledgments

Poonam Rajesh Prasad is grateful to BHU IOE (Seed Grant) & Science and Engineering Research Board (SERB), New Delhi for sponsoring research.

References

1. Kate, A., Sahu, L.K., Pandey, J., Mishra, M., Sharma, P.K., Green catalysis for chemical transformation: The need for the sustainable development. *Current Research in Green and Sustainable Chemistry*, 5, 100248, Elsevier, 2022.
2. Lokko, Y., Heijde, M., Schebesta, K., Scholtès, P., Biotechnology and the bioeconomy—Towards inclusive and sustainable industrial development. *New Biotechnol.*, 40, 5–10, Elsevier, 25 January 2018.
3. Virgen-Ortíz, J. J., Peirce, S., Tacias-Pascacio, V. G., Reuse of anion exchangers as supports for enzyme immobilization: Reinforcement of the enzyme-support multiinteraction after enzyme inactivation. *Process Biochem.*, 51, 10, 1391–1396, Elsevier, October 2016.
4. Ottone, C., Romero, O., Urrutia, P., Bernal, C., Illanes, A., Wilson, L., Piumetti, M., Bensaid, S. (Eds.), Nanostructured Catalysts for Environmental Applications, pp. 58934–9, Springer, Cham, 2021.
5. Sun, H., Zhang, H., Lui, E., Zhao, H., Biocatalysis for the synthesis of pharmaceuticals and pharmaceutical intermediates. *Bioorg. Med. Chem.*, 1–10, 2017.
6. Robinson, P.K. Enzymes: principles and biotechnological applications. *Essays Biochem.*, 59, 1–41, 2015.

7. Han, Y.S., Van der Heijden, R., Verpoorte, R. Chemical biotechnology – better enzymes for green chemistry. Editorial overview. *Curr. Opin. Biotechnol.*, 12, 6, 533–534, 1 December 2001.
8. Rowbotham, J.S., Ramirez, M.A., Lenz, O. *et al.*, Bringing biocatalyticdeuteration into the toolbox of asymmetric isotopic labelling techniques. *Nat. Commun.*, 11, 1454-60, 2020.
9. Alcalde, M., Ferrer, M., Plou, F.J., Ballesteros, A. Environmental biocatalysis: from remediation with enzymes to novel green processes. *Trends Biotechnol.*, 24, 6, 281–287, 2006.
10. Wu, S., Snajdrova, R., Moore, J.C., Baldenius, K., Bornscheuer, U.T., Biocatalysis: enzymatic synthesis for industrial applications. *Agnew. Chem. Int. Ed.*, 60, 1, 88–119, 2021.
11. Szymczak, T., Cybulska, J., Podlesny, M., Frąc, M., Various perspectives on microbial lipase production using agri-food waste and renewable product. *Agriculture*, 11, 6, 540, 2021.
12. Maddalwar, S., Kumar Nayak, K., Kumar, M., Singh, L., Plant microbial fuel cell: Opportunities, challenges, and prospects. *Bioresource Technol.*, 341, 125772, 2021.
13. Chandorkar, V., Chimurkar, A., Gomashe, A., Voltage production using metabolic activities of Azatobacter species and other soil microbial flora in rice field using microbial fuel cell and microbial solar cell technology. *Agric. Sci. Digest A Res. J.*, 38, 4, 2019.
14. Maddalwar, S. and Shanware, A., Growth curve analysis of rhizobium leguminosarum using voltage produced by microbial fuel cell. *Int. J. Life Sci. Sci. Res.*, 4, 6, 2111–2115, 2018.
15. Wang, Q., Qi, Y., Gao, X., Gong, L., Wan R., Lei, W., Wang, Z., Mao, J., Guan, H., Li, W., Walsh, P.J., Recent trends and developments in the asymmetric synthesis of profens. *Green Synth. Catal.*, 4, 89–103, 2023.
16. Dhall, H., Kumar, A., Mishra, K.A., Synthesis of some novel substituted phenyl β, β-diphenyl propanoate with analgesic and anti-inflammatory activity. *Curr. Bioact. Compd.*, 14, 26–32, 2018.
17. Landoni, F.M. and Soraci, A., Pharmacology of chiral compounds 2-arylpropionic acid derivatives. *Curr. Drug Metabol.*, 2, 37–51, 2001.
18. Schjerning, A.M., McGettigan, P., Gislason, G., Cardiovascular effects and safety of (non-aspirin) NSAIDs. *Nat. Rev. Cardiol.*, 17, 574–584, 2020.
19. Pope, J.E., Anderson, J.J., Felson, D.T., A meta-analysis of the effects of nonsteroidal anti-inflammatory drugs on blood pressure. *Arch. Intern. Med.*, 153, 477–484, 1993.
20. Straus, W., Ofman, J., Maclean, C., Morton, S., Roth, E., Berger, M., Shekelle, P., Assessment of the proportion of gastrointestinal perforations, ulcers and bleeds (PUBS) attributable to NSAIDs. *Gastroenterology*, 118, 3921–3924, 2000.
21. Gooch, K., Culleton, B.F., Manns, B.J., Zhang, J., Alfonso, H., Tonelli, M., Frank, C., Klarenbach, S., Hemmelgarn, B.R., NSAID use and progression of chronic kidney disease. *Am. J. Med.*, 120, 280.e1–280.e7, 2007.
22. Hull, S., Mathur, R., Dreyer, G., Yaqoob, M.M., Evaluating ethnic differences in the prescription of NSAIDs for chronic kidney disease: a cross-sectional survey of patients in general practice. *Br. J. Gen. Pract.*, 64, e448, 2014.
23. Levesque, L.E., Brophy, J.M., Zhang, B., The risk for myocardial infarction with cyclooxygenase-2 inhibitors: a population study of elderly adults. *Ann. Intern. Med.*, 142, I–45, 2005.
24. Chinthapalli, K., Recent trends and developments in the asymmetric synthesis of profens. *Br. Med. J. (Clin. Res. Ed.)*, 346, f3533, 2013.
25. Cashman, J. and McAnulty, G., Nonsteroidal anti-inflammatory drugs in perisurgical pain management: mechanisms of action and rationale for optimum use. *Drugs*, 49, 51–70, 1995.
26. Kawashima, M., Ogura, N., Akutsu, M., Ito, K., Kondoh, T., J., The anti-inflammatory effect of cyclooxygenase inhibitors in fibroblast-like synoviocytes from the human temporomandibular joint results from the suppression of PGE2 production. *Oral. Pathol. Med.*, 42, 499–506, 2013.
27. Hutt, A.J. and Caldwell, J., *Clin. Pharmacokinet.*, 9, 371–373, 1984.

28. Reichel, C., Brugger, R., Bang, H., Geisslinger, G., Brune, K., Molecular cloning and expression of a 2-arylpropionyl-coenzyme A epimerase: a key enzyme in the inversion metabolism of ibuprofen. *Mol. Pharmacol.*, 51, 576, 1997.
29. Brooks, P.M. and Day, R.O., Nonsteroidal antiinflammatory drugs—differences and similarities. *N. Engl. J. Med.*, 324, 1716–1725, 1991.
30. Knihinicki, R.D., Williams, K.M., Day, R.O., Chiral inversion of 2-arylpropionic acid non-steroidal anti-inflammatory drugs—1: *In vitro* studies of ibuprofen and flurbiprofen. *Biochem. Pharmacol.*, 38, 4389–4395, 1989.
31. Oliary, J., Tod, M., Nicolas, P., Petitjean, O., Caille, G., Pharmacokinetics of ibuprofen enantiomers after single and repeated doses in man. *Biopharm. Drug Dispos.*, 13, 337–344, 1992.
32. Piccolo, O., Spreafico, F., Visentin, G., Valoti, E., Alkylation of aromatic compounds with optically active lactic acid derivatives: synthesis of optically pure 2-arylpropionic acid and esters. *J. Org. Chem.*, 50, 3945–3946, 1985.
33. Noorduin, W.L., Vlieg, E., Kellogg, R.M., Kaptein, B., From Ostwald ripening to single chirality. *Angew. Chem. Int. Ed.*, 48, 9600–9606, 2009.
34. Walsh, P.J. and Kozlowski, M.C., *Fundamentals of Asymmetric Catalysis*, University Science Books, 2008.
35. Kourist, R., Domínguez de María, P., Miyamoto, K., Biocatalytic strategies for the asymmetric synthesis of profens–recent trends and developments. *Green Chem.*, 13, 2607–2618, 2011.
36. Friedfeld, M.R., Zhong, H., Ruck, R.T., Shevlin, M., Chirik, P.J., Cobalt-catalyzed asymmetric hydrogenation of enamides enabled by single-electron reduction. *Science*, 360, 888–893, 2018.
37. Friedfeld, M.R., Shevlin, M., Hoyt, J.M., Krska, S.W., Tudge, M.T., Chirik, P.J., Cobalt precursors for high-throughput discovery of base metal asymmetric alkene hydrogenation catalysts. *Science*, 342, 1076–1080, 2013.
38. Shevlin, M., Friedfeld, M.R., Sheng, H., Pierson, N.A., Hoyt, J.M., Campeau, L.C., Chirik, P.J., Nickel-catalyzed asymmetric alkene hydrogenation of α,β-unsaturated esters: high-throughput experimentation-enabled reaction discovery, optimization, and mechanistic elucidation. *J. Am. Chem. Soc.*, 138, 3562–3569, 2016.
39. Chen, J. and Lu, Z., Asymmetric hydrofunctionalization of minimally functionalized alkenes via earth abundant transition metal catalysis. *Org. Chem. Front.*, 5, 260–272, 2018.
40. Schotten, C., Howard, J.L., Jenkins, R.L., Codina, A., A continuous flow-batch hybrid reactor for commodity chemical synthesis enabled by inline NMR and temperature monitoring. *Tetrahedron*, Elsevier, 5, 2, 260–272, 2018.
41. Biffis, A., Centomo, P., Del Zotto, A., Zecca, M., Pd metal catalysts for cross-couplings and related reactions in the 21st century: a critical review. *Chem. Rev.*, 118, 2249–2295, 2018.
42. Moradi, W.A. and Buchwald, S.L., Palladium-catalyzed α-arylation of esters *J. Am. Chem. Soc.*, 123, 7996–8002, 2001.
43. Hama, T. and Hartwig, J.F., Palladium-catalyzed α-arylation of esters with chloroarenes. *Org. Lett.*, 10, 1549–1552, 2008.
44. Huang, Z., Liu, Z., Zhou, J.S., An enantioselective, intermolecular α-arylation of ester enolates to form tertiary stereocenters. *J. Am. Chem. Soc.*, 133, 15882–15885, 2011.
45. Li, B., Li, T., Aliyu, M.A., Li, Z.H., Tang, W., Enantioselective palladium-catalyzed cross-coupling of α-bromo carboxamides and Aryl Boronic acids. *Angew. Chem. Int. Ed.*, 58, 11355–11359, 2019.
46. Galdeano, C., Camps, P., Formosa, X., Synthesis, pharmacological evaluation,and molecular modeling of a novel family of 6-chlorotacrine-based dual binding site acetylcholinesterase inhibitors. *Org. Lett.*, 23, 16, 6439–6443, 2021.
47. Jin, M., Adak, L., Nakamura, M., Iron-catalyzed enantioselective cross-coupling reactions of α-chloroesters with Aryl Grignard reagents. *J. Am. Chem. Soc.*, 137, 7128–7134, 2015.

48. Mao, J., Liu, F., Wang, M., Wu, L., Zheng, B., Liu, S., Zhong, J., Bian, Q., Walsh, P.J., Cobalt-bisoxazoline-catalyzed asymmetric Kumada cross-coupling of racemic α-bromo esters with Aryl Grignard reagents. *J. Am. Chem. Soc.*, 136, 17662–17668, 2014.
49. Liu, F.. Bian, Q., Mao, J., Gao, Z., Liu, D., Liu, S., Wang, X., Wang, Y., Wang, M., Zhong, J., Asymmetric cross-coupling of racemic α-bromo esters with aryl Grignard reagents catalyzed by cyclopropane-based bisoxazolines cobalt complexes Tetrahedron. *Asymmetry*, 27, 663–669, 2016.
50. Affo, W., Ohmiya, H., Fujioka, T., Ikeda, Y., Nakamura, T., Yorimitsu, H., Oshima, K., Imamura, Y., Mizuta, T., Miyoshi, K., Cobalt-catalyzed trimethylsilylmethylmagnesium-promoted radical alkenylation of alkyl halides: a complement to the Heck reaction. *J. Am. Chem. Soc.*, 128, 8068–8077, 2006.
51. Tasker, S.Z., Standley, E.A., Jamison, T.F., Recent advances in homogeneous nickel catalysis. *Nature*, 509, 299–309, 2014.
52. Fu, G.C., Transition-metal catalysis of nucleophilic substitution reactions: A radical alternative to SN1 and SN2 processes. *ACS Cent. Sci.*, 3, 692–700, 2017.
53. Choi, J., Martin-Gago, P., Fu, G.C., Stereoconvergent arylations and alkenylations of unactivated alkyl electrophiles: catalytic enantioselective synthesis of secondary sulfonamides and sulfones. *J. Am. Chem. Soc.*, 136, 12161–12165, 2014.
54. Cordier, C.J., Lundgren, R.J., Fu, G.C., Enantioconvergent cross-couplings of racemic alkylmetal reagents with unactivated secondary alkyl electrophiles: catalytic asymmetric negishi α-alkylations of n-boc-pyrrolidine *J. Am. Chem. Soc.*, 135, 10946–10949, 2013.
55. Do, H.Q., Chandrashekar, E.R.R., Fu, G.C., Nickel/bis (oxazoline)-catalyzed asymmetric Negishi arylations of racemic secondary benzylic electrophiles to generate enantioenriched 1,1-diarylalkanes *J. Am. Chem. Soc.*, 135, 16288–16291, 2013.
56. Liang, Y. and Fu, G.C., Catalytic asymmetric synthesis of tertiary alkyl fluorides: Negishi cross-couplings of racemic α, α-dihaloketones. *J. Am. Chem. Soc.*, 136, 5520–5524, 2014.
57. Lou, S. and Fu, G.C., Nickel/bis (oxazoline)-catalyzed asymmetric Kumada reactions of alkyl electrophiles: cross-couplings of racemic α-bromoketones. *J. Am. Chem. Soc.*, 132, 1264–1266, 2010.
58. Oelke, A.J., Sun, J., Fu, G.C., Nickel-catalyzed enantioselective cross-couplings of racemic secondary electrophiles that bear an oxygen leaving group. *J. Am. Chem. Soc.*, 134, 2966–2969, 2012.
59. Saito, B. and Fu, G.C., Enantioselective alkyl–alkyl Suzuki cross-couplings of unactivated homobenzylic halides. *J. Am. Chem. Soc.*, 130, 6694–6695, 2008.
60. Kadunce, N.T. and Reisman, S.E., Nickel-catalyzed asymmetric reductive cross-coupling between heteroaryl iodides and α-chloronitriles. *J. Am. Chem. Soc.*, 137, 10480–10483, 2015.
61. Bigot, A., Williamson, A.E., Gaunt, M.J., Enantioselective α-Arylation of N-Acyloxazolidinones with Copper(II)-bisoxazoline Catalysts and Diaryliodonium Salts *J. Am. Chem. Soc.*, 133, 13778–13781, 2011.
62. Harvey, J.S., Simonovich, S.P., Jamison, C.R., MacMillan, D.W., Enantioselective α-arylation of carbonyls via Cu (I)-bisoxazoline catalysis. *J. Am. Chem. Soc.*, 133, 13782–13785, 2011.
63. Zhang, W., Wang, F., McCann, S.D., Wang, D., Chen, P., Stahl, S.S., Liu, G., Enantioselective cyanation of benzylic C–H bonds via copper-catalyzed radical relay. *Science*, 353, 1014–1018, 2016.
64. Wang, D., Zhu, N., Chen, P., Lin, Z., Liu, G., Enantioselective decarboxylative cyanation employing cooperative photoredox catalysis and copper catalysis *J. Am. Chem. Soc.*, 139, 15632–15635, 2017.
65. List, B., The direct catalytic asymmetric three-component Mannich reaction. *J. Am. Chem. Soc.*, 122, 9336–9337, 2000.

66. List, B., Pojarliev, P., Martin, H.J., Efficient proline-catalyzed michael additions of unmodified ketones to nitro olefins. *Org. Lett.*, 3, 2423–2425, 2001.
67. Allen, A.E. and MacMillan, D.W., Enantioselective α-arylation of aldehydes via the productive merger of iodonium salts and organocatalysis. *J. Am. Chem. Soc.*, 133, 4260–4263, 2011.
68. Huang, M., Zhang, L., Pan, T., Luo, S., Deracemization through photochemical E/Z isomerization of enamines. *Science*, 375, 869–874, 2022.
69. Wood, L., Poliakoff, M.,Anti-inflammatory Drugs Licence, P., *Nature*, 450, 810–8, 2007.
70. Bednar, D. and Damborsky, Computational design of enzymes for biotechnological applications. *J., Biotechnol. Adv.*, 47, 107696, 2021.
71. Schwarzwalder, G.M., Matier, C.D., Fu, G.C., Enantioconvergent cross-couplings of alkyl electrophiles: the catalytic asymmetric synthesis of organosilanes. *Angew. Chem. Int. Ed.*, 58, 3571–3574, 2019.
72. Binder, J.T., Cordier, C.J., Fu, G.C., Catalytic enantioselective cross-couplings of secondary alkyl electrophiles with secondary alkylmetal nucleophiles: negishi reactions of racemic benzylic bromides with achiral alkylzinc reagents. *J. Am. Chem. Soc.*, 134, 17003–17006, 2012.
73. Lundgren, R.J., Wilsily, A., Marion, N., Ma, C., Chung, Y.K., Fu, G.C., Catalytic asymmetric C–N bond formation: Phosphine-catalyzed intra-and intermolecular γ-addition of nitrogen nucleophiles to allenoates and alkynoates. *Angew. Chem. Int. Ed.*, 52, 2525–2528, 2013.
74. Wang, Z., Bachman, S., Dudnik Alexander, S., Fu Gregory, C., Nickel-catalyzed enantioconvergent borylation of racemic secondary benzylic electrophiles. *Angew. Chem. Int. Ed.*, 57, 14529–14532, 2018.

6

Application of Nanocatalysts in Greener Synthesis of Chemical Compounds

Karan Chaudhary[1,2] and Dhanraj T. Masram[1*]

[1]*Department of Chemistry, University of Delhi, Delhi, India*
[2]*Forensic Chemistry and Toxicology Laboratory, Department of Forensic Sciences, National Forensic Sciences University, Delhi, India*

Abstract

In the synthesis of organic compounds and transformation of organic compounds, nanocatalysts play an especially important role, as they provide a new mechanistic pathway for the transformation and synthesis of the desired organic compound. Over time, enormous research has been performed in the field of nanocatalysis and has led to the development of various catalytic methods that include biocatalysis, photocatalysis, organocatalysis, transition metal catalysis, and electrocatalysis. The most widely used catalytic system for the transformation and synthesis of the desired organic compound is metal-based catalysts. Furthermore, in green chemistry, catalysis is in itself a fundamental pillar, and the inclusion of green chemistry principles in the field of nanocatalysis has several benefits for the reduction of risk caused to our environment. As per green chemistry, it is designing a method or a process in which there is minimized use and minimum production of harmful and dangerous substances, avoiding the use of volatile organic solvents, and elimination of prolonged and challenging wasteful separation. Thus, a nontoxic, eco-friendly, economical, efficient, and simple approach used for the synthesis of nanomaterial or an organic compound is green synthesis. The greener synthesis of chemical compounds by utilization of nanocatalysts includes two major parts, which are the green synthesis of the nanocatalysts and the use of green reaction condition for obtaining the chemical compound. The ideal condition is the use of green synthesized nanocatalysts in greener reaction conditions for the synthesis of chemical compounds. Therefore, green chemistry with an objective of an increase in greener synthesis and processing is considered a valuable pathway toward environmental sustainability. The present chapter is based on the discussion of the development that has been made in recent years for the greener synthesis of chemical compounds using nanocatalysts. The focus of the present chapter is on three major ideas that are considered important in the green synthesis of chemical compounds that include (i) the synthesis of green nanocatalysts that comprises the use of compounds such as ionic liquids, ethylene glycol, plant extracts, and fruit extracts; (ii) the greener reaction condition such as the use of PEG, water, and ethanol as solvents or solvent-free condition for nanocatalysis; and (iii) the use of green nanocatalysts along with greener reaction conditions.

***Keywords*:** Nanocatalysis, green chemistry, environmental sustainability, green synthesis, organic transformation

**Corresponding author*: dhanraj_masram27@rediffmail.com

Mousumi Sen (ed.) Sustainable Green Catalytic Processes, (121–140) © 2024 Scrivener Publishing LLC

6.1 Introduction

In the present era of science, catalysis is one of the crucial areas that leads an innovative way to tackle the problems faced in the development of sustainable and environmentally friendly methods for chemical transformations. Based on green chemistry, catalysis is one of the fundamental pillars, as catalysis science is more energy-efficient and sustainable for carrying out chemical transformations.[1,2] The development of a more natural, environment-friendly, and economical catalytic system for a chemical transformation is the challenge of the present time. As per green chemistry, the chemical transformation processes where hazardous substances are neither used nor created are categorized as green chemical transformation. In green chemical transformations, the use of harsh reaction conditions such as the use of volatile organic solvents and toxic reagents is avoided.[1]

As per green chemistry, the chemical transformation process is designed in such a way that it prevents the use of harmful chemicals as well as minimizes the generation of chemical waste so that it does not deteriorate our environment. In terms of green chemistry, the 12 fundamental principles include atom economy, catalysis, designing safer chemicals, design for degradation, design for energy efficiency, inherently safer chemistry for accident prevention, less hazardous chemical syntheses, prevention, reduce derivatives, real-time analysis for pollution prevention, safer solvents and auxiliaries, and use of renewable feedstocks. Catalysis is one of the principles of green chemistry because catalyst use has successfully achieved several objectives of green chemistry.[1] As the use of catalysis leads to atom economy, energy efficiency, less hazardous chemical syntheses, and less waste generation, so, based on these, it is necessary for the present time to develop more sustainable and environment-friendly catalysts for chemical transformation. The use of catalysts aids in achieving the dual objectives of economic benefits and environmental protection.[2] Additional benefits associated with catalysis are the use of renewable feedstocks and the use of alternative solvents that are environmentally benign.[1]

Over time, enormous research has been performed in the field of catalysis. And it has led to the development of various catalytic methods that include biocatalysis, photocatalysis, organocatalysis, transition metal catalysis, and electrocatalysis. The most widely used catalytic system for the transformation and synthesis of the desired organic compound is metal-based catalysts because they have a notable contribution. The composition of these metal-based catalysts can be pure metal salts, oxides of metals, complexes of metal, and composites of metal with other types of materials.[3] Furthermore, catalysis by the use of a nanostructured catalytic system has become more important regarding its preparation and utilization for the transformation and synthesis of the desired organic compound. Catalyst activity mainly depends upon the surface area; thus, the smaller the size, the higher the surface area; therefore, the higher the catalytic activity.[4] So, nanocatalysts have particles whose size ranges between 1 and 100 nm, and due to these nano-dimensions, they possess unique properties including an exceptionally large surface area-to-volume ratio.[5,6] This particular property of enhanced surface area has an extensive advantage in catalyzing the reaction for the production of the desired organic compound. Other than this nanocatalyst had structural and electronic changes in comparison to that of bulk material, which results in greater selectivity in the catalytic process.[5] Furthermore, by altering the physical properties such as composition, shape, and morphology of the nanocatalyst, the activity and selectivity of the catalyst can be enhanced. But the challenge associated is how to alter and control these physical properties, which has been solved by solution-based synthesis of nanocatalysts by the use of the bottom-up approach.[7] Whereas for the synthesis of nanocatalysts by the bottom-up approach, harsh chemical and volatile organic solvents are used. Even though nanocatalysts produced are of desired properties, these are not economical and environmentally friendly.[8] Thus, development has been made and nanocatalysts are being prepared by a nontoxic, eco-friendly, economical, efficient, and simple approach, which is known as green synthesis of nanocatalysts. Green synthesis of nanocatalysts refers to the minimal use of toxic and harmful chemicals or the use of solvents that have been considered as green for the synthesis and processing of nanomaterials to be used as catalysts.[9,10] Over the time of two decades, a lot of effort and research activities have been observed for the development of greener catalysts with high selectivity, higher activity, and catalyst separation with lesser chemical waste generation.[11] Moreover, the greener synthesis of chemical compounds by utilization of nanocatalysts also includes the use of green reaction conditions for obtaining the chemical compound. The green reaction condition refers to the use of less harsh conditions, the use of solvents considered as green, and an environment-friendly process to perform the catalytic reaction for obtaining the chemical compound. Furthermore, the more ideal condition is the use of green synthesized nanocatalysts in greener reaction conditions for the synthesis of chemical compounds.[9] Therefore, green chemistry with an objective of an increase in greener synthesis and processing is considered a valuable pathway toward environmental sustainability. The focus of the present chapter is on three major ideas that are considered important in the green synthesis of chemical compounds that include (i) the synthesis of green nanocatalysts that comprises the use of compounds such as ionic liquids, ethylene glycol, plant extracts, and fruit extracts; (ii) the greener reaction condition such as the use of PEG, water, and ethanol as solvents or solvent-free condition for nanocatalysis; and (iii) the use of green nanocatalysts along with greener reaction conditions.

6.2 Green Strategies

6.2.1 Synthesis of Green Nanocatalysts

For the synthesis of green nanocatalysts, synthetic strategies involve the use of benign reagents in the matrix in which they are being synthesized, so that the use of hazardous substances can be eliminated in their synthetic process. Other than that, the use of biomolecules and carbohydrates as substrates in the synthesis of nanocatalysts improves their biological and green aspects. In the synthesis of $Fe_{3-x}Ti_xO_4$@TMSP@TBAAsp nanocatalyst, tetrabutylammonium asparaginate (TBAAsp), an amino acid ionic liquid, has been immobilized on the surface of $Fe_{3-x}Ti_xO_4$@TMSP.[12] Another catalyst, $NiFe_2O_4$@SiO_2@amino glucose, in this, a carbohydrate molecule that is 2-amino glucose has been used as a substrate in the synthesis of the catalyst.[13] A study showed that during catalyst synthesis, no toxic reducing agent was used, palladium nanoparticles were decorated on the magnetic lignin/chitosan microbeads while stirring in ethanol at an elevated temperature.[14] During the synthesis of a catalyst, Fe_3O_4@NFC-ImSalophCu, magnetic cellulose was chosen and synthesized as base material for the stabilization of the catalyst.[15] In a study, gold nanoparticles as a nanocatalyst have been synthesized by using sea cucumber extract that acted as a reducing as well as a stabilizing agent.[16] Another study shows that NiO nanoparticles have been synthesized by using marine macroalgae extract that acted as a reducing as well as a coating agent.[17] Green tea extract has also been used for the synthesis of green nanocatalyst Pd/Fe_3O_4. In this nanocatalyst, by the use of green tea extract and ultrasound irradiations, Pd nanoparticles have been immobilized *in situ* on the surface of Fe_3O_4 nanoparticles.[18] Other than these, molecules such as tryptophan and citric acid have been used as substrates in the synthesis of nanocatalysts, which makes them green nanocatalysts.[19, 20]

6.2.2 Greener Reaction Condition

There are alternate reaction conditions that are considered green such as performing reactions under microwave-assisted conditions, ultrasound-assisted conditions, solvent-free conditions, and the use of green solvents including water, ethanol, polyethylene glycol, and deep eutectic solvent during nanocatalysis for the transformation and synthesis of the desired organic compound.

Ultrasound-Assisted Synthesis Conditions: The use of ultrasonication method for the synthesis of organic compounds has been found efficient, as it offers mild reaction conditions, faster rate for organic reaction, high yields of products, waste minimization, lower production cost, energy efficiency, simple, clean approach compared to traditional methods. A multicomponent reaction for the synthesis of 3-substituted coumarins has been performed in one pot by Knoevenagel condensation of 1,3-dicarbonyl compound and o-hydroxy benzaldehyde by using $MgFe_2O_4$ as a nanocatalyst under 35-W power of ultrasonic irradiation at 45°C.[21] In the presence of $NiFe_2O_4$/geopolymer as catalyst for the synthesis of 2,4,5-trisubstituted imidazole derivatives by condensation reaction of ammonium acetate, benzil, and substituted aldehydes, the ultrasonication method was found to be more efficient.[22] In another study, the ultrasonication method found effective was utilization of Pd/Fe_3O_4 as nanocatalyst for catalyzing Suzuki–Miyaura coupling reaction for synthesizing biphenyl compounds.[18] For the synthesis of substituted 3-pyrrolin-2-one, in the presence of $CoFe_2O_4$@CA as nanocatalyst, the ultrasonication method was employed and excellent product yields were obtained when 100 W ultrasonic power was used.[19]

Microwave-Assisted Synthesis Conditions: The use of microwave irradiation technique has an advantage that organic transformation can be done in a shorter reaction time for which large reaction time and high reaction temperature were required in traditional synthesis method. For Suzuki–Miyaura coupling reactions between phenyl boronic acid and various aryl halides in the presence of Pd nanocatalyst on chitosan/cellulose composite, microwave radiation (400 W) at 50°C was used to obtain products in maximum yields.[23]

Solvent-Free Synthesis Conditions: Performing an organic reaction under solvent-free conditions offers enhanced efficiency and selectivity and prevents the use of volatile and toxic solvents. In a study, $Fe_{3-x}Ti_xO_4$@TMSP@TBAAsp nanoparticles were utilized as catalyst for the synthesis of 1,4-dihydropyrano[2,3-c]pyrazole derivatives. Under solvent-free conditions, maximum yields (86%–96%) of derivatives were obtained in lesser reaction time (15–60 min).[12] In another study, Fe_3O_4@SiO_2@Vanillin@Thioglycolic acid was applied as catalyst for carrying out the synthesis of azo-linked 5-amino-pyrazole-4-carbonitriles and 5-amino-bispyrazole-4-carbonitriles. Transformations were carried out under solvent-free conditions at room temperature in lesser time, and yields obtained were expeditiously high.[24]

Use of Green Solvents: Besides just being solvents for green reaction media, they can be involved in additional functions including the role of cocatalyst, promoters, costabilizers, and reductants. The use of green solvents help in eliminating toxic and volatile solvents used in traditional methods. In the presence of water as solvent, one-pot synthesis of spiro-oxindoles with fused chromenes was done by using Fe_3O_4/COS@β-CD-SO_3H nanocatalyst, and products were obtained in excellent yields in shorter reaction time (15–22 min) at 50°C.[25] In another study, ZnO@Fe_3O_4 was utilized as catalyst in a multicomponent reaction for the synthesis of 3,4-disubstituted isoxazol-5(4H)-ones. In this reaction, high yields were obtained when water was used as solvent.[26]

In the presence of ethanol as solvent, La-guanine@SBA-15 was used as catalyst for synthesis of 4,4'-(arylmethylene)-bis-(3-methyl-1-phenyl-1Hpyrazol-5-ols) and benzo [a] pyrano to obtain the best results.[27] On using ethanol as solvent, 4H-pyran derivatives were synthesized in excellent yields under shorter reaction time when halloysite nanotubes loaded with Fe_3O_4 nanoparticles were applied as nanocatalyst.[28] For the synthesis of 3-amino-1-aryl-5,10-dioxo-1H-pyrazolo[1,2-b]phthalazine-2-carbonitrile derivatives, $NiFe_2O_4$@TiO_2–ILPip was utilized as catalyst in a multicomponent reaction between aldehyde, phthalhydrazide, and malononitrile, high yields of products were obtained in the presence of PEG 400 as solvent.[29] In another study, the catalytic activity of Fe_3O_4@SiO_2@TiO_2-OSO_3H catalyst for the synthesis of spiroquinazolin-4(3H)-one derivatives and 2,3-dihydroquinazolin-4(1H)-one derivatives was exemplified in deep eutectic solvents. Derivative products were obtained in good yields at low temperature under short reaction time.[30]

6.3 Nanocatalysts for Green Synthesis of Organic Compounds

For the synthesis of biologically active tetrahydrobenzo[b]pyran compounds by greener pathway, Elhamifar *et al.*[31] have used a novel core shell structured material that is magnetic iron oxide-supported phenylsulfonic acid (Fe_3O_4@Ph-SO_3H) prepared by chemical modification of iron oxide nanoparticles. After complete physicochemical studies by various physicochemical techniques, Fe_3O_4@Ph-SO_3H was utilized as catalyst for the synthesis of tetrahydrobenzo[b]pyran compounds. Condensation of dimedone and benzaldehyde was chosen as model reaction, and for this catalysis reaction, several parameters were screened including temperature, catalyst amount, and solvent. The optimized condition was found to be 50°C temperature and 0.2 mol% catalyst amount along with water as solvent, which is eco-friendly and green solvent for efficient yield of the product. Furthermore, under same condition of 50°C, 0.2 mol% catalyst, and water as solvent, better yield for the product was obtained in ultrasonic media in very lesser time. Moreover, excellent yield of desired product was obtained in ultrasonic method when performed at room temperature. Series of derivative products were synthesized under these greener conditions in good to excellent yields. Without significant loss in efficiency and stability, catalyst was recoverable and recyclable up to nine times. The advantage of such catalyst was that excellent yields of tetrahydrobenzo[b]pyran compounds can be obtained in green conditions by ultrasonic method in less reaction time.[31]

Dihydropyrano[2,3-c]pyrazoles and their derivatives are among important organic compounds due to their therapeutic and pharmacological properties. Azarifar *et al.*[12] have synthesized titanomagnetite nanoparticles ($Fe_{3-x}Ti_xO_4$) onto which tetrabutylammonium asparaginate, an amino acid ionic liquid, was immobilized by using organosilane compound as spacer to obtain $Fe_{3-x}Ti_xO_4$@TMSP@TBAAsp nanoparticles. After complete physiochemical studies for the successful formation of the nanomaterial, it was used as catalyst for the synthesis of 1,4-dihydropyrano[2,3-c]pyrazole derivatives. For catalytic evaluation, model reaction chosen was between 3-methyl-1-phenyl-1H-2-pyrazol-5(4H)-one, malononitrile, and benzaldehyde, for which several parameters were screened including temperature, catalyst amount, and solvent. The optimum condition was found to be 100°C and 0.01 g catalyst when 1 mmol of each reaction was taken to obtain maximum yield in lesser time under solvent-free conditions. By using $Fe_{3-x}Ti_xO_4$@TMSP@TBAAsp as catalyst, several 1,4-dihydropyrano[2,3-c]pyrazole derivatives were obtained in high yields. Furthermore, for six consecutive runs, this catalyst was reusable without significant loss in catalytic activity. Such kind of catalyst that can lead to organic transformations under solvent-free conditions is environment-friendly and a greener method for the synthesis of organic products.[12] Zakeri *et al.*[32] prepared graphene oxide functionalized with phosphoric acid to be used as catalyst for the synthesis of pyrano[2,3-c]pyrazole derivatives. The reaction of 2 mmol of each reactant including hydrazine hydrate, ethyl acetoacetate, malonitrile, and aromatic aldehyde in the presence of catalyst was chosen as model reaction. For this multicomponent reaction, 5 mol% catalyst in water under reflux for 15 min was the optimized condition for obtaining products with high yield. Under optimized conditions, pyrano[2,3-c]pyrazole derivatives were also synthesized in high yields. This catalyst was reusable without catalytic activity loss for at least six runs. Simple procedure, high product yields, use of no toxic solvents, short reaction time, and water as solvent make this process a greener pathway for synthesis of pyrano[2,3-c]pyrazole derivatives.[32] A magnetic nanocatalyst composed of Fe_3O_4@chitosan and tannic acid bionanocomposite was utilized by Kamalzare *et al.*[33] for catalyzing the organic synthesis reaction of pyrano[2,3-c]pyrazole. One-pot four-component reaction between 4-chlorobenzaldehyde, malononitrile, ethyl acetate, and hydrazine hydrate was opted as model reaction for the optimization of reaction parameters. Parameters such as reaction temperature, solvent, and catalyst amount were optimized, and best results were obtained when reaction was performed at room temperature, ethanol was used as solvent, and 0.02 g was the catalyst amount. Several pyranopyrazole derivatives were synthesized under optimized conditions in excellent yield (86%–95%) under the reaction time of 15–45 min. Furthermore, without any significant loss in the catalytic action of the catalyst, it was reusable up to seven times. The use of green solvent such as ethanol, short reaction time, high product yield, and mild reaction condition show that this method is in agreement with the principle of green and sustainable chemistry.[33] Chen *et al.*[34] synthesized an organic-inorganic nanohybrid in which Ru^{III} was incorporated with CMC/Fe_3O_4. This Ru^{III}@ CMC/Fe_3O_4 was used as catalyst in the synthesis of pyrano[2,3-c] pyrazole and polyhydroquinoline derivatives. Reaction conditions were optimized for both the reactions by choosing suitable model reactions and found that, for both reactions, best results were obtained when ethanol was used as solvent, catalyst amount containing 0.5 mol% of Ru^{III}, and at specific temperatures. Under optimized conditions, the desired products were synthesized in excellent yields. The catalyst was easily separable with the help of an external magnet and found reusable for several runs without significant decrease in the catalytic performance. Advantages of this method that include excellent product yields, simple procedure, mild reaction conditions, high atom economy, energy efficiency, and use of green solvent such as ethanol show that this protocol follows the principle of green chemistry.[34] Mohtasham and Gholizadeh[35] have synthesized a nanocomposite material in which nanomesoporous silica obtained from horsetail plant is used to prepare aminated epibromohydrin-functionalized Fe_3O_4@SiO_2, further which has been used as support material for the immobilization of $H_3PW_{12}O_{40}$ to finally obtain Fe_3O_4@SiO_2-EP-NH-HPA. This Fe_3O_4@SiO_2-EP-NH-HPA was applied as nanocatalyst for the synthesis of pyrano[2,3-c]pyrazole derivatives. A reaction between 4-nitrobenzaldehyde, malononitrile, hydrazine hydrate, and acetoacetate was chosen as model reaction; various reaction parameters including solvent type, reaction temperature, and catalyst amount were optimized for best results. By using water as green solvent, 20 mg catalyst, and performing reaction at room temperature (25°C), product was obtained in excellent yield. Furthermore, under these optimized conditions, several different pyrano[2,3-c]pyrazole derivatives were obtained in good to excellent yield (89%–99%) under short reaction time (3–11 min). Furthermore, Fe_3O_4@SiO_2-EP-NH-HPA nanocatalyst being magnetic in nature was easily recoverable by use of an external magnet and was found to be reusable up to seven runs without significant loss in its catalytic activity. Advantages such as mild reaction conditions, shorter reaction time, high product yields, and use of water as green solvent show that this protocol is a greener method for the synthesis of pyrano[2,3-c]pyrazole derivatives.[35]

Nikoorazm *et al.*[27] synthesized La-guanine@SBA-15 in which chloro-functionalized SBA-15 is substrate on which lanthanum (III) complex with guanine ligand has been immobilized. This La-guanine@SBA-15 was used as catalyst for the synthesis of 4,4'-(arylmethylene)-bis-(3-methyl-1-phenyl-1Hpyrazol-5-ols) and benzo [a] pyrano [2, 3-c] phenazine derivatives. For both the organic transformation, it was found that best yields of derivative products were obtained when catalyst amount was 0.5 mol% w.r.t. La, greener solvent ethanol was used at reflux condition. Furthermore, without any significant loss in the catalytic activity, this catalyst was found to be reusable up to over five runs. This nanocatalyst has an advantage of carrying such organic transformation under safe, green, and eco-friendly conditions with minimal chemical waste.[27]

Nikpassand *et al.*[24] have prepared nanoparticles of silica-coated Fe_3O_4 functionalized with bis(thioglucolic acid)-vanillin (Fe_3O_4@SiO_2@Vanillin@Thioglycolic acid) and applied as catalyst for carrying out the synthesis of azo-linked 5-amino-pyrazole-4-carbonitriles and 5-amino-bispyrazole-4-carbonitriles via greener method. On optimization of reaction conditions, it was found that 0.1 g catalyst per 1 mmol of primary aldehyde used is enough to carry out the transformation under solvent-free condition at room temperature in lesser time. For synthesis of derivatives for both organic transformations, yields obtained were expeditiously high. Moreover, the catalyst was easily and quickly recoverable and found to be reusable up to six runs without significant loss of catalytic activity. Using such catalyst that can be used for obtained desired product in high yields, under solvent-free condition, and lesser chemical waste are greener methods for synthesizing organic compounds.[24] Hamidinasab *et al.*[29] prepared a bifunctional IL-based nanomaterial ($NiFe_2O_4$@TiO_2–ILPip) for catalyzing multicomponent reaction between aldehyde, phthalhydrazide, and malononitrile for the synthesis of 3-amino-1-aryl-5,10-dioxo-1H-pyrazolo[1,2-b]phthalazine-2-carbonitrile derivatives. When 1.5 mmol of malononitrile and 1 mmol of other two reactants were taken, the optimum condition for high yields of products was found to be

20 mg of catalyst, PEG 400 as green solvent, and heating at 90°C. Catalyst was easily separable by external catalyst and was reusable up to five runs without significant activity loss. On comparison with other catalysts for such reaction, this method was found more green and environment-friendly.[29]

For the synthesis of spirooxindoles with fused chromenes, Mohammadian and Akhlaghinia[25] synthesized novel green solid-phase acidic nanocatalyst Fe_3O_4/COS@β-CD-SO_3H. This nanocatalyst when applied for one-pot synthesis of spirooxindoles with fused chromenes, it was revealed that 0.02 g catalyst is efficient to produce products in excellent yields in shorter reaction time (15–22 min) at 50°C using water as green reaction medium. The present catalyst without significant loss in catalytic activity was reused for at least eight runs. This catalyst under greener solvent condition and mild reaction condition produced excellent product yields in less time, which is better than other catalysts used for the synthesis of spirooxindoles with fused chromenes.[25] Ghavidel *et al.*[36] has synthesized inexpensive magnetic nanoparticles coated with basic carbon (Fe_3O_4@C@ONa) for utilizing as nanocatalyst. They applied Fe_3O_4@C@ONa as catalyst for one-pot synthesis of 4H-chromene derivatives. Reaction between β-naphthol, dimedone, and salicylaldehyde was chosen as model reaction for optimization of the reaction conditions. Optimized conditions were found to be 1 mmol of each of the reactants, 30 wt% of Fe_3O_4@C@ONa catalyst treated with 17.5 M NaOH, 60°C temperature, and water as green solvent. Under such condition, excellent yields of 4H-chromene derivatives were obtained in 55–60 min. Furthermore, catalyst is magnetically separable and was found to be reusable for at least nine runs without suitable loss of catalytic activity. Therefore, using this catalyst under green conditions with advantage of easy catalyst recovery and reusability, excellent product yields, and less chemical wastage, provides greener method for the preparation of 4H-chromene derivatives.[36]

Ghomi and Akbarzadeh[21] have synthesized and characterized magnetic $MgFe_2O_4$ nanoparticles that were applied as nanocatalyst for the synthesis of 3-substituted coumarins via Knoevenagel condensation. Ultrasonic irradiation method was used to prepare 3-substituted coumarins from 1,3-dicarbonyl compound and o-hydroxy benzaldehyde at 45°C. For this reaction, other experimental conditions were optimized, and final conditions for obtaining excellent yield of products in shorter time (10–15 min) were 4 mol% $MgFe_2O_4$ catalyst amount, solvent-free condition, and 35-W power of ultrasonic irradiation when 2 mmol of o-hydroxy benzaldehyde and 3 mmol of 1,3-dicarbonyl compound were used. Catalyst is easily recoverable by use of an external magnet and found efficiently reusable up to six runs with significant loss of catalytic activity. Such high product yield under eco-friendly, solvent-free, and moderate reaction conditions makes this protocol a greener approach for synthesis of 3-substituted coumarins.[21]

Shanshak *et al.*[26] prepared core shell nanoparticles of ZnO@Fe_3O_4 that was utilized for catalyzing a multicomponent reaction between ethyl acetoacetate, hydroxylamine hydrochloride, and aromatic/heterocyclic aldehydes for the synthesis of 3,4-disubstituted isoxazol-5 (4H)-ones. On optimization, it was found that optimized conditions for obtaining high yields of 3,4-disubstituted isoxazol-5(4H)-ones products were as follows: ethyl acetoacetate was taken 2 mmol and other reactants as 1 mmol, 5 mg of catalyst, water as solvent at 70°C for 30 min. Moreover, magnetically separable catalyst was found to be reusable for six runs without significant loss of catalytic activity. Therefore, mild conditions, economical, atom efficiency, minimum chemical waste, reusability of catalyst, and water as solvent make this method a greener pathway for synthesizing 3,4-disubstituted isoxazol-5(4H)-ones.[26]

Sepahvand *et al.*[16] have prepared gold nanoparticles by a greener method for which sea cucumber extract has been used as reducing as well as stabilizing agent. These gold nanoparticles were applied as catalyst for the synthesis of polyhydroquinoline derivatives. When 1 mmol of reactants including aldehyde, dimedone, and 4-hydroxy coumarin and 2 mmol of ammonium acetate were taken, 1.5 mg catalyst amount under solvent-free condition at 60°C were optimized conditions to obtain derivative products in high yield in 50 min. The catalyst was found to be reusable up to five runs without significant loss in catalytic activity. Thus, green method for catalyst synthesis and greener protocol for the synthesis of polyhydroquinoline derivatives are sustainable and more environment-friendly approaches.[16]

To synthesize 4H-pyran derivatives via green method, Maleki *et al.*[28] used halloysite nanotubes loaded with Fe_3O_4 nanoparticles as magnetic nanocatalyst. A model reaction having 1 mmol of dimedone, malononitrile, and benzaldehyde was used for the optimization of reaction conditions for obtaining desired products in excellent yield. The optimized conditions were found to be 30 mg of the nanocatalyst, green solvent as ethanol, and 40°C temperature. Then, using these conditions, several 4H-pyran derivatives were synthesized in excellent yields under shorter reaction time. Moreover, the nanocatalyst was recoverable and was effectively reused for seven runs. Such mild reaction conditions for the conversion, high efficiency, use of green solvent, and eco-friendly nanocatalyst are the advantages that make this a green protocol for the synthesis of 4H-pyran derivatives.[28] For the synthesis of α-aminonitriles, Maleki *et al.*[37] prepared a hybrid nanocatalyst composed of $CuFe_2O_4$ nanoparticles supported on chitosan ($CuFe_2O_4$/chitosan). For a model reaction between trimethylsilyl cyanide, aniline, and 4-chlorobenzaldehyde, series of experiments were performed to optimize the reaction conditions including amount of catalyst and solvent. Best results were obtained when 20 mg of catalyst was used along with ethanol as solvent at room temperature for 15 min. Under these optimized conditions, different α-aminonitriles were synthesized in excellent yields (86%–94%). $CuFe_2O_4$/chitosan catalyst was recoverable and was reused for five consecutive runs without major loss in the catalytic property. The use of $CuFe_2O_4$/chitosan as catalyst for the synthesis of α-aminonitriles is efficient and a green method.[37]

Veisi *et al.*[18] used ultrasonication method for the preparation of Fe_3O_4 nanoparticles coated with green tea extract, and Pd nanoparticles were immobilized over it to obtain Pd/Fe_3O_4. Obtained Pd/Fe_3O_4 was utilized as nanocatalyst for catalyzing Suzuki–Miyaura coupling reaction for synthesizing biphenyl compounds via ultrasonication method. It was found that 35 mg of nanocatalyst was efficient to produce desired products in high yields in water–ethanol (1:1) as solvent and sonication at 40°C for desired time. Moreover, the nanocatalyst was recoverable and was found effective to catalyze the reaction up to six cycles without significant loss in the catalytic activity. Green synthesis of catalyst, ultrasonication method used for preparing the biphenyl product in high yields, and shorter reaction time by using the catalyst in green solvent are the advantages, proving this method greener for Suzuki–Miyaura coupling reaction.[18] Baran *et al.*[23] designed an environment-friendly catalyst to be utilized in Suzuki–Miyaura coupling reactions. Catalyst was composed of chitosan/cellulose composite on which palladium nanoparticles were stabilized. Then, this catalyst was utilized as catalyst for coupling reactions between phenyl boronic acid and various aryl halides. As a green method in which microwave radiation was used for performing the reactions. Various reaction parameters including catalyst amount, reaction duration, and base system were optimized using a model reaction between phenyl boronic acid and p-bromoanisole. And it was found that 0.004 mol% of catalyst, 5-min reaction time at 400 W at 50°C, and K_2CO_3 as base were the best conditions for maximum yields. A series of biphenyl products were synthesized in 55%–100% yield under optimized conditions. Furthermore, the catalyst was found to be recyclable for eight consecutive runs without significant loss in the catalytic activity. Green catalyst system and microwave method used for preparing the biphenyl product in high yields and shorter reaction time by using the catalyst are the advantages, proving this method greener for Suzuki–Miyaura coupling reaction.[23]

Rezaei and Mamaghani[38] have prepared a nanocatalyst that has been formed as iron oxide as core encapsulated by fluorapatite incorporated with cobalt (γ-Fe_2O_3@FAp@Co), which was applied in the synthesis of functionalized bis-triazole derivatives. For synthesizing bis-triazole derivatives in excellent yields, under optimized condition, a condensation reaction was performed between 2 mol of various aromatic aldehydes and 1 mol butane-1,4-diylbis(hydrazinecarbimidothioate) in the presence of 0.1 g catalyst while using ethanol as solvent at room temperature. The yield of product was 88%–95%, which was obtained under the reaction time of 60–80 min. Furthermore, the catalyst is recoverable by the use of an external magnet and was reused for seven cycles and was found that catalytic activity was unchanged. In the presence of inexpensive, eco-friendly, and reusable nanocatalyst, the bis-triazole derivative synthesis in high yields at room temperature in the presence of a green solvent provides a greener alternative method.[38]

To catalyze the synthesis of 1-substituted 1H-tetrazoles, Habibi *et al.*[39] designed and prepared the nanocatalyst of Fe_3O_4 that was functionalized by 5,10-dihydropyrido[2,3-b]quinoxaline-7,8-diol to which copper was complexed to finally achieve Fe_3O_4@Quindiol@Cu. For synthesis of various tetrazole compounds using Fe_3O_4@Quindiol@Cu catalyst, various parameters including solvent, reaction temperature, and catalyst amount were optimized through a model reaction of triethyl orthoformate, sodium azide, and aniline. Maximum yields for the product were obtained when reaction was performed under solvent-free condition having 0.01 g of catalyst at 100°C. Several tetrazole products were obtained in excellent yields (84%–93%) under the reaction time of 40–85 min. Furthermore, the catalyst is recoverable by the use of an external magnet and was reused for six consecutive runs and was found that catalytic activity was unchanged. Advantages of this method such as excellent product yields, solvent-free condition, reusability of catalyst, simple reaction procedure, and easy workup provide a greener pathway for the synthesis of 1-substituted 1H-tetrazoles.[39] Moeini *et al.*[20] prepared a nanocatalyst in which Ni was immobilized on tryptophan-coated Fe_3O_4 NPs (Fe_3O_4@tryptophan@Ni) and was used in the synthesis of 5-substituted 1H-tetrazoles. For the optimization of the catalysis process, a model reaction between sodium azide and malononitrile was chosen. The optimized condition obtained as poly(ethylene glycol) as green solvent and 0.08 g of catalyst at 120°C in which best results were obtained. Under these optimized conditions, various 5-substituted 1H-tetrazoles were prepared in excellent yields (88%–97%). The Fe_3O_4@tryptophan@Ni catalyst was easily separable from the reaction mixture with an external magnet and reusable for five runs without significant reduction in the catalytic activity. Easy and economical preparation of catalyst, excellent yields, and use of green solvent are advantages of this method, which makes it a greener method for the synthesis of 5-substituted 1H-tetrazoles.[20]

To synthesize substituted 3-pyrrolin-2-one, Ahankar *et al.*[19] have prepared the citric acid functionalized magnetic cobalt ferrite nanoparticles ($CoFe_2O_4$@CA). For the synthesis of substituted 3-pyrrolin-2-one in the presence of $CoFe_2O_4$@CA, ultrasonication method that is a greener pathway was employed. For best results, reaction parameters were investigated that included the amount of catalyst, solvent effect, and power of ultrasonic irradiation by choosing a model reaction between 4-chlorobenzaldehyde, diethyl acetylenedicaboxylate, and aniline. The best condition was found to be a green solvent ethanol as solvent, 100-W ultrasonic power, and 75 mg catalyst when 1 mmol of each reactant was taken. Under optimized conditions, substituted 3-pyrrolin-2-one was obtained in excellent yields (82%–92%) under shorter reaction time. The catalyst was easily separable with the help of an external magnet and found reusable for five consecutive runs without significant decrease in the catalytic performance. Use of green, reusable, and efficient catalyst, greener solvent like ethanol, along with the use of ultrasonication method, the present protocol is economical and environment-friendly.[19]

Akhlaghinia *et al.*[40] synthesized a nanocatalyst in which hydroxyapatite scaffold was taken and into which magnetic nanoparticles were incorporated (HAP/Fe_3O_4 NPs). This nanocatalyst was applied for the synthesis of biscoumarin derivatives. A reaction between 4-hydroxycoumarin (2 equivalents) and benzaldehyde (1 equivalent) was chosen as a model reaction, through which different parameters such as amount of catalyst, solvent, and reaction temperature were optimized. It was found that the product was obtained in best yield when reaction was performed with 0.008 g of catalyst at 90°C under solvent-free conditions. Several biscoumarin derivatives were synthesized under optimized condition in excellent yields (85%–98%) under specific time. The present catalyst was recoverable with the help of an external magnet and found to be reusable up to six runs with loss in catalytic activity under the optimized reaction condition. Various advantages of this protocol for the synthesis of biscoumarin derivatives were cheap and nontoxic catalyst, easy reaction procedure, reusability of catalyst, high conversion in shorter reaction time, and solvent-free condition, as all of these advantages show the agreement of this method with the green chemistry principles.[40]

Varzi and Maleki[41] utilized ZnS-$ZnFe_2O_4$ as a hybrid nanocatalyst for the synthesis of 2,4,5-triaryl-1H-imidazole derivatives. Through a model reaction between 4-chlorobenzaldehyde (1 mmol), benzil (1 mmol), and ammonium acetate (4 mmol), several reaction parameters were optimized, and it was found that the use of 2 mg catalyst and ethanol as solvent at 70°C under ultrasonication provided best results in shorter reaction time. Several 2,4,5-triaryl-1H-imidazole derivatives were synthesized under optimized conditions in good to excellent yields (51%–95%) under short reaction time. Furthermore, ZnS-$ZnFe_2O_4$ was recoverable and reusable up to six consecutive runs without significant loss of catalytic activity. This protocol for the synthesis of 2,4,5-triaryl-1H-imidazole derivatives is in good agreement with the green chemistry principles, as the catalyst had high efficiency, short reaction time, easy workup, high product yields, and use of ethanol that is a green solvent.[41]

Kooti *et al.*[42] have prepared a nanocomposite in which Cu(II) complex has been immobilized on the surface of CPTMS modified magnetically reduced graphene oxide (RGO/$CoFe_2O_4$@Cu(II)). Further for the synthesis of 1-amidoalkyl-2-naphthols, RGO/$CoFe_2O_4$@Cu(II) was employed as catalyst. To obtain the best results, different reaction parameters such as type of solvent, amount of catalyst, and reaction temperature were optimized using a model reaction between acetamide, 2-naphthol, and benzaldehyde. As per optimization, best conditions were the use of 0.03 g catalyst at 120°C under solvent-free condition. At these optimized conditions, several amidoalkylnaphthol derivatives were synthesized in excellent yields (85%–92%) under reaction time of 25–45 min. The RGO/$CoFe_2O_4$@Cu(II) catalyst was recoverable with the help of an external magnet and found to be reusable for at least five runs with no loss in its catalytic performance. The use of RGO/$CoFe_2O_4$@Cu(II) as catalyst for the synthesis of amidoalkylnaphthol derivatives is rapid, efficient, and a green method.[42]

Nanocatalysts for the green synthesis of organic compounds that have been discussed above along with numerous other nanocatalysts applied for the synthesis of organic compounds have been summarized in Table 6.1.

Table 6.1 Nanocatalysts for green synthesis of organic compounds.

SN	Catalyst	Organic transformation	Greener synthetic condition	Ref
1	Fe_3O_4@Ph-SO_3H	tetrahydrobenzo[b]pyrans and derivatives	water as solvent and ultrasonic method	31
2	$Fe_{3-x}Ti_xO_4$@ TMSP@ TBAAsp	1,4-Dihydropyrano[2,3-c]pyrazole derivatives	solvent-free condition	12
3	GO-PO_3H_2	pyrano[2,3-c]pyrazole derivatives	water as solvent	32
4	La-guanine@ SBA-15	4,4'-(arylmethylene)-bis-(3-methyl-1-phenyl-1Hpyrazol-5-ols) and benzo [a] pyrano [2, 3-c] phenazine derivatives	Ethanol as solvent	27
5	Fe_3O_4@SiO_2@ Vanillin@ Thioglycolic acid	azo-linked 5-amino-pyrazole-4-carbonitriles and 5-amino-bispyrazole-4-carbonitriles	solvent-free condition	24
6	$NiFe_2O_4$@ TiO_2–ILPip	3-amino-1-aryl-5,10-dioxo-1H-pyrazolo[1,2-b]phthalazine-2-carbonitrile derivatives	PEG 400 as green solvent	29
7	Fe_3O_4/COS@β-CD-SO_3H	spirooxindoles with fused chromenes	water as solvent	25
8	Fe_3O_4@C@ONa	4H-chromene derivatives	water as solvent	36
9	mica/Fe_3O_4	2-amino-4H-chromene derivatives	Ethanol as solvent	43
10	$MgFe_2O_4$	3-substituted coumarins	solvent-free condition and ultrasonic method	21
11	ZnO@Fe_3O_4	3,4-disubstituted isoxazol-5(4H)-ones	water as solvent	26

(Continued)

Table 6.1 Nanocatalysts for green synthesis of organic compounds. (*Continued*)

SN	Catalyst	Organic transformation	Greener synthetic condition	Ref
12	Gold nanoparticles	polyhydroquinoline derivatives	Green synthesis of catalyst and solvent-free condition for organic reaction	16
13	$NiFe_2O_4$@ SiO_2@amino glucose	bis(1,2-dihydro-4-hydroxy-2-oxoquinolin-3-yl)methanes	water as solvent	13
14	Pd NPs@Fe_3O_4/ lignin/chitosan	aryl nitrile and biaryl compounds.	Catalyst prepared without any toxic reagents	14
15	Fe_3O_4@NFC-ImSalophCu	1,4-disubstituted 1,2,3-triazoles	water as solvent	15
16	Fe3O4/HNTs	4H-pyran derivatives	Ethanol as solvent	28
17	Biosynthesized NiO NPs	Pyridopyrimidine derivatives	water as solvent	17
18	$NiFe_2O_4$/ geopolymer	2,4,5-trisubstituted imidazole derivatives	Ethanol as solvent	22
19	Pd/Fe_3O_4	biphenyl compounds by Suzuki–Miyaura coupling reaction	water–ethanol (1:1) as solvent and ultrasonic method	18
20	Cu-BPy@ Am SiO_2@ Fe_3O_4	polysubstituted oxazole by tandem oxidative cyclization of benzylamines and 1,3-dicarbonyls	H_2O_2 as oxidant	44
21	γ-Fe_2O_3@FAp@ Co	bis-triazole derivatives	Ethanol as solvent	38
22	nano–Fe_3O_4@ SiO_2–OSO_3H	metallopyrazinoporphyrazines	solvent-free condition for organic reaction	45
23	Fe_3O_4@ Quindiol@Cu	1-substituted 1H-tetrazoles	solvent-free condition for organic reaction	39

(*Continued*)

Table 6.1 Nanocatalysts for green synthesis of organic compounds. (*Continued*)

SN	Catalyst	Organic transformation	Greener synthetic condition	Ref
24	Pd nanocatalyst on chitosan/ cellulose composite	Suzuki–Miyaura cross-coupling reactions	solvent-free condition and under microwave radiation	23
25	Fe_3O_4@chitosan-tannic acid	pyrano[2,3-c]pyrazole derivatives	Ethanol as solvent	33
26	Ru^{III}@ CMC/ Fe_3O_4	pyrano[2,3-c]pyrazole and polyhydroquinoline derivatives	Ethanol as solvent	34
27	$CoFe_2O_4$@CA	substituted 3-pyrrolin-2-one	Ethanol as solvent and ultrasonic method	19
28	HAP/Fe_3O_4 NPs	biscoumarin derivatives	solvent-free condition	40
29	$NiFe_2O_4$@SiO_2-BPMN-Ni	biphenyl compounds by Suzuki–Miyaura coupling reaction	Water/EtOH (1:1) as solvent and ultrasonic method	46
30	Cu@imine/ Fe_3O_4 MNP	polysubstituted pyrroles and 1,2,4,5-tetrasubstituted imidazole derivatives	solvent-free condition	47
31	Fe_3O_4@SiO_2-EP-NH-HPA	pyrano[2,3-c]pyrazole derivatives	Water as solvent	35
32	ZnS-$ZnFe_2O_4$	2,4,5-triaryl-1H-imidazoles derivatives	Ethanol as solvent and ultrasonic method	41
33	Fe_3O_4@ tryptophan@ Ni	5-substituted 1H-tetrazoles	PEG as green solvent	20
34	RGO/$CoFe_2O_4$@ Cu(II)	1-amidoalkyl-2-naphthols	solvent-free condition	42
35	Fe_3O_4@SiO_2@ TiO_2-OSO_3H	spiroquinazolin-4(3H)-one derivatives and 2,3-dihydroquinazolin-4(1H)-one derivatives	Deep eutectic solvent	30
36	$CuFe_2O_4$/ chitosan	α-aminonitriles	Ethanol as solvent	37

(*Continued*)

Table 6.1 Nanocatalysts for green synthesis of organic compounds. (*Continued*)

SN	Catalyst	Organic transformation	Greener synthetic condition	Ref
37	Cu(II)-Fur-APTES/GO	Xanthene derivatives	Water/EtOH (1:1) as solvent	48
38	TiO_2@AAPTMS-rGO	2,4,5-triarylimidazoles	Ethanol as solvent	49
39	CC-TAPT-COP	quinazoline derivatives	Ethanol as solvent	50
40	La_2O_3/RGO	bis(indolyl)methanes	solvent-free condition	51
41	rGO-NiO-NCs	2, 4, 5-trisubstituted imidazole derivatives	Ethanol as solvent	52

6.4 Conclusion

Nanocatalysts play an especially important role in the synthesis of organic compounds, as a new mechanistic pathway is provided by the catalyst for the rapid transformation and synthesis of the desired organic compound in enhanced yields. Over time, there is a notable contribution of metal-based catalysts for the transformation and synthesis of the desired organic compound. As per green chemistry, a nontoxic, eco-friendly, economical, efficient, and simple approach used for the synthesis of nano material is green synthesis. Furthermore, if the use of catalyst in the organic transformation led to excellent product yields, mild conditions, economical, atom efficiency, simple procedure, energy-efficient, minimum chemical waste, reusability of catalyst, and use of green solvent, then it is said to be a greener pathway for syntheses of organic compounds as per the principles of green chemistry. Over time, a lot of progress has been made by the scientific community to apply green chemistry principle in catalysis, yet a lot of room is still there for further investigation and making catalysis process more sustainable and environment-friendly. Along with this development, one major problem that has to be resolved is removal of these nanocatalysts from the environment because their accumulation in the environment can lead to several other major problems.

References

1. Dandia, A., Parihar, S., Sharma, R., Rathore, K.S., Parewa, V., Nanocatalysis in Green Organic Synthesis, in: *Green Sustainable Process for Chemical and Environmental Engineering and Science*, pp. 71–103, Elsevier Science Publishing Co Inc, Amsterdam, Netherlands, 2020, https://doi.org/10.1016/B978-0-12-819539-0.00004-X.
2. Amrutkar, R.D., Bhalerao, S.S., Bhoir, A.S., Bhusare, R.H., Bodhare, S.S., Borse, J.N., Role of Catalyst in Organic Synthesis. *Curr. Trends Pharm. Pharma. Chem.*, *4*, 3, 115–119, 2022, https://doi.org/10.18231/J.CTPPC.2022.019.

3. Sutradhar, M., Metal-Based Catalysts in Organic Synthesis. *Catalysts*, *10*, 12, 1429, 2020, https://doi.org/10.3390/CATAL10121429.
4. Mostafa, A.M. and Menazea, A.A., Polyvinyl Alcohol/Silver Nanoparticles Film Prepared via Pulsed Laser Ablation: An Eco-Friendly Nano-Catalyst for 4-Nitrophenol Degradation. *J. Mol. Struct.*, *1212*, 128125, 2020, https://doi.org/10.1016/J.MOLSTRUC.2020.128125.
5. Varma, R.S., Nano-Catalysts with Magnetic Core: Sustainable Options for Greener Synthesis. *Sustain. Chem. Processes*, *2*, 1, 1–8, 2014, https://doi.org/10.1186/2043-7129-2-11.
6. Chaudhary, K. and Masram, D.T., Biological Activities of Nanoparticles and Mechanism of Action, in: *Model Organisms to Study Biological Activities and Toxicity of Nanoparticles*, pp. 19–34, Springer, Singapore, 2020, https://doi.org/10.1007/978-981-15-1702-0_2.
7. Polshettiwar, V. and Varma, R.S., Green Chemistry by Nano-Catalysis. *Green Chem.*, *12*, 5, 743–754, 2010, https://doi.org/10.1039/B921171C.
8. Varma, R.S., Greener Approach to Nanomaterials and Their Sustainable Applications. *Curr. Opin. Chem. Eng.*, *1*, 2, 123–128, 2012, https://doi.org/10.1016/J.COCHE.2011.12.002.
9. Narayanan, R., Synthesis of Green Nanocatalysts and Industrially Important Green Reactions. *Green Chem. Lett. Rev.*, 5, 4, 707–725, 2012, https://doi.org/10.1080/17518253.2012.700955.
10. Dikshit, P.K., Kumar, J., Das, A.K., Sadhu, S., Sharma, S., Singh, S., Gupta, P.K., Kim, B.S., Green Synthesis of Metallic Nanoparticles: Applications and Limitations. *Catalysts*, *11*, 8, 902, 2021, https://doi.org/10.3390/CATAL11080902.
11. Kalidindi, S.B. and Jagirdar, B.R., Nanocatalysis and Prospects of Green Chemistry. *ChemSusChem*, *5*, 1, 65–75, 2012, https://doi.org/10.1002/CSSC.201100377.
12. Azarifar, D., Badalkhani, O., Abbasi, Y., Amino Acid Ionic Liquid-Based Titanomagnetite Nanoparticles: An Efficient and Green Nanocatalyst for the Synthesis of 1,4-Dihydropyrano[2,3-c] Pyrazoles. *Appl. Organomet. Chem.*, *32*, 1, e3949, 2018, https://doi.org/10.1002/AOC.3949.
13. Aghazadeh, B. and Nikpassand, M., "2-Amino Glucose" as a Substrate for Synthesis of Magnetically Recoverable Nanocatalyst NiFe2O4@SiO2@amino Glucose for the Green Synthesis of Novel Bis (1,2-Dihydro-4-Hydroxy-2-Oxoquinolin-3-Yl)Methanes. *Carbohydr. Res.*, *483*, 107755, 2019, https://doi.org/10.1016/J.CARRES.2019.107755.
14. Baran, T. and Sargin, I., Green Synthesis of a Palladium Nanocatalyst Anchored on Magnetic Lignin-Chitosan Beads for Synthesis of Biaryls and Aryl Halide Cyanation. *Int. J. Biol. Macromol.*, *155*, 814–822, 2020, https://doi.org/10.1016/J.IJBIOMAC.2020.04.003.
15. Ghamari kargar, P., Bagherzade, G., Eshghi, H., Design and Synthesis of Magnetic Fe3O4@ NFC-ImSalophCu Nanocatalyst Based on Cellulose Nanofibers as a New and Highly Efficient, Reusable, Stable and Green Catalyst for the Synthesis of 1,2,3-Triazoles. *RSC Adv.*, *10*, 54, 32927–32937, 2020, https://doi.org/10.1039/D0RA06251K.
16. Sepahvand, M., Buazar, F., Sayahi, M.H., Novel Marine-Based Gold Nanocatalyst in Solvent-Free Synthesis of Polyhydroquinoline Derivatives: Green and Sustainable Protocol. *Appl. Organomet. Chem.*, *34*, 12, e6000, 2020, https://doi.org/10.1002/AOC.6000.
17. Moavi, J., Buazar, F., Sayahi, M.H., Algal Magnetic Nickel Oxide Nanocatalyst in Accelerated Synthesis of Pyridopyrimidine Derivatives. *Sci. Rep.*, *11*, 1, 6296, 2021, https://doi. org/10.1038/s41598-021-85832-z.
18. Veisi, H., Ghorbani, M., Hemmati, S., Sonochemical in Situ Immobilization of Pd Nanoparticles on Green Tea Extract Coated Fe3O4 Nanoparticles: An Efficient and Magnetically Recyclable Nanocatalyst for Synthesis of Biphenyl Compounds under Ultrasound Irradiations. *Mater. Sci. Eng.: C*, *98*, 584–593, 2019, https://doi.org/10.1016/J.MSEC.2019.01.009.
19. Ahankar, H., Ramazani, A., Ślepokura, K., Lis, T., Kinzhybalo, V., Magnetic Cobalt Ferrite Nanoparticles Functionalized with Citric Acid as a Green Nanocatalyst for One-Pot Three-Component Sonochemical Synthesis of Substituted 3-Pyrrolin-2-Ones. *Res. Chem. Intermed.*, *45*, 10, 5007–5025, 2019, https://doi.org/10.1007/S11164-019-03878-1.

20. Moeini, N., Tamoradi, T., Ghadermazi, M., Ghorbani-Choghamarani, A., Anchoring Ni (II) on Fe3O4@tryptophan: A Recyclable, Green and Extremely Efficient Magnetic Nanocatalyst for One-Pot Synthesis of 5-Substituted 1H-Tetrazoles and Chemoselective Oxidation of Sulfides and Thiols. *Appl. Organomet. Chem.*, *32*, 9, e4445, 2018, https://doi.org/10.1002/AOC.4445.
21. Ghomi, J.S. and Akbarzadeh, Z., Ultrasonic Accelerated Knoevenagel Condensation by Magnetically Recoverable MgFe2O4 Nanocatalyst: A Rapid and Green Synthesis of Coumarins under Solvent-Free Conditions. *Ultrason. Sonochem.*, *40*, 78–83, 2018, https://doi.org/10.1016/J.ULTSONCH.2017.06.022.
22. Hajizadeh, Z., Radinekiyan, F., Eivazzadeh-keihan, R., Maleki, A., Development of Novel and Green NiFe2O4/Geopolymer Nanocatalyst Based on Bentonite for Synthesis of Imidazole Heterocycles by Ultrasonic Irradiations. *Sci. Rep.*, *10*, 1, 11671, 2020, https://doi. org/10.1038/s41598-020-68426-z.
23. Yılmaz Baran, N., Baran, T., Menteş, A., Production of Novel Palladium Nanocatalyst Stabilized with Sustainable Chitosan/Cellulose Composite and Its Catalytic Performance in Suzuki-Miyaura Coupling Reactions. *Carbohydr. Polym.*, *181*, 596–604, 2018, https://doi.org/10.1016/J.CARBPOL.2017.11.107.
24. Nikpassand, M., Fekri, L.Z., Varma, R.S., Hassanzadi, L., Pashaki, F.S., Green Synthesis of Novel 5-Amino-Bispyrazole-4-Carbonitriles Using a Recyclable Fe3O4@SiO2@vanillin@thioglycolic Acid Nano-Catalyst. *RSC Adv.*, *12*, 2, 834–844, 2021, https://doi.org/10.1039/D1RA08001F.
25. Mohammadian, N. and Akhlaghinia, B., Magnetic Calcined Oyster Shell Functionalized with Taurine Immobilized on β-Cyclodextrin (Fe3O4/COS@β-CD-SO3H NPs) as Green and Magnetically Reusable Nanocatalyst for Efficient and Rapid Synthesis of Spirooxindoles. *Res. Chem. Intermed.*, *45*, 10, 4737–4756, 2019, https://doi.org/10.1007/ S11164-019-03860-X.
26. Shanshak, M., Budagumpi, S., Małecki, J.G., Keri, R.S., Green Synthesis of 3,4-Disubstituted Isoxazol-5(4H)-Ones Using ZnO@Fe3O4 Core–Shell Nanocatalyst in Water. *Appl. Organomet. Chem.*, *34*, 4, e5544, 2020, https://doi.org/10.1002/AOC.5544.
27. Nikoorazm, M., Khanmoradi, M., Mohammadi, M., Guanine-La Complex Supported onto SBA-15: A Novel Efficient Heterogeneous Mesoporous Nanocatalyst for One-Pot, Multi-Component Tandem Knoevenagel Condensation–Michael Addition–Cyclization Reactions. *Appl. Organomet. Chem.*, *34*, 4, e5504, 2020, https://doi.org/10.1002/AOC.5504.
28. Maleki, A. and Hajizadeh, Z., Magnetic Aluminosilicate Nanoclay: A Natural and Efficient Nanocatalyst for the Green Synthesis of 4H-Pyran Derivatives. *Silicon*, *11*, 6, 2789–2798, 2019, https://doi.org/10.1007/S12633-019-0069-4.
29. Hamidinasab, M., Bodaghifard, M.A., Mobinikhaledi, A., Green Synthesis of 1H-Pyrazolo[1,2-b] Phthalazine-2-Carbonitrile Derivatives Using a New Bifunctional Base–Ionic Liquid Hybrid Magnetic Nanocatalyst. *Appl. Organomet. Chem.*, *34*, 2, e5386, 2020, https://doi.org/10.1002/AOC.5386.
30. Maleki, A., Kari, T., Aghaei, M., Fe3O4@SiO2@TiO2-OSO3H: An Efficient Hierarchical Nanocatalyst for the Organic Quinazolines Syntheses. *J. Porous Mater.*, *24*, 6, 1481–1496, 2017, https://doi.org/10.1007/S10934-017-0388-Z.
31. Elhamifar, D., Ramazani, Z., Norouzi, M., Mirbagheri, R., Magnetic Iron Oxide/Phenylsulfonic Acid: A Novel, Efficient and Recoverable Nanocatalyst for Green Synthesis of Tetrahydrobenzo[b] Pyrans under Ultrasonic Conditions. *J. Colloid Interface Sci.*, *511*, 392–401, 2018, https://doi.org/10.1016/J.JCIS.2017.10.013.
32. Zakeri, M., Abouzari-lotf, E., Miyake, M., Mehdipour-Ataei, S., Shameli, K., Phosphoric Acid Functionalized Graphene Oxide: A Highly Dispersible Carbon-Based Nanocatalyst for the Green Synthesis of Bio-Active Pyrazoles. *Arabian J. Chem.*, *12*, 2, 188–197, 2019, https://doi.org/10.1016/J.ARABJC.2017.11.006.

33. Kamalzare, M., Ahghari, M.R., Bayat, M., Maleki, A., Fe3O4@chitosan-Tannic Acid Bionanocomposite as a Novel Nanocatalyst for the Synthesis of Pyranopyrazoles. *Sci. Rep.*, *11*, 1, 1–10, 2021, https://doi.org/10.1038/s41598-021-99121-2.
34. Chen, Y., Zhang, Z., Jiang, W., Zhang, M., Li, Y., RuIII@CMC/Fe3O4 Hybrid: An Efficient, Magnetic, Retrievable, Self-Organized Nanocatalyst for Green Synthesis of Pyranopyrazole and Polyhydroquinoline Derivatives. *Mol. Divers.*, *23*, 2, 421–442, 2019, https://doi.org/10.1007/S11030-018-9887-3.
35. Hosseini Mohtasham, N. and Gholizadeh, M., Nano Silica Extracted from Horsetail Plant as a Natural Silica Support for the Synthesis of H3PW12O40 Immobilized on Aminated Magnetic Nanoparticles (Fe3O4@SiO2-EP-NH-HPA): A Novel and Efficient Heterogeneous Nanocatalyst for the Green One-Pot Synthesis of Pyrano[2,3-c]Pyrazole Derivatives. *Res. Chem. Intermed.*, *46*, 6, 3037–3066, 2020, https://doi.org/10.1007/S11164-020-04133-8.
36. Ghavidel, H., Mirza, B., Soleimani-Amiri, S. A., A Novel, Efficient, and Recoverable Basic Fe3O4@C Nano-Catalyst for Green Synthesis of 4H-Chromenes in Water via One-Pot Three Component Reactions. *Polycycl. Aromat. Compd.*, 41, 3, 604–625, 2019, https://doi.org/10.1080/10406638.2019.1607413.
37. Maleki, A., Haji, R.F., Ghassemi, M., Ghafuri, H., Preparation and Application of a Magnetic Organic-Inorganic Hybrid Nanocatalyst for the Synthesis of α-Aminonitriles. *J. Chem. Sci.*, *129*, 4, 457–462, 2017, https://doi.org/10.1007/S12039-017-1253-Y.
38. Rezaei, I. and Mamaghani, M., Green Synthesis of Bis Pyrazole-Triazole and Azo-Linked Triazole Hybrids Using an Efficient and Novel Cobalt Nanocatalyst. *React. Kinet. Mech. Catal.*, *134*, 1, 385–400, 2021, https://doi.org/10.1007/S11144-021-02076-8.
39. Habibi, D., Pakravan, N., Arabi, A., Kaboudvand, Z., Preparation of Fe3O4@5,10-Dihydropyrido[2,3-b]Quinoxaline-7,8-Diol Copper Complex: A Capable Nanocatalyst for the Green Synthesis of 1-Substituted 1H-Tetrazoles. *Appl. Organomet. Chem.*, *32*, 1, e3988, 2018, https://doi.org/10.1002/AOC.3988.
40. Akhlaghinia, B., Sanati, P., Mohammadinezhad, A., Zarei, Z., The Magnetic Nanostructured Natural Hydroxyapatite (HAP/Fe3O4 NPs): An Efficient, Green and Recyclable Nanocatalyst for the Synthesis of Biscoumarin Derivatives under Solvent-Free Conditions. *Res. Chem. Intermed.*, *45*, 5, 3215–3235, 2019, https://doi.org/10.1007/S11164-019-03788-2.
41. Varzi, Z. and Maleki, A., Design and Preparation of ZnS-ZnFe2O4: A Green and Efficient Hybrid Nanocatalyst for the Multicomponent Synthesis of 2,4,5-Triaryl-1H-Imidazoles. *Appl. Organomet. Chem.*, *33*, 8, e5008, 2019, https://doi.org/10.1002/AOC.5008.
42. Kooti, M., Karimi, M., Nasiri, E.A., Novel Copper Complex Supported on Magnetic Reduced Graphene Oxide: An Efficient and Green Nanocatalyst for the Synthesis of 1-Amidoalkyl-2-Naphthol Derivatives. *J. Nanopart. Res.*, *20*, 2, 1–14, 2018, https://doi.org/10.1007/S11051-017-4107-0.
43. Maleki, A., Hajizadeh, Z., Valadi, K., Green and Eco-Friendly Mica/Fe3O4 as an Efficient Nanocatalyst for the Multicomponent Synthesis of 2-Amino-4H-Chromene Derivatives. *Green Chem. Lett. Rev.*, *14*, 1, 60–70, 2021, https://doi.org/https://doi.org/10.1080/17518253.2020.1860259.
44. Dutta, S., Sharma, S., Sharma, A., Sharma, R.K., Fabrication of Core-Shell-Structured Organic-Inorganic Hybrid Nanocatalyst for the Expedient Synthesis of Polysubstituted Oxazoles via Tandem Oxidative Cyclization Pathway. *ACS Omega*, *2*, 6, 2778–2791, 2017, https://doi.org/DOI: 10.1021/acsomega.7b00382.
45. Maleki, A., An Efficient Magnetic Heterogeneous Nanocatalyst for the Synthesis of Pyrazinoporphyrazine Macrocycles. *Polycycl. Aromat. Compd.*, 38, 5, 402–409, 2016, https://doi.org/10.1080/10406638.2016.1221836.

46. Kiani, F. and Naeimi, H., Ultrasonic Accelerated Coupling Reaction Using Magnetically Recyclable Bis (Propyl Molononitril) Ni Complex Nanocatalyst: A Novel, Green and Efficient Synthesis of Biphenyl Derivatives. *Ultrason. Sonochem.*, *48*, 267–274, 2018, https://doi.org/10.1016/J. ULTSONCH.2018.06.001.
47. Thwin, M., Mahmoudi, B., Ivaschuk, O.A., Yousif, Q.A., An Efficient and Recyclable Nanocatalyst for the Green and Rapid Synthesis of Biologically Active Polysubstituted Pyrroles and 1,2,4,5-Tetrasubstituted Imidazole Derivatives. *RSC Adv.*, *9*, 28, 15966–15975, 2019, https://doi.org/10.1039/C9RA02325A.
48. Subodh, Mogha, N.K., Chaudhary, K., Kumar, G., Masram, D.T., Fur-Imine-Functionalized Graphene Oxide-Immobilized Copper Oxide Nanoparticle Catalyst for the Synthesis of Xanthene Derivatives. *ACS Omega*, *3*, 11, 16377–16385, 2018, https://doi.org/10.1021/acsomega.8b01781.
49. Subodh, Chaudhary, K., Prakash, K., Masram, D.T., TiO2 Nanoparticles Immobilized Organo-Reduced Graphene Oxide Hybrid Nanoreactor for Catalytic Applications. *Appl. Surf. Sci.*, *509*, 144902, 2020, https://doi.org/10.1016/J.APSUSC.2019.144902.
50. Subodh, Prakash, K., Chaudhary, K., Masram, D.T.A., New Triazine-Cored Covalent Organic Polymer for Catalytic Applications. *Appl. Catal. A Gen.*, *593*, 117411, 2020, https://doi.org/10.1016/J.APCATA.2020.117411.
51. Mogha, N.K., Kirti, S., Masram, D.T., La2O3/Reduced Graphene Oxide Nanocomposite: A Highly Efficient, Reusable Heterogeneous Catalyst for the Synthesis of Biologically Important Bis(Indolyl)Methanes under Solvent Free Conditions. *J. Nanosci. Nanotechnol.*, *17*, 4, 2508–2514, 2017, https://doi.org/10.1166/JNN.2017.13441.
52. Kumar, G., Mogha, N.K., Kumar, M., Subodh; Masram, D. T. NiO Nanocomposites/RGO as a Heterogeneous Catalyst for Imidazole Scaffolds with Applications in Inhibiting the DNA Binding Activity. *Dalton Trans.*, *49*, 6, 1963–1974, 2020, https://doi.org/10.1039/ C9DT04416G.

7

Heterogeneous Photocatalysis: Recent Advances and Applications

Sher Mohd and Amjad Mumtaz Khan*

Department of Chemistry, Aligarh Muslim University, Aligarh, India

Abstract

Heterogeneous photocatalysis has gained considerable interest in recent years due to its promising potential for addressing challenges like environmental remediation, energy conversion, water splitting, CO_2 reduction, and organic synthesis. Recently, there have been several advances in heterogeneous photocatalysis, including the designing of novel photocatalysts, the optimization of photocatalytic processes, and the discovery of new applications, such as the use of nanoparticle-based photocatalysts, plasmonic photocatalysts, and 2D materials. Overall, recent advances in heterogeneous photocatalysis have opened new opportunities for using this technology to address a variety of energy and environmental related challenges. This process generally involves the use of a semiconductor material, typically titanium dioxide (TiO_2), which becomes energized by sunlight, resulting in the creation of electron-hole pairs. These pairs interact with surface molecules, leading to the formation of reactive oxygen species (ROS) and diverse chemicals. ROS assist in breaking down pollutants, splitting water into oxygen and hydrogen, and converting CO_2. Nanoparticle-based photocatalysts provide a notable advantage with their substantial surface-area to volume ratio, thereby elevating their photocatalytic performance. Plasmonic photocatalysts utilize surface plasmon resonance to boost their efficiency in converting CO_2, benefiting from their ability to absorb visible light, abundant in solar energy. Additionally, materials like graphene and transition-metal dichalcogenides, known as two-dimensional substances, have gained attention for their substantial surface areas and unique electronic properties, making them promising and versatile in various photocatalytic applications.

Heterogeneous photocatalysis stands as a highly promising technological innovation, garnering significant attention in the realm of environmental remediation. This advanced technique has been subjected to extensive research and exploration due to its immense potential in the purification of both air and water environments. This chapter gives a brief summary of recent advances in heterogeneous photocatalysis and its applications. Initially, it delves into the fundamental principles underpinning photocatalysis, encompassing a comprehensive examination of photocatalyst categories, the mechanisms governing photocatalysis, and the factors that influence its efficiency. Subsequently, it explores the diverse applications of heterogeneous photocatalysis, spanning pollutant degradation, CO_2 reduction, water splitting, and organic synthesis. Next, various applications of heterogeneous photocatalysis, such as pollutant degradation, CO_2 reduction, water splitting, and organic synthesis, are reviewed. The mechanisms and challenges associated with each application are discussed along with recent advances in enhancing the photocatalytic performance of catalysts.

**Corresponding author*: amjad.mt.khan@gmail.com

Mousumi Sen (ed.) Sustainable Green Catalytic Processes, (141–164) © 2024 Scrivener Publishing LLC

Finally, chapter discusses the future directions and challenges in heterogeneous photocatalysis highlighting the importance of heterogeneous photocatalysis in addressing global environmental and energy challenges.

Keywords: Nanomaterials, pollutants, photocatalysis, environmental remediation

7.1 Introduction

The use of light to accelerate chemical reactions is known as photocatalysis, and it is a fascinating phenomenon. Excitation of semiconductors produces electron–hole pairs that can initiate a number of redox processes at the surface of the material. The discovery of heterogeneous photocatalysis dates to the early 1970s when Fujishima and Honda demonstrated water splitting using a TiO_2 semiconductor under ultraviolet (UV) light. Since then, the discipline has had exponential growth with notable developments in comprehending the underlying concepts, investigating novel materials, and identifying a broad range of applications. With its roots in the study of catalysis, photocatalysis stands out as a highly promising technology in the pursuit of sustainable solutions. Heterogeneous photocatalysis offers a novel route toward ecologically friendly and sustainable processes by having the capacity to convert solar energy into a driving force for chemical changes. With the potential to transform solar energy into chemical energy, there are intriguing possibilities for tackling various global issues simultaneously, such as sustainable synthesis, environmental cleanup, and energy conversion [1]. This field has become much more interdisciplinary, encompassing chemical engineering, semiconductor physics surface sciences, photophysical chemistry, and materials science. Notably, there has been a growing recognition of the detrimental impacts of human activities on the environment, prompting an immediate call for the adoption of sustainable practices. The ecosystems of our planet's world and human well-being are under severe threat from rising greenhouse gas emissions, water pollution, and energy demand. The urgent need to address environmental contamination, the depletion of natural resources, and the escalating energy crisis has pushed scientists to develop novel and long-lasting solutions [2, 3]. Heterogeneous photocatalysis offers compelling potential for tackling these complex problems. Since then, there have been significant improvements in the knowledge of underlying concepts, the discovery of novel materials, and the disclosure of various applications. Scientific groups from all fields have come together to create creative solutions for a cleaner, greener, and more sustainable future as the globe struggles with these issues [4–6].

The fundamentals of photocatalysis are based on the physics of semiconductors, where photons of light drive electrons across the energy bandgap to produce electron–hole pairs. At the surface of the photocatalyst, these charge carriers are essential for a variety of redox processes [7]. The driving force behind this extraordinary process lies in the ability of the photoexcited charge carriers to participate in a series of electron transfer reactions with molecules adsorbed on the photocatalyst surface, leading to the transformation of energy and formation of valuable products. The various intriguing characteristics of semiconducting materials provide the basis of heterogeneous photocatalysis. These substances have distinctive band structures that specify their electrical and optical characteristics and control how well they function as photocatalysts [8, 9]. To fully realize the potential of photocatalysis, researchers have worked to build and optimize these materials to maximize their efficiency in light absorption, electron–hole pair production, and charge separation. A thorough understanding of the complex mechanics of the photocatalytic process is essential for the success of photocatalysis. Many reaction mechanisms have been uncovered by scientists over time, including the well-known Langmuir–Hinshelwood (L-H) mechanism and the more recent surface defect engineering process. The initial approach entails the binding of reactant molecules onto the surface of the photocatalyst, while the subsequent technique leverages surface imperfections to alter the catalytic activity and selectivity of the photocatalyst [10].

Photocatalysis possesses a broader impact that transcends surface-level reactions, encompassing the generation of reactive oxygen species (ROS), such as superoxide ions and hydroxyl radicals. The significant potential of photocatalysis in environmental remediation is illustrated by the key roles played by ROS in the photocatalytic destruction of numerous pollutants, organic compounds, and toxins [11]. Researchers have explored innovative materials as the field has developed to push the limits of photocatalysis, such as the design of semiconductor materials such as photocatalysts, metal–organic frameworks (MOFs), perovskite, and plasmonic photocatalysts. The versatility of heterogeneous photocatalysis is vividly showcased through its diverse applications, addressing some of the most critical challenges faced by society [12]. Water splitting is a key application that enables the creation of pure hydrogen fuel, which is a sustainable and safe energy source, from water. As a strategy to reduce greenhouse gas emissions by transforming carbon dioxide into useful fuels and chemicals and so promoting the circular economy with carbon, photocatalytic CO_2 reduction has attracted a great deal of interest [13, 14]. Photocatalysis has demonstrated remarkable effectiveness in the breakdown of contaminants, including emerging pollutants present in both water and air, as well as the decomposition of organic dyes and pharmaceutical compounds. An effective tool for safeguarding the environment and public health is the capacity to break down dangerous substances at the molecular level. As we explore diverse applications of photocatalysis, we inevitably encounter unique challenges and limitations. Understanding the mechanisms and addressing these challenges are critical for optimizing photocatalytic performance and facilitating practical implementation. The future of heterogeneous photocatalysis is promising, with the potential for groundbreaking discovery and transformative applications. This field will surely reach previously unimaginable heights as a result of ongoing research into specialized materials, cutting-edge reactor designs, and a fundamental comprehension of reaction dynamics. The integration of photocatalysis with other cutting-edge technologies has tremendous potential to amplify its impact and facilitate sustainable solutions. Their synergistic actions may enhance pollutant degradation, energy conversion, and overall photocatalytic efficiency [15, 16]. This chapter provides a comprehensive exploration of heterogeneous photocatalysis, spanning its fundamental principles, recent advancements in materials, and diverse applications. Figure 7.1 shows the various aspects of heterogeneous photocatalysis that are widely studied by researchers. By understanding the underlying principles, exploring recent advances, and uncovering potential applications, it intends to encourage deeper research and innovation in this captivating field, driving progress toward a cleaner and greener future by examining the complex interplay between light and matter.

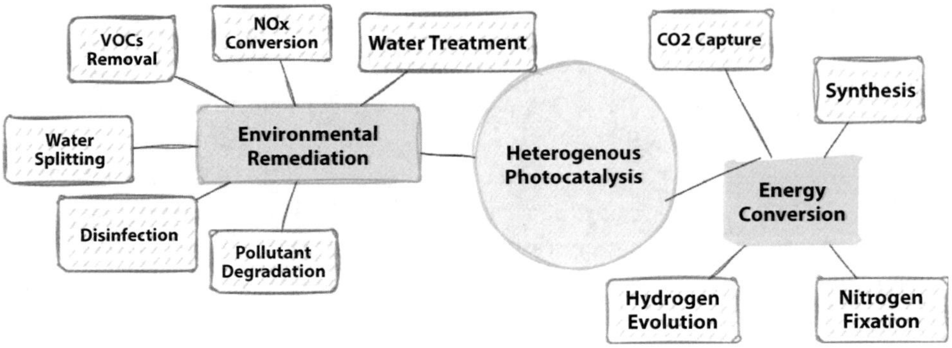

Figure 7.1 Multiple aspects of heterogeneous photocatalysis. (this figure has been adapted and redrawn from [17]).

7.2 Fundamental Principles of Photocatalysis

The versatile and promising process that leverages light energy to initiate chemical reactions on the surfaces of semiconducting materials is photocatalysis [17]. The fundamental principles behind this process are examined in this section, which also provides a full understanding of the key concepts and techniques involved, such as bandgap engineering, charge carrier generation, and separation, all of which are crucial for understanding and optimizing photocatalytic processes.

7.2.1 Bandgap Engineering

Semiconducting materials used as photocatalysts possess an energy range known as bandgap. The distinction between the valence band, where electrons reside in their lowest energy state, and the conduction band, where electrons exist in their highest energy state, is established through the presence of a bandgap. This bandgap serves as a barrier between these two states. When the photocatalyst absorbs photons with energies equal to or exceeding the bandgap's value, electrons undergo a transition from the valence band to the conduction band. This transition leads to the creation of electron–hole pairs, and the width of the bandgap plays a substantial role in shaping the photocatalytic performance of the material. Materials with narrow bandgaps tend to absorb visible light effectively, whereas those with wider bandgaps typically absorb UV light. Researchers can optimize the photocatalysts' ability to absorb light and thus boost the effectiveness of charge carrier generation by tuning the bandgap [18].

7.2.2 Charge Carrier Generation and Separation

The generation of electron–hole pairs, also known as charge carriers, plays a pivotal role in the photocatalytic process. While the efficient creation of charge carriers represents the initial phase, it is equally essential to follow this with their effective separation and measures to prevent recombination. If electrons and holes recombine quickly, the energy gained from light absorption is wasted, leading to lower photocatalytic efficiency [19]. To improve charge carrier separation, several methods have been used to solve this issue. Making heterojunctions or interfaces between various semiconducting materials is one strategy. By properly segregating the electrons and holes, these surfaces serve as charge sinks, reducing the likelihood of recombination [20].

7.2.3 Role of Photocatalytic Materials

The selection of a photocatalytic material holds significant importance in influencing the overall effectiveness and specificity of a photocatalytic procedure. Among the extensively investigated and widely adopted photocatalysts, titanium dioxide (TiO_2) stands out. It is preferred for its remarkable stability, nontoxic nature, and straightforward synthesis [21]. Nevertheless, owing to its wide bandgap, TiO2's photocatalytic capabilities are confined to the UV spectrum, restricting its utility in visible light applications [22]. Alternative semiconducting materials with various characteristics and bandgap engineering options have been investigated by researching this restriction. For visible light-active metal (ZnO), tantalum pentoxide (Ta_2O_5) and tungsten trioxide (WO_3) offer attainable possibilities. Additionally, non-oxides for visible light-active photocatalysis semiconductors, such as cadmium sulfide (CdS), cadmium telluride (CdTe), and bismuth vanadate ($BiVO_4$), have gained attention as promising visible light-active photocatalysts, offering tunable bandgaps and unique electronic properties [23, 24]. Recent development has facilitated the designing of hybrid and composite photocatalytic materials by combining several semiconducting substances or integrating photocatalysts with other beneficial components such as carbon-based substances or plasmonic nanoparticles.

7.3 Photocatalytic Mechanisms

Photocatalytic mechanisms in heterogeneous photocatalysis are the fundamental processes that occur during the interaction of light with photocatalytic materials, resulting in the production of charge carriers and the subsequent transformation of chemical species. These processes are essential for triggering photocatalytic reactions and have useful applications in the synthesis of chemical compounds, energy conversion, and rehabilitation of the environment [25]. Designing effective photocatalysts and maximizing their functionality require a thorough understanding of the complex functioning of these mechanisms. The key steps involved in the photocatalytic mechanism are as follows.

7.3.1 Light Absorption and Electron–Hole Pair Generation

The photocatalytic process initiates as photons are absorbed by the photocatalyst material. These photons transport energy corresponding to a specific light wavelength. When a photon possessing energy greater than the photocatalyst's bandgap strikes its surface, it gets assimilated by an electron, causing its transition from the valence band to the conduction band. Consequently, this event gives rise to a photogenerated electron–hole pair (e^-/h^+). The disparity in energy levels between the valence and conduction bands establishes the bandgap energy of the photocatalyst, ultimately governing the wavelength of light it can efficiently absorb [25, 26].

7.3.2 Charge Carrier Separation and Redox Reactions on the Photocatalyst Surface

When light is absorbed, it triggers the formation of electron–hole pairs (e-/h+) within the photocatalyst. The effectiveness of separating and maintaining the stability of these charge carriers is pivotal in determining the efficiency of photocatalysis. Once these photo-induced electrons and holes are successfully separated, they engage in redox reactions taking place on the photocatalyst's surface. The photoinduced holes (H^+) possess strong oxidizing properties and can interact with water or hydroxyl groups (OH-) located on the photocatalyst's surface, resulting in the production of hydroxyl radicals ($OH^•$). These radicals exhibit potent oxidative capabilities, playing indispensable roles in the degradation of pollutants and the synthesis of organic compounds [26, 27].

7.3.3 Reactive Oxygen Species Generation

The previously mentioned photocatalytic mechanism is fundamentally dependent on the production of reactive oxygen species. Within photocatalysis, hydroxyl radicals ($OH^•$) and superoxide radicals ($O_2^{•}$-) emerge as paramount ROS. These ROS contribute to the oxidation and degradation of pollutants and can play a crucial role in killing bacteria and other microorganisms in water and air treatment applications.

7.3.4 Langmuir–Hinshelwood and Surface Defect Engineering Mechanism

The crucial step in many photocatalytic reactions involves the adsorption of reactants onto the surface of the photocatalyst, a fundamental process in the Langmuir–Hinshelwood mechanism. In the L-H mechanism, reactants are absorbed on active sites, followed by subsequent reaction steps, leading to the formation of products. Properly designed catalysts with ample active sites and suitable reaction intermediates are crucial for achieving efficient photocatalytic performance. Understanding these mechanisms is essential for tailoring materials and optimizing their performance for specific applications, thereby advancing the potential of the field to address global challenges in energy and the environment [25, 28].

7.4 Factors Affecting Photocatalytic Efficiency

The photocatalytic reaction's efficiency and effectiveness are determined by a variety of variables that are controlled by the complex process of photocatalysis. To achieve optimal performance in photocatalytic systems and the creation of highly efficient photocatalysts, it is crucial to comprehend these underlying factors. The effectiveness of a photocatalytic substance is predominantly dictated by its semiconductor characteristics. Specifically, the bandgap energy (Eg) of the semiconductor plays a pivotal role in determining the range of wavelengths of light that it can absorb, thereby facilitating the generation of electron–hole pairs. When the bandgap is relatively narrow, the material can absorb visible light, resulting in a notable augmentation of its photocatalytic activity [29]. The photocatalytic effectiveness may experience a reduction due to the rapid recombination of electron–hole pairs when the bandgap is insufficiently wide. However, a wide bandgap may restrict visible light absorption, making the material ineffective for solar-powered applications.

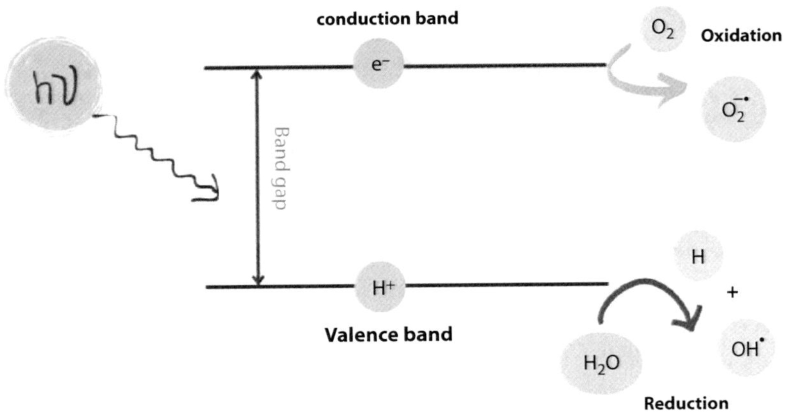

Figure 7.2 Mechanism for the generation of an electron-hole pair within a semiconductor-based photocatalyst upon exposure to light. (this figure has been adapted and redrawn from [30]).

Figure 7.2 displays the schematic scheme for the mechanism of generation of electron upon irradiation of light on the semiconductor photocatalyst.

The surface area and morphology of the photocatalysts significantly influence their photocatalytic activity. Increasing the surface area offers a greater number of active sites for the adsorption of reactants, enhancing the potential for molecules to engage with the photocatalyst surface, consequently expediting the reaction rate. Moreover, the configuration of the photocatalyst can influence the accessibility of reactive sites and the efficiency of light absorption. The controlled synthesis of well-defined morphologies can optimize light absorption and improve charge carrier separation, leading to enhanced photocatalytic efficiency. Doping stands as a commonly utilized method for modifying the band structure of a photocatalyst, facilitating improved charge separation, and expanding its capacity to absorb a broader spectrum of light. As an example, the introduction of nitrogen or sulfur into titanium dioxide (TiO_2) can effectively reduce the bandgap, thus enabling the absorption of visible light and consequently elevating its overall photocatalytic efficiency. Careful selection of dopants and their concentrations is crucial, as excessive doping may lead to recombination centers or unintended energy levels that hamper the photocatalytic efficiency.

The pH level within the reaction medium plays a significant role in influencing both the extent of ionization of the reactants and the surface charge characteristics of the photocatalyst, which also has an influence on photocatalytic effectiveness. Variations in pH can affect the stability of the photocatalyst as well as the pace of the surface reaction and reactant adsorption. To achieve the desired performance, it is essential to optimize the pH conditions for specific photocatalytic reactions. The rate of the surface reactions and the kinetics of the charge transfer processes are both significantly influenced by temperature, which is another important factor in photocatalysis. Extreme temperatures may lead to thermal degradation or reduced stability of the photocatalyst, emphasizing the need for careful temperature control during photocatalytic experiments. Moreover, the aggregation of photocatalyst particles can reduce the accessible surface area, thereby constraining the accessibility of reactants to the catalytic active sites. Aggregation often occurs during prolonged photocatalytic reactions or in complex reaction media. By optimizing semiconductor properties, surface area, doping, cocatalysts, pH, temperature, and stability, researchers can tailor photocatalysts for specific applications and contribute to the advancement of sustainable energy and environmental technologies [30, 31].

7.5 Recent Advances in Heterogeneous Photocatalysts

Recently, significant progress has been made in heterogeneous photocatalysis, leading to the emergence of innovative and highly effective photocatalytic materials. These developments are driven by the demand for effective and sustainable responses to a range of environmental and energy-related problems. They include a variety of substances and methods, each with special qualities and advantages for certain uses, which have significantly expanded the scope and practicality of photocatalysis as a sustainable technology. Researchers continue to unlock the potential of photocatalysts by customizing materials, improving their structures, and investigating cutting-edge ideas [21, 32]. This will lead to a greener and more sustainable future. Here, we explore three promising types of heterogeneous photocatalysts: plasmonic photocatalysts, perovskites, and metal–organic frameworks.

7.5.1 Metal–Organic Frameworks

MOFs represent a distinct category of highly porous crystalline materials formed by the combination of metal ions and organic ligands. This unique combination of metal nodes and organic linkers grants the ability to finely adjust MOF structures, thereby offering a diverse spectrum of properties and functions [33]. These characteristics have opened new avenues for their applications in photocatalysis. These inherent traits have ushered in novel opportunities for employing MOFs in the realm of photocatalysis. Indeed, researchers have achieved noteworthy progress in the integration of MOFs with diverse semiconductors, aiming to amplify both the efficacy of charge separation and light absorption processes. The high surface area and tunable structure of MOFs offer excellent opportunities for efficient interfacial charge transfer, leading to improved photocatalytic performance. This is reported in the incorporation of a cocatalyst, such as platinum or palladium nanoparticles, into a copper-based MOF. The MOF acts as a CO_2 adsorbent, capturing CO_2 molecules efficiently, whereas the cocatalyst facilitates the activation and reduction of CO_2 into useful hydrocarbons, such as methane or ethylene. These photocatalysts made from MOF have demonstrated promising CO_2 reduction results, providing an effective alternative for reducing greenhouse gas emissions and producing sustainable fuel. Recently, MOF engineering has enabled the development of photocatalysts modified with MOFs that can react with visible light, thereby expanding their solar absorption range. Researchers have functionalized titanium dioxide (TiO_2) photocatalysts with light-absorbing MOFs, enabling the use of visible light for photoexcitation. Photocatalysts that have been modified with MOFs (metal–organic frameworks) have demonstrated improved efficiency in harnessing solar energy for the purpose of photocatalytic pollutant degradation and water treatment applications, making them very desirable for environmental remediation [2, 34].

7.5.2 Perovskites

Recent research has focused on the bandgap and electrical structure of perovskites to boost their efficiency in certain photocatalytic applications. To alter the bandgap and improve light absorption in the desired spectral range, several doping techniques have been investigated. Due to their remarkable optoelectronic qualities and high charge carrier mobility, perovskite photocatalysts have emerged as feasible alternatives. The perovskite structure is typically composed of an ABX_3 where A and B are cations, an X is an anion. The unique crystal structure of perovskites allows facile charge transport, making them ideal materials for efficient photocatalysis. Perovskites like $LaFeO_3$ have demonstrated exceptional photocatalytic efficiency for the disintegration of dyes and pharmaceutical compounds. Perovskite materials possess a crucial capability to generate active oxygen species (commonly known as ROS) when exposed to light, which significantly contributes to the breakdown of pollutants in the environment [35].

Perovskite photocatalysts have shown remarkable potential for pollutant degradation; however, their stability under prolonged irradiation remains a challenge. To address this issue, scientists have devised approaches aimed at enhancing the durability of the substance under harsh photocatalytic conditions, such as using protective coatings or hybridizing perovskites with more stable materials.

7.5.3 Plasmonic Photocatalysts

Plasmonic photocatalysts have garnered significant interest in recent times owing to their remarkable ability to enhance light absorption and exploit localized surface plasmon resonances. They exhibit strong light–matter interactions, leading to enhanced light harvesting and hot electron generation, which can be employed for numerous photocatalytic applications. Plasmonic nanoparticles, such as gold or silver nanoparticles, are widely researched for their ability to enhance photocatalytic water splitting. The rational design of these nanostructures contributes to effective light-trapping and strong plasmonic effects. Gold nanoparticles immobilized on titania (Au/TiO_2) have demonstrated an increased capacity for converting CO_2 into carbon monoxide and methane when exposed to visible light irradiation [36]. The electric fields induced by localized surface plasmon resonance (LSPR) at the surface of nanoparticles played a crucial role in promoting the adsorption and activation of CO_2 molecules, thereby enhancing catalytic performance. Plasmonic photocatalysts have also demonstrated efficacy for pollutant degradation. Silver nanoparticle-decorated zinc oxide (Ag/ZnO) photocatalysts have exhibited exceptional photocatalytic performance in the elimination of organic dyes from wastewater [37].

Researchers have demonstrated the controlled synthesis of plasmonic photocatalysts with tailored size, shape, and composition to optimize their performance. Synergistic hybrid systems are produced by combining plasmonic nanoparticles with traditional photocatalytic materials. These hybrid structures combine the unique properties of plasmonic nanoparticles with the stability and catalytic activity of traditional semiconductors, resulting in significantly enhanced photocatalytic performance [38].

7.6 Applications of Heterogeneous Photocatalysis

Heterogeneous photocatalysis has emerged as a versatile and prospective technology offering a wide array of potential applications spanning from energy generation and environmental cleanup to healthcare and materials science, including antibacterial applications, self-cleaning

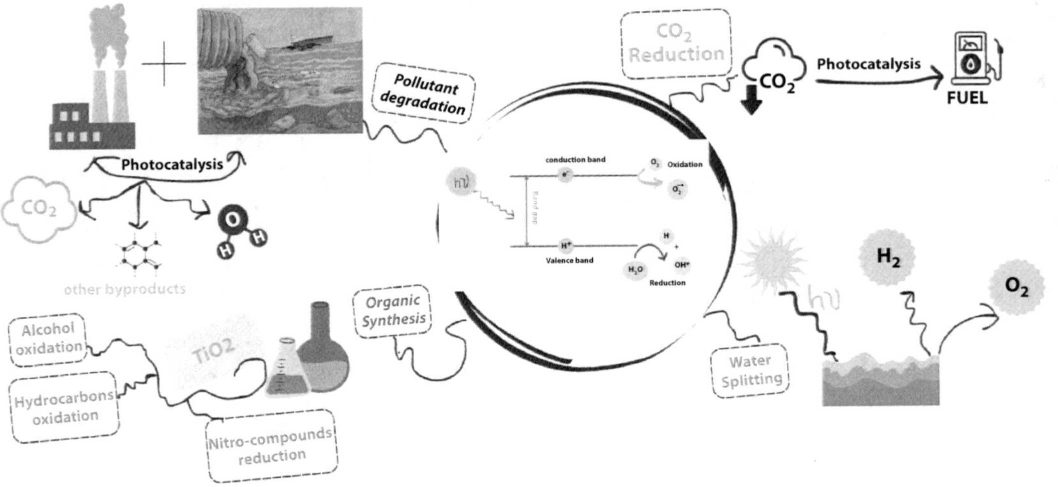

Figure 7.3 Illustration of different applications of heterogeneous photocatalysis. (this figure has been adapted and redrawn from [56].

coatings, air purification, photocatalytic sensors, environmental remediation, and energy storage. In Figure 7.3, various applications of heterogeneous photocatalysis are displayed. With the advancement of this field's study, the range of applications is expected to expand further, paving the way for a more sustainable and environmentally friendly future [1, 14]. The combination of photocatalysis with advanced materials and innovative designs continues to drive the exploration of new applications and optimize existing ones. The primary areas of interest in this section will encompass significant applications, such as water splitting, CO_2 reduction, the degradation of pollutants, and chemical synthesis.

7.6.1 Water Splitting

One of the crucial applications of photocatalysis is water splitting, which aims to harness solar energy to make clean sustainable hydrogen fuel. Heterogeneous photocatalysts have a crucial function in the process of water splitting, facilitating the efficient separation of photo-generated charge carriers, namely, holes and electrons. This separation is essential for driving the redox reactions required for the process [39]. The water-splitting process itself encompasses two half-reactions: the reduction of protons (H^+) to yield hydrogen gas (H_2) and the oxidation of water, resulting in the formation of oxygen (O_2) as its product.

The reaction occurring in water splitting can be expressed as follows:

(a) Hydrogen Evolution Reaction (HER):

$$4H^+ + 4\,e^- \rightarrow 2H_2$$

(b) Oxygen Evolution Reaction (OER):

$$2H_2O + h\nu \rightarrow O_2 + 4H^+ + 4\,e^-$$

The overall reaction is:

$$2H_2O + h\nu \rightarrow 2H_2 + O_2$$

In this process, the photocatalyst captures photons, leading to the formation of electron–hole pairs (e^-/h^+). The electrons generated by this process (e^-) play a crucial role in reducing protons (H^+) to form H_2 gas, while the photogenerated holes (h^+) are actively engaged in the oxidation of water, resulting in the production of protons and O_2 gas. Efficient charge separation and the prevention of electron–hole recombination are critical factors in achieving high water-splitting efficiency [40]. Titanium dioxide (TiO_2), metal oxides, and certain perovskite materials have shown promising ability for water splitting. The latest developments in water-splitting photocatalysts have focused on improving efficiency and stability. Researchers used cocatalysts and catalytic metal sites to accelerate the hydrogen evolution reaction while reducing electron–hole recombination. Furthermore, novel tandem photocatalytic devices that combine several semiconductors with complementary bandgap energies have demonstrated enhanced solar-to-hydrogen conversion efficiency [41].

7.6.2 CO_2 Reduction

Heterogeneous photocatalysts play a crucial role in initiating the activation of CO_2 molecules and facilitating their transformation into desired end products. Cutting-edge materials, including metal–organic frameworks, perovskites, and hybrid systems, have demonstrated significant promise in the realm of CO_2 reduction. The process of photocatalytic carbon dioxide (CO_2) conversion represents a viable approach to mitigate the presence of this greenhouse gas by converting it into valuable hydrocarbons and chemicals, thus contributing to the mitigation of climate change while simultaneously offering renewable resources [2, 42]. In the context of CO_2 reduction, the photocatalyst efficiently captures light energy, resulting in the generation of electron–hole pairs that are harnessed for the activation and reduction of CO_2 molecules. This intricate process mandates the efficient separation and orchestrated movement of charge carriers, ensuring the successful execution of the desired chemical transformations. Diverse heterogeneous photocatalysts, encompassing metal oxides, metal sulfides, and hybrid materials, have unveiled their potential in CO_2 reduction, channeling CO_2 into an array of end products, including carbon monoxide (CO), methane (CH_4), ethylene (C_2H_4), and formic acid (HCOOH).

For example, when exposed to visible light, Cu-MOF-74 has been employed as a photocatalyst to convert CO_2 into formate (HCOOH). The unique porous structure of MOFs provides ample sites for CO_2 adsorption, while the presence of copper sites enhances the selectivity for formate production. Furthermore, researchers have successfully synthesized Z-scheme g-C_3N_4/TiO_2 photocatalysts to enhance CO_2 reduction [43]. The staggered band alignment between g-C_3N_4 and TiO_2 facilitates efficient charge transfer, resulting in improved CO_2 conversion to hydrocarbons when exposed to visible light. Recent research efforts have been focused on augmenting the selectivity and efficiency of these photocatalysts for specific product formation.

7.6.3 Pollutant Degradation

These days, toxic pollutants are a major problem worldwide. Contaminants like petroleum oil, heavy metals, organic dyes, xenobiotics, hydrocarbons, pharmaceuticals, and other emerging pollutants are discharged into the environment and biota as a result of human actions. The increased awareness of environmental contamination has prompted scientists and researchers to devise and implement a variety of approaches that utilize the use of cutting-edge technology to identify and address this worldwide issue. Photocatalysis has demonstrated impressive effectiveness in the breakdown of various emerging pollutants, pharmaceuticals, and organic substances found in both water and air [44]. Within the process of photocatalytic degradation, illumination activates the photocatalyst, resulting in the creation of reactive oxygen species and electron–hole pairs. These ROS, which encompass hydroxyl radicals (OH•), superoxide radicals (O_2-•), and hydrogen peroxide (H_2O_2), play a vital role in the transformation of organic contaminants into less harmful by-products [45]. Simultaneously, the freed electrons and holes actively engage in redox reactions that facilitate the decomposition of organic compounds.

Titanium dioxide (TiO_2) stands out as a widely recognized photocatalyst due to its robust oxidative properties, which make it highly effective for breaking down pollutants. Conventional TiO_2 photocatalysis, on the other hand, is limited to UV light absorption, restricting its application in natural sunlight [46]. Researchers have placed their emphasis on modifying the attributes of the photocatalyst with the aim of broadening its light absorption capabilities into the visible spectrum. This endeavor has been undertaken to address and overcome this particular limitation effectively. Doping, cocatalyst integration, and surface functionalization have all been investigated as ways to optimize light consumption and photocatalytic activity. Methods for surface modification, including defect engineering, have been utilized to boost the surface reactivity of photocatalysts and enhance their effectiveness in degrading substances. Moreover, plasmonic photocatalysts have also gained attention for pollutant degradation due to their enhanced light absorption and hot electron generation properties. Plasmonic nanoparticles, such as gold or silver, can enhance light harvesting and promote charge transfer, leading to improved degradation rates under visible light [47, 48].

7.6.4 Organic Synthesis

With the ability to directly activate and functionalize bonds using visible light, photocatalysis has arisen as an eco-friendly and sustainable approach to chemical synthesis. This approach offers advantages over traditional chemical synthesis methods that rely on high temperatures and hazardous reagents [49]. It is an environmentally friendly method for performing chemical reactions using light energy. It involves the use of photocatalysts to initiate photochemical transformations of organic molecules, resulting in the formation of complex molecular structures and new chemical bonds [50]. A large number of photocatalysts, including metal oxides, organic photocatalysts, and carbon-based nanomaterials, have been employed in various organic transformations. Additionally, photoredox catalysis involves the use of photocatalysts to mediate electron transfer reactions.

In photocatalytic organic synthesis, the photocatalyst absorbs visible light, initiating photochemical reactions that activate organic molecules. The photocatalyst's excited state can participate in single-electron transfer reactions, generating radicals and initiating a cascade of chemical transformations. These photo-generated charge carriers can be involved in a variety of reactions, including single-electron transfer, energy transfer, and radical reactions. The excited state of the photocatalyst can transfer energy to nearby reactant molecules, leading to the activation of chemical bonds and subsequent transformations [51].

Photocatalysis has been widely applied to C-C bond formation reactions, enabling the construction of complex molecular frameworks. Cross-coupling of aryl halides with alkyl nucleophiles is facilitated by visible light. Researchers have used photocatalysts like iridium complexes to encourage the establishment of C-C bonds between distinct functional groups, providing an efficient and selective route to valuable organic compounds. Photocatalytic C-H functionalization reactions enable direct alteration of C-H bonds in organic compounds, eliminating the need for pre-functionalized starting materials [52]. For instance, ruthenium complexes have been used to activate C-H bonds in aromatic compounds, allowing for regioselective functionalization at sites that would otherwise be difficult to access. Additionally, asymmetric transformations have been used to generate enantioselective transformations by the development of chiral radical intermediates utilizing chiral photocatalysts such as chiral iridium complexes. This method has allowed the synthesis of complex chiral compounds with great enantioselectivity [50, 53].

7.7 Recent Advances in Enhancing Photocatalytic Performance

In the quest for more efficient and sustainable photocatalytic systems, researchers have focused on improving photocatalytic performance using novel methodologies. Recent breakthroughs in enhancing photocatalytic performance have pushed the field forward. These developments have paved the way for more effective and eco-friendly solutions across a range of applications, encompassing chemical synthesis, energy conversion, and environmental remediation. These developments are the result of a combination of novel material design, improved catalyst architectures, and a deeper understanding of fundamental photocatalytic processes [54]. Such advancements have expanded the capabilities of photocatalysis, making it a promising technology for addressing global challenges. Researchers are unlocking the full potential of photocatalysis to promote a more sustainable and cleaner future by combining novel materials design, precise engineering of catalyst structures, and new characterization techniques [55].

7.7.1 Cocatalyst Engineering for Charge Carrier Management

A fundamental obstacle within the realm of photocatalysis lies in the effective separation and movement of charge carriers (electrons and holes, represented as e^-/h^+), induced into excitation by photons. The latest developments are concentrating on cocatalyst engineering, which involves introducing cocatalysts onto photocatalytic surfaces to facilitate charge transfer and suppress charge recombination. Figure 7.4 shows the schematic display of loading of a cocatalyst on the surface of a photocatlyst to increase the active sites in order to significantly boost the overall photocatalytic performance. Cocatalysts, such as cocatalytic metallic nanoparticles (for instance, Pt and Rh), have been effectively incorporated onto semiconductor photocatalysts (such as TiO_2 and WO_3). This integration serves to amplify the efficiency of hydrogen evolution and oxygen evolution reactions in water-splitting applications [56]. These cocatalysts provide active sites for charge transfer, promoting efficient charge

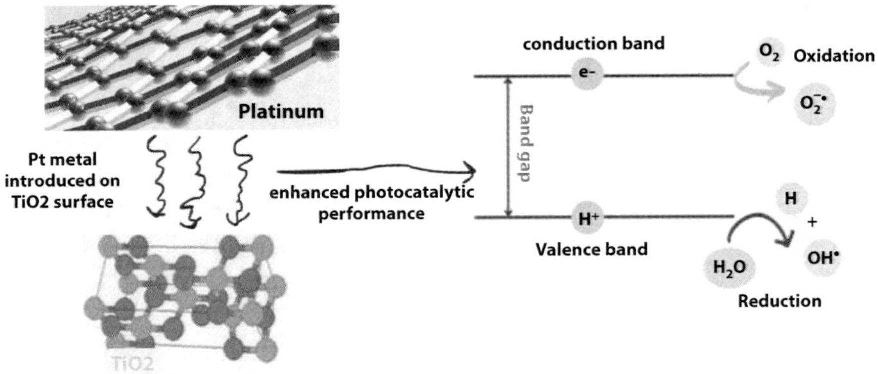

Figure 7.4 Platinum loaded onto TiO_2 to increase active sites for charge transfer. (this figure has been adapted and redrawn from [63]).

separation while minimizing the possibility of electron–hole recombination [57]. In addition, researchers developed a Pt-decorated $BiVO_4$ photocatalyst for solar water splitting [58]. The synergistic interaction between Pt cocatalysts and $BiVO_4$ significantly increased the efficiency of hydrogen synthesis by enhancing both charge separation and visible light absorption [59].

7.7.2 *In Situ* Photocatalyst Modification and Activation

In situ photocatalyst modification and activation have gained popularity as effective methods for increasing catalytic activity. During the process, the photocatalyst surface or structure is modified to generate active sites and facilitate certain reactions. Researchers recently discovered an *in situ* modification approach for improving CO_2 reduction over TiO_2-supported Au photocatalysts [36]. The surface of Au nanoparticles was selectively modified by exposing the photocatalyst to CO_2 during the reaction, resulting in improved CO_2 adsorption and enhanced photocatalytic CO_2 conversion to methane [60].

7.7.3 Surface Defect Engineering and Co-Doping

Surface defects have a substantial impact on catalytic activity and charge carrier dynamics in photocatalytic processes. Lately, researchers are using surface defect engineering for introducing controlled defects into photocatalytic materials to enhance their photocatalytic performance. It is easy to produce surface defects and improve charge carrier separation by incorporating two or more distinct materials into the photocatalyst lattice. This method has proven to be highly successful in fine-tuning the electronic and optical characteristics of photocatalysts and is alternatively referred to as co-doping [9, 61, 62]. Controlled annealing procedures are another successful strategy for surface defect engineering. High-temperature treatments in controlled atmospheres can induce oxygen vacancies, metal vacancies, or other lattice distortions on the surface of the photocatalyst. These defects may produce trap states that facilitate charge separation and promote specific photocatalytic reactions. This method has been extensively used with TiO_2, a renowned photocatalytic material [63]. Oxygen vacancies within TiO_2 can function as electron traps, effectively reducing the rate at which electron–hole recombination occurs on the surface. Scientists have documented enhanced photocatalytic performance in processes such as hydrogen generation and pollutant decomposition by intentionally creating controlled oxygen vacancies through thermal treatment methods [64].

Surface defect engineering and co-doping often lead to synergistic effects that amplify the photocatalytic activity beyond the individual contributions of defects or dopants. These phenomena encompass various advantages, including optimized charge separation, augmented reactant adsorption, and extended light absorption [65]. The incorporation of surface defects through co-doping has led to an increase in the absorption of visible light and an extension in the lifetime of photo-generated charge carriers. This has translated into significantly enhanced performance in the photocatalytic synthesis of nitrogen and sulfur co-doped TiO_2 materials designed for the degradation of pollutants under visible light. Nitrogen introduces additional energy levels within the bandgap, effectively expanding the range of visible light absorption, while carbon serves to improve charge carrier mobility. This cooperative synergy results in superior visible light-driven photocatalytic capabilities, positioning nitrogen and carbon co-doped TiO_2 as highly promising candidates for applications such as the degradation of organic pollutants [66, 67].

7.7.4 Advanced Characterization Techniques in Photocatalysis

Advances in characterization techniques have been critical in understanding the principles underlying improved photocatalytic performance. In recent years, the field of photocatalysis has been greatly enriched by advanced characterization techniques that provide previously unexplored insights into the complex mechanisms governing photocatalytic processes [68]. These techniques, often at the forefront of materials science and spectroscopy, have enabled researchers to elucidate the fundamental interactions and dynamics occurring at the nanoscale during photocatalytic reactions. In photocatalysis, SPM techniques have been used to visualize the distribution of charge carriers on photocatalyst surfaces, providing insights into charge transport and recombination processes. Additionally, atomic force microscopy-based Kelvin probe force microscopy (KPFM) has the capability to chart the localized work function of photocatalysts, exposing fluctuations in charge distribution that exert an impact on the photocatalytic performance, while SERS provides insights into the molecular interactions between reactants, intermediates, and catalyst surfaces. By monitoring changes in vibrational modes and shifts in Raman spectra, researchers can identify reaction pathways, adsorption geometries, and active sites involved in photocatalytic reactions [69].

Time-resolved spectroscopy stands as a powerful method for examining fleeting entities and events with exceptional time precision. Within the realm of photocatalysis, this approach proves indispensable for monitoring the durations of charge carriers (electrons and holes) as well as the reaction kinetics of intermediate species throughout the processes of photoexcitation and catalysis [70]. Femtosecond transient absorption spectroscopy (Fs-TA) allows researchers to follow the ultrafast dynamics of charge carrier generation, migration, and recombination. By probing the absorption spectra of the photocatalyst over femtosecond to nanosecond timescales after light excitation, researchers can precisely quantify the rates of charge separation and recombination, shedding light on factors influencing photocatalytic efficiency. Moreover, operando spectroscopy is the practice of observing chemical reactions in real-world operational settings. This approach offers valuable insights into the behavior of catalysts during catalytic processes. On the other hand, electron microscopy techniques, such as transmission electron microscopy (TEM) and scanning electron microscopy (SEM), allow scientists to visualize the nanoscale morphology, size, and structure of photocatalytic materials.

7.8 Prospects and Pioneering Challenges in Heterogeneous Photocatalysis

As heterogeneous photocatalysis continues to advance, it harbors the potential to tackle several significant global issues, encompassing energy generation, environmental restoration, and the sustainable synthesis of chemicals. However, several future directions and challenges lie ahead, shaping the trajectory of this dynamic and impactful area of research.

7.8.1 Future Directions

Future research in heterogeneous photocatalysis will likely focus on the tailored design of photocatalysts with specific properties optimized for target applications. In order to improve catalytic activity, material composition, charge separation, structure, light absorption, and surface properties must be customized. To do this, it will be crucial to use rational design methods supported by computational modeling and sophisticated characterization tools. The integration of multiple photocatalysts or catalytic functions within a single system holds promise for achieving more complex and efficient photocatalytic processes. Tandem and multifunctional photocatalytic systems can facilitate sequential reactions, such as multistep organic transformations or the coupling of water splitting and CO_2 reduction [71]. Optimizing energy transfer and charge carrier dynamics between various components will likely be the primary focus of research in this field. A promising direction in heterogeneous photocatalysis involves the production of earth-abundant and sustainable materials. While many successful photocatalysts rely on noble metals or rare elements, there is a growing emphasis on identifying and designing alternative materials that are both efficient and environmentally friendly. This includes exploring nonmetallic materials, such as nanomaterials based on carbon, semiconductors devoid of metals, and easily accessible metal oxides, to pave the way for scalable and cost-effective photocatalytic technologies [72, 73].

More importantly, with the advancements in technology after the innovation of AI, the integration of artificial intelligence (AI) and computational methods presents an exciting prospect for accelerating the discovery and design of new photocatalysts [74]. Machine learning algorithms and computational simulations can efficiently screen and predict the properties of potential photocatalytic materials, enabling researchers to identify promising candidates for experimental validation. This data-driven approach has the potential to revolutionize the process of materials discovery, significantly reducing the time and resources required for developing advanced photocatalysts.

Future efforts will concentrate on developing photocatalytic systems that mimic the intricate processes of natural photosynthesis, involving efficient light absorption, charge separation, and catalytic conversion. This direction holds the potential to provide sustainable and renewable energy sources [75, 76]. The development of multifunctional photocatalytic systems is an emerging trend that aims to simultaneously achieve multiple chemical transformations or applications. This includes designing photocatalysts capable of simultaneous water splitting and pollutant degradation or combining photocatalysis with other technologies, such as catalytic ozonation [77, 78]. By harnessing the synergistic effects of multiple processes, multifunctional systems could lead to enhanced overall performance and resource utilization. Innovations in plasmonic enhancement and nanoscale engineering offer promising possibilities for enhancing photocatalytic performance. It is expected to delve into the precise management of nanostructures and plasmonic nanoparticles to influence interactions between light and matter, resulting in enhanced light absorption and efficient charge separation. Such advancements hold the potential for creating exceptionally effective and adjustable photocatalysts applicable across a range of fields.

7.8.2 Challenges

The persistent problem of catalyst stability remains a matter of major concern in the field of heterogeneous photocatalysis, particularly for long-duration reactions or under difficult reaction conditions. Photocorrosion, agglomeration, and surface poisoning can lead to catalyst degradation over time [4]. Addressing these issues will require the development of stable photocatalyst architectures, protective coatings, and strategies to mitigate surface reactions. Maximizing the quantum efficiency of photocatalytic reactions remains a challenge, as high photocatalytic efficiency requires effective charge carrier dynamics, and a substantial portion of photogenerated charge carriers often recombine before participating in desired reactions. Performance can be constrained by charge carrier recombination, particularly in systems with short carrier lifetimes. Researchers will need to investigate novel approaches, such as cocatalyst engineering, defect control, and advanced heterostructures, to increase charge separation and lengthen charge carrier lifetimes. Achieving high selectivity in complex photocatalytic reactions, such as multi-electron transformations, remains a challenge. Understanding reaction mechanisms and controlling reaction pathways to selectively produce desired products are essential [79]. Further research efforts will involve elucidating reaction mechanisms through advanced spectroscopic and computational methods and designing catalysts that favor specific pathways.

The practical implementation of photocatalytic technologies on a large scale is a challenge that requires consideration of factors such as reactor design, catalyst immobilization, and scalability of synthesis methods. Researchers must bridge the gap between laboratory-scale demonstrations and real-world applications by developing cost-effective and scalable production methods. While photocatalysis offers the potential for environmentally friendly processes, the environmental impact and toxicity of photocatalytic materials must be thoroughly evaluated. Researchers need to assess the potential release of nanoparticles or by-products into the environment and develop strategies to minimize any negative effects. Heterogeneous photocatalysis offer potential to transform chemical synthesis, energy synthesis, and environmental remediation is quite promising [80]. To fully utilize the potential of photocatalysis, challenges with catalyst stability, charge carrier dynamics, selectivity, scalability, and environmental effect must be solved. Heterogeneous photocatalysis is positioned to be a key component of sustainable and clean technologies for a more energy- and environmentally conscious future with continuing interdisciplinary study and collaboration.

7.9 Conclusion

In this chapter, we have highlighted the transformative developments in fundamental principles, photocatalytic mechanisms, and materials engineering in the realm of photocatalysis, uncovering an extensive range of promising opportunities and groundbreaking achievements. Heterogeneous photocatalysis stands as a pillar of sustainable technology, offering innovative solutions to pressing environmental and energy challenges. The use of innovative materials like MOFs, perovskites, and plasmonic photocatalysts has shown the potential to transform solar-driven processes. These developments have not only deepened our understanding of photocatalysis but also propelled its practical application in various fields, including water splitting, CO_2 reduction, pollutant cleanup, and organic synthesis. In the future, collaborative interdisciplinary efforts and creative research will be crucial in addressing obstacles and advancing photocatalytic solutions. Continuous exploration, practical application, and improvement are guiding heterogeneous photocatalysis toward becoming a promising solution for a cleaner and more sustainable world.

References

1. Teoh, W.Y., Scott, J.A., Amal, R., Progress in heterogeneous photocatalysis: From classical radical chemistry to engineering nanomaterials and solar reactors. *J. Phys. Chem. Lett.*, *3*, 5, 629–639, **2012**, doi: 10.1021/JZ3000646/ASSET/IMAGES/MEDIUM/JZ-2012-000646_0007.GIF.
2. Basyach, P., Sonowal, K., Chetia, P., Kalita, L., Borthakur, S., Saikia, L., Photocatalytic conversion of CO2 into value added and renewable fuels over heterogeneous nanocatalysts: A green and environmental benign approach, in: *Heterogeneous Nanocatalysis for Energy and Environmental Sustainability*, vol. *2-2*, pp. 221–274, **2022**.
3. Oyetade, J.A., Machunda, R.L., Hilonga, A., Photocatalytic degradation of azo dyes in textile wastewater by Polyaniline composite catalyst-a review. *Sci. Afr.*, *17*, **2022**, doi: 10.1016/J.SCIAF.2022.E01305.
4. Kamat, P.V., TiO 2 nanostructures: Recent physical chemistry advances. *J. Phys. Chem. C*, *116*, 22, 11849–11851, **2012**, doi: 10.1021/JP305026H.
5. Yadav, S., Shakya, K., Gupta, A., Singh, D., Chandran, A.R., VarayilAanappalli, A. *et al.*, A review on degradation of organic dyes by using metal oxide semiconductors. *Environ. Sci. Pollut. Res.*, *30*, 28, 71912–71932, **2023**, doi: 10.1007/S11356-022-20818-6.
6. Mohd, S., Wani, A.A., Khan, A.M., ZnO/POA functionalized metal-organic framework ZIF-8 nanomaterial for dye removal. *Cleaner Chem. Eng.*, *3*, 100047, **2022**, doi: 10.1016/J.CLCE.2022.100047.
7. Zhang, J. and Nosaka, Y., Quantitative detection of OH radicals for investigating the reaction mechanism of various visible-light TiO2 photocatalysts in aqueous suspension. *J. Phys. Chem. C*, *117*, 3, 1383–1391, **2013**, doi: 10.1021/JP3105166.
8. Nguyen, M.A., Zahran, E.M., Wilbon, A.S., Besmer, A.V., Cendan, V.J., Ranson, W.A. *et al.*, Converting Light Energy to Chemical Energy: A New Catalytic Approach for Sustainable Environmental Remediation. *ACS Omega*, *1*, 1, 41–51, **2016**, doi: 10.1021/ACSOMEGA.6B00076.
9. Seriani, N., Pinilla, C., Crespo, Y., Presence of gap states at Cu/TiO2 anatase surfaces: Consequences for the photocatalytic activity. *J. Phys. Chem. C*, *119*, 12, 6696–6702, **2015**, doi: 10.1021/ACS.JPCC.5B00846.

10. Bian, Z., Tachikawa, T., Kim, W., Choi, W., Majima, T., Superior electron transport and photocatalytic abilities of metal-nanoparticle-loaded TiO2 superstructures. *J. Phys. Chem. C*, *116*, 48, 25444–25453, **2012**, doi: 10.1021/JP309683F.
11. Shu, J. and Tang, D., Recent Advances in Photoelectrochemical Sensing: From Engineered Photoactive Materials to Sensing Devices and Detection Modes. *Anal. Chem.*, *92*, 1, 363–377, **2020**, doi: 10.1021/ACS.ANALCHEM.9B04199.
12. Ganguly, P., Harb, M., Cao, Z., Cavallo, L., Breen, A., Dervin, S. *et al.*, 2D Nanomaterials for Photocatalytic Hydrogen Production. *ACS Energy Lett.*, *4*, 7, 1687–1709, **2019**, doi: 10.1021/ACSENERGYLETT.9B00940.
13. Trench, A.B., MacHado, T.R., Gouveia, A.F., Foggi, C.C., Teodoro, V., Sánchez-Montes, I. *et al.*, Rational Design of W-Doped Ag3PO4as an Efficient Antibacterial Agent and Photocatalyst for Organic Pollutant Degradation. *ACS Omega*, *5*, 37, 23808–23821, **2020**, doi: 10.1021/ACSOMEGA.0C03019.
14. Pirkanniemi, K. and Sillanpää, M., Heterogeneous water phase catalysis as an environmental application: A review. *Chemosphere*, *48*, 10, 1047–1060, **2002**, doi: 10.1016/S0045-6535(02)00168-6.
15. Maeda, K. and Domen, K., Photocatalytic water splitting: Recent progress and future challenges. *J. Phys. Chem. Lett.*, *1*, 18, 2655–2661, **2010**, doi: 10.1021/JZ1007966.
16. Sharma, P., Bano, A., Singh, S.P., Dubey, N.K., Chandra, R., Iqbal, H.M.N., Recent advancements in microbial-assisted remediation strategies for toxic contaminants. *Cleaner Chem. Eng.*, *2*, 100020, **2022**, doi: 10.1016/J.CLCE.2022.100020.
17. Herrmann, J.M., Fundamentals and misconceptions in photocatalysis. *J. Photochem. Photobiol. A: Chem.*, *216*, 2–3, 85–93, **2010**, doi: 10.1016/J.JPHOTOCHEM.2010.05.015.
18. Mahdi, Z. and El Hanandeh, A., Insight into copper and nickel adsorption from aqueous solutions onto carbon-coated-sand: Isotherms, kinetics, mechanisms, and cost analysis. *Cleaner Chem. Eng.*, *3*, 100045, **2022**, doi: 10.1016/j.clce.2022.100045.
19. Gillespie, P.N.O. and Martsinovich, N., Electronic Structure and Charge Transfer in the TiO2 Rutile (110)/Graphene Composite Using Hybrid DFT Calculations. *J. Phys. Chem. C*, *121*, 8, 4158–4171, **2017**, doi: 10.1021/ACS.JPCC.6B12506.
20. Zhang, H., Wang, W., Zhao, H., Zhao, L., Gan, L.Y., Guo, L.H., Facet-Dependent Interfacial Charge Transfer in Fe(III)-Grafted TiO2 Nanostructures Activated by Visible Light. *ACS Catal.*, *8*, 10, 9399–9407, **2018**, doi: 10.1021/ACSCATAL.8B02075.
21. Ibhadon, A.O. and Fitzpatrick, P., Heterogeneous Photocatalysis: Recent Advances and Applications. *Catalysts*, *3*, 1, 189–218, **2013**, doi: 10.3390/CATAL3010189.
22. Tachikawa, T., Fujitsuka, M., Majima, T., Mechanistic insight into the TiO2 photocatalytic reactions: Design of new photocatalysts. *J. Phys. Chem. C*, *111*, 14, 5259–5275, **2007**, doi: 10.1021/JP069005U.
23. Free Full-Text | Heterogeneous Photocatalysis A Promising Advanced Oxidation Process. *Catalysts-Mdpi,* 3, 1, 189–218, 2013, https://www.mdpi.com/2073-4344/3/1/189. Accessed 23.08.09.
24. Fujishima, A., Rao, T.N., Tryk, D.A., Titanium dioxide photocatalysis. *J. Photochem. Photobiol. C: Photochem. Rev.*, *1*, 1, 1–21, **2000**, doi: 10.1016/S1389-5567(00)00002-2.
25. Zhang, J. and Nosaka, Y., Mechanism of the OH radical generation in photocatalysis with TiO 2 of different crystalline types. *J. Phys. Chem. C*, *118*, 20, 10824–10832, **2014**, doi: 10.1021/JP501214M.
26. Kamat, P.V., Manipulation of charge transfer across semiconductor interface. A criterion that cannot be ignored in photocatalyst design. *J. Phys. Chem. Lett.*, *3*, 5, 663–672, **2012**, doi: 10.1021/JZ201629P.

27. Kamat, P.V. and Bisquert, J., Solar fuels. Photocatalytic hydrogen generation. *J. Phys. Chem. C*, *117*, 29, 14873–14875, **2013**, doi: 10.1021/JP406523W.
28. Wang, F., Jiang, Y., Lawes, D.J., Ball, G.E., Zhou, C., Liu, Z. *et al.*, Analysis of the Promoted Activity and Molecular Mechanism of Hydrogen Production over Fine Au-Pt Alloyed TiO2 Photocatalysts. *ACS Catal.*, *5*, 7, 3924–3931, **2015**, doi: 10.1021/ACSCATAL.5B00623.
29. Wei, Y., Han, S., Walker, D.A., Warren, S.C., Grzybowski, B.A., Enhanced photocatalytic activity of hybrid Fe2O3-Pd nanoparticulate catalysts. *Chem. Sci.*, *3*, 4, 1090–1094, **2012**, doi: 10.1039/C2SC00673A.
30. Ibhadon, A.O., Greenway, G.M., Yue, Y., Photocatalytic activity of surface modified TiO2/RuO2/SiO2 nanoparticles for azo-dye degradation. *Catal. Commun.*, *9*, 1, 153–157, **2008**, doi: 10.1016/J.CATCOM.2007.05.038.
31. Yu, J.C., Yu, J., Ho, W., Jiang, Z., Zhang, L., Effects of F- doping on the photocatalytic activity and microstructures of nanocrystalline TiO2 powders. *Chem. Mater.*, *14*, 9, 3808–3816, **2002**, doi: 10.1021/CM020027C.
32. Singh, S.K., Mishra, P.K., Upadhyay, S.N., Recent developments in photocatalytic degradation of insecticides and pesticides. *Rev. Chem. Eng.*, *39*, 2, 225–270, **2023**, doi: 10.1515/REVCE-2020-0074.
33. Bu, Y., Wang, K., Yang, X., Nie, G., Photoelectrochemical sensor for detection Hg2+ based on *in situ* generated MOFs-like structures. *Anal. Chim. Acta*, Elsevier, 1233, 340496, **2022**, doi: 10.1016/J.ACA.2022.340496.
34. Khan, M.A.M., Nain, P., Ahmed, J., Ahamed, M., Kumar, S., Characterization and photocatalytic performance of hydrothermally synthesized Cu-doped TiO2 NPs. *Opt. Mater.*, *133*, **2022**, doi: 10.1016/J.OPTMAT.2022.112983.
35. Maldonado Pérez, A.X. and Pérez Bueno, J.D.J., Solar lighting systems applied in photocatalysis to treat pollutants - A review. *Rev. Adv. Mater. Sci.*, *62*, 1, **2023**, doi: 10.1515/RAMS-2022-0293.
36. Tada, H., Mitsui, T., Kiyonaga, T., Akita, T., Tanaka, K., All-solid-state Z-scheme in CdS-Au-TiO2 three-component nanojunction system. *Nat. Mater.*, *5*, 10, 782–786, **2006**, doi: 10.1038/NMAT1734.
37. Strach, A., Dulski, M., Wasilkowski, D., Metryka, O., Nowak, A., Matus, K. *et al.*, Microwave Irradiation vs. Structural, Physicochemical, and Biological Features of Porous Environmentally Active Silver–Silica Nanocomposites. *Int. J. Mol. Sci.*, *24*, 7, **2023**, doi: 10.3390/IJMS24076632.
38. Wang, F., Wong, R.J., Ho, J.H., Jiang, Y., Amal, R., Sensitization of Pt/TiO2 Using Plasmonic Au Nanoparticles for Hydrogen Evolution under Visible-Light Irradiation. *ACS Appl. Mater. Interfaces*, *9*, 36, 30575–30582, **2017**, doi: 10.1021/ACSAMI.7B06265.
39. Kudo, A. and Miseki, Y., Heterogeneous photocatalyst materials for water splitting. *Chem. Soc. Rev.*, *38*, 1, 253–278, **2009**, doi: 10.1039/B800489G.
40. Kudo, A., Z-scheme photocatalyst systems for water splitting under visible light irradiation. *MRS Bull.*, *36*, 1, 32–38, **2011**, doi: 10.1557/MRS.2010.3.
41. Ryu, J. and Choi, W., Substrate-specific photocatalytic activities of TiO2 and multiactivity test for water treatment application. *Environ. Sci. Technol.*, *42*, 1, 294–300, **2008**, doi: 10.1021/ES071470X.
42. Domínguez-Espíndola, R.B., Arias, D.M., Rodríguez-González, C., Sebastian, P.J., A critical review on advances in TiO2-based photocatalytic systems for CO2 reduction. *Appl. Therm. Eng.*, *216*, **2022**, doi: 10.1016/J.APPLTHERMALENG.2022.119009.
43. Yan, S.C., Ouyang, S.X., Gao, J., Yang, M., Feng, J.Y., Fan, X.X. *et al.*, A Room-Temperature Reactive-Template Route to Mesoporous ZnGa2O4 with Improved Photocatalytic Activity in Reduction of CO2. *Angew. Chem.*, *122*, 36, 6544–6548, **2010**, doi: 10.1002/ANGE.201003270.
44. Rajeshwar, K., Osugi, M.E., Chanmanee, W., Chenthamarakshan, C.R., Zanoni, M.V.B., Kajitvichyanukul, P. *et al.*, Heterogeneous photocatalytic treatment of organic dyes in air and

aqueous media. *J. Photochem. Photobiol. C: Photochem. Rev.*, *9*, 4, 171–192, **2008**, doi: 10.1016/J.JPHOTOCHEMREV.2008.09.001.
45. Zhang, H., Guo, L.H., Zhao, L., Wan, B., Yang, Y., Switching oxygen reduction pathway by exfoliating graphitic carbon nitride for enhanced photocatalytic phenol degradation. *J. Phys. Chem. Lett.*, *6*, 6, 958–963, **2015**, doi: 10.1021/ACS.JPCLETT.5B00149.
46. Coleman, H.M., Chiang, K., Amal, R., Effects of Ag and Pt on photocatalytic degradation of endocrine disrupting chemicals in water. *Chem. Eng. J.*, *113*, 1, 65–72, **2005**, doi: 10.1016/J.CEJ.2005.07.014.
47. Xu, N., Shi, Z., Fan, Y., Dong, J., Shi, J., Hu, M.Z.C., Effects of particle size of TiO2 on photocatalytic degradation of methylene blue in aqueous suspensions. *Ind. Eng. Chem. Res.*, *38*, 2, 373–379, **1999**, doi: 10.1021/IE980378U.
48. Arabatzis, I.M., Antonaraki, S., Stergiopoulos, T., Hiskia, A., Papaconstantinou, E., Bernard, M.C. *et al.*, Preparation, characterization and photocatalytic activity of nanocrystalline thin film TiO2 catalysts towards 3,5-dichlorophenol degradation. *J. Photochem. Photobiol. A: Chem.*, *149*, 1–3, 237–245, **2002**, doi: 10.1016/S1010-6030(01)00645-1.
49. Nguyen, M.A., Bedford, N.M., Ren, Y., Zahran, E.M., Goodin, R.C., Chagani, F.F. *et al.*, Direct Synthetic Control over the Size, Composition, and Photocatalytic Activity of Octahedral Copper Oxide Materials: Correlation between Surface Structure and Catalytic Functionality. *ACS Appl. Mater. Interfaces*, *7*, 24, 13238–13250, **2015**, doi: 10.1021/ACSAMI.5B04282.
50. Gautam, A., Sk, S., Pal, U., Recent advances in solution assisted synthesis of transition metal chalcogenides for photo-electrocatalytic hydrogen evolution. *Phys. Chem. Chem. Phys.*, *24*, 35, 20638–20673, **2022**, doi: 10.1039/D2CP02089K.
51. Gong, T., Liu, R., Zuo, Z., Che, Y., Yu, H., Song, C. *et al.*, Metabolic Engineering of Pseudomonas putida KT2440 for Complete Mineralization of Methyl Parathion and γ-Hexachlorocyclohexane. *ACS Synth. Biol.*, *5*, 5, 434–442, **2016**, doi: 10.1021/ACSSYNBIO.6B00025.
52. Denny, F., Scott, J., Chiang, K., Teoh, W.Y., Amal, R., Insight towards the role of platinum in the photocatalytic mineralisation of organic compounds. *J. Mol. Catal. A: Chem.*, *263*, 1–2, 93–102, **2007**, doi: 10.1016/J.MOLCATA.2006.08.031.
53. Teoh, W.Y., Mädler, L., Amal, R., Inter-relationship between Pt oxidation states on TiO2 and the photocatalytic mineralisation of organic matters. *J. Catal.*, *251*, 2, 271–280, **2007**, doi: 10.1016/J.JCAT.2007.08.008.
54. Shan, Z., Clayton, D., Pan, S., Archana, P.S., Gupta, A., Visible light driven photoelectrochemical properties of Ti@TiO2 nanowire electrodes sensitized with core-shell Ag@Ag2S nanoparticles. *J. Phys. Chem. B*, *118*, 49, 14037–14046, **2014**, doi: 10.1021/JP504346K.
55. Svensson, F.G. and Österlund, L., One-Step Synthesis of Sulfate-Modified Titania Nanoparticles with Surface Acidic and Sustained Photocatalytic Properties via Solid-State Thermolysis of Titanyl Sulfate. *ChemCatChem*, *14*, 20, **2022**, doi: 10.1002/CCTC.202200682.
56. Nikitenko, S.I., Chave, T., Cau, C., Brau, H.P., Flaud, V., Photothermal Hydrogen Production Using Noble-Metal-Free Ti@TiO2 Core-Shell Nanoparticles under Visible-NIR Light Irradiation. *ACS Catal.*, *5*, 8, 4790–4795, **2015**, doi: 10.1021/ACSCATAL.5B01401.
57. Sher Shah, M.S.A., Park, A.R., Zhang, K., Park, J.H., Yoo, P.J., Green synthesis of biphasic TiO 2-reduced graphene oxide nanocomposites with highly enhanced photocatalytic activity. *ACS Appl. Mater. Interfaces*, *4*, 8, 3893–3901, **2012**, doi: 10.1021/AM301287M.
58. Hu, C., Chu, K., Zhao, Y., Teoh, W.Y., Efficient photoelectrochemical water splitting over anodized p-type NiO porous films. *ACS Appl. Mater. Interfaces*, *6*, 21, 18558–18568, **2014**, doi: 10.1021/AM507138B.
59. Eder, M., Courtois, C., Petzoldt, P., Mackewicz, S., Tschurl, M., Heiz, U., Size and Coverage Effects of Ni and Pt Co-Catalysts in the Photocatalytic Hydrogen Evolution from Methanol on TiO2(110). *ACS Catal.*, *12*, 15, 9579–9588, **2022**, doi: 10.1021/ACSCATAL.2C02230.

60. Wang, F., Jiang, Y., Gautam, A., Li, Y., Amal, R., Exploring the origin of enhanced activity and reaction pathway for photocatalytic H2 production on Au/B-TiO2 catalysts. *ACS Catal.*, *4*, 5, 1451–1457, **2014**, doi: 10.1021/CS5002948.
61. Fujishima, A., Zhang, X., Tryk, D.A., TiO2 photocatalysis and related surface phenomena. *Surf. Sci. Rep.*, *63*, 12, 515–582, **2008**, doi: 10.1016/J.SURFREP.2008.10.001.
62. Ibhadon, A.O., Greenway, G.M., Yue, Y., Falaras, P., Tsoukleris, D., The photocatalytic activity of TiO2 foam and surface modified binary oxide titania nanoparticles. *J. Photochem. Photobiol. A: Chem.*, *197*, 2–3, 321–328, **2008**, doi: 10.1016/J.JPHOTOCHEM.2008.01.010.
63. Henderson, M.A., A surface science perspective on TiO2 photocatalysis. *Surf. Sci. Rep.*, *66*, 6–7, 185–297, **2011**, doi: 10.1016/J.SURFREP.2011.01.001.
64. Umebayashi, T., Yamaki, T., Itoh, H., Asai, K., Band gap narrowing of titanium dioxide by sulphur doping. *Appl. Phys. Lett.*, *81*, 3, 454–456, **2002**, doi: 10.1063/1.1493647.
65. Kontos, A.G., Pelaez, M., Likodimos, V., Vaenas, N., Dionysiou, D.D., Falaras, P., Visible light induced wetting of nanostructured N-F co-doped titania films. *Photochem. Photobiol. Sci.*, *10*, 3, 350–354, **2011**, doi: 10.1039/C0PP00159G/.
66. Murase, T., Irie, H., Hashimoto, K., Visible light sensitive photocatalysts, nitrogen-doped Ta2O 5 powders. *J. Phys. Chem. B*, *108*, 40, 15803–15807, **2004**, doi: 10.1021/JP047874I.
67. Asahi, R., Morikawa, T., Ohwaki, T., Aoki, K., Taga, Y., Visible-light photocatalysis in nitrogen-doped titanium oxides. *Science*, *293*, 5528, 269–271, **2001**, doi: 10.1126/SCIENCE.1061051.
68. Cai, T., Liu, Y., Wang, L., Dong, W., Zeng, G., Recent advances in round-the-clock photocatalytic system: Mechanisms, characterization techniques and applications. *J. Photochem. Photobiol. C: Photochem. Rev.*, *39*, 58–75, **2019**, doi: 10.1016/J.JPHOTOCHEMREV.2019.03.002.
69. Mu, C., Lv, C., Meng, X., Sun, J., Tong, Z., Huang, K., *In Situ* Characterization Techniques Applied in Photocatalysis: A Review. *Adv. Mater. Interfaces*, *10*, 3, 2201842, **2023**, doi: 10.1002/ADMI.202201842.
70. Gao, Y., Nie, W., Wang, X., Fan, F., Li, C., Advanced space- and time-resolved techniques for photocatalyst studies. *Chem. Commun.*, *56*, 7, 1007–1021, **2020**, doi: 10.1039/C9CC07128H.
71. Loddo, V., Bellardita, M., Camera-Roda, G., Parrino, F., Palmisano, L., Chapter- 1 Heterogeneous Photocatalysis A Promising Advanced Oxidation Process, in: *Current Trends and Future Developments on (Bio-) Membranes*, pp. 1–43, Elsevier, 2018, doi: 10.1016/B978-0-12-813549-5.00001-3.
72. Tahir, M.B., Iqbal, T., Rafique, M., Rafique, M.S., Nawaz, T., Sagir, M., Nanomaterials for photocatalysis, in: *Nanotechnology and Photocatalysis for Environmental Applications*, pp. 65–76, Elsevier, 2020, doi: 10.1016/b978-0-12-821192-2.00005-x.
73. Thakur, N., Singh, S.B., Anshuman, Use of Photocatalyst in Self-Cleaning Constructions Material: A Review, pp. 117–132, **2022**, doi: 10.1007/978-981-16-9744-9_8.
74. Priyanga, G.S., Pransu, G., Krishna, H., Thomas, T., Discovery of Novel Photocatalysts Using Machine Learning Approach, in: *Machine Learning for Advanced Functional Materials*, pp. 233–261, **2023**.
75. Yu, H., Zhu, J., Qiao, R., Zhao, N., Zhao, M., Kong, L., Facile Preparation and Controllable Absorption of a Composite Based on PMo12/Ag Nanoparticles: Photodegradation Activity and Mechanism". *ChemistrySelect*, *7*, 2, **2022**, doi: 10.1002/SLCT.202103668.
76. Zhang, T., Lei, W., Liu, P., Rodriguez, J.A., Yu, J., Qi, Y. *et al.*, Organic Pollutant Photodecomposition by Ag/KNbO3 Nanocomposites: A Combined Experimental and Theoretical Study. *J. Phys. Chem. C*, *120*, 5, 2777–2786, **2016**, doi: 10.1021/ACS.JPCC.5B11297.
77. Zhang, H., Guo, L.H., Wang, D., Zhao, L., Wan, B., Light-induced efficient molecular oxygen activation on a Cu(II)-grafted TiO2/graphene photocatalyst for phenol degradation. *ACS Appl. Mater. Interfaces*, *7*, 3, 1816–1823, **2015**, doi: 10.1021/AM507483Q.

78. Wang, F., Ho, J.H., Jiang, Y., Amal, R., Tuning Phase Composition of TiO2 by Sn4+ Doping for Efficient Photocatalytic Hydrogen Generation. *ACS Appl. Mater. Interfaces*, *7*, 43, 23941–23948, **2015**, doi: 10.1021/ACSAMI.5B06287.
79. Schmitt-Kopplin, P., Hertkorn, N., Schulten, H.R., Kettrup, A., Structural changes in a dissolved soil humic acid during photochemical degradation processes under O2 and N2 atmosphere. *Environ. Sci. Technol.*, *32*, 17, 2531–2541, **1998**, doi: 10.1021/ES970636Z.
80. Mills, A., Davies, R.H., Worsley, D., Water purification by semiconductor photocatalysis. *Chem. Soc. Rev.*, *22*, 6, 417–425, **1993**, doi: 10.1039/CS9932200417.

8

Role of Biocatalysis-Biotransformations in Sustainable Chemistry

Devashish Singh, Surya Prakash Verma and Poonam Rajesh Prasad*

Department of Chemistry, Institute of Science, Banaras Hindu University, Varanasi, India

Abstract

Based on the principles and metrics of green chemistry and sustainable development, biocatalysis is both a green and sustainable technology. This is largely a result of the spectacular advances in molecular biology and biotechnology achieved in the past two decades. Protein engineering has enabled the optimization of existing enzymes and the invention of entirely new biocatalytic reactions that were previously unknown in nature. It is now eminently feasible to develop enzymatic transformations to fit predefined parameters, resulting in processes that are truly sustainable by design. This approach has successfully been applied, for example, in the industrial synthesis of active pharmaceutical ingredients. In addition to the use of protein engineering, other aspects of biocatalysis engineering, such as substrate, medium, and reactor engineering, can be utilized to improve the efficiency and cost-effectiveness and, hence, the sustainability of biocatalytic reactions. Furthermore, immobilization of an enzyme can improve its stability and enable its reuse multiple times, resulting in better performance and commercial viability. Consequently, biocatalysis is being widely applied in the production of pharmaceuticals and some commodity chemicals. Moreover, its broader application will be further stimulated in the future by the emerging biobased economy. Chemical synthesis, which primarily uses fossil resources including solvents, organic molecules, inorganic catalysts, and inert gases, has been widely used in the manufacturing of chemicals. Several of these chemical reactions take place in harsh environments and produce wasteful by-products that are undesired. Green chemistry, sometimes referred to as sustainable chemistry, is a viable substitute for conventional chemical synthesis that does not have a detrimental effect on the environment. This method is based on biocatalysis, in which cells or enzymes are utilized for a more environmentally friendly conversion of chemicals into the required products (*in vitro* biocatalysis or whole-cell biocatalysis). The high catalytic activity of the enzymes and their chemo-, regio-, and stereospecificity for substrates and catalytic processes are utilized in biocatalysis. Since enzymes function, they also comply with green chemistry principles. Biocatalysis is the use of organic elements, such as cells or enzymes, to catalyze chemical processes, and this transformation of one chemical to another form is biotransformation. It is a green and sustainable technology that is based on the ideas and standards of green chemistry and sustainable development. An enzyme can be immobilized to increase its stability and allow for many uses, leading to higher performance and increased economic viability.

***Keywords*:** Biocatalysis, biotransformations, sustainable chemistry, biobased economy, chiral pharmaceutical intermediate, KREDs, green chemistry, bioeconomy

**Corresponding author*: poonamrp.chem@bhu.ac.in; poonam11a86@gmail.com

Mousumi Sen (ed.) Sustainable Green Catalytic Processes, (165–182) © 2024 Scrivener Publishing LLC

8.1 Introduction

Utilizing natural materials, such as biologically sourced enzymes or whole cells, to expedite chemical processes is referred to as "biocatalysis." With broad implementation and more sophisticated application, biocatalysis has witnessed enormous growth in the pharmaceutical industry over the past few decades. The formation of alcohol from fermentation and the making of cheese by the way milk proteins break down are only two examples of the many events that enzymes play a crucial part in catalyzing. The understanding of the structure and functional functions of enzymes has improved recently as a result of improvements in scientific study, which have also increased the stability, activity, sustainability, and substrate specificity of enzymes. Currently, several pharmaceutical, chemical, culinary, and agro-based sectors have used various types of biocatalytic processes. Reaction kinetics and stability are crucial considerations for both single-step and multistep reactions in biocatalytic systems, since they are comparable to conventional processes in many aspects. Identification of the target reaction, the finding of the biocatalyst, its characterization, engineering, and process modeling are the first steps in the process of biocatalysis. Future biocatalysis will have a precise impact on our ability to use enzymes to catalyze chemical reactions in industrial processes, such as the production of drugs, flavors, fragrances, electronic chemicals, and polymers—chemicals that directly affect almost every aspect of our lives. Greener, cheaper, and more sustainable chemical manufacturing is made possible by using biocatalysis as a standard method. without being consumed during the reaction Typically, catalysts are used to speed up transformation processes without depleting them during the reaction. The application of a catalyst to speed up a transformation reaction is known as a catalysis reaction in such a process[1] of biocatalysis in water-free ILs, suspensions of Candida antarctica lipase B (CaLB) in anhydrous 1-butyl-3-methylimidazolium hexafluorophosphate ([bmim][PF6]) and 1-butyl-3-methylimidazolium tetrafluoroborate ([bmim][BF4]) were shown to catalyze transesterifications. Chemocatalysis and biocatalysis are two subcategories of it. Chemocatalysis can be an organic reaction that takes place in challenging circumstances, uses little Additionally, it produces poisonous and dangerous waste streams, all of which have a negative effect on the environment.[2] Chemical synthesis has a number of distinct difficulties and boundaries, including a requirement for organic solvents, a number of step reactions, the need for transition metal cofactor protection/deprotection loops, and hazardous reaction products.[3] By bypassing the thermodynamic obstacles through microbial cascade biotransformations, the increasing complexity of industrial pipelines could be avoided. This viewpoint holds that the innovative application in the area of catalysis is the possibility for microbial biotransformation and the significance of bio-catalysis Enantio- and regioselectivity of known as compounds can be repurposed using biocatalysis to increase new reactivity. To reduce the waste of industrial nonselective hazardous chemicals, this might be accomplished using complete intact organisms or isolated enzymes. In addition, the pharmaceutical industry has observed enormous growth in the adoption and application of biocatalysis over the past 10 years due to its use becoming more widespread and sophisticated. As an enabling synthetic technology, enzyme catalysis provides a number of significant advantages, comprising the propensity to extraordinarily enhance chemo-, regio-, and enantioselectivity in chemical interactions[4]. The ability to evolve enzymes is an essential component in the enormous expansion of biocatalytic implementation in the pharmaceutical industry.[5] This is essential to increasing the range of wild-type enzyme activity and capacity for the challenging procedure environments required to make syntheses economically and environmentally feasible, as well as for expanding their range toward non-natural substrates.[6] Additionally, as a result of significant industrial–university cooperation, scholarly advancements in the field, and internal enzyme discovery initiatives, the range of the biocatalysis tool kit keeps expanding.[7] Biocatalytic reactions can be made to be truly green processes by having high utilization factors for biocatalysts (high substrate:enzyme ratios), which can be achieved by lowering process mass intensity (PMI)[8]. A green and sustainable technology called biocatalysis is based on the tenets and benchmarks for sustainable development and green chemistry. This is primarily due to the amazing progress made in molecular biology and biotechnology during the past 20 years.[9] Bioengineering of proteins has made it possible to improve already-existing enzymes and create brand-new biocatalytic processes that were previously unheard of in nature. The development of enzymatic transformations to meet predetermined criteria is now eminently achievable, leading to the development of processes that are really sustainable by design.[10] In addition to using protein engineering, biocatalysis engineering also makes use of engineering of the substrate, medium, and reactor to increase the efficacy and cost-efficiency of biocatalytic processes and, as a result, their sustainability. An enzyme's stability and capacity to be reused repeatedly might both be improved by immobilization, leading to enhanced performance and increased economic viability. As a result, biocatalysis is widely used in the manufacturing of medications as well as certain other commodity chemicals. Furthermore, the expanding biobased economy will encourage its wider adoption in the future. Chemical synthesis, which primarily uses fossil resources including solvents, organic molecules, inorganic catalysts, and inert gases, has been widely applied in the manufacturing of chemicals. Several of these chemical reactions are carried out in harsh environments and produce wasteful by-products that are undesired. However, green chemistry, referred to as sustainable chemistry, is a viable substitute for conventional chemical synthesis that does not have a detrimental effect on the environment. This technique is based on biocatalysis, in which cells or enzymes are utilized for a more environmentally friendly conversion of chemicals into the required products (*in vitro* biocatalysis or whole-cell biocatalysis). The high catalytic activity of the enzymes and their chemo-, regio-, and stereospecificity for substrates and catalytic processes are utilized in biocatalysis. Since enzymes function, they also comply with green chemistry principles. Using biocatalysis involves organic elements, such as cells or enzymes, to catalyze chemical processes, and this transformation of one chemical to another form is biotransformation. It is a sustainable and green technology that is according to the ideas and standards of development that is sustainable and green chemical. An enzyme can be immobilized to increase its stability and allow for many uses, leading to higher performance and increased economic viability. Mostly, enzymes are biocompatible, biodegradable, and nonhazardous. Given that they are produced from inexpensive resources that are renewable and readily available. Combining numerous enzyme-catalyzed levels in a single pot reduces the time for required to complete the reaction, resource-intensive, and expensive separation techniques. This is one of the more modern biocatalytic capabilities that has been put into practice on a large scale.[11–13] As a result, the process mass intensity (PMI) of synthetic pathways can be greatly decreased by these cascade processes.[8] When time and effort are invested in focused process chemistry optimization for scale-up and enzyme engineering methodologies, biocatalysis has a long history of having an impact on modern development.[14] However, as the range of applications for biocatalysis expands, so does the impact of this technology on earlier discovery areas, which are characterized by large numbers of trial compounds, fast growing cycles, and high abrasion.[15]

Table 8.1 Difference between catalysis and biocatalysis.[16, 17]

Catalysis	Biocatalysis
1. Catalysis is nonbiodegradable and nonrenewable. 2. It produces products that are toxic and environment-polluting in nature. 3. It increases protection and deprotection step, cost incurring in nature. 4. It requires organic solvent. 5. Resources of catalysis are undersupply.	1. Biocatalysis is eco-friendly and sustainable.[16] 2. It has developed into a technique that is appealing to industry into mainstream enantioselective synthesis organic synthesis.[17] 3. Biocatalysis requires water as a solvent, environmentally benign. 4. It is economical and renewable. 5. Reduced production time. 6. Toxicity level is low in comparison to catalysis.

8.2 Principle of Biocatalysis

According to Anastas and Horvath, 12 well-known principles of green chemistry serve as the foundation for these new scientific philosophies and can conceptually be summed up as follows:

1. Instead of remediation, prioritize waste prevention
2. Atomic effectiveness
3. Less detrimental/toxic compounds
4. Designing safer products
5. Auxiliaries and harmless solvents
6. Design for energy efficiency
7. Utilize renewable sources of raw materials and solvents
8. Reduced steps of protection and deprotection during synthesis
9. Instead of thermal processes, biocatalytic or catalytic processes
10. Design the products to degrade without difficulties with pollution
11. Effective analytical techniques for preventing and detecting pollution
12. Naturally safer techniques

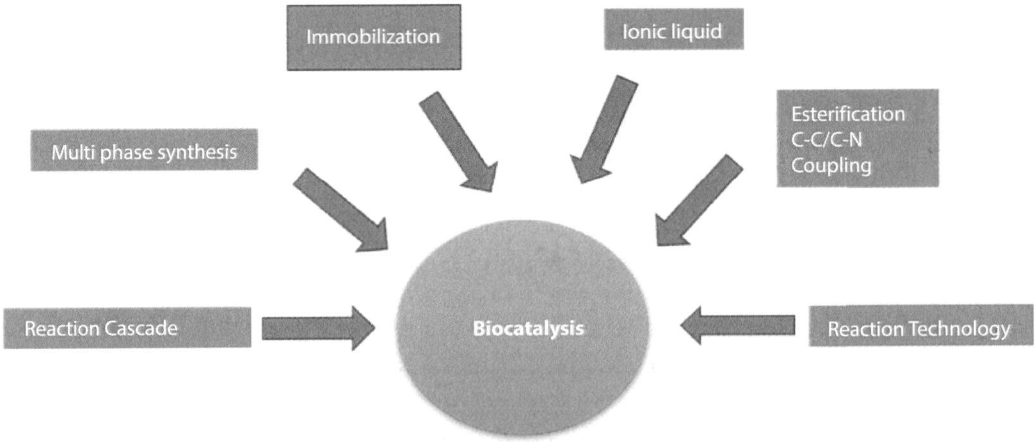

8.3 Recent Development in Biocatalysis

Due to its unmatched selectivity and efficiency, biocatalysis is a highly prized enabling technique under pharmaceutical research and growth. It could be open up single synthetic pathways to complicated chiral motifs. With broader adoption and more sophisticated application, biocatalysis has witnessed enormous growth in the pharmaceutical industry during the last one decade.[18]

Profens (2-aryl-propionic acids) are among the most significant nonsteroidal anti-inflammatory medications (NSAIDs) used to treat inflammation produced by tissue injury. Examples include ibuprofen, naproxen, ketoprofen, flurbiprofen, and ketoprofen. Although the ability of profens' (S)-enantiomer to inhibit cyclooxygenase (COX) is primarily responsible for their anti-inflammatory action, these medications are also sold as racemates. icrobial lipases, particularly those produced by fungi like *Candida antarctica* and *Candida rugosa*, in addition to fungi like *Mucor javanicus* and *Rhizomucor miehei*, can resolve profen racemates in an enantioselective manner. For the purpose to achieve optically pure (S)-acids, racemate resolution can be carried out using one of two enzymatic methods (Scheme 8.1), either an enantioselective esterification of 2-aryl-propionic acid racemates in organic media catalyzed by microbial lipases that favor microorganism lipases catalyzes the (R)-isomer or an enantioselective hydrolysis of 2-aryl-propionate ester racemates in aqueous environments that prefers the (S)-isomer. Both methods have been effectively used in the productive synthesis of (S)-ibuprofen, (S)-naproxen, (S)-ketoprofen, and (S)-flurbiprofen.

At GSK, different biocatalytic methods for chiral 3-substituted cyclohexanones were enabled so that many enzymatic actions could be applied to access the essential components of medicinal chemistry. To accept the (R)- or (S)-, 3-nitrile or methyl-3-carboxylate molecules, enantio-complementary options were generated: (R)-2, (S)-2, (R)-4, and (S)-4. Ene-reductases (EREDs) were used for these one-step reactions to reduce the cyclohexanone derivatives 1 and 3 (Scheme 8.2) asymmetrically, while lipases or nitrilases were applied to resolve the equivalent nitrile rac-4 or racemic ester rac-2 precursors (Scheme 8.3). A reaction pH of 7 was crucial for the ERED-catalyzed synthesis of (S)-2 in order to reduce the

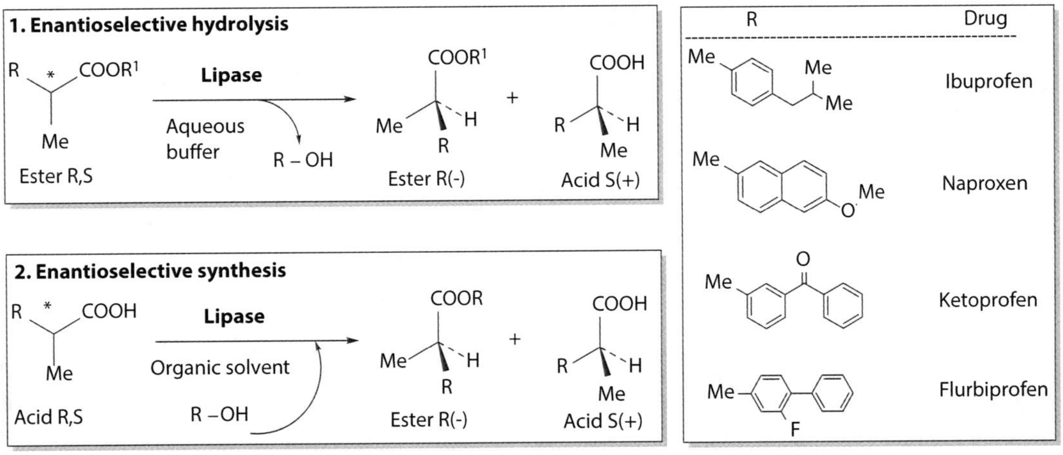

Scheme 8.1 Resolution of profens catalyzed by lipase.[19]

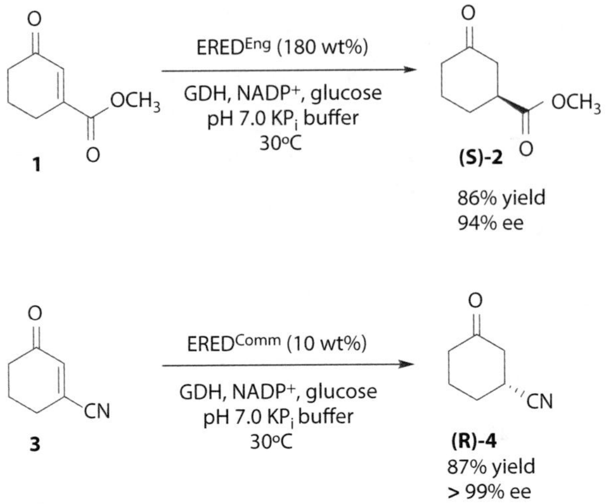

Scheme 8.2 Chiral 3-substituted cyclohexanone building synthesis through ERED catalysis.[20]

development at pH 7.5, a disproportionation by-product was seen. Further pH control was required due to the employment of the glucose cofactor recycling mechanism of GDH in order to neutralize the equivalent of gluconic acid that was produced throughout the reaction. The nitrile analog was the starting point for the development of a comparable procedure. To provide material for medicinal chemistry studies, each of these methods was effectively scaled up to multigram quantities (Scheme 8.2). Ene-reductases were used by Merck & Co. researchers to create -chiral carboxylic acids for the creation of pharmaceuticals (Scheme 8.3). In this study, a Prozomix ERED that is readily accessible on the market was found to have outstanding enantioselectivity when reducing a group of aliphatic and aromatic acrylic acids. For NADPH recycling, a phosphite dehydrogenase was utilized, which was advantageous because it eliminated the requirement for pH regulation and monitoring,

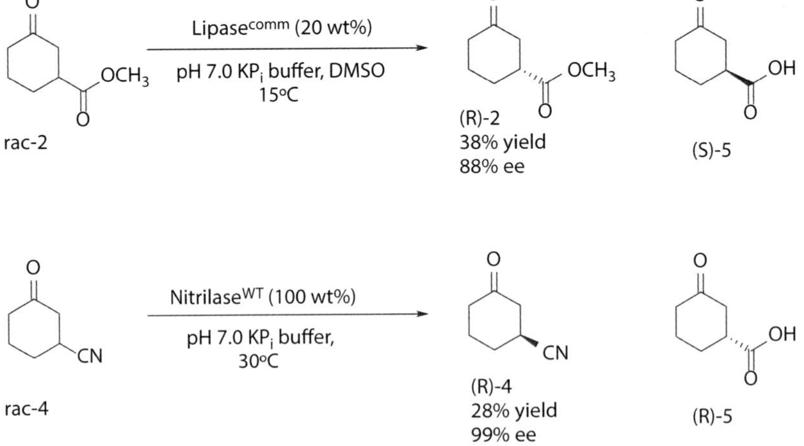

Scheme 8.3 Biocatalytic techniques using lipase and nitrilase to synthesize chiral 3-substituted cyclohexanone building.[21]

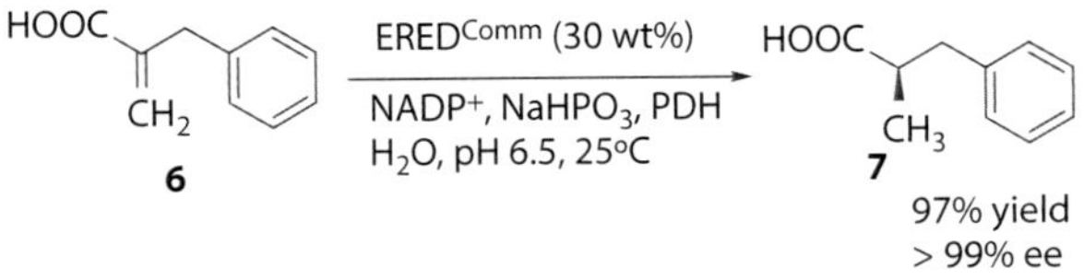

Scheme 8.4 Acrylic acid (6) asymmetrically reduced with ERED to produce a chiral acid product (7).[22]

in contrast to more conventional gluconic acid is a by-product of GDH/glucose recycling. The characterization for a number of acrylic acid substrates, ERED to carry this transition is an addition to the instrument of biocatalysts because unusual ERED action on unsaturated acids. With substrate **6**, the new method was tested on a gram scale to provide excellent yields of enantiopure **7** (**Scheme 8.4**). GSK researchers additionally created hydrolase resolutions to the enantio-complementary derivatives (R)-2 and (S)-4 (Scheme 8.3) and ERED-catalyzed methods to the fundamental elements of medicinal chemistry **(S)-2** and **(R)-4** (*vide supra*, Scheme 8.2). From screening, a wild-type nitrilase and a commercial lipase were chosen to transfer multigram quantities of each item quickly. The absence of great selectivity in any one enzyme class that can deliver enantiomers of every building block in high purity evolved to the several biocatalytic approaches to these molecules.

For the manufacture of a chiral amine functional group of compounds **9** applicable in medicinal chemistry preparation, the Pfizer team used a commercial Codexis transaminase (Scheme 8.5).

Ipatasertib, an Akt inhibitor, was created by Genentech and Roche using a KRED approach (Scheme 8.6). The extremely diastereoselective reduction required to access intermediate compound **11** and reproduce the cofactor NADPH as a terminal reductant from i-PrOH. could be carried out by a commercially available enzyme from Codexis. This method had previously used a mechanism for recycling GDH and glucose cofactors, but just one enzyme system made it more straightforward to work. The method was applied as a slurry-to-slurry reaction at high substrate loading to overwhelm some reported substrate deprivation and was demonstrated to produce high yield. Due to the strong diastereoselectivity of the enzymatic process and the difficulties in eliminating residual Ru, this method was selected over an alternate Ru-catalyzed asymmetric transfer hydrogenation procedure.

An ADH procedure was applied to obtain a chiral hydroxy ester that is necessary to produce a gamma secretase inhibitor, which is required. Ac-LEcta ADH enzyme is selected for the huge-scale enablement of this transformation after Pfizer researchers examined 190 commercially available enzymes for it. The application of i-PrOH as a terminal reductant for these procedures was considered because of how straightforward it was to work, but

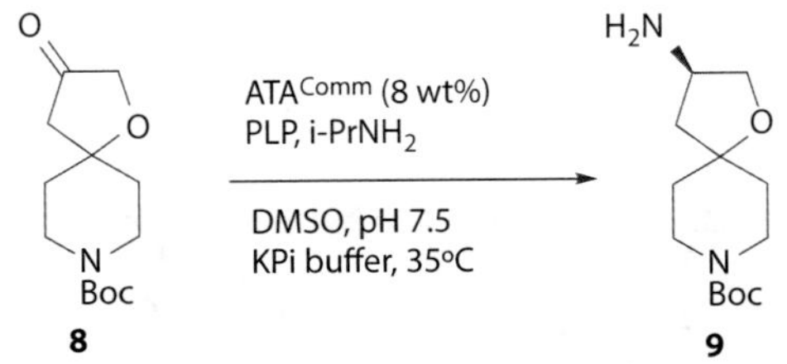

Scheme 8.5 Asymmetric synthesis of the chiral pharmaceutical intermediate using transaminase catalysis.[23]

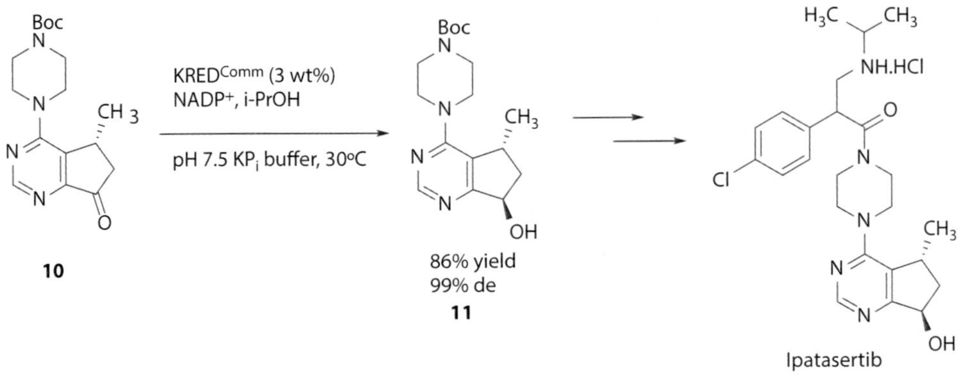

Scheme 8.6 The Akt kinase inhibitor ipatasertib is synthesized using an asymmetric KRED process to access a chiral alcohol intermediate (11).[24]

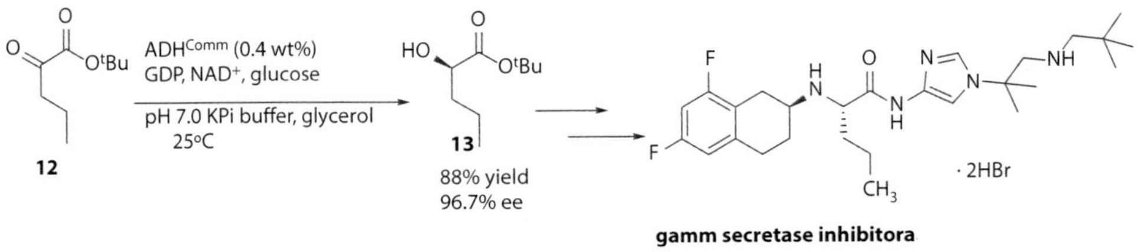

Scheme 8.7 Obtaining a hydroxy ester intermediate (13) via the asymmetric KRED process to produce a gamma secretase inhibitor.[25]

because the method was hampered by enzyme instability at high i-PrOH concentrations and the emergence of an unidentified contaminant. instead carried out using a GDH/glucose Co-factor recycling system where base titration controls pH is used. To produce the desired product **13** in a good yield and high enantiomeric excess, a high amount of substrate and a small amount of enzyme were used to conduct the reaction (Scheme 8.7).

Researchers at Novartis explained an enzyme discovery project to find out a novel proline hydroxylase (PHs) to access hydroxylated proline molecules as part of attempts creating synthetic approaches to essential pharmacological the foundations for studying medicinal chemistry. These enzymes were recognized by applying genomic mining to a multi-milligram the preparative synthesis scale for various enantiomers and regioisomers of hydroxyproline. They were then separated as the Fmoc derivatives and were chromatographically purified (Scheme 8.8).

Asymmetric sulfoxidation was developed by AstraZeneca *et al.* to generate a chiral sulfoxide compound of the kinase inhibitor. Compound **21** was accomplished *via uncontrolled m-CPBA oxidation* of sulfide **20** chromatography, too. An enzyme activity for the sulfoxidation was observed after screening a variety of commercially accessible Codexis Baeyer–Villiger monooxygenases Baeyer–Villiger monooxygenases (BVMOs) using KRED with i-PrOH for recycling NADPH. When specific process chemistry optimization was applied to the best BVMO result from screening, it was discovered that the mass transport of oxygen was crucial to the increased scale-up achievement of the reaction with outstanding diastereoselectivity and good yield (Scheme 8.9).

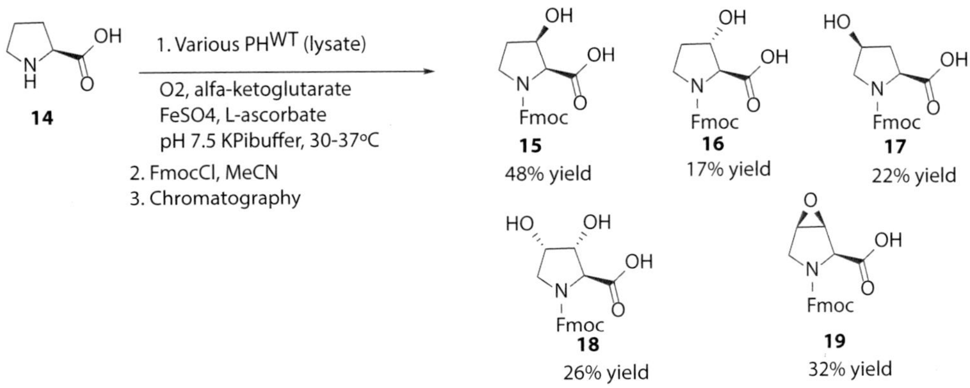

Scheme 8.8 Asymmetric oxidations catalyzed by proline hydroxylase to access various hydroxylated proline motifs for medicinal chemistry applications.[26]

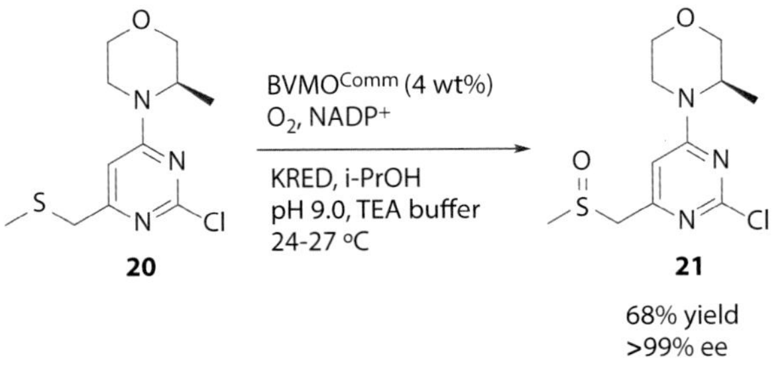

Scheme 8.9 Asymmetric sulfoxidation catalyzed by BVMO for obtaining access to a derivative (21) of the experimental kinase inhibitor.[27]

The Pfizer team invented a one-pot sequential cascade for the preparation of disubstituted chiral tetrahydropyran **24** (Scheme 8.10). This procedure substituted a five-step synthesis that includes distinct procedures for lipase resolution and transaminase in the early medicinal chemistry phase. A two-enzyme solution was created, since it was not possible to find a transaminase that could carry out the direct examination and resolution of the

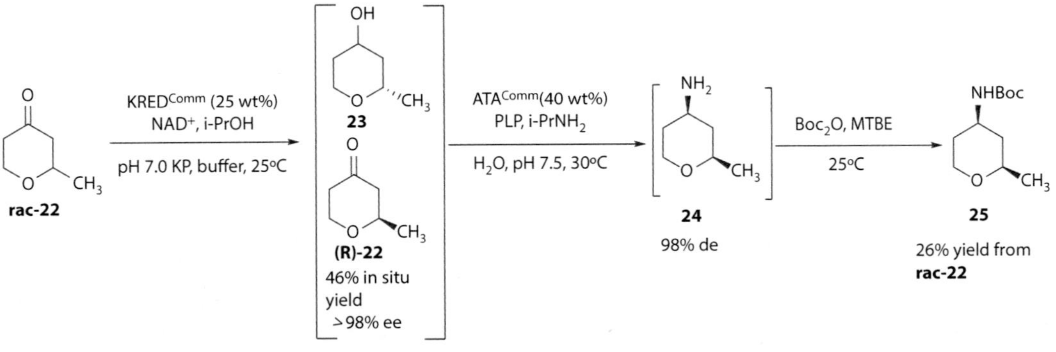

Scheme 8.10 Chiral pharmaceutical intermediate synthesis via KRED-ATA cascade.[28]

ketone Rac-22 at the same time. Rac-22 with a KRED was the starting material, which was resolution of the racemic ketone is initial stage of the cascade. As long as the necessary stereo distinction of the remote chiral methyl center was accomplished, the KRED's absolute selectivity for setting the chiral alcohol stereo-center is not crucial. After an acidification-pH adjustment, the reaction mixture was then immediately telescoped into the subsequent transamination process to produce amine **24,** removal of the protein catalyst from the first step and celite filtration. Once accomplished, the enzyme was once more eliminated using acidification-basification, through filtration of celite, followed by distillation of extra i-PrNH2. Lastly, isolation of the end product **25** was made possible by an *in situ* Boc protection. The choice in commercially accessible (c-LEcta and Codexis) enzymes for this procedure was made following a screening of KREDs and transaminases at home and from vendors at the gram scale and during initial development.

To make a racemic mixture into an enantiopure alcohol (R)-26, chemists at Merck and Co. developed a deracemization procedure (Scheme 8.11). Two designed KREDs with distinct preferred cofactors were used in this cascade to work orthogonally and enable the transformation with major conversion and enantiomeric excess. The initial KRED, this was fueled by the NADH oxidase (NOx) turnover of NAD, specifically oxidized the alcohol's (S)-enantiomer to the corresponding ketone **27**. The required (R)-enantiomer was subsequently obtained by selectively reducing the ketone using a second KRED, and NADPH was recycled using a phosphite dehydrogenase (PDH)/phosphite system. These enzymatic processes were successfully operated in parallel by applying two enzymes with orthogonal cofactor preferences.

According to Codexis in partnership with Merck, a two-enzyme cascade from L-lysine **28** to the synthesis of (2S,5S)-hydroxy-pipecolic acid 30 was described. A lysine cyclodeaminase

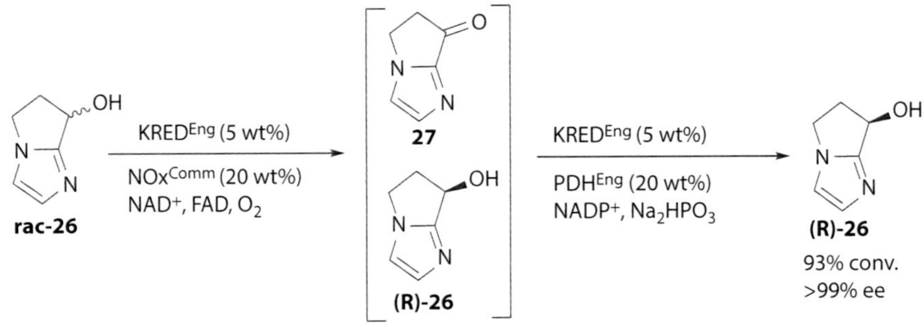

Scheme 8.11 Racemic alcohol rac-26 cascade deracemization by two engineered KREDs operating with orthogonal.[29]

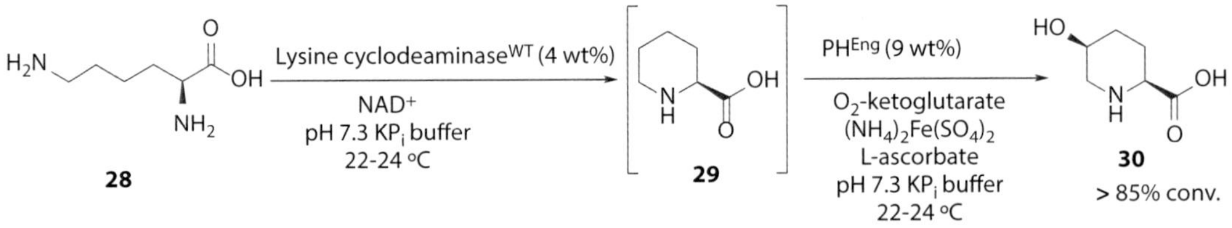

Scheme 8.12 Using a wild-type lysine cyclodeaminase and an engineered proline hydroxylase, cascade synthesis of (2S,5S)-hydroxy-pipecolic acid 30 from L-lysine 28.[30]

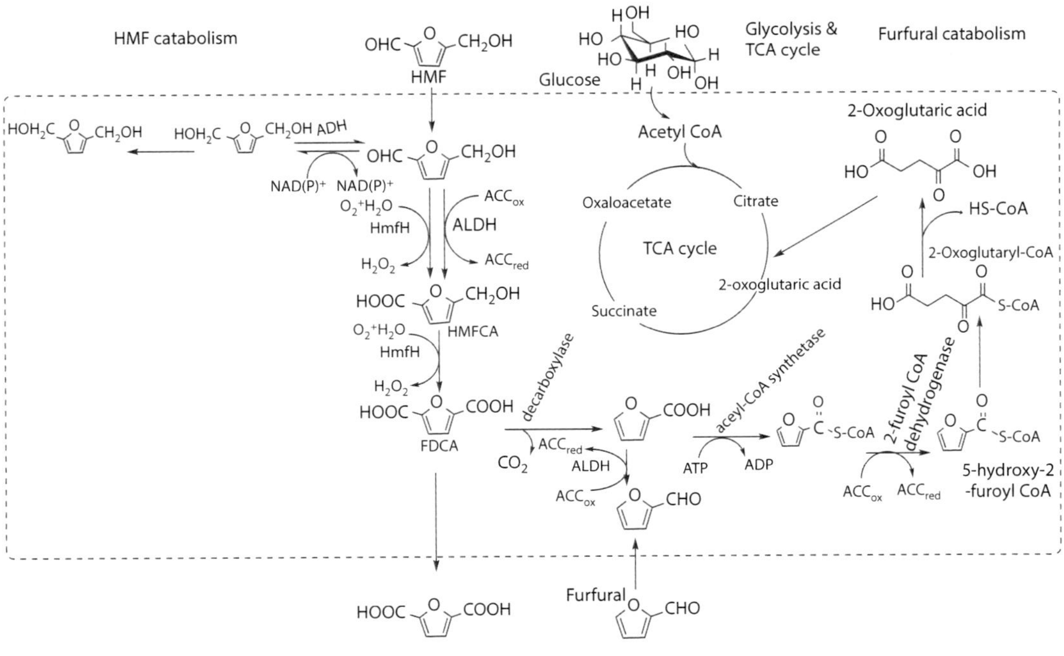

Scheme 8.13 A description of the HMF and furfural-related metabolic pathways.[31–38]

catalyst was used in the first phase of the cascade to produce the equivalent pipecolic acid **29**. Since L-lysine was the enzyme's original substrate and it was already a highly active biocatalyst, it was used in this transformation. Proline hydroxylase (PH), which is created with increased reactivity and selectivity of pipecolic acid, was used to complete the next step. The metal cofactor $(NH_4)_2Fe(SO_4)_2$ (Mohr's salt), which is required for the reaction because the proline hydroxylase (PH) was Fe^{2+}-dependent, was added. Ascorbic acid was discovered to improve the hydroxylase activity, which supports the maintenance or regeneration of the active enzyme-Fe^{2+} catalytic form. In order to provide the O_2 required for the oxidation, air was also sparged into the reaction (Scheme 8.12).

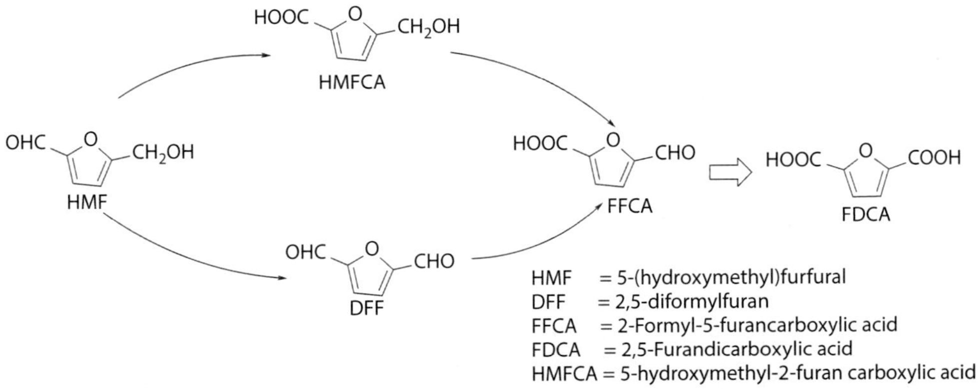

Example: HMF oxidation pathway to FDCA.
Using metabolically active whole cells, where the respiratory chain reoxidizes decreased cofactors, is thought to be the simpler approach for synthetic reactions that need cofactors and less costly choice cofactors versus stoichiometric cofactor reoxidants. Additionally, endogenous catalases found in whole cells catabolize H_2O_2, a typical inhibitory consequence of enzymatic oxidations. Whole cells also operate as a barrier against reactive substrates and products as well as the stress of aeration. Aldehydes, like HMF and furfural, In general, reactive oxygen species (ROS) and their inhibition of key enzymes in primary metabolism cause significant damage to proteins, nucleic acids, and organelles. They are also severe inhibitors of fermentation. Since these aldehydes are harmful, many species have evolved "generic" ALDHs that change them into less harmful acid/alcohol types (such as HMF alcohol/HMFCA). Aspergillus, Serratia liquefaciens, Serratia marcescens, Ureibacillus thermosphaericus, Cupriavidus basilensis, Acetobacter rancens, and Pseudomonas aeruginosa are just a few of the organisms that have been found to convert HMF to HMF acid.[39]

An investigation after that identified the genes in charge of the metabolic pathway used to break down the hazardous substances HMF and furfural (Scheme 8.13). It starts with HMF oxidation to HMFCA by either nonspecific dehydrogenases or by HMF/furfural oxidoreductase (HmfH), which is associated with the hmfFGH′H cluster in C. basilensis. The subsequent oxidation of HMFCA to FDCA requires HmfH, which it does. The HMF and furfural catabolic pathways converge at this point as FDCA is decarboxylated to produce 2-furoic acid by decarboxylases linked with hmf FG (whose disruption caused the buildup of HMF acid and FDCA when the microbe was cultured in the presence of HMF). Furfural is broken down into 2-furoic acid in the furfural catabolic pathway by the action of a dehydrogenase with an unidentified cofactor. Following that, 2-furoic acid is transformed progressively by acyl-CoA synthetase and a molybdenum-dependent 2-furoyl-CoA dehydrogenase produces 5-hydroxy-2-furoyl-2-furoyl-CoA and 2-furoyl-CoA. Finally, 5-hydroxy-2-furoyl-CoA is converted into 2-oxoglutarate and CoA by combining spontaneous keto-enol tautomerizations, delactonization, and thioester hydrolysis. The TCA cycle can include 2-oxoglutaric acid for metabolic energy and the manufacture of numerous important compounds, such as amino acids.

Several organisms, including Cupriavidus, Methylovorus, and Burkholderia, have contributed HMF catabolic genes to the industrially significant bacteria Pseudomonas putida S12 possesses an intrinsic tolerance to chemical stimuli and is active with respect to HMF (e.g., it has endogenous ALDHs that oxidize HMF to its corresponding carboxylic acid). Since glucose is transformed to gluconate and 2-ketogluconate in this case, which may interfere with pH and process control, glycerol was chosen as the carbon (C) source. The GMC oxidoreductase family, which contains FAD, includes the enzyme HmfH. Furfural, HMF, HMF acid, and furfural alcohol are all oxidized by it, proving that it can react with both alcohol and aldehyde moieties. It generates H_2O_2 by using molecular oxygen as an electron acceptor rather than NAD^+, making it a genuine oxidase. In the methylotrophic bacteria Methylovorus, the hmfH gene, which produces the HMF oxidase (HMFO) enzyme, has a homolog. One such investigation detected bacteria in soil from a biofuel pilot plant's acid pretreatment spirits disposal location. Although an analysis of the gene sequences involved in the previously described HMF catabolic pathway using BLAST only identifies a small number of bacteria, and the activity of Bacteria are not the only organisms that may oxidize HMF to FDCA. Low product recovery and the requirement for a constant source of carbon feed are problems with whole-cell biotransformation. Because filamentous fungus can utilize a wider variety of carbon sources may be preferable than using bacteria. Filamentous fungi are unique in their ability to use Almost all organisms can ferment six-carbon sugars (like glucose) as well as five-carbon sugars (like xylose and arabinose). These sugars are plentiful in appealing and affordable feedstocks like agricultural residues (such as corn stover, straw, and sugar cane bagasse) and used hydrolysates. The most difficult to decompose lignocellulosic materials, like wood, can be efficiently broken down by filamentous fungus and the enzymes they produce. Additionally, fungus can survive low pH levels (most bacteria prefer neutral pH), which is useful for infection prevention and downstream processing.

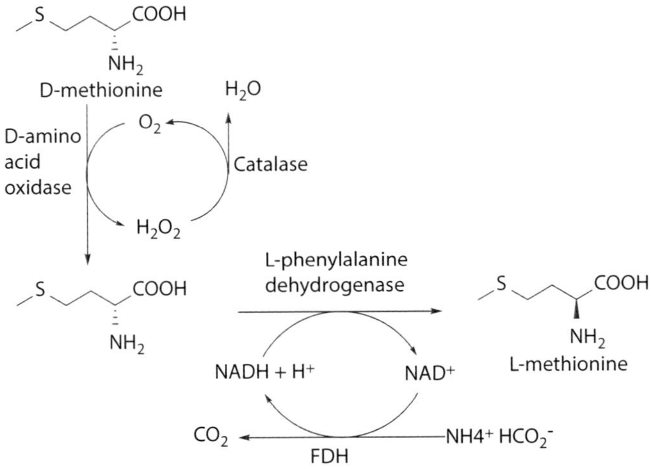

Another significant instance is the deracemization of D and L-methionine to L-methionine, which is accomplished by converting D-methionine to L-methionine using a four-enzyme cascade that includes catalase, format dehydrogenase, L-amino acid dehydrogenase, and D-amino acid oxidase.[40]

8.4 Future in Biocatalysis

Biocatalysis offers a desirable alternative in the context of worries about the environmental effects of chemical manufacturing. Each year, the US Environmental Protection Agency gives out five honors in the Challenge for the Presidential Green Chemistry Award. The nominations place a focus on the 12 Principles of Green Chemistry, which take into account worker safety, renewable feedstock use, and environmental considerations. Since 2000, biocatalysis has garnered 16 prizes, whether it uses entire cells or enzyme technology. Biocatalysts are produced using renewable resources, are biodegradable, and are nontoxic. Their high selectivities make reaction setups simpler and result in better product yields. Because they normally operate at room temperature, atmospheric pressure, and neutral pH, biocatalytic reactions are also secure. Therefore, it is not unexpected that biocatalysis receives so many honors[1].

Table 8.2 Green chemistry and biocatalysis.[1]

Green chemistry principle	Biocatalysis
• Energy efficiency	• greatly decreased waste
• Waste minimization	• greater "atom" and "step"-efficiency
• Atom-based economy	• often low-toxic
• Less dangerous synthetics	• not relevant (procedure or product)
• Creating safer products	• most usually done in water
• Safer auxiliaries and solvents	• modest environmental and energy efficiency
• Renewable sources of fuel	• Recycling enzymes is possible.
• Diminished derivatization	• avoids the protective/deprotective stages
• Catalysis	• Catalysts are enzymes
• Create a degradable design	• Applicability in biocatalytic processes
• Instantaneous analysis	• Usefulness in biocatalytic processes
• Innately safer methods	• conditions are benign and secure

In conclusion, biocatalysis is an incredibly sustainable and green technology. In addition to being advantageous for the environment, biotechnology advancements have made it relatively affordable.[41]

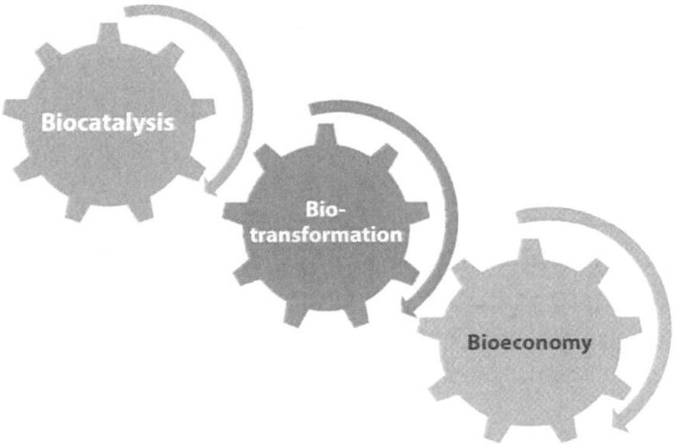

5. The discovery that organic solvents increase the thermal stability of lipases like toluene than in water, made by the 1984 study of Zaks and Klibanov, resulted in the understanding that enzymes may be widely used in chemical synthesis. In addition, the expanding demand for efficient techniques for the industrial manufacture of chiral medicines' pure enantiomers[42] created a chance for the widespread use of highly enantioselective biocatalytic techniques[43]. Continued discoveries and advancements in the fields of protein engineering and enzyme discovery have the potential to expand the range of chemicals available and the biocatalytic transformations available to synthetic chemists. For many years to come, the function of biocatalysis in the pharmaceutical business appears to be on track for an exhilarating expansion in scope and influence[18]. With fascinating possibilities for machine learning-guided evolution as well as the ability to diversify enzyme libraries using techniques like ancestral sequence reconstruction, the complexity of enzyme evolution is still increasing[44].

Improvements to high-throughput platforms like microfluidic workflows[45] and expedited screening protocols[46] have the potential to revolutionize the way that biocatalysis development is carried out because the pace of enzyme engineering is still a development constraint. In addition, intriguing developments in newly constructed proteins[47, 48] and enzymes that integrate noncanonical amino acids[49, 50] continue to be made, which could also broaden the range of biocatalysts accessible as possible ancestors for evolution. These developments go beyond the wild-type enzyme sequences found in nature. The tendency toward increasingly sophisticated multienzymatic syntheses is also noticeable as the area develops and the pharmaceutical industry's ambitions for biocatalysis grow. These computer-aided synthesis techniques can help create single- and multistep biocatalytic transformations for complex compounds and continue to promote the use of biocatalysis throughout the synthetic chemistry community. The successful fusion of flow chemistry and biocatalysis also holds considerable promise.[51] However, due to the difficulties dealing with heterogeneous reactions, which are commonly connected with biocatalysis and managing enzymes that recycle cofactor needs, it has not yet been widely adopted in the pharmaceutical business. As a result, the field of flow biocatalysis still has a tonne of room for expansion.[52]

8.5 Conclusion

In summary, biocatalysis is essential to the development of sustainable chemistry. Given both the constant growth in the world's population and the evolving conditions imposed by climate change, it is becoming increasingly essential to develop and use, as a practice, sustainable production techniques to be able to facilitate the establishment of a healthy and sustainable bioeconomy. Biofuels and bioproducts are among the best approaches to lower greenhouse gas emissions and combat global warming while still providing for humanity's energy needs. There are many advantages to biocatalysis in this situation. By using a catalyst that is biocompatible, biodegradable, and generated from renewable resources, reactions are carried out under benign conditions, avoiding the problems with scarcity and product contamination brought on by using catalysts made of precious metals as catalysts. Processes are also extremely selective and step-economic, which results in improved product quality and less waste formation. Briefly put, compared to traditional chemical processes, biocatalytic processes are more cost- and environmentally friendly as well as sustainable. According to N. N. Semenov, a renowned Russian chemist and Nobel laureate who wrote about the chemistry of the future 40 years ago: "By applying the ideas of biochemistry, chemical science may solve the energy crisis, make industrial production infinitely more efficient, and provide mankind with wings."

Acknowledgments

Poonam Rajesh Prasad is grateful to BHU IOE(Seed Grant) & Science and Engineering Research Board (SERB) New Delhi for sponsoring research.

References

1. Sheldon Orcid, R.A. and Woodley, J.M., Role of Biocatalysis in Sustainable Chemistry. *Chem. Rev.*, *118*, 2, 801–838, **2018**.
2. Krogan, N.J. *et al.*, A SARS-CoV-2 protein interaction map reveals targets for drug repurposing. *Nature*, *583*, 459–468, **2020**.
3. Ayano, Y., Min-Sun, K., Satoshi, A., The effects of self-construals, self-criticism, and self-compassion on depressive symptoms. *Pers. Individ. Differ.*, 68, 65–70, October **2014**.
4. Bell, E.L., Finnigan, W., France, S.P., Green, A.P., Hayes, M.A., Hepworth, L.J., Lovelock, S.L., Niikura, H., Osuna, S., Romero, E., Ryan, K.S., Turner, N.J., Flitsch, S.L., Biocatalysis. *Nat. Rev. Methods Primers*, *1*, 1–21, **2021**.
5. Martinengo, B., Diamanti, E., Uliassi, E., Bolognesi, M.L., Harnessing the 12 green chemistry principles for sustainable antiparasitic drugs: Toward the one health approach. 6, 40126–40141, **2024**.
6. De Marco, B.A., Rechelo, B.S., Tótoli, E.G., Kogawa, A.C., Salgado, H.R.N., Evolution of Green Chemistry and Its Multidimensional Impacts: A Review. *Saudi Pharm. J.*, *27*, 1–8, **2019**.
7. Intasian, P., Prakinee, K., Phintha, A., Trisrivirat, D., Weeranoppanant, N., Wongnate, T., Chaiyen, P., Enzymes, *in Vivo* Biocatalysis, and Metabolic Engineering for Enabling a Circular Economy and Sustainability. *Chem. Rev.*, *121*, 10367–10451, **2021**.
8. Sherer, E.C., Bagchi, A., Kosjek, B., Maloney, K.M., Peng, Z., Robaire, S.A., Sheridan, R.P., Metwally, E., Campeau, L.-C., Driving Aspirational Process Mass Intensity Using Simple Structure-Based Prediction. *Org. Process Res. Dev.*, *26*, 1405–1410, **2022**.
9. Sheldon, R.A., The: E Factor 25 Years on: The Rise of Green Chemistry and Sustainability. *Green Chem.*, *19*, 18–43, **2017**.
10. Campos, K.R., Coleman, P.J., Alvarez, J.C., Dreher, S.D., Garbaccio, R.M., Terrett, N.K., Tillyer, R.D., Truppo, M.D., Parmee, E.R., The Importance of Synthetic Chemistry in the Pharmaceutical Industry. *Science*, *363*, 1–8, **2019**.
11. France, S.P., Hepworth, L.J., Turner, N.J., Flitsch, S.L., Constructing Biocatalytic Cascades: In Vitro and *in Vivo* Approaches to de Novo Multi-Enzyme Pathways. *ACS Catal.*, *7*, 710–724, **2017**.
12. Schrittwieser, J.H., Velikogne, S., Hall, M., Kroutil, W., Artificial Biocatalytic Linear Cascades for Preparation of Organic Molecules. *Chem. Rev.*, *118*, 270–348, **2018**.
13. McIntosh, J.A. and Owens, A.E., Enzyme Engineering for Biosynthetic Cascades. *Curr. Opin. Green Sustainable Chem.*, *29*, 100448, **2021**.
14. Truppo, M.D., Biocatalysis in the Pharmaceutical Industry: The Need for Speed. *ACS Med. Chem. Lett.*, *8*, 476–480, **2017**.
15. Goodwin, N.C., Morrison, J.P., Fuerst, D.E., Hadi, T., Biocatalysis in Medicinal Chemistry: Challenges to Access and Drivers for Adoption. *ACS Med. Chem. Lett.*, *10*, 1363–1366, **2019**.
16. Antar, M., Lyu, D., Nazari, M., Shah, A., Zhou, X., Smith, D.L., Biomass for a sustainable bioeconomy: An overview of world biomass production and utilization. *Renew. Sustain. Energy Rev.*, 139, 110691–110708, April 2021.
17. Sheldon, R.A., Biocatalysis and Green Chemistry, in: *Green Biocatalysis*, Wiley Online Library, 56, 1–15, 2016.
18. France, S.P., Lewis, R.D., Martinez, C.A., The Evolving Nature of Biocatalysis in Pharmaceutical Research and Development. *JACS Au*, *3*, 3, 715–735, **2023**.
19. Hadi, T., Dıaz-Rodrıguez, A., Khan, D., Morrison, J.P., Kaplan, J.M., Gallagher, K.T., Schober, M., Webb, M.R., Brown, K.K., Fuerst, D., Snajdrova, R., Roiban, G.-D., Identification and Implementation of Biocatalytic Transformations in Route Discovery: Synthesis of Chiral 1,3-Substituted Cyclohexanone Building Blocks. *Org. Process Res. Dev.*, *22*, 871–879, **2018**.

20. An, C., Shaw, M.H., Tharp, A., Verma, D., Li, H., Wang, H., Wang, X., Enantioselective Enzymatic Reduction of Acrylic Acids. *Org. Lett.*, *22*, 8320–8325, **2020**.
21. Pellisa, A., Cantoneb, S., Ebertc, C., Gardossic, L., Evolving biocatalysis to meet bioeconomy challenges and opportunities. *New Biotechnol.*, *40*, 154–169, **2018**.
22. Kohrt, J.T., Dorff, P.H., Burns, M., Lee, C., O'Neil, S.V., Maguire, R.J., Kumar, R., Wagenaar, M., Price, L., Lall, M.S., Application of Flow and Biocatalytic Transaminase Technology for the Synthesis of a 1-Oxa-8-Azaspiro-[4.5]-Decan-3-Amine. *Org. Process Res. Dev.*, *26*, 616–623, **2022**.
23. Han, C., Savage, S., Al-Sayah, M., Yajima, H., Remarchuk, T., Reents, R., Wirz, B., Iding, H., Bachmann, S., Fantasia, S.M., Scalone, M., Hell, A., Hidber, P., Gosselin, F., Asymmetric Synthesis of Akt Kinase Inhibitor Ipatasertib. *Org. Lett.*, *19*, 4806–4809, **2017**.
24. Burns, M., Martinez, C.A., Vanderplas, B., Wisdom, R., Yu, S., Singer, R.A., A Chemoenzymatic Route to Chiral Intermediates Used in the Multikilogram Synthesis of a Gamma Secretase Inhibitor. *Org. Process Res. Dev.*, *21*, 871–877, **2017**.
25. Tassano, E., Moore, C., Dussauge, S., Vargas, A., Snajdrova, R., Discovery of New Fe(II)/α-Ketoglutarate-Dependent Dioxygenases for Oxidation of l-Proline. *Org. Process Res. Dev.*, *26*, 1996–2003, **2022**.
26. Goundry, W.R.F., Adams, B., Benson, H., Demeritt, J., McKown, S., Mulholland, K., Robertson, A., Siedlecki, P., Tomlin, P., Vare, K., Development and Scale-up of a Biocatalytic Process to Form a Chiral Sulfoxide. *Org. Process Res. Dev.*, *21*, 107–113, **2017**.
27. Burns, M., Bi, W., Kim, H., Lall, M.S., Li, C., O'Neill, B.T., Ketoreductase / Transaminase, One-Pot, Multikilogram Biocatalytic Cascade Reaction. *Org. Process Res. Dev.*, *25*, 941–946, **2021**.
28. Moore, J.C., Liang, J., Penfield, J., Nazor, J., Dellas, N., Mitchell, V., Duan, D., Farasat, I., Rodriguez-Granillo, A., Murphy, G., Marshall, N., Ketoreductase Polypeptides and Polynucleotides, **2020**.
29. Nazor, J., Osborne, R., Liang, J., Vroom, J., Zhang, X., Entwistle, D., Voladri, R., Garcia, R.D., Moore, J.C., Grosser, S., Kosjek, B., Truppo, M., Biocatalysts and Methods for Hydroxylation of Chemical Compounds, **2017**.
30. Leatham, G.F., Forrester, I.T., Mishra, C., Enzymes from Solid Substrates. *Enzymes in Biomass Conversion, ACS Symposium Series; ACS*, pp. 95–110, **1991**.
31. McKenna, S.M., Leimkühler, S., Herter, S., Turner, N.J., Carnell, A.J., Enzyme cascade reactions: synthesis of furandicarboxylic acid (FDCA) and carboxylic acids using oxidases in tandem. *Green Chem.*, *17*, 6, 3271–5, **2015**.
32. De Bont, J., Ruijssenaars, H., Werij, J., Fungal Production of FDCA, **2016**.
33. Zhang, X.-Y., Zong, M.-H., Li, N., Whole-cell biocatalytic selective oxidation of 5-hydroxymethylfurfural to 5-hydroxymethyl-2-furancarboxylic acid. *Green Chem.*, *19*, 19, 4544–51, **2017**.
34. Rajesh, R., Godan, T.K., Pandey, A., Binod, P., Whole cell based biocatalytic production of 2, 5-furan dicarboxylic acid. *Indian J. Exp. Biol.*, 56, 493–497, **2018**.
35. Rajesh, R.O., Godan, T.K., Rai, A.K., Sahoo, D., Pandey, A., Binod, P., Biosynthesis of 2, 5-furan dicarboxylic acid by Aspergillus flavus APLS-1: Process optimization and intermediate product analysis. *Bioresour. Technol.*, *284*, 155–60, **2019**.
36. Yang, Z.-Y., Wen, M., Zong, M.-H., Li, N., Synergistic chemo/biocatalytic synthesis of 2,5-furandicarboxylic acid from 5-hydroxymethylfurfural. *Catal. Commun.*, *139*, 105979, **2020**.
37. Crigler, J., Eiteman, M.A., Altman, E., Characterization of the Furfural and 5-Hydroxymethylfurfural (HMF) Metabolic Pathway in the Novel Isolate Pseudomonas putida ALS1267. *Appl. Biochem. Biotechnol.*, *190*, 3, 918–30, **2020**.
38. Findrik, Z. and Vasić-Racki, D., Biotransformation of D-Methionine into L-Methionine in the Cascade of Four Enzymes. *Biotechnol. Bioeng.*, *98*, 956–967, **2007**.
39. Bornscheuer, U.T., Huisman, G.W., Kazlauskas, R.J., Lutz, S., Moore, J.C., Robins, K., Engineering the thirdwave of biocatalysis. *Nature*, *485*, 185–194, **2012**.

40. FDA's Policy Statement for the Development of New Stereoisomeric Drugs. *Chirality*, *4*, 338–340, **1992**.
41. Sheldon, R.A., *Chirotechnology, Industrial Synthesis of Optically Active Compounds*, Marcel Dekker, New York, **1993**.
42. Livada, J., Vargas, A.M., Martinez, C.A., Lewis, R.D., Ancestral Sequence Reconstruction Enhances Gene Mining Efforts for Industrial Ene Reductases by Expanding Enzyme Panels with Thermostable Catalysts. *ACS Catal.*, *13*, 2576–2585, **2023**.
43. Stucki, A., Vallapurackal, J., Ward, T.R., Dittrich, P.S., Droplet Microfluidics and Directed Evolution of Enzymes: An Intertwined Journey. *Angew. Chem. Int. Ed.*, *60*, 24368–24387, **2021**.
44. Kempa, E.E., Galman, J.L., Parmeggiani, F., Marshall, J.R., Malassis, J., Fontenelle, C.Q., Vendeville, J.-B., Linclau, B., Charnock, S.J., Flitsch, S.L., Turner, N.J., Barran, P.E., Rapid Screening of Diverse Biotransformations for Enzyme Evolution. *JACS Au*, *1*, 508–516, **2021**.
45. Huang, P.-S., Boyken, S.E., Baker, D., The Coming of Age of de Novo Protein Design. *Nature*, *537*, 320–327, **2016**.
46. Lovelock, S.L., Crawshaw, R., Basler, S., Levy, C., Baker, D., Hilvert, D., Green, A.P., The Road to Fully Programmable Protein Catalysis. *Nature*, *606*, 49–58, **2022**.
47. Zhao, J., Burke, A.J., Green, A.P., Enzymes with Noncanonical Amino Acids. *Curr. Opin. Chem. Biol.*, *55*, 136–144, **2020**.
48. Burke, A.J., Lovelock, S.L., Frese, A., Crawshaw, R., Ortmayer, M., Dunstan, M., Levy, C., Green, A.P., Design and Evolution of an Enzyme with a Non-Canonical Organocatalytic Mechanism. *Nature*, *570*, 7760, 219–223, **2019**.
49. Rocha, R.A., Speight, R.E., Scott, C., Engineering Enzyme Properties for Improved Biocatalytic Processes in Batch and Continuous Flow. *Org. Process Res. Dev.*, *26*, 1914–1924, **2022**.
50. Fessner, N.D., Badenhorst, C.P.S., Bornscheuer, U.T., Enzyme Kits to Facilitate the Integration of Biocatalysis into Organic Chemistry – First Aid for Synthetic Chemists. *ChemCatChem*, 14, 156–166, **2022**.
51. Patel, R.N., Biocatalysis for synthesis of pharmaceuticals. *Bioorg. Med. Chem.*, *26*, 1252–1274, **2018**.
52. Devine, P.N., Howard, R.M., Kumar, R., Thompson, M.P., Truppo, M.D., Turner, N.J., Extending the application of biocatalysis to meet the challenges of drug development. *Nat. Rev. Chem.*, *2*, 409–421, **2018**.

9

Synthesis and Functionalization of Natural Products with Light-Driven Reactions

Kanchanbala Sahoo, Gitanjali Mishra and Barla Thirupathi*

Department of Chemical Sciences, Indian Institute of Science Education and Research Berhampur, Transit Campus, Govt. ITI Building, Ganjam-District, Berhampur, Odisha, India

Abstract

By replicating the photosynthesis that occurs in plants by nature, chemists have given significant attention to the use of photochemical syntheses in laboratory settings. Over the past few decades, visible light-driven photochemical synthesis has become a powerful synthetic tool in organic synthesis, with the notable feature of green and sustainable chemistry. Notable, a variety of photoactivation modes, including photo-SET (single-electron transfer), PCET (proton-coupling electron transfer), direct photoexcitation of substrates, reagents, or their complex, etc., have triggered several upgraded or unprecedented chemical transformations under mild conditions. Though visible light-driven photochemical synthesis has the potential to address significant issues, in organic synthesis, it can also play a pivotal role in the total synthesis of several structurally diverse biologically important natural products. The utility of natural products has been expanding their application in several fields like human and veterinary medicine, agriculture, food, and the cosmetic industry. Additionally, natural compounds are frequently employed as starting points for drug discovery and development. Accordingly, the synthesis of various analogs with enhanced efficacy, potency, safety, and purity has been developed. Natural products also provide the crucial starting point for many other concepts and directions in research, advancement, and the pharmaceutical industry. Recently, several such natural products have been synthesized under photochemical reaction conditions. Photochemical conditions include household bulbs or easily accessible LEDs under mild conditions. The widespread application of visible light-mediated photocatalysis evolved as a new chemical tool for various C-C bond formation reactions. In this chapter, we have highlighted the visible light-driven photocatalysis process employed in the total synthesis and functionalization of natural products.

***Keywords*:** Functionalization, light-driven reactions, natural products, photochemical reactions, total synthesis

**Corresponding author*: thirupathibarla@iiserbpr.ac.in; bthirupathi56@gmail.com

Mousumi Sen (ed.) Sustainable Green Catalytic Processes, (183–224) © 2024 Scrivener Publishing LLC

9.1 Introduction

By replicating the photosynthesis happening in plants by nature, synthetic chemists have given significant attention to the use of photochemical syntheses in laboratory settings.[1] The visible light-driven photochemical synthesis has become a powerful synthetic tool in organic synthesis, with the notable feature of green and sustainable chemistry over the past few decades.[2–5] The main concern in a visible light-driven reaction is to use an appropriate photocatalyst to avoid unwanted side reactions and product decomposition. Additionally, the absorbance energy of the photocatalyst should not match with the energy absorbed by the reagents or the products. Basically, a wavelength having greater than 400 nm is compatible for visible light-driven reactions.[5] In this context, numerous photocatalysts (PCs) have been employed in the total syntheses and functionalizations of natural products and as shown in Figure 9.1 (PCI-PCXVI).

A variety of photoactivation modes, including photo-SET (single-electron transfer), PCET (proton-coupling electron transfer), direct photoexcitation of substrates, reagents, or their complex, etc., have triggered several upgraded or unprecedented chemical transformations under mild conditions.[1] However, the initial step of the photoredox catalysis involves an excitation of the reactant that is thermodynamically feasible. In order to understand the feasibility of the reaction, it should follow the Rehm–Weller equation.[6]

$$\Delta G_{SET} = E^{ox}_{1/2}(D) - E^{red}_{1/2}(A) - E^{*}(D \text{ or } A) + \Delta E_{Coulombic}$$

Equation 9.1 Rehm–Weller equation

ΔG_{SET} = Gibb's free energy in single-electron transfer (SET) process,

$E^{ox}_{1/2}(D)$ = Oxidation potential of the compound, which donates electron

$E^{red}_{1/2}(A)$ = Reduction potential of the compound, which accepts electron

$E^{*}(D \text{ or } A)$ = Excited state energy of the compound including both donor and accepter ones

$\Delta E_{Coulombic}$ = Interaction between charged ions in the dielectric constant of the solvent used in the reaction

In a general photoredox cycle, in the presence of electron donor compound or electron accepter, the photocatalyst forms $PC^{\bullet+}/PC^{\bullet-}$ as the intermediate, which then goes to triplet state to form PC^{3} with the aid electron accepter compound/electron donor compound. The whole event occurred through single-electron transfer process. It is important to make sure that in single-electron transfer process, excited state of the photocatalyst should be thermodynamically feasible and the given photocatalyst has a suitable redox potential to form corresponding $PC^{\bullet-}$ or $PC^{\bullet+}$ intermediate and, with an appropriate reagent, it returns to the ground state and completes the catalytic process as shown in Figure 9.2.[7]

Over the last two decades, visible light-driven organic photochemical synthesis has attracted wide research interest from academic and industrial communities due to its features of green and sustainable chemistry. Though visible light-driven photochemical synthesis has the potential to address significant issues, in organic synthesis, it can also play a key step in the total synthesis of several structurally diverse biologically important natural products. Due to its energy-conserving and environmentally friendly features, visible light-induced photoredox catalysis has undergone rapid progress in the synthesis and functionalization of

fac-Ir(ppy)$_3$
PC-I

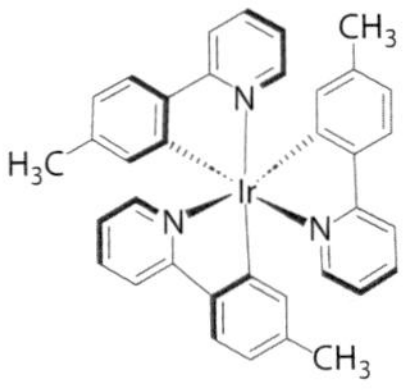

fac-Ir(mppy)$_3$
PC-II

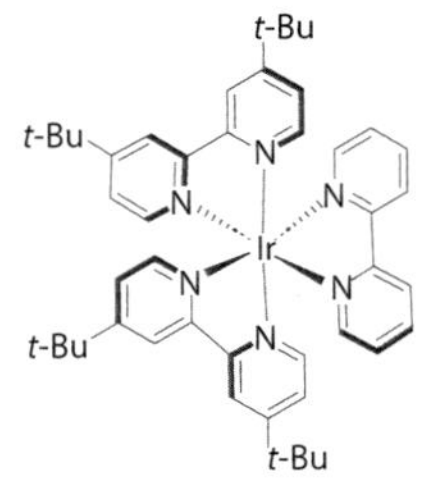

Ir[(dtbpy)(ppy)$_2$]PF$_6$
PC-III

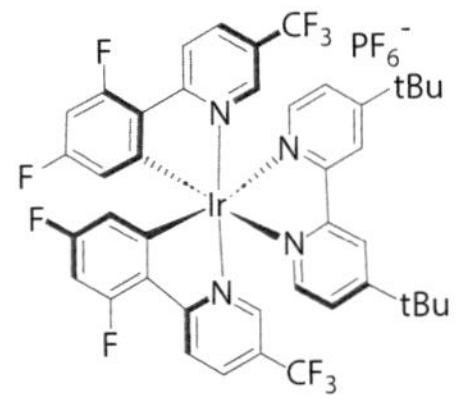

Ir[dF (CF$_3$)(dtbpy)(ppy)$_2$]PF$_6$
PC-IV

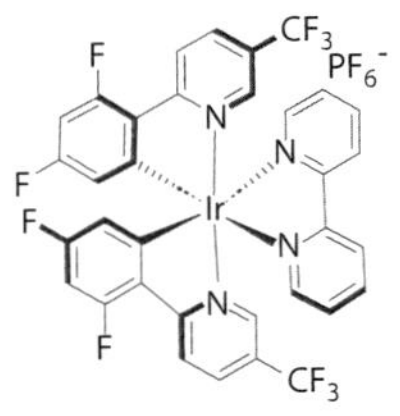

Ir[dF (CF$_3$)(bpy)(ppy)$_2$]PF$_6$
PC-V

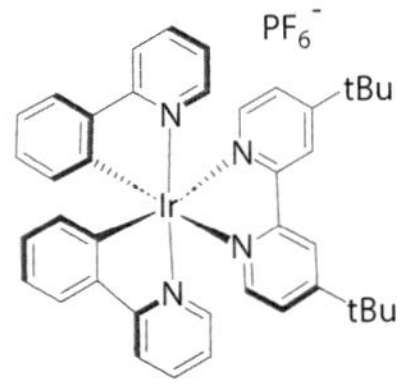

Ir[(dtbpy)(ppy)$_2$]PF$_6$
PC-VI

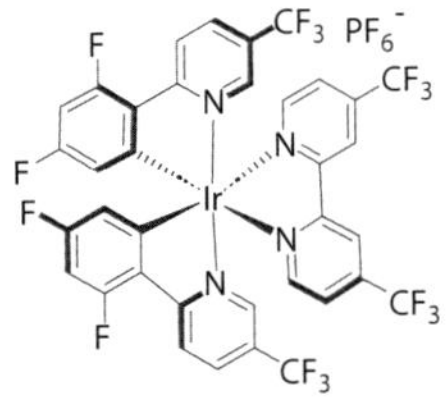

Ir[dF(CF$_3$)(ppy)$_2$d(CF$_3$)(bpy)]PF$_6$
PC-VII

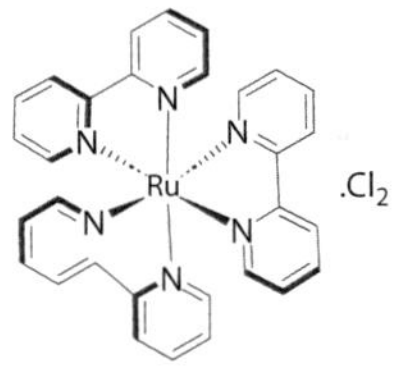

Ru(bpy)$_3$Cl$_2$
PC -VIII

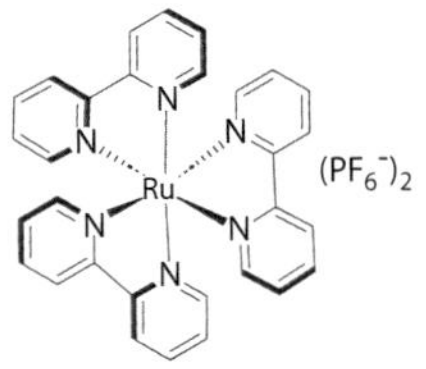

Ru(bpy)$_3$(PF$_6$)$_2$
PC -IX

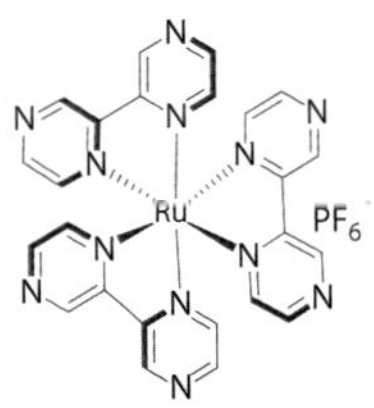

Ru(bpz)$_3$PF$_6$
PC -X

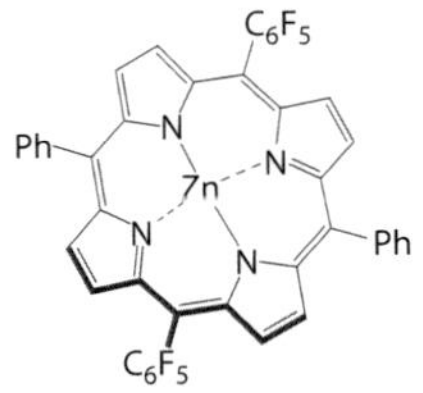

Zinc(II)porphyrin
PC-XI

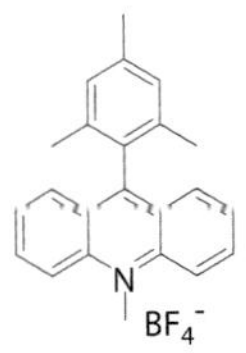

9-Mes-10-Meacridinium
PC-XII

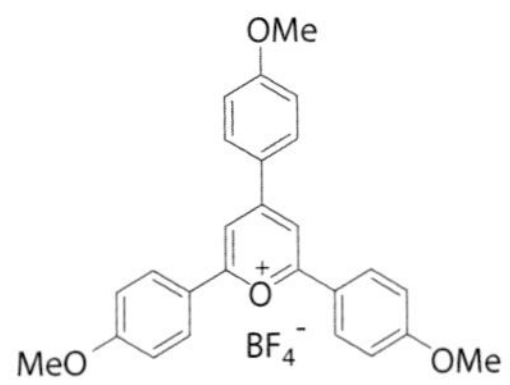

2,4,6-Tris(4-MeOPh)pyrylium
PC-XIII

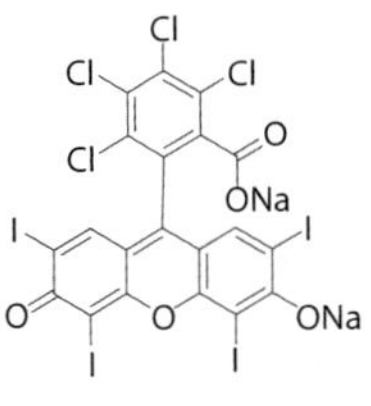

Rose Bengal
PC-XIV

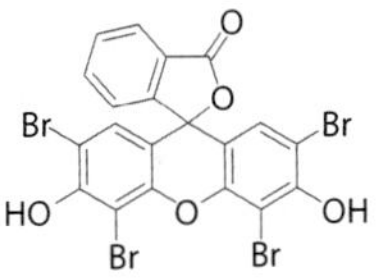

Eosin
PC-XV

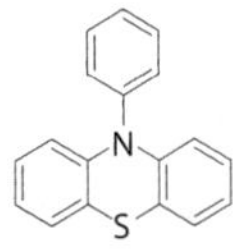

10-Phenylphenothiazine
PC-XVI

Figure 9.1 Few of the selective photocatalysts (PCs).

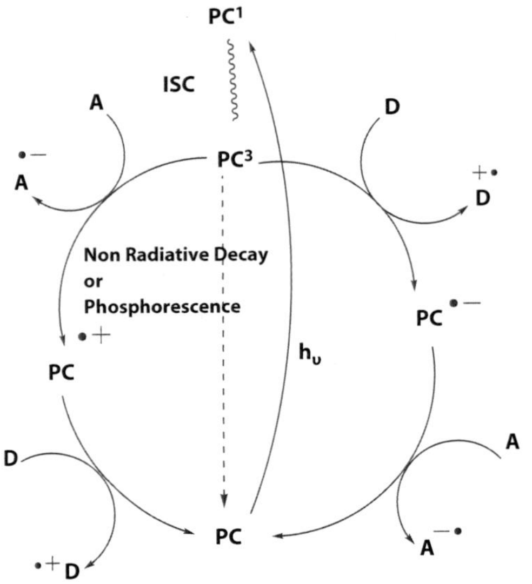

Figure 9.2 General photoredox cycle.

natural products. Natural products give the origin of stereochemistry, optical activity, regioselectivity, chirality, and many other concepts and directions within science, development, and industry in a scope, which is indispensable.[8] The utility of natural products has been expanding their application in several fields like human and veterinary medicine, agriculture, food, and the cosmetic industry.[8–12] Additionally, natural compounds are frequently employed as starting points for drug discovery and development. Accordingly, the synthesis of various analogs with enhanced efficacy, potency, safety, and purity has been developed. Natural products also provide the crucial starting point for many other concepts and directions in research, advancement, and the pharmaceutical industry. Recently, several such natural products have been synthesized under photochemical conditions. Photochemical conditions include household bulbs or easily accessible LEDs under mild conditions. The widespread application of visible light-mediated photocatalysis evolved as a new chemical tool for various C-C bond formation reactions.

9.2 Visible Light-Driven Total Synthesis of Natural Products

In this chapter, we have highlighted the visible light-driven photocatalysis processes, which have been employed in the total synthesis and functionalization of natural products. The first part of the chapter describes the total synthesis of natural products by utilizing a light-driven reaction as one of the key steps in their total synthesis. Natural products are arranged in alphabetical order, and the second part of the chapter describes the functionalization of natural products by utilizing visible light-driven photocatalysis in an alphabetical order.

9.2.1 Total Synthesis of (–)-Actinophyllic Acid

Actinophyllic acid (**5**) was isolated from the leaves of *Alstonia actinophylla* by Carroll *et al.*[13] in 2005. It is a monoterpenoid indole alkaloid and found as an effective

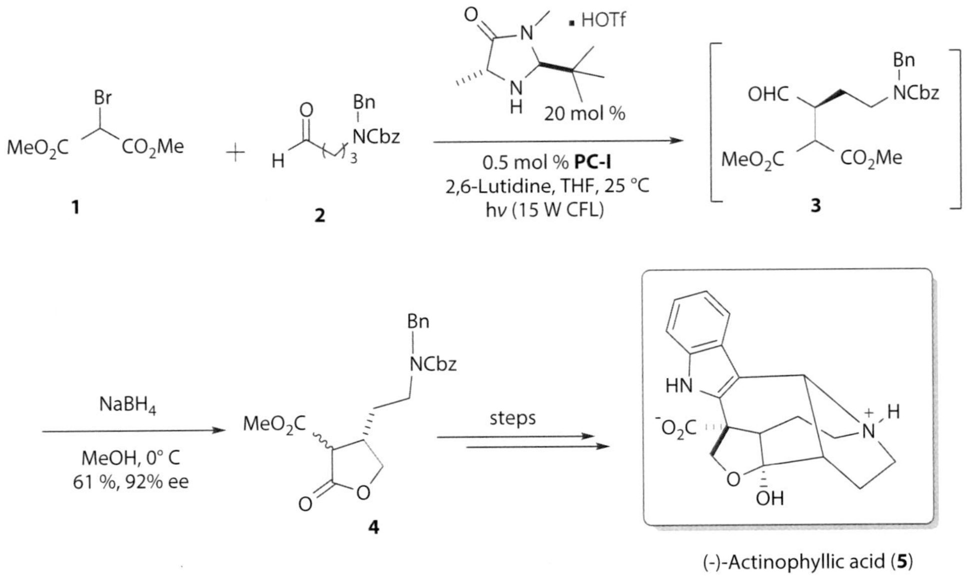

Scheme 9.1 Total synthesis of (–)-actinophyllic acid (**5**).

carboxypeptidase-U inhibitor. Structurally, actinophyllic acid has a novel hexacyclic ring that includes an azabicyclo[3.3.2]decanene scaffold, a pyrrolidine ring, a hemiketal moiety including five contiguous stereo centers, and a quaternary carbon center. In 2018, Xue *et al.*[14] achieved the formal total synthesis of (–)-actinophyllic acid. Authors have achieved this total synthesis by employing the photoredox strategy with organocatalyst developed by Nicewicz and MacMillan[15] as shown in Scheme 9.1. Accordingly, the dimethyl bromomalonate (**1**) was treated with aldehyde **2** to form intermediate **3**, which was further reduced to produce butyrolactone (**4**) in 61% yield and 92% ee. This intermediate **4** was further converted to the desired natural product (–)-actinophyllic acid (**5**). Notably on a gram scale, the yield and enantioselectivity of this photocatalytic asymmetric alkylation were readily maintained.

9.2.2 Total Synthesis of (±)–Aglacin B/C

Wang *et al.*[16] isolated aglacins from the methanol extract of the stem bark of *Aglaia cordata* collected in Indonesia. Aglacins belong to aryl tetralin cyclic ether lignans. This family of

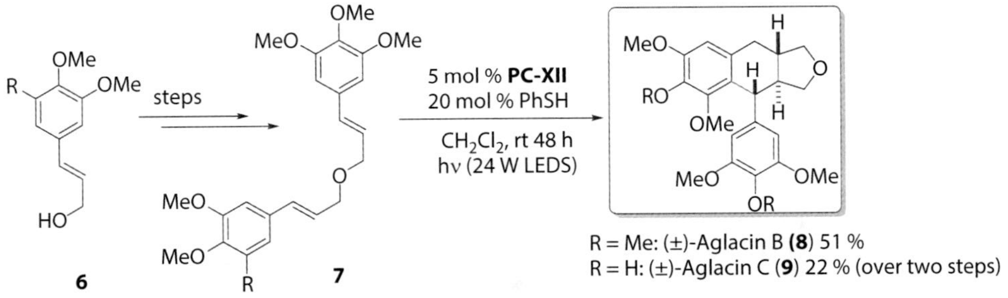

Scheme 9.2 Total synthesis of (±)–aglacin B (**8**) and (±)–aglacin C (**9**).

natural lignans exhibits a broad spectrum of biological activities, such as antiviral, insecticidal, anticancer, antifungal, anti-inflammatory, neurotoxic, and immunosuppressive activities.[17–19] In 2020, Xiang *et al.*[20] successfully applied the Diels–Alder reaction and photoredox reaction to prepare aryl tetralin cyclic ether of lignans. The authors used the styrene (**6**) that undergoes [2+2] cycloaddition with an oxidizing photocatalyst, thiophenol as cocatalyst, provided the desired natural product aglacin B (**8**) and aglacin C (**9**), with good yield and excellent diastereoselectivity as shown in Scheme 9.2.

9.2.3 Total Synthesis of (1*R*, 3*S*)-Albucidin

A novel nucleoside phytotoxin, albucidin (**13**), is a potent herbicide, and it was isolated in 2009 from *Streptomyces albus* subsp. *chlorinus* by Hahn *et al.*[21] It exhibits herbicidal activity against a broad spectrum of weeds.[21] In 2016, Zhang *et al.*[22] synthesized the oxetane core of the natural product, albucidin (**13**), by employing visible light irradiation. This synthesis starts from readily available *D*-xylose. In seven additional steps, *D*-xylose was advanced to epinoroxetanocin (**10**) as shown in Scheme 9.3. Subsequently, compound (**10**) was converted to iodine intermediate **11**, which underwent benzyl deprotection to form the intermediate **12**. The treatment of intermediate **12** with photocatalyst, *N,N*-diisopropylethylamine, and *p*-toluenethiol in the combination of acetonitrile and DMF resulted in the natural product (1*R*, 3*S*)-albucidin (**13**) in 82% yield.

9.2.4 Total Synthesis of (–)-Aspergillide A

In 2008, Kito *et al.*[23] isolated aspergillides A–C from the marine fungus *Aspergillus ostianus*. It showed an important cytotoxic activity against lymphocytic leukemia cell lines.

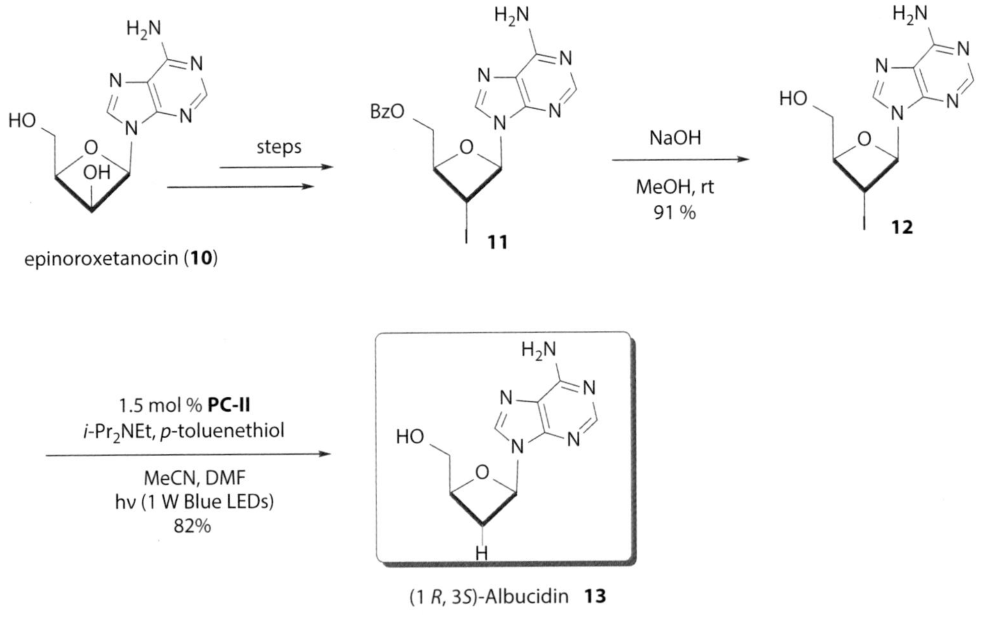

Scheme 9.3 Total synthesis of (1*R*, 3*S*)-albucidin (**13**).

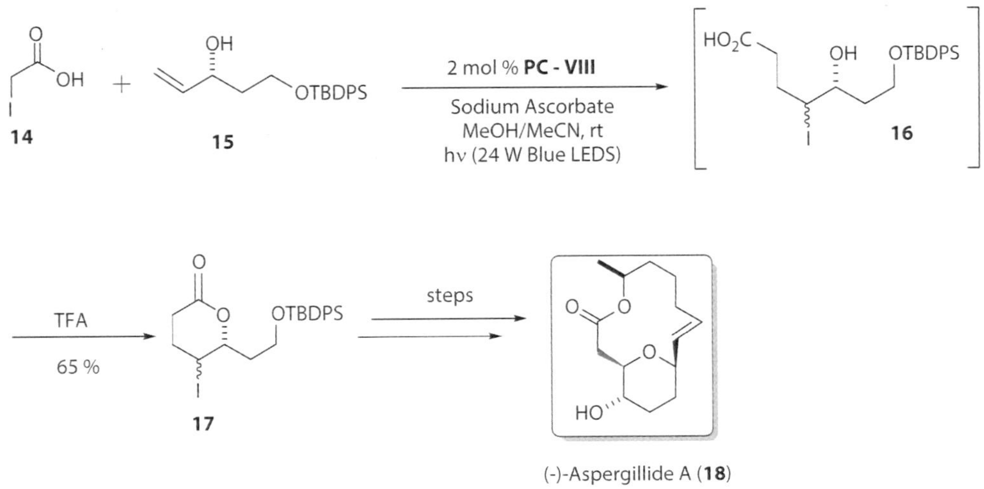

Scheme 9.4 Total synthesis of (–)-aspergillide A (**18**).

Aspergillides have an interesting trisubstituted tetrahydropyran ring that is encased in a 12-membered macrolactone that contains an *E*-alkene, a hydroxy group, and four stereogenic centers. In 2019, Mateus-Ruiz and Cordero-Vargas[24] utilized photoredox catalysis for the concise total synthesis of (–)-aspergillide A (**18**). Iodoacetic acid (**14**) was allowed to react with allylic alcohol **15** through an atom transfer radical addition (ATRA) followed by acid-catalyzed lactonization to form hydropyranone **17** under visible light irradiation as shown in Scheme 9.4. Subsequently, (–)-aspergillide A (**18**) was synthesized by using the intermediate **17**.

9.2.5 Total Synthesis of (–)-Batrachotoxin

Batrachotoxin (**26**) is a highly toxic steroidal alkaloid isolated from the skin of *Phyllobates* poison-dart frogs in South America.[25] It has been used as an important tool for studying ion transport, as it specifically stabilizes voltage-gated sodium (NaV) channels in an active open form. In 2020, Guo *et al.*[26] reported the total synthesis of (–)-batrachotoxinin A. This protocol starts by using *cis*-decalin derivative **19**, which was converted to tricyclic α-bromoketone **21**. The intermediate **21** was allowed to be treated with diketone **22**, in the presence of Ru-catalyst (**PC-VIII**), as a photocatalyst under the irradiation of visible light furnished the intermediate **25** in 70% yield as shown in Scheme 9.5. Authors anticipated that the C–C bond-forming in compound **25** is generated from the coupling radical intermediate **23** and enamine intermediate **24** in the presence of photoredox catalysis. Intermediate **25** was again employed in the total synthesis of (–)-batrachotoxin (**26**).

9.2.6 Total Synthesis of (–)-Burshernin

Bursehernin (**33**) is a lignan compound extracted from many plants such as *Bursera simaruba* Sarg, *Macrococculus pomiferus*, *Linum meletonis*, *Bursera fagaroides*, and *Geranium thunbergii*.[27] It showed biological activity against many cancer cells such as breast, nasopharyngeal, colon, and prostate cancer.[28] In 2015, MacMillan *et al.*[29] accomplished

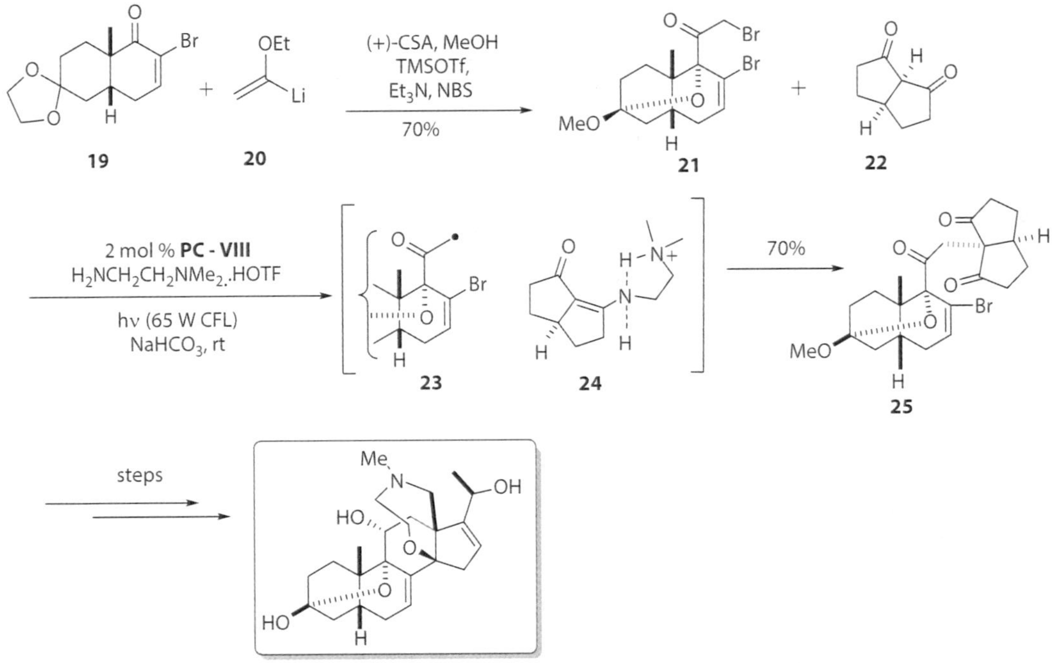

Scheme 9.5 Total synthesis of (−)-batrachotoxin (**26**).

the enantioselective total synthesis of cytotoxic lignan (−)-bursehernin (**33**) employing photoredox-organocatalysis. They suggested that the primary radical **30** is initially furnished from bromoacetonitrile **28**, which then combines with intermediate **31** to generate compound **32** as shown in Scheme 9.6. The natural product (−)-bursehernin (**33**) was finally synthesized using the advanced intermediate **32** in 84% yield.

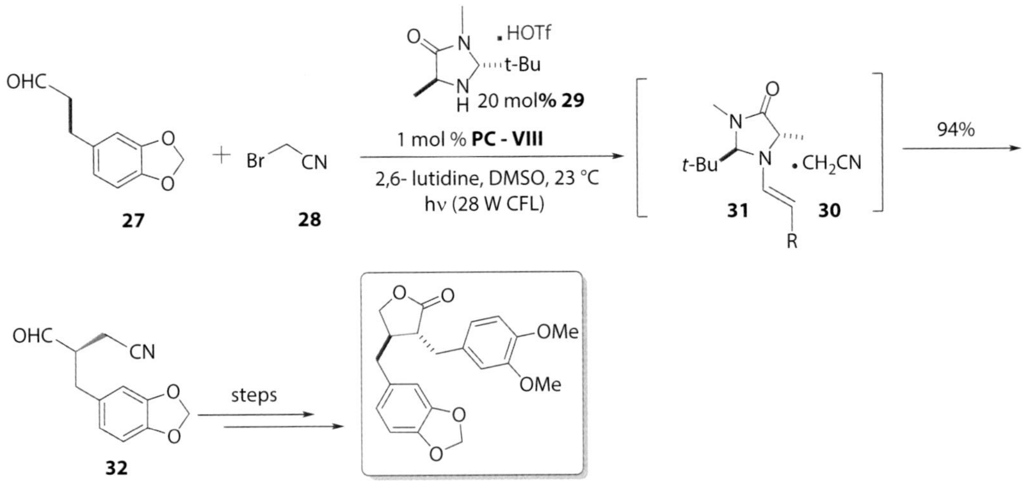

Scheme 9.6 Total synthesis of (−)-bursehernin (**33**).

9.2.7 Total Synthesis of (±)-Cannabiorcicyclolic Acid

In 2010, Qin *et al.*[30] isolated (±)-cannabiorcicyclolic acid from the leaves of *R. anthopogon*. Lu and Yoon[31] effectively synthesized (±)-cannabiorcicyclolic acid using photocatalysis strategy in 2012. Accordingly, the chromene precursor **36** was prepared from phenol **34** and citral (**35**) in 54% yield under a base-catalyzed condensation reaction as shown in Scheme 9.7. Thereafter, Ir-catalyzed photocycloaddition of intermediate **36** produced the cyclobutane-cycloadduct that was followed by saponification furnished the natural product (±)-cannabiorcicyclolic (**37**) in **97**% yield (Scheme 9.7).

9.2.8 Total Synthesis of (+)-Cephalosporolides E and F

In 1985, Ackland *et al.*[32] isolated cephalosporolides from the fermentation of the fungus *C. aphidicola*, ACC 3490. Structurally, cephalosporolides are 5,5-spiroketal-fused *γ*-lactone rings. Recently, Xochicale-Santana *et al.*[33] successfully completed the total synthesis of cephalosporolide F using *D*-glucose (**38**) as the starting material. The intermediate octono-1,4-lactone (**40**) derivative was initially prepared from Meldrum's acid (**39**) and the *D*-glucose (**38**) under microwave irradiation conditions. Under exploiting visible light photoredox conditions, compound **41** generates radical precursor **42** in the presence of photocatalyst with Hantzsch ester as the hydrogen atom transfer reagent at low temperature. Accordingly, radical precursor **42** was exclusively converted into the natural product (+)-cephalosporolide F (**45**) in 80% yield as shown in Scheme 9.8. The formation of the radical cation **44** intramolecularly combined by the hydroxyl group leads to the stereoselective formation of the spiroketal center of (+)-cephalosporolide F (**45**) and is thought to be enabled by the destabilization of Pauli repulsions of compound **43** under non-thermodynamic conditions.

9.2.9 Total Synthesis of (±)-Cermizine

Lycopodium alkaloid cermizine C was initially identified in 2004 from the club moss *Lycopodium cernuum* by Morita *et al.*[34] In 2016, Benimana *et al.*[35] achieved the concise

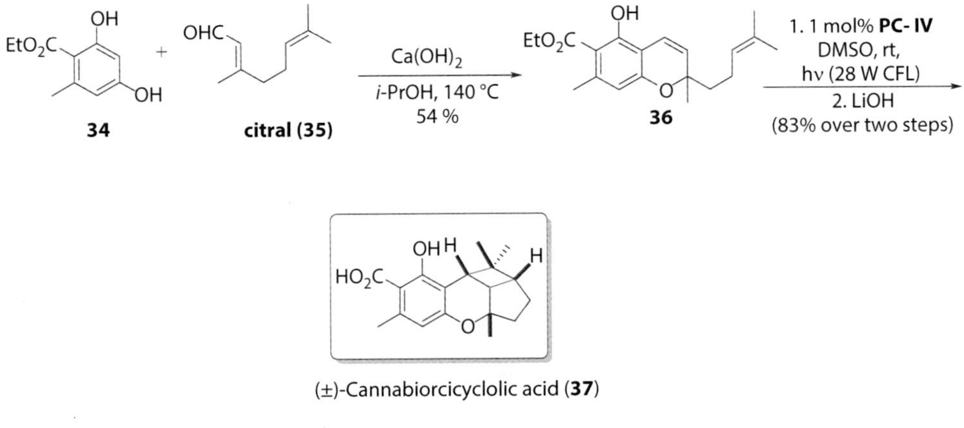

Scheme 9.7 Total synthesis of (±)-cannabiorcicyclolic (**37**).

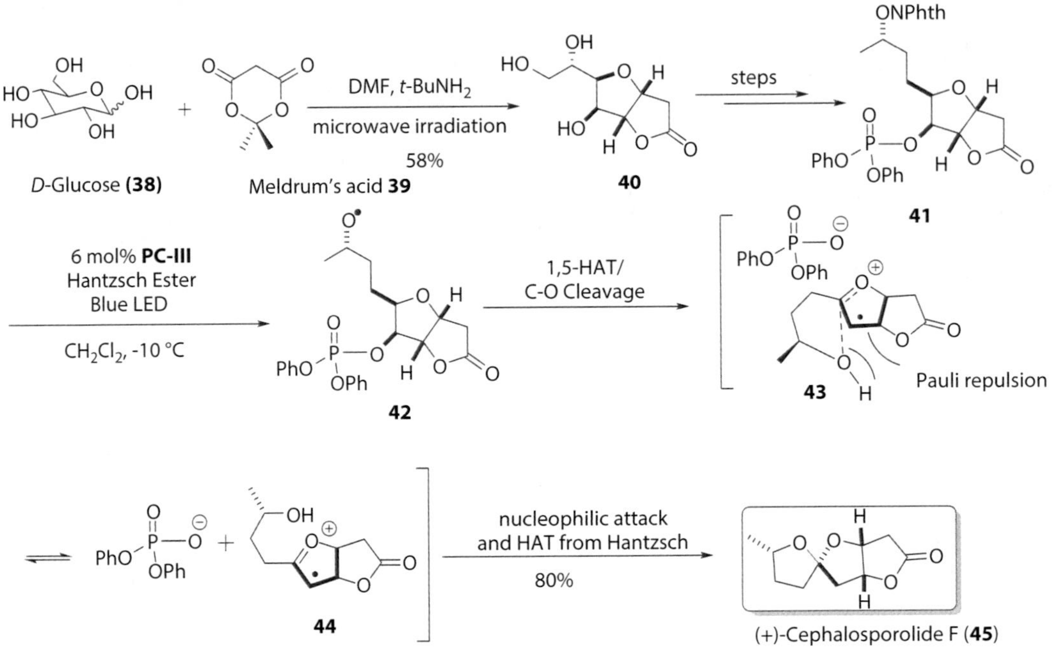

Scheme 9.8 Total synthesis of (+)-cephalosporolide F (**45**).

total synthesis of the (±)–5-*epi*-cermizine (**51**). In order to find the desired natural product, authors synthesized azacyclic molecule **50** by the intramolecular reactions between iminium ion **49** and alkene **47**. Accordingly, the intermediate allylsilane **48** was prepared from the inexpensive starting material *N*-alkyl piperidine **46** after three steps. Exploiting photoredox catalysis, the iminium intermediate **49** is formed from allylsilane nucleophile **48**, which was the key step in the total synthesis of (±)–*epi*-cermizine (Scheme 9.9). Finally, hydrogenation of azacyclic **50** furnished the desired natural product (±)–*epi*-cermizine (**51**) in 52% yield.

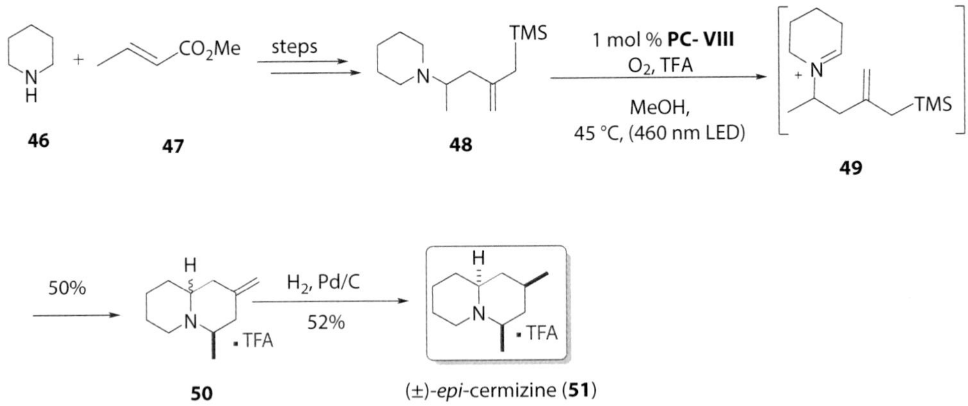

Scheme 9.9 Total synthesis of (±)–*epi*-cermizine (**51**).

9.2.10 Total Synthesis of (–)-Chromodorolide B

Chromodorolides were isolated from nudibranches of the genus *Chromodoris* and the encrusting sponges on which these nudibranches potentially feed.[36–38] Chromodorolide B contains 10 contiguous stereocenters, embedded with 6-acetoxy-2,7- dioxabicyclo[3.2.1] octan-3-one and 7-acetoxy-2,8-dioxabicyclo-[3.3.0]octan-3-one rings. Chromodorolides displayed distinct biological activities like antitumor, nematocidal, and antimicrobial activities.[37, 39, 40] Tao *et al.*[36] reported the total synthesis of (–)-chromodorolide B (**60**) by applying bimolecular radical addition/cyclization/fragmentation (ACF) cascade reaction. This protocol starts with (*S*)-enedione (**52**) that is elaborated to *trans*-hydrindanone (**53**). Intermediate **53** was further converted to acetonide NHPI ester **54** in a few steps. Compound **54** was treated with (*S*)-chlorobutenolide (**55**) and photocatalyst in THF solution under the irradiation of visible light, followed by dechlorination resulting in a mixture of **59**. The required natural product (–)-chromodorolide B (**60**) was finally synthesized from the intermediate **59** in few additional steps as shown in Scheme 9.10.

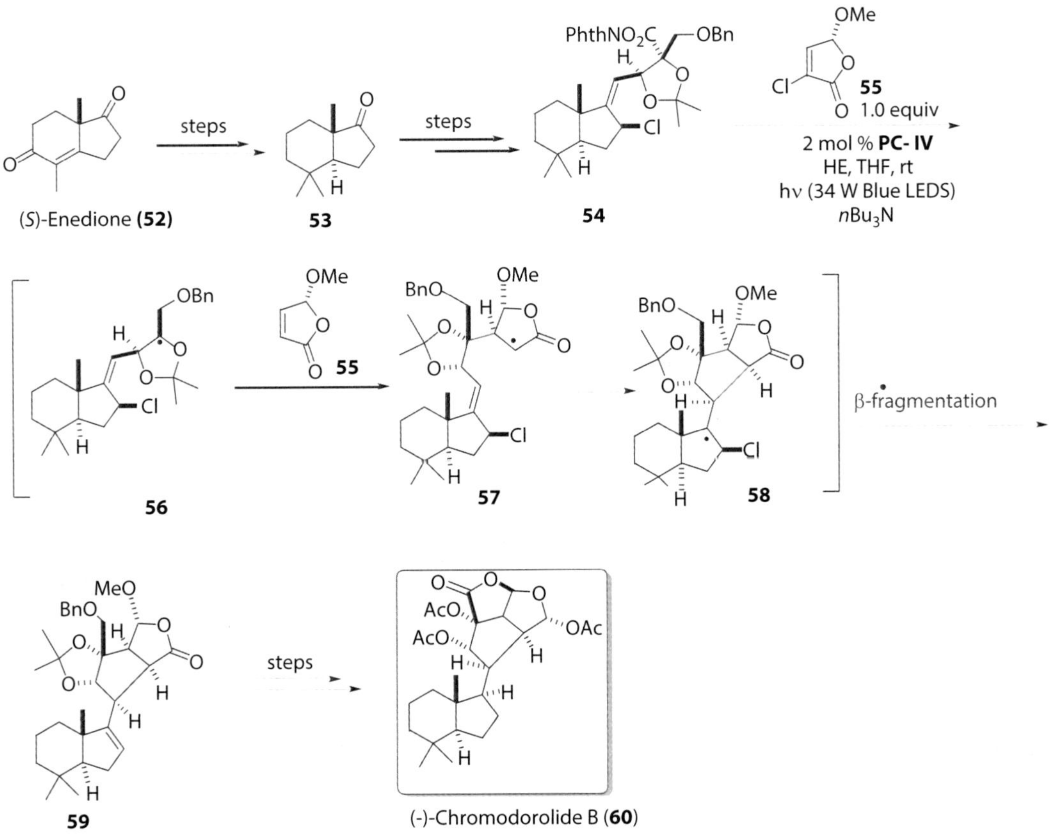

Scheme 9.10 Total synthesis of (–)-chromodorolide B (**60**).

9.2.11 Total Synthesis of (–)-Coerulescine

In 1998, Anderton *et al.*[41] isolated coerulescine from *Pharalis coerulescens.* Structurally, coerulescine belongs to the spirooxindole family. Creating a stereogenic center at quaternary carbon has been quite challenging, which makes it difficult to achieve the enantioselective total synthesis of (–)-coerulescine (**65**). In 2016, Chen *et al.*[42] utilized a photocatalysis reaction for the total synthesis of (–)-coerulescine (**65**). Accordingly, compound **61** was subjected to photocatalysis reaction to produce radical species **62**, which underwent an intramolecular 1,5 hydrogen atom transfer process to give radical species **63**, and finally the spirooxindole **64** was obtained. The natural product (–)-coerulescine (**65**) was finally synthesized by using the intermediate **64** as shown in Scheme 9.11.

9.2.12 Total Synthesis of (±)-Crispine A

Crispine A (**70**) is a tricyclic tetrahydroisoquinoline alkaloid isolated from the welted thistle, *Carduus crispus.*[43] Orejarena Pacheco *et al.*[44] successfully utilized photocatalysis reaction for the construction of the natural product crispine A (**70**). Accordingly, they prepared pyrrolidine **68** from the commercially available homoveratrylamine (**66**) in a two-step procedure. Then, compound **68** was treated with trimethylsilyl cyanide (TMSCN), and photocatalyst under visible light irradiation conditions furnished α-aminonitrile **69** in 78% yield. Compound **69** was allowed to react with $AgBF_4$ and at -78°C generated an iminium ion, followed by Pictet–Spengler cyclization of **69** in the presence of trifluoroacetic acid (TFA) produced natural product (±)-crispine A (**70**) in 91% yield as shown in Scheme 9.12.

9.2.13 Total Synthesis of (±)-Danshenspiroketallactones

Danshenspiroketallactone is a mono-benzannulated 5,5-spiroketal that was isolated from *Danshen*, which is used for cardiac arrest.[45, 46] In 2019, Deng *et al.*[47] successfully employed

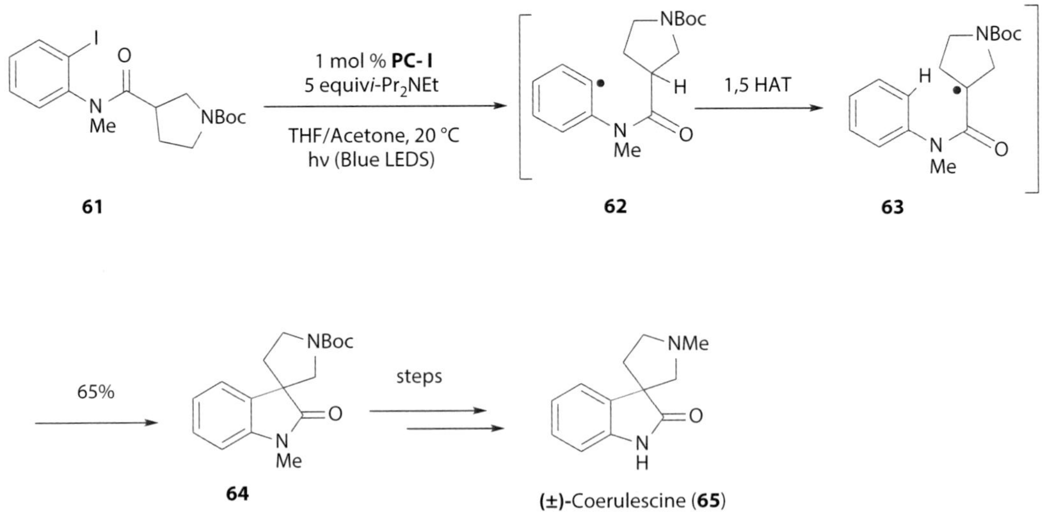

Scheme 9.11 Synthesis of (–)-coerulescine (**65**).

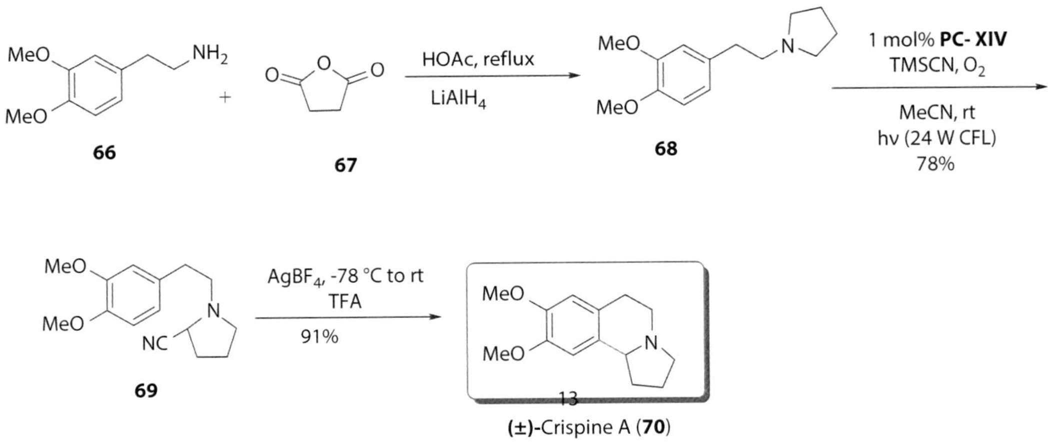

Scheme 9.12 Total synthesis of (±)–crispine A (**70**).

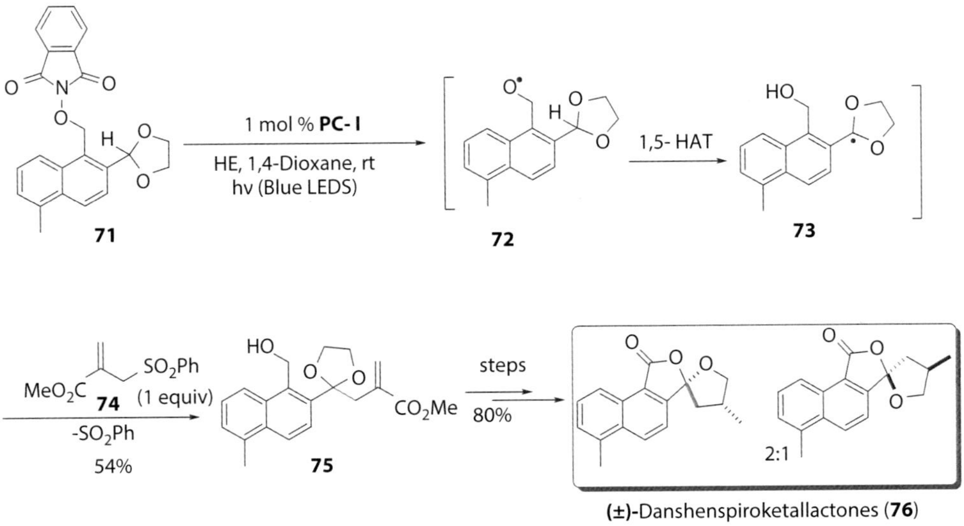

Scheme 9.13 Total synthesis of (±)–danshenspiroketallactone (**76**).

a well-suited photoctalyst and an external reagent Hantzsch ester (HE) as a hydrogen atom transfer reagent (HAT) to generate benzyloxy radical **72** and dioxolanyl radicals **73**, which is the key step to produce the intermediate **75**. The intermediate **75** was further elaborated to provide the natural product (±)–danshenspiroketallactone (**76**) in 80% yield over three steps (Scheme 9.13).

9.2.14 Total Synthesis of Daurioxoisoporphine C

Daurioxoisoporphine C (**81**) was first isolated in 2001 from the rhizomes of *Menispermum dauricum* by Yu *et al.*[48] In 2014, Zhang *et al.*[49] developed the total synthesis of natural product daurioxoisoporphine C by employing photoredox catalysis strategy. Accordingly, they

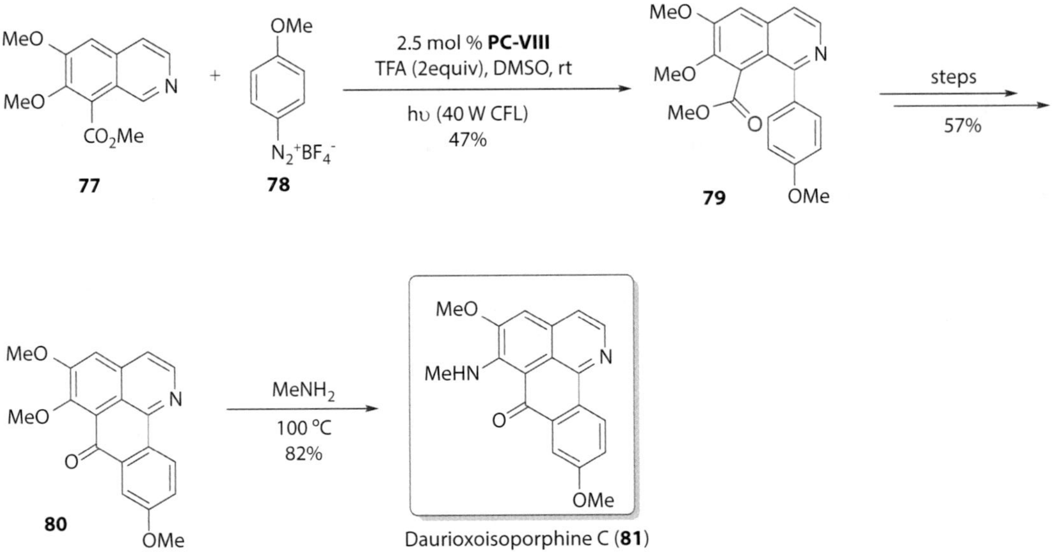

Scheme 9.14 Total synthesis of daurioxoisoporphine C (**81**).

prepared compound **79** from the reaction between isoquinoline derivative **77** and diazonium salt **78** under photoredox condition. The advanced intermediate **79** was further elaborated to give the compound **80**, which on treatment with methylamine at 100 °C furnished the natural product daurioxoisoporphine C (**81**) in 82% yield (Scheme 9.14).

9.2.15 Total Synthesis of Drimentine F

Drimentine (**87**) is a pyrroloindoline alkaloid that shows potent biological activities like antibacterial, anticancer, antifungal, and anthelmintic properties.[50, 51] In 2013, Sun *et al.*[51]

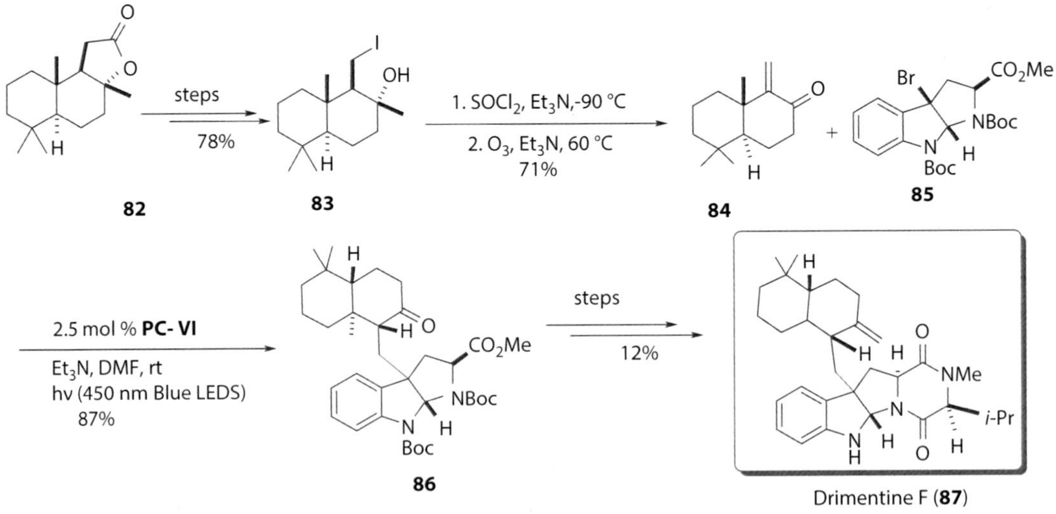

Scheme 9.15 Total synthesis of drimentine F (**87**).

reported total synthesis of drimentines A, F, and G using photoredox strategy. Accordingly, they used (+)–sclareolide (**82**) as the starting material, which produced α-methylene *trans*-decalone **84** in five steps. Treatment of intermediate **84** with bromopyrrolindoline **85** and photocatalyst in the presence of visible light furnished the compound **86** in 87% yield. Compound **86** was advanced to complete the concise total synthesis of drimentine F (**87**) in 12% yield by performing three additional steps (Scheme 9.15).

9.2.16 Total Synthesis of (±)–Epiraikovenal

Epiraikovenal (**93**) was isolated in 1995 from strains SB8, GA8, and 39W of *Euplotes raikovi*, collected from the Thyrrenian and Adriatic coasts of Italy and the Pacific Ocean, by Snider and Lu[52] and Guella *et al.*[53] Hurtley *et al.*[54] utilized late-stage functionalization of 1,3-dienes to accomplish a concise total synthesis of epiraikovenal. Authors had synthesized triene intermediate **90** from aldehyde **88** by applying a Horner–Wadsworth–Emmons olefination reaction. Next, intermediate **91** is converted to compound **90** by exploiting photoredox-catalyzed intermolecular [2+2] cycloadditions. Finally, the intermediate **91** is allowed to cross-metathesis with enal **92** producing the racemic epiraikovenal (**93**) in 42% yield (Scheme 9.16).

9.2.17 Total Synthesis of (+)–Flavisiamine F

Flavisiamine F (**99**) is an alkaloid isolated from the leaves of *Kopsia* by Gallagher and Magnus[55] in 1983. Tong *et al.*[56] reported total synthesis of (+)–flavisiamine F (**99**) by employing a photoredox catalysis strategy. They achieved Fischer indolization with compounds **94** and **95** to produce compound **96**, which was expanded in eight additional steps to give dihydrocarboline allylic alcohol **97** as shown in Scheme 9.17. To perform the deiodination process, compound **97** was treated under visible light irradiation to obtain the advanced intermediate **98**, which was then subjected to complete the enantioselective synthesis of (+)–flavisiamine F (**99**) in two steps (Scheme 9.17).

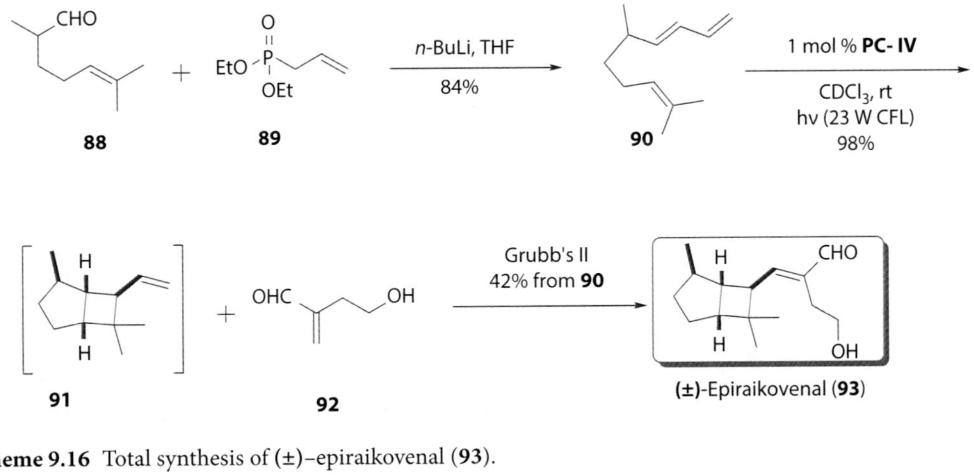

Scheme 9.16 Total synthesis of (±)–epiraikovenal (**93**).

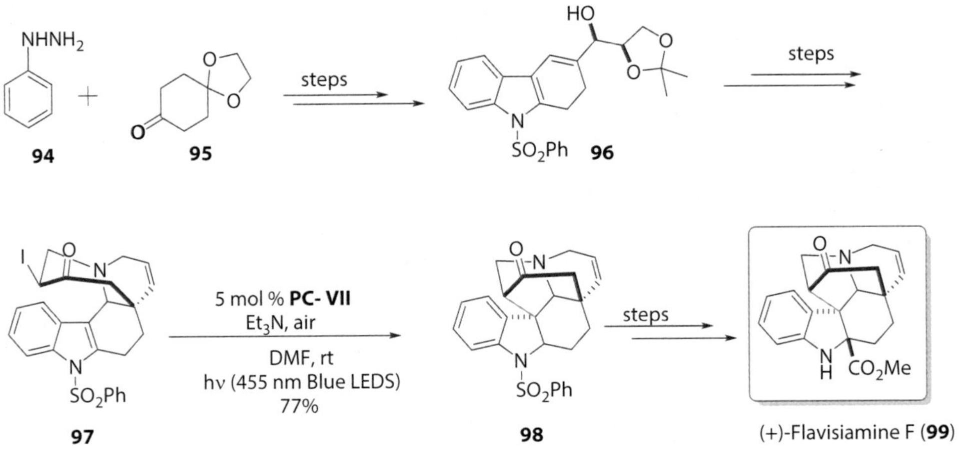

Scheme 9.17 Total synthesis of (+)–flavisiamine F (**99**).

9.2.18 Total Synthesis of (–)-Flustraminol

Flustraminols are marine alkaloids, which were obtained from the bryozoan *Flustra foliacea*.[57–59] Wu *et al.*[60] developed the late-stage synthesis of pyrroloindoline derivatives to construct (–)–flustraminol B by employing photoredox strategy. The synthesis started from the 6-bromoindole acetic acid (**100**), which was subjected to produce compound **101** in 75% yield in two steps. The compound **101** is then subjected to photocatalytic amidyl radical **102** generation with organophotocatalyst, followed by intramolecular cyclization, producing the radical intermediate **103**. The intermediate was then treated with O_2 and produced the corresponding alcohol **104** (Scheme 9.18). The consequent reduction of the amide group on **104** completed the total synthesis of (–)–flustraminol B (**105**) in 87% yield.

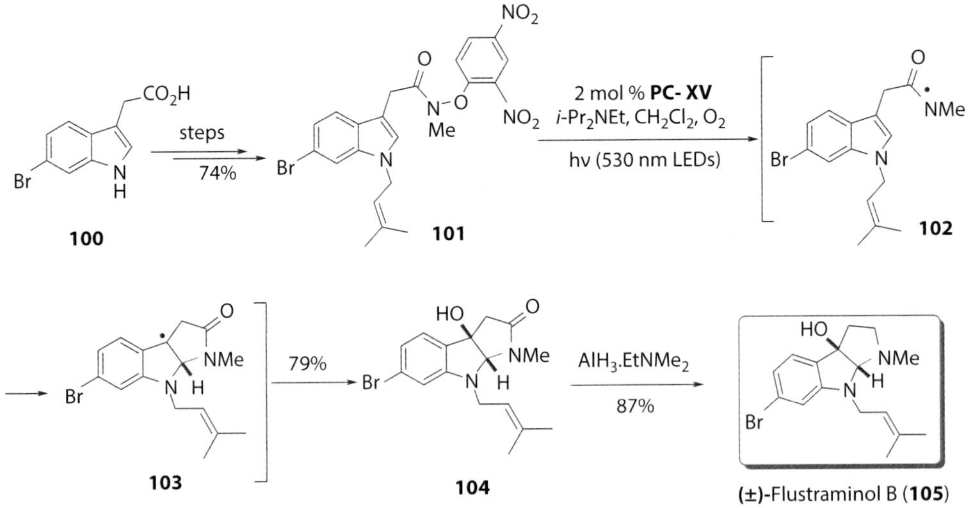

Scheme 9.18 Total synthesis of (–)–flustraminol B (**105**).

9.2.19 Total Synthesis of (+)–GB22

(+)-GB22 is from galbulimima alkaloid family; it was isolated from the extraction of *Galbulimima belgravena* plant.[61, 62] Burdge *et al.*[63] successfully achieved the total synthesis of natural product (+)–GB22 by employing a dual nickel catalyst and photocatalyst in 2019. Siloxycyclopropanol **107**, which was prepared from ketone **106**, and aryl bromide **108** were irradiated with visible light in the presence of 20 mol% photocatalyst and 10 mol% Ni–catalyst to give corresponding ketone derivative **111** in 48% yield as shown in Scheme 9.19. Subsequently, (+)–GB22 (**112**) was prepared from the intermediate **111**.

9.2.20 Total Synthesis of (+)–Gliocladin C

Gliocladin C (**119**) was first isolated from a strain of *Gliocladium roseum* OUPS-N132 by Usami *et al.*[64] in 2004. It exhibits significant cytotoxic activity against the murine P388 lymphocytic leukemia cell line.[64, 65, 67] In 2011, Furst *et al.*[66] developed visible light-driven total synthesis of (+)–gliocladin C (**119**). Initially, the ester derivative **113** was converted to bromopyrrolidinoindoline (**114**) in three steps. Compound **114** and 2-formylindole (**115**) was treated with Ru- catalyst as the photocatalyst, and *n*-Bu_3N as the reductive quencher compound produced the intermediate **117** in 82% yield. Intermediate **117** undergoes Rh-promoted deformylation and furnished the compound **118**. Subsequently, (+)–gliocladin C (**119**) was synthesized in 86% yield from the intermediate **118** as shown in Scheme 9.20.

9.2.21 Total Synthesis of Hamigeran B

Hamigeran B (**127**) is an antiviral natural product isolated from the marine sponge *Hamigera tarangaensis.*[68] It shows antiviral activity against herpes and polioviruses. Recently, Okanishi *et al.*[69] effectively utilized a radical route to construct the tricyclic molecular structure of hamigeran B. Irradiation of visible light plays a pivotal role to achieve the total synthesis of hamigeran B (**127**). For instance, compound **120** was subjected to photoredox-catalyzed

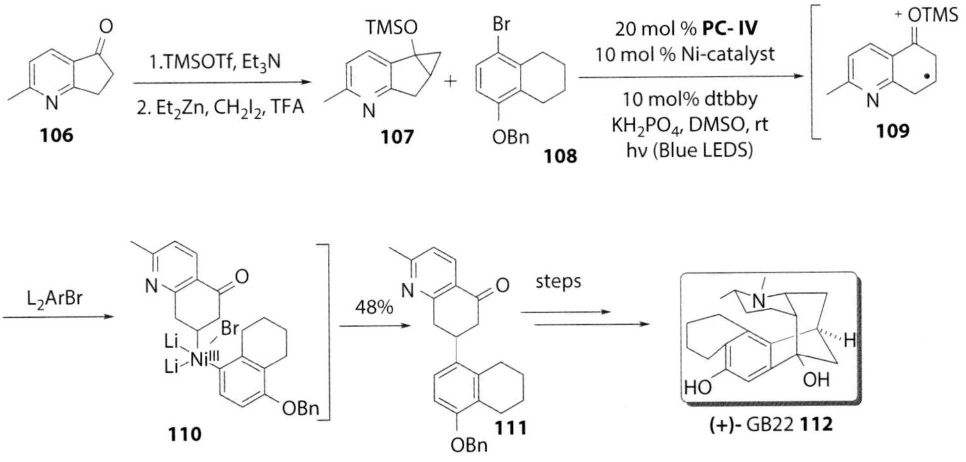

Scheme 9.19 Total synthesis of (+)–GB22 (**112**).

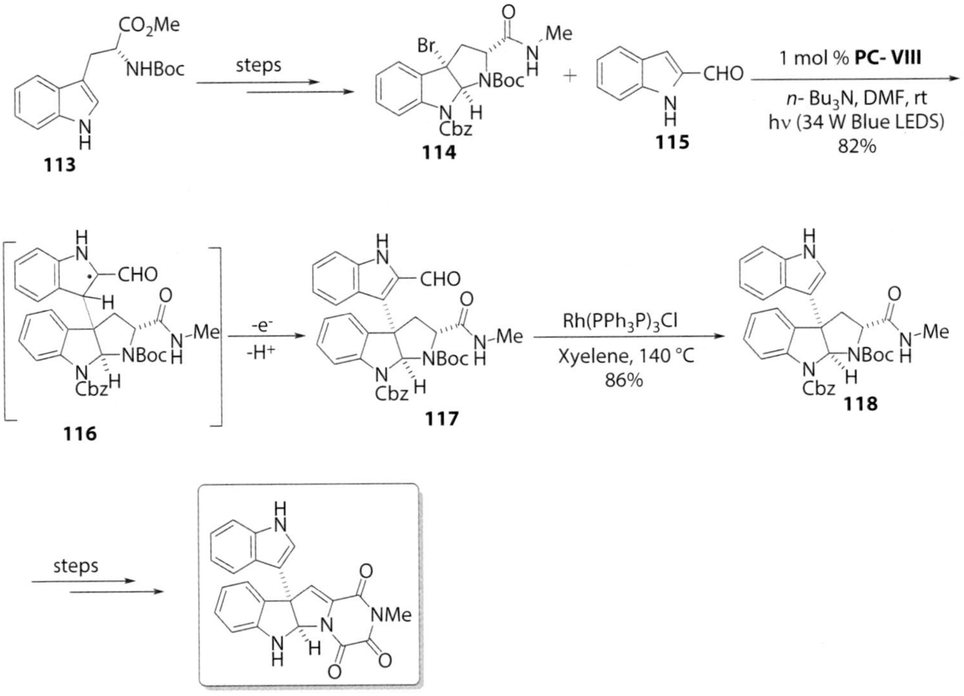

(+)-Gliocladin C (**119**)

Scheme 9.20 Total synthesis of (+)–gliocladin C (**119**).

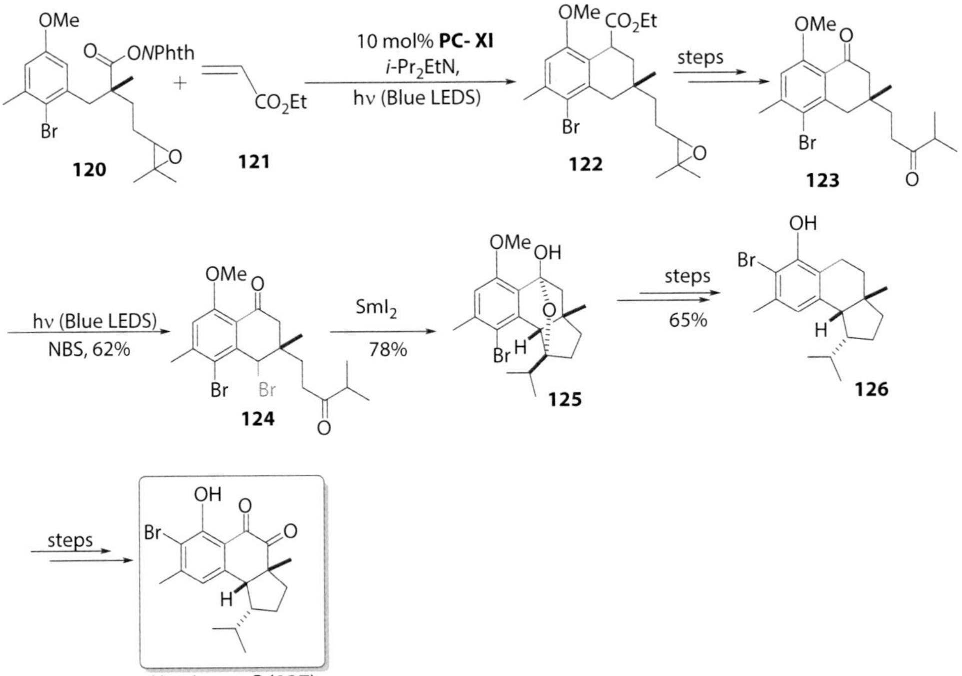

Hamigeran B (**127**)

Scheme 9.21 Total synthesis of hamigeran B (**127**).

cyclization to produce tetralin derivative **122**. Subsequently, in the absence of catalyst bromination that occurred in compound **123** using NBS under visible light irradiation, then it underwent cyclization to produce the intermediate **125** (Scheme 9.21). Finally, the natural product hamigeran B (**127**) was obtained from the intermediate **125**.

9.2.22 Total Synthesis of (±)–Heitziamide A

Heitziamide A (**131**) is an amide-containing natural product, and it was isolated from an African medicinal shrub *Fagara heitzii*.[70] Late-stage coupling of myrcene (**129**) and styrene **128** successfully accomplished the total synthesis of heitziamide A, which was reported by Lin *et al.*[71] in 2011. The authors found that the Diels–Alder adduct **130** was produced from compounds **128** and **129** under irradiation of visible light using 2 mol% Ru-catalyst as the photocatalyst as shown in Scheme 9.22. Intermediate **130** was elaborated to the total synthesis of (±)–heitziamide A (**131**) over five additional steps.

9.2.23 Total Synthesis of (±)–Hongoquercin A

In 2017, Yang *et al.*[72] developed the visible light-driven concise total synthesis of (±)–hongoquercin A (**136**). This synthesis started from farnesol (**133**), compound **134** in two steps. Using visible light and organophotocatalyst, the intermediate **134** produced tetracyclic product **135** in 60% yield (Scheme 9.23). This advanced intermediate **135** was elaborated to (±)–hongoquercin A (**136**) in four steps.

9.2.24 Total Synthesis of Indotertine A

Sun *et al.*[51] reported total synthesis of indotertine A from drimentine F in 2013. We have discussed the total synthesis of drimentine F (**87**) using photoredox catalysis in an earlier

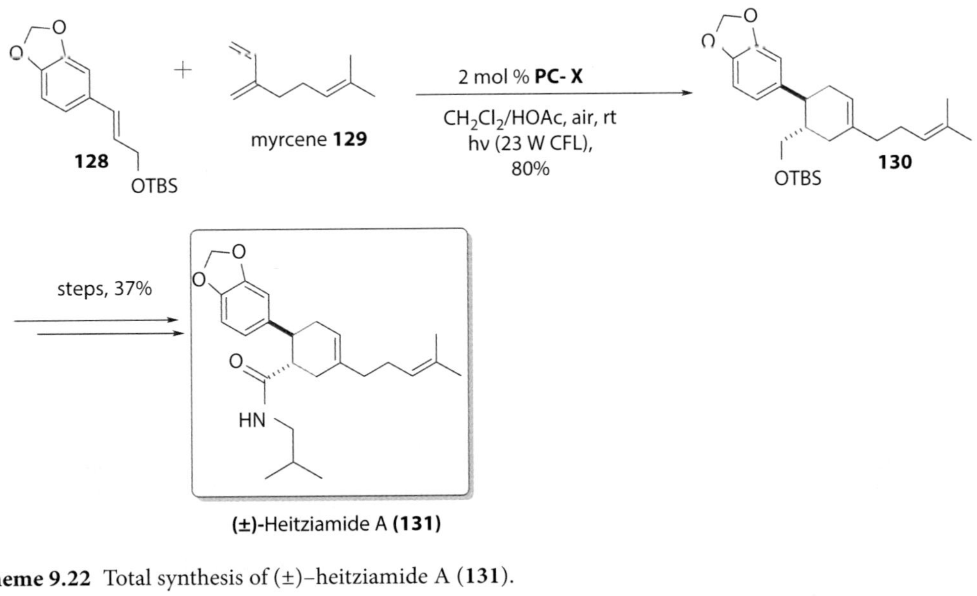

Scheme 9.22 Total synthesis of (±)–heitziamide A (**131**).

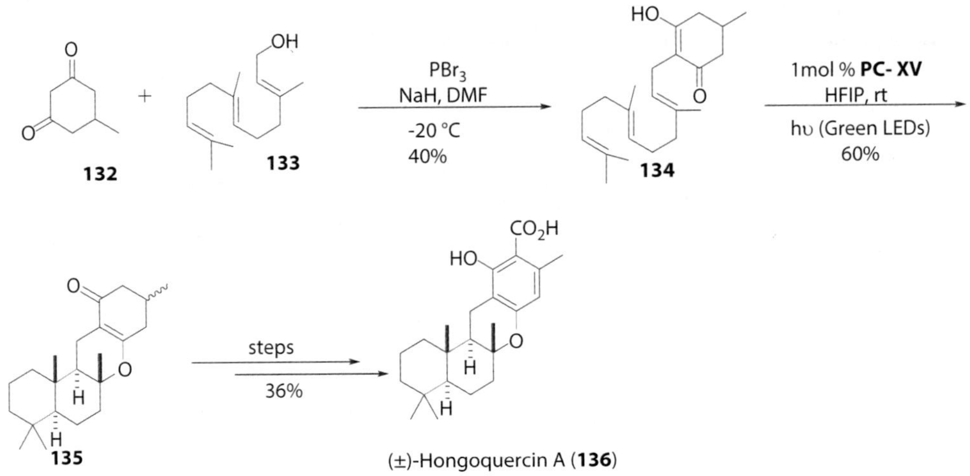

Scheme 9.23 Total synthesis of (±)-hongoquercin A (**136**).

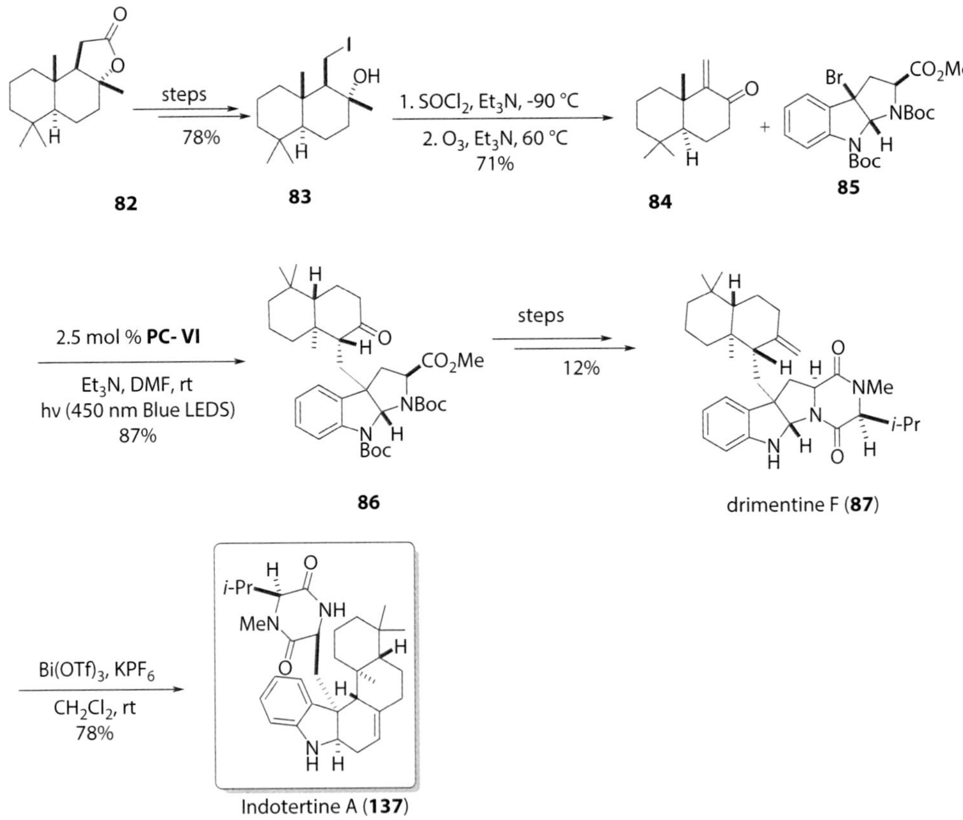

Scheme 9.24 Total synthesis of indotertine A (**137**).

section as shown in Scheme 9.15. The compound drimentine F (**87**) was treated with bismuth triflate under acidic conditions that afforded indotertine A (**137**) in 78% yield (Scheme 9.24).

9.2.25 Total Synthesis of (–)-Kadsulignan E

In 2021, Huang and Lumb[73] reported the total synthesis of (–)-kadsulignan (**143**) using visible light photoredox catalysis *via* oxidative radical cyclization process. Accordingly, they prepared the intermediate **139** prepared from 3,4-dimethoxy-5-hydroxybenzaldehyde over 21 steps. Using visible light irradiation, followed by decarboxylation, the intermediate phenoxy methyl radical intermediate **140** is produced from intermediate **139** as shown in Scheme 9.25. The intermediate **140** undergoes 5-exo-trig cyclization followed by a one-electron oxidation reaction resulting in pentadienyl carbocation **142**. Consequently, the intermediate **142** was trapped by H_2O, and loss of methanol produced *p*-spirodienone (–)-kadsulignan E (**143**).

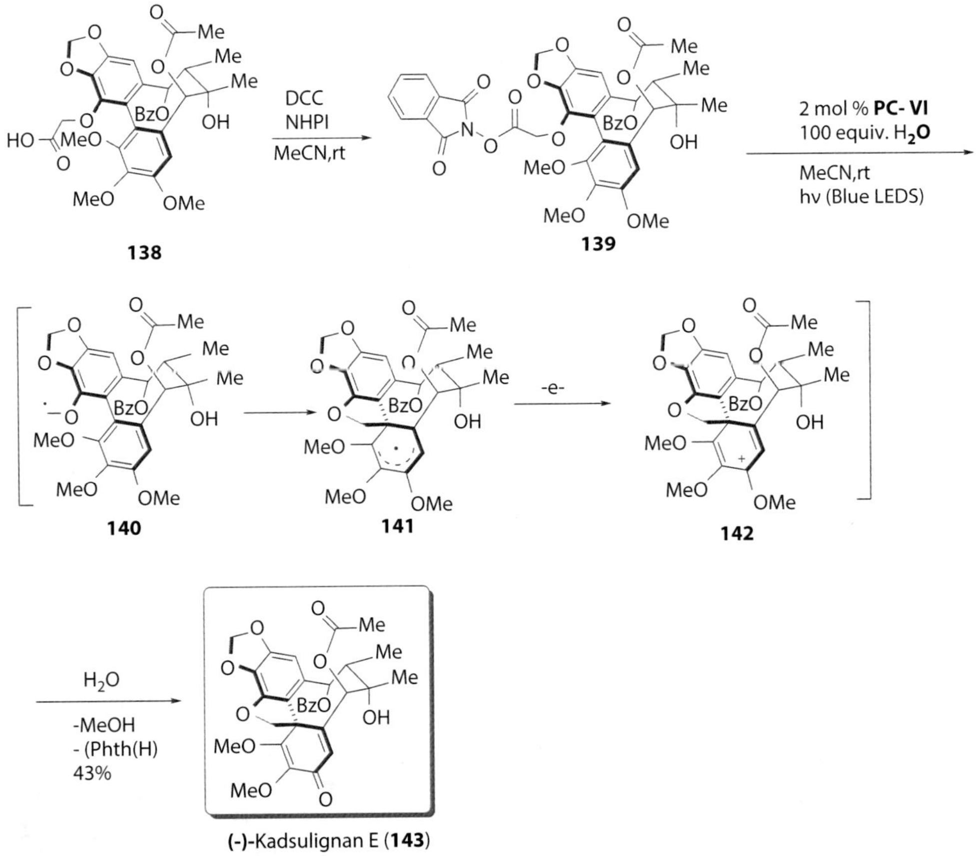

Scheme 9.25 Total synthesis of (–)-kadsulignan E (**143**).

9.2.26 Total Synthesis of (±)–Leptosperol B

Leptosperol B (**148**) is a cinnamoyl phloroglucinol–sesquiterpenoid hybrid isolated from *Leptospermum scoparium*.[74] *L. scoparium* is an herbal medicine, which was applied to cure urinary and skin disorders. It is obtained from manuka honey, which is found in New Zealand. In a seminal report in 2020, Gu *et al.*[74] achieved the total synthesis of leptosperol B (**148**) using visible light irradiation in the presence of photocatalyst. Starting from (–)–menthone (**144**), (–)–cadina-1,4-diene which **145** was prepared in six steps. Natural product leptosperol B (**148**) is produced upon irradiation of (–)–cadina-1(6),4-diene **145** by exploiting De Mayo reaction as shown in Scheme 9.26. The compound **146** was treated with a photocatalyst, and anhydrous DCM solution provided the key intermediate **147**, which was further transformed into the desired product leptosperol B.

9.2.27 Total Synthesis of (–)–Macfarlarndin C

In 2020, Allred *et al.*[75] reported the preparation of diterpenoid Macfarlarndin C (**155**). They successfully utilized visible light photoredox catalysis to synthesize the key intermediate **154** from the intermediates **152** and **153** using 1 mol% of Ir-photocatalyst as shown in Scheme 9.27. The compound **154** was finally elaborated to provide the required natural product macfarlarndin C (**155**).

9.2.28 Total Synthesis of (+)–Monomorine I

Monomorine I (**160**) is a bicyclic amine that was isolated from the trail pheromone of *Monomorium pharaonis*[76], and it first isolated by Ritter *et al.*[76] McManus *et al.*[77] employed a late-stage alkylation of carbamate to synthesize the natural product monomorine I. They found that *Boc* proctated 2-methylpiperidine (**156**) treated with photocatalyst in dichloromethane-generated α-carbamyl radical cation **158**[77] as shown in Scheme 9.28. This radical undergoes a Giese coupling reaction with enone **157** to produce **159** in a 54% yield.

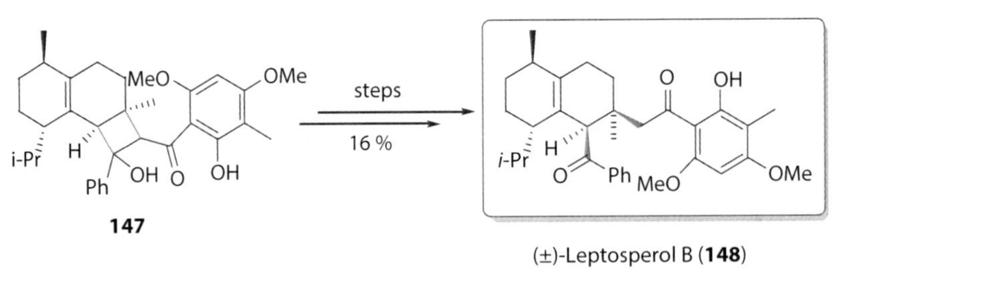

Scheme 9.26 Total synthesis of (±)–leptosperol B (**148**).

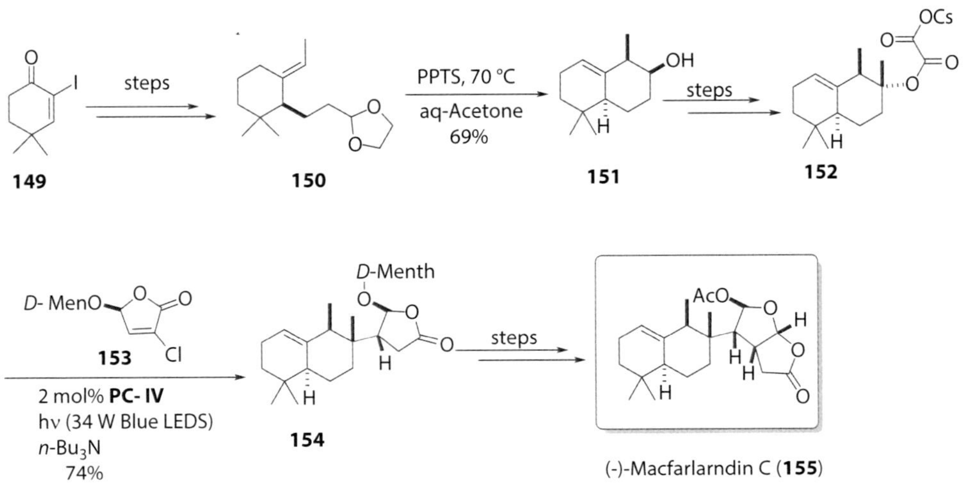

Scheme 9.27 Total synthesis of (–)-macfarlarndin C (**155**).

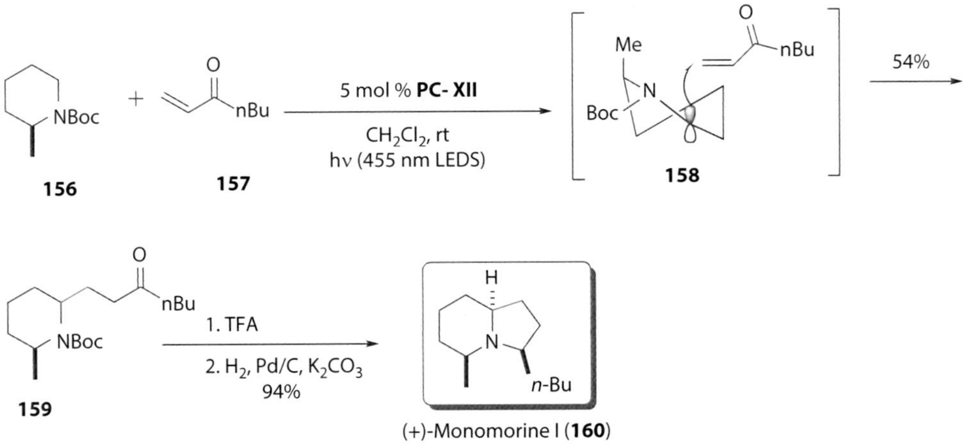

Scheme 28 Total synthesis of (+)-monomorine I (**160**).

Subsequent deprotection followed by reductive amination afforded the natural product monomorine I (**160**) in 51% yield.

9.2.29 Total Synthesis of (±)-Norruspoline

In a seminal report, in 2014, Noble and MacMillan[78] achieved visible light-driven total synthesis of pyrrolidine alkaloid (±)-norruspoline (**164**) by applying photoredox catalysis. This three-step synthesis begins with the coupling of α-amino acids (**162**) and compound **161** in the presence of photocatalyst (**PC-VI**) in 1,4-dioxane at 50°C, furnishing the radical intermediate **163** *via* a catalytic oxidative decarboxylation reaction as shown in Scheme 9.29. Consequently, treatment of the radical intermediate **163** with TFA produced the natural product (±)-norruspoline (**164**) in 76% yield.

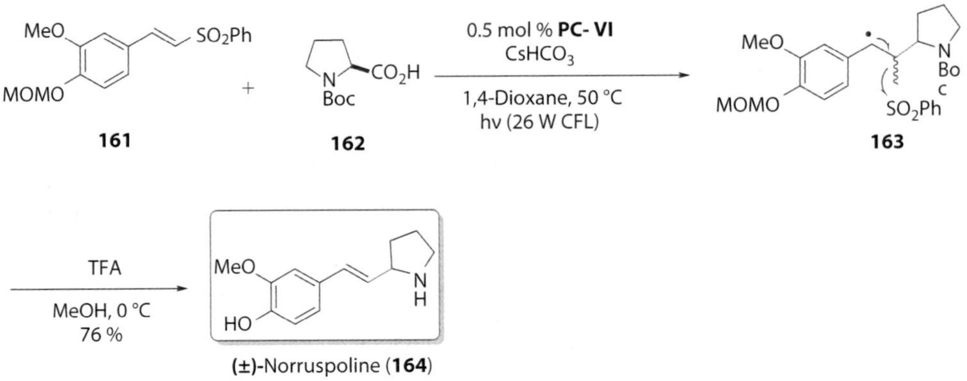

Scheme 9.29 Total synthesis of (±)-norruspoline (**164**).

9.2.30 Total Synthesis of (±)-Oxycodone

Oxycodone is an opium alkaloid that is used as a pain killer. In a seminal report, Park and Chen[79] achieved the total synthesis of oxycodone (**168**). The compound **165** was elaborated to give the advanced intermediate **166**, which upon visible light irradiation provided the compound **167** as shown in Scheme 9.30. This compound **167** was further applied in order to get the compound oxycodone (**168**).

9.2.31 Total Synthesis of (–)-Pavidolide B

Pavidolide B is a tetracyclic diterpenoid, and it was isolated for the first time in 2012 from the marine soft coral *Sinularia pavida* by Shen *et al.*[80] It shows highly selective inhibitory activity against several human promyelocytic leukemia cell lines. The enantioselective synthesis of pavidolide B (**173**) was reported by Zhang *et al.*[81] in 2017. Accordingly, the vinylcyclopropane carboxylic ester **169** and enantiopure hydroxy cyclohexanone **170** undergo Mitsunobu coupling to form intermediate **171**. The crucial intramolecular [3+2]

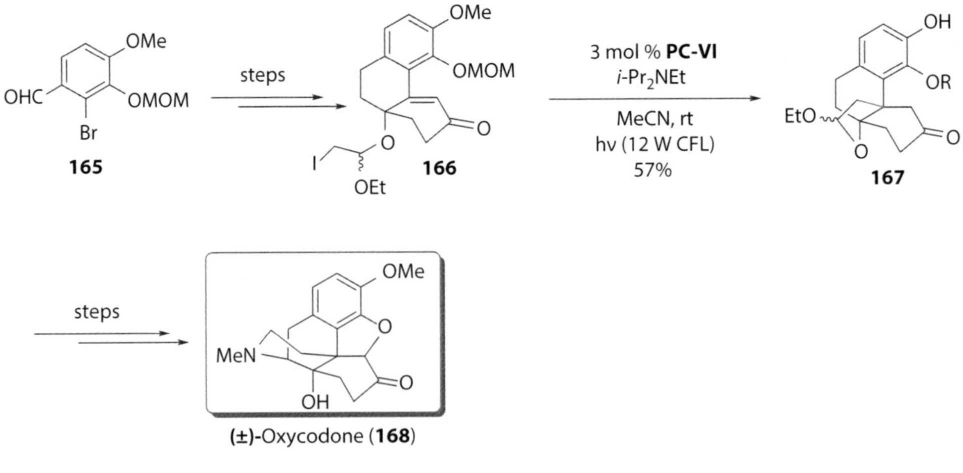

Scheme 9.30 Total synthesis of (±)-oxycodone (**168**).

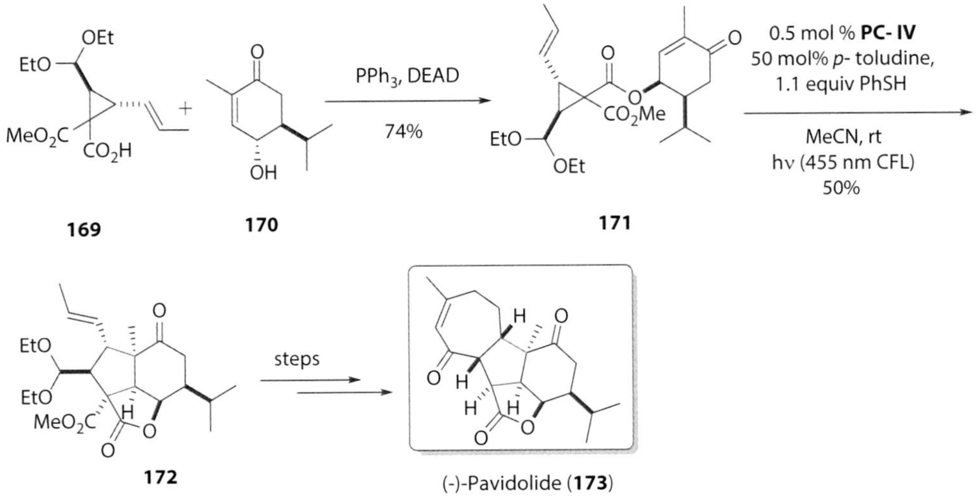

Scheme 9.31 Total synthesis of (−)-pavidolide B (**173**).

cycloaddition of **171** produced compound **172** using thiophenol and AIBN employing visible light irradiation as shown in Scheme 9.31. The compound **172** was elaborated to the synthesis of (−)-pavidolide B (**173**).

9.2.32 Total Synthesis of (±)-Pellucidin A

Pellucidin A (**175**) was isolated from the stem bark of *Pachypodanthium staudtii* by Ngadjui *et al.*[82] in 1989. By reducing the number of writhes, pellucidin A showed an antinociceptive effect against chemically induced pain in mice.[83] In 2013, Riener and Nicewicz[84] synthesized lignin natural products (±)-pellucidin A by applying visible light photocatalysis. Structurally, it contains a cyclobutane ring that can be achieved by [2+2] cycloadditions of electron-rich styrenes, i.e., *2,4,5*-trimethoxystyrene **174** as shown in Scheme 9.32. The natural product pellucidin A (**175**) was produced upon irradiation of compound **174** in the presence of a 3 mol% photocatalyst (**PC-XIII**) in acetone at low temperature.

9.2.33 Total Synthesis of Pentachloropseudilin

Pseudilin is a marine natural product, and it exhibits antibiotic, antitumor, and phytotoxic activities.[85–87] Pentachloropseudine was isolated from Actinoplanes by Cavalleri *et al.*[88] in 1978.

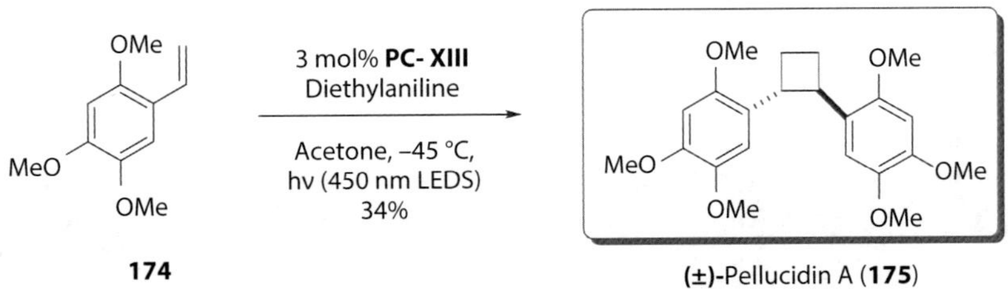

Scheme 9.32 Total synthesis of (±)-pellucidin A (**175**).

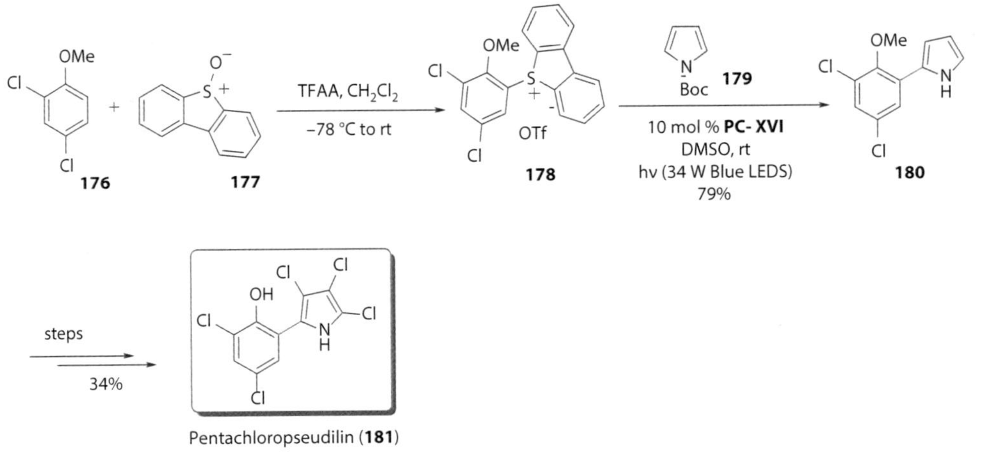

Scheme 9.33 Total synthesis of pentachloropseudilin (**181**).

The key step in the preparation of pentachloropseudilin is the coupling of two aryl radicals, which was achieved by Aukland *et al.*[89] using photoredox catalysis. They prepared triaryl-sulfonium salt **178** from arene **176** and dibenzothiophene *S*-oxide (**177**). Then, the coupling of sulfonium salt **178** and *N*-Boc-pyrrole (**179**) produced 2-aryl pyrrole **180** in 79% yield by employing the organic photocatalyst using visible light as shown in Scheme 9.33. The compound **180** was utilized further to complete the enantioselective synthesis of penta-chloropseudilin **181**.

9.2.34 Total Synthesis of (–)-Polyoxamic Acid

Polyoxamic acid is a natural nucleoside peptide antibiotic isolated from polyoxins.[90, 91] It was a polyhydroxy α-amino acid with three consecutive chiral centers. The preparation of poly-oxamic acid (**189**) was accomplished by Matsuoka *et al.*[92] in 2020 by employing visible light photoredox catalysis. This synthesis was started from the commercially available *L*-glucose (**182**).[92] The pivotal step in this chemical transformation involves radical intermediate using hydrogen atom transfer. Subjecting *N*-alkoxyphthalamide precursor **184** in the presence of photocatalyst under visible light irradiation generated benzylidene acetal **188** in 63% yield. They suggested the following transformation proceeds through three steps, i.e., β-scission, 1,5- hydrogen atom transfer process, and cyclization as shown in Scheme 9.34. Compound **188** was elaborated to the total synthesis of (–)-polyoxamic acid (**189**).

9.2.35 Total Synthesis of (±)-Protolichesterinic Acid

Protolichesterinic acid (**196**) was isolated from *Cetraria islandical* and *Cladonia papillaria.*[93] It shows antifungal activity and inhibitory activity toward colon carcinoma.[94–96] In 2014, Zeller *et al.*[97] reported the total synthesis of protolichesterinic acid (**196**). In this chemical transformation, the substituted styrene styrene **190** upon visible light irradiation generated the radical species **192**. Subsequently, the radical species **192** intermolecularly treated with

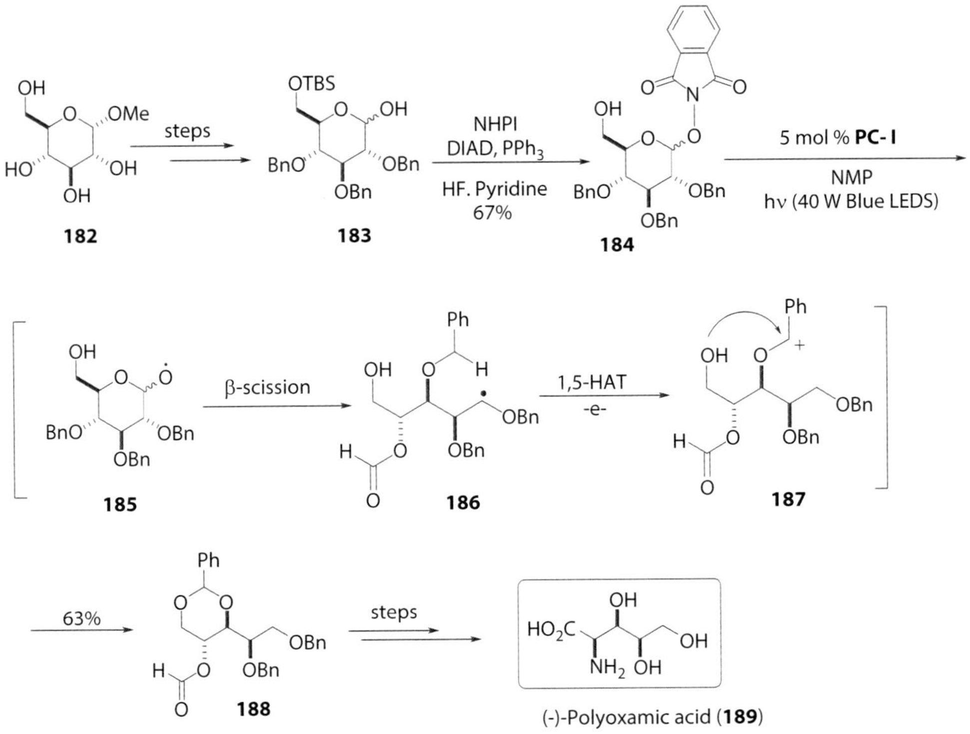

Scheme 9.34 Total synthesis of (–)-polyoxamic acid (**189**).

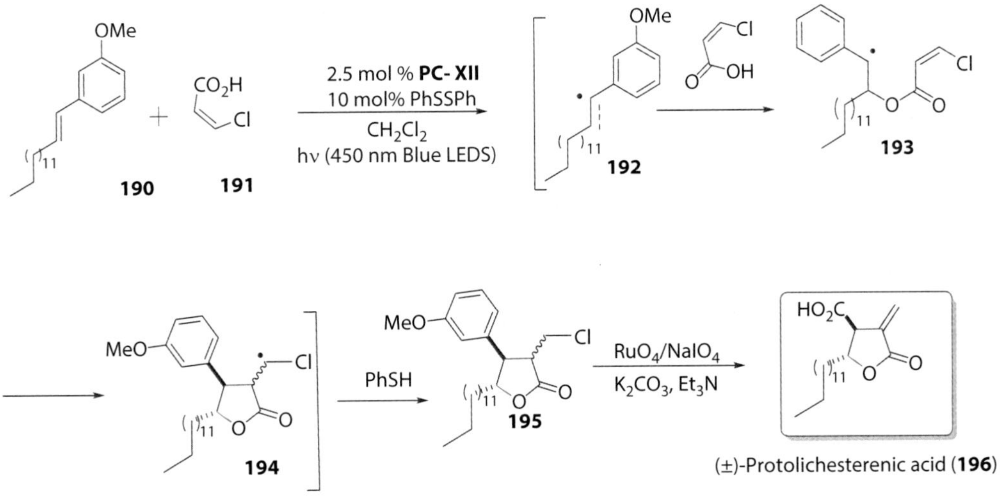

Scheme 9.35 Total synthesis of (±)-protolichesterinic acid (**196**).

acid compound **191** followed by cyclization and on treatment with thiophenol gave compound **195**. Compound **195** then underwent oxidation with a $RuO_4/NaIO_4$ system followed by dechlorination under basic conditions to produce the natural product protolichesterinic acid (**196**) as shown in Scheme 9.35.

9.2.36 Total Synthesis of (–)-Pseudotabersonine

Pseudotabersonine (**199**), a member of the pandoline alkaloids, was isolated from *Pandaca caducifolia* in 1975.[98] In 2014, Beatty and Stephenson achieved the total synthesis of pseudotabersonine starting from commercially available (+)-catharanthine by employing a photoredox catalysis reaction as a key step.[99] In the presence of light and photocatalyst, (+)-catharanthine (**197**) generated radical cation intermediate by breaking C_{16}–C_{21} bond. The radical species was then quenched with trimethylsilyl cyanide and formed 93% α-aminonitrile derivative (**198**) as shown in Scheme 9.36. Refluxing the intermediate **198** with trifluoracetic acid in toluene produced the desired natural product (–)-pseudotabersonine (**199**) in 90% yield.

9.2.37 Total Synthesis of (±)-Sceptrin

Sceptrin is the first dimeric pyrrole–imidazole alkaloid discovered in isolated form isolated from the Caribbean sponge Agelas sceptrum in 1981 by Faulkner, Clardy, and co-workers.[100] In 2020, Nguyen and Jamison accomplished total synthesis of (±)-sceptrin (**206**).[101] The synthesis starts with *Boc*-protected propargylamine **200**. Suzuki–Miyaura cross-coupling was used to prepare the cycloaddition precursor **202**. The compound **202** treated

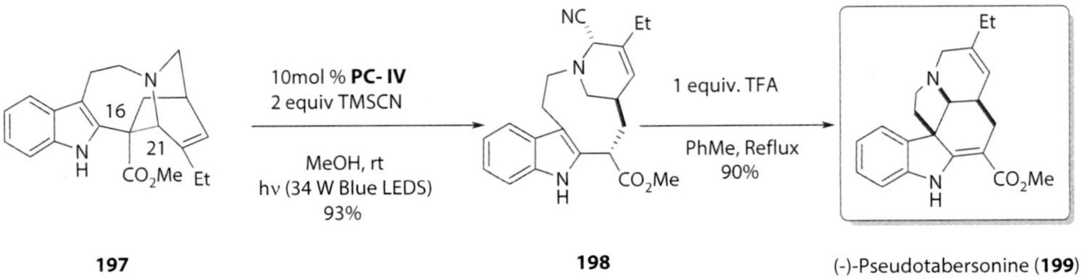

Scheme 9.36 Total synthesis of (–)-pseudotabersonine (**199**).

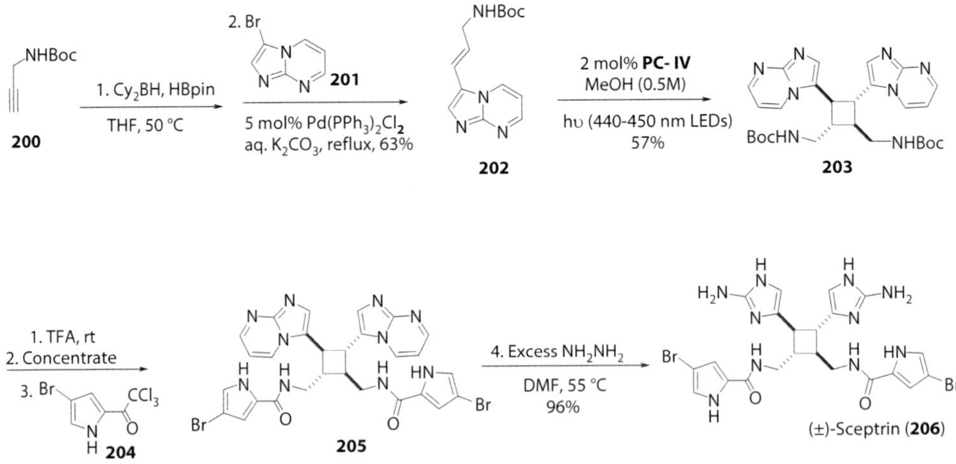

Scheme 9.37 Total synthesis of sceptrin (**206**).

with photocatalyst in methanol solution furnished 57% yield of its corresponding dimer **203** under visible light as shown in Scheme 9.37. The intermediate **203** was then converted to (±)–sceptrin (**206**) in 96% yield after three additional steps.

9.2.38 Total Synthesis of Tjipanazoles B and D

Tjipanazoles are the indolocarbazole alkaloids that were isolated from *Tolypothrix tjipanasensis*.[102] Recently, Tang *et al.*[103] introduced late-stage functionalization of indolocarbazole products to synthesize the natural product tjipanazoles B and D. This chemical transformation involves radical cyclization and dimerization using photoredox catalysis. This sequence leads to the formation of indolocarbazole product **208** from *N*-allenyl-2-iodoaniline **207** as shown in Scheme 9.38. Subsequent functional group manipulations on **208** completed the total synthesis of tjipanazole B (**209**) and tjipanazole D (**210**).

9.2.39 Total Synthesis of (±)–Tylophorine

Tylophorine is an alkaloid found in *T. indica* and exhibits distinct biological activities like antitumor and antifeedant.[104] In 2019, Zhang *et al.*[105] synthesized quaternary carbon-centered benzoindolizidines by exploiting photoredox catalysis in order to get the natural product tylophorine. The core starting material **213** is produced from coupling compounds **211** and **212** after two additional steps. The intermediate **213** upon visible light irradiation produced compound **217** as shown in Scheme 9.39. They found that the reaction proceeds through the formation of benzamidyl radical **214** followed by cyclization. Consequently, they successfully synthesized the natural product tylophorine (**218**) by reduction of intermediate **217**.

9.2.40 Total Synthesis of (–)–Vincorine

Vincorine is an akuammiline alkaloid isolated from *Vinca minor* L.[106] Structurally, it contains a bicyclic framework, with a hexahydro pyrrolo [2,3,b] indole skeleton. Stanek *et al.*[107]

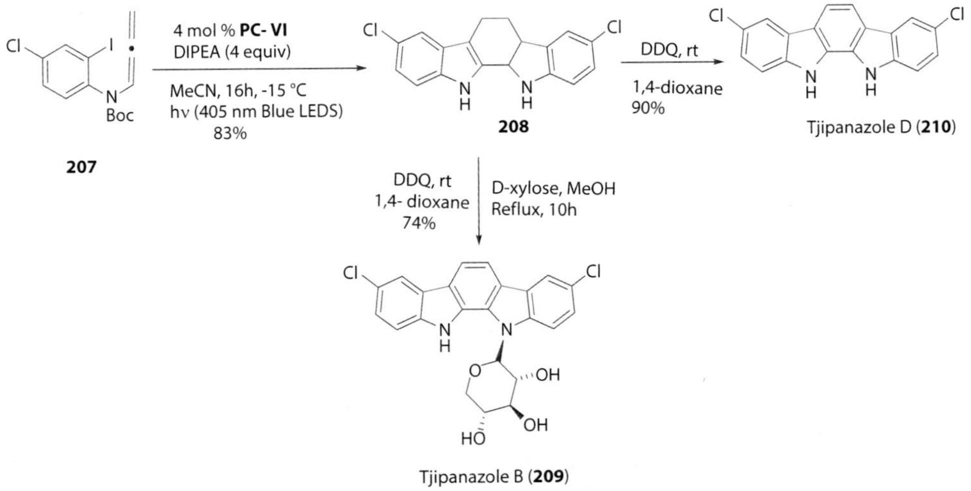

Scheme 9.38 Total synthesis of tjipanazole B (**209**) and tjipanazole D (**210**).

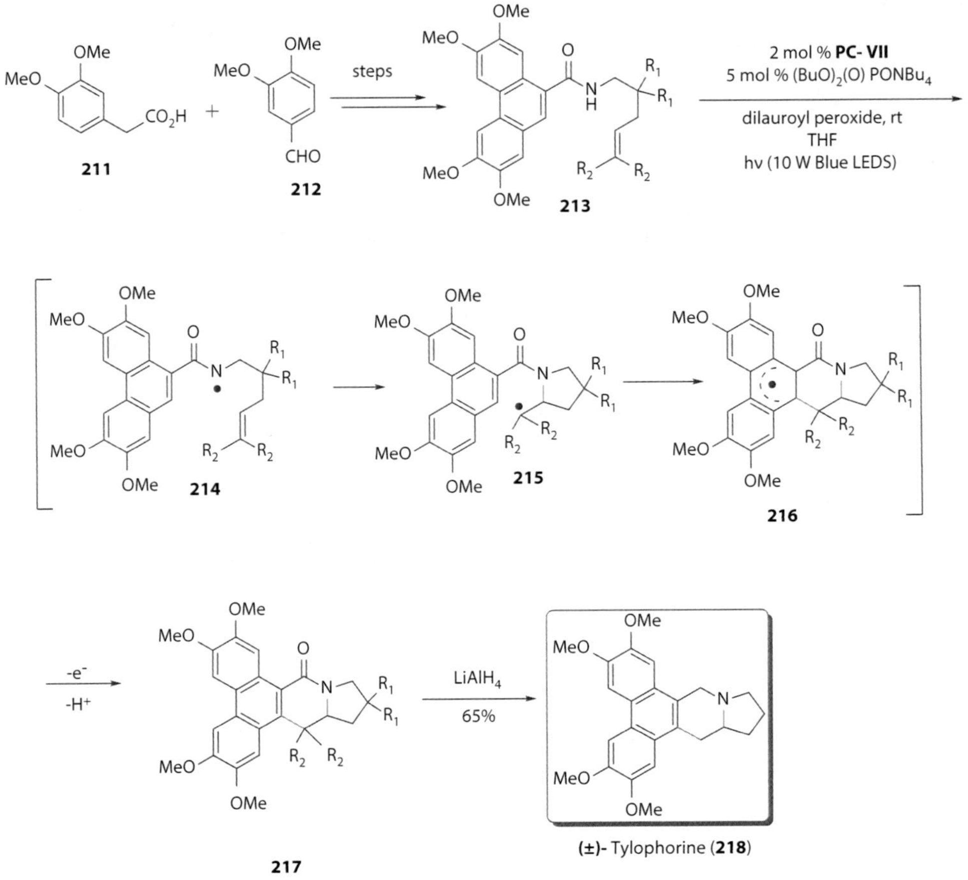

Scheme 9.39 Total synthesis of (±)-tylophorine (**218**).

reported total synthesis of the akuammiline alkaloid (–)-vincorine under visible light irradiation. The synthesis started from 5-methoxy indole derivative **219**, which undergoes oxygenation with photocatalyst under visible light irradiation, and produced compound **220**. Compound **221** is obtained from the compound **220** under nitro aldol condensation

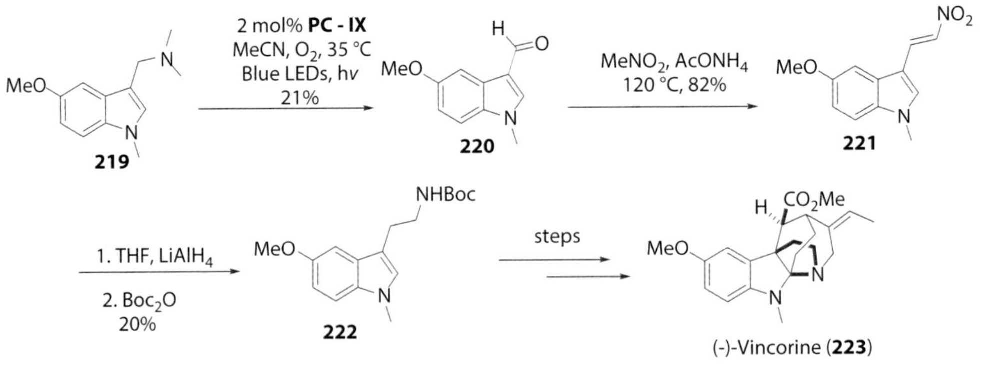

Scheme 9.40 Synthesis of (–)-vincorine (**223**).

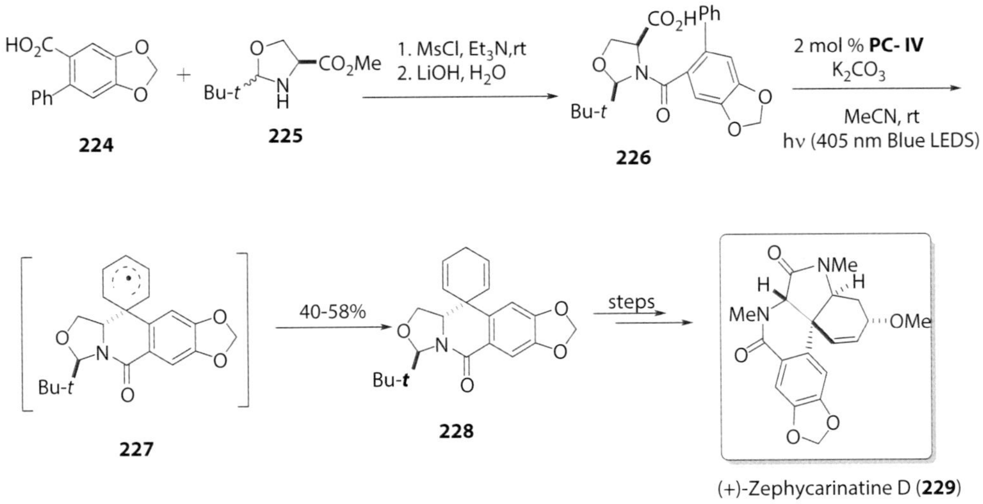

Scheme 9.41 Total synthesis of (+)–zephycarinatine D (**229**).

reaction as shown in Scheme 9.40. The compound **220** underwent nitration, followed by *Boc* protection and furnished the compound **222** in total 20% yield, which finally elaborated to produce (–)–vincorine.

9.2.41 Total Synthesis of (+)–Zephycarinatine D

Zephycarinatines belong to a plicamine family of amayllidaceae alkaloids.[108] In 2020, Takeuchi *et al.*[109] achieved the total syntheses of (+)–zephycarinatines C and D *via* reductive radical *ipso*-cyclization. The crucial step in this stereo-controlled preparation of alkaloids is the synthesis of spirocyclic ring with a quaternary carbon stereocenter. Initially, they synthesized an amide derivative **226** from the compound **224** and compound **225**. The compound **226** was allowed to react with photocatalyst and K_2CO_3 in acetonitrile solution under visible light irradiation resulting in the compound **228** *via* the formation of spirodienyl radical intermediate **227** as shown in Scheme 9.41. The compound **228** was then used to produce the desired natural product (+)–zephycarinatine D (**229**).

9.3 Visible Light-Driven Functionalization of Natural Products

Chemists started employing photoredox-catalyzed late-stage functionalization of natural products recently. Functionalization of natural products with photoredox catalysis are quite few when compared to total synthesis. However, we have described a few such processes involving photoredox-mediated functionalization of natural products in the following section in alphabetical order.

9.3.1 Functionalization of Aminoestrone

In 2014, Xie *et al.*[110] reported trifluoromethylation of anilines using Togni reagent (**231**) using photoredox catalysis. Applying this protocol, they were able to functionalize

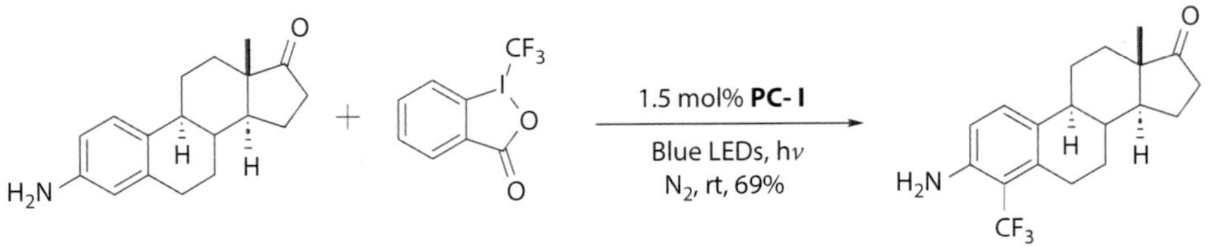

3-Aminoestrone (**230**) Togni Reagent (**231**) 3-Aminoestrone-CF_3 (**232**)

Scheme 9.42 Functionalization of aminoestrone (**230**).

3-aminoestrone (**230**), resulting in a corresponding 3-aminoestrone-CF_3 (**232**) motif in 69% yield as shown in Scheme 9.42.

9.3.2 Functionalization of Caffeine

Baar and Blechert[111] achieved heterogeneous catalyzed trifluoromethylation of arenes utilizing mesoporous graphitic carbon nitride (mpg-CN) as a photocatalyst under visible light irradiation. Caffeine (**233**) was treated with trifluoro sulfonyl chloride under standard conditions, which afforded the trifluoromethylated caffeine (**234**) in 75% yield as shown in Scheme 9.43. In 2015, Beatty *et al.*[112] also developed a Ru-catalyzed trifluoromethylation reaction by utilizing a pyridine *N*-oxide-mediated photoredox catalysis. Accordingly, the trifluoromethylated caffeine (**234**) compound was prepared in 54% yield[112] as shown in Scheme 9.44.

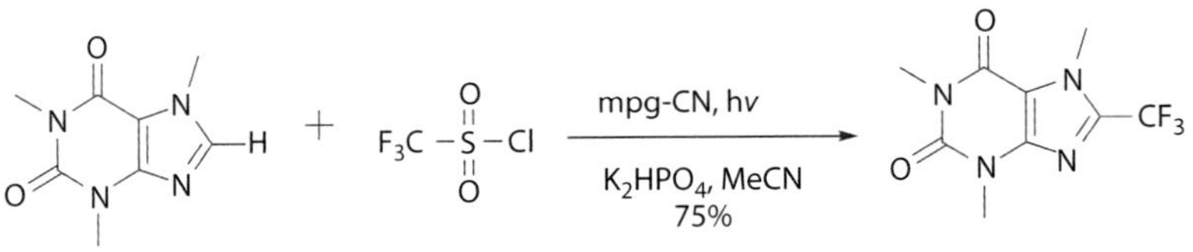

Caffeine-CF_3 (**234**)

Caffeine (**233**)

Scheme 9.43 Functionalization of caffeine.

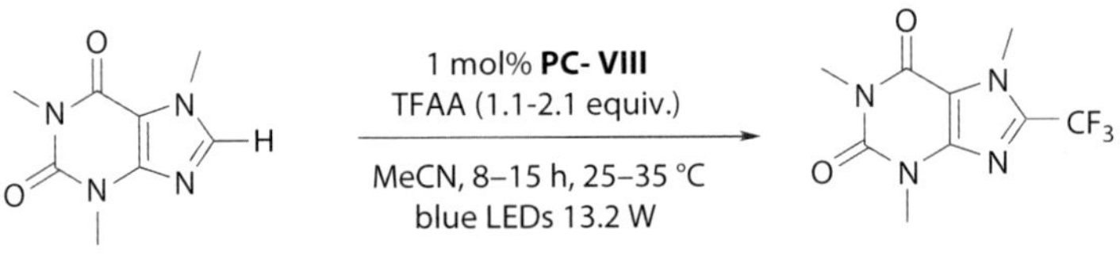

Caffeine (**233**) Caffeine-CF_3 (**234**)

Scheme 9.44 Functionalization of caffeine.

9.3.3 Functionalization of Cedrol

More than 50 sesquiterpenes containing the seco-prezizaane skeleton have been isolated from plants of the *Illicium* genus, which includes pseudoanisatin (**237**), majucin (**240**), and jiadifenoxolane A (**241**). In 2016, Hung *et al.*[113] reported the total synthesis of pseudoanisatin. This synthesis started from the natural product cedrol (**235**). Cedrol initially treated with PIDA under visible light irradiation produced compound (**236**). The intermediate (**236**) was elaborated to give the natural product pseudoanisatin (**237**) in 11 steps as shown in Scheme 9.45.

They further attempted to synthesize other seco-prezizaane skeletons like majucin (**240**) and jiadifenoxolane A (**241**) using cedrol (**235**) as the starting material after successfully achieving the total synthesis of pseudoanisatin (**237**).[114] They also exploited visible light irradiation throughout the experiments like before. For instance, the compound **238** was generated from **236** upon visible light radiation, producing intermediate **239**. Subsequently, they synthesized majucin (**240**) and jiadifenoxolane A (**241**) as shown in **Scheme 9.46.**

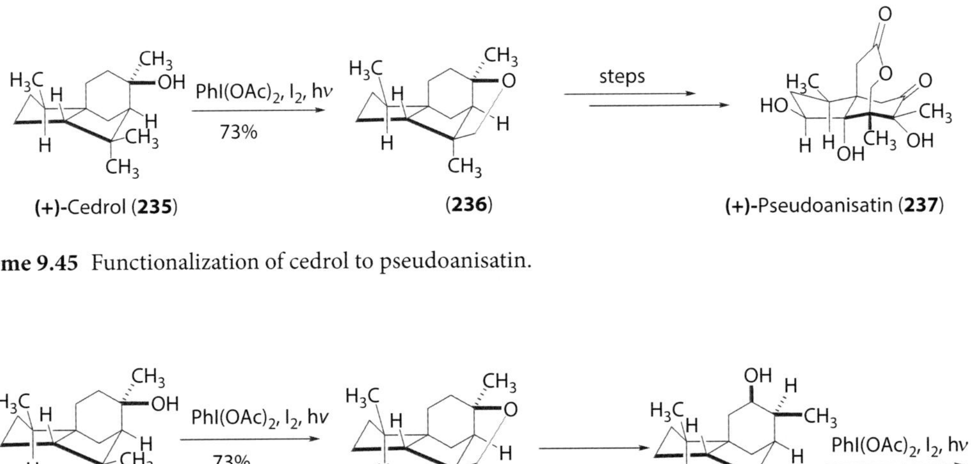

Scheme 9.45 Functionalization of cedrol to pseudoanisatin.

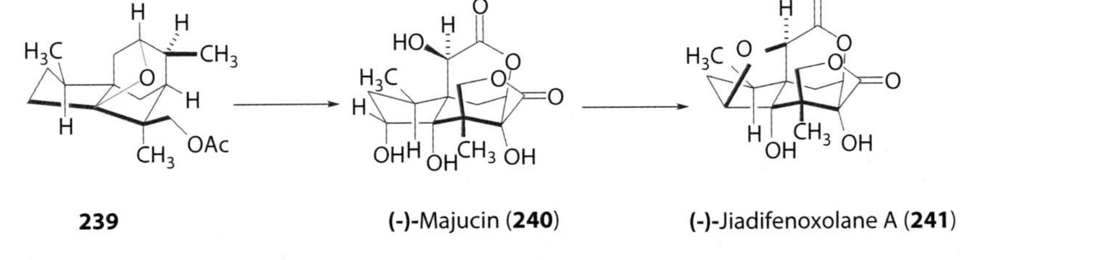

Scheme 9.46 Functionalization of cedrol to majucin and jiadifenoxolane A.

9.3.4 Functionalization of Methylvanillin

Methylvanillin is a flavoring agent found in chocolate and coffee. In 2011, Nagib and MacMillan[115] developed photoredox-catalyzed trifluoromethylation of arenes and heteroarenes. In this context, methylvanillin (**242**) on treatment with CF_3SO_2Cl and K_2HPO_4 in the presence of Ru-catalyst produced trifluoromethylated methylvanillin (**243**) in 82% yield as shown in Scheme 9.47.

9.3.5 Functionalization of Steviol to (–)–Isoatisiane (247)

Cherney *et al.*[116] reported the total synthesis of the *atisian*e motif from *ent-kaurane* (–)–steviol. Steviol is a diterpene that was first isolated from the plant *Stevia rebaudiana.*[117] This synthesis begins from steviol (**244**) that underwent functional group manipulation to produce the intermediate **245**. Treatment of the intermediate **245** with I_2 and PIDA in the presence of visible light irradiation followed by hydrolysis produced iodinated compound **246**. Compound **246** was later used to synthesize (–)–isoatisiane (**247**) as shown in Scheme 9.48.

9.3.6 Functionalization of Tocopherol

As we discussed in the earlier sections (Scheme 9.42), trifluoromethylation of free anilines was achieved by Xie *et al.*,[110] and they successfully applied this procedure for late-stage functionalization of tocopherol. Then, tocopherol (**248**) was treated with Togni reagent in

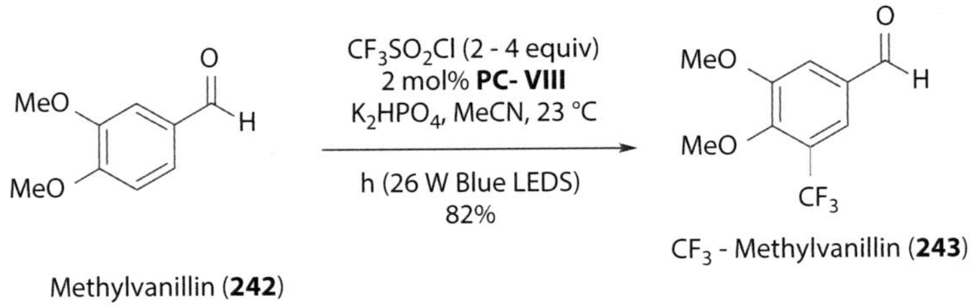

Scheme 9.47 Functionalization of methylvanillin (**242**).

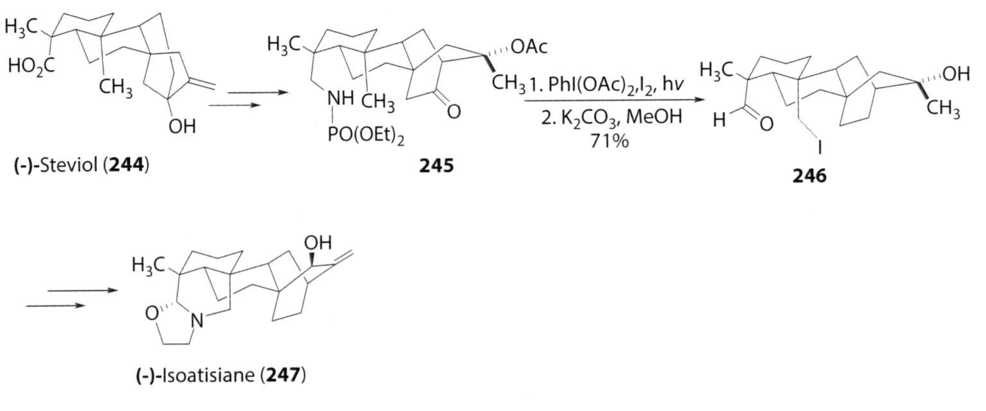

Scheme 9.48 Functionalization of steviol to (–)–isoatisiane (**247**).

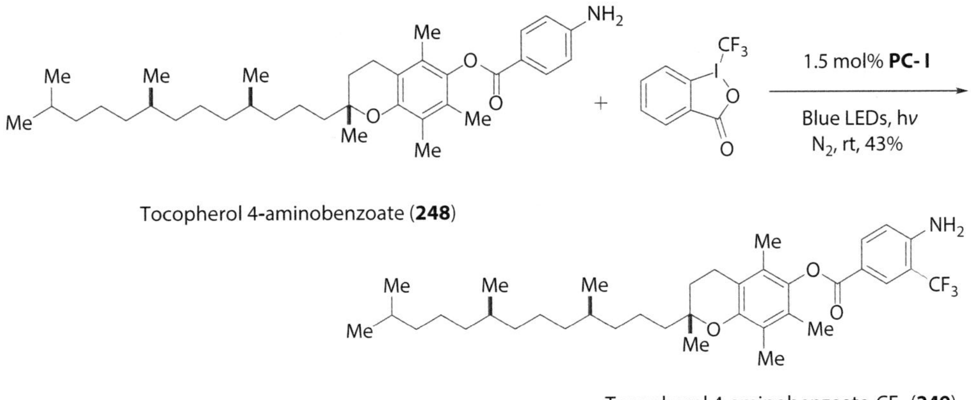

Scheme 9.49 Functionalization of tocopherol 4-aminobenzoate (**248**).

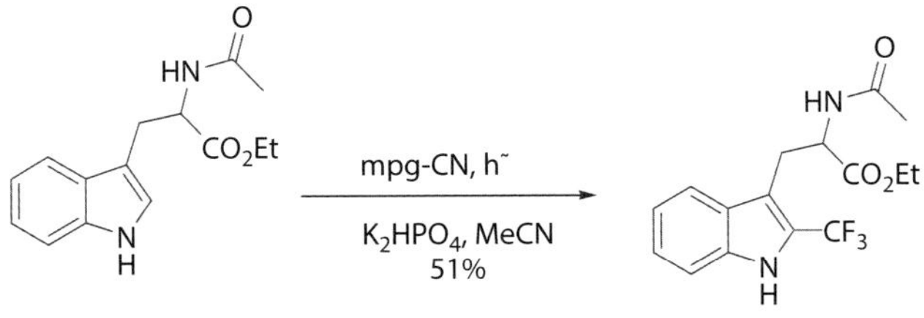

Scheme 9.50 Functionalization of tryptophan (**250**).

the presence of Ir-photocatalyst under a nitrogen atmosphere, resulting in its corresponding -CF_3-substituted product (**249**) in 43% yield as shown in Scheme 9.49.

9.3.7 Functionalization of Tyropotophan

Baar and Blechert[111] achieved heterogeneous catalyzed trifluoromethylation of arenes using mesoporous graphitic carbon nitride (mpg-CN) as a photocatalyst under visible light irradiation. Tryptotophane (**250**) on treatment with trifluoro sulfonyl chloride under standard condition afforded the tryptotophane-CF_3 (**251**) in 51% yield as shown in Scheme 9.50.

9.4 Conclusion

This chapter describes the total synthesis and functionalization of natural products with light-driven reactions. We attempted to disclose the important photoredox catalysis process that is extensively used for the total synthesis of natural products of their functionalization. Because it has influenced other fields of chemistry, such as chemical biology, medicinal chemistry, biology and biotechnology, etc. The widespread application of visible light-mediated photocatalysis has been evolving as a new chemical tool for various C-C bond formation reactions. Upcoming directions could be the use of these photocatalytic approaches immensely because of their atom economical, green, and efficient characteristics.

Acknowledgements

This work was financially supported by the Department of Science and Technology, SERB, New Delhi (Grant No. CRG/2022/003143) and the Indian Institute of Science Education and Research Berhampur. KS acknowledge IISER Berhampur for a research fellowship. G. M. thank CSIR, New Delhi, India, for the financial assistance in the form of fellowships.

References

1. Chen, Y., Lu, L.Q., Yu, D.G., Zhu, C.J., Xiao, W.J., Visible light-driven organic photochemical synthesis in China. *Sci. China Chem.*, *62*, 24, **2019**.
2. Liu, Q., and Wu, L.Z., Recent advances in visible-light-driven organic reactions. *Natl. Sci. Rev.*, *4*, 359, **2017**.
3. Wei, Y., Zhou, Q.Q., Tan, F., Lu, L.Q., Xiao, W.J., Visible-light-driven organic photochemical reactions in the absence of external photocatalysts. *Synth.*, *51*, 3021, **2019**.
4. Kaur, N., Photochemical reactions as key steps in five-membered *N*-heterocycle synthesis. *Synth. Commun.*, *48*, 1259, **2018**.
5. Chen, J., Cen, J., Xu, X., Li, X., The application of heterogeneous visible light photocatalysts in organic synthesis. *Catal. Sci. Technol.*, *6*, 349, **2016**.
6. Rehm, D. and Weller, A., Kinetics of fluorescence quenching by electron and H-Atom transfer. *Isr. J. Chem.*, *8*, 259, **1970**.
7. Pitre, S.P. and Overman, L.E., Strategic use of visible-light photoredox catalysis in natural product synthesis. *Chem. Rev.*, *122*, 1717, **2022**.
8. Drasar, P.B. and Khripach, V.A., Growing importance of natural products research. *Molecules*, *25*, 14, **2020**.
9. Hong, J., Natural product synthesis at the interface of chemistry and biology. *Chem. - Eur. J.*, *20*, 10204, **2014**.
10. Bach, T. and Hehn, J.P., Photochemical reactions as key steps in natural product synthesis. *Angew. Chem. - Int. Ed.*, *50*, 1000, **2011**.
11. Butler, M.S., The role of natural product chemistry in drug discovery. *J. Nat. Prod.*, *67*, 2141, **2004**.
12. Baran, P.S., Natural product total synthesis: As exciting as ever and here to stay. *J. Am. Chem. Soc.*, *140*, 4751, **2018**.
13. Carroll, A.R., Hyde, E., Smith, J., Quinn, R.J., Guymer, G., Forster, P.I., Actinophyllic acid, a potent indole alkaloid inhibitor of the coupled enzyme assay carboxypeptidase U/hippuricase from the leaves of *Alstonia actinophylla* (Apocynaceae). *J. Org. Chem.*, *70*, 1096, **2005**.
14. Xue, F., Lu, H., He, L., Li, W., Zhang, D., Liu, X.Y., Qin, Y., Formal total syntheses of (−)- and (+)-actinophyllic acid. *J. Org. Chem.*, *83*, 754, **2018**.
15. Nicewicz, D.A. and MacMillan, D.W.C., Merging photoredox catalysis with organocatalysis: The direct asymmetric alkylation of aldehydes. *Science*, *322*, 77, **2008**.
16. Wang, B.-G., Ebel, R., Wang, C.-Y., Wray, V., Proksch, P., New methoxylated aryltetrahydronaphthalene lignans and a norlignan from Aglaia cordata. *Tetrahedron Lett.*, *43*, 5783, **2002**.

17. Tran, T.D., Pham, N.B., Booth, R., Forster, P.I., Quinn, R.J., Lignans from the Australian endemic plant austrobaileya scandens. *J. Nat. Prod.*, *79*, 1514, **2016**.
18. Smith, P.J., Soues, S., Gottlieb, T., Falk, S.J., Watson, J.V., Osborne, R.J., Bleehen, N.M., Etoposide-induced cell cycle delay and arrest-dependent modulation of DNA topoisomerase H in small-cell lung cancer cells. *Br. J. Cancer*, *70*, 914, **1994**.
19. Xu, M., Hou, M., He, H., Gao, S., Asymmetric total synthesis of Aglacins A, B, and E. *Angew. Chem. - Int. Ed.*, *60*, 16655, **2021**.
20. Xiang, J.C., Wang, Q., Zhu, J., Radical-cation cascade to aryltetralin cyclic ether lignans under visible-light photoredox catalysis. *Angew. Chem. - Int. Ed.*, *59*, 21195, **2020**.
21. Hahn, D.R., Graupner, P.R., Chapin, E., Gray, J., Heim, D., Gilbert, J.R., Gerwick, B.C., Albucidin: A novel bleaching herbicide from Streptomyces albus subsp. chlorinus NRRL B-24108. *J. Antibiot. (Tokyo)*, *62*, 191, **2009**.
22. Zhang, H., Liu, P.F., Chen, Q., Wu, Q.Y., Seville, A., Gu, Y.C., Clough, J., Zhou, S.L., Yang, G.F., Synthesis and absolute configuration assignment of albucidin: A late-stage reductive deiodination by visible light photocatalysis. *Org. Biomol. Chem.*, *14*, 3482, **2016**.
23. Kito, K., Ookura, R., Yoshida, S., Namikoshi, M., Ooi, T., Kusumi, T., New cytotoxic 14-membered macrolides from marine-derived fungus Aspergillus ostianus. *Org. Lett.*, *10*, 225, **2008**.
24. Mateus-Ruiz, J.B. and Cordero-Vargas, A., Stereoselective total synthesis of Aspergillide A: A visible light-mediated photoredox access to the trisubstituted tetrahydropyran core. *J. Org. Chem.*, *84*, 11848, **2019**.
25. Tokuyama, T., Daly, J.W., Witkop, B., The structure of batrachotoxin, a steroidal alkaloid from the colombian arrow poison frog, phyllobates aurotaenia, and partial synthesis of batrachotoxin and its analogs and homologs. *J. Am. Chem. Soc.*, *91*, 3931, **1969**.
26. Guo, Y., Guo, Z., Lu, J.T., Fang, R., Chen, S.C., Luo, T., Total synthesis of (−) batrachotoxinin A: A local-desymmetrization approach. *J. Am. Chem. Soc.*, *142*, 3675, **2020**.
27. Rattanaburee, T., Thongpanchang, T., Wongma, K., Tedasen, A., Sukpondma, Y., Graidist, P., Anticancer activity of synthetic (±)-kusunokinin and its derivative (±)-bursehernin on human cancer cell lines. *Biomed. Pharmacother.*, *117*, 109115, **2019**.
28. Rojas-Sepulveda, A.M., Mendieta-Serrano, M., Mojica, M.Y.A., Salas-Vidal, E., Marquina, S., Villarreal, M.L., Puebla, A.M., Delgado, J.I., Alvarez, L., Cytotoxic podophyllotoxin type-lignans from the steam bark of Bursera fagaroides var. fagaroides. *Molecules*, *17*, 9506, **2012**.
29. Welin, E.R., Warkentin, A.A., Conrad, J.C., MacMillan, D.W.C., Enantioselective α-alkylation of aldehydes by photoredox organocatalysis: Rapid access to pharmacophore fragments from β-Cyanoaldehydes. *Angew. Chem. - Int. Ed.*, *54*, 9668, **2015**.
30. Qin, C., Mei, Y., Zhou, X., Ping, A., Huang, S., Five monoterpenes from leaves of Rhododendron anthopogon. *Zhongguo Zhong Yao Za Zhi*, *35*, 2568, **2010**.
31. Lu, Z. and Yoon, T.P., Visible light photocatalysis of [2+2] styrene cycloadditions by energy transfer. *Angew. Chem. - Int. Ed.*, *51*, 10329, **2012**.
32. Ackland, M.J., Hanson, J.R., Hitchcock, P.B., Ratcliffe, A.H., Structures of the cephalosporolides B-F, a group of C_{10} lactones from cephalosporium aphidicola. *J. Chem. Soc., Perkin Trans.*, *1*, 843, **1985**.
33. Xochicale-Santana, L., Cortezano-Arellano, O., Frontana-Uribe, B.A., Jimenez-Perez, V.M., Sartillo-Piscil, F., The Stereoselective total synthesis of the elusive cephalosporolide F. *J. Org. Chem.*, *88*, 4880, **2023**.
34. Morita, H., Hirasawa, Y., Shinzato, T., Kobayashi, J., New phlegmarane-type, cernuane-type, and quinolizidine alkaloids from two species of Lycopodium. *Tetrahedron*, *60*, 7015, **2004**.

35. Benimana, S.E., Cromwell, N.E., Meer, H.N., Marvin, C.C., Visible light photoredox and Polonovski-Potier cyclizations for the synthesis of (±)-5-epi-cermizine C and (±)-epimyrtine *Tetrahedron Lett.*, *57*, 5062, **2016**.
36. Tao, D.J., Slutskyy, Y., Overman, L.E., Total Synthesis of (−)-Chromodorolide B. *J. Am. Chem. Soc.*, *138*, 2186, **2016**.
37. Rungprom, W., Chavasiri, W., Kokpol, U., Kotze, A., Garson, M.J., Bioactive chromodorolide diterpenes from an aplysillid sponge. *Mar. Drugs*, *2*, 101, **2004**.
38. Katavic, P. L., Jumaryatno, P., Hooper, J. N. A., Blanchfield, J. T., Garson, M. J., Oxygenated terpenoids from the australian sponges coscinoderma matthewsi and dysidea sp., and the nudibranch chromodoris albopunctata. *Aust. J. Chem.*, *65*, 531, **2012.**
39. Morris, S.A., Silva, E.D.D., Andersen, R.J., Chromodorane diterpenes from the tropical dorid nudibranch Chromodoris cavae. *Can. J. Chem.*, *69*, 768, **1991**.
40. Uddin, M.H., Hossain, M.K., Nigar, M., Roy, M.C., Tanaka, J., New Cytotoxic Spongian-class rearranged diterpenes from a marine sponge. *Chem. Nat. Compd.*, *48*, 412, **2012**.
41. Anderton, N., Cockrum, P.A., Colegate, S.M., Edgar, J.A., Flower, K., Vit, I., Willing, R.I., Oxindoles from phalaris coerulescens. *Phytochemistry*, *48*, 437, **1998**.
42. Chen, J.Q., Wei, Y.L., Xu, G.Q., Liang, Y.M., Xu, P.F., Intramolecular 1,5-H transfer reaction of aryl iodides through visible-light photoredox catalysis: a concise method for the synthesis of natural product scaffolds. *Chem. Commun.*, *52*, 6455, **2016**.
43. Zhang, Q., Tu, G., Zhao, Y., Cheng, T., Novel bioactive isoquinoline alkaloids from Carduus crispus. *Tetrahedron*, *58*, 6795, **2002**.
44. Orejarena Pacheco, J.C., Lipp, A., Nauth, A.M., Acke, F., Dietz, J.P., Opatz, T., A highly active system for the metal free aerobic photocyanation of tertiary amines with visible light: Application to the synthesis of tetraponerines and crispine A. *Chem. - Eur. J.*, *22*, 5409, **2016**.
45. Chorley, D.F., Chen, J.L.Y., Furkert, D.P., Sperry, J., Brimble, M.A., Total synthesis of danshenspiroketallactone. *Synlett*, *8*, 128, **2012**.
46. Zhou, L., Zuo, Z., Chow, M.S.S., Danshen: An overview of its chemistry, pharmacology, pharmacokinetics, and clinical use. *J. Clin. Pharmacol.*, *45*, 1345, **2005**.
47. Deng, Y., Nguyen, M.D., Zou, Y., Houk, K.N., Smith, A.B., Generation of dithianyl and dioxolanyl radicals using photoredox catalysis: Application in the total synthesis of the danshenspiroketallactones via radical relay chemistry. *Org. Lett.*, *21*, 1708, **2019**.
48. Yu, B.W., Meng, L.H., Chen, J.Y., Zhou, T.X., Cheng, K.F., Ding, J., Qin, G.W., Cytotoxic oxoisoaporphine alkaloids from Menispermum dauricum. *J. Nat. Prod.*, *64*, 968, **2001**.
49. Zhang, J., Chen, J., Zhang, X., Lei, X., Total syntheses of menisporphine and daurioxoisoporphine C enabled by photoredox-catalyzed direct C–H arylation of isoquinoline with aryldiazonium salt. *J. Org. Chem.*, *79*, 10682, **2014**.
50. Che, Q., Zhu, T., Qi, X., Mandi, A., Kurtan, T., Mo, X., Li, J., Gu, Q., Li, D., Hybrid isoprenoids from a reeds rhizosphere soil derived actinomycete streptomyces sp. CHQ-64. *Org. Lett.*, *14*, 3438, **2012**.
51. Sun, Y., Li, R., Zhang, W., Li, A., Total synthesis of indotertine A and drimentines A, F, and G. *Angew. Chem. - Int. Ed.*, *52*, 9201, **2013**.
52. Snider, B.B. and Lu, Q., Syntheses Of (±)-Raikovenal, (±)-Preraikovenal, and (±)- Epiraikovenal. *Synth. Commun.*, *27*, 1583, **1997**.
53. Guella, G., Pietra, F., Dini, F., From epiraikovenal, an instrumental niche-exploitation sesquiterpenoid of some strains of the marine ciliated protist euplotes uaikovi, to an unusual intramolecular tele-Dienone-Olefin [2+2] photocycloaddition *Helv. Chim. Acta*, *78*, 1747, **1995**.
54. Hurtley, A.E., Lu, Z., Yoon, T.P., [2+2] Cycloaddition of 1,3-Dienes by visible light photocatalysis. *Angew. Chem. - Int. Ed.*, *53*, 8991, **2014**.

55. Gallagher, T. and Magnus, P., Synthesis of (±)-Kopsanone and (±)- 10,22-Dioxokopsane, heptacyclic indole alkaloids. *J. Am. Chem. Soc.*, *105*, 2086, **1983**.
56. Tong, X., Shi, B., Liang, K., Liu, Q., Xia, C., Enantioselective total synthesis of (++)-flavisiamine F via late-stage visible-light-induced photochemical cyclization. *Angew. Chem. - Int. Ed.*, *58*, 5443, **2019**.
57. Peters, L., Konig, G.M., Terlau, H., Wright, A.D., Four new bromotryptamine derivatives from the marine bryozoan flustra foliacea. *J. Nat. Prod.*, *65*, 1633, **2002**.
58. Carle, J.S. and Christophersen, C., Marine alkaloids. 3. Bromo-Substituted alkaloids from the marine bryozoan flustra foliacea, Flustramine C and Flustraminol A and B. *J. Org. Chem.*, *46*, 3440, **1982**.
59. Julian, P.L. and Pikl, J., Studies in the indole series. III. On the synthesis of physostigmine. *J. Am. Chem. Soc.*, *57*, 539, **1935**.
60. Wu, K., Du, Y., Wei, Z., Wang, T., Synthesis of functionalized pyrroloindolines via a visible-light-induced radical cascade reaction: Rapid synthesis of (±)-flustraminol B. *Chem. Commun.*, *54*, 7443, **2018**.
61. Bhattacharyya, D., The galbulimima alkaloidsd-A new frontier in alkaloid synthesis. *Tetrahedron*, *67*, 5525, **2011**.
62. Rinner, U., Lentsch, C., Aichinger, C., Syntheses of galbulimima alkaloids. *Synthesis (Stuttg)*, 22, 3763, **2010**.
63. Burdge, H.E., Oguma, T., Kawajiri, T., Shenvi, R., Concise synthesis of GB22 by endo-selective siloxycyclopropane arylation. *chemRxiv*, *22*, 1, **2019**.
64. Usami, Y., Yamaguchi, J., Numata, A., Gliocladins A – C and glioperazine; Cytotoxic dioxo- or trioxopiperazine metabolites from a *gliocladium sp.* separated from a sea hare. *Heterocycles*, *63*, 1123-1129, **2004**.
65. Boyer, N. and Movassaghi, M., Concise total synthesis of (+)-gliocladins B and C. *Chem. Sci.*, *3*, 1798, **2012**.
66. Furst, L., Narayanam, J.M.R., Stephenson, C.R.J., Total synthesis of (+)-Gliocladin C enabled by visible-light photoredox catalysis. *Angew. Chem. - Int. Ed.*, *50*, 9655, **2011**.
67. Sun, M., Hao, X.Y., Liu, S., Hao, X.J., Formal synthesis of (+)-gliocladin C. *Tetrahedron Lett.*, *54*, 692, **2013**.
68. Wellington, K.D., Cambie, R.C., Rutledge, P.S., Bergquist, P.R., Chemistry of sponges. 19. novel bioactive metabolites from hamigera tarangaensis. *J. Nat. Prod.*, *63*, 79, **2000**.
69. Okanishi, Y., Ishikawa, T., Jinnouchi, T., Hayashi, S., Takanami, T., Aoyama, H., Yoshimitsu, T., Radical-based route to functionalized tetralin: Formal total synthesis of (±)-Hamigeran B. *J. Org. Chem.*, *88*, 1085, **2023**.
70. Mbaze, L. M. a., Lado, J.A., Wansi, J.D., Shiao, T.C., Chiozem, D.D., Mesaik, M.A., Choudhary, M.I., Lacaille-Dubois, M.A., Wandji, J., Roy, R., Sewald, N., Oxidative burst inhibitory and cytotoxic amides and lignans from the stem bark of Fagara heitzii (Rutaceae). *Phytochemistry*, *70*, 1442, **2009**.
71. Lin, S., Ischay, M.A., Fry, C.G., Yoon, T.P., Radical cation diels-alder cycloadditions by visible light photocatalysis. *J. Am. Chem. Soc.*, *133*, 19350, **2011**.
72. Yang, Z.B., Li, S.J., Luo, S.Z., Total synthesis of (±)-Hongoquercin A via visible-light-mediated organocatalytic polyene cyclization. *Huaxue Xuebao*, *75*, 351, **2017**.
73. Huang, Z. and Lumb, J.P., Mimicking oxidative radical cyclizations of lignan biosynthesis using redox-neutral photocatalysis. *Nat. Chem.*, *13*, 24, **2021**.
74. Gu, J.H., Wang, W.J., Chen, J.Z., Liu, J.S., Li, N.P., Cheng, M.J., Hu, L.J., Li, C.C., Ye, W.C., Wang,L., Leptosperols A and B, two cinnamoylphloroglucinol–sesquiterpenoid hybrids from leptospermum scoparium: Structural elucidation and biomimetic synthesis. *Org. Lett.*, *22*, 1796, **2020**.

75. Allred, T.K., Dieskau, A.P., Zhao, P., Lackner, G.L., Overman, L.E., General access to concave-substituted cis-Dioxabicyclo[3.3.0]octanones: Enantioselective total syntheses of macfarlandin C and dendrillolide A. *J. Org. Chem.*, *85*, 15532, **2020**.
76. Ritter, F.J., Rotgans, I.E.M., Talman, E., Verwiel, P.E.J., Stein, F., 5-methyl-3-Butyl-Octahydroindolizine, a novel type of pheromone attractive to pharaoh's ants (Monomorium pharaonis (L.). *Experientia*, *29*, 530, **1973**.
77. McManus, J.B., Onuska, N.P.R., Nicewicz, D.A., Generation and alkylation ofα-carbamyl radicals via organic photoredox catalysis. *J. Am. Chem. Soc.*, *140*, 9056, **2018**.
78. Noble, A. and MacMillan, D.W.C., Photoredox α-vinylation of α-amino acids and n-aryl amines. *J. Am. Chem. Soc.*, *136*, 11602, **2014**.
79. Park, K.H. and Chen, D.Y.K., A desymmetrization-based approach to morphinans: application in the total synthesis of oxycodone. *Chem. Commun.*, *54*, 13018, **2018**.
80. Shen, S., Zhu, H., Chen, D., Liu, D., van Ofwegen, L., Proksch, P., Lin, W., Pavidolides A–E, new cembranoids from the soft coral Sinularia pavida. *Tetrahedron Lett.*, *53*, 5759, **2012**.
81. Zhang, P.P., Yan, Z.M., Li, Y.H., Gong, J.X., Yang, Z., Enantioselective total synthesis of (−)-Pavidolide B. *J. Am. Chem. Soc.*, *139*, 13989, **2017**.
82. Ngadjui, B.T., Lontsi, D., Ayafor, J.F., Sondengam, B.L., Pachypophyllin and Pachypostaudins A and B: Three bisnorlignans from pachypodanthium staustii. *Phytochemistry*, *28*, 231, **1989**.
83. Santos Queiroz, A.P., Costa Freitas, M.C., Silva, J.R.A., Lima, A.B., Sawada, L., Martins Monteiro, R.F., Albuquerque de Freitas, A.C.G., Loureiro Maues, L.A., Arruda, A.C., Silva, M.N., Ferraz Maia, C.S., Fontes-Junior, E.A., do Nascimento, J.L.M., Arruda, M.S.P., Bastos, G.N.T., Pellucidin A promotes antinociceptive activity by peripheral mechanism inhibiting COX-2 and NOS: *In vivo* and *in silico* study. *PLoS One*, *15*, 1, **2020**.
84. Riener, M. and Nicewicz, D.A., Synthesis of cyclobutane lignans via an organic single electron oxidant–electron relay system. *Chem. Sci.*, *4*, 2625, **2013**.
85. Preller, M., Chinthalapudi, K., Martin, R., Knolker, H.J., Manstein, D.J., Inhibition of myosin ATPase activity by halogenated pseudilins: A structure-activity study. *J. Med. Chem.*, *54*, 3675, **2011**.
86. Kunfermann, A., Witschel, M., Illarionov, B., Martin, R., Rottmann, M., Hoffken, H.W., Seet, M., Eisenreich, W., Knolker, H.J., Fischer, M., Bacher, A., Groll, M., Diederich, F., Pseudilins: Halogenated, allosteric inhibitors of the non-mevalonate pathway enzyme IspD. *Angew. Chem. - Int. Ed.*, *53*, 2235, **2014**.
87. Martin, R., Jager, A., Bohl, M., Richter, S., Fedorov, R., Manstein, D.J., Gutzeit, H.O., Knolker, H.J., Total synthesis of pentabromo- and pentachloropseudilin, and synthetic analogues—Allosteric inhibitors of myosin ATPase. *Angew. Chem. - Int. Ed.*, *48*, 8042, **2009**.
88. Cavalleri, B., Volpe, G., Tuan, G., Berti, M., Parenti, F., A chlorinated phenylpyrrole antibiotic from actinoplanes. *Curr. Microbiol.*, *1*, 319, **1978**.
89. Aukland, M.H., Šiaučiulis, M., West, A., Perry, G.J.P., Procter, D.J., Metal-free photoredox-catalysed formal C–H/C–H coupling of arenes enabled by interrupted Pummerer activation. *Nat. Catal.*, *3*, 163, **2020**.
90. Naider, F., Shenbagamurthi, P., Steinfeld, A.S., Smith, H.A., Boney, C., Becker, J.M., Synthesis and biological activity of tripeptidyl polyoxins as antifungal agents. *Antimicrob. Agents Chemother.*, *24*, 787, **1983**.
91. Isono, K., Asahi, K., Suzuki, S., Studies on polyoxins, antifungal antibiotics. XIII.' The structure of polyoxin *J. Am. Chem. Soc.*, *91*, 7490, **1969**.
92. Matsuoka, T., Inuki, S., Miyagawa, T., Oishi, S., Ohno, H., Total synthesis of (+)-polyoxamic acid via visible-light-mediated photocatalytic β-Scission and 1,5-Hydrogen atom transfer of glucose derivative. *J. Org. Chem.*, *85*, 8271, **2020**.

93. Bloomer, J. L., Eder, W. R., and Hoffman, W.F., The biosynthesis of (+)-protolichesterinic acid. *Chem. Commun.*, *120*, 354, **1968**.
94. Goel, M., Dureja, P., Rani, A., Uniyal, P.L., Laatsch, H., Isolation, characterization and antifungal activity of major constituents of the himalayan lichen Parmelia reticulata Tayl. *J. Agric. Food Chem.*, *59*, 2299, **2011**.
95. Brisdelli, F., Perilli, M., Sellitri, D., Piovano, M., Garbarino, J.A., Nicoletti, M., Bozzi, A., Amicosante, G., Celenza, G., Cytotoxic activity and antioxidant capacity of purified lichen metabolites: An *in vitro* study. *Phyther. Res.*, *27*, 431, **2013**.
96. Kumar KC, S. and Muller, K., Lichen metabolites. 2. Antiproliferative and cytotoxic activity of gyrophoric, usnic, and diffractaic acid on human keratinocyte growth. *J. Nat. Prod.*, *62*, 821, **1999**.
97. Zeller, M.A., Riener, M., Nicewicz, D.A., Butyrolactone synthesis via polar radical crossover cycloaddition reactions: Diastereoselective syntheses of methylenolactocin and protolichesterinic acid. *Org. Lett.*, *16*, 4810, **2014**.
98. Zeches, M., Debray, M.M., Ledouble, G., Le Men-Olivier, L., Le Men, J., Alcaloïdes du Pandaca caducifolia. *Phytochemistry*, *14*, 1122, **1975**.
99. Beatty, J.W. and Stephenson, C.R.J., Synthesis of (−)-Pseudotabersonine, (−)-Pseudovincadifformine, and(+)-Coronaridine enabled by photoredox catalysis in flow. *J. Am. Chem. Soc.*, *136*, 10270, **2014**.
100. Walker, R.P., John Faulkner, D., Van Engen, D., Clardy, J., Sceptrin, an antimicrobial agent from the sponge agelas sceptrum. *J. Am. Chem. Soc.*, *103*, 6772, **1981**.
101. Nguyen, L.V. and Jamison, T.F., Total synthesis of (±)-Sceptrin. *Org. Lett.*, *22*, 6698, **2020**.
102. Bonjouklian, R., Smitka, T.A., Doolin, L.E., Molloy, R.M., Debono, M., Shaffer, S.A., Moore, R.E., Stewart, J.B., Patterson, G.M.L., Tjipanazoles, new antifungal agents from the blue-green alga tolypothrix tjipanasensis. *Tetrahedron*, *47*, 7739, **1991**.
103. Tang, J., Ren, L., Li, J., Wang, Y., Hu, D., Tong, X., Xia, C., Photochemical synthesis of indolocarbazoles through tandem indolization/dimerization/mannich cyclization from allenes. *Org. Lett.*, *24*, 3582, **2022**.
104. Lakkakula, Sasanala, S., Seher Lavanya, Y., Alavilli, G., Swamy, H., Kumara, M., *Plant-derived Bioactives: Production, Properties and Therapeutic Applications*, Springer, p. 61, **2020.** ISBN: 978-981-15-1760-0.
105. Zhang, C., Wang, Y., Sang, Y., Gao, H., Sun, Y., Sun, X., Yang, Y., He, M., Yang, Z., Zhan, L., Yu, Z.-X., Rao, Y., Synthesis of quaternary carbon-centered benzoindolizidinones via novel photoredox-catalyzed alkene aminoarylation: Facile access to tylophorine and analogues. *CCS Chem.*, *1*, 352, **2019**.
106. Yasui, Y., Kinugawa, T., Takemoto, Y., Synthetic studies on vincorine: Access to the 3a,8a-dialkyl-1,2,3,3a,8,8ahexahydropyrrolo[2,3-b]indole skeleton. *Chem. Commun.*, *28*, 4275, **2009**.
107. Stanek, F., Pawłowski, R., Mlynarski, J., Stodulski, M., Visible-light-mediated α-oxygenation of 3-(N,N-Dimethylaminomethyl)-Indoles to aldehydes. *Eur. J. Org. Chem.*, *2018*, 6624, **2018**.
108. Zhan, G., Zhou, J., Liu, J., Huang, J., Zhang, H., Liu, R., Yao, G., Acetylcholinesterase inhibitory alkaloids from the whole plants of zephyranthes carinata. *J. Nat. Prod.*, *80*, 2462, **2017**.
109. Takeuchi, H., Inuki, S., Nakagawa, K., Kawabe, T., Ichimura, A., Oishi, S., Ohno, H., Total synthesis of zephycarinatinesvia photocatalytic reductive radical ipso-cyclization. *Angew. Chem. - Int. Ed.*, *59*, 21210, **2020**.
110. Xie, J., Yuan, X., Abdukader, A., Zhu, C., Visible-light-promoted radical C–H trifluoromethylation of free anilines. *Org. Lett.*, *16*, 1768, **2014**.
111. Baar, M. and Blechert, S., Graphitic carbon nitride polymer as a recyclable photoredox catalyst for fluoroalkylation of arenes. *Chem. - Eur. J.*, *21*, 526, **2015**.

112. Beatty, J.W., Douglas, J.J., Cole, K.P., Stephenson, C.R.J., A scalable and operationally simple radical trifluoromethylation. *Nat. Commun.*, *6*, 6, **2015**.
113. Hung, K., Condakes, M.L., Morikawa, T., Maimone, T.J., Oxidative entry into the illicium sesquiterpenes: Enantiospecific synthesis of (+)-Pseudoanisatin. *J. Am. Chem. Soc.*, *138*, 16616, **2016**.
114. Condakes, M.L., Hung, K., Harwood, S.J., Maimone, T.J., Total syntheses of (−)-Majucin and (−)-Jiadifenoxolane A, complex majucin-type illicium sesquiterpenes. *J. Am. Chem. Soc.*, *139*, 17783, **2017**.
115. Nagib, D.A. and Macmillan, D.W.C., Trifluoromethylation of arenes and heteroarenes by means of photoredox catalysis. *Nature*, *480*, 224, **2011**.
116. Cherney, E.C., Lopchuk, J.M., Green, J.C., Baran, P.S., A unified approach to ent-atisane diterpenes and related alkaloids: Synthesis of (−)-Methyl atisenoate, (−)-Isoatisine, and the hetidine skeleton. *J. Am. Chem. Soc.*, *136*, 12592, **2014**.
117. Curi, A.B.R., Alvarez, M., Bazotte, R.B., Botion, L.M., Godoy, J.L., Effect of Stevia rebaudiana on glucose tolerance in normal adult humans. *Braz. J. Med. Biol. Res.*, *19*, 771, **1986**.

10

Metrics of Green Chemistry and Sustainability

Ramuel John I. Tamargo*, Hannah Shamina O. Cosiñero, Don Nelson C. Potato and Apraile Hope P. Dumrigue

Advanced Materials and Organic Synthesis Laboratory, Department of Chemical Engineering, University of the Philippines, Diliman, Quezon City, Philippines

Abstract

This chapter tackles various metrics of green chemistry and sustainability as well as its tools and applications. For the first part of the chapter, mass-based metrics, which simplify the assessment of the greenness of chemical processes by solely considering mass as a factor, are discussed. This chapter discusses a sample chemical process to enhance comprehension and distinction between these mass-based metrics, accompanied by demonstrations of how mass-based metrics are computed. On the other hand, as opposed to mass-based metrics which are commonly classified as univariate analysis metrics, impact-based metrics allow the combination of mass and environmental impacts to assess green chemistry and sustainability of chemical reactions and chemical processes. The last section of this chapter discusses the development of eco-friendly chemical products and processes guided by principles and metrics like Eco-Footprint and Environmental Factors. Analytical chemistry emphasizes eco-friendly procedures, leading to the development of metrics like the 12 principles of Green Analytical Chemistry (GAC) and tools such as NEMI and GAPI. Meanwhile, in the pharmaceutical industry, metrics like PMI, E-Factor, and EI improve sustainability, aided by tools like the Green Solvent Selection Tool. Organic synthesis uses metrics like carbon efficiency and mass intensity to assess eco-friendliness. Moreover, Life cycle assessment (LCA) comprehensively evaluates environmental impacts throughout a product's life cycle.

Introduction

In an era where the delicate balance between humanity and the planet is more critical than ever, the paradigm has shifted to "Green Chemistry" and "Sustainability." These two transformative concepts are intertwined and represent a collective endeavor to reframe our approach to industry, innovation, and existence itself.

Green chemistry and sustainability are related concepts that both aim to promote environmentally friendly practices, but they focus on different aspects of achieving a more sustainable world. Like two sides of a coin, they complement each other, with green chemistry offering the tools and methodologies to transform chemical processes with the environment in mind, while sustainability embraces a holistic vision, encompassing economic vitality, ecological integrity, and social equity toward balanced progress while striving to safeguard our planet's vitality for generations yet to come.

Green chemistry is a specific approach within the broader framework of sustainability. It focuses on improving chemical processes and products to reduce their environmental impact by using

**Corresponding author*: ritamargo@up.edu.ph; ramueljohn.tamargo@gmail.com

Mousumi Sen (ed.) Sustainable Green Catalytic Processes, (225–258) © 2024 Scrivener Publishing LLC

innovative approaches that are inherently safer and more eco-friendly. Green chemistry principles include the 12 Principles of Green Chemistry, which provide a framework for making chemical processes and products more sustainable.

The application and interpretation of these principles will vary depending on the level of importance of several factors involved. Thus, we need to qualify and quantify these into metrics that serve as qualitative and quantitative tools to evaluate and measure the environmental impacts of chemical products, processes, and practices. Importantly, these powerful tools will be harnessed by scientists and engineers for better decision-making to satisfy the functional requirements and needs of the industry with safety and sustainability in mind. More so, it will also lead to better appreciation and acceptance by the wider society of sustainable chemistry.

While sustainability is a broader concept that encompasses a wide range of environmental, social, and economic issues, as it is not limited to the field of chemistry but applies to various aspects of human activities, this section will focus on the scientific metrics that we use in green chemistry as sustainable chemistry.

For this chapter, let us embark on a journey through the verdant realms of chemistry, where molecules hold the promise of a sustainable tomorrow, and explore the expansive landscapes of sustainability, where the well-being of generations yet to come guides our every step.

***Keywords*:** Green chemistry, sustainability, mass-based metrics, impact-based metrics, life cycle assessment

10.1 Green Metrics

Green metrics are a tool in green chemistry that is fundamental to sustainability. It helps the academe and industries in decision-making, assessing and improving the environmental and sustainability performance of chemical processes and products. The green metric supports the broader principle of green chemistry, which aims to reduce chemical innovations' environmental and health impact while promoting more sustainable and responsible practices.

10.1.1 Mass-Based Metrics

Starting from 1990, with the emerging attention to sustainable development, various metrics have been created to evaluate process impacts. The primary purpose of metrics is to facilitate comparisons when there are multiple economically viable approaches for manufacturing a product with the goal of finding the one that has the least environmental impact.

The first and simplest green metrics introduced for the chemical industry and organic synthesis were the environmental factor (E-factor; measured in kg/kg) and atom economy (AE; expressed as a percentage), both focusing on mass considerations. Subsequently, mass-based metrics such as reaction mass efficiency (RME), effective mass yield (EMY), process mass intensity (PMI), and others have been devised for specific chemistry domains. These mass-based metrics, listed in Table 10.1, can be divided into two categories: one that signifies a percentage of the ideal green process, akin to atom economy, and another that relies on kg/kg, akin to the E-factor.

One major limitation of mass-based metrics is their inability to distinguish between more or less environmentally harmful waste products. Using mass-based metrics, the production of less waste implies a greener process. However, it could be less environmentally friendly if the waste generated is more detrimental to the environment. This limitation suggests that using mass-based metrics alone may be inadequate for determining the greener synthetic method.

Despite this, mass-based metrics offer a significant advantage in terms of simplicity where they can be calculated using readily available data with minimal assumptions. This is particularly beneficial for companies that manufacture a wide array of products, as mass-based metrics might be the only feasible choice for monitoring overall reductions in the environmental impact across the company.

This section will briefly discuss percentage yield and its shortcomings in measuring the "greenness" of a chemical process that motivated the development of other green chemistry metrics. The driving motivation of this section is to present an overview of the different mass-based metrics used in green chemistry.

10.1.1.1 Percentage Yield

Percentage yield can be calculated by dividing the actual quantity of the desired product by the theoretical yield. The motivation of this metric is that not all reactants in a chemical

Table 10.1 Mass-based metrics categorized as analogs of either E-factor or atom economy.

E-factor and its analogs (kg/kg): Focused on waste minimization
Environmental factor (E-factor) Mass intensity (MI) Process mass intensity (PMI)
Atom economy and its analogs (%): Focused on the incorporation of reactants into the product
Atom economy (AE) Reaction mass efficiency (RME) Mass productivity (MP) Effective mass yield (EMY) Carbon economy (CE)

process are converted into products due to several factors such as the reversible nature of some reactions and losses in the process such as loss due to an undesired side reaction. To assess these losses, one must experimentally measure and calculate the actual yield using the following equation:

$$\text{Percentage Yield} = \frac{\text{Actual mass of product}}{\text{Theoretical mass of product}} \times 100 \tag{10.1}$$

The theoretical mass of the product can be calculated using the stoichiometric ratio of reactants. For any given reaction A + B –→ P, where A is the limiting reagent:

$$\text{Theoretical yield} = [A] \times \left[\frac{P}{A}\right] \times [\text{Formula weight of P}] \tag{10.2}$$

where [A] is the limiting reagent in moles and [P/A] is the stoichiometric ratio of the desired product to the limiting reagent.

The percentage yield can be affected by chemical equilibrium, and introducing an excess of one or more reactants can enhance the overall yield. This approach may not necessarily align with green practices, as it implies that a significant quantity of the excess reactant remains unused, resulting in waste production. To assess the utilization of excess reactants, one can compute the excess reactant factor (ERF) as follows:

$$\text{ERF} = \frac{\text{Reactants' stoichiometric mass} + \text{reactants' excess mass}}{\text{Reactants' stoichiometric mass}} \tag{10.3}$$

An ERF with a value of 1 means that no excess was added to the reaction and the reaction proceeds under normal stoichiometric conditions. On the other hand, a factor with a value far greater than 1 may signify a large amount of waste chemicals [1].

Since yield does not consider excess reactants, catalysts, and solvents, it may not be a reliable metric for measuring the overall efficiency of a process. This led to the development of other metrics [2].

10.1.1.2 Atom Economy

The concept of atom economy as a metric for the greenness of a chemical transformation was introduced in 1991 by Barry M. Trost with the primary motivation to maximize the incorporation of reactant atoms into final products [3]. AE is a theoretical value obtained from an ideal chemical reaction with a 100% percentage yield and assumes the precise use of stoichiometric amounts of reactants while excluding substances like solvents and additional chemicals not included in the stoichiometric equation.

$$\text{Atom Economy (AE)} = \frac{\text{Molecular mass of desired product}}{\Sigma \text{ Molecular masses of reactants}} \times 100 \tag{10.4}$$

The strength of AE is its applicability to both single-stage and multistage reactions without the need for experimentation. This feature makes it an invaluable tool for swiftly forecasting and assessing the waste quantities produced in different pathways to a specific target molecule [4].

An example of the computation of the percent AE for producing the desired product I from a generic multistage reaction can be illustrated as follows:

$$A + B - \rightarrow C + D$$
$$C + E - \rightarrow F + G$$
$$F + H - \rightarrow I + J$$

$$\text{Atom Economy (AE)} = \frac{\text{molecular mass of I}}{\Sigma \text{ molecular masses of A, B, E, and H}} \times 100 \tag{10.5}$$

In accordance with the law of conservation of mass, which asserts that the combined mass of the reactants equals the combined mass of the products, the total molecular masses of A, B, E, and H should be equivalent to the total molecular masses of I, D, G, and J. Since the desired product is only I, the atoms of D, G, and J are considered wastes or by-products. A low atom economy signifies more wastes generated during the reaction, resulting in a higher cost of disposal and, thus, less green.

To illustrate the determination of AE, an example presented by Curzons and coworkers [2, 5, 11] for the esterification of benzyl alcohol and p-toluenesulfonyl chloride is discussed in this section. In the example, 10.81 g (0.10 mol, 108.14 g/mol) of benzyl alcohol and 21.9 g (0.115 mol, 190.64 g/mol) of p-toluenesulfonyl chloride are added together in the presence of 500 g toluene as solvent and 15 g of triethylamine as reagent or base to form 23.6 g (0.09 mol, 262.32 g/mol) of benzyl-4 toluene sulfonate ester.

$$C_6H_5CH_2OH + CH_3C_6H_4SO_2Cl \xrightarrow[N(CH_2CH_3)_3]{C_6H_5CH_3} C_{14}H_{14}O_3S + HCl$$

$$AE = \frac{262.32}{108.14 + 190.64} \times 100 \tag{10.6}$$

$$AE = 88\% \tag{10.7}$$

From the above example, it can be observed that wastes, solvents, and reagents are not included in the calculation of atom economy. Furthermore, the formation of HCl as a by-product resulted in an atom economy of less than 100% [5].

10.1.1.3 *Environmental Factor*

Proposed by Roger Sheldon in the early 1990s, the environmental factor is calculated as the total mass in kilograms of all wastes generated in a process per kilogram of a product wherein the term waste is defined as "anything that is not the desired product" [6–8].

$$\text{Environmental Factor (E-factor)} = \frac{\text{Total mass of waste in kg}}{\text{Mass of product in kg}} \times 100 \qquad (10.8)$$

E-factor was designed as a simple and fast metric to quantify the amount of waste produced in industrial processes. E-factors for individual steps are additive and can be computed for single-step or multistep reactions. A higher E-factor indicates more waste and a larger environmental impact, while a lower E-factor means using fewer materials and correlates with reduced manufacturing and waste disposal expenses. The ideal E-factor is 0, aligning with the primary principle of green chemistry that emphasizes preventing wastes over dealing with them afterward. Considering 0 as the ideal value represents the ultimate goal of waste-free manufacturing facilities [6].

Table 10.2 was derived and published by Sheldon in 1992 which challenged different chemical industries, especially fine chemical and pharmaceutical segments, to reduce waste generation during their processes. This resulted in a heightened understanding of waste issues and enabled exploring solutions to address them [3].

While the E-factor is a flexible measure applicable across different chemical industry sectors, it has a limitation in that it does not account for the hazards and environmental risks associated with the produced waste. It treats all waste types equally. However, considering the environmental impact of generated waste is crucial. Therefore, it is important to complement the E-factor with other metrics to address this limitation [6].

Table 10.2 E-factors for different chemical industries [6].

Industry sector	Annual production, tonnes	E-factor, kg waste/kg product
Oil refining	10^6 - 10^8	<0.1
Bulk chemicals	10^4 - 10^6	<1–5
Fine chemistry	10^2 - 10^4	5–50
Pharmaceuticals	10^1 - 10^3	25 to >100

10.1.1.4 *Effective Mass Yield*

The effective mass yield metric, introduced by Hudlicky and colleagues in 1999 [9], is defined as the percentage obtained by dividing the mass of the desired product by the total mass of all non-benign materials employed in the synthesis. Hudlicky defined benign as "those by-products, reagents or solvents that have no environmental risk associated to them, for example, water, low-concentration saline, dilute ethanol, autoclaved cell mass, etc." In mathematical terms, EMY can be expressed as follows:

$$\text{Effective Mass Yield (EMY)} = \frac{\text{Mass of products}}{\text{Mass of non-benign reagents}} \times 100 \quad (10.9)$$

This metric considers that not all chemicals have an equivalent impact and permits a weighted assessment focusing solely on the harmful waste generated, emphasizing the benefits of employing environmentally friendly solvents and reagents. However, one of its drawbacks arises from potential misconceptions in defining what qualifies as a "non-benign" reagent [7, 10].

10.1.1.5 *Reaction Mass Efficiency*

Reaction mass efficiency is the percentage of the actual mass of reactants in the final product that can be calculated by dividing the mass of the product by the sum of the masses of reactants that appears in the balanced equation [5, 11].

$$\text{Reaction Mass Efficiency (RME)} = \frac{\text{Actual mass of desired product}}{\text{Total mass of reactants}} \times 100 \quad (10.10)$$

Curzons and colleagues [5, 11] introduced RME to address the shortcomings of the atom economy metric of not considering the excess of reactants to maximize the yield and selectivity while maintaining its emphasis on reactants and reaction efficiency. Eventually, they recognized that RME accounts for yield, atom economy, and stoichiometry and can be mathematically written as follows:

$$\text{RME} = \text{Yield} \times \text{Atom Economy} \times \frac{1}{\text{Stoichiometric factor, SF}} \quad (10.11)$$

where the stoichiometric factor (SF) is a term to measure excess reactant.

The above equation can be expanded using a generic equation where product P is formed by reacting excess reactant B with reactant A in the presence of reagents, solvents, and catalysts, A + B –→ P:

$$RME = \frac{z}{x} \times \frac{MW_P}{MW_A + MW_B} \times \frac{1}{1 + \frac{(y-x)MW_B}{x(MW_A + MW_B)}} \tag{10.12}$$

where x, y, and z are the number of moles of reactants A and B and product P, respectively, while MW_A, MW_B, and MW_P are the molecular weights of A, B, and P, respectively.

For the same example presented by Curzons and colleagues for the esterification of benzyl alcohol and p-toluenesulfonyl chloride presented in Section 10.1.1.2, Dicks and Hent [11] calculated RME as follows:

$$C_6H_5CH_2OH + CH_3C_6H_4SO_2Cl \xrightarrow[N(CH_2CH_3)_3]{C_6H_5CH3} C_{14}H_{14}O_3S + HCl$$

$$RME = \frac{23.6}{10.81 + 21.9} \times 100 \tag{10.13}$$

$$RME = 72\% \tag{10.14}$$

With an AE of 88% obtained from Section 10.1.1.2, Dicks and Hent [11] computed RME as follows:

$$Yield = \frac{0.09}{0.10} \tag{10.15}$$

$$Yield = 0.90 \tag{10.16}$$

$$SF = 1 + \frac{(0.115 - 0.10) \times 190.64}{0.10 \times (108.14 + 190.64)} \tag{10.17}$$

$$SF = 1.10 \tag{10.18}$$

$$RME = 0.9 \times 88 \times \frac{1}{1.10} \tag{10.19}$$

$$RME = 72\% \tag{10.20}$$

Despite the significant advantage of this metric over the others, the definition of RME still does not include all materials involved in the chemical reaction such as reagents, solvents, and catalysts. As a result, Andraos and Sayed [12] proposed a generalized RME formula that included the material recovery parameter (MRP) to account for catalysts, solvents, and workup or purification materials. Mathematically, the generalized RME can be defined as follows:

$$\text{Generalized RME} = \text{Yield} \times \text{AE} \times \frac{1}{\text{SF}} \times \text{MRP} \tag{10.21}$$

The formula can be expanded further by defining the MRP to get the final generalized RME formula expressed as follows:

$$\text{Generalized RME} = \epsilon \times \text{AE}\frac{1}{\text{SF}} \times \left[\frac{1}{1 + \frac{\epsilon \times \text{AE} \times (c + s + \omega)}{SF \times m_{cp}}} \right] \tag{10.22}$$

where ε is the reaction yield; AE and SF are the atom economy and stoichiometric factor, respectively; c, s, and ω are the masses of catalyst, solvent, and all other post-reaction materials, respectively; and m_{cp} is the mass of the actual target product [12].

10.1.1.6 Mass Intensity and Process Mass Intensity

Mass intensity (MI) is a metric defined as the mass ratio of all input materials, excluding water, to the final product. This can be represented mathematically as follows:

$$\text{Mass Intensity (M)} = \frac{\text{Total mass in reaction vessel(excluding water) (kg)}}{\text{Mass of product (kg)}} \tag{10.23}$$

Yield, stoichiometry, solvent, and all reagents used, with the exception of water, are all accounted in MI calculation. Because water does not carry a significant impact on the environment, it has been excluded from mass calculations. This also avoids skewness of mass data in many processes.

To illustrate how MI is calculated, the esterification of benzyl alcohol and p-toluenesulfonyl chloride discussed from Section 10.1.1.2 can be used as follows [2]:

$$C_6H_5CH_2OH + CH_3C_6H_4SO_2Cl \xrightarrow[N(CH_2CH_3)_3]{C_6H_5CH_3} C_{14}H_{14}O_3S + HCl$$

$$\text{Mass Intensity (MI)} = \frac{10.81 + 21.9 + 500 + 15}{23.6} \tag{10.24}$$

$$\text{MI} = 23.2 \text{ g/g or } 23.2 \text{ kg/kg} \tag{10.25}$$

The term MI was adopted by the Green Chemistry Institute Pharmaceutical Round Table, relabeling it as process mass intensity, to establish environmental standards for pharmaceutical companies. Unlike MI, the PMI calculation includes evaluating the mass ratio of all materials used in a process, encompassing reactants, reagents, catalysts, solvents, and any other substances, in relation to the final product. This concept evolved from the idea of reaction MI and can be represented mathematically as follows:

$$\text{Process Mass Intensity (PMI)} = \frac{\text{Total mass in reaction vessel(including water) (kg)}}{\text{Mass of product (kg)}} \tag{10.26}$$

It can also be related to the E-factor, where:

$$\text{PMI} = \text{E-factor} + 1 \tag{10.27}$$

For a PMI with a value of 1, it means that the total mass of reagents is equal to the mass of product and thus considered to be ideal [2, 7, 10].

10.1.1.7 Mass Productivity

Mass productivity (MP) is the percentage reciprocal of mass intensity that was proposed for easier comprehension in the business world, as it highlights resource utilization. As such, it is a metric more valued in the business industry than the mass intensity [7].

$$\text{Mass Productivity (MP)} = \frac{1}{\text{MI}} \times 100 \tag{10.28}$$

$$\text{Mass Productivity (MP)} = \frac{\text{Mass of product (kg)}}{\text{Total mass in reaction vessel (kg)}} \times 100 \tag{10.29}$$

For the previous esterification process with an MI of 23.2 kg/kg presented in Section 10.1.1.6, MP can be calculated to be 4.3%, which means that 4.3% of the total process input is converted to the desired product.

10.1.1.8 Carbon Efficiency

Carbon efficiency (CE) is used to estimate the percentage of carbon in the reagents used that remains in the final desired product. Yield and stoichiometry are included in the calculation [2].

$$\text{Carbon Efficiency (CE)} = \frac{\text{Mass of carbon in the product}}{\text{Total mass of carbon present in reactant}} \times 100 \tag{10.30}$$

As an example, for a generic reaction A + B —> P , CE can be calculated as follows:

$$CE = \frac{\text{moles of P} \times \text{no. of carbons in P} \times 100}{(\text{moles of A} \times \text{no. of carbons in A}) + (\text{moles of B} \times \text{no. of carbons in B})} \tag{10.31}$$

Calculation of the CE can be illustrated using the esterification process presented in previous sections as follows [2]:

$$C_6H_5CH_2OH + CH_3C_6H_4SO_2Cl \xrightarrow[N(CH_2CH_3)_3]{C_6H_5CH_3} C_{14}H_{14}O_3S + HCl$$

$$CE = \frac{0.9 \times 14}{(0.10 \times 7) + (0.115 \times 7)} \times 100 \tag{10.32}$$

$$CE = 83.7\% \tag{10.33}$$

10.1.2 Impact-Based Metrics

Sustainability and green chemistry are multifaceted concepts that cannot be completely assessed through mass-based metrics only. Green chemistry metrics have been classified as univariate and multivariate in certain literature; most mass-based metrics are classified as univariate analysis metrics, meanwhile, multivariate analysis metrics involve mass-based metrics combined with environmental impact metrics [4, 13]. Impact-based assessments also include mass metrics and are generally considered a better approach to green chemistry assessment. To assess the "greenness" of a chemical, reaction, or a process, information on mass may not be enough, as an example, 1 gram of an environmentally friendly chemical may be assessed the same as 1 gram of a hazardous chemical. But their effects are entirely different, especially in the environmental sense [14].

In this section, impact-based metrics will be discussed with a certain focus on the Eco-Scale, toxicological hazard metrics, global hazard metrics, and the Eco-Footprint. The Eco-Scale is a green metric that assesses the quality of the experimental procedure based on six categories including product yield, experimental cost, safety of procedure, chosen conditions, and difficulty of workup or purification. Eco-Scale is usually used on laboratory or industrial scale chemical reactions and chemical productions. Toxicological hazard metrics involve the measurement of toxicity to humans, animals, and the environment. For humans, toxicity hazards are measured through the human toxicity by ingestion potential (INGTP) and human toxicity by inhalation potential (INHTP). For the environment, air quality, water quality, and soil quality are the common focus of measurements. Global hazard metrics are commonly associated with the global impacts such as global warming, ozone depletion, acidification, and many others. The Eco-Footprint or Ecological Footprint (EF) measures the demand on the ecosystem of certain industrial processes. EF measures the ability of the ecosystem to provide natural resources and absorb wastes from chemical reactions, processes, or even big projects.

10.1.2.1 The Eco-Scale

The Eco-Scale is a novel metric tool designed by Van Aken *et al.* [15] in 2006 that uses a 0–100 scale, wherein a score of 0 represents a totally failed reaction and a score of 100 represents the ideal reaction. An ideal reaction is defined as follows: "Compound A + Compound B = Compound C (100% yield) with minimal operator risk and minimal environmental impact" [15].

The score starts with 100 points and penalty points according to six general parameters that influence the quality of the reaction process and conditions are deducted. The six parameters include (1) product yield, (2) experimental costs, (3) safety of procedure, (4) conditions and technical setup, (5) chosen temperature/time, and (6) workup and purification. The Eco-Scale can be calculated as follows:

$$\text{Eco-Scale} = 100 - \text{sum of individual penalties} \tag{10.34}$$

Penalty points from 0 to 10 were assigned based on the environmental and greenness of each parameter (Table 10.3).

Table 10.3 Assigned penalty points for Eco-Scale calculation from Van Aken *et al.* [15] (doi:10.1186/1860-5397-2-3 an Open Access article).

Category	Penalty points
1. Yield	(100 - % yield)/2
2. Price of reaction components to obtain mmol of end product	
Inexpensive (< $10)	0
Expensive (> $10 and < $50)	3
Very expensive (> $50)	5
3. Safety (based on the hazard warning symbols)	
N: dangerous to environment	5
T: toxic	5
F: highly flammable	5
E: explosive	10
F+: extremely flammable	10
T+: extremely toxic	10

(*Continued*)

Table 10.3 Assigned penalty points for Eco-Scale calculation from Van Aken *et al.* [15] (doi:10.1186/1860-5397-2-3 an Open Access article). (*Continued*)

Category	Penalty points
4. Technical setup	
Common setup	0
Instruments for controlled addition of chemicals	1
Unconventional activation technique (microwave irradiation, ultrasound or photochemical activation, etc.)	2
Pressure equipment, > 1 atm	3
Any additional special glassware	1
Inert gas atmosphere	1
Glove box	3
5. Temperature and time	
Room temperature, < 1 h	0
Room temperature, < 24 h	1
Heating, < 1 h	2
Heating, > 1 h	3
Cooling to 0°C	4
Cooling, < 0°C	5
6. Workup and purification	
None	0
Cooling to room temperature	0
Adding solvent	0
Simple filtration	0
Removal of solvent with bp < 150°C	0
Crystallization and filtration	1
Removal of solvent with bp > 150°C	2
Solid-phase extraction	2
Distillation	3
Sublimation	3
Liquid-liquid extraction (includes drying of solvent with dessicant and filtration of dessicant	3
Classical chromatography	10

The Eco-Scale metric favors chemical processes that have higher yield, lower costs, safer reaction conditions, and easier purification. A standard Eco-Scale value can be set by individual researchers or research institutions. In any case, the Eco-Scale can be used to calculate whether a chemical process is "greener" than alternatives. An example of Eco-Scale calculation and comparison between two reaction processes is adapted from the publication Van Aken *et al.* [15] in 2006 (Table 10.4).

The calculated total penalty points of the reduction to nitrobenzene to aniline reaction is 36, having an Eco-Scale score of 64 points. Meanwhile, the calculated total penalty points of the oxidation of benzyl chloride to benzoic acid reaction is 22, and the Eco-Scale score is 78 points. Comparing the Eco-Scale scores of both reactions, the oxidation of benzyl chloride

Table 10.4 An example of Eco-Scale comparison of (1) reduction to nitrobenzene to aniline and (2) oxidation of benzyl chloride to benzoic acid reactions.

Reduction of nitrobenzene to aniline		**Oxidation of benzyle chloride to benzoic acid**	
Parameter	Penalty	Parameter	Penalty
1. Yield: 90%	5	1.Yield: 87%	6
2. 5% Pt/C, 0.3 g	3	2. H_2O_2 (30%, 4.1mL, 36mmol)	0
		Na+WO_4 $2H_2O$ (66mg, 0.2 mmol)	0
		[(Octyl]$_3$NMe]HSO_4 (93mg, 0.2mmol)	0
		Molecular sieves 4Å(100mg)	0
3. Nitrobenzene (T,N)	10	3. Benzyl Chloride (T)	5
MeOH (T,F)	10		
5% Pt/C (F)	5		
4. Basic glassware, stirring	0	4. Addition of H_2O_2	1
5. Room temperature, 1 h	0	5. 90°C, 10h	3
6. Filtration of the catalyst	0	6. Extraction with AcOEt (3 × 10mL)	3
Removal of MeOH	0	Washing with Aq.$Na_2S_2O_4$	3
Addition of $CHCl_3$	0	Drying over $MgSO_4$	0
Washing with aq. NaCl	3	Removal of AcOEt	0
Removal of $CHCl_3$	0	Crystallization from hexanes	1
Total Penalty Points	36	Total Penalty Points	22

to benzoic acid reaction can be considered "greener" than reduction of nitrobenzene to aniline reaction.

10.1.2.2 Toxicological Hazard Metrics

The presence of substantial quantities of toxic chemicals in areas with human and animal populations can produce toxicological hazards. Although the human-caused release of toxicological hazards is illegal and controlled, accidental spills can still happen. To ensure sustainability and greenness, it is important to assess the potential toxicological hazards from chemicals, reactions, or processes. This section will discuss four metrics on toxicological hazards divided into toxicity to humans (ingestion and inhalation), air toxicity (AT), water ecotoxicity, and bioaccumulation.

10.1.2.2.1 Human Toxicity by Ingestion Potential and Human Toxicity Inhalation Potential

Toxic substances are harmful to humans when ingested or inhaled at high doses and concentrations. To assess these, factors such as human toxicity by ingestion potential (INGTP) and human toxicity inhalation potential (INHTP) are used. A simple equation to calculate the INGTP and INHTP is shown below [13]:

$$\mathrm{INGTP} = \frac{C^{w}}{LD_{50}} \tag{10.35}$$

$$\mathrm{INHTP} = \frac{C_{a}}{LC_{50}} \tag{10.36}$$

where C_w is the concentration of the compound in water, LD50 is the estimated lethal dose to 50% of the population of test animals, Ca is the concentration of the emitted chemical, and LC50 is the estimated lethal concentration that is expected to kill 50% of the group of test animals at single exposure. The INGTP and INHTP are usually used in life cycle analyses.

10.1.2.2.2 Air Toxicity

Some chemical reactions, especially on large industrial scales, can contribute greatly to air pollution. Air pollution imposes health risks not just for human populations but for all living things. Countries have different air pollution regulations by assigning values for acceptable values of gas or vapor concentrations in the air. To calculate the air toxicity, Y_{AT}, the following equation is used [16]:

$$Y_{AT} = \frac{X_{AT(T)}}{X_{AT}} \tag{10.37}$$

where X_{AT} is the calculated air toxicity from the annual life cycle emissions, and $X_{AT(T)}$ is the target emission value published by regulatory boards. The value for Y_{AT} is the air toxicity score in which lower values represent cleaner air and higher values represent hazardous air. Projects should aim to have a greener assessment by having lower toxicity scores, Y_{AT}.

10.1.2.2.3 Water Ecotoxicity

The assessment of water ecotoxicity is also important especially for aquatic ecosystems where human food comes from and where most of marine life reside. Industrial companies release toxic and lethal substances into water bodies, and regulation of these substances is important. The equation to calculate the water ecotoxicity score, $Y_W E$, is as follows [16]:

$$Y_{WE} = \frac{X_{WE(T)}}{X_{WE}} \tag{10.38}$$

where X_{WE} is the life cycle emission of 1,4-DCB per capita per year, and $X_{WE(T)}$ is the target emissions to freshwater systems per capita per year. The upper target value of the water ecotoxicity score is calculated as follows:

$$YWE(T) = WE_{ref} \times \alpha_{WE} \tag{10.39}$$

where WE_{ref} is the global annual per capita of 1,4-1DCB emissions to freshwater systems, and α_{WE} is the adjustment factor. The assessment of water ecotoxicity is relevant especially in areas with rich marine life and countries that are surrounded by waters.

10.1.2.2.4 Bioaccumulation

Bioaccumulation is the increase in concentration of chemicals inside a biological organism (e.g., fish) over time. Chemicals such as polychlorinated biphenyls (PCBs), dichlorodiphenyltrichloroethane (DDT), dioxins, and mercury are commonly observed in fishes. The bioaccumulation score, BAP, can be approximated by the octanol-water partition coefficient, log P or log K_{ow}. Some papers set log K_{ow} value of <3.5 to have low bioaccumulation potential, values of 3.5–4.3 as moderate bioaccumulation potential, and >4.3 values as high bioaccumulation potential [13].

10.1.2.3 Global Hazard Metrics

Even though toxicological hazard metrics also involve environmental impacts, these are mostly affecting certain areas and communities. In this subsection, potential large-scale and global impacts to the environment will be discussed. Factors such as smog formation potential (SFP), ozone depletion potential (ODP), global warming potential (GWP), acidification potential (AP), abiotic depletion potential (ADP), and environmental persistence (EP) are included. The higher the global hazard metric scores are, the more dangerous it is for the environment.

10.1.2.3.1 Acidification Potential

There are several compounds that are known as potential precursors of acid rain, which include sulfur dioxide (SO_2), nitrogen oxides (NOx), nitrogen monoxide (NO), and dinitrogen oxide (N_2O). The ability of a gaseous chemical or compound to generate acid as precursors to acid rain can be measured through the acidification potential (AP). The AP is usually characterized relative to the mass of SO_2 or SO_2-equivalence. The equation used to calculate the AP score is as follows [17]:

$$Y_{AP} = \frac{X_{AP(T)}}{X_{AP}} \tag{10.40}$$

where X_{AP} is the calculated acidification potential (concentration of SO_2) in the local environment, and $X_{AP(T)}$ is the standard set by the environmental protection agency (EPA). The X_{AP} is calculated as follows:

$$X_{AP} = SO_{2,0+} \frac{SO_2}{Area_{community} \times MH_{SO2}} \times \frac{\tau_{SO2}}{8760} \tag{10.41}$$

where $SO_{2,0}$ is the background concentration, SO_2 is the annualized life cycle emissions, τSO_2 is the residence time, and MH_{SO2} is the vertical mixing height of sulfur dioxide (SO_2).

10.1.2.3.2 Ozone Depletion Potential

The ozone layer protects the Earth's surface and its inhabitants from long exposure to the sun's ultraviolet (UV) rays. Recent technological advancements had significant contributions to the ozone layer. One example is the chlorofluorocarbons or CFCs emitted from old equipment that have a long residence time of 46–1,700 years. The ODP score is calculated using the following equation [17]:

$$Y_{ODP} = \frac{X_{ODP(T)}}{X_{ODP}} \tag{10.42}$$

where XODP represents the actual annual CFC-11 emissions per capita, and the XODP(T) is the upper limit of the CFC-11 emissions per capita. As suggested by Hacatoglu *et al.* 2014 [17], the XODP(T) can be calculated using the following equation:

$$X_{ODP(T)} = \frac{O_3}{k_{Cl-O3} - f_{CFC-11 \times nCl} \times POP_{world} \times t_{Sust}} \times \alpha_{ODP} \tag{10.43}$$

where $k_{Cl\text{-}O3}$ is the relationship between the concentration of stratospheric chlorine and ozone depletion, f_{CFC-11} is the fate factor for CFC-11 when emitted from the Earth's surface, POP_{world} is the world population, n_{Cl} is the number of chlorine atoms in a CFC-11 molecule, and t_{Sust} is the time for sustainability assessment (usually from 5 years to infinity).

10.1.2.3.3 Smog Formula Potential

Smog air is formed due to particulate matter trapped beneath ground-level ozone. Smog is visually manifested as a haze composed of a mixture of smoke, gases, and other particles. These matters are usually from vehicle exhaust and industrial gas emissions from highly urban and industrialized areas. In countries with warmer climates, smog is usually common especially in the summer season, and smog only lasts for a short period of time. The SFP equation is as follows [17]:

$$Y_{SFP} = \frac{X_{SFP(T)}}{X_{SFP}} \tag{10.44}$$

where X_{SFP} is the calculated concentration of ground-level ozone (O_3), and $X_{SA(T)}$ is the upper threshold for ground-level ozone set by the Environmental Protection Agency standards.

$$X_{SFP} = O_{3,0} + \frac{O_3}{Area_{community} \times MH_{O3}} \times \frac{\tau_{O3}}{8760} \tag{10.45}$$

where $O_{3,0}$ is the background concentration, O_3 is the annualized life cycle emissions, τO_3 is the residence time, and MH_{O3} is the vertical mixing height of O_3.

10.1.2.3.4 Global Warming Potential

Greenhouse gas emissions are the main culprits to global climate change and global warming, which include carbon dioxide (CO_2), methane (CH_4), and chlorofluorocarbons (CFCs). When these gases are trapped within the atmosphere, global consequences emerge and some of these can be catastrophic in the long run. Global warming potential is a metric to measure the impact of different gases on the atmosphere. It measures the amount of absorbed energy when 1 ton of gas is released in the atmosphere over a certain period. The GWP equation is as follows [17]:

$$Y_{GWP} = \frac{X_{GWP(T)}}{X_{GWP}} \tag{10.46}$$

where X_{GWP} is the actual greenhouse gas emissions in a 100-year period, and $X_{GWP(T)}$ is the target value for the time period, usually the minimum acceptable greenhouse gas emission value.

10.1.2.3.5 Abiotic Depletion Potential

The abiotic depletion potential is a metric on the use of nonrenewable sources for energy production that measures each element's risk of depletion relative to antimony (Sb). AD is a factor that is part of the life cycle analysis (which will be discussed in the next section of this chapter). The equation used to calculate the ADP score is as follows [17]:

$$Y_{ADP} = \frac{X_{ADP(T)}}{X_{ADP}} \tag{10.47}$$

where X_{ADP} is the life cycle use of antimony (Sb) and its equivalents per capita per year, and $X_{ADP(T)}$ is the annual sustainable antimony allocation. The $X_{ADP(T)}$ can be calculated using the following equation:

$$X_{ADP(T)} = \frac{R_{Sb}}{POP_{world} \times t_{Sust}} \times \alpha_{ADP} \tag{10.48}$$

where R_{Sb} is the value for the antimony recoverable reserves, POP_{world} is the world population count, and t_{Sust} is the timescale considered during the metric assessment.

10.1.2.3.6 Environmental Persistence

The persistence of an organic chemical in the environment is a function of biodegradability for the majority of organic chemicals released to soil and water. The general equation for the environmental persistence (EP) of a substance in the environment is as follows:

$$EP = \frac{C^0}{k} \tag{10.49}$$

where C_0 is the initial concentration of the substance in the environment, and k is the rate constant for the degradation or removal of the substance. Generally, for the substance to be greener, the degradation constant k should be higher. In consequence, a lower EP score is considered greener and healthier for the environment. This equation is a simplified version, and other factors such as chemical properties, environmental conditions, and presence of other substances can be added to further refine the assessment of the environmental persistence. Another way to calculate persistence is through estimation of the aerobic biodegradation lifetime using the Boethling Index equation:

$$\text{Boethling Index} = 3.199 - 0.0021\text{MW} + {}^{\Sigma}(a_n + f_n) \tag{10.50}$$

where MW is the molecular weight of the substance, a_n is the number of functional groups, and f_n is the type of functional group.

10.1.2.4 Eco-Footprint

The ecological footprint (eco-footprint) is a measure of impact assessment that provides a simple, comprehensive, and effective tool to evaluate the environmental impact of production and consumption. Eco-footprint (EF) was introduced by Rees and Wackernagel [18] in the 1990s as a measure of an ecosystem's ability to absorb the consumer waste and replenish for the resources utilized in the production of goods and rendering of services in a distinct area without deterioration. Lower EF values, in terms of global hectare (gha) per person, mean that the process is environmentally friendly. Specific EF for individual factors is being used in recent literature, such factors include nitrogen footprint, chemical footprint, water footprint, energy footprint, land footprint, carbon footprint, material footprint, and phosphorus footprint. This metric is also used in manufacturing companies to assess their impacts to the environment [8].

10.1.2.5 Summary of Impact-Based Metrics

Impact-based metrics do not involve evaluation using mass, and most are included in the life cycle analysis (LCA). This subsection summarized some of the impact-based measures that are well-studied by green chemists in recent years. It should be noted that not all impact-based metrics currently being used in green chemistry are listed in this chapter due to the flexible and dynamic nature of these types of metrics. The summary of impact-based metrics discussed in this chapter is as follows:

1. Eco-Scale categories are the Yield, Price of Reaction Components, Safety, Technical Setup, Temperature/Time, and Workup and Purification;
2. Toxicological Hazard Metric categories are Human Toxicity by Ingestion Potential (INGTP), Human Toxicity by Inhalation Potential (INHTP), Air Toxicity (AT), Water Ecotoxicity (WE), and Bioaccumulation (ACCU);
3. Global Hazard Metric categories are Acidification Potential (AP), Ozone Depletion Potential (ODP), Smog Formation Potential (SFP), Global Warming Potential (GWP), Abiotic Depletion Potential (ADP), and Persistence (PER);
4. Eco-Footprint categories are Chemical Footprint, Material Footprint, Energy Footprint, Land Footprint, Water Footprint, Carbon Footprint, Nitrogen Footprint, and Phosphorus Footprint;

10.2 Tools and Applications of Green Metrics

As the world faces the challenge of transitioning towards sustainability, the shifting requires suitable key performance indicators to screen technologies and effectively develop a more optimal pathway. For the past decades, as mentioned above, green chemistry has been guided by 12 general principles and associated metrics, like Eco-Footprint and Environmental Factor, which are the typical metrics used in industries and research [19]. They offer a plan for making chemical products and processes more green and environmentally friendly. However, Eco-Footprint and Environmental Factors are single-factor metrics showing practical limitations [8]. Consequently, the emerging trend is to shift from single-factor metrics to versatile multifactor indicators in any experimental work. Recently, there have been existing eco-friendly analytical methods that comply with green chemistry principles used by researchers to design and implement chemical processes and products that minimize environmental and health impacts, employing various tools for assessing and measuring green metrics. Depending on its application, several green metrics and tools are developed to evaluate the effectiveness and efficiency of the method's eco-friendliness and greenness. This section will discuss different green metrics distinguished by a different degree of sophistication in different applications and their proper tools to help design and optimize processes and products toward sustainable growth.

10.2.1 Green Metrics Applied in Analytical Chemistry

Green analytical chemistry (GAC) focuses on controlling more eco- and environmentally friendly analytical procedures and cognizance of health and safety issues during the activities. In evaluating how environmentally friendly an analytical method is, factors like amounts of materials, producing minimal waste, using less energy, having few processes, making things smaller, and automating processes are some of the criteria that should be taken into account. The greenness of analytical procedures is a multivariate complex parameter that is not easily quantifiable and requires dedicated tools. In analytical chemistry, the analytical process refers to determining certain substances. It differs from the industrial processes mainly by its scale, and emissions are low but are more dispersed, making it very difficult to control [8]. Compared to industrial processes, analytical chemistry uses a small scale of solvents and reagents to produce the outcome of the analytical process. This means that materials like solvents, reagents, energetic inputs, and others contribute to waste generation. The volume of waste generated during analytical processes can be reduced by controlling the quantity of the materials and energetic inputs. The final product of analytical chemistry results from the analytical measurement, contrary to green chemistry metric systems, mainly applied in industrial processes, usually referring to the mass of the product [20]. Traditional sustainable metrics used by green chemistry are unsuitable for green analytical chemistry [10]. Given all these contrary, in 2013, 12 principles of GAC were formulated to measure the greenness of analytical methods. These principles have different aspects of the analytical procedure and suggest the greenest solution for each element, as shown in Figure 10.1.

Thus, it is clear that a different approach should be made to green analytical chemistry, which requires the development of several metrics to measure the degree of greenness of analytical methodologies. To date, the environmental impact of analytical methods/procedures and their alternatives has been appropriately evaluated using the following GAC approaches, as shown in Figure 10.2.

As shown in Figure 10.2, many metrics show growing popularity; however, this section will specifically concentrate on a few widely used metrics in green analytical chemistry, such as Analytical Eco-Scale, Green Analytical Procedure Index (GAPI), Analytical GREEnness (AGREE), and National Environmental Methods Index (NEMI).

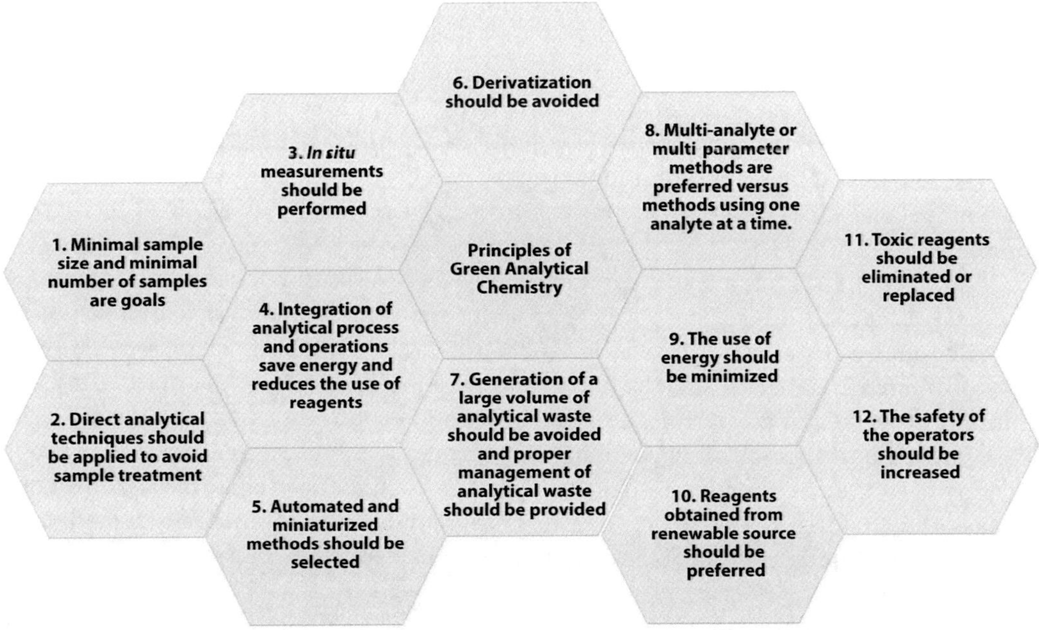

Figure 10.1 The 12 principles of green analytical chemistry.

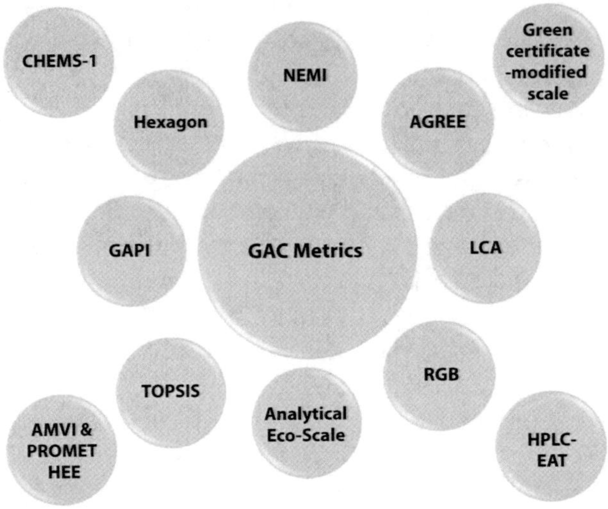

Figure 10.2 Metrics that can be employed in GAC.

10.2.1.1 National Environmental Methods Index

The initial method for evaluating the greenness of an analytical method was based on the National Environmental Methods Index. NEMI, established in 2002, offers a searchable database containing a wide range of techniques, protocols, statistical and analytical methods, and processes. NEMI is a simple pictogram and easy tool consisting of four fields stating if hazardous or corrosive reagents are used or if the procedure generates significant waste. Specific criteria must be met in each category to be considered environmentally friendly. The first category is considered green when none of the chemicals used in the analysis are on the list of persistent, bio-accumulative, and toxic substances. The second requirement is that the chemicals used should not be listed on the Toxic Release Inventory (TRI) and Resource Conservation and Recovery Act (RCRA) hazardous waste lists. The third condition is that it avoids highly corrosive conditions when the pH levels fall between 2 and 12. The fourth and final condition for being labeled green is that the total waste produced during the analysis is less than 50 grams [8, 20–23]. An example of a NEMI pictogram is presented in Figure 10.3.

Moreover, NEMI has a significant advantage as a tool for assessing the greenness of the analytical methods because it is easy for users to understand. The pictogram provides enough information about the environmental impacts of operations. However, the tools have drawbacks, such as the information being too general and the time-consuming processes of preparing the NEMI symbol, especially when dealing with non-standard chemicals. An improvement was suggested by de la Guardia and Armenta, proposing a color-coded three-degree scale for each field—red for environmentally unfriendly analysis, yellow for moderate, and green for environmentally friendly analysis, as shown in Figure 10.4. This modification makes the NEMI procedure assessment more quantitative [22].

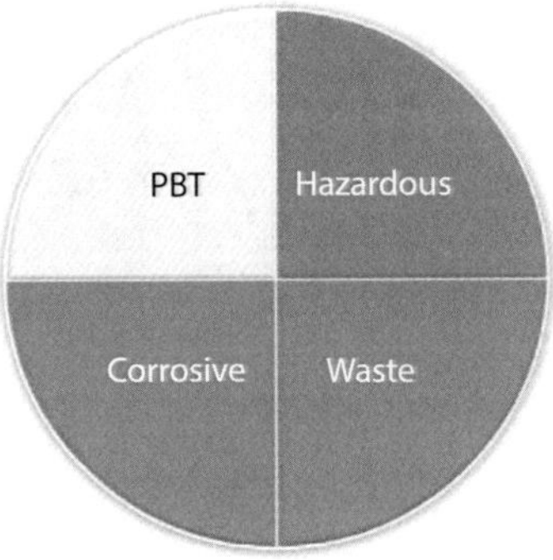

Figure 10.3 NEMI pictogram. The field is filled with green, which means that the criteria are met.

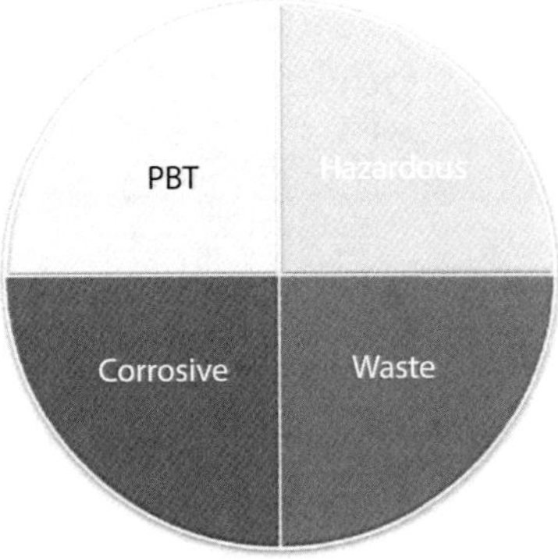

Figure 10.4 Modified NEMI pictogram. A three-degree scale for the greenness assessment.

10.2.1.2 Green Analytical Procedure Index

This metric tool evaluates the environmental aspects of the entire analytical method. GAPI offers general and qualitative information. This tool uses five pentagrams to represent different stages of the analytical process, which are 1) preparing the sample, 2) using reagents and compounds, 3) collecting the sample, 4) using instrumentation, and 5) the type of general methods used. Each pentagram is divided into sections corresponding to different parameters; for instance, the sample preparation considers factors like scale extraction, the type of solvent and reagent, and additional treatment. The GAPI pictogram includes a total of 15 parameters, which are color-coded as green, yellow, and red [10, 14, 21, 22], as shown in Figure 10.5, to indicate the severity of the impact, GAPI rules and criteria are used, as shown in Figure 10.6.

The five pentagrams of GAPI assess different aspects of the analytical methodology.

1. Sampling: The first pentagram evaluates the sampling. If the sample is collected inline, it is considered green, while on-line or at-line collection is

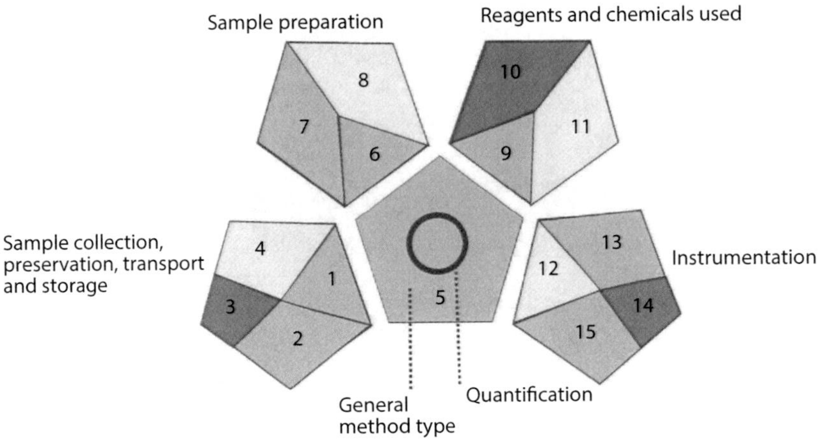

Figure 10.5 Description of different pentagrams of the GAPI pictogram.

Pictogram Name	Subpart number	Name of the subpart	Colour selection by the actual conditions		
			Green	Yellow	Red
Sample sourcing	1	Collection	In-line	Online	Offline
	2	Preservation	None	Chemical or Physical	Both (physical and chemical
	3	Transport	None	Required	
	4	Storage	None	Normal conditions	Special condition
Method type	5	Direct/Indirect	No preparation	Simple preparation	Required extraction
Sample preparation	6	Scale of extraction	Nano	Micro	Macro
	7	Solve/reagents	None	Green solvents/reagents	Non-green solvents/reagents
	8	Additional treatments	None	Simple(cleaning and solvent removal)	Advanced (derivatization)
Reagents and chemicals used	9	Amount	<10 mL	10-100 mL	>100 mL
	10	NFPA health hazard value	0 to 1	2 to 3	4 or above
	11	NFPA flammability or instability	0 to 1	2 to 3	4 or above
Instrumentation	12	Energy usage	<0.1 KWh per sample	<1.5 KWh per sample	>1.5 KWh per sample
	13	Occupational hazard	None	-	Vapors to atmosphere
	14	Waste	<1 mL	1-10 mL	>10 mL
	15	Waste treatment	Recycling	Degradation	No treatment
Symbol	O	Present	If the method is used for qualitative and quantitative purposes		
		Absent	If the method used for only qualitative		

Figure 10.6 GAPI rules and criteria for assessing the analytical method.

yellow, and off-line collection is red. On the other hand, preservation, transportation, and storage are also assessed in this pentagram.

2. Method Type: The second pentagram focuses on the type of method. If no sample preparation is needed, it is considered green; simple preparation is yellow; and extraction is red.
3. Sample Preparation: The third pentagram has three fields related to sample preparation. The scale of extractions (nano, micro, and macro), the nature of

the solvents used, and additional treatments are assessed. Green, yellow, and red are used to indicate the level of their impacts.

4. Solvents/Reagents: The fourth pentagram evaluates the volume/amount, health hazards, and safety hazards of solvents/reagents. Green, yellow, and red are used to represent their impacts.
5. Instrument and Waste: The fifth and final pentagram considers per sample of energy consumption by the instrument, occupational hazards, waste generation, and waste treatment. Green, yellow, and red signify the environmental impacts.

The presence of a circle in the second pentagram indicates if the method is used for qualification only or for both qualification and quantification. A circle means it is used for both, while no circle indicates qualification only.

10.2.1.3 Analytical GREEnness

Analytical GREEnness is one of the attractive green metric tools that aid in developing a green analytical methodology that circulates around the 12 principles of GAC. This metric tool has been developed to allow comprehensiveness and flexibility of inputs and simplicity and clarity of outputs. The tool expresses the analytical method in the AGREE score by a circle-shaped diagram. Each input principle transforms into a score ranging from 0 to 1, and the final score, placed in the middle, is obtained from the final assessment of the 12 principles. Using this tool, a pictogram, shown in Figure 10.7, offers the result where the user designates the scale values [20]. All 12 Green Analytical Chemistry principles are rated on a scale based on factors like sample treatment, sample size, energy use, and operator safety. Each aspect is color-coded on a scale from 0 to 1, with 1 indicating high eco-friendliness (dark green) and 0 representing the least eco-friendly (red).

The effectiveness of each principle is shown using a simple color scale (red, yellow, green) to indicate its performance. Additionally, the importance of each principle is represented by

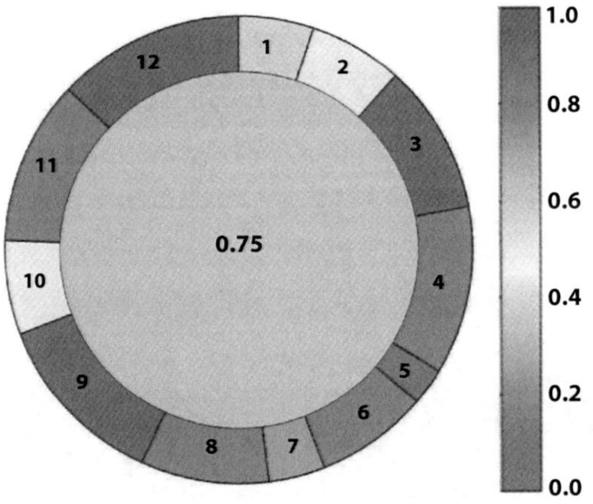

Figure 10.7 Sample pictogram of AGREE tool in color scale.

the width of its corresponding section. In this way, one can easily judge the greenness of a method in terms of the 12 GAC principles and the overall procedure [14, 21, 22, 24].

10.2.1.4 Analytical Eco-Scale

In 2012, the Eco-Scale was the most common tool used to assess the greenness of analytical methods. This comprehensive approach, an ideal green analysis, corresponds to a score of 100 [22]. The aspect that decreases the greenness of the process, like toxic reagents, waste generation, or high energetic demand, is subtracted from the ideal reaction with a value of 100 [14]. After subtracting, the number of remaining points will be used by the user to determine whether the procedure is ideally green or acceptable [20]. The points subtracted will be based on the penalty point criteria shown in Table 10.5 [21]. Ideally, the higher the score, the greener and more economical the organic preparation, as shown in Figure 10.8 [25]. Penalty points related to the chemicals used are readily accessible because hazard pictograms are displayed on chemical containers, and safety sheets contain this information. This scoring system is easy to understand and allows straightforward comparisons between different analytical methods. According to the analytical Eco-Scale, an analytical method can be considered "ideally environmentally friendly" if it meets these criteria: 1. The solvents or reagents do not pose any physical, environmental, or health risks; 2. The energy consumption per sample is less than 0.1 kWh; 3. No waste is produced.

The Analytical Eco-Scale has several advantages, including easy-to-read results, making it simple to compare different analytical methods, and considering various aspects of environmental impact. However, a significant drawback is that it does not provide information about the specific environmental hazards. The final score does not explain why an analytical procedure has a certain environmental impact like using solvents, reagents, workplace safety concerns, or waste generation. Therefore, the Eco-Scale score cannot help pinpoint the weaknesses in a procedure for improvement [22, 23].

10.2.2 Green Metrics in Pharmaceutical Development

As the pharmaceutical industry is committed to developing novel lifesaving medicines that reproducibly deliver the desired compound with acceptable quality, cost, and speed at a sufficient scale, metrics are a beneficial tool for enhancing sustainability and efficiency, providing a powerful opportunity for holistic process understanding and the ability to apply resources for the highest impact strategically [8, 26]. Therefore, the power of using green chemistry metrics to pharmaceutical processes encompasses improved chemical and pharmaceutical process performance potential and implies a significant competitive

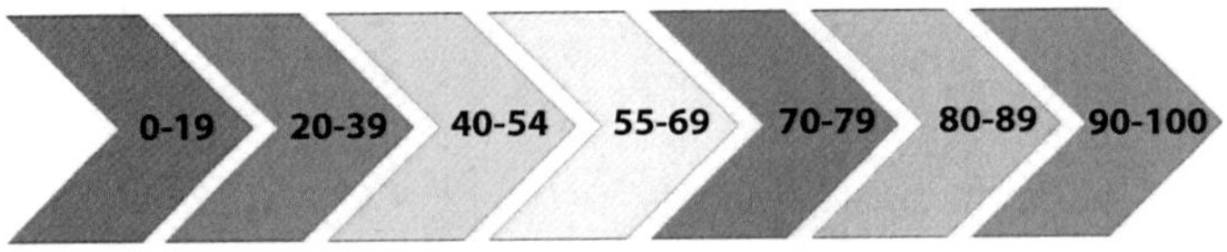

Figure 10.8 Color scale for evaluation of green analytical chemistry using Eco-Scale metrics.

Table 10.5 Penalty points applied to calculate the final analytical Eco-Scale score.

Names	**Penalty points calculation criteria**
Chemicals Reagents	Step 1: Check the Material Safety Data Sheet (MSDS) symbols. If the symbol is labeled "DANGER," multiply each pictogram by 2. If it is labeled "WARNING," count 1 point for each symbol. Step 2: Depending on the volume of the chemical or reagents used: • If it is less than 10 mL, add 1 penalty point • If it is between 10 mL and 100 mL, add 2 penalty points • If it is more than 100 mL, add 3 penalty points. Step 3: Multiply the penalty points from Step 1 by the penalty points from Step 2 to get the final penalty points for the specific chemical or reagent.
Instruments	Penalty points for energy consumption are determined this way: If energy usage is below 0.1 kWh per sample, you get 0 penalty points. If it is under 1.5 kWh per sample, you receive 1 penalty point. If it exceeds 1.5 kWh per sample, you get 2 penalty points.
Waste	The penalty points are determined by whether vapors are released into the environment. If there are no vapors released, the penalty is 0. If vapors are released, the penalty is 3. When it comes to the total waste produced, the penalty is assigned based on the amount of waste: 1 point for less than 1 mL of waste, 2 points for 1–10 mL, and 5 points for more than 10 mL.
Management	The penalty points are assigned based on how the waste is managed. If the waste can be recycled, there is no penalty (0 points). If it undergoes degradation, there is a penalty of 1 point. If it is treated through passivation, the penalty is 2 points. And if the waste is not treated at all, it gets a penalty of 3 points.

advantage [27]. The American Chemical Society Green Chemistry Institute Pharmaceutical Roundtable (ACS GCI PR) has adopted process mass intensity as its mass metric because it aligns with sustainability by focusing on the increase of resource efficiency and by having a direct correlation with upstream supply chain life cycle impacts [28, 29]. Over the years, the most used metrics in the pharmaceutical industry are the mass metrics [2, 30], as shown in Table 10.6.

The list of the metrics shown in Table 10.6 is easy to calculate and understand, provided the input data are available. However, there are identified disadvantages to using these metrics—for example, atom economy excludes yield, stoichiometry, solvents, and

Table 10.6 Most used metrics in the pharmaceutical industry [2, 28].

Metrics	Calculation	Definition
Process Mass Intensity (PMI)	$PMI = \frac{\sum m(Inputs)}{m(Products)}$	Number of mass units of process inputs needed to produce one mass unit of product
E-Factor (E-Factor)	$E-Factor = \frac{p\, m_{(Waste)}}{m_{(Products)}}$	Ratio of waste mass to product mass
Energy Intensity (EI)	$EI = \frac{p\ _{(EnergyUsed)}}{m_{(Products)}}$	Energy expanded to produce one mass unit of product
Atom Economy (AE)	$AE = \frac{m_{W(Product)}}{p\, m_{W(Reactants)}}$	Atomic percent of reactants incorporated in to product
Effective Mass Yield (EMY)	$EMY = \frac{m_{Product}}{m_{(non-benign\ inputs)}}$	Percent mass of product to non benign inputs
P Mass Efficiency = Mass pro-	Mass percent of the inputs inductivity $ME = \frac{m_{Product}}{p\, m_{inputs})} = \frac{1}{PMI}$	Corporated into the final product
Reaction Mass Efficiency (RME)	$RME = \frac{m_{Product}}{p\,(m_{reactants})}$	Mass percent of the reactants incorporated in to the product

Note: mw: molecular weight, m: mass, E: energy

catalysts. Reaction mass efficiency excludes solvents and other materials. Initially, E-factor and PMI did not consider water in their calculations because it could lead to very high results, making comparisons challenging. However, they have now been updated to include water, reflecting the growing significance of biological systems in pharmaceuticals and the impact of water scarcity. While E-factor and PMI are easy to calculate, they do not address the intricacies of the process. Recognizing these limitations, efforts have been made to create new metrics and tools to bridge these gaps [31].

10.2.2.1 Green Solvent Selection Tool

This tool lets you pick solvents based on their key characteristics. Solvents that are close together on the PCA map have similar qualities, while others are quite different. The program is accessible and can be downloaded to access information about solvent properties, including physical characteristics, environmental impact, safety, and health data. Although it does not align with the 12 green principles, it is handy for evaluating environmental friendliness without instruments. However, using it for assessing an analytical method is challenging [21].

10.2.2.2 Green Chemistry Innovation Scorecard Calculator or Innovation Green Aspiration Level

The proposed approach allows us to compare an ideal process for producing a specific pharmaceutical agent with the average commercial process regarding environmental impact [10, 14]. The Green Chemistry Innovation Scorecard Calculator tool takes a different approach, focusing on reducing waste and considering process mass intensity. It is a web calculator developed by the IQ Consortium, the ACS GCI Pharmaceutical Roundtable, and academic experts. This tool demonstrates how innovations in green chemistry and engineering can reduce the amount of waste generated when making large quantities of active pharmaceuticals. It helps compare different processes and their waste reduction efforts effectively. While it does not strictly follow the 12 green principles, it can be valuable for drug synthesis and manufacturing [30].

10.2.2.3 Process Mass Intensity-Environmental Life Cycle Assessment Tool

The Process Mass Intensity-Environmental Life Cycle Assessment (PMI-LCA) tool is designed for estimating the process mass intensity and conducting an environmental life cycle assessment for small-molecule Active Pharmaceutical Ingredients (APIs). It can be customized to suit various linear and convergent processes. While it is a streamlined tool and not exhaustive, it is "fit for purpose" when comparing and measuring progress toward more eco-friendly and sustainable manufacturing practices. It mainly focuses on four categories: Mass Net, Energy, Global Warming Potential, and Water Depletion, with two additional options for Acidification and Eutrophication [21].

10.2.2.4 Biopharma Process Mass Intensity Tool

The Biopharma Process Mass Intensity metric is a standard method for measuring the amount of water, raw materials, and supplies needed to produce 1 kg of a biological drug substance (API). This metric provides a common way to assess biologics production, enabling industry-wide comparisons, greater transparency in process development, and a fair means of evaluating different processes [21].

10.2.3 Green Metrics Applied to Organic Synthesis

Applying green metrics in organic synthesis aims to promote sustainable process interventions [32, 33]. To achieve this, designing the synthesis of new compounds requires using alternative green materials and processes. The primary purpose of applying metrics in organic synthesis is to obtain clear, simple, and fast information about the greenness of the organic process. They also enable us to predict how a specific change in a synthesis path, such as the elimination or replacement of a solvent, would influence its environmental impact [34]. This section will provide several metrics used to assess the greenness of organic syntheses with their short characteristics.

The first metric, carbon efficiency, is employed to calculate the portion of carbon from the initial reagents in organic synthesis that ends up in the final product. It is a way to assess the eco-friendliness of organic synthesis by focusing solely on carbon content. The equation of the metric is shown in Equations 10.20 and 10.21.

The second metric is the effective mass yield: this metric calculates the percentage of the final product found in all of the substances and materials used during organic synthesis. It does not include reagents with minimal or negligible environmental impact, such as sodium chloride or acetic acid, in the calculation. The equation of the metric is shown in Equation 10.10.

The third metric is mass intensity, which considers how efficiently a reaction proceeds, along with factors like stoichiometry, the volume of solvents, all reagents, and other substances used in the synthesis. In an ideal synthesis, where the total input mass equals the product mass, this metric has a value of 1. The equation of the metric is shown in Equation 10.16.

The fourth metric is reaction mass efficiency, essentially the opposite of the E-factor described by Sheldon. RME is a more straightforward way to identify the most environmentally friendly or the least friendly reactions within an industrial process or synthesis. This metric effectively reduces waste at both the fundamental and overall levels. The equation of the metric is shown in Equation 10.15.

The fifth metric is atom utilization (AU), which offers a quick and easy way to assess how environmentally friendly a process or a specific reaction is regarding waste production. It measures the percentage of the mass of the final product compared to the combined mass of all products (the final product and any by-products) obtained during the synthesis. AU can be stated as follows:

$$\%AU = \frac{(\text{mass of the final product})}{(\text{total mass of all the substances produced})} \tag{10.51}$$

The sixth metric is the environmental impact parameter for solvents and catalysts (f). It considers the actual masses of materials used in the process. It equals 0 only if all of the materials (like solvents and catalysts) used in the process or a specific synthesis step are recycled, recovered, or eliminated. Otherwise, it has a non-zero value. F can be stated as follows:

$$\frac{P_{\text{mass of reaction and postreaction solvents and materials+mass of catalyst used}}}{f = \text{mass of the final product}} \tag{10.52}$$

The final metric is the stoichiometric factor, used when one or more reagents are used in excess during a synthesis. For stoichiometric reactions, SF equals 1. If the reaction is non-stoichiometric, SF is greater than 1. SF can be stated as follows:

$$Sf = 1 + (AE)\frac{P_{\text{mass of excess reagents/chemicals}}}{\text{expected product mass at 100\% yield}} \tag{10.53}$$

10.3 Life Cycle Assessment

In the field of sustainable development, life cycle (LCA) assessment is a comprehensive approach that is widely used in decision-making for industries and organizations that aims to promote and ensure sustainability by evaluating the environmental impacts and resource usage associated with their product or process throughout its entire life cycle. LCA involves tracking all of the materials and energy involved, from raw material extraction to distribution, use, and the final disposal of waste, which is often referred to as "cradle-to-grave." As a tool, it prevents "problem shifting," where solving one environmental issue creates new problems elsewhere in the life cycle or environment. However, due to the lack of specific time and location data, LCA can only assess "potential impacts" rather than real observed ones [35–38].

Moreover, the ISO 14040 standard outlines four main stages for LCA studies: (1) Defining Goals and Scope, (2) Analyzing the Life Cycle Inventory, (3) Assessing Impact, and (4) Interpreting the Life Cycle. These stages are shown in Figure 10.9. LCA is typically an iterative process. Problems in data collection or new information can require adjustments in earlier decisions. Therefore, revisiting earlier stages and making necessary changes are often useful and essential.

Stage 1: In this stage, the purpose and extent of the product life cycle assessment are determined. It outlines what part of the product's life cycle will be assessed and why. This stage clarifies the study's objectives, defines the system under examination (its boundaries, function, data requirements, etc.), and sets the criteria for comparing systems.

Stage 2: This step involves analyzing the flow of materials and energy within the product system. The goal is to identify and measure energy consumption, raw materials, and emissions to the environment (air, water, soil, solid waste) that could have environmental

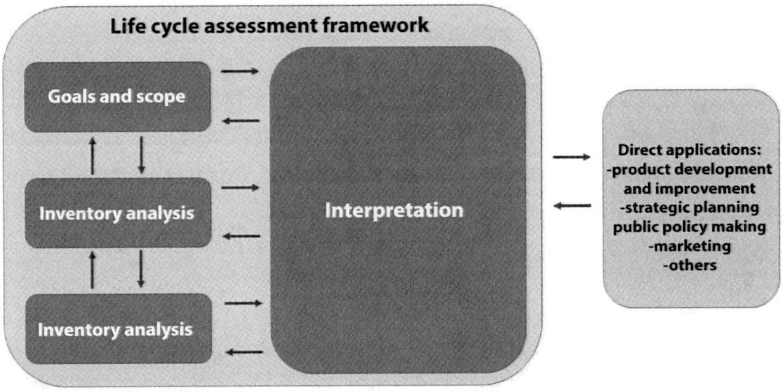

Figure 10.9 Stages of an LCA according to EN ISO 14040 [35].

impacts. The environmental significance of these substances is assessed in the next phase. The results are then discussed in the interpretation phase.

Stage 3: The details gathered from the inventory analysis are used for impact assessment. This phase breaks down the results for various impact categories and evaluates their importance through normalization and weighting. The life cycle impact assessment involves four steps:

- Classification: Grouping environmental interventions from the inventory analysis into different impact categories based on their expected environmental effects.
- Characterization: Weighing the different substances contributing to the same environmental impact, resulting in aggregated outcomes in specific units of measurement.
- Normalization: Relating the system's environmental profile to a broader dataset or context.
- Weighting: Reducing the environmental profile of multiple indicators to a single impact score using weighting factors based on subjective value judgments. This simplifies decision-making by converting various criteria into a numerical environmental impact score.

Stage 4: The final stage involves interpreting the life cycle assessment. It includes a critical review, determining data sensitivity, and presenting results. This stage ensures the consistency of assumptions and data quality concerning the study's objectives and scope.

10.4 Conclusions

The introduction of Green Metrics has played a pivotal role, having been extensively examined and applied, underscoring their critical contribution to advancing sustainable development. Measuring progress facilitates the evaluation of eco-friendly processes, simplifies the identification of critical stages, and allows for comparisons between various approaches. In this chapter, it is important to note that although numerous metrics have been introduced across different fields, many have seen limited use for specific one-time purposes. This is due to the constant emergence of new metrics, often based on existing ones with minor alterations, indicating a lack of consensus in their application. As exemplified in this chapter, some metrics center on aspects such as mass, while others focus on energy, waste, and emissions, serving various applications in fields like analytical chemistry, pharmaceuticals, and organic synthesis. Nevertheless, yields, costs, and social impact carry substantial significance. Therefore, only metrics that comprehensively evaluate all process parameters can effectively and accurately gauge the progress in adopting environmentally responsible practices.

References

1. Andraos, J., Unification of reaction metrics for green chemistry: applications to reaction analysis. *Org. Process Res. Dev.*, 9, 2, 149–163, 2005.
2. Constable, D.J.C., Curzons, A.D., Cunningham, V.L., Metrics to 'green' chemistry—which are the best? *Green Chem.*, 4, 6, 521–527, 2002.
3. Dicks, A.P. and Hent, A., The e factor and process mass intensity, in: *Green Chemistry Metrics: A Guide to Determining and Evaluating Process Greenness*, vol. 45–67, 2015.
4. Sheldon, R.A., Metrics of green chemistry and sustainability: past, present, and future. *ACS Sustain. Chem. Eng.*, 6, 1, 32–48, 2018.
5. Curzons, A.D., Constable, D.J.C., Mortimer, D.N., Cunningham, V.L., So you think your process is green, how do you know?—using principles of sustainability to determine what is green–a corporate perspective. *Green Chem.*, 3, 1, 1–6, 2001.
6. Sheldon, R.A., Bode, M.L., Akakios, S.G., Metrics of green chemistry: Waste minimization. *Curr. Opin. Green Sustain. Chem.*, 33, 100569, 2022.
7. J. Glaser, Green chemistry metrics. *Clean Technol. Environ. Policy*, 11, 371–374, 2009, doi: 10.1007/s10098-009-0264-x.
8. Tobiszewski, M., Marć, M., Gałuszka, A., Namieśnik, J., Green chemistry metrics with special reference to green analytical chemistry. *Molecules*, 20, 6, 10928–10946, 2015.
9. Hudlicky, T., Frey, D.A., Koroniak, L., Claeboe, C.D., Brammer Jr., L.E., Toward a 'reagent- free' synthesis: Tandem enzymatic and electrochemical methods for increased effective mass yield (EMY). *Green Chem.*, 1, 2, 57–59, 1999.
10. M. Espino, F. J. V. Gomez, J. Boiteux, M. de los Ángeles Fernández, and M. F. Silva, "Green Chemistry Metrics," in *Comprehensive Foodomics*, vol. 2, Berlin, Chausseestraße 20, Germany: ResearchGate, 2020, pp. 825–833. doi: 10.1016/B978-0-08-100596-5.22822-9.
11. A. P. Dicks and A. Hent, "Atom Economy and Reaction Mass Efficiency," in *Green Chemistry for Sustainability*, SpringerBriefs, 2015, pp. 17–44. doi: 10.1007/978-3-319-10500-0_2.
12. Andraos, J. and Sayed, M., On the use of " green" metrics in the undergraduate organic chemistry lecture and lab to assess the mass efficiency of organic reactions. *J. Chem. Educ.*, 84, 6, 1004, 2007.
13. Mercer, S.M., Andraos, J., Jessop, P.G., Choosing the greenest synthesis: a multi- variate metric green chemistry exercise. *J. Chem. Educ.*, 89, 2, 215–220, 2012.
14. Martínez, J., Cortés, J.F., Miranda, R., Green chemistry metrics, a review. *Processes*, 10, 7, 1274, 2022.
15. K. Van Aken, L. Strekowski, and L. Patiny, "EcoScale, a semi-quantitative tool to select an organic preparation based on economical and ecological parameters," *Beilstein J. Org. Chem.*, 2, 1–7, 2006, doi: 10.1186/1860-5397-2-3
16. I. Dincer and A. Abu-Rayash, "Sustainability modeling," in *Energy Sustainability*, Cambridge Massachusetts: Academic Press, 2020, pp. 119–164. doi: 10.1016/b978-0-12-819556-7.00006-1.
17. Hacatoglu, K., Dincer, I., Marc, A., Rosen. A new model to assess the environmental impact and sustainability of energy systems. *J. Clean. Prod.*, 103, 211–218, 2014.
18. Wackernagel, M. and Rees, W., *Our Ecological Footprint: Reducing Human Impact on the Earth*, New Society Publishers, Philadelphia, 1996.
19. Tulus, V., Pérez-Ramírez, J., Guillén-Gosálbez, G., Planetary metrics for the absolute environmental sustainability assessment of chemicals. *Green Chem.*, 23, 24, 9881–9893, 2021.
20. Pena-Pereira, F., Wojnowski, W., Tobiszewski, M., Agree—analytical greenness metric approach and software. *Anal. Chem.*, 92, 14, 10076–10082, 2020.

21. Kowtharapu, L.P., Katari, N.K., Muchakayala, S.K., Marisetti, V.M., Green metric tools for analytical methods assessment critical review, case studies and crucify. *TrAC, Trends Anal. Chem.*, 166, 117196, 2023.
22. Sajid, M. and Płotka-Wasylka, J., Green analytical chemistry metrics: A review. *Talanta*, 238, 123046, 2022.
23. Tobiszewski, M., Metrics for green analytical chemistry. *Anal. Methods*, 8, 15, 2993–2999, 2016.
24. J. Silva, V. Bühl, F. Iaquinta, and M. Pistón, "Should we think about green or white analytical chemistry? Case study: Accelerated sample preparation using an ultrasonic bath for the simultaneous determination of Mn and Fe in beef," *Heliyon*, 9, 10, 209267, 2023, doi: 10.1016/j.heliyon.2023.e20967.
25. Gałuszka, A., Migaszewski, Z.M., Konieczka, P., Namieśnik, J., Analytical eco-scale for assessing the greenness of analytical procedures. *TrAC, Trends Anal. Chem.*, 37, 61–72, 2012.
26. M. Shi, X. Zheng, N. Zhang, Y. Guo, M. Liu, and L. Yin, "Overview of sixteen green analytical chemistry metrics for evaluation of the greenness of analytical methods," *TrAC - Trends Anal. Chem.*, 166, June, 117211, 2023, doi: 10.1016/j.trac.2023.117211
27. Tucker, J.L., The hidden power and competitive advantage of applying green chemistry metrics. *Curr. Opin. Green Sustain. Chem.*, 35, 100606, 2022.
28. A. P. Dicks and A. Hent, *Green Chemistry Metrics: A Guide to Determining and Evaluating Process Greenness.* Springer Cham Heidelberg New York Dordrecht London, 2015. [Online]. Available: http://www.springer.com/series/10045
29. Rose, H.B., Kosjek, B., Armstrong, B.M., Robaire, S.A., Green and sustainable metrics: Charting the course for green-by-design small molecule api synthesis. *Curr. Opin. Green Sustain. Chem.*, 5, 100324, 2022.
30. Wuitschik, G., Olbrich, M., Hildbrand, S., Roche's approach to green metrics in drug development. *Curr. Res. Green Sustain. Chem.*, 5, 100293, 2022.
31. Lund, C. *et al.*, Green metrics in pharmaceutical development. *Curr. Res. Green Sustain. Chem.*, 33, 100564, 2022.
32. Banitaba, S.H., Safari, J., Khalili, S.D., Ultrasound promoted one- pot synthesis of 2-amino-4, 8-dihydropyrano [3, 2-b] pyran-3-carbonitrile scaffolds in aqueous media: A complementary 'green chemistry'tool to organic synthesis. *Ultrason. Sonochem.*, 20, 1, 401–407, 2013.
33. Schäfer, H.J., Contributions of organic electrosynthesis to green chemistry. *C.R. Chim.*, 14, 7-8, 745–765, 2011.
34. Loupy, A., Solvent-free microwave organic synthesis as an efficient procedure for green chemistry. *C.R. Chim.*, 7, 2, 103–112, 2004.
35. Kralisch, D., Ott, D., Gericke, D., Rules and benefits of life cycle assessment in green chemical process and synthesis design: a tutorial review. *Green Chem.*, 17, 1, 123–145, 2015.
36. Tufvesson, L.M., Tufvesson, P., Woodley, J.M., Börjesson, P., Life cycle assessment in green chemistry: overview of key parameters and methodological concerns. *Int. J. Life Cycle Assess.*, 18, 431–444, 2013.
37. Muñoz Ortiz, I. Life Cycle Assessment as a Tool for Green Chemistry: Application to Different Advanced Oxidation Processes for Wastewater Treatment. PhD Dissertation, Universitat Autònoma de Barcelona, Barcelona, Spain, 2006.
38. Chang, D., Lee, C.K.M., Chen, C.-H., Review of life cycle assessment towards sustainable product development. *J. Clean. Prod.*, 83, 48–60, 2014.

11

Biocatalysis and Biobased Economy

Gyanendra Kumar[1,2], Nitanshu Dhama[2], Rohit Yadav[2] and Dhanraj T. Masram[2*]

[1]Swami Shraddhanand College, University of Delhi, Alipur, Delhi, India
[2]Department of Chemistry, University of Delhi, Delhi, India

Abstract
In this chapter, authors discussed about the biobased economy as the best approach to convert waste biomass. Like that agricultural forestry by-products and food supply chain waste convert into liquid fuels as well as common chemicals and biopolymers. However, carbon-based nanocomposites have been utilized as biocatalysts, which are environmentally friendly. Carbon based biocomposites and renewable biomass are two of the most challenging economic and technological issues of the 21st century. Increased global energy utilization and the resulting carbon emissions need a transition toward the application of clean and sustainable energies. Because of its renewable nature and high energy content, biomass is a promising resource for energy generation. The liquid fuel produced by thermal cracking of biomass seems to have the opportunity to be blended with conventional petroleum products. The usage of pyro-oil in place of petroleum fuels helps to minimize greenhouse gas emissions and global warming. Every technology's success is determined by its accessibility of use and economic feasibility. This is the first evaluation of its sort to shed light on the energy and exergy evaluation of catalytic pyrolysis. As an alternative, exploration toward biocatalysts is now happening because of their several benefits, including milder reaction conditions, recyclability, selectivity, and biodegradability. Biocatalyst research has been going on for decades because the pharmaceutical and fine chemical industries have needed it. Today, these changes could help biobased chemistry by making it easier to develop helpful new ideas. Enzymes are biocompatible, biodegradable, and almost entirely harmless. Enzymes are helpful tools for processing renewable feedstock and adding benefit to a wide range of building blocks and monomers in the chemistry sector. This is because enzymes are very selective and catalytically active. The use of enzymes to transform a broader range of renewable feedstocks will increase due to computational approaches.

Keywords: Biocatalysis, biobased economy, enzymes, computational approaches, biocompatible

11.1 Introduction of Biocatalysis and Biobased Economy

In the scientific and technological field of biocatalysts, the chemical reactions are carried out by using enzymes as natural catalysts. These organisms are bacteria, fungi, plants, and animals as the source of these biocatalysts, which are essential for catalyzing a variety of chemical reactions in biological systems [1]. It is an integral part of chemical synthesis and

**Corresponding author*: dhanraj_masram27@rediffmail.com

Mousumi Sen (ed.) Sustainable Green Catalytic Processes, (259–272) © 2024 Scrivener Publishing LLC

provides access to a broad range of selective transformations that are often inaccessible with traditional chemical catalysts [1(a)]. Biocatalysis has gained significant attention and importance in various industries due to its numerous advantages over traditional chemical processes. which are different key ideas to take into account while adopting biocatalysis enzymes are extremely effective and selective catalysts that may accelerate chemical reactions under benign settings, such as neutral pH levels and moderate temperatures. They frequently show extraordinary specificity, which causes fewer undesirable by-products and less waste to be produced. When compared to conventional chemical processes, biocatalytic processes often use fewer harsh chemicals, produce less hazardous waste, and consume less energy. For these reasons, they are regarded as being ecologically benign. Whereas enzymes are extremely useful in the synthesis of complex compounds, including pharmaceutical intermediates, they may be tuned for specific reactions due to their great substrate to the target sites. Biocatalysis both in academics and in the chemical and pharmaceutical fields has become a significant part of current organic synthesis. The result has been partly attributed to a fast increase in the available spectrum of chemical reactions, which has been made feasible by enhanced enzyme discovery tools in addition to high-through laboratory evolution methods for biocatalyst optimization [2]. It is a green and sustainable technology that can be used in a wide range of industrial applications, from the production of active pharmaceutical ingredients to the synthesis of fine chemicals and biofuels. Biocatalysis offers several advantages as catalysis like mild reaction conditions, high selectivity, and the ability to use renewable resources as starting materials [3]. Moreover, several sources of biocatalysts that can be used in chemical processes: Sources of biocatalysts available commercially: Chemical sellers frequently provide purified enzymes. Moreover, enzymes can be modified to perform new tasks in synthetic chemistry by emulating natural biosynthetic processes. In this strategy, enzymes that catalyze processes resembling the desired synthetic transformation are found, and their activity and selectivity are subsequently enhanced by engineering. Whereas many bacterial strains that produce enzymes that can be used for biocatalysis like *E. coli* are often used in biotechnology for their ability to express various enzymes. Yeast, such as Saccharomyces cerevisiae, can produce enzymes like alcohol dehydrogenase and various hydrolases. Additionally, fungi like Aspergillus niger produce enzymes like cellulases and amylases. It is essential to take into account aspects like substrate specificity, reaction conditions, stability, and affordability when choosing a biocatalyst for a certain application. Biocatalysts are often screened and improved by researchers to satisfy the demands of a particular process.

The adoption of a biobased economy signifies a substantial change in the way we produce, use, and perceive diverse goods and resources. To address the needs of business, agriculture, and society, a biobased economy relies on renewable biological resources and environmentally friendly procedures. It is a shift from the conventional economy based on fossil fuels and attempts to lessen environmental impact, increase sustainability, and open up economic prospects. There are versatile ideas of a biobased economy that are explained in the following manner: Transitioning to renewable resources from fossil fuels is called "energy transition." Here are some strategies of switching from fossil fuels to more environmentally friendly energy sources.

A biobased economy turns biomass, comprising crops, forestry products, agricultural by-products, and organic waste, into a key resource. Using procedures including fermentation, enzymatic conversion, and chemical synthesis, it is transformed into a variety of biobased goods. The biobased economy helps to combat climate change and preserve natural resources by lowering greenhouse gas emissions and dependency on nonrenewable resources. Moreover, the biobased economy is based on three main pillars such as green biotechnology, white biotechnology, and thermochemical conversion [4]. The use of genetic engineering and other biotechnological technologies in "green" biotechnology aims to increase the sustainability and productivity of crops and other biomass sources. In white biotechnology, biomass is transformed into a variety of products, including biofuels, bioplastics, and other chemicals, with the use of biocatalysts such as enzymes and microbes. Biomass may be transformed into biofuels, charcoal, and other products by thermochemical conversion, which uses heat and other kinds of energy. There are many promising future aspects to this emerging field. One key area of growth is in the development of new biobased products and materials, such as bioplastics, biofuels, and biocomposites, which have the potential to replace traditional fossil-based products and reduce greenhouse gas emissions. The incorporation of circular economy ideas with biobased production systems is another crucial factor that can assist in cutting waste and improving resource efficiency. In areas with a strong agricultural or forestry industry, the biobased economy is also anticipated to assist rural development and offer new job possibilities [5]. The sustainability of biobased production systems and possible conflicts with food production are two difficulties that must be resolved, though.

11.2 Carbon-Based Biocomposites

Carbon-based biocomposites are substances that incorporate carbon-based elements with biodegradable or biocompatible substances to produce a composite substance with special qualities. Due to their potential for lightweight, robust, and ecologically friendly uses, these materials are including aerospace, automotive, construction, and biomedicine. These materials have also been used in the fields of biomedicine, energy storage, and environmental remediation [6]. The use of biocomposites can help reduce the environmental impact of products by lowering energy demands and greenhouse gas emissions [7, 8].

11.2.1 Carbon Fibers

It is a lightweight and high-strength material, which is used in sporting goods, automotive, and aerospace. It is produced by weaving together very fine strands of carbon atoms to form a fabric. The fabric is then soaked with a resin heated and compressed to cure, producing a stiff resilient structure. Carbon fibers are added to the composite to improve its mechanical properties, such as strength, stiffness, and fatigue resistance.

11.2.2 Biocompatible Matrix

The choice of matrix material depends on the specific application and the properties required for the desired tissue regeneration. It can be manufactured from natural polymers such as cellulose, starch, or proteins, as well as synthetic biodegradable polymers such as polylactic acid (PLA) or polyhydroxyalkanoates (PHAs).

Recently, different types of carbon-based biocomposite have been synthesized by researchers, for example, Soni *et al.* [8] synthesized a chitosan carbon-based biocomposite by following several steps. First, chitosan is dissolved in an acidic solution to form a chitosan solution [8]. Then, activated carbon is added to the chitosan solution and mixed thoroughly to form a homogeneous mixture. The mixture is then cast into a mold and allowed to dry. Once the composite has dried, it is cut into the desired shape and size. The resulting chitosan carbon-based biocomposite is a porous material that can be used for a variety of applications, including water treatment, tissue engineering, and drug delivery. The ratio of chitosan to activated carbon and the synthesis conditions may both be changed to customize the qualities of the biocomposite [8a, 8b]. The resultant biocomposite has a large surface area, is biodegradable, and is biocompatible, making it a desirable material for several biomedical applications.

According to Nicora *et al.* [6], synthesized graphene oxide-based biocomposite scaffolds provide a promising potential for use in bone tissue engineering applications. The hydrothermal technique with microwave assistance was used to create the hydroxyapatite/carbon-based biocomposite scaffolds. The biocomposite scaffolds were then made by homogeneously dispersing the powders in collagen/chitosan solution-based polymer gels. The resultant liquid was sonicated for 30 minutes before being uniformly disseminated by magnetic stirring. Overall, this study provides a foundation for the development of new and improved biocomposite materials for bone tissue engineering. The use of biocomposites as raw materials for sustainable products can help reduce the environmental impact of products by lowering energy demands and greenhouse gas emissions [7]. Utilizing the carbon footprint indicator will help you determine the environmental costs associated with biobased materials and how using biocomposites in product development will affect those costs. Carbon-based biocomposites have emerged as promising materials with a wide range of applications. These materials offer several advantages, as follows:

- **Sustainability:** The environmental impact of and reliance on fossil fuels are decreased by carbon-based biocomposites' use of renewable resources.
- **Improved Mechanical Characteristics:** They are excellent for a variety of structural applications because they frequently demonstrate increased strength, stiffness, and durability compared to their separate elements.
- **Biodegradability:** Numerous carbon-based biocomposites degrade naturally, minimizing waste and adverse environmental effects at the end of their useful lives.

However, difficulties still exist in simplifying production methods, ensuring uniform quality, and resolving these composites' long-term behavior. Carbon-based biocomposites are still being improved via research and development to realize their full potential in environmentally friendly and high-performance applications.

11.3 Waste Biomass

The term "waste biomass" describes biological materials that have lost their original value and are frequently thrown away as garbage. The sources of these materials, which can include a wide variety of organic matter, include forestry, agricultural, and municipal operations. There are several types of waste biomass such as *Residues From Agriculture*: Straw, maize stalks, rice husks, and other crop leftovers, as well as by-products like bagasse from sugarcane processing. *Woodland Waste*: It includes resources left over from handling wood trees, such as old dyewood, aging wood and bark, and effluent treatment leftovers. *Industrial Waste:* Food processing waste and wastewater sludge are examples of organic waste produced by industry operations. Waste biomass provides potential for resource use and sustainability as well as environmental issues because it can contribute to landfills and greenhouse gas emissions. Biomass is found in many varieties like forestry, agriculture, and agro-industry as well as crop residues. While these various biomass sources offer a free source of fuel in rural regions, they are also to blame for air pollution. The quantity of air supply for combustion, fuel type, combustion equipment, fuel characteristics, and climatic circumstances all have a significant role in global air pollution. The growing interest in recycling waste biomass into useful goods through a variety of techniques has emerged in recent years, including the following:

- Production of Bioenergy: Renewable energy sources are provided by biofuel production from waste biomass, such as biogas, bioethanol, and biodiesel.
- Biomaterials: Reducing dependency on fossil fuels is possible by using waste biomass as a raw material for bioplastics, biobased chemicals, and other sustainable products.
- Soil Modification: By adding organic matter to the soil through composting, soil fertility and structure can be improved.

Motghare *et al.* [9] have used diverse waste biomass for energy applications. The study investigated several biomass species from the Maharashtra (India) area, including agricultural and forest wastes (garden waste). It is based on moisture content, ultimate analysis, and heating value for energy-efficient cooking appliances using solid biomass fuel Cotton waste, soybean, rice straw, wheat straw, forest wastes, and coconut coir that biomass were chosen. According to the results of the proximate analysis, cotton waste and leaves have a high volatile matter content and a low ash level, which suggest that they have the potential to replace fossil fuels. All biomass samples had almost identical levels of carbon (C), according to elemental analysis, although cotton waste had the highest carbon content and the lowest sulfur concentration. According to the study, biomass has the potential to be a source of sustainable energy for rural regions.

Sheldon [10] focused on the role of catalysis in both chemical as well as biological. In developing sustainable technologies for converting waste biomass into fuels, biobased materials, and commodity chemicals. Utilizing a variety of processes such as chemical, physical, and biological methods, waste lignocellulosic biomass can be transformed into biobased materials. One way to extract the lignocellulose biomass into its constituents, e.g., cellulose, hemicellulose, and lignin, is through chemistry methods such as hydrolysis or pyrolysis. For the production of biobased materials, e.g., biofuels, plastics, and biochemicals, these components can be further processed. In addition, conversion of the lignocellulosic biomass into biobased products can be achieved by using biochemical methods like fermentation and enzyme hydrolysis. This means using microbes such as bacteria and fungi to break down complex carbohydrates found in biomass into simpler sugars that can then be used as a feedstock to produce biobased materials. Overall, sustainable and efficient technologies that can maximize the value of biomass while minimizing waste and environmental impact are needed to convert waste lignocellulosic biomass into biobased materials.

Donar *et al.* [11] had described the importance of generating carbonaceous materials from waste biomasses as a renewable and environmentally friendly method in their paper. Through hydrothermal carbonization, agricultural waste biomass-based hydrocars have been produced. The feedstocks, which were initially dried and crushed into tiny particles, were hazelnut shell and olive waste. These particles were then put into an autoclave made of stainless steel together with deionized water at a solid-to-liquid ratio of 1:10. After a 6-hour hydrothermal carbonization procedure at various temperatures (180°C, 200°C, and 220°C), the hydrocar products were obtained by cooling and filtering. The research discovered that the hydrocars' carbon content rose with temperature and that they included a range of functional groups, including hydroxyl, carbonyl, carboxyl, ether, and ester groups. The hydrocars also showed spherical/semispherical particle shape and a semi-graphitized amorphous carbonaceous structure. Overall, the study demonstrated the potential of hydrothermal carbonization as a sustainable and efficient method for producing carbonaceous materials from waste biomasses. In comparison to nonrenewable energy sources like coal, oil, and natural gas, it is, therefore, a sustainable option. Waste biomass also contributes to a decrease in greenhouse gas emissions. The carbon dioxide that was initially taken in from the atmosphere by the plants during photosynthesis is released when waste biomass is utilized as fuel. Unlike burning fossil fuels, which adds more carbon to the atmosphere, this results in a carbon-neutral cycle. The use of waste biomaterial also encourages recycling and waste reduction. It promotes resource efficiency and lessens the need for trash disposal, which helps to create a cleaner environment. In conclusion, waste biomass is an important resource that may have a big impact on environmental sustainability and trash management. It provides a carbon-neutral and renewable energy source, lessens the need for waste disposal, and encourages resource efficiency. However, implementing the proper regulations and technology will be necessary to fully utilize the potential of waste biomass. To overcome the difficulties involved with its collection, processing, and usage, investments must be made in infrastructure, research and development, and public awareness. Also, the potential of waste biomass as a source of renewable energy due to conversion efficiency and capability to produce and consume CO_2 gas [12].

11.4 Enzymes as Catalytically Active

Enzymes are tiny proteins that function as catalysts in biological processes, enabling things to proceed considerably more quickly than they otherwise could. They also contribute significantly to metabolism by giving cells the energy they require for cellular development and repair. Enzymes are crucial for life because they speed up the chemical processes that keep it alive. This article will examine the function of enzymes and their significance to living things. They serve as biological catalysts by reducing the activation energy needed for chemical processes in living organisms to proceed. Enzymes are vital for numerous biological activities, such as cellular respiration, digestion, and DNA replication. It is appreciations to their catalytic activity because of their extreme specificity [13]. Enzyme Synthesis: Protein synthesis is a process used by cells to create enzymes. In this procedure, DNA is converted into messenger RNA (mRNA), which is then translated into a protein [14]. As proteins, enzymes are synthesized in the cell using the same basic process as other proteins. Once the protein is fully synthesized, it undergoes further modifications to become an active enzyme. These modifications can include the addition of chemical groups, the formation of disulfide bonds, and its final 3-dimensional structure. Now we discuss some research articles regarding enzymes as catalytic active material.

Lin *et al.* discuss the synthesis of catalytically active nanomaterials [13], which are man-made, and it is catalytic function proved by natural enzymes. These nanoparticles develop through the design and development of nanomaterials based on carbon, metal, and metal oxides, which are categorized depending on their structural properties. These nanomaterials are created using a variety of techniques, including electrochemical deposition, sol-gel synthesis, hydrothermal synthesis, chemical reduction, and hydrothermal synthesis. The specific nanomaterial and the purpose for which it is designed will determine the synthesis process to be used. For example, graphene oxide could potentially be used as a modulator to significantly increase the catalytic activity at neutral pH. Which has an extensive range of applications in biological systems. Researchers can provide a simple but extremely effective method to enhance the catalytic performance of artificial enzymes or achieve the catalytic processes by including modulators in such enzymes [15].

Another researcher explored the concept of catalytic promiscuity, which refers to the ability of enzymes to catalyze more than one distinct chemical transformation [14]. They also talk about how many enzymes' broad substrate specificity has contributed significantly to the development of biocatalysis over the past 20 years, particularly in chemical synthesis. The a/b-hydrolase-fold superfamily and the enolase superfamily are two examples of individual proteins having various catalytic functions that are discussed in the study. Overall, the chapter highlights the importance of proteins, particularly enzymes, as catalysts in biocatalysis.

11.5 Immobilization of Enzymes in Biocatalysts

Immobilization of enzymes is a process of fixing the enzyme molecules onto a support material, such as a polymer, a gel, or a solid matrix, to form a biocatalyst. The immobilization process can improve the stability, reusability, and selectivity of the enzyme, making it more suitable for industrial applications. There are three main methods of immobilization: adsorption, encapsulation, and cross-linking. Encapsulation includes trapping the enzyme inside a polymer or gel matrix, whereas adsorption involves the physical attachment of the enzyme to the support material. Covalent bonds are created between the enzyme molecules and the support substance during cross-linking. The intended application determines the method of immobilization to be used. Cross-linked enzyme aggregates (CLEAs) and mesoporous silicas are a few of the innovative supports that have been used in enzyme immobilization recently [13].

Now, recombinant DNA technology, which entails the introduction of a gene encoding the required enzyme into a host organism, such as bacteria or yeast, can be used to create the enzymes employed in biocatalysis. The enzyme is then produced by growing the host organism in a bioreactor under carefully regulated conditions. Using several techniques like adsorption, covalent binding, or entrapment, the resultant enzyme can be purified and immobilized. The characteristics of the enzyme and the intended application determine the method of immobilization to use. Another method for immobilization is protein-coated microcrystals (PCMCs), which are a novel method of enzyme immobilization on an inorganic support. PCMCs are developed by mixing in an aqueous solution of the enzyme through a concentrated solution of a salt, such as potassium sulfate, an amino acid. This method's ability to dehydrate enzyme molecules using a process that largely maintains active enzyme conformations and reduces denaturation is a significant benefit. The PCMCs can be isolated, stored, or utilized in an organic solvent solution. They dissolve in an aqueous medium to release the free enzyme.

Catalytic Activity of Enzymes: Enzymes differ from other biological molecules due to their catalytic activity. Enzymes can speed up chemical reactions by lowering the activation energy required for the reaction. The active site of an enzyme is where it binds to the reactant molecules, occasionally referred to as substrates. The area of the enzyme that is complementary to the substrate in shape and charge is known as the active site. Due to their ability to attach entirely to the proper substrates, enzymes may catalyze particular reactions—thanks to their specificity [16]. When a substrate attaches to the enzyme's active site, the enzyme changes its conformation, bringing the substrate molecules closer together and orienting them in a way that helps the process. The term "induced fit model" is frequently used to describe this conformational shift. The enzyme then catalyzes the reaction by either creating new bonds or breaking existing ones in the substrate.

Once the reaction is finished, the products are released from the active site, allowing the enzyme to attach to a different substrate molecule. Enzymes may be utilized again since the processes they catalyzed do not consume them. Due to this characteristic, enzymes are extremely effective catalysts that may speed up hundreds of processes per second. In conclusion, enzymes are catalysts of biological reactions, essential for the functioning of living organisms. They are synthesized through the process of protein synthesis and regulated to ensure proper metabolic function. Understanding the synthesis, regulation, and catalytic activity of enzymes is crucial for comprehending the intricate workings of biological systems. This knowledge can also be applied in various fields, including medicine, agriculture, and biotechnology, where enzymes are used to catalyze specific reactions or to develop new drugs and therapies.

11.6 Biopolymer

Biopolymers are macromolecules made up of repeating subunits of smaller molecules called monomers. They are naturally occurring materials that are found in all living organisms. Biopolymers are increasingly being used in biocatalysis because of exceptional properties, such as the following:

- High Selectivity and Specificity: Biopolymers may be made to identify and catalyze certain reactions, which can aid in increasing yield and product purity.
- Mild Reaction Conditions: Biopolymers frequently function as catalysts in processes that may be catalyzed at low temperatures and pressures, which can save energy and cut down on waste.

- Sustainability: Biopolymers are often manufactured from environmentally friendly sources because they are renewable resources.

Enzymes are examples of biopolymers that are used in biocatalysts. They are highly selective and specific and can often catalyze reactions under mild conditions. Enzymes are used in an extensive range of biocatalytic processes, such as the production of pharmaceuticals, biofuels, and food additives [17]. There are some examples of biopolymers being used in biocatalysts:

- Polymer Synthesis by Enzymes: Numerous polymers, such as polyesters, polyamides, and polysaccharides, may be produced using enzymes. This provides a viable and safe substitute for traditional polymer synthesis techniques, which frequently involve the use of toxic chemicals and solvents [18].
- Biocatalysts in the Food Industry and Pharmaceutical Industry: Enzymes are frequently utilized in food operations to enhance the nutritional content, safety, and overall quality of food items. For instance, cheese, yogurt, bread, and beer are all made with the help of enzymes. In the pharmaceutical sector, enzymes are also utilized to make a range of medications and drug intermediates. For instance, enzymes are utilized in the synthesis of statins, antibiotics, and other drugs [19].

Biopolymers can be produced utilizing a variety of techniques, including the following:

- Chemical Synthesis: This entails the synthesis of a polymer by the reaction of monomers. Large amounts of biopolymers like polylactic acid (PLA) and polyhydroxy butyrate (PHB) are frequently produced chemically [17].
- Biochemical Synthesis: Biopolymers are synthesized in this manner by living organisms like bacteria or yeast. Biopolymers that are challenging or impossible to chemically synthesize are frequently produced by biochemical synthesis. For instance, the biopolymer cellulose is created by bacteria and plants.
- Technology Based on Recombinant DNA: To develop organisms that can manufacture certain biopolymers, genetic engineering is used in this process. Biopolymers having particular characteristics, such as biopolymers that are biodegradable or may create therapeutic proteins, are frequently produced using recombinant DNA technology [20].

Honarkar and Barikani [21] discuss chitosan, which is a biopolymer that is derived from chitin. It is the second most abundant natural biopolymer in the world. Chitosan has unique properties that make it useful in an extensive range of applications, such as drug delivery, biomedicine, water treatment, and food packaging [21]. Some of the applications of chitosan are biomedicine, membranes, drug delivery systems, etc.

Biopolymers are now used in many biocatalytic processes, such as the synthesis of drugs, biofuels, and food additives. We may anticipate seeing much more creative and useful uses of biopolymers in biocatalytic processes as research in this field develops. Biopolymers are employed as catalysts as well as to create new kinds of biocatalytic supports and immobilization matrices. These supports can make biocatalysts more stable and effective while also making it simpler to recover and reuse them [22]. Overall, biopolymers offer several advantages for biocatalysts.

11.7 Catalytic Applications in Biocatalysts

The use of enzymes or other biological molecules to accelerate chemical processes is known as catalytic applications in biocatalysts. Since enzymes are extremely focused catalysts, they can speed up one specific reaction without interfering with others that could be taking place in the same system. They are therefore perfect for use in the manufacturing of intricate compounds, such as those found in medications and fine chemicals [23, 24]. Some common catalytic applications in biocatalysts are as follows:

11.7.1 Use of Enzymes to Synthesize Chiral Compounds

Chiral compounds are composed of up to two or more mirror-image forms of a molecule, such as a left and a right hand. It is frequently crucial to synthesize the right enantiomer of the substance, since many medicines and other significant compounds are chiral. Enantioselective catalysis is the selective synthesis of one enantiomer over the other using enzymes.

11.7.2 Use of Enzymes to Catalyze Reactions Under Mild Conditions

Many conventional chemical processes call for high pressures and temperatures, which may be costly and energy-consuming. Enzymes may catalyze the same processes under much more benign circumstances, which can save money and have a less negative effect on the environment.

Kazlauskas [25] highlights the potential of enzyme catalysis to provide sustainable and efficient solutions for a wide range of chemical reactions and introduced the term catalytic promiscuity. Catalytic promiscuity refers to a single active site's capacity to catalyze several chemical reactions, each of which may include a different functional group and/or follow a different path for forming and breaking bonds. On the other hand, conventional enzyme catalysis generally uses a single substrate and a single reaction route. As an example, the carboxyl ester hydrolase lipase B from *Candida antarctica* (CAL-B) may also catalyze the creation of a carbon-carbon bond in the aldol addition of hexanal in cyclohexane. This highlights the possibility of catalyzing novel reactions with already-in-use enzymes. In this chapter, the authors discussed the use of protein engineering to enhance new catalytic activities in existing enzymes. This approach can be used to create enzymes with specific catalytic properties that are not found in nature. For example, the success in adding and enhancing catalytic activities by protein engineering, such as the catalytic antibody that catalyzed both decarboxylation and ester hydrolysis. Thus, the chapter also emphasizes the relevance of catalytic promiscuity in extending the spectrum of reactions that enzymes may catalyze as well as the promise of enzyme catalysis to offer effective and long-lasting solutions for a variety of chemical processes. Pharmaceuticals, fine chemicals, biofuels, and goods for bioremediation are just a few examples of the many items that may currently be produced using biocatalytic applications [26]. Some specific examples of catalytic applications in biocatalysts are as follows:

- Pharmaceutical Manufacture: Pharmaceuticals including statins, antibiotics, and painkillers are all made using enzymes. For instance, the antibiotic *penicillin G* is made using the enzyme *penicillin G acylase*.
- Production of Fine Chemicals: A wide range of fine compounds, including flavors, perfumes, and food additives, are also made with enzymes. For instance, the taste ester isoamyl acetate is made by the enzyme *lipase*.
- Bioremediation: Environmental contaminants can be removed using enzymes. For instance, the enzyme *laccase* is employed to break down the prevalent form of environmental pollution known as polycyclic aromatic hydrocarbons (PAHs).

Overall, catalytic applications in biocatalysts offer many advantages over traditional chemical synthesis methods, making them a more sustainable, efficient, and selective way to produce chemicals and clean up the environment [27].

11.8 Computational Approaches in Biocatalyst

In the subject of biocatalysis, computational methods are becoming more and more significant [28]. These methods can be utilized to:

- **Construct Novel Biocatalysts:** The development of novel enzymes with desirable catalytic features, such as improved activity, selectivity, or stability, can be aided by computational approaches. The structure and operation of enzymes may be simulated, and machine learning methods can be used to foretell the consequences of amino acid alterations.
- **Improve on Current Biocatalysts:** Computational methods can also be used to optimize existing biocatalysts for specific applications. This can be done by looking for mutations that enhance an enzyme's activity, stability, or solubility or by creating novel cofactors or substrates.
- **Recognize the Biocatalytic Mechanisms:** The mechanics of biocatalysis can also be understood using computational approaches. This may be done by computing the electrical structure of enzyme active sites and modeling the interactions between enzymes and their substrates.

According to Romero-Rivera *et al.*, research papers in computational sciences have led to novel refined computational methods, which describe the biocatalyst machinery in detail [28]. In biocatalysis research, valuable computational methods like statistical analysis, molecular modeling, and multivariate chemometrics are frequently used to provide sensible guidelines. However, direct experimental design, and steer clear of costly and time-consuming trials. The following are the most popular computational strategies in biocatalysis:

- **Molecular Docking:** Using this method, one can predict how an enzyme will react with a ligand (such as a substrate or an inhibitor). The most likely ligands to bind to the enzyme with high affinity and specificity can be found by screening huge libraries of candidate ligands using molecular docking [29].
- **Simulations of Molecular Dynamics:** The movement of atoms and molecules over time is simulated using molecular dynamics (MD) simulations. The structure, kinetics, and interactions of enzymes with substrates and other ligands may all be studied using MD simulations [29].
- **Calculations Based on Quantum Mechanics:** The electronic structure of molecules and atoms is determined by quantum mechanics (QM) calculations. The mechanics of enzyme catalysis may be studied using QM calculations, and novel enzymes with desired catalytic features can be created.

Computational techniques are critical in biocatalysis, from understanding the underlying principles of enzyme action to designing and modifying enzymes for specific purposes. The creation of novel biocatalytic processes for various kinds of goods, including medicines, biofuels, and fine chemicals, is advancing because of the integration of computational and experimental methodologies [30]. Here are some concrete instances of how computational methods have been successfully used in biocatalysis: developing and creating enzymes using computational techniques for the generation of biofuels like ethanol and biodiesel, the production of novel enzyme-based catalysts for the manufacture of drugs and other acceptable compounds, and the application of computational techniques to find fresh biocatalysts from nature, such as enzymes that can break down waste or create beneficial compounds [28]. The evolution of enzymes has been significantly accelerated by recent developments in computational enzyme design, which have also fundamentally altered the process of enzyme engineering. Computational enzyme design will gather momentum and see substantial achievements in the future with the development of enhanced experimental methodologies [31]. Overall, computational approaches are essential for the development of new biocatalytic processes and the improvement of existing ones. Future developments in computational approaches should lead to even more creative and significant uses of biocatalysts.

11.9 Conclusion and Future Prospects

Utilizing waste biomass helps create a more sustainable and resource-efficient future by reducing its negative effects on the environment. Researchers and industries continue to explore innovative ways to harness the potential of waste biomass for various applications. The sciences of biocatalysis and the biobased economy are both emerging quickly and have the potential to completely change how we generate and use materials. While the biobased economy is based on using renewable biological resources to manufacture chemicals and materials, biocatalysis employs enzymes and other biological catalysts to carry out chemical processes. Both of these industries have the potential to lessen our dependency on fossil fuels and create more ecologically friendly and sustainable processes and products. Pharmaceuticals, food additives, and biofuels are just a few of the high-value compounds and materials that might be produced *via* biocatalysis. Prospects for biocatalysis and the biobased economy are highly positive. These technologies are likely to continue to have a bigger and bigger part in the world economy as they advance and become more affordable. The following are some particular future prospects for catalysis, including biocatalysis and the biobased economy: Pharmaceuticals, food additives, and biofuels are just a few of the high-value materials and compounds that will be produced *via* biocatalysis. The development of biocatalysis will be used to create novel and improved processes for the production of currently used chemicals and materials. The novel and environmentally friendly techniques for the generation of energy and also with the help of the combination of computational and experimental research in biocatalysis offers. Consequently, the possibility to greatly boost the level of knowledge of a biocatalytic scheme with cost-effective and reduced experimental efforts.

References

1. Robinson, P.K., Enzymes: principles and biotechnological applications. *Essays Biochem.*, 59, 1–41, 2015, doi: 10.1042/bse0590001. (a). Torrelo, G., Hanefeld, U., Hollmann, F., Biocatalysis. *Catal. Lett.*, 145, 309–345, 2015, https://doi.org/10.1007/s10562-014-1450-y.
2. Bell, E.L., Finnigan, W., France, S.P., Green, A.P., Hayes, M.A., Hepworth, L.J., Lovelock, S.L., Niikura, H., Osuna, S., Romero, E., Ryan, K.S., Turner, N.J., Flitsch, S.L., Biocatalysis. *Nat. Rev. Methods Prim.*, 1, 1–21, 2021, https://doi.org/10.1038/s43586-021-00044-z.
3. Sheldon, R.A. and Woodley, J.M., Role of Biocatalysis in Sustainable Chemistry. *Chem. Rev.*, 118, 801–838, 2018, https://doi.org/10.1021/acs.chemrev.7b00203.
4. Vanholme, B., Desmet, T., Ronsse, F., Rabaey, K., Van Breusegem, F., Towards a carbon-negative sustainable bio-based economy. 4, 1–17, 2013, https://doi.org/10.3389/fpls.2013.00174.
5. Ladu, L. and Blind, K., ScienceDirect Overview of policies, standards and certifications supporting the European bio-based economy. *Curr. Opin. Green Sustain. Chem.*, 8, 30–35, 2017, https://doi. org/10.1016/j.cogsc.2017.09.002.
6. Nicoara, A.I., Neacsu, I.A., Ene, V.L., Vasile, B.S., Ficai, A., Andronescu, E., hydroxyapatite/carbon based biocomposite scaffolds as prospective materials for bone tissue engineering. *U.P.B. Sci. Bull and Series B.*, 81, 107-120, 2019.
7. Correa, J.P., Montalvo-Navarrete, J.M., Hidalgo-Salazar, M.A., Carbon footprint considerations for biocomposite materials for sustainable products: A review. *J. Clean. Prod.*, 208, 785–794, 2019, https://doi.org/10.1016/j.jclepro.2018.10.099.
8. Kumar, G. and Masram, D.T., Sustainable Synthesis of MOF-5@GO Nanocomposite for Efficient Removal of Rhodamine-B from Water. *ACS Omega*, 6, 14, 9587–9599, 2021, doi: 10.1021/acsomega.1c00143. (a) Kumar, G., Chaudhary, K., Mogha, N.K., Kant, A., Masram, D.T., Extended-release of metronidazole drug using chitosan/graphene oxide bio-nanocomposite beads as drug carrier. *ACS Omega*, 2021, doi: 10.1021/acsomega.1c02422. (b). Soni, U., Bajpai, J., Kumar, S., Bajpai, A.K., Journal of Water Process Engineering Evaluation of chitosan-carbon based biocomposite for efficient removal of phenols from aqueous solutions. *J. Water Process Eng.*, 16, 56–63, 2017, https://doi.org/10.1016/j.jwpe.2016.12.004.
9. Motghare, K.A., Rathod, A.P., Wasewar, K.L., Labhsetwar, N.K., Comparative study of different waste biomass for energy application. *Waste Manage.*, 47, 40–45, 2016, https://doi.org/10.1016/j.wasman.2015.07.032.
10. Sheldon, R.A., Green chemistry, catalysis and valorization of waste biomass, *J. Mol. Catal. A Chem.*, 422, 3–12, 2016, https://doi.org/10.1016/j.molcata.2016.01.013.
11. Sınag, A., Preparation and characterization of agricultural waste biomass based hydrochars. 183, 366–372, 2016, https://doi.org/10.1016/j.fuel.2016.06.108.

12. Iakovou, E., Karagiannidis, A., Vlachos, D., Toka, A., Malamakis, A., Waste biomass-to-energy supply chain management: A critical synthesis. *Waste Manage.*, 30, 1860–1870, 2010, https://doi.org/10.1016/j.wasman.2010.02.030.
13. Sheldon, R.A. and van Pelt, S., Enzyme immobilisation in biocatalysis: Why, what and how. *Chem. Soc. Rev.*, 42, 6223–6235, 2013, https://doi.org/10.1039/c3cs60075k.
14. Bornscheuer, U.T. and Kazlauskas, R.J., Catalytic promiscuity in biocatalysis: Using old enzymes to form new bonds and follow new pathways. *Angew. Chem. Int. Ed.*, 43, 6032–6040, 2004, https://doi.org/10.1002/anie.200460416.
15. Lin, Y., Ren, J., Qu, X., Catalytically active nanomaterials: A promising candidate for artificial enzymes, *Acc. Chem. Res.*, 47, 1097–1105, 2014, https://doi.org/10.1021/ar400250z.
16. Sun, H., Zhou, Y., Ren, J., Qu, X., Carbon Nanozymes: Enzymatic Properties, Catalytic Mechanism, and Applications. *Angew. Chem. Int. Ed.*, 57, 9224–9237, 2018, https://doi.org/10.1002/ anie.201712469.
17. Díez-Pascual, A.M., Synthesis and applications of biopolymer composites. *Int. J. Mol. Sci.*, 20, 2321, 2019, https://doi.org/10.3390/ijms20092321.
18. Van De Velde, K. and Kiekens, P., Ein Jahrhundert Bahnelektroindustrie, in: *Eb - Elektrische Bahnen*, vol. 99, p. 483, 2001.
19. Kaplan, D.L., Introduction to Biopolymers from Renewable Resources. *Biopolym. Renew. Resour.*, 22, 1–29, 1998, https://doi.org/10.1007/978-3-662-03680-8_1.
20. Schnepp, Z., Biopolymers as a flexible resource for nanochemistry. *Angew. Chem. Int. Ed.*, 52, 1096–1108, 2013. https://doi.org/10.1002/anie.201206943.
21. Honarkar, H. and Barikani, M., Applications of biopolymers I: chitosan. 140, 1403–1420, 2009, https://doi.org/10.1007/s00706-009-0197-4.
22. Reddy, N., Reddy, R., Jiang, Q., Crosslinking biopolymers for biomedical applications. *Trends Biotechnol.*, 33, 1–8, 2015, https://doi.org/10.1016/j.tibtech.2015.03.008.
23. Ran, N., Zhao, L., Chen, Z., Tao, J., Recent applications of biocatalysis in developing green chemistry for chemical synthesis at the industrial scale. *Green Chem.*, 10, 361–37, 2008, https://doi.org/10.1039/b716045c.
24. Singh, S., Catalytically Active Nanomaterials: Artificial Enzymes of Next Generation. *Nanosci. Technol. Open Access*, 5, 1–6, 2017, https://doi.org/10.15226/2374-8141/5/1/00150.
25. Kazlauskas, R.J., Enhancing catalytic promiscuity for biocatalysis, *Curr. Opin. Chem. Biol.*, 9, 195–201, 2005, https://doi.org/10.1016/j.cbpa.2005.02.008.
26. Khalid, F.E., Lim, Z.S., Sabri, S., Gomez-Fuentes, C., Zulkharnain, A., Ahmad, S.A. Bioremediation of Diesel Contaminated Marine Water by Bacteria: A Review and Bibliometric Analysis. *J. Mar. Sci. Eng.*, 2021, 9, 155. https://doi.org/10.3390/jmse9020155.
27. Zaks, A. and Dodds, D.R., Application of biocatalysis and biotransformations to the synthesis of pharmaceuticals. *Drug Discov. Today*, 2, 513–531, 1997, https://doi.org/10.1016/S1359-6446(97)01078-7.
28. Romero-Rivera, A., Garcia-Borràs, M., Osuna, S., Computational tools for the evaluation of laboratory-engineered biocatalysts. *Chem. Commun.*, 53, 284–297, 2017, https://doi.org/10.1039/C6CC06055B.
29. Khan, A.Z., Bilal, M., Rasheed, T., Iqbal, H.M.N., Advancements in biocatalysis: From computational to metabolic engineering. *Cuihua Xuebao/Chinese J. Catal.*, 39, 1861–1868, 2018, https:// doi.org/10.1016/S1872-2067(18)63144-4.
30. Braiuca, P., Ebert, C., Basso, A., Linda, P., Gardossi, L., Computational methods to rationalize experimental strategies in biocatalysis. *Trends Biotechnol.*, 24, 419–425, 2006, https://doi.org/10.1016/j.tibtech.2006.07.001.
31. Li, X., Zhang, Z., Song, J., Computational enzyme design approaches with significant biological outcomes: Progress and challenges. *Comput. Struct. Biotechnol. J.*, 2, e201209007, 2012, https://doi.org/10.5936/csbj.201209007.

12

Chemistry and Technology Innovation to Advance Green and Sustainable Chemistry

Prabitha Prabhakaran[1], Sakshi Bhardwaj[2], Bhawna Chopra[3], Ashwani K. Dhingra[3]* and Madhur Kant[4]

[1]*Department of Pharmaceutical Sciences, JSS College of Pharmacy, Mysore, India*
[2]*Department of Life Sciences, Altem Technologies, Bangalore, India*
[3]*Guru Gobind Singh College of Pharmacy, Yamuna Nagar, Haryana, India*
[4]*Amity Institute of Applied Sciences, Amity University, Noida, Uttar Pradesh, India*

Abstract

The designing of chemicals and processes that minimize manufacturing of compounds that are detrimental to people, plants, animals, and the environment is known as "green chemistry" or "sustainable chemistry" or "circular chemistry." Many conditions can be used to achieve the main goals of green chemistry, including the design of molecules, materials, products and processes that are more naturally safer and efficient. Some principles of green chemistry include the following: the prevention of waste, atom economy, less hazardous chemical synthesis, safer chemicals need to be designed, avoiding the auxiliary substances, energy requirements should be minimized, processes should be conducted at ambient temperature and pressure whenever possible, the use of catalysis should be controlled, chemical products should be designed so that they do not pollute the environment, and real-time in-process monitoring and control before hazardous substances form. Additionally, various other factors are also being considered such as chemical yield, cost of reaction components, safety of chemical handling, reduced by products, hardware requirements, energy profile and ease of products workup and purification. The well-known and integrated application of green chemistry, is in the drug discovery and development process through computational methods, where we may optimize chemical entities with high throughput screening to find required molecule. This process encompasses the concepts of green chemistry, which have discovered possibilities for pharmaceutical goods, as the development of novel synthesis techniques and the production of new drugs with reduced resource consumption and waste output. According to the United States Environmental Protection Agency (EPA), green chemistry is "the design of chemical products and processes that decrease the formation of harmful pollutants." Green chemistry, which offers a framework for innovation, assessment, and the pursuit of better practices, is thereby linked to computational approaches. Because these computational methods using machine learning and artificial intelligence algorithms to optimize chemical entities to be faster, cheaper, less toxic, and require less waste and energy while increasing yield and efficiency, they are more environmentally friendly and sustainable when it comes to innovation. These computational methods deploy various approaches for optimizing and designing chemical entities like fragment-based drug designing, structure-based drug designing, ligand-based drug designing, ADMET, TOPKAT, SAR, and QSAR. These methods have the unique

**Corresponding author*: ashwani1683@gmail.com

Mousumi Sen (ed.) Sustainable Green Catalytic Processes, (273–300) © 2024 Scrivener Publishing LLC

ability to map polypharmacological properties across numerous chemical libraries with different drug targets. The variety of computational methods that offer a solid foundation integrated with green chemistry will be covered in this book chapter.

Keywords: Green chemistry, sustainable chemistry, circular chemistry, SBDD, LBDD, ADMET

12.1 Introduction

The designing of chemicals and processes that minimize or eliminate the use or manufacturing of compounds that are detrimental to people, animals, plants, and the environment is known as "green chemistry" or "sustainable chemistry" or "circular chemistry." The objective of green and sustainable chemistry is to create and use chemical methods and outcomes that are socially conscious, commercially viable, and ecologically beneficial. It entails the use of guidelines and procedures that limit the consumption and production of dangerous materials and waste, protect the environment, and lessen the environmental impact of chemical procedures and end products [1]. In order to develop more environmentally friendly chemical processes and products that satisfy societal demands while minimizing their negative effects on the environment, green and sustainable chemistry is an interdisciplinary field that integrates knowledge from chemistry, physics, engineering, biology, environmental science, and other disciplines. Green chemistry is a rapidly evolving and interdisciplinary field of study nowadays [2]. The principles of green chemistry include avoiding auxiliary substances, preventing trash, atom economy, less dangerous chemical synthesis, safer chemical design, minimizing energy requirements, conducting processes at ambient pressure and temperature whenever possible, controlling the use of catalysis, designing chemical products to ensure they do not pollute the environment, and real-time in-person monitoring. Other considerations are made in addition to trying to quantify a sustainable chemical process like the chemical yield, expense of reaction components, safety when handling chemicals, number of waste products, hardware necessities, energy profile, and ease of product workup along with purification [3–5].

As a result, in order to proceed from green chemistry to sustainable chemistry, one must develop systems thinking skills; otherwise, solutions based on green chemistry are unlikely to be long-lasting. The goal of shedding light on green synthesis methods was to draw attention to new ecologically sound techniques for creating unique chemical entities that are significant in the pharmaceutical and pharmacological domains. It is generally established that the creation of new and improved medications has a significant environmental cost due to the expanding population and harm from industrial pharmaceutical manufacture. The development of environmentally friendly technology is therefore one of the most well-liked research areas at every level of the drug discovery process. Although there is resistance to altering the established practices, eco-friendly practices may encourage advancement in this direction because of the financial rewards they offer [6].

When thinking about a chemical product's entire life cycle, the following problems are crucial to consider workup processes and solvent selection strategies should be established or enhanced as a crucial component of reaction design [7]. According to this perspective, the process design *ex ante*, the preceding reaction screening, or the computational design of chemicals become increasingly important. These techniques, which are also outlined in the green chemistry postulates, help identify bottlenecks and minimize pointless tests, which ultimately result in cheaper costs. Particularly in terms of computational designing of chemicals, such technology is connected to rising computer performance and capacity as well as expanding Internet connectivity to databases.

A few years ago, computer-aided drug design was thought to be an impossibly difficult task, but today, it is a reality that incorporates a forecast of potential by-products in addition to the medication's qualities and toxicity. The ability to predict all effects and results for the environment and health is provided by toxicodynamics and toxicokinetics analysis. This is a further advantage in the comprehensive examination of green processes and products. Chemistry is the cornerstone of designing and delivering medications and vaccines that prevent and treat disease; therefore, for the pharmaceutical sector, technological advancement starts in the lab. With computational designing of chemicals, as a part of green chemistry, every effort is made during the drug development process to attempt and utilize fewer resources, produce less waste, and conduct safe operations. As part of green chemistry, computational chemical design makes every effort to try and use fewer resources, produce less waste, and conduct safe operations throughout the drug development process [8].

Greener technologies will be necessary to manage the full chemical life cycle, from chemical manufacture and supply chains to use and end of life of finished products. This move toward a sustainable future will necessitate the use of these technologies [9]. Chemistry has produced many useful products, including pharmaceuticals, cosmetics, food items, nanoparticles, liquid crystals, polymers, paints, biomolecules, and agrochemicals. In order to produce useful chemical molecules in a shorter span of time with less toxicity, computational techniques are heavily utilized in each sector [10]. With the development of computational techniques, we may also address significant issues like reusability, toxicity, and global impact by molecularly building the drug and selecting the kinds of reaction circumstances and reactions we want to occur [11, 12]. These reasonings make green chemistry a cutting-edge chemical philosophy. Its ideas motivate the development of novel methods and raw materials that reduce the consumption and production of dangerous substances, where the advancement of computational approaches comes into the picture to get the desired candidate. Pharmaceutical companies are more motivated to adopt environmentally friendly and sustainable practices as the cost of getting a medicinal molecule to market rises. Historically, medicinal chemists have attempted to use several green strategies to enhance the drug development process.

To make current procedures or chemicals more ecologically friendly, it becomes essential to update them occasionally. With the development of powerful computer technology, it is now possible to create predictive algorithms that will enable more sustainable methods of performing medicinal chemistry. In any pharmaceutical process chemistry, algorithms are used to calculate the process mass intensity (PMI) that aids in attenuating product sustainability. Designing more active compounds in a more environmentally friendly manner is made possible by cheminformatics tools. Modern drug development procedures depend on predictive algorithms for molecular characteristics including solubility, lipophilicity, and ADME (absorption, distribution, metabolism, and excretion). Modern pharmaceutical chemistry is now stronger and more effective—thanks to these eco-friendly and sustainable tools. In any pharmaceutical process chemistry, algorithms are used to calculate the process mass intensity that contributes to improved product sustainability.

Chemical informatics technologies like the computational tools assist in creating more active compounds in a more environmentally friendly manner. Modern drug development processes depend heavily on predictive algorithms for molecular characteristics including solubility, lipophilicity, and ADME (absorption, distribution, metabolism, and excretion). Modern pharmaceutical chemistry is now stronger and more effective—thanks to these environmentally friendly and sustainable techniques. The possibility for creating enzymatic cascades for the effective, step-economical, and selective synthesis of target molecules has been increased by recent developments in computer-aided synthesis design and the extended enzyme toolbox for biocatalysis [13, 14]. However, automation of drug discovery, artificial intelligence (AI), and big data to increase speed and efficiency decrease the number of test animals and the burden on humans and the environment [15]. In comparison to laborious serial compound synthesis, it provides the possibility of exploring broader areas of the chemical space within a particular period when innovative approaches to designing of new scaffolds are combined with it. Recent developments in AI have been helpful in planning chemical synthesis of challenging target compounds. Compound activity data *in vivo*, in animal models, and in human patients all increase at previously unheard-of rates, providing priceless large data for deep learning (DL) methodologies for activity and bioavailability prediction. DL architectures have recently been used to make predictions about the solubility of active pharmacological ingredients in water, and they have shown great promise in data-rich disciplines like image analysis and natural language processing [16–18]. Future drug discovery tool kits are probably going to include big data approaches more prominently.

12.2 Computational Chemistry Methods in Green and Sustainable Drug Design and Development

Drug discovery and development, especially the production of environmentally friendly and sustainable drugs, now heavily rely on computational chemistry approaches. These techniques can be utilized to increase productivity, reduce failure, and make the overall drug discovery process more environmentally friendly. Here are a few computational chemistry techniques for green and sustainable drug development:

12.2.1 High-Throughput Virtual Screening

A prevalent phase in the drug discovery process is high-throughput virtual screening (HTVS), which enables quick, inexpensive screening of far larger chemical libraries than is practical in experimental research (Figure 12.1) [19]. Effective computational and experimental techniques are used to speed up the lead optimization and identification procedure. The chemical molecules are small potentially drug-like compounds that have been further optimized to operate as a therapeutic medication against a certain disease. They can modify the activity of the target proteins. For quickly identifying drug leads, high-throughput virtual screening remains the preferred technique.

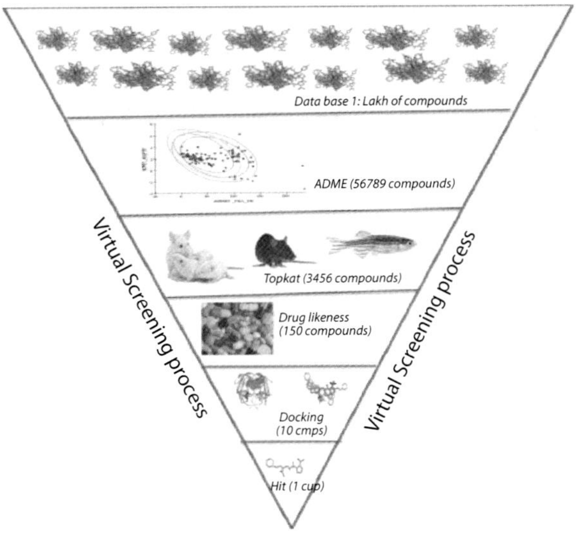

Figure 12.1 Virtual screening process.

These HTVS methods can quickly and more affordably identify the lead molecule among millions of other molecules than traditional high throughput screening (HTS) methods. HTVS is a computational screening technique that is frequently used to examine *in silico* collections of compound libraries in order to determine the target receptors' affinity for certain library compounds [20]. Various databases are freely available to get the libraries of the chemical compounds for desired purposes. The databases include Swiss Target Prediction [21], Uniprot [22], PDB [23], NCBI [24], IUBMB ExplorEnz Enzyme [25], Expasy [26], Rhea [27], BRENDA [28], CheMBL [29], Pharos [30], KEGG [31], Metacyc [32], and BindingDB [33]. By foreseeing the physicochemical characteristics and toxicity of potential medication candidates, high-throughput virtual screening can also be utilized to choose ecologically friendly compounds. Any virtual screening approach seeks to identify a set of high-performing compounds—here, computational "hits" with the lowest docking scores—within a much wider search space. By keeping some virtual filters like pharmacokinetic and pharmacodynamic, the chemical components can be eliminated. The selected and optimized components from the series further provided for the molecular modeling, docking, and simulation.

12.2.2 Molecular Docking

In order to anticipate interactions between chemical compounds and target proteins, computer methods are used in molecular modeling techniques. It can aid in the creation of molecules with greater binding affinities and selectivity, which can increase medicinal effectiveness and lower toxicity. Based on the interactions between tiny and large molecules (Figure 12.2), these molecular docking techniques are utilized to better comprehend the fundamental biological processes. The physicochemical properties of molecules can be predicted using molecular modeling in green and sustainable medication design, which can

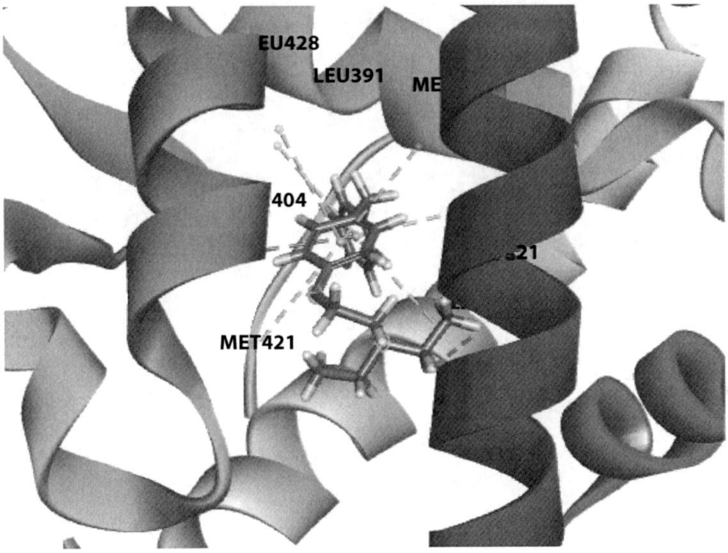

Figure 12.2 Molecular docking pose of a ligand into the binding cavity of the protein.

aid in the selection of ecologically friendly solvents and lessen the necessity for animal testing [34, 35].

The two fundamental processes in the docking procedure are prediction of the ligand conformation in addition to its orientation and position within these sites (often referred to as pose) of target molecule and evaluation of the binding affinity of receptor–ligand complex. The efficiency of docking is greatly improved by being aware of the binding site prior to docking operations. Numerous software are available to carry out molecular docking of the selected ligand and target molecules that include Discovery Studio, Schrodinger, AutoDock, Flex X, GOLD, Argus lab, MedChem Studio, UCSF Chimera, and PyMOL [36].

12.2.3 Molecular Dynamics and Simulation

A computer simulation method is called molecular dynamics (MD), and simulation examines the actual motions of atoms and molecules (Figure 12.3). A glimpse of the dynamic "evolution" of the system is provided by allowing the atoms and molecules to interact for a predetermined amount of time. The atomic level behavior of drug molecules and target proteins can be studied using molecular dynamics simulations. These simulations can aid in the formulation of more stable and bioavailable medications by revealing information about the dynamic behavior of drug molecules in solution. The thermodynamics and kinetics of protein–ligand interaction can also be predicted using molecular dynamics simulations, which can aid in the development of compounds with higher binding affinities [37]. The energy terms discussed above have been parameterized according to quantum-mechanical simulations and experimental (for example, spectroscopic) data in order to simulate the actual behavior of real molecules in motion. This tuning entails choosing the most desirable partial atomic charges for computing electrostatic interaction energies, choosing the

Figure 12.3 Representation of molecular dynamics and simulation of targeted protein.

appropriate van der Waals atomic radii, etc. The springs describing chemical bonds and atomic angles should have the ideal stiffness and lengths.

These variables, when taken as a whole, are referred to as a "force field" because they outline the role of the many atomic forces that control molecular motion. In molecular dynamics simulations, several force fields are frequently employed, including CHARMM [38], GROMOS [39], AMBER [40, 41], and NAMD [42].

12.2.4 Quantitative Structure–Activity Relationship

It is a technique for computer or mathematical modeling that identifies connections between biological processes and the structural characteristics of chemical molecules. The fundamental idea is that diverse biological activities result from varied structural characteristics [43, 44]. Physical-chemical properties are referred to as structural properties, and biological activities are referred to as pharmacokinetic properties including absorption, distribution, metabolism, excretion, and toxicity. Being an *in silico* methodology, quantitative structure–activity relationship (QSAR) modeling aids in the prioritization of many compounds in terms of their intended biological activities, greatly reducing the total number of prospective chemicals that need to be tested in *in vivo* tests. It is possible to predict the activity of novel pharmacological compounds using QSAR models (Figure 12.4), which eliminates the need for costly and time-consuming experimental testing. By foreseeing drug molecule toxicity and choosing compounds with enhanced biodegradability, QSAR models can be utilized to design medications that are more ecologically friendly and sustainable. In order

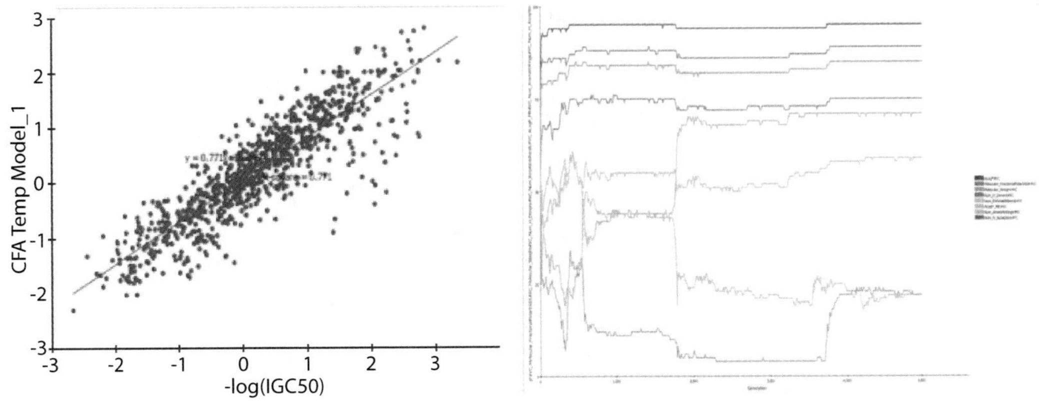

Figure 12.4 QSAR model generation.

to anticipate QSAR, machine learning (ML) techniques have been used. These predictions have been carried out using Bayesian neural networks [45] and linear regression models [46]. Quantitative structure–property relationships (QSPRs) are used in conjunction with this concept when a chemical feature is modeled as the response variable [47]. In the area of QSPR, various chemical molecule characteristics or behaviors have been studied. Additionally, a mathematical relationship, or QSAR, can be found when physicochemical properties or structures are expressed in numerical form. The mathematical equation can subsequently be used to forecast the modeled response of different chemical structures [48] after being rigorously validated [49].

For risk management, (Q)SAR models have been employed [50]. Due to a better understanding of the science behind it that allows one to create mathematical (QSAR) models in order to precisely forecast the biological impact or potency of novel or untested compounds, QSAR is becoming more and more popular among clinicians and researchers in related fields. In lead optimization and virtual screening, QSAR has the capacity to prioritize concepts.

12.3 Green Chemistry Principles in Computational Drug Design

Computational drug design can be used to design and develop ecologically friendly medicines using the principles of green chemistry. Incorporating green chemistry ideas into computational drug design can take several different forms like the development of nontoxic solvents, atom economy, reduction in waste, renewable resources, biodegradability, energy efficiency, and toxicity reduction (Figure 12.5). We will be discussing each of them one by one:

12.3.1 Use of Nontoxic Solvents

Computational techniques can be used to forecast the characteristics and solubility of medicinal molecules in different solvents. This can aid in choosing nontoxic and ecologically friendly solvents for the synthesis of drugs, cutting down on the usage of hazardous solvents

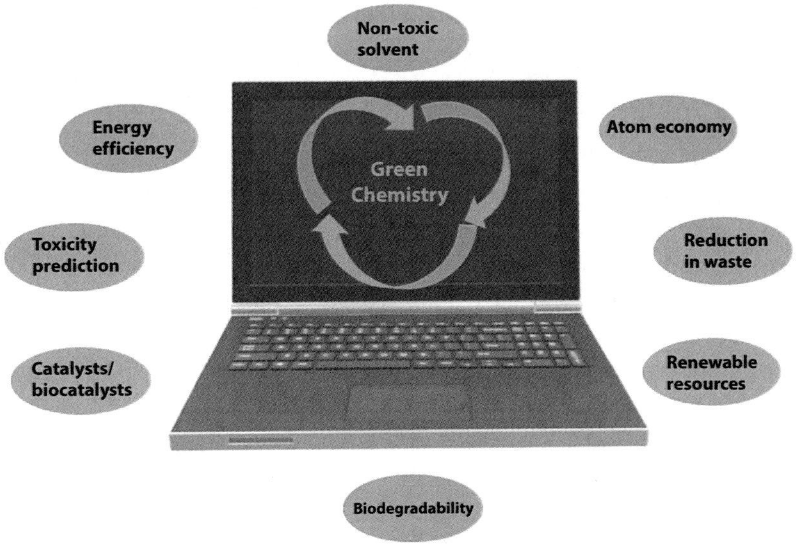

Figure 12.5 Computational approaches in green chemistry.

and generating less waste. The best technique to use solvents while synthesizing pharmaceuticals is using computational methods, which can be used to optimize solvent utilization. By choosing solvents with high boiling points, low toxicity, and low volatility, one can reduce the amount of solvent used and minimize solvent waste. Drug production procedures can be designed using computational techniques that enable solvent recovery and reuse. This can lessen the demand for new solvents and lower the amount of hazardous waste produced. Nevertheless, drug compounds that are compatible with biobased solvents, which are made from renewable resources like biomass, can be created using computational approaches. This can lessen the need for nonrenewable resources and lessen the negative effects of medicine production on the environment [51]. This has been an established preoccupation, and molecules used as solvents, such as benzene, carbon disulfide, and carbon tetrachloride, have been utilized at first and then replaced as new substances that were more effective or less dangerous became available. Other solvents that are no longer widely used in the industry are hexane (neurotoxic), nitromethane (explosion risk), diisopropyl ether (peroxide formation), diethyl ether (due to its flammability and propensity to form peroxides), and ethylene glycol dimethyl ether (teratogenicity).

12.3.2 Atom Economy

Atom economy is essentially about molecular waste prevention. It is better if the atom economy is higher, since atoms that are not used in the final product are deemed to be wasted. It is an essential concept in green chemistry that quantifies the effectiveness of a chemical reaction by estimating the proportion of atoms that are transferred from the starting ingredients to the final product. It is preferable to have an efficient atom economy, since it reduces the amount of waste produced during the process of synthesis of a molecule. A reaction's percent yield has historically been used to assess its efficiency [52]. Since chemists frequently must carry out numerous reaction steps, a step economy approach may be used to enhance the process's overall effectiveness. By reducing the number of reaction steps, chemists can view a smaller number of syntheses as being more environmentally friendly.

In addition to enhancing the atom and step economies, computer-aided organic synthesis (CAOS) is a helpful computational tool that can also optimize the reaction's underlying mechanism. To avoid producing undesired products and enhance reaction networks, CAOS employs a retrosynthetic methodology [53]. Chemical processes in drug manufacturing can be optimized using computational techniques to increase atom economy. Computational approaches can assist in determining the most productive reaction conditions and improving the yield of the intended product by foretelling the outcomes of chemical reactions. New chemical processes with greater atom economies can be created using computational approaches. Computational drug design can reduce waste production and boost the effectiveness of drug synthesis by creating processes that employ fewer starting materials or produce fewer by-products. Drug compounds that have greater stability and are less prone to degradation can be created using computational approaches. Because fewer starting materials are required and less waste is produced, this can increase the efficacy of drug synthesis. Drug synthesis procedures can be designed using computational approaches to have lesser steps, which can increase atom economy. Computational techniques can increase drug synthesis efficiency and decrease waste production by minimizing the total number of intermediates and lowering the number of reaction steps [54].

12.3.3 Reduction in Waste

Actually, chemical reactions can be optimized using computational techniques to reduce waste production. This can be done by choosing reaction conditions that generate fewer waste products or by creating molecules that are more stable and resistant to degradation. The number of physical experiments needed in the drug development process can be reduced by performing virtual screening for drug candidates using computational approaches. This lessens the amount of waste produced by experiments, including the usage of chemical solvents, reagents, and lab supplies. In order to reduce the number of compounds that need to be synthesized and evaluated, computational methods can be utilized for rational drug design. Drug designers can choose the most promising candidates for further development by applying computational approaches to anticipate the properties of drug molecules, reducing the waste produced during the drug discovery process. To reduce the amount of dangerous chemicals released into the environment, computational approaches can be utilized to find nontoxic solvents that can be used in the drug synthesis process. It is possible to create catalytic reactions using computational methods that produce less waste and use fewer stoichiometric reagents. This may result in less waste being produced when pharmaceuticals are being created. Drug synthesis procedures, including conditions for reaction, pressure, temperature, and reactant concentrations, can be optimized using computational approaches. This can reduce waste production and increase medication synthesis efficiency [55].

12.3.4 Renewable Resources

The goal of green chemistry is to create chemical processes that use as few nonrenewable resources as possible while maximizing the use of sustainable and renewable resources, such as biomass, water, and sunlight. The usage of biomass feedstocks in drug production can be found and optimized using computational approaches. Computational drug designing can lessen the reliance on nonrenewable resources in drug synthesis by anticipating the qualities of biomass feedstocks and creating synthetic routes that utilize these feedstocks effectively. Drug synthesis techniques that leverage renewable energy sources, such solar and wind power, can be designed using computational methods. Drug synthesis's carbon footprint can be decreased by maximizing the use of renewable energy sources—thanks to the computational drug design. Green solvents are nontoxic biodegradable solvents that are made from renewable resources. Computational drug designing can lessen the negative effects of drug manufacturing on the environment by using green solvents. Designing biocatalytic reactions that make use of renewable resources, including enzymes and microorganisms, can be done using computational approaches. By using fewer nonrenewable resources and producing less hazardous waste, biocatalytic processes have the potential to lessen the environmental impact of medication manufacturing [56].

The conversion of biomass into useful products involves a variety of chemical processes, including those involving mono- and disaccharides, oils, and terpenes, as well as the chemistry of the by-products of the fermentation of glucose. Carbohydrates are mostly used outside of food in the manufacturing of ethanol, surfactants, lactic acid, d-sorbitol sweetener, furfural, and medications like penicillin. Defatted soy flour can be cross-linked with plant fibers like flax, hemp, or bamboo to create biodegradable composite materials that are mostly made of soy protein and carbohydrates. In the building business, eco-friendly composites show potential as a replacement for composites that are not biodegradable, such as polypropylene or glass [57]. Chemically altered soy flour has been commercialized as a wood glue, replacing toxic urea-formaldehyde resins that are frequently utilized in plywood and particleboard. This product was developed to mimic the sticky proteins found in mussels [58, 59].

For instance, the paper and pulp sector produces a substantial amount of lignin as a waste product. It has been used as fuel for a long time at industrial locations. In recent years, it has come to light that it can be used to make things like dispersing agents, additives, and raw materials to produce other substances like vanillin or humic acid [60, 61].

12.3.5 Biodegradability

The term "biodegradability" describes a substance's capacity to degrade naturally into safe by-products such as carbon dioxide, water, and biomass. The development of safer alternatives via computational drug designing and the principles of green chemistry are increasingly being included into industrial research and development, but up until now, safer has typically meant less hazardous. But it is equally important to consider biodegradability when designing a product. Chemicals that do not degrade through biodegradation are still present in the biota and can have harmful consequences, some of which may not be immediately apparent or expected. Even more of a worry are persistent compounds that bioaccumulate because they can reach concentrations in creatures that first pass the acute toxicity test but ultimately have long-term or other unanticipated hazardous effects.

As a result of the larger and more diverse and, consequently, catabolically more adaptable microbial populations that these environments tend to support, biodegradation is more likely to occur in nutrient-rich environments. Furthermore, nutrients are unlikely to limit the growth of degrading microorganisms, should there be any present in that specific location. Furthermore, the rate of biodegradation is associated with temperature; therefore, biodegradation is likely to occur more quickly at higher temperatures, which fall within the range of temperatures considered conventional for environmental conditions. Less visibly, in comparison with laboratory screening procedures, biodegradation typically occurs more quickly in aquatic environments at lower substrate concentrations. The second point is difficult to imagine as having any direct benefit for chemical design, although it might provide test data comparisons in some context [62].

The tendency to adsorb surrounding solid phases and water solubility, which are often seen as being on opposite sides of the same coin, are two generalizations that could be significant in chemical design. Substances that are soluble in water biodegrade more quickly than those that are not. The latter have a strong tendency to adsorb to solid phases, and for very small compounds, sorption/sequestration is frequently linked to slower rates of biodegradation. The molecular structure of a chemical affects the reactions that it might go through once within the cell. Even though there have been hundreds of different transformations described in the literature, practically all of them may be categorized as oxidative, hydrolytic, reductive, or conjugative. The impact of chemical composition on biodegradability has given rise to a lot of generalizations. These deal with substituent position, the impact of different substituent groups or substructures, and the frequency with which a particular substituent appears in a molecule. The computational methods could be useful in this case to deal with a tiny molecule's biodegradability by constructing the molecules with various substructures and, of course, by increasing their water solubility.

The usage of biodegradable starting materials in drug manufacturing can be found and optimized using computational approaches. Computational drug designing can lessen the negative effects of drug synthesis on the environment and increase the likelihood that the final pharmaceuticals will be biodegradable by starting with biodegradable components. Drug delivery systems that are biodegradable and minimize the environmental impact of drug use can be created using computational methods. Drug delivery methods that are biodegradable assist in ensuring that the drug gets broken down to harmless compounds in the environment while reducing the quantity of trash produced after use. The metabolic destiny of drug molecules in the biological system can be predicted using computational approaches, which can also be used to pinpoint the biodegradable metabolites. Computational drug designing can lessen the negative effects of drug use on the environment by emphasizing the development of medications that result in biodegradable metabolites and ensuring that the metabolites are converted into harmless chemicals in the environment [63].

12.3.6 Energy Efficiency

Designing chemical processes with a minimal energy footprint and a lower carbon footprint is the goal of green chemistry. Energy efficiency is the term used to describe how effectively energy is used in chemical processes, which can lessen the environmental impact of chemical synthesis. The characteristics of a drug molecule can be predicted using computational approaches, and the most energy-efficient synthesis routes can be found. Computational drug designing can aid in lowering the energy required for drug synthesis by predicting the characteristics of drug molecules and creating synthetic methods that use less energy. The conditions for reaction in the production of medicines can be improved using computational techniques to utilize less energy. Computational drug design can lower the energy needed for drug production and increase energy efficiency by maximizing reaction conditions such as pressure, temperature, and catalysts. The energy needed for solvent recovery can be decreased and the environmental impact of drug synthesis can be reduced by using green solvents using computational approaches. Computational tools can lower the amount of energy needed for drug production and its carbon footprint by maximizing the usage of renewable energy sources [64].

12.3.7 Toxicity Prediction

The term "toxicity reduction" refers to the process of making medications and their by-products less toxic, which can enhance the safety and environmental effects of developing and using new drugs. The toxicity of pharmacological compounds can be predicted using computational methods based on their chemical structure. Computational drug designing can assist in prioritizing the creation of pharmaceuticals that are less hazardous to people and the environment by anticipating the toxicity of therapeutic molecules at the design stage. The usage of safe starting materials in medication manufacturing can be determined and optimized using computational approaches. The toxicity of the finished medication product and its by-products can be decreased by using safer starting ingredients in computational drug design. In order to reduce the production of harmful by-products during the manufacturing of drugs, reaction conditions can be optimized using computational approaches. Computational drug designing can decrease the creation of hazardous by-products and increase the safety of medication development and administration by optimizing reaction conditions including temperature, pressure, and catalysts. Designing biodegradable drug delivery systems that lessen the toxicity of pharmaceuticals and their by-products can be done using computational approaches. Drugs and their by-products will be broken down into harmless compounds with the help of biodegradable drug delivery systems, which can also lower the amount of harmful chemicals that humans and the environment are exposed to [65]. To minimize the dangers to both individuals and the environment, toxicity testing is necessary. However, sustainable practices—the cornerstones of green practices—call for a decrease in both animal testing and the chemicals used in toxicity testing. The foundation of the Green Toxicology approach is front-loading toxicity evaluations using intelligent and predictive testing strategies, which implies conducting safety assessments before regulatory or market release. This study discusses the several benefits that Green Toxicology offers in terms of the usage and application of Green Chemistry. By utilizing fresh and cutting-edge predictive computational toxicological methods and techniques, the Green Toxicology principles provided by Maertens *et al.* [66] offer a framework for creating chemicals that are safer for people and the environment. In addition to enhancing predictive toxicology, the use of *in silico* techniques helps prioritize substances, offers insight into upcoming toxicity testing, and reduces late-stage failures in chemical design (i.e., test early, create safety). All of these advantages support the use of green toxicology in the creation of new chemicals and goods that are beneficial to both the environment and consumers. For the advancement of Green Toxicology efforts in product development, *in silico* methods like TOPKAT are beneficial in high-throughput toxicity predictions of vast chemical datasets [67].

12.3.8 Catalysts and Biocatalysts

The efficiency and selection of chemical reactions can be increased by catalysts and biocatalysts, which can also lower the energy and material requirements for drug manufacturing. A broader variety of enzymes for catalyzing chemical reactions are becoming available to us because of developments in the field of genomics. Enzymes with the capacity to carry out nonnatural chemistry can also be engineered. This implies that over the coming years, the range of chemical reactions that we can carry out using biocatalysis will be much greater than it is now, providing the prospect of many more sustainable chemical reactions in drug development. In terms of green chemistry, biocatalysis has many advantages over chemocatalysis, including mild reaction conditions (physiological, pH, and temperature), the use of environmentally friendly catalysts (enzymes) and solvents (typically water), high catalytic activity, and good regio- and chemo-selectivity for multifunctional molecules. Therefore, the introduction of enzymes can eliminate the necessity for pointless protection/deprotection stages as well as functional group activation [68]. Whereas the biocatalysts are having enzyme-based applications for the conversion of molecular substrate into various naturally occurring and synthetic chemicals [69]. Enzymes have a high degree of substrate recognition and catalyzed reaction specificity. As a result, they are a prime candidate for the development and advancement of sustainable chemical processes. Additionally, this specificity, sustainability, low toxicity, and high efficiency can all be easily attained with the aid of computational tools [70, 71]. An effective biocatalyst must be sufficiently compatible with several characteristics, including thermostability, catalytic capacity, substrate specificity, operational stability in turbulent flow regimes, dangerous solvents, and substrate inhibition [72]. Due to easier availability to enzymes and the capacity to modify those enzymes to match the needs of industrial processes, the usage of biocatalysis in the pharmaceutical sector is continuing to grow and computational techniques are giving a boost to it.

12.4 Case Studies in Green and Sustainable Drug Design Using Computational Approaches

There are various studies that have been done for green and sustainable drug designing using computational tools. Here, we are going to discuss a few case studies:

12.4.1 Green Synthesis of Artemisinin-Based Antimalarial Drugs

The disease of malaria, which affects millions of people worldwide, is one of the most effectively treated with artemisinin-based antimalarial medications. However, the manufacture of artemisinin and its by-products is linked to several environmental and social problems, such as the use of hazardous solvents, excessive energy consumption, and a limited supply of raw materials. By promoting the use of ecologically friendly and sustainable drug synthesis techniques, green synthesis of artemisinin-based antimalarial medicines seeks to address these problems. Artemisinin is traditionally produced by extracting it from the leaves of the *Artemisia annua* species using harmful solvents like hexane or petroleum ether. However, environmentally friendly extraction techniques have been created to lessen the usage of hazardous solvents and increase the output of artemisinin, including extraction with supercritical fluid and microwave-assisted extraction. Enzymes like cytochrome P450 and artemisinic acid 9-hydroxylase have been used to study the biocatalytic production of artemisinin and its derivatives. Drug synthesis can potentially use less energy and resources with the help of biocatalytic synthesis, which also has the potential to produce fewer harmful by-products. Semi-synthesis process used to alter a natural product through chemical processes to create a derivative with better qualities.

Using techniques like ozonolysis and peroxidation, the semi-synthesis of artemisinin and its derivatives has been studied. Drug synthesis can be made more effective overall and with less starting material by using semi-synthesis. A study was conducted by Kojom *et al.* [73], where they have emphasized the possibility of discovering and developing new antimalarial medications based on green synthesis procedures by highlighting the great antiplasmodial potential of metal nanoparticles made using green synthesis techniques. They produced nanoparticles (NPs) using plant leaves. Fuentes *et al.* [74] from the Max-Planck Institute in Germany achieved success in expressing all of the enzymes involved in the artemisinin biosynthesis pathway inside the chloroplast of tobacco plants by employing a novel synthetic biology strategy. They utilized their combinatorial super transformation of transplastomic recipient lines (COSTREL) technique to simultaneously insert the genes for the artemisinin complete production pathway into the chloroplast. Because foreign genes can be easily accommodated by homologous recombination when they are transported, the chloroplast is a good place for the expression of foreign genes.

They also inserted a few auxiliary genes into the tobacco plant's nucleus; their precise purpose is unknown, but it has been hypothesized that they are crucial for controlling the artemisinin production pathway. Following transplastomy, the species of plant were cultivated on spectinomycin to make them homoplasmic (all copies of the plastid DNA include the genes for artemisinin production). The plants produced the Nt-AO3-1 phenotype, in which artemisinic acid builds up during the leaf's life. A further improvement was made to this genotype, which was given the designation Nt-AO3-CS180, and it was able to acquire 120.4 ± 42 mg per kg of fresh weight (FW). The correlation investigations of numerous transplastomic lines revealed that the high levels of artemisinin in phenotypes are caused by the presence of the genes for the two enzymes aldehyde dehydrogenase (ALDH1) and 1-deoxy-d-xylulose-5-phosphate reductoisomerase (DXR). The amount of artimisinic acid increases with the expression of these two genes.

The highest generating artemisinic acid phenotypic plant lines only saw a 13% reduction in biomass when compared to wild-type plants, and there is no harmful effect on the chloroplast of leaves.

12.4.2 Design of Green Inhibitors for Cancer Treatment

Since cancer is one of the main causes of death in the world, research efforts are heavily focused on finding novel and efficient therapies. Utilizing inhibitors, which are compounds with the ability to specifically bind to and stop the function of proteins that are essential for the growth and proliferation of cancer cells, is one method of treating cancer. Utilizing ecologically responsible and sustainable techniques for antagonist synthesis and testing is essential to the development of environmentally friendly inhibitors for the treatment of cancer. Natural materials have proven a significant source of possible cancer treatment inhibitors, including plant extracts and marine creatures. Many natural compounds can be synthesized utilizing green chemistry techniques and have been discovered to have a strong inhibitory effect against cancer cells. To create inhibitors with high selectivity and potency against certain cancer targets, computational techniques like molecular docking and molecular dynamics simulations can be applied. Utilizing fewer resources and materials can minimize the requirement for experimental testing. On other hand, biodegradable inhibitors are made to dissolve into harmless compounds once they have served their purpose. These inhibitors support sustainable growth by reducing the buildup of hazardous residues in the environment.

Due to their capability to facilitate the body's immune system's ability to identify and combat cancer cells, immune checkpoint inhibitors (ICIs) represent a major advance in immunotherapy [75]. The immune response is tightly controlled by the PD-1/PD-L1 immune checkpoint axis [76]. Drugs that target the PD-1/PD-L1 axis therefore could inhibit their interaction, allowing the immune system to continue preparing an attack against the cancer cells. In comparison to peptide- and particularly Ab-based therapies, small molecule-based immunotherapies have benefits including a higher oral bioavailability, overall lower costs, and a relative shorter half-life, which results in less possible toxicity and unwanted effects [77]. In 2015, Bristol-Myers Squibb (BMS) revealed a set of drugs containing a biphenyl group as the first small molecules with a promising inhibitory activity against PD-L1 [78]. However, Guzik *et al.* [79] discovered the binding mechanism when they found that the BMS compounds caused the PD-L1 protein to dimerize. Structure-based drug design [79] was made clear by the discovery of two co-crystal structures of PD-L1 in association with small-molecule inhibitors. In a PD-1/PD-L1 complex using BMS inhibitors of the PD-L1 protein, Mittal *et al.* [80] used molecular docking experiments and found that the residues Tyr56, Asp122, and Lys124 are essential for ligand binding to the PD-L1 protein.

12.4.3 Discovery of Green Peptidomimetic Drugs

Peptidomimetics are compounds that closely resemble the structure and functional characteristics of peptides but are more durable and have superior pharmacokinetic characteristics. Drugs that are peptididomimetic have been created for a variety of therapeutic purposes, including the treatment of inflammation, cancer, and infectious disorders. Natural items have been a great source of prospective peptidomimetic medications, including marine creatures and plant extracts. Numerous natural substances that can be synthesized utilizing green chemistry techniques have been found to have strong medicinal efficacy. Peptidomimetic medicines with excellent selectivity and potency against targets can be created using computational techniques like molecular docking and molecular dynamics simulations. Utilizing fewer resources and materials can minimize the requirement for experimental testing. After performing their intended purpose, biodegradable peptidomimetics are made to disintegrate into innocuous compounds. These medications can encourage sustainable growth and help the environment become less polluted. When compared to antibodies or small molecules, peptidomimetics based on endogenous peptides offer intriguing lead compounds since peptides are frequently ligands for endogenous receptors. Normally, endogenous precursor peptides are broken down to produce metabolites that are physiologically active. Therapeutics based mostly on the active form of endogenous peptides can be developed by taking advantage of this natural processing [81]. Giovanna *et al.* [82] outline some of the most recent initiatives in the design and synthesis of peptidomimetic compounds, as well as their biological assessment toward a panel of targets. By single-residue scanning of a parent sequence or scaffold [83], computational and rational design, or phenotypic and/or genotype-linked display from phage [84] or yeast [85], unnatural amino acid-containing peptidomimetics can be created.

12.5 Technology Innovations in Computational Green and Sustainable Drug Design

Drugs with lesser toxicity, greater efficacy, and a smaller negative impact on the environment are designed using green and sustainable drug design techniques. Integrating sustainable drug discovery techniques provides more creative and workable solutions that reduce their negative effects on the environment. The need for new drugs and treatments will not go away; therefore, it is critical that the research and development process is ethical and environmentally sustainable. In the long run, this could result in better health outcomes, as it will enable researchers to explore and develop medicines while also minimizing environmental harm. Pharmaceutical businesses are more motivated to adopt environmentally friendly and sustainable practices as the cost of getting a medicinal molecule to market rises. Historically, medicinal chemists have attempted to use several green strategies to enhance the drug development process. Sometimes making a process more ecologically friendly calls for changing it or replacing it with a new synthetic approach. Methods included choosing catalysts and/or solvents or substituting sustainable reaction conditions for conventional reaction steps. Predictive algorithms can be used to produce more sustainable methods of performing medicinal chemistry—thanks to the development of powerful computer technologies. Designing more active compounds in a more environmentally friendly manner is made possible by cheminformatics tools. Modern drug development procedures depend on predictive algorithms for molecular characteristics including solubility, lipophilicity, and ADME. These green and ecological techniques have made modern pharmaceutical chemistry stronger and more potent (Figure 12.6).

The qualities of therapeutic molecules, such as stability, solubility, and bioactivity, which are essential for drug design, have been predicted using machine learning algorithms. The goal of machine learning, a branch of artificial intelligence, is to develop a statistical model for a given dataset in order to solve real-world issues [86]. This has made it possible for researchers to more rapidly and precisely identify potential medication candidates, hence minimizing the need for costly and time-consuming experimental testing. These days, AI technologies, in particular, the deep learning paradigms, show enormous potential in drug design because of their excellent generalization and feature extrication power [87]. These predictive analytics algorithms could significantly speed up the development of sustainable chemicals. The DL technique has been shown to have enormous potential in the field of drug creation—thanks to the aforementioned advantages in data mining. The DL paradigms often include deep neural networks (DNNs), recurrent neural networks (RNNs), convolutional neural networks (CNNs), restricted Boltzmann machines (RBNs), and autoencoders [88]. In order to link trained feature representations of molecular descriptors to

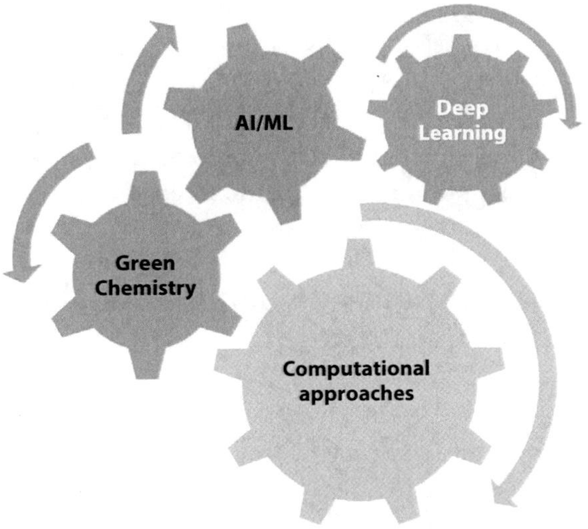

Figure 12.6 Integrated green chemistry with modern computational techniques.

the intended chemical, physical, or biological properties of catalysts, predictive QSPR models are frequently employed in conjunction with deep learning-based generative models.

Deep learning has recently shown the capacity to make property predictions that are comparable to calculations using density functional theory (DFT), outperforming previous machine learning techniques in this area as well [89]. The drug development process is additionally hampered by the complexity and volume of data generated by genomes, proteomics, microarray analysis, and clinical trials. Innovative methods like artificial intelligence and machine learning are essential in the research and discovery of new drugs (Figure 12.7). Researchers may now conduct extensive simulations and molecular modeling studies to create medications with increased efficacy and lower toxicity—thanks to advancements in high-performance computers. As a result, fewer chemicals need to be examined experimentally and new therapeutic candidates can now be identified using virtual screening techniques. Collaborative drug design approaches have been facilitated by the ease with which researchers can now access and share vast amounts of data and computational resources—thanks to cloud computing. Additionally, this has made it possible to create Web-based platforms and tools for drug creation, which makes the procedure more accessible to researchers with constrained computational capabilities.

A rapidly developing technology, i.e., quantum computing, has the power to completely alter medication development. More precise simulations and modeling of molecular systems are possible—thanks to quantum computers' ability to carry out complex calculations considerably faster than conventional computers. Researchers may be able to create medications with previously unheard-of accuracy and efficacy because of this [90]. It is anticipated that deep learning techniques will greatly enhance the efficiency and precision of AI-based catalyst design for sustainable chemistry applications as a result of advancements in deep learning algorithms, more potent architectures of generative models, and increased availability of experimental and computational chemistry data. Nevertheless, chemical processes in drug manufacturing can be optimized using machine learning.

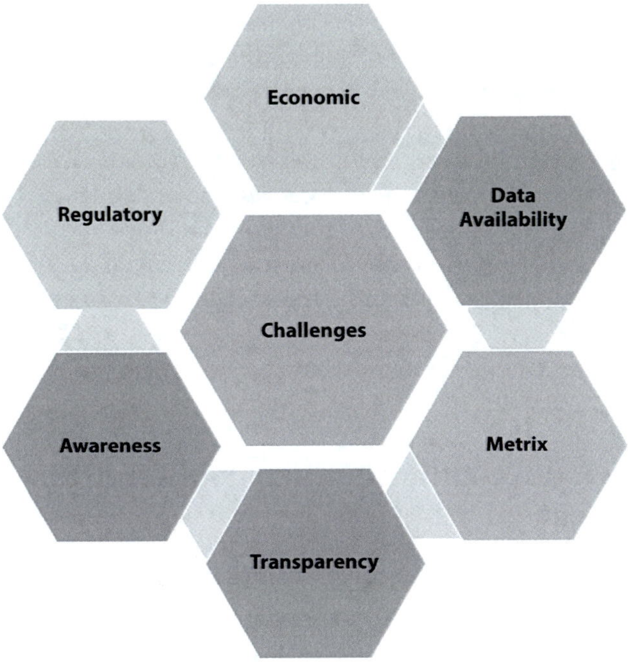

Figure 12.7 Challenges and limitations in computational green and sustainable drug design.

Machine learning algorithms can pinpoint the circumstances of chemical reactions that result in high yields and little waste by analyzing big datasets of chemical reactions. This permits the creation of pharmaceuticals with greater purity and potency while reducing the environmental effect of drug manufacturing. In order to find prospective therapeutic candidates, machine learning algorithms can also be employed to virtually analyze huge databases of chemicals. In order to determine which molecules are most likely to have a certain biological impact, these algorithms can be trained using datasets of known active substances. This makes the process of discovering new drugs more effective and long-lasting by lowering the number of molecules that must be synthesized and tested experimentally. In order to maximize the utilization of physical and logical resources, cloud computing integrates already-existing technologies and paradigms. A cloud is a kind of parallel and distributed system, according to Buyya *et al.* [91], consisting of a collection of interconnected and computerized technology that are dynamically allocated and displayed as one or more unified computing resources based on service-level agreements negotiated between the service provider and customers [92].

The organizational level needs to implement green computing practices to support and uphold the underlying optimization methodologies and technologies suggested by researchers. The uniqueness of green computing approaches rests in their diversity, taking into account both greedy and evolutionary algorithms as well as low-level processor, memory, and network aspects for system optimization [93].

Researchers can now more easily and affordably do extensive simulations and analyses because of cloud computing platforms that provide shared access to computational resources like CPUs, GPUs, and storage. As a result, less costly hardware and software architecture is required, making drug design accessible to researchers with little funding. The virtual screening of huge databases of substances using cloud computing also enables researchers to swiftly and effectively find prospective therapeutic candidates. This results in a more efficient and cost-effective drug development process by lowering the number of molecules that must be synthesized and tested in experiments. These platforms provide cutting-edge data management solutions that let researchers store, exchange, and analyze massive amounts of data. This lowers the chance of data loss and makes it possible for researchers to run intricate studies on huge datasets. The advantages of cloud computing, it has additionally facilitated the creation of Internet-based tools and platforms for drug design, facilitating researchers' access to computational resources and analysis functions without the need for specialized gear or software.

12.6 Challenges and Limitations in Computational Green and Sustainable Drug Design

The development of therapeutics that are sustainable and kind to the environment is the goal of the burgeoning discipline of "green and sustainable drug design." Even if there is a lot of opportunity for improvement in this area, drug development still has a negative impact on the environment. Computational techniques can significantly cut down on the time and cost of the drugs' development. Generating potential target proteins for drug screening and design, bioinformatics can help in identifying the pathophysiological roles of drug-able genes from a big amount of genomic data [94]. The structural and thermodynamic properties of target proteins can be obtained using multidisciplinary biomolecular simulations, which may be useful for identifying drug binding sites and clarifying the mechanisms behind therapeutic action. Although various computational techniques are used in the design of medicines, many challenging conditions have also been observed. A crucial phase in the development of drugs is the identification and confirmation of biological targets with therapeutic application. Identifying the chemical space that is physiologically relevant is one of the most important concerns. Studying polypharmacology and attempting to foresee how a drug candidate would interact with each of its potential target partners are also very important. In computational biology, network-based methods are frequently employed to identify targets and interactions of novel medicinal compounds [95]. In these models, the nodes in the networks stand in for drugs, diseases, or gene products, while the edges reflect their interactions or linkages. These include protein–protein interactions, transcriptional networks, and drug–drug, drug–target, drug–disease, disease–disease, gene–disease, and drug–disease interactions [96].

However, there are still a number of issues and restrictions that must be resolved, which include, but are not limited to, increasing the effectiveness of virtual screening, increasing the quantity and caliber of online computational resources, advancing the field of computational chemogenomics, enhancing the design of drugs targeting a variety of molecular targets, enhancing the predictive power of toxicity models and side effects, and enhancing collaboration with other disciplines to maximize the search for bioactive compounds. A significant primary problem is the absence of drug-disease indication data. The missing indications may be compromised as real negatives by the computational biology's predictive power for repurposing drug candidate development, or they may be ignored in training. It is difficult for *in silico* techniques to assess findings because there is not a genuinely superior dataset for therapeutic repositioning. As a result, standard performance metrics like specificity, sensitivity, and precision are used to evaluate the usefulness of computerized drug repurposing algorithms. Last but not least, current computational techniques are frequently/potently one-sided (for example, drug- or disease-centric).

A bioactive substance that has been demonstrated to be safe and efficient in biological systems is believed to have a variety of therapeutic purposes, according to mounting scientific evidence. Given the growing popularity of this "simple" method of repositioning, there is an urgent need for more effective and organized computational methods to deal with and first organize the available genomic and pharmacological knowledge repositories for portrayal and information revelation and then use these datasets and example disclosure tools to predict the potential new uses for existing medications [97]. Therefore, creating individualized phenotypic validation procedures in addition to speeding up the repurposing of established medications represents a significant challenge. In order to foster intellectual cooperation between government organizations, professionals, and institutes, access to datasets that are already available should also be increased. To facilitate the advancement of AI algorithms, which necessitates addressing several fundamental technological, legal, and maybe ethical issues, it ought to be encouraged to contribute validated data [98]. The dearth of information regarding pharmaceuticals' effects on the environment is one of the main obstacles to designing computationally green and sustainable drugs.

This makes it challenging to precisely forecast a drug's environmental impact and to design eco-friendly long-lasting alternatives. With a thorough understanding of a medication's chemical structure, mode of action, and manufacturing process, computational models that can precisely forecast the environmental impact of the drug are currently being developed. Nevertheless, these models still face complex challenges. On the other side, regulatory guidelines on how to assess a drug's environmental impact are currently lacking. This makes it challenging for researchers to create eco-friendly long-lasting medications that adhere to regulatory standards. Also, it can be more costly to develop sustainable and green pharmaceuticals than it is to develop conventional drugs. This is because environmentally friendly and sustainable pharmaceuticals frequently call for the employment of more expensive and ecologically hazardous components and manufacturing techniques [99]. Nevertheless, environmental data are not sufficiently reported throughout the development of new drugs and during clinical trials.

During drug development or after the drug is in the market, many pharmaceutical corporations fail to reveal how their products affect the environment. The fragmented and scattered data sources are the biggest challenge to collect and analyze the data for computational modeling. Another crucial component of computational green and sustainable drug design is reproducibility and validation. Validation is the procedure of ascertaining the accuracy and dependability of a computational model, whereas reproducibility is the capacity to duplicate the output of a computational model. Regulatory and legal obstacles are also some major obstacles in computational green and sustainable medication design. The approaches employed in model development and validation need to be rigorous and transparent, and on the other hand, the scientific community, including academic scientists, the pharmaceutical industry, regulatory bodies, and HTA/payers, needs to accept models more widely as a valuable source of evidence. The techniques employed for this need to be thoroughly explained and widely accepted. We believe that because these mechanistic *in silico* models are unique, diverse, and relatively new to the drug development environment, we are currently in the learning process of determining the best techniques for model verification, validation, and certification. It is anticipated that as the use of these kinds of models grows, model evaluation techniques will develop in number and performance, much like other quantitative tools that have been around longer and are more well-recognized (such as statistical methods, popPK, and PK/PD models).

12.7 Future Directions and Conclusion

The process of finding new drugs and developing them takes a long time and costs a lot of money. The advancement of computational techniques facilitates the investigation of medications. The prediction of prospective compounds and treatments against fatal diseases has greatly benefited from computer-aided drug design, nucleic acid-based therapeutics, and the application of AI. But at the same time, they come with some difficulties. When creating AI applications for medication repurposing that could improve efficiency, it is necessary to have a long-term perspective due to the complex and regulated aspect of drug research. Indeed, whenever enabled by the volume of data, AI-related techniques are gaining significant attention in the areas of molecular property and behavior prediction [100]. The advantages of artificial intelligence have already been felt in several different areas of computational chemistry, with an emphasis on the early discovery environment [101, 102]. On-target and off-target impact predictions in computational toxicology [103, 104] that are constructed in the family of AI-based algorithms for target prediction based only on ligand chemical data serve as one particularly pertinent example. On the structure-based side, recently developed and described algorithms include delta-learning [105], deep learning-based 3D pocket mapping [106], and AI rescoring. The prioritization performance provided by the traditional scoring methodologies has not been matched by these most recent methods [107].

Quantum mechanics techniques, on the other hand, are becoming more and more practical [108]. Although the use of AI to anticipate QM properties is well-established in the CADD sector [109, 110], it is conceivable that the widespread computation of such traits will become commonplace with the development of quantum computing. Hence, by lowering costs, boosting productivity, and minimizing environmental effect, computational approaches for environmentally friendly and sustainable drug design have the potential to completely transform the drug discovery process. The combination of machine learning and artificial intelligence can be used to find new targets, forecast drug toxicity, and improve therapeutic formulations for increased efficacy and less environmental impact.

The development of extensive databases of chemical and environmental data can help to increase the precision of computational models and make it easier to create pharmaceutical products that are more environmentally friendly. The adoption of computational approaches in green and sustainable drug design can be facilitated by collaboration between researchers, pharmaceutical companies, regulatory agencies, and other stakeholders. This can also help to develop standardized protocols and improve the validation and reproducibility of computational models. In order to minimize the usage of dangerous reagents and minimize waste in the drug discovery process, new green and sustainable technologies like biocatalysis and flow chemistry are being developed. To ensure that the development of sustainable medications is equitable and beneficial to all populations, social responsibility and justice for the environment factors should be incorporated into the drug discovery process. The use of a life cycle approach to medication development, which takes the environment's impact on pharmaceutical goods into account through every phase of their development, can aid in locating chances for improvement and reducing environmental impact. Computational methods for green and sustainable drug design and development have the potential to advance and aid in the creation of pharmaceuticals that are more environmentally and socially conscious by following these new directions.

References

1. Mutlu, H. and Barner, L., Getting the Terms Right: Green, Sustainable, or Circular Chemistry? *Macromol. Chem. Phys.*, 223, 13, 2200111, 2022.
2. Constable, D.J., Green and sustainable chemistry – the case for a systems-based, interdisciplinary approach. *iScience*, 24, 103489, 2021.
3. Khan, I., Hou, F., Le, H.P., The impact of natural resources, energy consumption, and population growth on environmental quality: Fresh evidence from the United States of America. *Sci. Total Environ.*, 754, 142222, 2021.
4. Marteel, A.E., Davies, J.A., Olson, W.W., Abraham, M.A., Green chemistry and engineering: Drivers, Metrics, and Reduction to Practice. *Annu. Rev. Environ. Resour.*, 28, 401–428, 2023.
5. Dhage, S.D. and Shisodiya, K.K., Application of Green Chemistry in Sustainable Development. *Int. Res. J. Pharm.*, 4, 7, 1–4, 2013.
6. Allais, F., Brogi, S., Castro, G.R., Vidali, V.P., Editorial: Advances in green synthesis for drug discovery. *Front. Chem.*, 11, 1166887, 2023.
7. Escribà, G.M., Green chemistry strategies for drug discovery. *Green Process. Synth.*, 4, 6, 509–510, 2015.
8. Guo, M., Nowakowska-Grunt, J., Gorbanyov, V., Egorova, M., Green technology and sustainable development: assessment and green growth frameworks. *Sustainability*, 12, 16, 6571, 2020.
9. Desai, M., Njoku, A., Nimo-Sefah, L., Comparing Environmental Policies to Reduce Pharmaceutical Pollution and Address Disparities. *Int. J. Environ. Res. Public Health*, 19, 14, 8292, 2022.
10. Ravichandran, S., Green chemistry–a potential tool for chemical synthesis. *Int. J. Chem. Tech Res.*, 2, 2191, 2010.
11. Clark, J.H., Luque, R., Matharu, A.S., Green Chemistry, Biofuels, and Biorefinery. *Annu. Rev. Chem. Biomol. Eng.*, 3, 183–207, 2012.
12. Sindhu, R., Verma, A., Sharma, D., Gupta, S., Arora, S., Application of green chemistry in pharmaceutical chemistry and day today life. *Arch. Med. Pharm. Sci. Res.*, 1, 39–44, 2017.
13. Finnigan, W., Hepworth, L.J., Flitsch, S.L., Turner, N.J., RetroBioCat as a computer-aided synthesis planning tool for biocatalytic reactions and cascades. *Nat. Catal.*, 4, 98–104, 2021.
14. Sheldon, R.A. and Woodley, J.M., Role of biocatalysis in sustainable chemistry. *Chem. Rev.*, 118, 801–838, 2018.
15. Hunter, J., Adopting AI is essential for a sustainable pharma industry. *Drug Discovery World Winter*, 69–71, 2017.
16. Boobier, S., Hose, D.R.J., Blacker, A.J., Nguyen, B.N., Machine learning with physicochemical relationships: solubility prediction in organic solvents and water. *Nat. Commun.*, 11, 5753, 2020.

17. Cui, Q., Lu, S., Ni, B., Zeng, X., Tan, Y., Chen, Y.D., Zhao, H., Improved prediction of aqueous solubility of novel compounds by going deeper with deep learning. *Front. Oncol.*, 10, 121, 2020.
18. Evelien, W., Christophe, F., Bartosz, W., Per, L., Eelko, H., Thomas, B., Serge, V.C., Nicolas, W., Michal, M., Luke, R.O., Gerrit, J.P., Bart, D.S., Sustainability in drug discovery. *Med. Drug Discov.*, 12, 100107, 2021.
19. Zhang, S., Kumar, K., Jiang, X., Wallqvist, A., Reifman, J., DOVIS: An implementation for high-throughput virtual screening using AutoDock. *BMC Bioinf.*, 9, 126, 2008.
20. Subramaniam, S., Mehrotra, M., Gupta, D., Virtual high throughput screening (vHTS)–a perspective. *Bioinf.*, 3, 1, 14–17, 2008.
21. David, G., Aurélien, G., Matthias, W., Antoine, D., Olivier, M., Vincent, Z., Swiss Target Prediction: a web server for target prediction of bioactive small molecules. *Nucleic Acids Res.*, 42, W1, W32–W38, 2014.
22. Coudert, E., Gehant, S., de Castro, E., Pozzato, M., Baratin, D., Neto, T., Sigrist, C.J.A., Redaschi, N., Bridge, A., UniProt Consortium. Annotation of biologically relevant ligands in UniProtKB using ChEBI. *Bioinf.*, 39, 793, 2023.
23. Berman, H.M., Westbrook, J., Feng, Z., Gilliland, G., Bhat, T.N., Weissig, H., Shindyalov, I.N., Bourne, P.E., The Protein Data Bank. *Nucleic Acids Res.*, 28, 1, 235–42, 2000.
24. Li, W., O'Neill, K.R., Haft, D.H., DiCuccio, M., Chetvernin, V., Badretdin, A., Coulouris, G., Chitsaz, F., Derbyshire, M.K., Durkin, A.S., Gonzales, N.R., Gwadz, M., Lanczycki, C.J., Song, J.S., Thanki, N., Wang, J., Yamashita, R.A., Yang, M., Zheng, C., Marchler-Bauer, A., Thibaud-Nissen, F., RefSeq: expanding the Prokaryotic Genome Annotation Pipeline reach with protein family model curation. *Nucleic Acids Res.*, 49, D1, D1020–8, 2021.
25. McDonald, A.G., Boyce, S., Tipton, K.F., ExplorEnz: the primary source of the IUBMB enzyme list. *Nucleic Acids Res.*, 37, D593–D597, 2009.
26. Gasteiger, E., Gattiker, A., Hoogland, C., Ivanyi, I., Appel, R.D., Bairoch, A., ExPASy: the proteomics server for in-depth protein knowledge and analysis. *Nucleic Acids Res.*, 31, 3784–3788, 2003.
27. Bansal, P., Morgat, A., Axelsen, K.B., Muthukrishnan, V., Coudert, E., Aimo, L., Hyka-Nouspikel, N., Gasteiger, E., Kerhornou, A., Neto, T.B., Pozzato, M., Blatter, M.C., Ignatchenko, A., Redaschi, N., Bridge, A., Rhea, the reaction knowledgebase in 2022. *Nucleic Acids Res.*, 50, D1, D693–D700, 2022.
28. Chang, A., Jeske, L., Ulbrich, S., Hofmann, J., Koblitz, J., Schomburg, I., Neumann-Schaal, M., Jahn, D., Schomburg, D., BRENDA, the ELIXIR core data resource in 2021: new developments and updates. *Nucleic Acids Res.*, 49, D1, D498–D508, 2021.
29. Mendez, D., Gaulton, A., Bento, A.P., Chambers, J., De Veij, M., Félix, E., Magariños, M.P., Mosquera, J.F., Mutowo, P., Nowotka, M., Gordillo-Marañón, M., Hunter, F., Junco, L., Mugumbate, G., Rodriguez-Lopez, M., Atkinson, F., Bosc, N., Radoux, C.J., Segura-Cabrera, A., Hersey, A., Leach, A.R., ChEMBL: towards direct deposition of bioassay data. *Nucleic Acids Res.*, 47, D1, D930–D940, 2019.
30. Keith, J.K., Timothy, K.S., Stephen, L.M., Jeremy, J.Y., Vincent, T.M., Vishal, B.S., Dac-Trung, N., Lars, J.J., Dušica, V., Stephan, C.S., Jayme, H., Karlie, R.S., Ajay, P., Cristian, G.B., Jeremy, S.E., Ewy, A.M., Tudor, I.O., Pharos 2023: an integrated resource for the understudied human proteome. *Nucleic Acids Res.*, 51, D1, D1405–D1416, 2023.
31. Kanehisa, M., Furumichi, M., Sato, Y., Kawashima, M., Ishiguro-Watanabe, M., KEGG for taxonomy-based analysis of pathways and genomes. *Nucleic Acids Res.*, 51, D587–D592, 2023.
32. Caspi, R., Altman, T., Billington, R., Dreher, K., Foerster, H., Fulcher, C.A., Holland, T.A., Keseler, I.M., Kothari, A., Kubo, A., Krummenacker, M., Latendresse, M., Mueller, L.A., Ong, Q., Paley, S., Subhraveti, P., Weaver, D.S., Weerasinghe, D., Zhang, P., Karp, P.D., The MetaCyc Database of metabolic pathways and enzymes and the BioCyc collection of Pathway/Genome Databases. *Nucleic Acids Res.*, 42, 1, D459–D471, 2014.

33. Liu, T., Lin, Y., Wen, X., Jorissen, R.N., Gilson, M.K., Binding DB: a web-accessible data-base of experimentally determined protein-ligand binding affinities. *Nucleic Acids Res.*, 35, D198–201, 2007.
34. McConkey, B.J., Sobolev, V., Edelman, M., The performance of current methods in ligand-protein docking. *Curr. Sci.*, 83, 845–855, 2002.
35. Meng, X.Y., Zhang, H.X., Mezei, M., Cui, M., Molecular docking: a powerful approach for struc- ture-based drug discovery. *Curr. Comput. Aided Drug Des.*, 7, 2, 146–57, 2011.
36. Bradley, F., Ashcroft, D.M., Noyce, P.R., Integration and differentiation: a conceptual model of general practitioner and community pharmacist collaboration. *Res. Soc. Adm. Pharm.*, 8, 1, 36–46, 2012.
37. Lamoureux, G., Harder, E., Vorobyov, I.V., Roux, B., MacKerell, A.D., A polarizable model of water for molecular dynamics simulations of biomolecules. *Chem. Phys. Lett.*, 418, 1, 245–249, 2006.
38. Brooks, B.R., Bruccoleri, R.E., Olafson, B.D., States, D.J., Swaminathan, S., Karplus, M., CHARMM - a program for macromolecular energy, minimization, and dynamics calculations. *J. Comput. Chem.*, 4, 187–217, 1983.
39. Christen, M., Hünenberger, P.H., Bakowies, D., Baron, R., Bürgi, R., Geerke, D.P., Heinz, T.N., Kastenholz, M.A., Kräutler, V., Oostenbrink, C., Peter, C., Trzesniak, D., van Gunsteren, W.F., The GROMOS software for biomolecular simulation: GROMOS05. *J. Comput. Chem.*, 26, 1719–1751, 2005.
40. Wang, J., Wolf, R.M., Caldwell, J.W., Kollman, P.A., Case, D.A., Development and testing of a general amber force field. *J. Comput. Chem.*, 25, 1157–1174, 2004.
41. Durrant, J.D. and McCammon, J.A., Molecular dynamics simulations and drug discovery. *BMC Biol.*, 9, 71, 2011.
42. Phillips, J.C., Braun, R., Wang, W., Gumbart, J., Tajkhorshid, E., Villa, E., Chipot, C., Skeel, R.D., Kale, L., Schulten, K., Scalable molecular dynamics with NAMD. *J. Comput. Chem.*, 26, 1781–1802, 2005.
43. Verma, J., Khedkar, V.M., Coutinho, E.C., 3D-QSAR in drug design-a review. *Curr. Top. Med. Chem.*, 10, 1, 95–115, 2010.
44. Kwon, S., Bae, H., Jo, J. *et al.*, Comprehensive ensemble in QSAR prediction for drug discovery. *BMC Bioinform.*, 20, 521, 2019.
45. Pradeep, P., Povinelli, R.J., White, S., Merrill, S.J., An ensemble model of qsar tools for regulatory risk assessment. *J. Cheminformatics.*, 8, 1, 48, 2016.
46. Luco, J.M. and Ferretti, F.H., QSAR based on multiple linear regression and PLS methods for the antihiv activity of a large group of hept derivatives. *J. Chem. Inf. Comput. Sci.*, 37, 2, 392–401, 1997.
47. Nantasenamat, C., Isarankura-Na-Ayudhya, C., Prachayasittikul, V., Advances in computational methods to predict the biological activity of compounds. *Expert Opin. Drug Discov.*, 5, 7, 633–54, 2010.
48. Tropsha, A., Best Practices for QSAR Model Development, Validation, and Exploitation. *Mol. Inform.*, 29, 6–7, 476–488, 2010.
49. Ruusmann, V., Sild, S., Maran, U., QSAR DataBank repository: open and linked qualitative and quantitative structure–activity relationship models. *J. Cheminformatics*, 7, 32, 2015.
50. Fioravanzo, E., Bassan, A., Pavan, M., Mostrag-Szlichtyng, A., Worth, A.P., Role of *in silico* genotoxicity tools in the regulatory assessment of pharmaceutical impurities. *SAR QSAR Environ. Res.*, 23, 3–4, 257–277, 2012.
51. Winterton, N., The green solvent: a critical perspective. *Clean Techn. Environ. Policy*, 23, 2499–2522, 2021.

52. Barry, M.T., The Atom Economy-A Search for Synthetic Efficiency. *Science*, 254, 1471–1477, 1991.
53. Erythropel, H.C., Zimmerman, J.B., de Winter, T.M., Petitjean, L., Melnikov, F., Lam, C.H., Lounsbury, A.W., Mellor, K.E., Janković, N.Z., Tu, Q., Pincus, L.N., Falinski, M.M., Shi, W., Coish, P., Plata, D.L., Anastas, P.T. The Green ChemisTREE: 20 years after taking root with the 12 principles. *Green Chem.*, 20, 1929–1961, 2018.
54. Szymkuc, S., Gajewska, E.P., Klucznik, T., Molga, K., Dittwald, P., Startek, M., Bajczyk, M., Grzybowski, B.A., Computer-Assisted Synthetic Planning: The End of the Beginning. *Angew. Chem. Int. Ed. Engl.*, 55, 5904–5937, 2016.
55. Carola, C., Pierre, J., Annika, M., Claus, J., Christian., D., Sergio, V., Dante, R., Rossella, F., Clemens, Z., Antonello, M., Green Med Chem: the challenge in the next decade toward eco-friendly compounds and processes in drug design. *Green Chem.*, 25, 2109, 2023.
56. Melero, J.A., Iglesias, J., Garcia, A., Biomass as renewable feedstock in standard refinery units. Feasibility, opportunities and challenges. *Energy Environ. Sci.*, 5, 7393–7420, 2012.
57. Marta, B., Oksana, Z., Xiang-Fa, W., Green Synthesis of Soy Protein Nanocomposites: Effects of Cross-Linking and Clay Nanoparticles on the Mechanical Performance. *ACS Omega*, 7, 7, 5883–5893, 2022.
58. Chen, N., Lin, Q., Zheng, P., Rao, J., Zeng, Q., Sun, J., A sustainable bio-based adhesive derived from defatted soy flour and epichlorohydrin. *Wood Sci. Technol.*, 53, 801–817, 2019.
59. Samson, D.O., Mat Jafri, M.Z., Hashim, R., Sulaiman, O., Aziz, M.Z.A., Yusof, M.F.M., Shukri, A., Rhizophora spp. Particleboards incorporating defatted soy flour bonded with NaOH/IA-PAE: Towards a water equivalent phantom material. *Radiat. Phys. Chem.*, 176, 109057, 2020.
60. Kazzaz, A.E., Feizi, Z.H., Fatehi, P., Grafting strategies for hydroxy groups of lignin for producing materials. *Green Chem.*, 21, 5714–5752, 2019.
61. Österberg, M., Sipponen, M.H., Mattos, B.D., Rojas, O.J., Spherical lignin particles: A review on their sustainability and applications. *Green Chem.*, 22, 2712–2733, 2020.
62. Boethling, R.S., Elizabeth, S., David, D., Designing Small Molecules for Biodegradability. *Chem. Rev.*, 107, 6, 2207–2227, 2007.
63. Stefano, N., Achille, A., Carlotta, P., Lorenzo, G., Nadia, B., Marilena, C., Matteo, B., Critical Assessment of the Sustainability of Deep Eutectic Solvents: A Case Study on Six Choline Chloride-Based Mixtures. *ACS Omega*, *7*, 51, 47449–47461, 2022.
64. Philip, D.H. and Richard, A.F., Electropolishing of Aluminum at Room Temperature Using a Green DES of Choline Chloride and Propylene Glycol. *ACS Omega*, 8, 13, 11801–11805, 2023.
65. Crawford, S.E., Hartung, T., Hollert, H., Mathes, B., van Ravenzwaay, B., Steger-Hartmann, T., Studer, C., Krug, H.F., Green Toxicology: a strategy for sustainable chemical and material development. *Environ. Sci. Eur.*, 29, 1, 16, 2017.
66. Maertens, A., Anastas, N., Spencer, P.J., Stephnes, M., Goldberg, A., Hartung, T., Food for thought: Green Toxicology. *ALTEX Altern. Anim. Exp.*, 31, 243–249, 2014.
67. Nagendran, S., Balasubramaniyan, S., Irfan, N. Virtually screened novel sulfathiazole derivatives as a potential drug candidate for methicillin-resistant *Staphylococcus aureus* and multi-drug-resistant tuberculosis. *J. Biomol. Struct. Dyn.*, 41, 11, 5086-5095, 2023.
68. Wells, A.S., Biocatalysis for Medicinal Chemistry, in Green and Sustainable Medicinal Chemistry: Methods, Tools and Strategies for the 21st Century Pharmaceutical Industry, ch. 15, in: *Royal Society of Chemistry*, L. Summerton, H.F. Sneddon, L.C. Jones, J.H. Clark, (Eds.), pp. 180–191, Cambridge, UK, 2016.
69. Hughes, G. and Lewis, J.C., Introduction: Biocatalysis in industry. *Chem. Rev.*, 118, 1–3, 2018.
70. Woodley, J.M., Accelerating the implementation of biocatalysis in industry. *Appl. Microbiol. Biotechnol.*, 103, 12, 4733–4739, 2019.
71. Sheldon, R.A. and Brady, D., The limits to biocatalysis: Pushing the envelope. *Chem. Commun.*, 54, 6088–6104, 2018.

72. Grigoras, A.G., Catalase immobilization—A review. *Biochem. Eng. J.*, 117, 1–20, 2017.
73. Kojom, F.L.P., Eya'ane, M.F., Eboumbou, M.C.E., Ntoumba, A.A., Ngaha, N.M.I., Belle, E.K.P., Ayong, L., Lehman, L.G., A systematic review on anti-malarial drug discovery and antiplasmodial potential of green synthesis mediated metal nanoparticles: overview, challenges and future perspectives. *Malar. J.*, 18, 337, 2019.
74. Fuentes, P., Zhou, F., Erban, A., Karcher, D., Kopka, J., Bock, R., A new synthetic biology approach allows transfer of an entire metabolic pathway from a medicinal plant to a biomass crop. *eLife*, 5, e13664, 2016.
75. Bagchi, S., Yuan, R., Engleman, E.G., Immune Checkpoint Inhibitors for the Treatment of Cancer: Clinical Impact and Mechanisms of Response and Resistance. *Annu. Rev. Pathol.*, 16, 223–249, 2021.
76. Dobosz, P., Stepien, M., Golke, A., Dzieciatkowski, T., Challenges of the Immunotherapy: Perspectives and Limitations of the Immune Checkpoint Inhibitor Treatment. *Int. J. Mol. Sci.*, 23, 2847, 2022.
77. DiFrancesco, M., Hofer, J., Aradhya, A., Rufinus, J., Stoddart, J., Finocchiaro, S., Mani, J., Tevis, S., Visconti, M., Walawender, G., DiFlumeri, J., Fattakhova, E., Patil, S.P., Discovery of small-molecule PD-1/PD-L1 antagonists through combined virtual screening and experimental validation. *Comput. Biol. Chem.*, 102, 107804, 2022.
78. L.S. Chupak and X. Zheng, Bristol-Myers Squibb Company Assignee. Compounds Useful as Immunomodulators. *International Patent.* WO034820 Al, 2015.
79. Guzik, K., Zak, K.M., Grudnik, P., Magiera, K., Musielak, B., Torner, R., Skalniak, L., Domling, A., Dubin, G., Holak, T.A., Small-Molecule Inhibitors of the Programmed Cell Death-1/Programmed Death-Ligand 1 (PD-1/PD-L1) Interaction via Transiently Induced Protein States and Dimerization of PD-L1. *J. Med. Chem.*, 60, 5857–5867, 2017.
80. Mittal, L., Tonk, R.K., Awasthi, A., Asthana, S., Targeting cryptic-orthosteric site of PD-L1 for inhibitor identification using structure-guided approach. *Arch. Biochem. Biophys.*, 713, 109059, 2021.
81. Anthony, L. and Freda, P.U., From somatostatin to octreotide LAR: evolution of a somatostatin analogue. *Curr. Med. Res. Opin.*, 25, 2989–2999, 2009.
82. Giovanna, L.P., Simona, D.M., Maria, D.R., Peptidomimetics: An Overview of Recent Medicinal Chemistry Efforts toward the Discovery of Novel Small Molecule Inhibitors. *J. Med. Chem.*, 65, 11, 7438–7475, 2022.
83. Gentilucci, L., De Marco, R., Cerisoli, L., Chemical modifications designed to improve peptide stability: incorporation of non-natural amino acids, pseudo-peptide bonds, and cyclization. *Curr. Pharm. Des.*, 16, 318–3203, 2010.
84. Galan, A., Comor, L., Horvatić, A., Kuleš, J., Guillemin, N., Mrljaka, V., Bhide, M., Library-based display technologies: where do we stand? *Mol. Biosyst.*, 12, 2342–2358, 2016.
85. Gietz, R.D., Yeast two-hybrid system screening. *Methods Mol. Biol.*, 313, 345–371, 2006.
86. Burkov, A., *The Hundred-Page Machine Learning Book*, vol. 1, pp. 3–5, Andriy Burkov, Canada, 2019.
87. Ma, J., Sheridan, R.P., Liaw, A., Dahl, G.E., Svetnik, V., Deep neural nets as a method for quantitative structure-activity relationships. *J. Chem. Inf. Model.*, 55, 263–274, 2015.
88. Sarkar, C., Das, B., Rawat, V.S., Wahlang, J.B., Nongpiur, A., Tiewsoh, I., Lyngdoh, N.M., Das, D., Bidarolli, M., Sony, H.T., Artificial Intelligence and Machine Learning Technology Driven Modern Drug Discovery and Development. *Int. J. Mol. Sci.*, 24, 3, 2026, 2023.
89. Jha, D., Choudhary, K., Tavazza, F., Liao, W.K., Choudhary, A., Campbell, C., Agrawal, A., Enhancing materials property prediction by leveraging computational and experimental data using deep transfer learning. *Nat. Commun.*, 10, 1–12, 2019.
90. Arindam, C., Computational methods and tools for sustainable and green approaches in drug discovery edited by Bimal Krishna Banik, in: *Advances in Green and Sustainable Chemistry:*

Green Approaches in Medicinal Chemistry for Sustainable Drug Design, pp. 965–988, Elsevier, 2020.

91. Buyya, R., Yeo, C.S., Venugopal, S., Broberg, J., Brandic, Y., Cloud computing and emerging IT platforms: Vision, hype, and reality for delivering computing as the 5th utility. *Future Gener. Comput. Syst.*, 25, 599–616, 2009.
92. Radu, L.D., Green Cloud Computing: A Literature Survey. *Symmetry*, 9, 12, 295, 2017.
93. Shuja, J., Gani, A., Ahmad, R.W., Ahmed, A.I.A., Siddiqa, A., Nisar, A., Khan, S.U., Zomaya, A., Greening Emerging IT Technologies: Techniques and Practices. *JISA*, 8, 9, 2017.
94. Moult, J., Fidelis, K., Kryshtafovych, A., Schwede, T., Tramontano, A., Critical assessment of methods of protein structure prediction (CASP)—Round XII. Proteins: Structure. *Funct. Bioinf.*, 86, 7–15, 2018.
95. Schadt, E.E., Molecular networks as sensors and drivers of common human diseases. *Nature*, 461, 7261, 218–223, 2009.
96. Chen, H., Zhang, H., Zhang, Z., Cao, Y., Tang, W., Network-based inference methods for drug repositioning. *Comput. Math. Methods Med.*, 2015, 2015.
97. Chang, K., Balachandar, N., Lam, C., Yi, D., Brown, J., Beers, A., Rosen, B., Rubin, D.L., Kalpathy-Cramer, J., Distributed deep learning networks among institutions for medical imaging. *J. Am. Med. Inform. Assoc.*, 25, 8, 945–954, 2018.
98. Char, D.S., Shah, N.H., Magnus, D., Implementing machine learning in health care' addressing ethical challenges. *NEJM*, 378, 11, 981–983, 2018.
99. Prasad, S., Srivastava, A., Singh, N., Singh, H., Saluja, R., Kumar, A., Yadav, N., Qidwai, T., Present and future challenges in therapeutic designing using computational approaches. Computational Approaches for Novel Therapeutic and Diagnostic Designing to Mitigate SARS-CoV-2. *Infection*, 2022, 489–505, 2022.
100. Ayatunga, M.K., Xie, W., Ruder, L., Schulze, U., Meier, C., AI in small-molecule drug discovery: A coming wave? *Nat. Rev. Drug Discov.*, 21, 175–176, 2022.
101. Boniolo, F., Dorigatti, E., Ohnmacht, A.J., Saur, D., Schubert, B., Menden, M.P., Artificial intelligence in early drug discovery enabling precision medicine. *Expert Opin. Drug Discov.*, 16, 991–1007, 2021.
102. Cavasotto, C.N. and Di Filippo, J.I., Artificial intelligence in the early stages of drug discovery. *Arch. Biochem. Biophys.*, 698, 108730, 2020.
103. Füzi, B., Mathai, N., Kirchmair, J., Ecker, G.F., Toxicity prediction using target, interactome, and pathway profiles as descriptors. *Toxicol. Lett.*, 381, 20–26, 2023.
104. Hao, Y. and Moore, J.H., TargetTox: A Feature Selection Pipeline for Identifying Predictive Targets Associated with Drug Toxicity. *J. Chem. Inf. Model.*, 61, 5386–5394, 2021.
105. Atz, K., Isert, C., Böcker, M.N.A., Jiménez-Luna, J., Schneider, G., Δ-Quantum machine-learning for medicinal chemistry. *Phys. Chem. Chem. Phys.*, 24, 10775–10783, 2022.
106. Pu, L., Govindaraj, R.G., Lemoine, J.M., Wu, H.-C., Brylinski, M., DeepDrug3D: Classification of ligand-binding pockets in proteins with a convolutional neural network. *PloS Comput. Biol.*, 15, e1006718, 2019.
107. Graff, D.E., Shakhnovich, E.I., Coley, C.W., Accelerating high-throughput virtual screening through molecular pool-based active learning. *Chem. Sci.*, 12, 7866–7881, 2021.
108. Pozzan, A., QM Calculations in ADMET Prediction. *Quantum Mech. Drug Discov.*, 21, 14, 285–305, 2020.
109. Isert, C., Atz, K., Jiménez-Luna, J., Schneider, G., QMugs, quantum mechanical properties of drug-like molecules. *Sci. Data*, 9, 273, 2022.
110. Böselt, L., Thürlemann, M., Riniker, S., Machine Learning in QM/MM Molecular Dynamics Simulations of Condensed-Phase Systems. *J. Chem. Theory Comput.*, 17, 2641–2658, 2021.

13

Green Chemistry: The Emergence of a Transformative Framework

Priyanka Chaudhary[1,2*], **Rapelly Venkatesh**[3,4] **and Reena Singh**[5]

[1]*Amity Institute of Applied Sciences, Amity University, Noida, Uttar Pradesh, India*
[2]*Department of Chemistry, University of Lucknow, Lucknow, Uttar Pradesh, India*
[3]*Pondicherry University, Puducherry, India*
[4]*Institute of Chemistry, Academia Sinica, Taipei, Taiwan*
[5]*Sunbeam College of Women, Bhagwanpur, Varanasi, India*

Abstract

Green chemistry can be defined as the design of chemical products and routes that reduce the generation of dangerous substances. The progressive transformation and distinct discipline has led to substantial environmental assistances, modernization, and a reinforced economy. More than a Triennial, the scientific community has been contributing to achieve the environmental benign and sustainability goals through the improvement and the progress of the different synthetic routes. The remarkable milestone has been set with the green chemistry principles. Since they play an important role in ensuring that our next generation of chemicals, resources, and energy is more sustainable than the current generation. The most appealing theory is Green Chemistry, which is the utilization of a set of principles that reduces or eliminates the use or generation of hazardous substances in the structured design, construction, and purposes of chemical products. The Green Chemistry's concept was formulated in the early 1990s, which was the drive for the scientists to achieve them by facing different challenges and resolving them. The problems faced in conventional synthesis are the complexity of the reaction methods, slow reaction rates, hazardous and use of large quantity of solvents and catalysts, and completion time. The concept of designing new alternative synthetic routes while keeping in mind the conscious aspects such as impact of chemicals and by-products on human health, animals, environment, and sustainability. This chapter includes the various research on synthetic transformations since recent years that has been built on the pillar of Green Chemistry. The reusability of the catalyst, innocuous by-products, and biodegradability would be an important aspect in the design of the synthetic organic products. The impacts of green chemistry on pharmaceutical industries, environment, population, and chemical industries are some of the important aspects to be taken care for the sustainable environment.

This emphasizes some advances made in green chemistry research areas that can be considered as an emergence of a transformative framework:

(i) Synthetic routes with catalysts in stoichiometric amounts with the higher selectivity of the chemistry showcasing its advancement
(ii) Solvent-free syntheses or alternative environmental benign solvents

**Corresponding author*: pchaudhary1@amity.edu; priyankaciitbhu@gmail.com

Mousumi Sen (ed.) Sustainable Green Catalytic Processes, (301–328) © 2024 Scrivener Publishing LLC

(iii) Overcoming the conventional methods by switching to microwave, ball milling, and photochemical synthesis
(iv) Preventing the usage of toxic chemicals, use of alternative chemicals, or designing for lessened eco-friendly disclosure

***Keywords*:** Green chemistry, solvent-free, environmentally benign, microwave synthesis, photochemical synthesis, toxic chemical alternatives

13.1 Introduction

The accomplishments of green chemistry in recent years have shown a great emergence of a unified field that is remarkable.[1] It is significant to diagnose the scientific innovations that have been made through outstanding research that has provided the wider organization needed to drive the field.

The foundation building blocks of green chemistry have built up the central of this field.[2] Therefore, the essential understanding to find the reach of green chemistry and the challenges and opportunities of future synthetic chemistry lies where the necessary steps should be accepted to follow sustainability in the environment.[3]

There are important terms such as E-factor[4] and atom economy,[5] which are used to calculate the environmental factor and conversion efficiency of the reaction. Efficiency could also be calculated in relation of an "E-factor," or the amount of discarded materials generated per kilogram of the desired product. Reaction mass efficiency (RME) considers the measurement of atom economy along with the stoichiometry and percentage yield of all steps.[6] These systems of measurement have been efficaciously useful in commercial platform in industries to invent greener developments, for example, in medicinal product

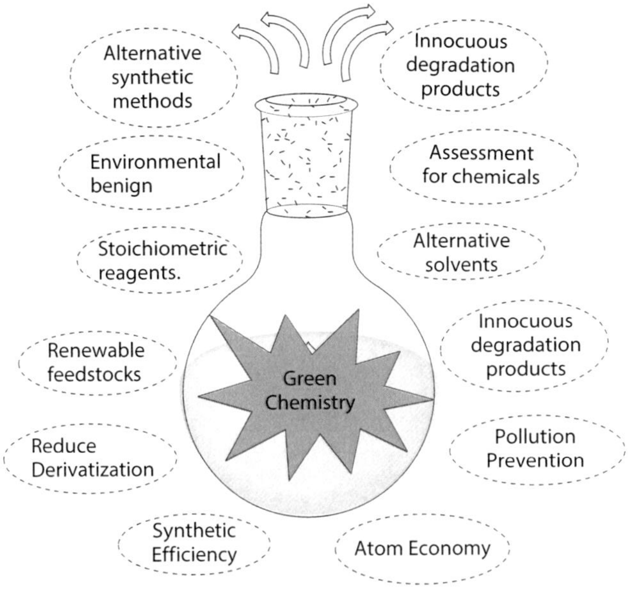

Figure 13.1 Green chemistry: the emergence of a transformative framework.

formation. RME calculations and atom economy classically do not interpret for usage of toxic chemicals, or solvents, which might comprise the initial basis of generation of waste in the syntheses. The basis of green chemistry involves the challenges in describing the security and waste in relation of environmental impact for sustainability.[7]

Formula:

$$\text{Atom Economy} = \left[\frac{\text{Molecular weight of desired product}}{\text{Molecular weight of all reactants}}\right] \text{X } 100\%$$

Green chemistry has made incredible advancement over a period and has facilitated the route and its analysis to synthesize the molecule practically with simplicity to obtain its characterization and understand the complexity of the molecules and complex molecular systems.[8] Since they have emerged to be boon, the green synthetic routes are applied that lead to more efficiency and selectivity for the construction of compounds that address new enhanced properties in making novel products.[9] Green chemistry aims were the important advances. They have appeared as a structured design framework that would adventure the capabilities of chemists and molecular designers. They mention the concerns of the essential effect of chemicals on the condition of living beings, the ecosystem, and sustainability.[10] The effort of the green chemistry team, universally, has developed in substantial development in demonstration of the methodology, this strategy agenda, should be practiced (Figure 13.1). The chemists followed the 12 Principles of Green Chemistry to provide clean, ecologically benevolent approaches that are balanced for an extensive period.[11] The summary of some research progresses in green chemistry is emphasized here:

13.2 Synthetic Routes with Catalysts in Stoichiometric Amounts with the Higher Selectivity of the Chemistry Showcasing Its Advancement

The efficient synthetic approaches[12] were vital or essential to form the complex molecular arrangements that involve the chemical reactions. This would involve the pathways that have selectivity such as chemoselectivity,[13] regioselectivity,[14] diastereoselectivity,[15] and enantioselectivity,[16] and inexpensive with the atom economy.[17] The different approaches Includes the simple combination of two or more construction blocks with other reactants that are required in stoichiometric amounts of catalysts with the highest degree of atom economy.[18]

The chiral α-sulfenylated ketones[19] are useful construction blockades. During research, all of the chemists are curious for their syntheses, but still, there are various drawbacks and restrictions with their syntheses. Therefore, the need of the hour was reported recently by Samec *et al.* where a novel two-step method that consists of palladium-mediated hydrothiolation of propargylic alcohols pursued by an enantioselective rhodium isomerization of allylic alcohols.[20] The key step in the following reaction was the isomerization reaction for the formation of ketones in their enantioenriched (er) form.[21] The importance of the methodology is the induction of high yields of high enantioselectivity, high atom economy, and with no waste produced. Later, the mechanism for the isomerization reaction was also proposed with a rhodium hydride-enone intermediate.[22] The authors mentioned that it is a greener approach to synthesize racemic α-sulfenylated ketones *via* formed intermediates of sulfenylated allylic alcohols.[23]

As the ligand, catalyst, and solvent screenings were optimized, therefore, the study of substrate scope has been investigated. The substrates with electron-donating showed a high enantioselectivity (~88:12), whereas the electron-withdrawing groups showed almost no conversion (Scheme 13.1). The halogen-containing reactant had slightly lowered the enantioselectivity (~82:18). All of the reactants containing C-thioether moiety in substrates also furnished desired products with quantitative yields.

The proposed mechanism is mentioned in Scheme 13.2. Initially, the Rh complex coordinates with the substrate that led to the formation of the alkoxide intermediate **III** with the elimination of HCl. In this study, there are some substrates containing a thioether group, which is a good ligand for the transition metals. Subsequently, there are chances that intermediate **II** could be formed as off-cycle species. Later, Rh-enone-hydride intermediate **IV** is formed *via* the β-hydride elimination. In this step, the chirality of the substrate would be lost, allowing a stereoconvergent progression. Following, this intermediate **III** undergoes 1,2-insertion with the formation of intermediate **VI**. This will form the enolate intermediate **VII** through tautomerization of **VII**. Lastly, protonation took place enantioselectively and the enolate furnishes the chiral ketone with catalyst regeneration.[24]

Therefore, it can be concluded that the authors demonstrated the asymmetric isomerization of allylic alcohols which form chiral α-sulfenylated ketones,[26] and they are used for

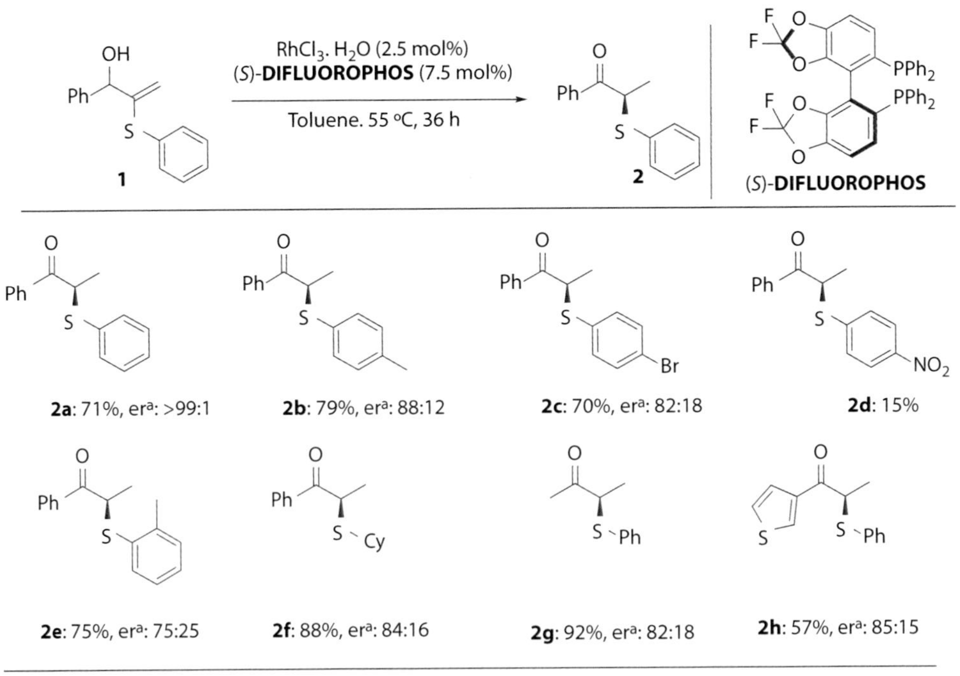

[a]Determined by HPLC.

Scheme 13.1 Substrate scope of the rhodium-mediated isomerization of secondary allylic alcohols.

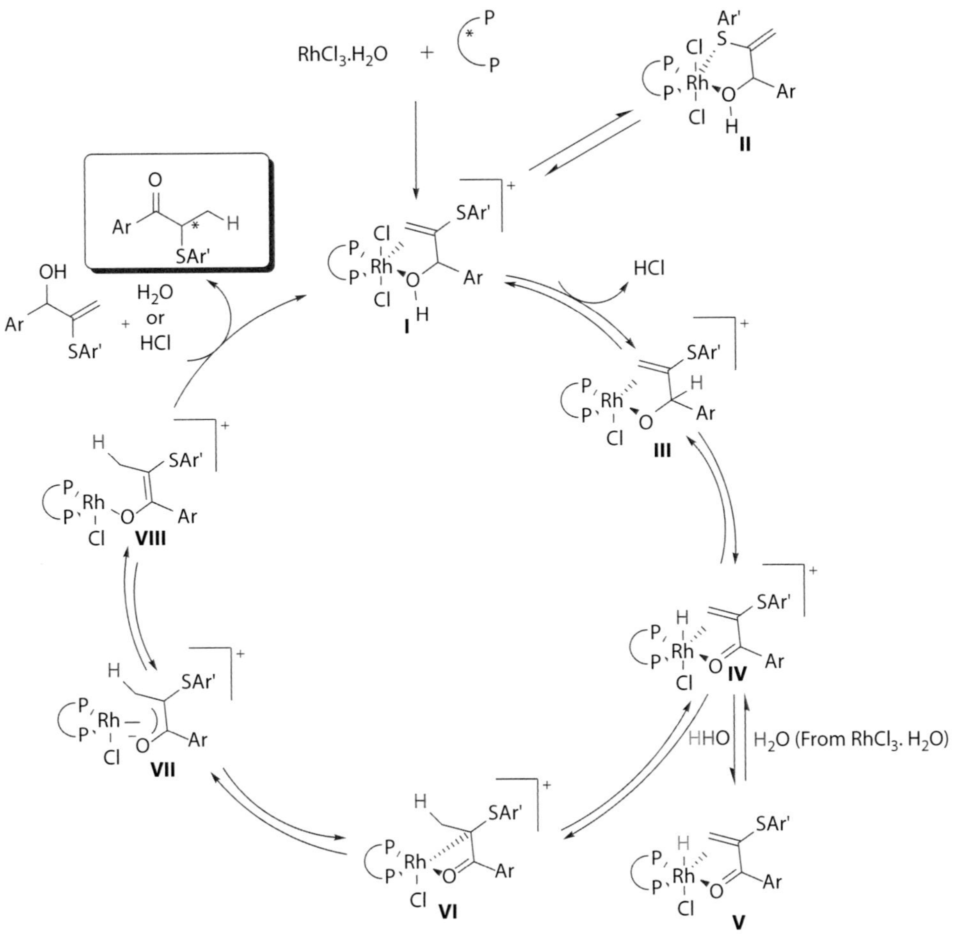

Scheme 13.2 Plausible mechanism for the rhodium-catalyzed isomerization reaction.

the construction of vital building blocks.[25] In some cases, the good enantioselectivities were obtained (87:13 er and >99:1 for a few substrates). They have also discussed that the methodology is limited for allylic alcohols containing arylthioether groups and aryl ketones, which shows high atom economy with enantio-riched selectivity.

13.3 Solvent-Free Syntheses or Alternative Environmental Benign Solvents

Solvent-free organic reactions[27] under conditions suggest improved discernment and productivity, ease of exploitation, and frequently prevent noxious and volatile solvents.[28] These advances are relevant to a broad range of reactions such as cleavage,[29] condensation,[30] cyclization,[31] oxidation,[32] and reduction reactions,[33] and to the rapid parallel amalgamation of a collection of biologically active molecules.[34] The further short discussion events that a variation of organic reactions, which are conventionally conducted in solvent avenues, can be conducted more cost-effectively in the absence of solvents.[35] The synthetic organic chemists will absolutely renew their undertaking to get additional reactions into the fold up of the solvent-free synthetic approach.

13.3.1 Metal-Free, Acid-Free, and Solvent-Free Reaction

In this context, recently, Chaudhary *et al.*[36] had reported an effective synthesis of *N*-nitrosamines of secondary amines under solvent-free conditions using tert-butyl nitrite (TBN) (Scheme 13.3). This advance of an acid-free and metal-free nitrosating reagent[37] that worked proficiently under mild conditions remains a significant approach. With the growth in environmentally friendly responsiveness in accordance with green chemistry, there is also an insistent requirement to progress eco-friendly methods for the formulation of numerous acceptable compounds. The TBN as an acid-free reagent has many benefits like commercial availability in cheap price, quite easy handing, and great solubility in most of the regular solvents.

At the outset, N-methylaniline (**1a**) was used for the investigation of N-nitrosation reaction using tert-butyl nitrite (TBN). The reactions were executed at room temperature in different protic and aprotic solvents. The best optimized standard reaction conditions of the N-nitrosation were solvent-free conditions using one equivalent of tert-butyl nitrite at room temperature within 5 minutes. N-nitrosation of several secondary amines was explored using TBN under solvent-free conditions as mentioned in Scheme 13.3. The various amines such as diaryl amines, N-alkyl N-aryl amines, and dialkyl amines were investigated. All of the substrates were well-tolerated, and functionality remained sustainable under the reaction conditions. In general, the reactivity of several categories of secondary (2°) amines was observed in the subsequent sequence where aryl amines were the most reactive, then benzyl amines and later alkyl amines. The mechanism of the above reaction has not been mentioned in their research but based on their following research on TBN believed that the research has been expected to be free-radical mechanism.[38]

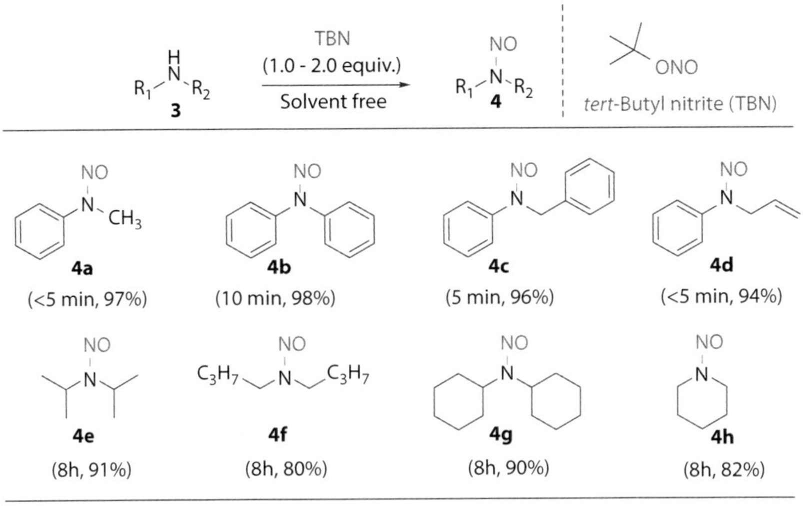

Scheme 13.3 N-nitrosation of secondary amines using tert-butyl nitrite under solvent-free conditions.

Therefore, the versatile reagent tert-butyl nitrite has supported the solvent-free N-nitrosation reaction in organic synthesis. Several aryl, benzyl, and alkyl secondary amines were nitrosated in excellent yields under mild conditions under solvent-free conditions that reduced the time and labor for the synthesis.

13.3.2 Catalyst-Free, Additive-Free, and Solvent-Free Reaction

Li *et al.*[39] and Sarkar *et al.*[40] in 2023 has disclosed the accessible method for the synthesis of aryl-comprising trisubstituted alkenes *via* direct alkylation of alkenes under solvent-free and catalyst-free (SFCF) conditions (Scheme 13.4). Such synthesis has expanded extensive consideration in the area of green chemistry and sustainable improvement due to its ability to reduce waste, preserve resources, and modernize strategies. The reaction was optimization under various conditions such as temperature, reaction time, and reactant ratio. Here, 1,1-diphenylethylene (1 equiv.) and (1-chloroethyl)benzene (1.1 equiv.) were considered as model substrates for the optimization where the reaction gave 90% isolated yield at 80°C for 3 h under air. The standard conditions were applied to various substituted alkenes and haloalkyl benzene. Remarkably, when asymmetric alkenes are used as substrates, a mixture of Z/E alkylation products is attained, and the enhances the stereoselectivity by using steric effects.[41]

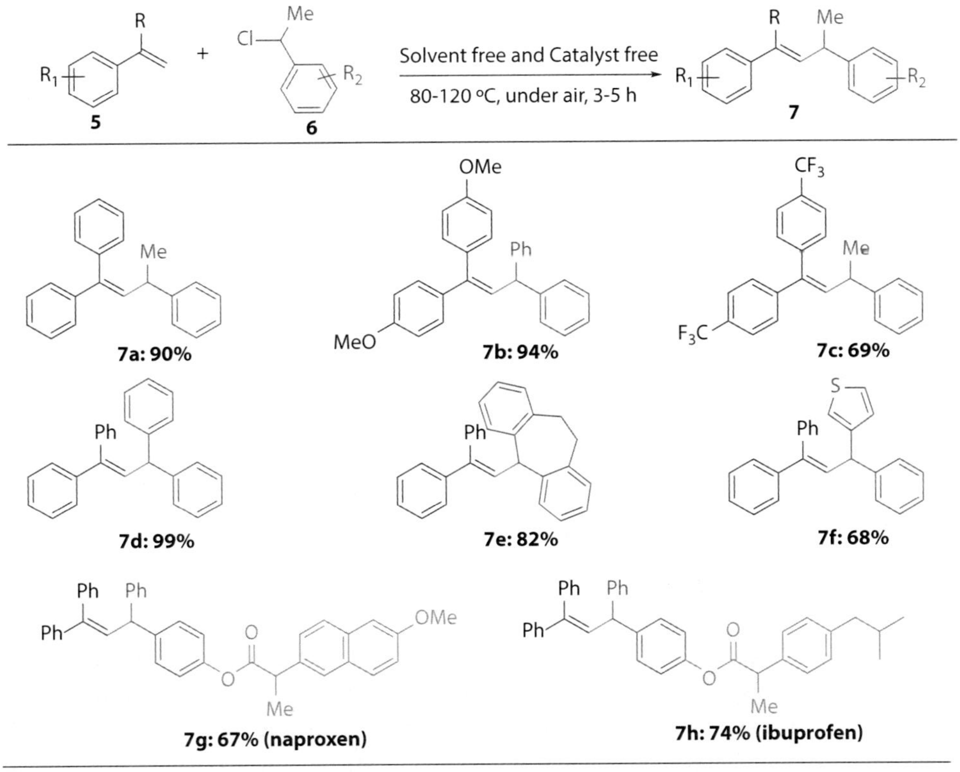

Scheme 13.4 Scope of benzyl halides and alkenes under solvent-free and catalyst-free reaction.

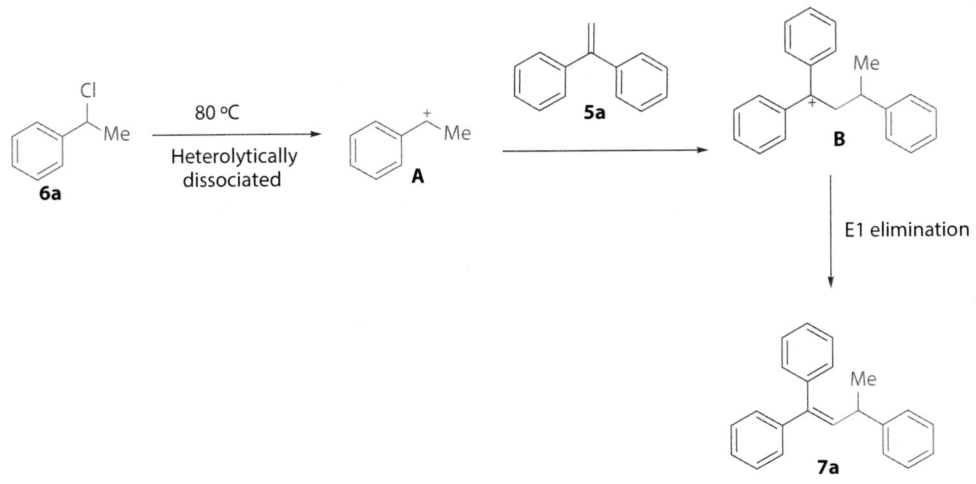

Scheme 13.5 The proposed mechanism for the synthesis of aryl-containing trisubstituted alkene.

The electron-donating and electron-withdrawing groups on substrates were compatible, whereas the introduction of EWG such as trifluoromethyl group on aromatic ring reduces the yields that increases the energy for heterolytic cleavage of the C-X bond. The corresponding products were obtained in high yields. Also, the heterocyclic rings such as thienyl were also sustainable under standard reaction conditions. The additional merits of the methodology were proven by the synthesis of a variation of medicine-comprising reactants, such as ibuprofen (**5g**) and naproxen (**5h**). Besides this, the authors have also performed the gram-scale syntheses under the solvent-free conditions with excellent yields. They have also calculated the E-factor,[4] carbon efficiency,[42] and atom economy[5, 18] as 0.18, 100%, and 91%, respectively, proving the approach's superiority.

To propose the mechanism, the mechanistic studies were investigated such radical trapping that proves that the reaction takes place via ionic reaction.[43] To prove it, cation-exchange experiment was attempted. In the given Scheme 13.5, **6a** undergoes heterolytically dissociation of C-Cl bond at 80°C that gave the benzyl cation **A** that is later captured by **5a** to form another cationic intermediate **B**. Conclusively, the corresponding compound **7a** is formed *via* the E1 elimination process. This procedure displays an exceptional atom economy, step economy, and up-front procedure, supporting the principles of green chemistry.[44]

13.4 Overcoming the Conventional Methods by Switching to Microwave, Ball Milling, and Photochemical Synthesis

13.4.1 Microwave-Accelerated Reactions

An unconventional energy source is microwave irradiation that has been used for a diversity of organic conversions.[45] The rapid synthetic conversions were enabled using microwave irradiation under solvent-free conditions enabled at moderate pressure. Thus, novel chemical reactions with features such as ease of operation, improved reaction rates, and greater yields. The microwave-assisted method comprises the syntheses of a broad variation of industrially noteworthy complexes and intermediates, specifically enones,[46] imines,[47] enamines,[48] and nitroalkenes[49]. The solvent-free reactions in microwave irradiation were preparation of ionic liquids that are characterized as catalysts.[50] Moreover, there is the possibility that microwave irradiation reactions could be attempted with water as a reaction medium that has been proven to be a greener chemical approach.[51] For example, the tertiary amines were synthesized with the aqueous N-alkylation of amines, synthesis of N-azacycloalkanes,[52] which play an important role for the building blocks in natural products[53] and pharmaceuticals.[54] These microwave-assisted, greener, and expeditious chemical transformations circumvent the need for multistep processes that use expensive metal catalysts, proceed under mild reaction conditions, and minimize or eliminate the formation of by-products.[55]

13.4.1.1 *Synthesis of Benzo[f][1,4]oxazepine-2H-Chromene and 1,2-Dihydroquinoline-Fused Polycyclic Nitrogen Heterocycles*

For the wide applications in biological and medicinal field of small molecules such as quinazolinones, chromenes, quinolines, and benzoxazepines, only a few reports were accessible in the literature where their fused forms such as quinazolinone-fused benzoxazepines,[56] quinoline-fused chromenes,[57] and 1,6-naphthyridines.[58] Nowadays, the thriving partition of chemical synthesis, chemo-divergent synthesis, leads to the formation of the molecules with complexity and their high structural range employs general substrates.[59] The substrate- or reagent-based strategies have been used in the area of drug discovery is basically a diversity-oriented synthesis (DOS) which is an exciting and proficient approach to achieving such scaffold.[60] Diversity-oriented synthesis (DOS) directs to initiate structural diversity in an efficient manner. The strategy for the reagent-based application led to the miscellaneous products from common substrates in the existence of diverse reagents, and the reactant-based application accomplishes the compounds initiating from prefunctionalized reactants under reaction conditions.

The reaction was optimized where the strategic reaction that takes place with the synthesis of oxazepino[5,4-b]quinazolin-9-ones[61] was examined by exposing a variation of O-propargylated 2-hydroxybenzaldehydes (**11**) and 2-aminobenzamides[62] (**10**) under microwave irradiation and conventional heating conditions (Scheme 13.6). The standard reaction conditions were set as $CuBr_2$ catalyst, potassium carbonate (K_2CO_3) in DMF. Therefore, the impact of the different functionalities such as electron-donating and electron-withdrawing substitutions was investigated and was found to deliver the desired products smoothly with quantitative yields. Later, the conclusion was drawn from the comparison between microwave irradiation and conventional heating conditions through the yield of the products for the synthesis of oxazepino-[5,4-b]quinazolin-9-ones (**12**). Furthermore, the range of diversity-oriented synthetic approach was explored where the reaction between O/N-propargylated 2-hydroxy/aminobenzaldehydes (**8**) and 2- aminobenzyl alcohols (**9**). The substrates were well-tolerated with sensitive functional groups as well. Overall, the microwave-aided reactions were clean, fast, and accomplished within 15 minutes at 90°C, although the same reactions completed in 4 h at 120°C under conventional heating. All of the compounds' structures were established by ^1H, ^{13}C NMR studies[63] and single-crystal XRD.[64]

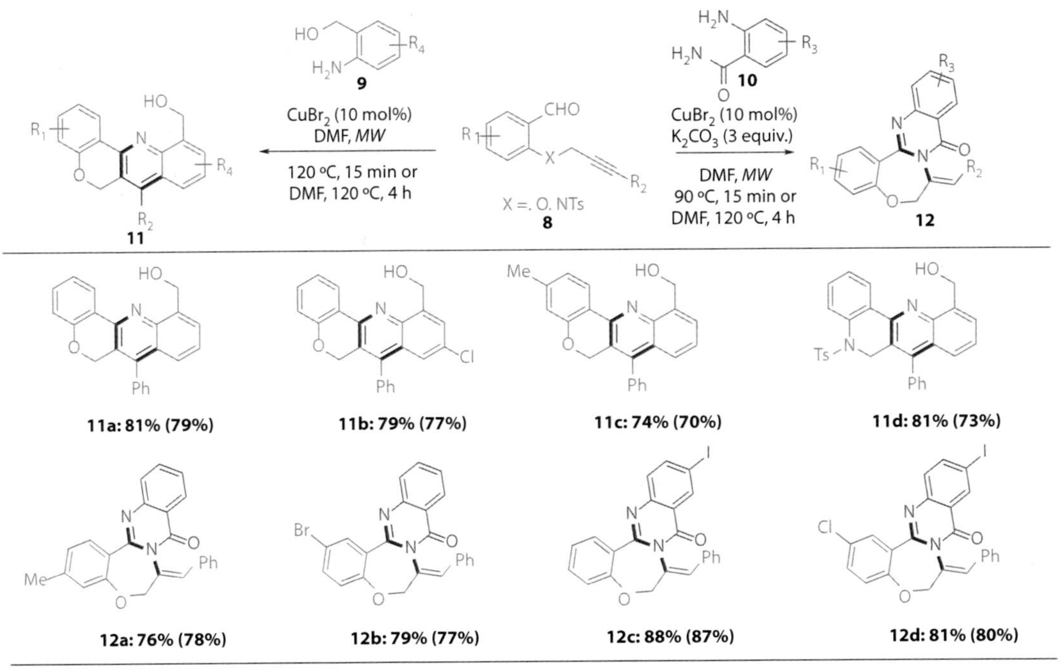

Scheme 13.6 Scope of the synthesis of oxazepino[5,4-b]quinazolin-9-ones, 6h-chromeno[4,3-b]quinolines, and dibenzo[b,h][1,6]naphthyridines.

The authors attempted the control experiments that have led to the proposed mechanistic pathway for the synthesis of oxazepino[5,4- b]quinazolin-9-ones (Scheme 13.7). Initially, aldehyde **8** and the amine **11** formed an imine as **I'** in the presence of the CuBr$_2$ catalyst, which led 6-exo-trig cyclization to furnish 2,3-dihydroquinazolin-4(1H)-one **II'** that on the oxidation in the presence of air (Scheme 13.7) led to the formation of the intermediate **III'**. Later, the 1,3-proton shift followed by 7-exo-dig cyclization furnished the desired products

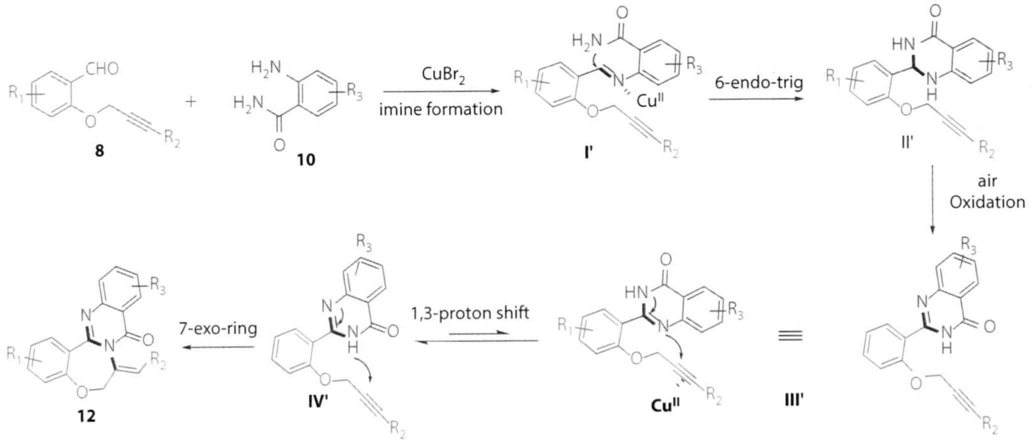

Scheme 13.7 CuBr$_2$-catalyzed diversity-oriented synthetic routes to access substituted oxazepino[5,4-b] quinazolin-9-ones.

9 through the formation of the intermediate of **IV'**. The obtained product is formed since the presence of the amino group enhances the nucleophilicity of the amide N-group that assisted the formation of the product.

The proposed mechanism for the formation of dibenzo[b,h][1,6]naphthyridines and 6H-chromeno[4,3-b]quinolines is mentioned in Scheme 13.8. Like the earlier reaction, the aldehyde and 2-aminobenzyl alcohol to form the imine intermediate **I"** that led to the *in situ* consequent Povarov-type [4 + 2] cycloaddition in the presence of $CuBr_2$ catalyst that afforded the desired products **11** and **11'**. The alcoholic functionality supports the active coordination of the Cu (II) with the imine nitrogen (N) that increases the e-mandate hetero Diels–Alder reaction inversely. The poor nucleophilicity of the hydroxy group does not lead to the competitive endo/exo-trig cascade.

13.4.1.2 Ball Milling Approach

Mechanochemistry stipulates a different method of powerful force to amend the energy site of organic chemical reactions and accompaniments predictable solution-based procedures.[65, 66] The previous era has viewed the stimulating developments of mechanical organic synthesis in a wide range of covalent bond-formation reactions.[67–69] This has enhanced the less usage of high quantity of organic solvents and minimalized ecological effects that have also shortened the time of reactions and improved yields or selectivity.[70–73] Interestingly, mechanochemistry research has discovered the friendly connections of mechanical force with the chemical synthesis and reactivity of cysteine disulfide bridge reduction by resources of single-molecule force spectroscopy, molecular force probes, and computational studies.[74–77] Specifically, it was recommended that kinetics of the thiol-disulfide exchange reaction in proteins might be structured by mechanical force through the lengthening of bond. This mechanical force encouraged the conformational distortion of a disulfide that could complicatedly modify the cleavage of disulfide S–S setup.[78, 79]

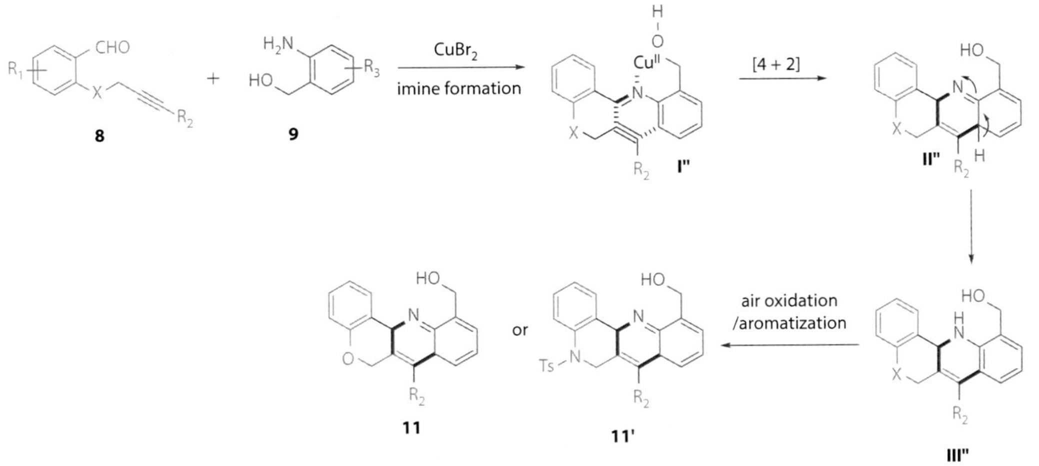

Scheme 13.8 $CuBr_2$-catalyzed diversity-oriented synthetic routes to various substituted 6h-chromeno[4,3-b] quinolines.

Qin *et al.*[80] in 2020 have discovered the first mechanical synthesis that demonstrates an electrophilic C-H sulfenylation of indoles under iodine catalysis. They have reported the first electrophilic C–H sulfenylation by amalgamation of iodine catalysis and ball milling activation that has not only enhanced S–S bond polarity but also promoted an aerobic oxidation, which finally interprets the high atom-economic harness of the sulfur reagent under metal-free benevolent conditions.[80] This mechanochemistry had directed to the reactivity of disulfides concerning arene electrophilic addition. The unfunctionalized indole **13** (0.5 mmol) was taken as a model substrate and treated with dialkyl/diaryl disulfide **14** (0.25 mmol) under ball milling which was optimized using different catalysts with various solvents and additives where Iodine (I_2) with DMF and silica (SiO_2) gave the best results. The reaction conditions were applied to various substrates by varying indole and substituted groups of the disulfide substrates (Scheme 13.9). All of the substrates showed a great tolerance capacity toward the standard reaction conditions along with the mechanical force. Also, the facile sulfenylation of tryptophol produced the compound **15g** in a synthetically useful yield (67%).

The plausible mechanism for the reaction has been proposed (Scheme 13.10). Diaryldisulfide **14a** reacts with molecular iodine (I_2) that generated the reactive sulfenyl iodide (R_2SI). The reactive intermediate has led to the electrophilic substitution at C3 position of indole. This led to the release of the hydrogen iodide (HI) that is oxidized by molecular oxygen (O_2)[81] from the air that led to the regeneration of iodine. The important role in the above reaction is mainly of activation of mechanical force that lies in the polarization of S–S bonds and the advancement of aerobic oxidation of HI.

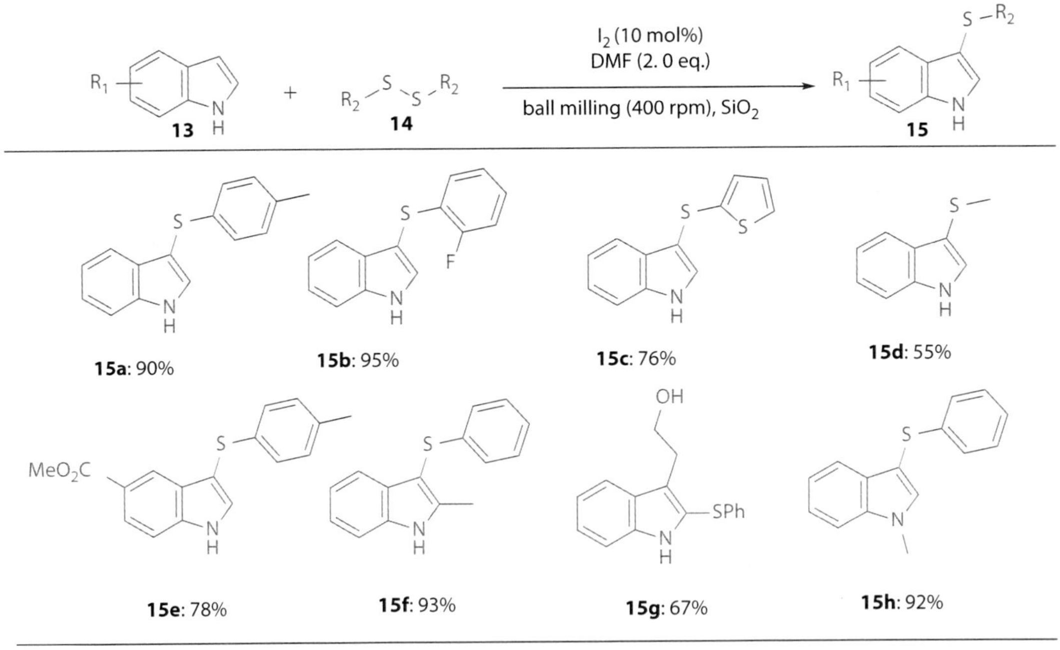

Scheme 13.9 Substrate scope with respect to disulfide and indole.

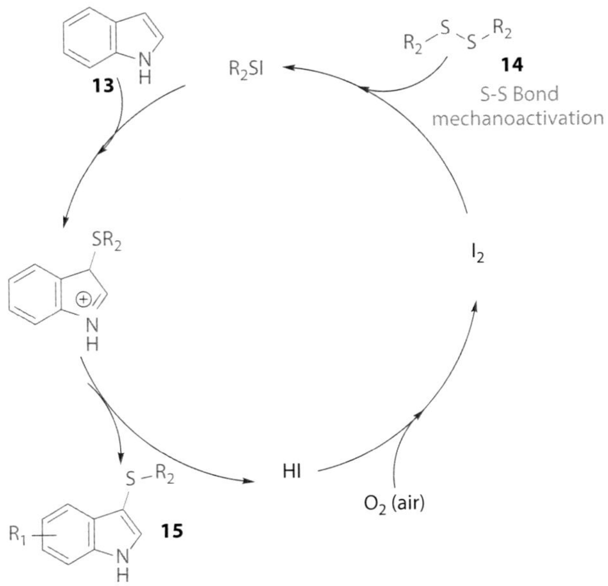

Scheme 13.10 Plausible reaction mechanism for the catalytic electrophilic C–H sulfenylation of indoles with disulfides.

13.4.1.3 Photochemical Green Approach

Photocatalysis has been lately advanced as an efficient and greener approach for the establishment of drug derivatization through visible light-mediated catalysis that is advancement within the synthetic methodology for the investigation of versatility and utility across the broad spectrum of the chemical structures. Hence, the green, alternative, safe, and efficient method for the photocatalysis was investigated over the high-energy UV light and the critical chemicals that provided a great catalytic platform for green synthesis.[81] There has been, nowadays, availability of new green methodologies based on photocatalysts that have been developed for their robustness on demonstrative functionalities of pharmaceutical compounds. This signifies an embrace and simple screening to study the effectiveness of a visible light-mediated process, eventually important to the superior approval of such types of transformations transversely the pharmaceutical and fine chemical industries.[82]

13.4.1.3.1 Photoredox Catalysis: Carbon-Carbon Bond Formation in the Presence of Organic Dye Rose Bengal

Xhang *et al.* disclosed an efficient and direct Rose Bengal (RB)-mediated conversion of different heterocycles to difluoro methylated heterocycles that have great scope as anticancerous agents[83]. This investigated methodology does not involve use of any poisonous or costly metal catalyst or an external oxidant that clearly indicates the advantages over the conventional methods. The authors have actively used Hu's reagent sodium difluoromethane sulfonate (CF_2HSO_2Na)[84, 85] that is used as a radical precursor for heterocycle functionalization in the presence of photocatalyst, i.e., Rose Bengal, to produce difluoroethylation product[86] (Scheme 13.11). The optimized reaction conditions revealed tremendous compatibility with an extensive range of various substituted heterocycles including pyridines,

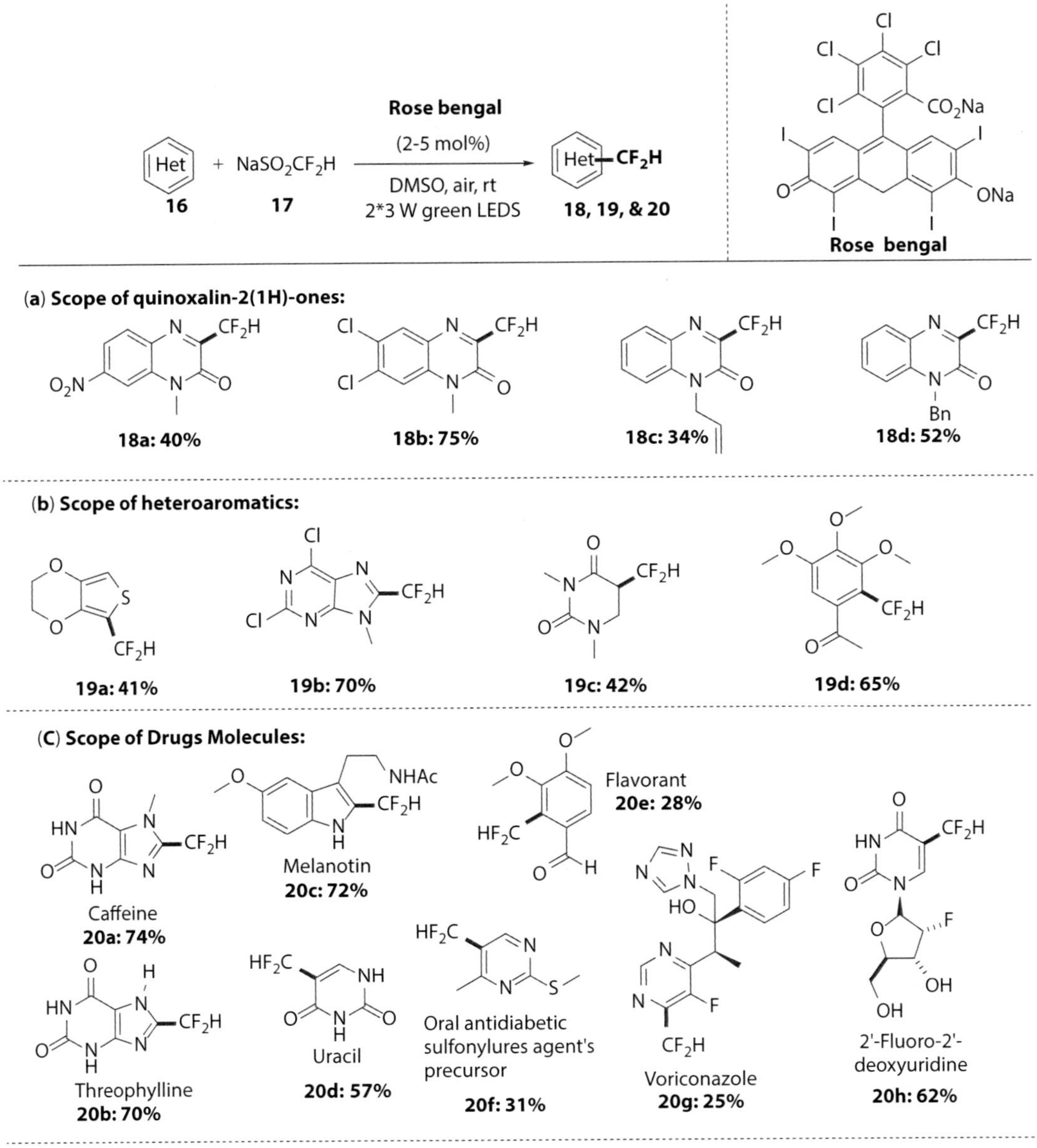

Scheme 13.11 Substrate scope of direct C-H difluoromethylation of quinoxaline-2(1H)-ones, hetero-aromatics, and late-stage functionalization of drug molecules.

pyrazoles, pyrazines, quioxalines, and more to be explored later. They have examined the substrate scope of bioactive molecules with optimized conditions and obtained quantitative yields of the desired products. The synthesis of different difluoromethylated caffeine derivatives, nucleoside derivatives was observed with good yields, along with these, several difluoromethylated drug compounds comprising free secondary -NH groups and other bioactive heteroarenes (Scheme 13.11). This methodology has been found to have great substrate scope that makes this transformation as a brilliant means for advancement of drug library.

The scope of the substrates was detailed investigated with the optimized conditions a variety of quinoxalin-2(1H)-ones, heterocycles, and bioactive drug molecules. The reactions went smoothly, and the desired products formed with quantitative yields. The methodology was well-tolerated with different functionalities such as hydroxyl (-OH), amine (-NH), or aldehyde (-CHO) with the reaction conditions. Moreover, the viability of the methodology was found to proceed well with voriconazole, uracil, melatonin, flavorant, and sulfonylureas, giving the difluoromethylated products with good yields.

The contributors of this protocol have also investigated control experiments to propose the plausible mechanism. The detailed investigation of the mechanism has been studied through a possible mechanism that was anticipated (Scheme 13.12). The absorption of visible light led the photocatalyst RB is excited into RB* and a single electron is transferred from CF_2HSO_2Na to RB*, which furnished CF_2H radical and generated an $RB^{\bullet -}$ radical anion. This photocatalytic cycle was finished with the oxidation of $RB^{\bullet -}$ in the presence of molecular oxygen along with the RB and $O_2^{\bullet -}$. Later, the addition of CF_2H radical to **18e** led to intermediate **A**, which underwent a 1,2-H shift to generate radical intermediate **B**. The loss of a hydrogen atom from the intermediate **B** to $O_2^{\bullet -}$ which afforded the desired product.

Therefore, they have developed a practical approach for difluoromethylation reactions with cheap reagent sodium difluoromethane sulfonate (CF_2HSO_2Na) that produces $^{\cdot}CF_2$ radical precursor under standard conditions. This reaction involves the catalytic system of CF_2HSO_2Na in DMSO with organic photocatalysis, *i.e.*, Rose Bengal (2–5 mol%), which gives a viable route for the preparation of different classes of heterocycles. Outstandingly, 2′-deoxy-5-difluoromethyluridine (F_2TDR) that is a difluoromethylation product significantly can inhibit the proliferation of cancer cells, such as HepG-2 and MCF-7.

13.4.1.3.2 Photoredox Catalysis: Carbon-Oxygen Bond Formation in the Presence of Organic Dye Eosin Y

In 2020, Inoa *et al.*[87] has reported a visible light-mediated Csp^3 –H activation that is a highly efficient benzylic hydroperoxidation. The reaction was found to be feasible using metal-free organic dye catalyst (eosin Y) and an economical blue LED with a molecular oxygen as an oxidant. The extensive investigation of various metal and metal-free catalysts, eosin Y was found to be suitable to serve the purpose.[88] Zhao and Wang[89] have already explored and reported that the neutral eosin Y-derived photoexcited states can be served as photoacids.

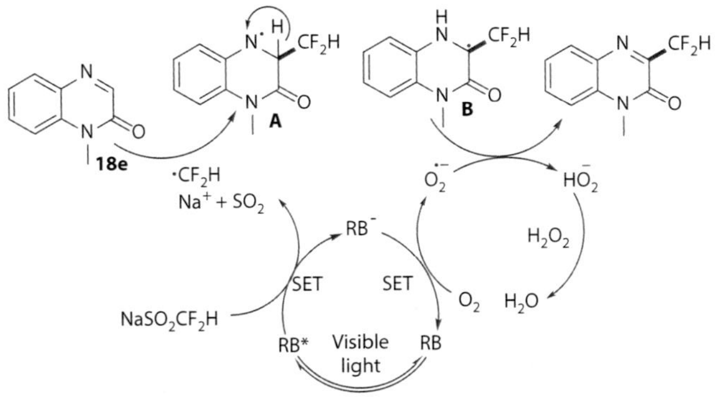

Scheme 13.12 Mechanism of direct C-H difluoromethylation of heterocycles.

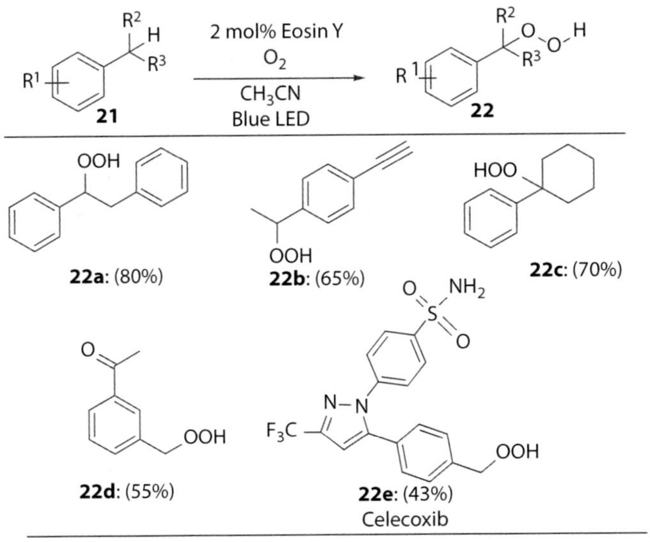

Scheme 13.13 Substrate scope of benzylic hydroperoxidation.

It was also believed that the reaction undergoes through direct HAT catalysts.[90] Also, it was observed that a range of 1°, 2°, and 3° hydroperoxides as well as benzyl, silyl, and acyl peroxides were smoothly synthesized with optimum yields, and great functionality tolerance in substrates was also found (Scheme 13.13). Remarkably, the approach demonstrated a wide approach for late-stage functionalization for drug derivatization, for example, modification of the benzylic C–H bond on the anti-inflammatory drug, celecoxib[91] (**22e**), has been reported.

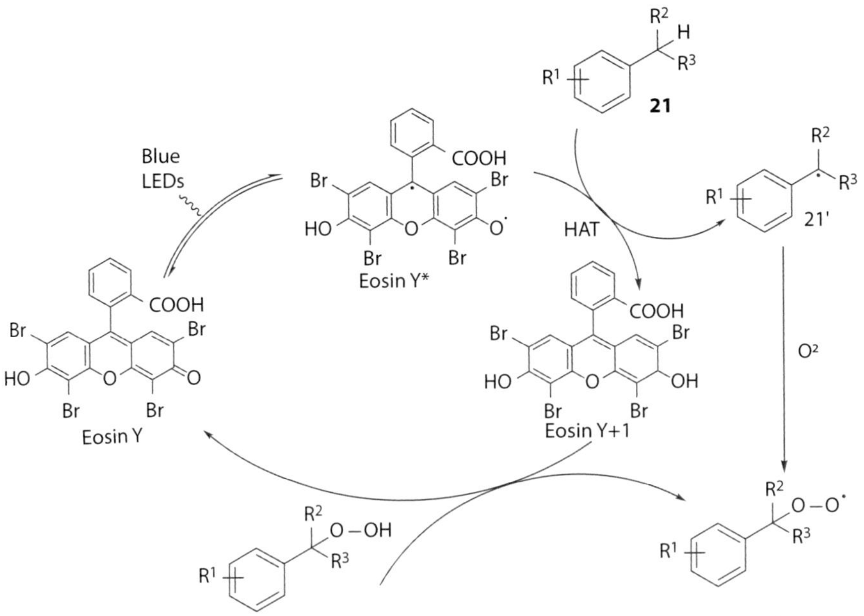

Scheme 13.14 Mechanism of benzylic hydroperoxidation.

The peroxides containing the hydroperoxides and endoperoxides are present in different natural compounds that are isolated from plants, fungi, endophytes, and other organisms.[92] Such compounds possess antibacterial, antimalarial, and cytotoxic also.[93] Especially the hydroperoxides have been found to have significant usefulness in materials science chemistry also.

The description of the mechanism was mentioned in Scheme 13.14. Eosin Y goes to the excited state eosin Y* upon light irradiation led to the removal of a hydrogen atom from the respective benzylic C–H bond to furnish a benzylic radical **21** with eosin Y-H. The molecular oxygen then traps radical **21'** to produce a peroxy radical. Finally, a retro-HAT from eosin Y–H to peroxy radical forms hydroperoxide product **22** and reestablishes eosin Y into the catalytic cycle. After Wu's development, a great advancement toward the Eosin Y as a direct HAT catalyst has been published in the work.[94]

13.5 Preventing the Usage of Toxic Chemicals, Use of Alternative Chemicals

Since the last decade, researchers have established approaches that are known as alternatives valuation to produce a wide-ranging method to recognizing better chemicals, resources, and developments. Experts use the alternatives valuation procedure to assist and circumvent "regrettable substitutions"—the auxiliary of an identified noxious chemical with alternative that verifies to be correspondingly or more harmful to human health or the environment. For example, the chemical used in butter for flavoring the microwave popcorn is diacetyl that disables and causes irreversible lung disease in employees. Therefore, an alternative chemical was used to substitute the diacetyl to serve the humanity for the safety.[95]

13.5.1 Thiourea Dioxide Mediated Reduction of N-Nitroso Functionality

Chaudhary *et al.*[96] recently reported the reduction of aryl-N-nitrosamines to respective hydrazines under metal-free condition using reductant thiourea dioxide (TDO) under aqueous medium to provide high yields of corresponding products. Thiourea dioxide could be prepared from thiourea that is a very stable crystalline material using hydrogen peroxide (Scheme 13.15).[97] Thiourea dioxide is available in a cheaper price commercially, which could be handled and stored well. Additionally, the reduction of TDO produces only harmless by-products such as urea and sodium sulfite that is one of its important advantages. TDO is eco-friendly in nature, and it has been used as a reducing agent in the paper, textile, and leather industries.[98] The researchers had the aim focused for the development of

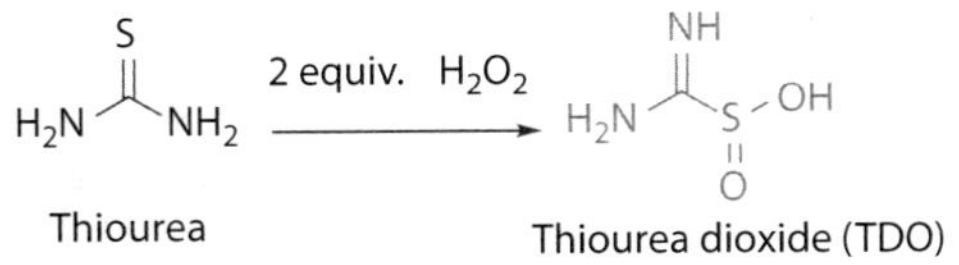

23

Scheme 13.15 Preparation of thiourea dioxide.

green and efficient approaches for organic synthesis. Herein, they described the practical approach under metal-free condition in aqueous medium for the reduction of aryl-N-nitrosamines into desired hydrazines using thiourea dioxide (TDO).[96] It was the first green approach revealed for the reduction of aryl-N-nitrosamines.

The authors described the synthesis of N-nitrosamines that undergo reaction under the standard conditions of TDO, aqueous solution of sodium hydroxide at 50°C. The reactions went through reduction smoothly without any electronic effect that provided the corresponding hydrazines with excellent yields. Later, under the optimized conditions, substrates such as N-nitroso-N-benzylanilines, N-methyl/ethyl/iso-propyl (or in general, alkyl groups), and phenyl anilines were also efficaciously giving the corresponding hydrazines in excellent yields (Scheme 13.16). All of the synthesized hydrazines were found to be stable in the alkaline medium.

The reduction plausible mechanism of TDO that was proposed for the reduction of N-nitrosamine is shown in Scheme 13.17. Initially, the TDO reacts with NaOH that gives sulfoxylic acid, and it has a strong role in the reduction reaction (Scheme 13.17, Step 1).[98, 99] The formed acid undergoes addition reaction with nitroso moiety (-NO) to furnish an intermediate **A** with the release the sulfur dioxide (SO_2). Then, intermediate **A** is heated to provide a partially reduced hydroxylamine derivative **B** (Scheme 13.17, Step 2). Later, one more molecule of sulfoxylic acid reacts with intermediate **B** that would furnish the respective hydrazine through intermediate **C** as mentioned in Scheme 13.17.

Later, the authors had established methodology for the synthesis of α-substituted aryl hydrazines from desired N-nitroso compounds using industrial green reductant thiourea

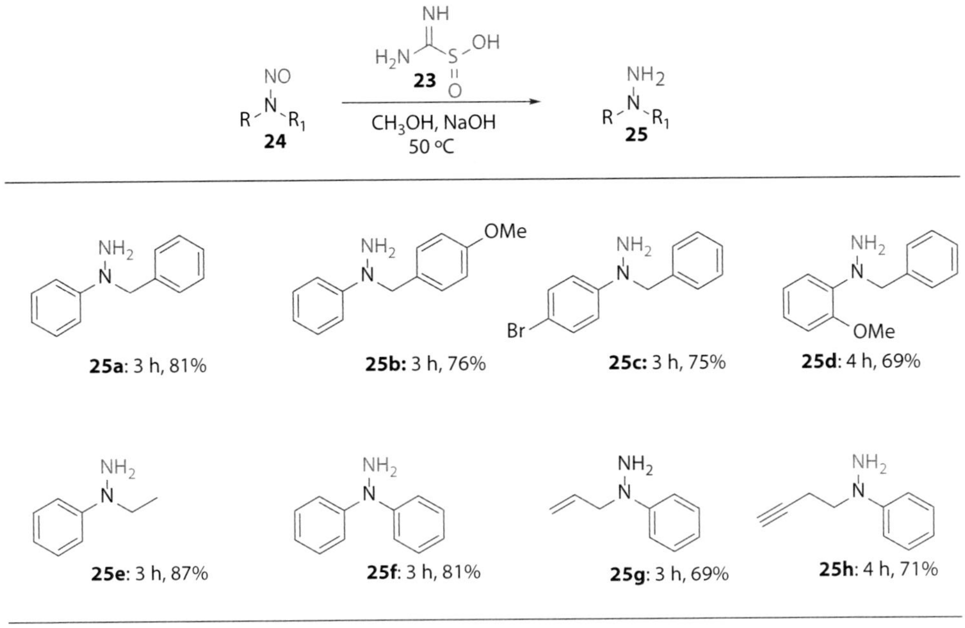

Scheme 13.16 Reduction of N-nitrosamines using TDO.

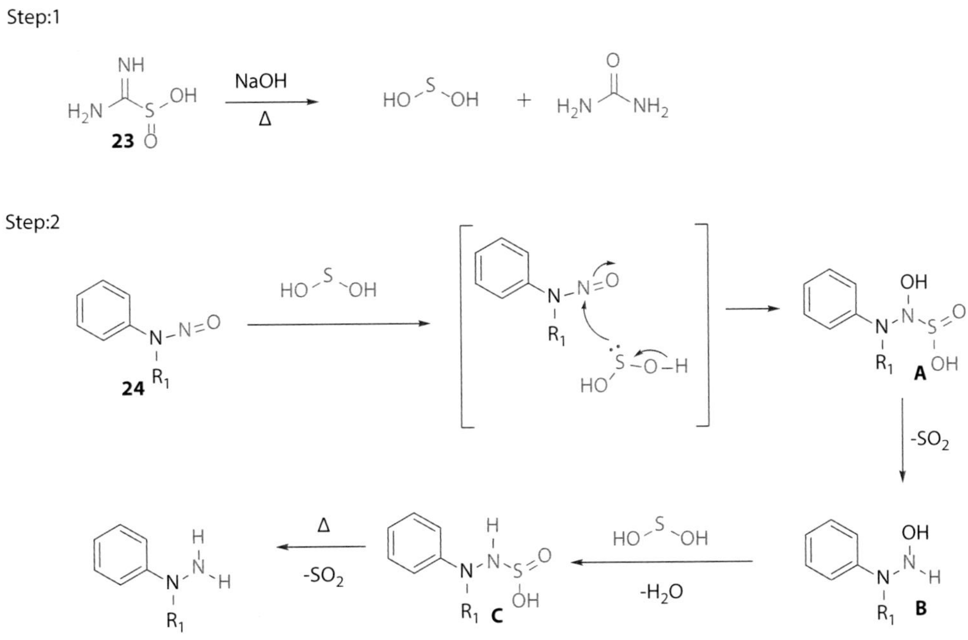

Scheme 13.17 The proposed mechanism for the reduction of N-nitrosamines using TDO.

dioxide. TDO was found to be the most efficient sulfur-containing reducing agent and also delivers a high to excellent yields of the products. Additionally, all of the functional groups such as aryl halides, olefin, and alkyne were tolerant under the reduction reaction. Harmless and mild reagent, suitable procedure, and high yield made the protocol as striking for the syntheses of aryl hydrazines from easy reactants.

Conclusion

Overall, there are many benefits of following the principles of green chemistry that validate its power and potential that would be highly beneficial to the living beings and the sustainability of the environment. The nascent research of the green chemistry has already begun, and the encouraging results in the research have been always inspiring for the scientific community and industrial realms. This concept has been exploring every phase and scope of the potential accomplishments in green chemistry. Nowadays, the perceptions have been vividly and unambiguously established and concluded the progression of current years. The word-"green" shows an implication of the new, fresh that basically related to the environment. The guidelines of Green Chemistry is the strong foundation for different innovative ideas applicable in synthetic chemistry followed by scientists to succeed in present and also, has promising future. The discipline of green chemistry demonstrated that the innovation of thorough experts of each region are producing then interesting and new chemistry that commands the drive of this transformation in theory and practice as well.

References

1. (a) Anastas, P.T. and Warner, J.C., *Green Chemistry: Theory and Practice*, Oxford University Press, Oxford, 1998. (b) Trost, B.M., The Atom Economy—A Search for Synthetic Efficiency, *Science*, 254, 5037, 1471–1477, 1991. (c) Wender, P.A. and Ternansky, R.J., Synthetic studies on arene-olepin cycloadditions-VII:1 a three-step total synthesis of (±)-silphinene, *Tetrahedron Lett.*, 26, 2625–2628, 1985. (d) Bertz, S.H. and Sommer, T.J., Rigorous mathematical approaches to strategic bonds and synthetic analysis based on conceptually simple new complexity indices, *Chem. Commun.*, 2409–2410, 1997.
2. Sarkar, A., Santra, S., Kundu, S.K., Hajra, A., Zyryanov, G.V., Chupakhin, O.N., Charushin, V.N., Majee, A., A decade update on solvent and catalyst-free neat organic reactions: a step forward towards sustainability, *Green Chem.*, 18, 4475–4525, 2016.
3. Gawande, M.B., Bonifácio, V.D.B., Luque, R., Branco, P.S., Varma, R.S., Solvent-free and catalysts-free chemistry: a benign pathway to sustainability, *ChemSusChem*, 7, 24–44, 2014.
4. Sheldon, R.A., Atom efficiency and catalysis in organic synthesis, *Pure Appl. Chem.*, 72, 12331246, 2000.
5. Constable, D.J.C., Curzons, A.D., Cunningham, V.L., Metrics to 'green' chemistry—which are the best?, *Green Chem.*, 4, 521527, 2002.
6. Dunn, P.J., Galvin, S., Hettenbach, K., The development of an environmentally benign synthesis of sildenafil citrate (Viagra™) and its assessment by Green Chemistry metrics, *Green Chem.*, 6, 4348, 2004.
7. Herrchen, M. and Klein, W., Use of the life-cycle assessment (LCA) toolbox for an environmental evaluation of production processes, *Pure Appl. Chem.*, 72, 12471252, 2000.
8. Debecker, D.P., Hii, K.K., Moores, A., Rossi, L.M., Sels, B., Allen, D.T., Subramaniam, B., Shaping Effective Practices for Incorporating Sustainability Assessment in Manuscripts Submitted to ACS Sustainable Chemistry & Engineering: Catalysis and Catalytic Processes. *ACS Sustain. Chem. Eng.*, 9, 14, 4936–4940, 2021.
9. Webster, K., A Circular Economy is about the Economy. *Circ. Econ. Sustain.*, 1, 115–126, 2021.
10. Pansera, M., The Origins and Purpose of Eco-Innovation. *Glob. Environ.*, 4, 128–155, 2016.
11. Erythropel, H.C., Zimmerman, J.B., de Winter, T.M., Petitjean, L., Melnikov, F., Lam, C.H., Lounsbury, A.W., Mellor, K.E., Jankovic, N.Z., Tu, Q.S., Pincus, L.N., Falinski, M.M., Shi, W.B., Coish, P., Plata, D.L., Anastas, P.T., The Green ChemisTREE: 20 years after taking root with the 12 principles. *Green Chem.*, 20, 9, 1929–1961, 2018.
12. Garrett, R.L., *Designing Safer Chemicals: Green Chemistry for Pollution Prevention*, S.C. DeVito, and R.L. Garrett, (Eds.), pp. 2–15, American Chemical Society, Washington, DC, 1996.
13. DeVito, S.C., DeVito, S.C., Garrett, R.L., *Designing Safer Chemicals: Green Chemistry for Pollution Prevention*, pp. 16–59, American Chemical Society, Washington, DC, 1996.
14. Yang, N., Sheng, X., Ti L., Jia, H., Ping, Qi., Li, N., Ball-milling as effective and economical process for biodiesel production under Kraft lignin activated carbon stabilized potassium carbonate, *Bioresour. Technol.*, 369, 128379, 2023.
15. Shearouse, Wi.C., Kortea C. M., Mack J., A two-step ball milling method synthesizes and purifies α,β-unsaturated esters, *Green Chem.*, 13, 598-601, 2011.
16. Williams, M. T. J., Morrill, L. C., Browne D. L., Mechanochemical Organocatalysis: Do High Enantioselectivities Contradict What We Might Expect?, 15(2), e202102157, 2022.
17. Tundo, P. and Selva, M., The Chemistry of Dimethyl Carbonate, *Acc. Chem. Res.*, 35, 706–716, 2002.
18. (a) Mc Ghee, W.D., Paster, M., Riley, D., Ruettimann, K., Solodar, J., Waldman, T., *Green Chemistry: Designing Chemistry for the Environment*, P.T. Anastas, and T.C. Williamson, (Eds.), American Chemical Society, Washington, DC, 1996. (b)Thomas, J.M., Raja, R., Lewis, D.W.,

Single-Site Heterogeneous Catalysts, *Angew. Chem., Int. Ed.*, 44, 64566482, 2005. (c) Thomas, J.M. and Raja, R., The advantages and future potential of single-site heterogeneous catalysts, *Top. Catal.*, 40, 3-17, 2006. (d) Bruggink, A., Schoevaart, R., Kieboom, T., Concepts of nature in organic synthesis: cascade catalysis and multistep conversions in concert, *Org. Process Res. Dev.*, 7, 622–640, 2003.

19. (a) Chauhan, P., Mahajan, S., Enders, D., Organocatalytic Carbon–Sulfur Bond-Forming Reactions. *Chem. Rev.*, 114, 8807, 2014. (b) Yu, J.-S., Huang, H.-M., Ding, P.-G., Hu, X.-S., Zhou, F., Zhou, J., Catalytic Enantioselective Construction of Sulfur-Containing Tetrasubstituted Carbon Stereocenters. *ACS Catal.*, 6, 5319, 2016. (c) Wang, W. and Zhang, X., C–X. Bond Formation: Organocatalytic α-Sulfenylation and α-Selenenylation, in: *Comprehensive Chirality*, vol. 6, p. 518, 2012. (d) Wladislaw, B., Marzorati, L., Di Vitta, C., Sulfanylation Reactions of Carbonyl Compounds and Carboxylic Acid Derivatives Employing Sulfur Electropfflles. A Review. *Org. Prep. Proced. Int.*, 39, 447, 2007. (e) Zhao, X., Shen, J., Jiang, Z., Catalytic Asymmetric α-Sulfenylation: A New and Efficient Pathway to Access Chiral C-S Bonds. *Mini-Rev. Org. Chem.*, 11, 424, 2014.
20. (a) Tanaka, K. and Fu, G.C., A Versatile New Catalyst for the Enantioselective Isomerization of Allylic Alcohols to Aldehydes: Scope and Mechanistic Studies. *J. Org. Chem.*, 66, 8177, 2001. (b) Mantilli, L., Gérard, D., Torche, S., Besnard, C., Mazet, C., Iridium-Catalyzed Asymmetric Isomerization of Primary Allylic Alcohols. *Angew. Chem., Int. Ed.*, 48, 5143, 2009. (c) Mantilli, L. and Mazet, C., Expanded Scope for the Iridium-Catalyzed Asymmetric Isomerization of Primary Allylic Alcohols Using Readily Accessible Second-Generation Catalysts. *Chem. Commun.*, 46, 445, 2010. (d) Quintard, A., Alexakis, A., Mazet, C., Access to High Levels of Molecular Complexity by One-Pot Iridium/Enamine Asymmetric Catalysis. *Angew. Chem., Int. Ed.*, 50, 2354, 2011. (e) Li, J.-Q., Peters, B., Andersson, P.G., Highly Enantioselective Asymmetric Isomerization of Primary Allylic Alcohols with an Iridium–N,P Complex. *Chem. - Eur. J.*, 17, 11143, 2011. (f) Biswas, S., Watile, R.A., Samec, J.S.M., Tandem Pd/Au-Catalyzed Route to α-Sulfenylated Carbonyl Compounds from Terminal Propargylic Alcohols and Thiols. *Chem. -Eur. J.*, 20, 2159–2163, 2014.
21. (a) Arai, N., Sato, K., Azuma, K., Ohkuma, T., Enantioselective Isomerization of Primary Allylic Alcohols into Chiral Aldehydes with the Tol-Binap/dbapen/Ruthenium(II) Catalyst. *Angew. Chem., Int. Ed.*, 52, 7500, 2013. (b) Cahard, D., Gaillard, S., Renaud, J.-L., Asymmetric Isomerization of Allylic Alcohols. *Tetrahedron Lett.*, 56, 6159, 2015.
22. (a) Liebeskind, L.S. and Srogl, J., Heteroaromatic Thioether– Boronic Acid Cross-Coupling under Neutral Reaction Conditions. *Org. Lett.*, 4, 979, 2002. (b) Arambasic, M., Majhail, M.K., Straker, R.N., Neuhaus, J.D., Willis, M.C., A Rhodium-Catalysed Sonogashira- Type Coupling Exploiting C–S Functionalisation: Orthogonality with Palladium-Catalysed Variants. *Chem. Commun.*, 55, 2757, 2019.
23. (a) Yus, M., Gutiérrez, A., Foubelo, F., Polylithiation of Thioethers: A Versatile Route for Polyanionic Synthons. *Tetrahedron*, 57, 4411, 2001. (b) Matsumoto, K., Kozuki, Y., Ashikari, Y., Suga, S., Kashimura, S., Yoshida, J., Electrophilic Substitution Reactions Using an Electrogenerated ArS(ArSSAr)+ Cation Pool As an ArS+ Equivalent. *Tetrahedron Lett.*, 53, 1916, 2012.
24. (a) Shirakawa, S., Tokuda, T., Kasai, A., Maruoka, K., Design of Chiral Bifunctional Quaternary Phosphonium Bromide Catalysts Possessing an Amide Moiet. *Org. Lett.*, 15, 3350, 2013. (b) Deng, Q.-H., Rettenmeier, C., Wadepohl, H., Gade, L.H., Copper–Boxmi Complexes as Highly Enantioselective Catalysts for Electrophilic Trifluoromethylthiolations. *Chem. - Eur. J.*, 20, 93, 2014. (c) Cui, L., You, Y., Mi, X., Luo, S., Catalytic Enantioselective α-Sulfenylation of β-Keto-carbonyls by Chiral Primary Amines. *Org. Chem. Front.*, 5, 2313, 2018.

25. (a) Wang, C., Yang, X., Loh, C.C.J., Raabe, G., Enders,Organocatalytic, D., Asymmetric Synthesis of 3-Sulfenylated N-Boc-Protected Oxindoles. *Chem. - Eur. J.*, 2012, 18, 11531. (b) Cui, L., You, Y., Mi, X., Luo, S., Catalytic Enantioselective α-Sulfenylation of β-Ketocarbonyls by Chiral Primary Amines. *Org. Chem. Front.*, 5, 2313, 2018. (c) Jiao, L., Bu, L., Ye, X., Zhao, X., Jiang, Z., Catalytic Asymmetric Conjugate Addition and Sulfenylation of Diarylthiazolidin-2,4-diones. *J. Org. Chem.*, 81, 9620, 2016.
26. Gupta, S., Chaudhary, P., Seva, L., Sabiah, S., Kandasamy, J., Bio-based green solvent for the catalyst free oxidation of arylboronic acids into phenols, *RSC Adv.*, 5, 89133–89138, 2015.
27. Namieoenik, J. and Wardencki, W., Solventless sample preparation techniques in environmental analysis. *J. High Resolut. Chromatogr.*, 23, 297-303, 2000.
28. (a) Adsul, M.G., Varma, A.J., Gokhale, D.V., Lactic acid production from waste sugarcane bagasse derived cellulose, *Green Chem.*, 9, 58–62, 2007. (b) Finkenstadt, V.L., Liu, L., Willett, J.L., Evaluation of poly (lactic acid) and sugar beet pulp green composites, *J. Polym. Environ.*, 15, 1–6, 2007.
29. Cintas, P. and Luche, J.-L., Green chemistry. The sonochemical approach, *Green Chem.*, 1, 115–125, 1999.
30. (a) Macquarrie, D.J., Nazih, R., Sebti, S., KF/natural phosphate as an efficient catalyst for synthesis of 2′-hydroxychalcones and flavanones, *Green Chem.*, 4, 56-59, 2002. (b) Raston, C.L. and Scott, J.L., *Green Chem.*, 2, 49–52, 2000.
31. McCarroll, A.J. and Walton, J.C., Chemoselective, solvent-free aldol condensation reaction, *Angew. Chem., Int. Ed.*, 40, 2224–2248, 2001.
32. Trost, B.M., Frederiksen, M.U., Rudd, M.T., Ruthenium-catalyzed reactions—a treasure trove of atom-economic transformations, *Angew. Chem., Int. Ed.*, 44, 66306666, 2005.
33. (a) Chisem, I.C., Chisem, J., Rafelt, J.S., Macquarrie, D.J., Clark, J.H., Utting, K.A., Environmentally friendly liquid phase oxidation: enhanced selectivity in the aerial oxidation of alkyl aromatics, epoxidations and the Baeyer–Villiger oxidation using novel silica supported transition metal ions, *J. Chem. Technol. Biotechnol.*, 74, 923–930, 1999. (b) McCarroll, A.J. and Walton, J.C., Programming organic molecules: Design and management of organic syntheses through free-radical cascade processes, *Angew. Chem., Int. Ed.*, 40, 2224–2248, 2001.
34. Banerjee, D., Markley, A.L., Yano, T., Ghosh, A., Berget, P.B., Minkley, E.G., Jr., Khetan, S.K., Collins, T.J., "Green" oxidation catalysis for rapid deactivation of bacterial spores, *Angew. Chem., Int. Ed.*, 45, 3974–3977, 2006.
35. Freire, M.G. and Coutinho, J.A.P., Editorial overview: a closer look on sustainable solvents and processes. *Curr. Opin. Green Sustain. Chem.*, 18, A4–A6, 2019.
36. Chaudhary, P., Gupta, S., Muniyappan, N., Sabiah, S., Kandasamy, J., An efficient synthesis of N-nitrosamines under solvent, metal and acid free conditions using tert-butyl nitrite, *Green Chem.*, 18, 2323–2330, 2016.
37. (a) Taniguchi, T., Yajima, A., Ishibashi, H., Oxidative Nitration of Alkenes with tert-Butyl Nitrite and Oxygen, *Adv. Synth. Catal.*, 353, 2643–2647, 2011. (b) Wu, X.F., Schranck, J., Neumann, H., Beller, M., Convenient and mild synthesis of nitroarenes by metal-free nitration of arylboronic acids, *Chem. Commun.*, *47*, 12462–12463, 2011. (c) Kilpatrick, B., Heller, M., Arns, S., Chemoselective nitration of aromatic sulfonamides with tert-butyl nitrite, *Chem. Commun.*, 49, 514–516, 2013. (d) Maity, S., Naveen, T., Sharma, U., Maiti, D., Stereoselective Nitration of Olefins with tBuONO and TEMPO: Direct Access to Nitroolefins under Metal-free Conditions *Org. Lett.*, 15, 3384–3387, 2013. (e) Manna, S., Jana, S., Saboo, T., Maji, A., Maiti, D., Synthesis of (E)-nitroolefins via decarboxylative nitration using t-butylnitrite (t-BuONO) and TEMPO, *Chem. Commun.*, 49, 5286–5288, 2013. (f) Liu, Y., Zhang, J.L., Song, R.J., Qian, P.C., Li, J.H., Cascade Nitration/Cyclization of 1,7-Enynes with tBuONO and H2O: One-Pot Self-Assembly of Pyrrolo[4,3,2-de]quinolinones, *Angew. Chem. Int. Ed. Engl.*, 53, 9017–9020, 2014.

(g) Majhi, B., Kundu, D., Ahammed, S., Ranu, B.C., tert-Butyl Nitrite Mediated Regiospecific Nitration of (E)-Azoarenes through Palladium-Catalyzed Directed C-H Activation, *Chem-Eur. J.*, 20, 9862–9866, 2014. (h) Shen, T., Yuan, Y.Z., Jiao, N., Metal-free nitro-carbocyclization of activated alkenes: a direct approach to synthesize oxindoles by cascade C–N and C–C bond formation, *Chem. Commun.*, 50, 554–556, 2014. (i) Deng, G.B., Zhang, J.L., Liu, Y.Y., Liu, B., Yang, X.H., Li, J.H., Metal-free nitrative cyclization of N-aryl imines with tert-butyl nitrite: dehydrogenative access to 3-nitroindoles, *Chem. Commun.*, 51, 1886–1888, 2015. (j) Hao, X.H., Gao, P., Song, X.R., Qiu, Y.F., Jin, D.P., Liu, X.Y., Liang, Y.M., Metal-free nitro-carbocyclization of 1, 6-enynes with t BuONO and TEMPO, *Chem. Commun.*, 51, 6839–6842, 2015. (k) Ji, Y.F., Yan, H., Jiang, Q.B., Effective Nitration of Anilides and Acrylamides by tert-Butyl Nitrite, *Eur. J. Org. Chem.*, 2015, 9, 2051–2060, 2015. (l) Yang, X.H., Ouyang, X.H., Wei, W.T., Song, R.J., Li, J.H., Nitrative Spirocyclization Mediated by TEMPO: Synthesis of Nitrated Spirocycles from N-Arylpropiolamides, tert-Butyl Nitrite and Water, *Adv. Synth. Catal.*, 357, 1161–1166, 2015. (m) Koley, D., Colon, O.C., Savinov, S.N., Chemoselective Nitration of Phenols with tert-Butyl Nitrite in Solution and on Solid Support, *Org. Lett.*, 11, 4172–4175, 2009. (n) Dutta, U., Maity, S., Kancherla, R., Maiti, D., Aerobic Oxynitration of Alkynes with tBuONO and TEMPO, *Org. Lett.*, 16, 6302–6305, 2014.

38. Murphy, J.A., Free radicals in synthesis. Clean reagents affording oxidative or reductive termination, *Pure Appl. Chem.*, 72, 1327–1334, 2000.
39. Li, M.-Y., Li, J., Gu, A., Nong, X.-M., Zhai, S., Yue, Z.-Y., Feng, C.-G., Liu, Y., Lin, G.-Q., Solvent-free and catalyst-free direct alkylation of alkenes, *Green Chem.*, 25, 7073–7078, 2023.
40. (a) A decade update on solvent and catalyst-free neat organic reactions: a step forward towards sustainability, Sarkar, A., Santra, S., Kundu, S.K., Hajra, A., Zyryanov, G.V., Chupakhin, O.N., Charushin, V.N., Majee, A., *Green Chem.*, 18, 4475–4525, 2016. (b) Gawande, M.B., Bonifácio, V.D.B., Luque, R., Branco, P.S., Varma, R.S., Solvent-free and catalysts-free chemistry: a benign pathway to sustainability, *ChemSusChem*, 7, 24–44, 2014.
41. Rossi, R.A., Pierini, A.B., Peñéñory, A.B., Nucleophilic substitution reactions by electron transfer, *Chem. Rev.*, 103, 71–168, 2003.
42. Schaub, T., Efficient Industrial Organic Synthesis and the Principles of Green Chemistry. *Chem. Eur. J.*, 27, 6, 1865–1869, 2021.
43. (a) Abraham, M.H., *Tetrahedron Lett.*, 11, 5233–5236, 1970. Solvent effects on the free energies of ion-pairs, and of transition states in an SN1 and an SN2 reaction, (b) Yue, H.-L., Wei, W., Li, M.-M., Yang, Y.-R., Ji, J.-X., sp3-sp2 C-C Bond Formation via Brønsted Acid Trifluoromethanesulfonic Acid-Catalyzed Direct Coupling Reaction of Alcohols and Alkenes, *Adv. Synth. Catal.*, 353, 3139–3145, 2011.
44. Silvestri, R., Artico, M., De Martino, G., Ragno, R., Massa, S., Loddo, R., Murgioni, C., Loi, A.G., La Colla, P., Pani, A., Synthesis, Biological Evaluation, and Binding Mode of Novel 1-[2-(Diarylmethoxy)ethyl]-2-methyl-5-nitroimidazoles Targeted at the HIV-1 Reverse Transcriptase, *J. Med. Chem.*, 45, 1567–1576, 2002.
45. (a) Martina, K., Cravotto, G., Varma, R.S., Impact of Microwaves on Organic Synthesis and Strategies toward Flow Processes and Scaling Up, *J. Org. Chem.*, 86, 20, 13857–13872, 2021. (b) Kappe, C.O., Controlled Microwave Heating in Modern Organic Synthesis, *Angew. Chem. Int. Ed.*, 43, 46, 6250–6284, 2004.
46. (a) Adhikari, A., Bhakta, S., Ghosh, T., Synthesis and characterisation of an integratively self-sorted [Fe4L6]8+ tetrahedron, *Tetrahedron*, *126*, 133085–133089, 2022. (b) Kakati, D. and Sarma, J.C., Microwave assisted solvent free synthesis of 1,3-diphenylpropenones, *Chem. Cent J.*, 5, 8, 2011, doi: 10.1186/1752-153X-5-8. (c) Mukku, N., Davanagere, P.M., Chanda, K., Maiti, B., [CMMIM][BF4–] Ionic Liquid-Catalyzed Facile, One-Pot Synthesis of Chromeno[4,3-d]

pyrido[1,2-a]pyrimidin-6-ones: Evaluation of Their Photophysical Properties and Theoretical Calculations, *ACS Omega*, 5, 43, 28239–28248, 2020.
47. (a) Souza, F.B., Pimenta, D.C., Stefani, H.A., Microwave-assisted one-pot three-component synthesis of imine 1,2,3-triazoles, *Tetrahedron Lett.*, 57, 14, 1592–1596, 2016.
(b) Das, S., Das, V.K., Saikia, L., Thakur, A.J., Nano rod-shaped and reusable basic Al2O3catalyst for N-formylation of amines under solvent-free conditions: A novel, practical and convenient 'NOSE' approach, *Green Chem. Lett. Rev.*, 5, 3, 457–474, 2012.
48. Varma, R.S., Dahiya, R., Kumar, S., Clay catalyzed synthesis of imines and enamines under solvent-free conditions using microwave irradiation, *Tetrahedron Lett.*, 38, 12, 2039–2042, 1997.
49. Varma, R.S., Dahiya, R., Kumar, S., Microwave-assisted Henry reaction: Solventless synthesis of conjugated nitroalkenes, *Tetrahedron Lett.*, 38, 29, 5131–5134, 1997.
50. Pârvulescu, V., II and Hardacre, C., Catalysis in Ionic Liquids, *Chem. Rev.*, 107, 6, 2615–2665, 2007.
51. Gawande, M.B., Shelke, S.N., Zboril, R., Varma, R.S., Microwave-assisted chemistry: synthetic applications for rapid assembly of nanomaterials and organics, *Acc. Chem. Res.*, *47*, 4, 1338–1348, 2014.
52. (a) Jua, Y. and Varma, R.S., Aqueous N-alkylation of amines using alkyl halides: direct generation of tertiary amines under microwave irradiation, *Green Chem.*, 6, 219–221, 2004. (b) Rathi, A.K., Gawande, M.B., Zboril, R., Varma, R.S., Microwave-assisted synthesis – Catalytic applications in aqueous media, *Coord. Chem. Rev.*, 291, 68–94, 2015.
53. Newman, D.J. and Cragg, G.M., Natural Products as Sources of New Drugs over the Nearly Four Decades from 01/1981 to 09/2019, *J. Nat. Prod.*, 83, 3, 770–803, 2020.
54. David, B., Wolfender, J.-L., Dias, D.A., The pharmaceutical industry and natural products: historical status and new trends, *Phytochem. Rev.*, 14, 299–315, 2014.
55. Varma, R.S., Green Chemistry with Microwave Energy, in: *Encyclopedia of Sustainability Science and Technology*, 2012, ISBN : 978-0-387-89469-0.
56. Kim, D.Y., Dao, P.D.Q., Cho, C.S., Recyclable Magnetic Cu-MOF-74-Catalyzed C(sp2)-N Coupling and Cyclization under Microwave Irradiation: Synthesis of Imidazo[1,2-c]quinazolines and Their Analogues, *ACS Omega*, 3, 12, 17456–17465, 2018.
57. Mal, S. and Jana, M., Recent Advances on Synthesis of Fused Chromeno-Quinoline Heterocycles under Transition Metal-Free Condition, *ChemistrySelect*, 7, 6, e202103525, 2022.
58. Devadoss, T., Sowmya, V., Bastati, R., Synthesis of 1,6-Naphthyridine and Its Derivatives: A Systematic Review, *ChemistrySelect*, 6, 15, 3610–3641, 2021.
59. Muthukrishnan, I., Karuppasamy, M., Vachan, B.S., Rajput, D., Subbiah, N., Maheswari, U., Sridharan, V., Chemodivergent synthesis of functionalized methanodibenzo[b,f][1,5]diazocin-13-ylmethanones and tetrahydroquinolines via solvent-dependent AB2 and A2B2 multicomponent annulation reactions, *Org. Chem. Front.*, 7, 1616–1625, 2020.
60. Galloway, W., Isidro-Llobet, A., Spring, D., Diversity-oriented synthesis as a tool for the discovery of novel biologically active small molecules, *Nat. Commun.*, 1, 80, 2010.
61. Zhang, Y., Liu, J., Wang, X.-S., An efficient synthesis of 16H-dibenzo[2,3:6,7][1,4]oxazepino[5,4-b]quinazolin-16-ones via an Ullmann reaction catalyzed by CuI, *Org. Biomol. Chem.*, 16, 10, 1679–1685, 2018.
62. Huang, K.H., Veal, J.M., Fadden, R.P., Rice, J.W., Eaves, J., Strachan, J.-P., Barabasz, A.F., Foley, B.E., Barta, T.E., Wei. Ma, M.A., Silinski, M., Hu, J.M., Partridge, A., Scott, L.G., DuBois, T., Freed, P.M., Steed, A.J., Ommen, E.D., Smith, P.F., Hughes, A.R., Woodward, G.J., Hanson, W.S., Mc Call, C.J., Markworth, L., Hinkley, M., Jenks, L., Geng, M., Lewis, J., Otto, B., Pronk, K., Verleysen, Hall, S.E., A Systematic Protocol for the Characterization of Hsp90 Modulators, *J. Med. Chem.*, 52, 14, 4288–4305, 2009.

63. (a) Qing, W., Chunxia, J., Jianxin, G., Wenxue, G., 1H NMR and 13C NMR Studies of Oil from Pyrolysis of Indonesian Oil Sands, *Energy Fuels*, 30, 3, 2478–2491, 2016. (b) Fallot, L.S.B., Suresh, R.R., Fisher, C.L., Salmaso, V., O'Connor, R.D., Kaufman, N., Gao, Z.-G., Auchampach, J.A., Jacobson, K.A., 1H-Imidazo[4,5-c]quinolin-4-amines: novel non-xanthine adenosine antagonists, *J. Med. Chem.*, 65, 22, 15238–15262, 2022.
64. Zhang, J.-P., Liao, P.-Q., Zhou, H.-L., Lina, R.-B., Chen, X.-M., Single-crystal X-ray diffraction studies on structural transformations of porous coordination polymers, *Chem. Soc Rev.*, 43, 5789–5814, 2014.
65. Do, J.-L. and Friščić, T., Mechanochemistry: A Force of Synthesis, *ACS Cent. Sci.*, 3, 1, 13–19, 2017.
66. Wang, G.-W., Mechanochemical organic synthesis, *Chem. Soc Rev.*, 42, 7668–7700, 2013.
67. Mateti, S., Mathesh, M., Liu, Z., Tao, T., Ramireddy, T., Glushenkov, A.M., Yang, W., Chen, Y., II, Mechanochemistry: A force in disguise and conditional effects towards chemical reactions, *Chem. Commun.*, 57, 1080–1092, 2021.
68. Covalent Organic Frameworks: Chemistry beyond the Structure, Kandambeth, S., Dey, K., Banerjee, R., *J. Am. Chem. Soc*, 141, 5, 1807–1822, 2019.
69. Yu, J., Gaedke, M., Schaufelberger, F., Dynamic Covalent Chemistry for Synthesis and Co-conformational Control of Mechanically Interlocked Molecules, *Eur. J. Chem.*, 26, 8, e202201130, 2023.
70. Neuman, K. and Nagy, A., Single-molecule force spectroscopy: optical tweezers, magnetic tweezers, and atomic force microscopy. *Nat. Methods*, 5, 491–505, 2008. https://doi.org/10.1038/nmeth.1218.
71. Ingenbosch, K.N., Vieyto-Nuñez, J.C., Ruiz-Blanco, Y.B., Mayer, C., Hoffmann-Jacobsen, K., Sanchez-Garcia, E., Effect of Organic Solvents on the Structure and Activity of a Minimal Lipase, *J. Org. Chem.*, 87, 3, 1669–1678, 2022.
72. Piras, C. C., Fernández-Prieto, S., De Borggraeve, W. M., Ball milling: a green technology for the preparation and functionalisation of nanocellulose derivatives, *Nanoscale Adv.*, 2019, 1, 937-947.
73. Shuai, L. and Luterbacher, J., Organic Solvent Effects in Biomass Conversion Reactions, *ChemSusChem*, 9, 2, 133–155, 2016.
74. Wiita, A.P., Perez-Jimenez, R., Walther, K.A., Grater, F., Berne, B.J., Holmgren, A., Sanchez-Ruiz, J.M., Fernandez, J.M., Probing the chemistry of thioredoxin catalysis with force, *Nature*, 450, 124–127, 2007.
75. Ainavarapu, S.R.K., Wiita, A.P., Dougan, L., Uggerud, E., Fernandez, J.M., Single-molecule force spectroscopy measurements of bond elongation during a bimolecular reaction, *J. Am. Chem. Soc*, 130, 6479–6487, 2008.
76. Garcia-Manyes, S., Liang, J., Szoszkiewicz, R., Kuo, T.L., Fernandez, J.M., Quaternary stereocentres via catalytic enantioconvergent nucleophilic substitution reactions of tertiary alkyl halides, *Nat. Chem.*, 1, 236–242, 2009.
77. Iozzi, M.F., Helgaker, T., Uggerud, E., Influence of external force on properties and reactivity of disulfide bonds., *J. Phys. Chem. A*, 115, 2308–2315, 2011.
78. Popa, I., Kosuri, P., Alegre-Cebollada, J., Garcia-Manyes, S., Fernandez, J.M., Force dependency of biochemical reactions measured by single molecule force-clamp spectroscopy, *Nat. Protoc.*, 8, 1261–1276, 2013.
79. (a) Dopieralski, P., Ribas-Arino, J., Anjukandi, P., Krupicka, M., Marx, D., Unexpected mechanochemical complexity in the mechanistic scenarios of disulfide bond reduction in alkaline solution, *Nat. Chem.*, 9, 164–170, 2017. (b) Dopieralski, P., Ribas-Arino, J., Anjukandi, P., Krupicka, M., Kiss, J., Marx, D., The Janus-faced role of external forces in mechanochemical disulfide bond cleavage, *Nat. Chem.*, 5, 685–691, 2013.

80. Qin, J., Zuo, H., Ni, Y., Yu, Q., Zhong, F., Catalytic Electrophilic C–H Sulfenylation of Indoles with Disulfides under Ball Milling, *ACS Sustain. Chem. Eng.*, 8, 33, 12342–12347, 2020.
81. Holleman, A.F. and Wiberg, E., *Inorganic Chemistry*, vol. 371, pp. 432–433, Academic Press, San Diego, 2001.
82. (a) Ji, P., Zhang, Y., Wei, Y., Huang, H., Hu, W., Mariano, P.A., Wang, W., Visible-Light-Mediated, Chemo- and Stereoselective Radical Process for the Synthesis of C-Glycoamino Acids. *Org. Lett.*, 21, 9, 3086–3092, 2019. (b) Cuadros, S., Horwitz, M.A., Schweitzer-Chaput, B., Melchiorre, P., A visible-light mediated three-component radical process using dithiocarbamate anion catalysis. *Chem. Sci.*, 10, 5484–5488, 2019.
83. Zhang, W., Xiang, X.X., Chen, J. *et al.*, Direct C–H difluoromethylation of heterocycles via organic photoredox catalysis. *Nat. Commun.*, 11, 638, 2020.
84. He, Z., Tan, P., Ni, C., Hu, J., Fluoroalkylative aryl migration of conjugated N-arylsulfonylated amides using easily accessible sodium Di- and monofluoroalkanesulfinates. *Org. Lett.*, 17, 1838–1841, 2015.
85. (a) Sakamoto, R., Kashiwagi, H., Maruoka, K., The direct C–H difluoromethylation of heteroarenes based on the photolysis of hypervalent iodine(III) reagents that contain difluoroacetoxy ligands. *Org. Lett.*, 19, 5126–5129, 2017. (b) Tung, T.T., Christensen, S.B., Nielsen, J., Difluoroacetic acid as a new reagent for direct C-H difluoromethylation of heteroaromatic compounds. *Chemistry*, 23, 18125–18128, 2017. (c) Fujiwara, Y. *et al.*, A new reagent for direct difluoromethylation. *J. Am. Chem. Soc*, *134*, 1494–1497, 2012.
86. (a) Sessler, C.D. *et al.*, CF2H, a hydrogen bond donor. *J. Am. Chem. Soc*, 139, 9325–9332, 2017.
87. Inoa, J., Patel, M., Dominici, G., Eldabagh, R., Patel, A., Lee, J., Xing, Y., Benzylic Hydroperoxidation via Visible-Light-Induced Csp3–H Activation. *J. Org. Chem.*, 85, 6181–6187, 2020.
88. (a) Hari, D.P. and Konig, B., Synthetic applications of eosin Y inphotoredox catalysis. *Chem. Commun.*, 50, 6688–6699, 2014. (b) Srivastava, V. and Singh, P.P., Eosin Y catalysed photoredox synthesis: a review. *RSC Adv.*, 7, 31377–31392, 2017. (c) Yan, D.M., Chen, J.R., Xiao, W.J., New Roles for Photoexcited Eosin Y in Photochemical Reactions. *Angew. Chem., Int. Ed.*, 58, 378–380, 2019.
89. Zhao, G. and Wang, T., Stereoselective Synthesis of 2-Deoxyglycosides from Glycals by Visible-Light-Induced Photoacid Catalysis. *Angew. Chem., Int. Ed.*, 57, 6120–6124, 2018.
90. Fan, X., Rong, J., Wu, H., Zhou, Q., Deng, H., Tan, J., Xue, C., Wu, L., Tao, H., Wu, J., Eosin Y as a Direct Hydrogen-Atom Transfer Photocatalyst for the Functionalization of C-H Bonds. *Angew. Chem., Int. Ed.*, 57, 8514–8518, 2018.
91. (a) Scholtz, C. and Riley, D.L., Improved batch, and flow syntheses of the nonsteroidal anti-inflammatory COX-2 inhibitor celecoxib. *React. Chem. Eng.*, 6, 138–146, 2021. (b) Dembitsky, V.M., Astonishing Diversity of Natural Peroxides as Potential Therapeutic Agents. *J. Mol. Genet. Med.*, 9, 1, 2014. (c) Norris, M.D. and Perkins, M.V., Structural diversity and chemical synthesis of peroxide and peroxide-derived polyketide metabolites from marine sponges. *Nat. Prod. Rep.*, 33, 861–880, 2016.
92. (a) Cheng, J.-K. and Loh, T.-P., Copper-and Cobalt-Catalyzed Direct Coupling of sp3 α-Carbon of Alcohols with Alkenes and Hydroperoxides. *J. Am. Chem. Soc*, 137, 42–45, 2015. (b) Kong, D.-L., Cheng, L., Yue, T., Wu, H.-R., Feng, W.-C., Wang, D., Liu, L., Cobalt-Catalyzed Peroxidation of 2-Oxindoles with Hydroperoxides. *J. Org. Chem.*, 81, 5337–5344, 2016.
93. Chaudhari, M.B., Moorthy, S., Patil, S., Bisht, G.S., Mohamed, H., Basu, S., Gnanaprakasam, B., Iron-Catalyzed Batch/Continuous Flow C-H Functionalization Module for the Synthesis of Anticancer Peroxides. *J. Org. Chem.*, 83, 1358–1368, 2018.

94. Fan, X., Rong, J., Wu, H., Zhou, Q., Deng, H., Tan, J., Xue, C., Wu, L., Tao, H., Wu, J., Eosin Y as a Direct Hydrogen-Atom Transfer Photocatalyst for the Functionalization of C-H Bonds. *Angew. Chem., Int. Ed.*, 57, 8514–8518, 2018.
95. (a) Chen, Q. and Beckman, E.J., One-step synthesis of N,N′-dialkyl-p-phenylenediamines, *Green Chem.*, 9, 802–808, 2007. (b) Danciu, T., Beckman, E.J., Hancu, D., Cochran, R.N., Grey, R., Hajnik, D.M., Jewson, J., Direct Synthesis of Propylene Oxide with CO2 as the Solvent, *Angew. Chem., Int. Ed.*, 42, 1140–1142, 2003.
96. Chaudhary, P., Gupta, S., Sureshbabu, P., Sabiah, S., Kandasamy, J., A metal free reduction of aryl-N-nitrosamines to the corresponding hydrazines using a sustainable reductant thiourea dioxide, *Green Chem.*, 18, 6215–6221, 2016.
97. Makarov, S.V., Horvath, A.K., Silaghi-Dumitrescu, R., Gao, Q., Recent Developments in the Chemistry of Thiourea Oxides, *Chem. Eur. J.*, 20, 14164–14176, 2014.
98. Gao, W.Y., Qi, W.J., Lai, J.P., Qi, L.M., Majeed, S., Xu, G.B., Thiourea dioxide as a unique eco-friendly coreactant for luminol chemiluminescence in the sensitive detection of luminol, thiourea dioxide and cobalt ions, *Chem. Commun.*, 51, 1620–1623, 2015.
99. Makarov, S.V., Recent trends in the chemistry of sulfur-containing reducing agents, *Russ. Chem. Rev.*, 70, 885–895, 2001.

14

Sustainable Therapeutic Approaches with Nanophotocatalyst

Rajarshi Sarkar[1], Dripta De Joarder[1] and Dilip K. Maiti[2*]

[1]Department of Chemistry, School of Advanced Sciences, VIT-AP University, Andhra Pradesh, India
[2]Department of Chemistry, University of Calcutta, Kolkata, India

Abstract

Light as a treatment of numerous diseases predates modern history. Ancient civilizations originated in different parts of the world in different times such as Indian, Egyptian, and Chinese discovered the benefits of sunlight to treat various skin diseases, such as vitiligo, psoriasis, and skin cancer. In Europe, sunlight was used to cure different deadly and debilitating conditions such as tuberculosis, scurvy, rickets, edema, rheumatism, and paralysis in the 18th and 19th centuries. The potential of light-based treatment was realized at the beginning of the 20th century when Dr. Niels Finsen utilized light from a carbon arc lamp for the treatment of skin tuberculosis lupus vulgaris. Phototherapy is generally described as the use of light to treat various diseases, however, photochemotherapy involves administration of a chemical agent that can be sensitized and subsequently activated to alleviate the condition at the tissue where the agent is localized. Seminal work by Tappeiner and Jodlbauer revealed that the presence of oxygen is absolutely imperative for photosensitization process in 1907, and they proposed the name "photodynamic action" to characterize this phenomenon. Post this, photochemotherapy has continued to grow and evolve and its application in therapeutics was expanded to many ailments including tumors and cancers and viral, bacterial, and parasitic diseases. In this context, photonanotechnology is an emerging division of nanotechnology that focuses on the design and synthesis of photoactive nanomaterials that are capable to release therapeutic agents or generate functional effects in response to light stimulation. Despite its wide use in the field, the term photonanotechnology is still vaguely defined partly because it encompasses a number of interconnected interdisciplinary areas of research and development. These nanomaterials capable of phototherapeutics have been playing an increasingly crucial role as a novel clinical treatment protocol and *state-of-the-art* scaffolding in numerous biomedical utilizations for imaging, therapy, and diagnostic devices. Owing to their rapid growth and increasing impacts, it is of paramount interest and importance to address and understand their design protocols, synthetic methodologies, and potential practical applications. This book chapter will aim to discuss the principles and functions of photonanotechnology thoroughly. The authors will discuss major areas of its clinical impacts based upon therapeutic and imaging methodologies comprising photocontrolled drug delivery and photodynamic therapy (PDT), as well as discuss unmet needs to treat other deadly diseases such as cancers and bacterial infections.

**Corresponding author*: dkmchem@caluniv.ac.in

Mousumi Sen (ed.) Sustainable Green Catalytic Processes, (329–342) © 2024 Scrivener Publishing LLC

Keywords: Nanophotocatalyst, photodynamic therapy, cancer therapeutics, drug delivery, sustainable

14.1 Introduction

Nanotechnology is one of the most sought-after technologies of the modern era. It applies the principles of nanoscience to develop technology for various human needs.[1] Nanotechnology has found application in energy, healthcare, electronics, and other industrial sectors.[1, 2] Photocatalysis is a process in which the chemical reaction is carried out and accelerated by the activation of a catalyst known as photocatalyst by light.[3–7] Since light is used for activation, it is an eco-friendly process, and during the reaction, an electron–hole pair is generated post the activation of the semiconducting substance by light, and that in turn generates various reactive free radicals such as hydroxyl radicals and reactive oxygen species (ROS).[8] Generally, photocatalysis can be classified into two main categories—homogeneous and heterogeneous. In case of homogeneous photocatalysis, the reactant and semiconductor belong to the same phase; on the contrary, in case of heterogeneous catalysis, the semiconductor and reactant belong to different phases.[9] Photocatalyst semiconductors conduct electricity by absorbing photons from irradiated light of specific wavelength at room temperature and resulting in excitation of electrons from valence bond to conduction band producing a hole/electron pair.[8] During the process of photocatalysis, the excited electron reduces the acceptor moiety; on the other hand, the donor molecule is oxidized by the hole simultaneously.[8] Photocatalysis is advantageous for various reasons such as eco-friendly, sustainability, and ease of scaling-up and implementation.[10–12] Numerous photocatalysts have been designed and synthesized by using carbon dots, MXenes, covalent organic frameworks (COFs), metal organic frameworks (MOFs), graphene, metals, and metal oxides, etc., for energy conversion, abating pollution, CO_2 reduction, hydrogen production, detoxification, and purification of water, etc.[13–23] However, usage of photocatalysts in therapeutics is comparatively recent. In this regard, various metal oxide nanoparticles and their derivatives have displayed remarkable optical, electronic, photocatalytic, and biological properties, and among them, TiO_2 nanoparticles have widely been used for biomedical applications.[23–27] Therefore, we take it on ourselves to summarize the recent progress of use of photocatalytic materials in the biomedical field including cancer therapeutics, tissue engineering, drug delivery, and biosensing in this book chapter.

14.2 Cancer Therapeutics

Cancer has become a silent pandemic of modern-day society. Some of the common treatment protocols of cancer around the world are surgery, radiotherapy, chemotherapy, gene delivery, and immunotherapy.[24, 28–33] However, these protocols have various severe diverse side effects, and therefore, there is a huge requirement for safe, targeted, efficient cancer treatment.[24, 27] Nanocatalytic medicines are expected to increase the efficacy of cancer treatment and reduce the harmful side effects as compared to current treatment, *viz.*, radiotherapy and chemotherapy.[34–37] Photodynamic therapy or PDT using nanophotocatalyst has shown tremendous potential in cancer treatment as it is noninvasive and targeted.

Here, a specific wavelength of light is used to activate the photosensitizer molecule and thereby transmit energy to the oxygen molecule present in the surroundings. However, the clinical applicability of these PDT treatments is impeded by lack of oxygen and hydrogen peroxide near the cancer cells, microenvironment around the tumor growth, and of course due to chemical properties of the photosensitizers.[38] Hence, numerous studies have been conducted to overcome these challenges, *e.g.*, a unique nanocatalyst made up of indocyanine green and calcium peroxide supported with manganese silicate, whose surface was modified with lauric acid. When applied, this H_2O_2/O_2 self-supplying nanomaterial has shown to impede the tumor growth by ROS formation both *in vivo* and *in vitro*.[6] Qin *et al.*[39] have reported a novel therapeutic material by incorporating catalase enzyme with lactate oxidase onto Fe_3O_4 nanoparticle/indocyanine green gels. This nanomaterial resulted in a significant increase in the level of intracellular ROS, hence causing fatal damages to the cancer cells. The PDT process results in the production of very toxic singlet oxygen from the tissue oxygen that kills the cancer cell in the vicinity. Free oxygen radicals generated from the oxygen molecule are usually represented as ROS, which can be of various types such as hydroxyl radical (OH·), superoxide radical ($O_2^{\cdot-}$), hydrogen peroxide, and singlet oxygen ($_1O^2$). Therefore, PDT is highly dependent on the presence of oxygen to impart any cytotoxic effects and that coupled with limited aqueous solubility of nanophotocatalysts diminishes the scope of application of PDT.[27, 40] To overcome these challenges, various new protocols are being developed including nanophotocatalysts with the possibility that they will become viable anticancer therapeutic protocols of the future.[27, 41, 42] Lin *et al.*[43] have developed a photosensitizer composed of polymer-encapsulated carbonized hemin nanoparticles.

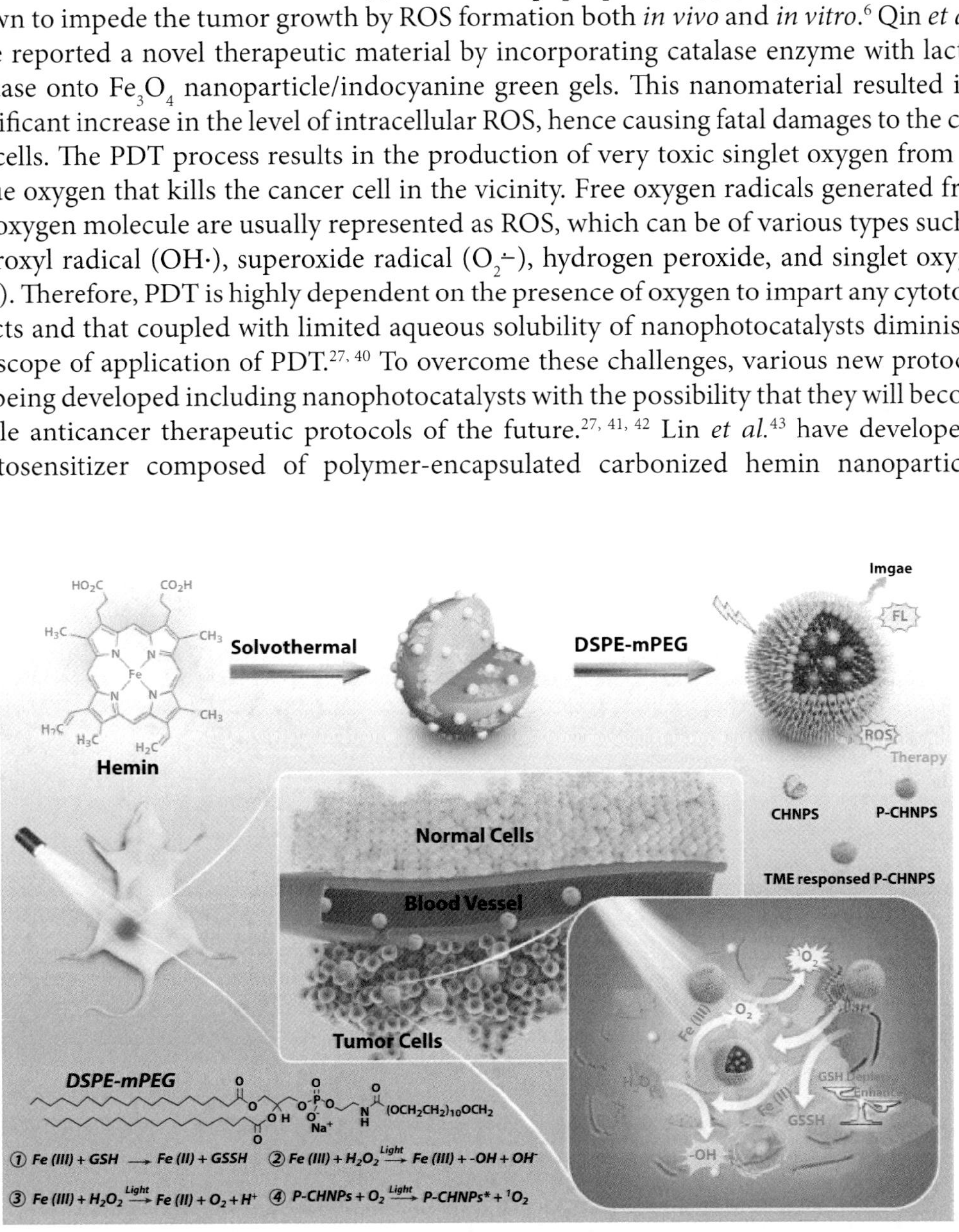

Figure 14.1 Carbonized hemin nanoparticle encapsulated in polymer for PDT to increase oxidative stress around the tumor cells. Copyright 2022 Springer Nature.

However, despite presence of hemin and its efficacy in catalysis, the overall amplification of oxidative stress in the tumor and cancer cells is limited. To improve the oxidative stress of this nanomaterial, solvothermal treatment of hemin was performed by Lin *et al.*[43] that exhibited greater production of ROS, glutathione depletion, hypoxia relief was observed both *in vitro* and *in vivo* in PDT (Figure 14.1).

Gong *et al.*[44] used carbon dots anchored on single-atom gold to amplify ROS concentration at the mitochondria of the cancer cells and thereby increasing the efficiency of the cancer treatment. Wang *et al.*[45] utilized a self-assembled nanoenzyme with an active site consisting of single-atom ruthenium installed within the MOF $Mn_3[Co(CN)_6]_2$. They introduced a self-assembled PDT single-atom nanozyme containing single-atom ruthenium (Ru) as the active catalytic site anchored in the MOF $Mn_3[Co(CN)_6]_2$ with encapsulated chlorine6 (Ce6), serving as a catalase-like nanozyme for oxygen formation (Figure 14.2). This OxgeMCC-r nanozymes are employed for targeted therapy against cancer and tumors

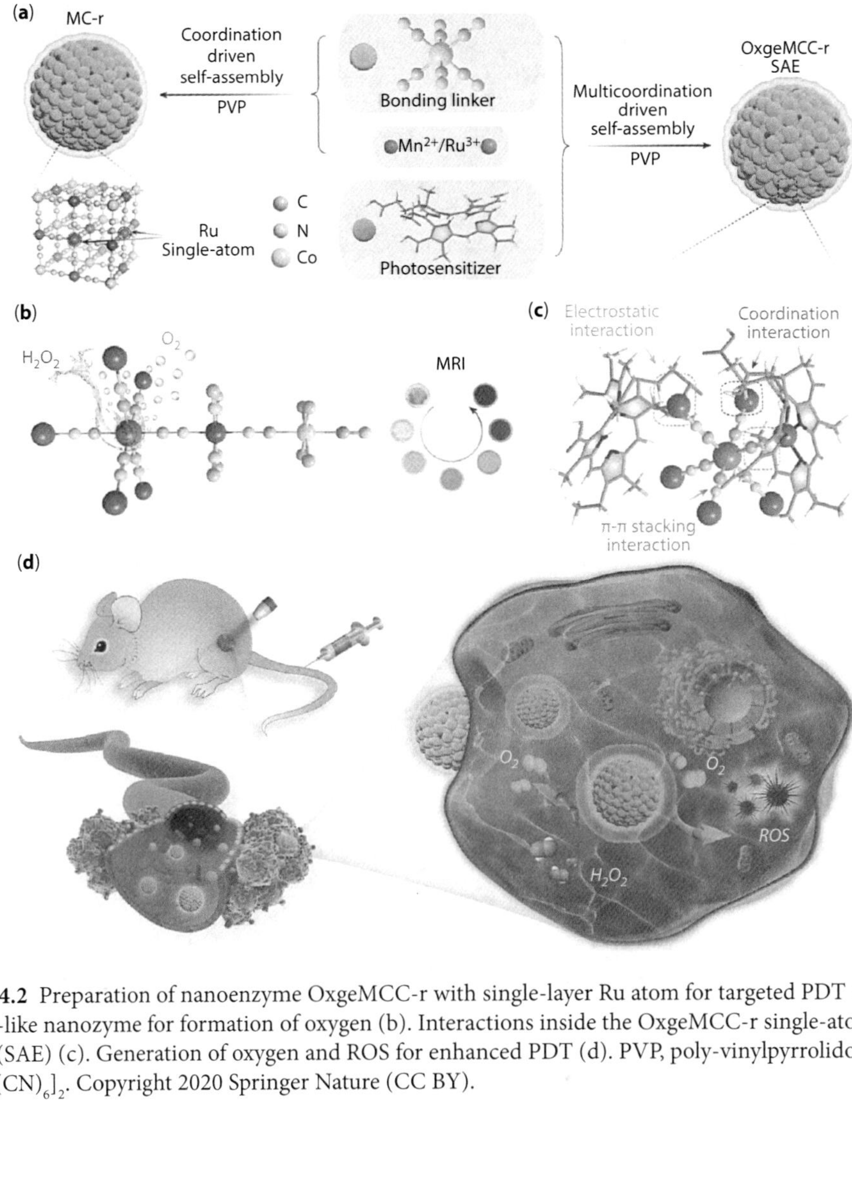

Figure 14.2 Preparation of nanoenzyme OxgeMCC-r with single-layer Ru atom for targeted PDT (a). Catalase-like nanozyme for formation of oxygen (b). Interactions inside the OxgeMCC-r single-atom enzyme (SAE) (c). Generation of oxygen and ROS for enhanced PDT (d). PVP, poly-vinylpyrrolidone; MC, $Mn_3[Co(CN)_6]_2$. Copyright 2020 Springer Nature (CC BY).

due to their highly efficient single-atom catalytic sites. The results obtained revealed that nanozymes with such good biocompatibility can be applied to decompose intracellular H_2O_2 into O_2 to alleviate hypoxia in solid tumors, thereby enhancing tumor formation into ROS, and therefore, apoptotic cell death may occur. Recently, organo-iridium (Ir) III complexes have emerged as very promising agents for the treatment of cancer, acting by converting the reduced nicotinamide adenine dinucleotide (NADH) to NAD+ via the photooxidation and single-electron transfer (SET) pathways that lead to cell cancer mortality.[46]

Furthermore, ROS generated by the photocatalytic activity of Ir(III) have been shown to have photocytotoxic effects in the mitochondria and lysosomes of drug-resistant cancer cells, whereas they have no toxic effects on benign cells, even in the long term. However, despite the promising results of using Ir(III) for cancer treatment, this concept faces several challenges in achieving favorable phototherapy for clinical use, e.g., designing Ir(III) photosensitizers, which are not only biocompatible but also have strong light absorption ability.[47] Semiconductor metal oxides such as ZnO, TiO_2, CdO, NiO, and other metal complexes have been widely used to treat cancer for their superior chemical stability, continuous absorption band, and catalytic efficiency.[48] The shape and size of the materials determine the transport characteristics of phonons and photons, and their surface-to-volume ratio can be increased by reducing their size to the nanoscale, leading to the exposure of new properties and the creation of various novel photocatalytic materials. As a result, numerous nanocomposite materials have been explored for their potential in the treatment of cancer. Nano-sized particles of NiO, TiO_2, CuO, and ZnO are part of metal oxide nanophotocatalysts, which play their role through four main mechanisms, including 1) increased mobility and communication with the cell membrane due to the small size and specific shape of the nanostructures, leading to improved cellular uptake of these nanomaterials; 2) avoiding the capture of NPs by macrophages—thanks to the surface modification of NPs using hydrophilic polymers; 3) increased intracellular drug concentration in cancer cells; 4) chemotherapeutic drugs trapped or bio-conjugated in NPs to increase drug loading efficiency.[24] Another interesting approach to reduce the cytotoxic effects of chemotherapy using PDT is usage of near-infrared (NIR) photocatalysts. NIR light has many important advantages over UV and visible light due to its greater tissue penetration capacity along with lower phototoxicity. NIR photocatalysts play a very crucial role via generating the oxidation holes and H_2, by oxidizing glutathione that is overexpressed in cancer cells in a controlled manner. Due to glutathione decomposition and concomitant hydrogen production, cancer cells become energy-deficient and an imbalance in the redox state in the cell is observed. This results in DNA damage and eventually apoptosis of the tumor cells.[49] In a study, Z-scheme SnS1.68–WO2.41 nanocatalysts were prepared for NIR photocatalytic formation of oxidation holes and hydrogen molecules to evaluate hole/hole combination therapy (Figure 14.3). Hydrogen molecules produced and the consumption of GSH in turn inhibit the energy of cancer cells and destroy the redox balance in the tumor, thereby causing aggregated DNA damage and inducing hypertrophy and apoptosis of tumor cells. However, utilizing NIR light for photocatalytic hydrogen generation also appears to be challenging due to the small energy bandgap and low oxidation potential of NIR light. Therefore, developing catalysts with high oxidation capacity with expanded energy range to overcome current shortcomings is necessary in the future.

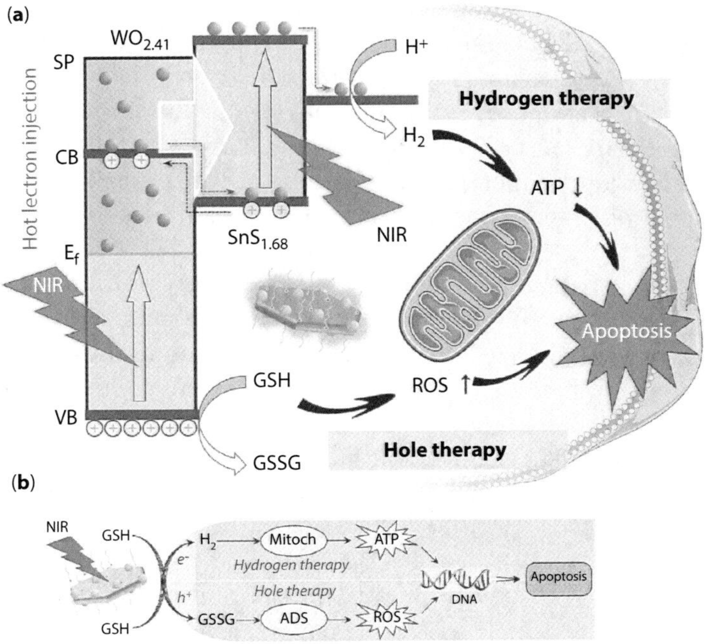

Figure 14.3 (a, b) The combined hole/hydrogen therapy protocol using nanocatalysts Z-scheme $SnS_{1.68}$–$WO_{2.41}$ under NIR irradiation. Copyright 2021 Springer Nature.

14.3 Photocatalysis and Drug Delivery

The need to develop effective low-toxicity delivery systems is in high demand, and over the past decade, scientists have focused on improving long-term performance, stability, and controllability of the distribution system.[50–54] Many nanomaterials and therapeutic molecules are being developed as drug carriers to increase the efficacy of drugs and minimize their side effects by delivering them to specific disease sites.[55–57] Encapsulating many drugs and therapeutic agents containing photosensitizers in nanomaterials is a promising approach to increase the therapeutic efficacy of drugs.[58] TiO_2 nanotube (TiNT) is a fascinating drug delivery material owing to its ordered morphology, superior biocompatibility, and outstanding photocatalytic activity. Self-ordered TiNTs have many –OH groups and internal spaces, which make these materials excellent medicinal goods. Xu *et al.*[59] developed TiNT with regular intertube spacing, which was produced by irrigation and served as a photocatalyst for controlled drug delivery. In this study, the inner and outer walls of TiNTs showed different chemical modifications. For example, the biocompatible inner wall of the TiNT was suitable for binding with antibiotics, while the outer wall was used for capturing NIR light. The platinum nanoparticles used in this study emitted luminescence in the visible region due to NIR radiation. The absorption of visible light by Pt NPs promotes 5d electrons (e^{-1}) to the TiO_2 conduction band through the Schottky barrier at the Pt/TiO_2 heterojunction. The promoted e^{-1} at the TiO_2 conduction band reacts with O_2 to produce O_2 radicals. On the other hand, H_2O reacts with the positive holes on the surface of Pt nanoparticle to produce OH radicals. The ROS released in the process damage bacterial membranes, thereby enabling the antibiotics to penetrate bacterial cells and accumulate in the cytoplasm in high concentrations within a short period of time (2 hours) and cause loss of bacterial function. Zhao *et al.*[60] in a similar work synthesized amphiphilic Au/TiO_2 NT consisting of a hydrophilic lower layer containing antibiotic ampicillin sodium and a hydrophobic upper layer for collecting NIR light and to prevent the release of loaded antibiotic. Controllable drug release was achieved based on the photocatalytic activity of the NIR-activated top layer (Figure 14.4).[60] Furthermore, this drug delivery system when protected by the hydrophobic "cap" in the top layer exhibits nominal cytotoxicity to healthy skin cells and therefore shows great promise for biomedical implants suitable for drug delivery. Incorporation of photosensitizers (PSs) and chemotherapeutic drugs into nanocarriers is a feasible approach to combine photodynamic therapy (PDT) and chemotherapy to enhance the efficacy of the anticancer medicines. Sensitive chemical bonds such as thioketals, vinyldithioether, and aminoacrylates can be easily broken irreversibly by singlet oxygen. This makes the combination of a photosensitizer (PS) and a singlet oxygen reactive (SOR) polymer nanocarrier an excellent substitute for light-controlled drug release. Scientists have designed a light-driven

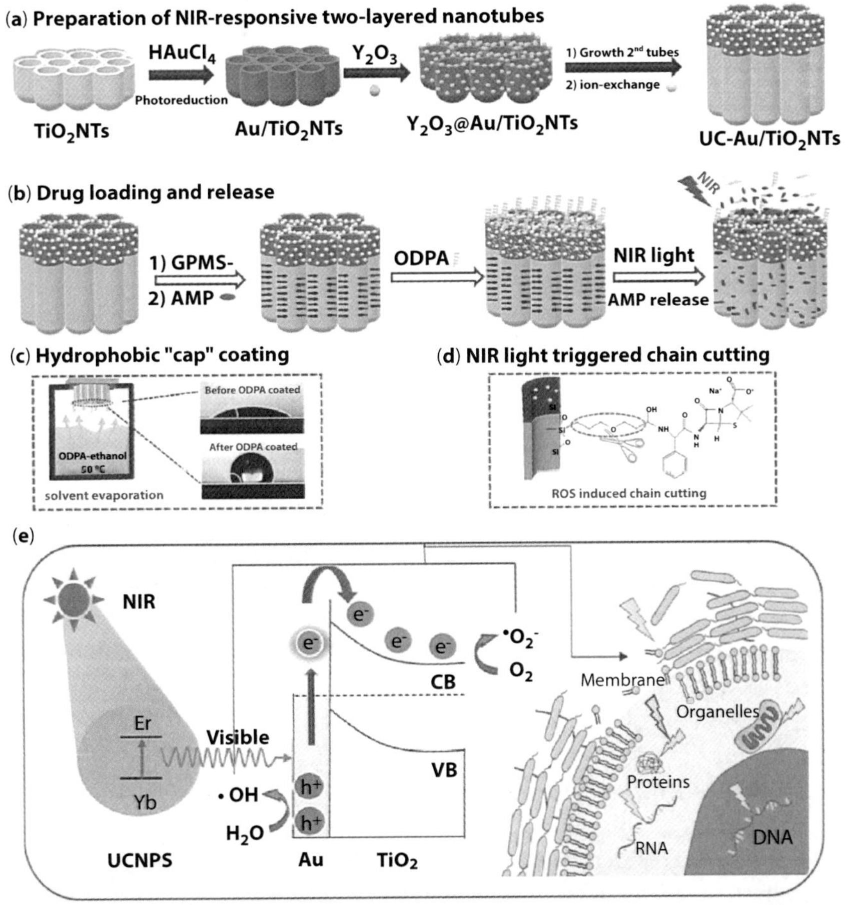

Figure 14.4 Synthesis of upconversion-Au/TiO_2 nanotubes (a). Controlled drug release (the drug loading and releasing) (b). The process of hydrophobic "Cap" coating (c), and NIR light triggered chain cutting (d). The related photocatalysis mechanisms under NIR light irradiation (e). Copyright 2020 American Chemical Society.

O_2-sensitive drug delivery vehicle where SOR-NP was utilized to co-load photosensitizer BDP and paclitaxel (PTX) as a chemotherapy drug. The composite nanomedicine SOR-NPBDP/PTX showed high oxygen content, improved water solubility, and greater stability. The release of PTX from SOR-NPBDP/PTX increased after the exposure to light and subsequent O_2 production, which then led to degradation of aminoacrylate groups and dissociation of NP. The results indicate that upon laser irradiation, SORNPBDP/PTX effectively reduces tumor growth by combining PDT with chemotherapy, while causing no notable side effects, as so it can be used as a safe and effective photodynamic-chemical combination therapy.[58] Additionally, Zhang *et al.*[61] fabricated a multifunctional drug delivery system by combining the conjugated polymer poly(fluorene-co-vinylene) (PFV), the prodrug doxorubicin modified with a phenylboronic acid ester group (BDOX), and RGD peptide is internalized. The hydrophobic inner core was functionalized with PFV chains, which have a high affinity for BDOX and generate significant amounts of ROS upon irradiation of light. The prodrug BDOX is then activated by ROS and that leads to effective release of drug resulting in dual PDT/chemotherapy. Therefore, this light-activated and ROS-responsive nanomaterial enables controlled drug release, enhanced tumor cell targeting, and effective chemotherapy/PDT synergism. Encapsulation of photosensitizers in nanomaterials is a viable approach to increase the therapeutic effects of drugs and offers new perspectives for the development of controlled drug release. Yet, despite the benefits of using photocatalysts for drug delivery, their application still faces multiple challenges mostly due to concerns about the adverse effects of ROS present and photocatalyst toxicity on metabolism and the immune system. Therefore, the development of nanocarriers containing therapeutic agents and photosensitizers with minimal side effects should be considered.[27]

14.4 Challenges and Perspectives

Despite increased inquisitiveness in the use of photocatalysts in medical usages, universal adoption of the same confronts significant difficulties. Light harvesting, surface reaction, and separation and transfer of photogenerated charge carriers are the four major elements influencing photocatalyst catalytic performance.[62] Light absorption is related to the configuration of photocatalysts' energy band. However, most current photocatalysts have a narrow scope of light absorption because of their broad bandgap, which reduces photocatalytic efficacy. The re-coalescence of the majority of electrons and holes created in the bulk phase of photocatalysts is another fundamental challenge in their use, as they can only participate in reactions at the photocatalyst's surface. Furthermore, the issue of photocatalyst corrosion in the planned reaction is not trivial. To address these challenges, researchers proposed various strategies for modifying the chemical and physical characterization of photocatalysts in order to accelerate surface reactions while increasing the range of light absorption and decreasing photogenerated charge carrier recombination.[63] When working with nanophotocatalysts, several key considerations should be considered. It is crucial to emphasize that although nanophotocatalysts have important properties like nontoxicity, they still require extensive study to enhance their efficacy. In general, a dynamic approach for increasing overall productivity is the use of diverse constituents for designing specific interfaces in composite systems to enable competent interactions.[64–68] Due to increased bacterial resistance to various antibiotics, photocatalysts have recently been introduced as an alternative for antibiotic therapy following the implantation of scaffolds and devices, and studies have indicated promising results in terms of antibacterial properties of photocatalysts including TiO_2 and their efficacy in the prevention of tissue infection.[69] In terms of the photocatalytic reaction technique, the generation of a significant amount of ROS that generates antibacterial activity may be harmful to the implant's surrounding cells and tissues.[70] Furthermore, the biocompatibility of implant materials should also be considered in clinical application. Hence, it is imperative to conduct additional studies to understand the biocompatibility and cytotoxicity of ROS generated via photocatalytic process. It is critical to evaluate the existence of endogenous photosensitizers (e.g., *Porphyromonas gingivalis*) as deadly photosensitization might happen even in the absence of photocatalytic material coating. When triggering such microorganisms, it is critical to use materials whose activated visible light wavelength is in accord with the bacterium's absorption band.[70] Several constraints of using single semiconductors for practical applications in biosensor design, such as their broad bandgap, have been mentioned. Other drawbacks of employing semiconductors include quick electron-hole re-coalescence and reduced visible light absorption efficacy. To address these complications, designer heterojunctions composed of two different semiconductors with uneven band structures have been put forward. Additionally, there are some limitations in terms of mobility and the high cost of such equipment; therefore, there will be an increasing demand for the design and production of affordable and easily portable biosensors.[71] Another significant barrier to widespread use of these types of catalysts is the lack of an acceptable criterion for evaluating catalyst activity. Most studies now normalize photocatalyst activity using their surface area or mass; however, due to the nonlinear association between the photocatalyst's surface area/mass and the desired product production rate, this technique appears to be ineffective.[72] Moreover, there is a wide range of typical light sources, illumination distances, and reactors employed in diverse research to induce photocatalytic activity. As a result, comparing photocatalytic activity of catalysts in different research based on their intended productivity yield is futile.[73] Recently, researchers have addressed this issue by comparing photocatalytic activity in various systems using apparent quantum yield (AQY) rather than traditional yield, and this appears to be a promising trend toward the creation of standardized AQY computation methods.[74] Regardless of recent breakthroughs in the use of photocatalysts for various laboratory applications, there is still a significant discordance between benchtop and market for these materials. This disparity could be attributed to the fact that researchers are primarily interested in studying the most feasible approaches to utilize these materials, despite the fact that there are no significant limitations or related ecological concerns. However, the commercialization of a product is dependent on a number of uncontrollable circumstances, and the production of optimal conditions for the catalyst's activity is not as steady and managed as in the laboratory.[75] As a consequence, there is a high requirement for cheap, practical, and sturdy large-scale manufacturable photocatalysts for using devices in commercial applications. Furthermore, a well-designed reactor not only improves reaction efficiency and economic benefits but also reduces waste of catalyst and energy, making it critical to optimize and improve the reactor's system design to achieve the highest photocatalytic efficiency.[73]

14.5 Conclusion

Various photocatalysts with high activity/functionality, better physicochemical stability, and superior biocompatibility have been developed, with potential applications in the biomedical field. These compounds can also be used to manage biomedical waste. Although nanophotocatalysts have clinical potential, however, their usage remains limited. Biological problems, large-scale production, biocompatibility and safety, and other factors all contribute to the difficulties in transferring nanophotocatalysts to clinical usage. This technique can be used to oxidize and kill tumor cells with a minimal risk of reinfection, holding significant potential for effective cancer cell treatment. Consequently, it has cleared the way for researchers to investigate novel therapeutic options for treating cancer, allowing them to overcome the obstacles of systemic toxicity and medication resistance. Furthermore, they have various advantages, such as benign nature, low cost, and higher absorption efficiency, which make them great candidates for photodynamic therapeutics. Various standard physicochemical techniques have been introduced to create nanophotocatalysts, but they may have impediments such as the use of unsafe chemicals, arduous methods, expensive instruments, and time-consuming reactions. The development of cost-effective and eco-friendly manufacturing strategies that can minimize rigor and complexity of traditional processes is in great demand. Poor stability and selectivity, separation difficulty, fast charge recombination, low migration potential of photogenerated electrons/holes, and low visible light utilization are some of the critical issues that need to be solved for commercial application and industrial production of nanophotocatalysts.

References

1. Bayda, S. *et al.*, The history of nanoscience and nanotechnology: from chemical–physical applications to nanomedicine. *Molecules*, 25, 112, 2019.
2. Karimi-Maleh, H. *et al.*, State-of-art Advances on Removal, Degradation Electrochemical Monitoring of 4-aminophenol Pollutants in Real Samples: A Review. *Environ. Res.*, 222, 115338, 2023.
3. Alizadeh, E. and Baseri, H., Photocatalytic degradation of sumatriptan succinate by ZnO, Fe doped ZnO and TiO2-ZnO nanocatalysts. *Mater. Chem. Horiz.*, 1, 7–21, 2022.
4. Candish, L. *et al.*, Photocatalysis in the life science industry. *Chem. Rev.*, 122, 2907–2980, 2022.
5. Fang, Y. *et al.*, Photocatalysis: an overview of recent developments and technological advancements. *Sci. China Chem.*, 63, 149–181, 2020.
6. Liu, C. *et al.*, An open source and reduce expenditure ROS generation strategy for chmodynamic/photodynamic synergistic therapy. *Nat. Commun.*, 11, 1735, 2020.
7. Thongam, D.D. and Chaturvedi, H., Advances in nanomaterials for heterogeneous photocatalysis. *Nano Express*, 2, 012005, 2021.
8. Ameta, R. *et al.*, Chapter 6 Photocatalysis, in: *Advanced Oxidation Processes for Waste Water Treatment*, pp. 135–175, 2018.
9. Widegren, J.A. and Finke, R.G., A review of the problem of distinguishing true homogeneous catalysis from soluble or other metal-particle heterogeneous catalysis under reducing conditions. *J. Mol. Catal. Chem.*, 198, 317–341, 2003.
10. Iravani, S., Nanophotocatalysts against viruses and antibiotic-resistant bacteria: recent advances. *Crit. Rev. Microbiol.*, 48, 67–82, 2022.

11. Khatami, M. and Iravani, S., Green and eco-friendly synthesis of nanophotocatalysts: an overview. *Comments Mod. Chem.*, 41, 133–187, 2021.
12. Nasrollahzadeh, M. *et al.*, Green-synthesized nanocatalysts and nanomaterials for water treatment: current challenges and future perspectives. *J. Hazard Mater.*, 401, 123401, 2021.
13. Fardood, S.T. *et al.*, Green synthesis and characterisation of ZnMn2O4 nanoparticles for photocatalytic degradation of Congo red dye and kinetic study. *Micro Nano Lett.*, 14, 986–991, 2019.
14. Iravani, S. and Varma, R.S., MXene-based composites as nanozymes in biomedicine: a perspective. *Nano Micro Lett.*, 14, 213, 2022.
15. Zhang, X. *et al.*, A MXene-based bionic cascaded-enzyme nanoreactor for tumor phototherapy/enzyme dynamic therapy and hypoxia-activated chemotherapy. *Nano Micro Lett.*, 14, 22, 2022.
16. Chen, H. *et al.*, Engineering covalent organic frameworks as heterogeneous photocatalysts for organic transformations. *Angew. Chem.*, 61, e202204938, 2022.
17. Yuan, G. *et al.*, MOF-COF composite photocatalysts: design, synthesis, and mechanism. *Cryst. Growth Des.*, 22, 893–908, 2022.
18. Zeng, Q. *et al.*, Hyperpolarized Xe NMR signal advancement by metal-organic frame-entrapment in aqueous solution. *Proc. Natl. Acad. Sci. U.S.A.*, 117, 17558–17563, 2020.
19. Zhang, Y. *et al.*, Application of MOFs and COFs for photocatalysis in CO2 reduction, H2 generation, and environmental treatment. *Energy Chem.*, 4, 100078, 2022.
20. Herrmann, J.-M., Heterogeneous photocatalysis: fundamentals and applications to the removal of various types of aqueous pollutants. *Catal. Today*, 53, 115–129, 1999.
21. Rodenas, P. *et al.*, Quantum dot based heterostructures for unassisted photoelectrochemical hydrogen generation. *Adv. Energy Mater.*, 3, 176–182, 2013.
22. Sudhagar, P. *et al.*, Exploring graphene quantum dots/TiO2 interface in photoelectrochemical reactions: solar to fuel conversion. *Electrochim. Acta*, 187, 249–255, 2016.
23. Akter, T. *et al.*, Rapid adsorption mechanism of methylene blue onto a porous mixed Ti-Nb oxide. *Mater. Chem. Horiz.*, 1, 49–67, 2022.
24. Kannan, K. *et al.*, Nanostructured metal oxides and its hybrids for photocatalytic and biomedical applications. *Adv. Colloid Interface Sci.*, 281, 102178, 2020.
25. Mehdizadeh, A. *et al.*, Preparation of a new magnetic nanocomposite for the removal of dye pollution from aqueous solutions: synthesis and characterization. *Mater. Chem. Horiz.*, 1, 23–34, 2022.
26. Tomas-Gamasa, M. and Mascarenas, J.L., TiO2-Based photocatalysis at the interface with biology and biomedicine. *Chembiochem.*, 21, 294–309, 2020.
27. Yi, H. and Cheng, Z., A literature review on high-performance photocatalysts for sustainable cancer therapy. *Crystals*, 11, 1241, 2021.
28. Ali, E.S. *et al.*, Targeting cancer cells with nanotherapeutics and nanodiagnostics: current status and future perspectives. *Semin. Cancer Biol.*, Elsevier, 69, 52–68, 2021.
29. Ashrafizadeh, M. *et al.*, Nanotechnological approaches in prostate cancer therapy: integration of engineering and biology. *Nano Today*, 45, 101532, 2022.
30. Ashrafizadeh, M. *et al.*, (Nano) platforms in bladder cancer therapy: challenges and opportunities. *Bioeng. Transl. Med.*, 8, e10353, 2023.
31. Lu, L. *et al.*, New insights into natural products that target the gut microbiota: effects on the prevention and treatment of colorectal cancer. *Front. Pharmacol.*, 13, 964793, 2022.
32. Mirzaei, S. *et al.*, Employing siRNA tool and its delivery platforms in suppressing cispla- tin resistance: approaching to a new era of cancer chemotherapy. *Life Sci.*, 277, 119430, 2021.

33. Tian, Y. *et al.*, Crosstalk between 5-methylcytosine and N6-methyladenosine machinery defines disease progression, therapeutic response and pharmacogenomic landscape in hepatocellular carcinoma. *Mol. Cancer*, 22, 1–25, 2023.
34. Fan, K. *et al.*, *In vivo* guiding nitrogen-doped carbon nanozyme for tumor catalytic therapy. *Nat. Commun.*, 9, 1–11, 2018.
35. Huo, M. *et al.*, Tumor-selective catalytic nanomedicine by nanocatalyst delivery. *Nat. Commun.*, 8, 1–12, 2017.
36. Lin, H. *et al.*, Nanoparticle-triggered in situ catalytic chemical reactions for tumour-specific therapy. *Chem. Soc. Rev.*, 47, 1938–1958, 2018.
37. Yang, B. *et al.*, Nanocatalytic medicine. *Adv. Mater.*, 31, 1901778, 2019.
38. Fan, W.P. *et al.*, Overcoming the Achilles' heel of photodynamic therapy. *Chem. Soc. Rev.*, 45, 6488–6519, 2016.
39. Qin, X. *et al.*, Peroxisome inspired hybrid enzyme nanogels for chemodynamic and photodynamic therapy. *Nat. Commun.*, 12, 5243, 2021.
40. Imberti, C. *et al.*, New designs for phototherapeutic transition metal complexes. *Angew. Chem. Int. Ed.*, 59, 61–73, 2020.
41. Karges, J. *et al.*, Metal dipyrrin complexes as potential photosensitizers for photodynamic therapy. *Inorg. Chim. Acta*, 505, 119482, 2020.
42. McKenzie, L.K. *et al.*, Transition metal complexes as photosensitisers in one-and two-photon photodynamic therapy. *Coord. Chem. Rev.*, 379, 2–29, 2019.
43. Lin, L., Pang, W., Jiang, X., Light amplified oxidative stress in tumor microenvironment by carbonized hemin nanoparticles for boosting photodynamic anticancer therapy. *Light Sci. Appl.*, 11, 47, 2022.
44. Gong, N. *et al.*, Carbon-dot-supported atomically dispersed gold as a mitochondrial oxidative stress amplifier for cancer treatment. *Nat. Nanotechnol.*, 14, 379–387, 2019.
45. Wang, D. *et al.*, Self-assembled single-atom nanozyme for enhanced photodynamic therapy treatment of tumor. *Nat. Commun.*, 11, 357, 2020.
46. Huang, H. *et al.*, Targeted photoredox catalysis in cancer cells. *Nat. Chem.*, 11, 1041–1048, 2019.
47. Huang, C. *et al.*, *In-vitro* and *in-vivo* photocatalytic cancer therapy with biocompatible iridium (III) photocatalysts. *Angew. Chem.*, 133, 9560–9565, 2021.
48. Hassanzadeh-Afruzi, F. *et al.*, Magnetic nanocomposite hydrogel based on Arabic gum for remediation of lead (II) from contaminated water. *Mater. Chem. Horiz.*, 1, 107–122, 2022.
49. Zhao, B. *et al.*, Photocatalysis-mediated drug-free sustainable cancer therapy using nanocatalyst. *Nat. Commun.*, 12, 1–11, 2021.
50. Gao, Y. *et al.*, Polysaccharide-based supramolecular drug delivery systems mediated via host-guest interactions of cucurbiturils. *Chin. Chem. Lett.*, 32, 949–953, 2021.
51. Huang, L. *et al.*, Rational design of nanocarriers for mitochondria-targeted drug delivery. *Chin. Chem. Lett.*, 33, 4146–4156, 2022.
52. Shen, Z. *et al.*, pH-and H2O2-sensitive drug delivery system based on sodium xanthate:dual-responsive supramolecular vesicles from one functional group. *Chin. Chem. Lett.*, 33, 4563–4566, 2022.
53. Yang, J. *et al.*, Molecular-scale drug delivery systems loaded with oxaliplatin for supra-molecular chemotherapy. *Chin. Chem. Lett.*, 32, 729–734, 2021.
54. Zhang, Q. *et al.*, A brief review of polysialic acid-based drug delivery systems. *Int. J. Biol. Macromol.*, 230, 123151, 2023.
55. Harandi, H. *et al.*, Nanoliposomal formulation of pistachio hull extract: preparation, characterization and anti-cancer evaluation through Bax/Bcl2 modulation. *Mol. Biol. Rep.*, 49, 2735–2743, 2022.

56. Poustforoosh, A. *et al.*, Recent advances in Bio-conjugated nanocarriers for crossing the Blood-Brain Barrier in (pre-) clinical studies with an emphasis on vesicles. *J. Control. Release.*, 343, 777–797, 2022.
57. Zhao, X. *et al.*, Preparation of photocatalytic and antibacterial MOF nanozyme used for infected diabetic wound healing. *ACS Appl. Mater. Interfaces*, 14, 18194–18208, 2022.
58. Yang, D.-C. *et al.*, Singlet oxygen-responsive polymeric nanomedicine for light controlled drug release and image-guided photodynamic–chemo combination therapy. *ACS Appl. Mater. Interfaces*, 13, 33905–33914, 2021.
59. Xu, Y. *et al.*, Engineering tailorable TiO2 nanotubes for NIR-controlled drug delivery. *Nano Res.*, 14, 4046–4055, 2021.
60. Zhao, J. *et al.*, NIR light-driven photocatalysis on amphiphilic TiO2 nanotubes for controllable drug release. *ACS Appl. Mater. Interfaces*, 12, 23606–23616, 2020.
61. Zhang, Z. *et al.*, ROS-Responsive and active targeted drug delivery based on conjugated polymer nanoparticles for synergistic chemo-/ therapy. *J. Mater. Chem. B*, 9, 2240–2248, 2021.
62. Han, C. *et al.*, Gold nanorods-based hybrids with tailored structures for photoredox catalysis: fundamental science, materials design and applications. *Nano Today*, 27, 48–72, 2019.
63. Wu, H. *et al.*, Metal–organic framework decorated cuprous oxide nanowires for long- lived charges applied in selective photocatalytic CO2 reduction to CH4. *Angew. Chem. Int. Ed.*, 60, 8455–8459, 2021.
64. Chen, M. *et al.*, A novel Z-scheme AgBr/Pg-C3N4 heterojunction photocatalyst: excellent photocatalytic performance and photocatalytic mechanism for ephedrine degradation. *Appl. Catal. B Environ.*, 266, 118614, 2020.
65. Khademi, D. *et al.*, A novel γ-BMO@ BMWO Z-Scheme heterojunction for promotion photocatalytic performance: nanofibers thin film by Co-axial-electrospun. *Environ. Res.*, 219, 115154, 2023.
66. Mehrkhah, R. *et al.*, Antibacterial evaporator based on wood-reduced graphene oxide/titanium oxide nanocomposite for long-term and highly efficient solar-driven wastewater treatment. *Ind. Eng. Chem. Res.*, 62, 4573–4586, 2023.
67. Zargazi, M. *et al.*, New highly efficient 2D/1D HAp/g-C3N4 photocatalyst thin film insight into crystal orientation and C-vacancy effects. *Chemosphere*, 303, 135079, 2022.
68. Zuo, G. *et al.*, Direct Z-scheme TiO2–ZnIn2S4 nanoflowers for cocatalyst-free photocatalytic water splitting. *Appl. Catal. B Environ.*, 291, 120126, 2021.
69. Moon, K.-S. *et al.*, Visible light-mediated sustainable antibacterial activity and osteo-functionality of Au and Pt multi-coated TiO2 nanotubes. *Materials*, 14, 5976, 2021.
70. Leelanarathiwat, K. *et al.*, The antibacterial activity of hydroxyapatite-tryptophan complex with gray titania by photocatalysis using LED diodes. *Int. J. Oral. Maxillofac. Implants*, 35, 265, 2020.
71. Qiu, Z. and Tang, D., Nanostructure-based photoelectrochemical sensing platforms for biomedical applications. *J. Mater. Chem. B*, 8, 2541–2561, 2020.
72. Albero, J. *et al.*, Photocatalytic CO2 reduction to C2+ products. *ACS Catal.*, 10, 5734–5749, 2020.
73. Xu, Y.-J., Promises and challenges in photocatalysis. *Front. Cataly.*, 1, 6, 2021.
74. Wang, C. *et al.*, Co and Pt dual-single-atoms with oxygen-coordinated Co–O–Pt dimer sites for ultrahigh photocatalytic hydrogen evolution efficiency. *Adv. Mater.*, 33, 2003327, 2021.
75. Tang, X. *et al.*, Self-cleaning and de-pollution efficacies of photocatalytic architectural membranes. *Appl. Catal. B Environ.*, 281, 119260, 2021.

15

Chemistry for Catalytic Conversion of Biomass/Waste Into Green Fuels

Poulami Hota and Dilip K. Maiti*

Department of Chemistry, University of Calcutta, Kolkata, India

Abstract

Sustainable development of modern society largely depends on the improvement of biorefinery platforms, that is, the development of green renewable fuels that reduce environmental pollution and meet the global energy demand. Biofuels generated via catalytic conversion of biomass and waste gain major importance in this respect. Synthesis of hydrocarbon fuels from biomass involves the formation of carbon-carbon bonds between intermediates derived from biomass, leading to the generation of hydrocarbon having high molecular weight. Depending on the nature of feedstocks, biofuels are classified into three main categories. Biofuels generated from edible food crops such as coconut, wheat, and sugarcane are termed first-generation biofuels. Whereas agricultural, industrial, and municipal solid waste-derived biofuels are known as second-generation biofuels. Basically, lignocellulosic materials present in wood, straw, and biomass residues are responsible for second-generation biofuel production. The third-generation biofuel uses algae-based feedstock that is largely utilized for biodiesel production due to its high potential for lipid production. The utilization of biomass leftovers and trash as primary resources for biofuels is a potential approach for reducing waste disposal environmental issues by converting wastes into valuable biofuels. The most common type of biomass obtained in such leftovers and trash is lignocellulosic biomass, which acts as abundant feedstock for second-generation biofuel production. It has three main fractions: lignin, hemicellulose, and cellulose. In case of cellulose and hemicellulose, they are hydrolyzed into sugars and upgraded to renewable green fuel by either concerted or stepwise process. Whereas lignin initially undergoes depolymerization to produce chemicals such as alcohols, aldehydes, and ketones, followed by transformation into hydrocarbon fuels. Catalysis allows for the effective breaking of C-C and C-O bonds, as well as the elimination of oxygen content. The present book chapter mainly focuses on the generation of second-generation biofuels such as bioethanol, biodiesel, and syngas via catalytic conversion of lignocellulosic biomass and briefly discusses the chemistry behind the process. Different chemical condensation reactions such as aldol condensation, acetalization, etherification that are used for upgradation of biomass are discussed. This book chapter will give an idea about the chemistry behind the biofuel generation process and will be helpful in designing robust strategies for large-scale production of green fuels derived from biomass and waste for industrial applications.

***Keywords*:** Green renewable fuels, second-generation biofuel, lignocellulosic biomass, gasification, pyrolysis, liquefaction, hydrolysis

**Corresponding author*: dkmchem@caluniv.ac.in

Mousumi Sen (ed.) Sustainable Green Catalytic Processes, (343–376) © 2024 Scrivener Publishing LLC

15.1 Introduction

The development of the modern world is inextricably linked to the utilization of energy resources that led to an increase in fossil fuel consumption. Until now, nonrenewable fuel petroleum is used as the main source of energy in the transportation sector. Since the stock is limited, extensive utilization of petroleum in the transportation sector reduces its worldwide supply and makes it very expensive. In addition to this, the combustion of petroleum derivatives produces greenhouse gases that cause global warming. Depletion of fossil fuels and increased fuel consumption coupled with global warming lead to the development of clean technologies to meet the global energy demand. To minimize the fossil fuel dependency, technologies powered by renewable energy resources such as solar energy, hydrogen fuel cells, and biofuels are all being actively researched, but commercialization of these new technologies needs time and modification of existing infrastructure to support cutting-edge technologies. As per the report by the International Renewable Energy Agency (IRENA), by the year 2050, approximately 90% of the world's electricity can and should come from renewable energy. Among different renewable energy resources, biofuels generated from plants gain major attention due to their similarity to the widely used liquid fuels. Biofuels can be implemented in the transportation sector without prior changes in the infrastructure and internal combustion engine, making them promising fossil fuel alternatives. In addition, biorefining of renewable biomass-derived fuels generates fuels and valuable chemicals as occur in the petroleum industry. Biomass is a generic term attributed to the organic substances obtained from plants and animals. Depending on the nature of biomass sources, biofuels are classified into three main categories.[1–4] Biofuels generated from edible food crops such as coconut, wheat, and sugarcane are termed first-generation biofuels. Whereas agricultural, industrial, and municipal solid waste-derived biofuels are known as second-generation biofuels. First-generation biomass is mainly starch-based feedstocks having glucose polysaccharides bonded via α glycosidic linkage (ex-amylase, amylopectin), which on hydrolysis produces sugar monomers that can be directly utilized in biorefineries for the generation of biofuel. However, first-generation biofuel is obtained from edible crops, which increases the cost of food and thereby conflicts with the need for primary resources to feed the world.[5] Lignocellulosic materials present in wood, straw, and biomass residues are responsible for second-generation biofuel production. [6] Such biomass needed several pretreatment steps to recover the fermentable sugars and additional processing steps to obtain biofuel. Therefore, operational costs will be high in this case. The third-generation biofuel uses algae-based feedstock that is largely utilized for biodiesel production due to its high potential for lipid production.[7] Different types of biofuels such as biodiesel, bioethanol, syngas, and biohydrogen may be produced from algae-based feedstock. Apart from these categories, municipal solid waste, industrial effluent, and sewage sludge may also be used for biofuel production. [8, 9] The utilization of biomass leftovers and trash as primary resources for biofuels is a potential approach for reducing waste disposal environmental issues by converting waste into valuable biofuels. The most common type of biomass obtained in such leftovers and trash is lignocellulosic biomass, which acts as abundant feedstock for second-generation biofuel production.[10] It has three main fractions: lignin, hemicellulose, and cellulose. The polymeric structure of cellulose consists of glucose units joined via beta (1,4)-glycosidic linkage, and hemicellulose contains various hexose

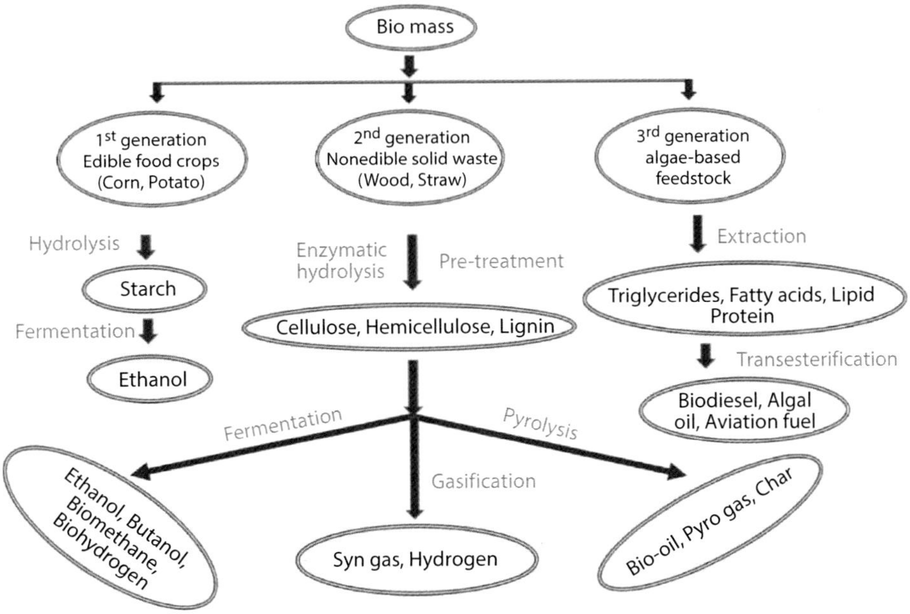

Figure 15.1 Classification of biofuels based on biomass feedstock.

units such as glucose, mannose and pentose units, xylose, and arabinose. Lignin has cross-linked aromatic polymeric structures having three major phenolic units, p-hydroxyphenyl, guaiacyl, and syringyl moiety, which make lignocellulosic biomass hydrophobic. In case of cellulose and hemicellulose, they are hydrolyzed into sugars and upgraded to renewable green fuel by either concerted or stepwise process. Whereas lignin initially undergoes depolymerization to produce chemicals such as alcohols, aldehydes, and ketones, followed by transformation into hydrocarbon fuels. Catalysis allows for the effective breaking of C-C and C-O bonds and the elimination of oxygen content. The present book chapter mainly focuses on the generation of second-generation biofuels such as bioethanol, biodiesel, and syngas via catalytic conversion of lignocellulosic biomass and briefly discusses the chemistry behind the process. Classification of biofuels depending on the biomass feedstock is given in Figure 15.1.

15.2 Lignocellulosic Biomass

Lignocellulosic biomass is a widespread and sustainable carbon source with prospective applications in the production of biofuels. Lignocellulosic biomass is composed of three biopolymers, 25%–35% hemicellulose, 40%–50% cellulose, and 15%–20% lignin (Figure 15.2).[11] These biomass fractions must be isolated before acting as abundant feedstock for biofuel generation. The isolation process proceeds through a series of pretreatment steps including physical methods such as milling, comminuting, and steam and chemical treatment such as acidic and alkaline hydrolysis. In the upcoming sections, the composition of each biomass fraction and the production of biofuels from every feedstock are elaborately discussed.

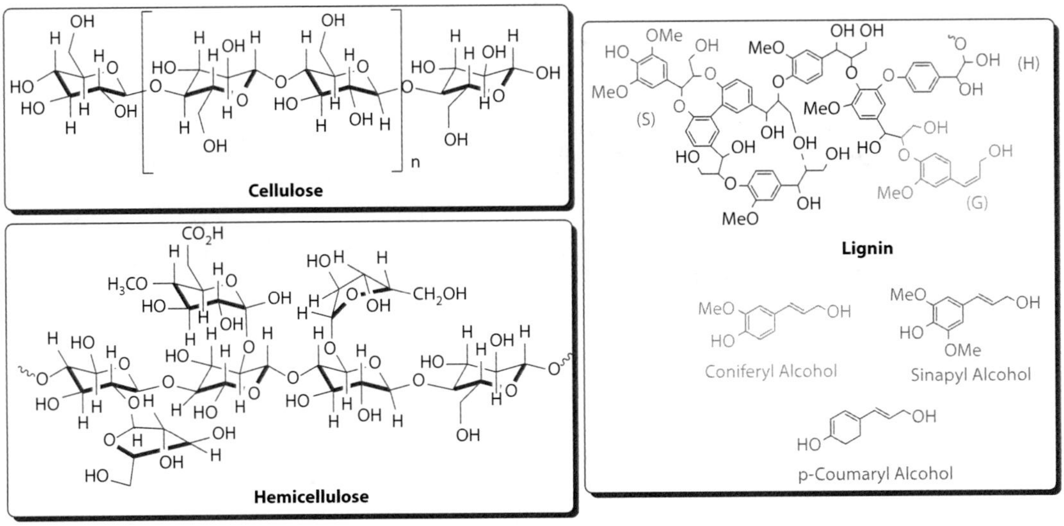

Figure 15.2 Chemical structure of cellulose, hemicellulose, and lignin fraction of lignocellulosic biomass.

15.2.1 Cellulose

Cellulose is the most abundant natural polymer on earth. It is a complex carbohydrate [generic formula $(C_6H_{10}O_5)_n$] containing glucose units joined together via β-glycosidic bonds that makes them resistant toward hydrolysis. Uniform distribution of hydroxyl group on the linear structures enables cellulose to parallelly align producing nanofibrils, which on further aggregation produce macrofibers (size in the millimeter range), held together via hydrogen bonding. Separation of cellulose from the lignocellulosic part is carried out via several pretreatment steps such as milling and physical/chemical treatments that on subsequent hydrolysis produces glucose monomers. The hydrolysis of cellulose is considered more difficult than the hemicellulose fraction. Maximum glucose yield (90%) was reported to be obtained via enzymatic hydrolysis followed by biomass pretreatment.[12, 13] Strong mineral acids such as sulfuric acid can also be used for cellulose hydrolysis; however, such a process is accompanied by the production of degradation products such as hydroxymethylfurfural (HMF), levulinic acid, and insoluble humins.[14]

15.2.2 Hemicellulose

After cellulose, hemicellulose is the second most abundant natural polymer found on earth. Hemicellulose consists of five principal sugar monomers, D-glucose, D-xylose, D-galactose, L-arabinose, and D-mannose. The structure (linear or branched) of hemicellulose is highly influenced by the presence of principal sugar moiety in its β-(1,4)-linked backbone chain. The primary hemicelluloses present in the plant cell walls are xylan, arabinoxylan, xyloglucan, glucomannan, galactoglucomanna, and arabinoglucuronoxylan. In lignocellulosic biomass, hemicellulose is bonded to lignin and cellulose is entangled with hemicellulose.[15] Separation of constituent polymer fractions is carried out via several pretreatment steps involving physical methods such as steam explosion, hot water treatment, or combination of physical and chemical methods such as dilute acid hydrolysis. During enzymatic depolymerization or dilute acid (H_2SO_4) hydrolysis of hemicellulose, β-(1,4)-glycosidic bonds break to generate soluble sugars (xylose monomers) that on yeast or bacteria-mediated fermentation produces bioethanol, and dehydration of the xylose monomers generates furfural. Among the pretreatment steps, enzymatic hydrolysis and fermentation step involved in lignocellulosic biofuel production, enzymatic hydrolysis is the slowest and is the rate-limiting step.[16–18]

15.2.3 Lignin

Along with cellulose and hemicellulose, lignin is one of the major components in lignocellulosic biomass. Lignin contains a methoxylated phenylpropane system that is responsible for water transportation and structural rigidity of plants. These phenyl propane units are cross-linked via different types of bonds, such as carbon-oxygen-carbon (C-O-C) and carbon-carbon (C-C) bonds generating rigid lignin matrix. Three aromatic alcohols, p-coumaryl alcohol, coniferyl alcohol, and sinapyl alcohol, are involved in the formation of such a structure. Different phenolic substructures are produced from these monolignols such as guaiacyl termed as G-units obtained from coniferyl alcohol, p-hydroxyphenyl named as H-units obtained from coumaryl alcohol, and syringyl named as S-units obtained from sinapyl alcohol as shown in Figure 15.2. Each chemical structure incorporates specific properties to lignin. Lignins consisting of G-units in majority are usually softwood lignins, whereas different proportion of G- and S-units results in hardwood lignins. Lignin encircles the hemicellulose and cellulose fractions. Therefore, to synthesize lignin-derived biofuels and chemicals, depolymerization of the lignin seal via pretreatment is necessary. Different physical and chemical techniques such as pyrolysis, kraft, organosov, steam explosion, and alkaline oxidation are used for the efficient generation of usable lignins from lignocellulosic matrix via depolymerization.[19, 20]

15.3 Conventional Approach for the Generation of Liquid Fuels From Lignocellulosic Biomass

Generation of biofuel from lignocellulosic biomass is accompanied by oxygen removal from the biomass feedstock followed by carbon-carbon bond formation via coupling reactions, leading to the formation of final hydrocarbons of desired molecular weight utilizing the least amount of hydrogen from an external source as occur in case of steam reforming of petroleum. Different thermochemical routes such as gasification combined with Fischer–Tropsch (FT) synthesis, pyrolysis, liquefaction, and hydrolysis methods such as aqueous-phase reforming and catalytic dehydration adopted for the generation of biofuel are briefly discussed in the following section. Figure 15.3 represents the mostly used methods for biomass processing.[4]

In thermochemical approaches such as pyrolysis, gasification, and liquefaction, by applying high temperature and pressure, biomasses are thermally degraded to upgradable intermediates such as bio-oils by pyrolysis and synthesis gas (syngas), which is a mixture of CO and H_2 by gasification. Production of fuel-range hydrocarbons involves thermal processing along with subsequent catalytic upgradation processes (Fischer–Tropsch synthesis,

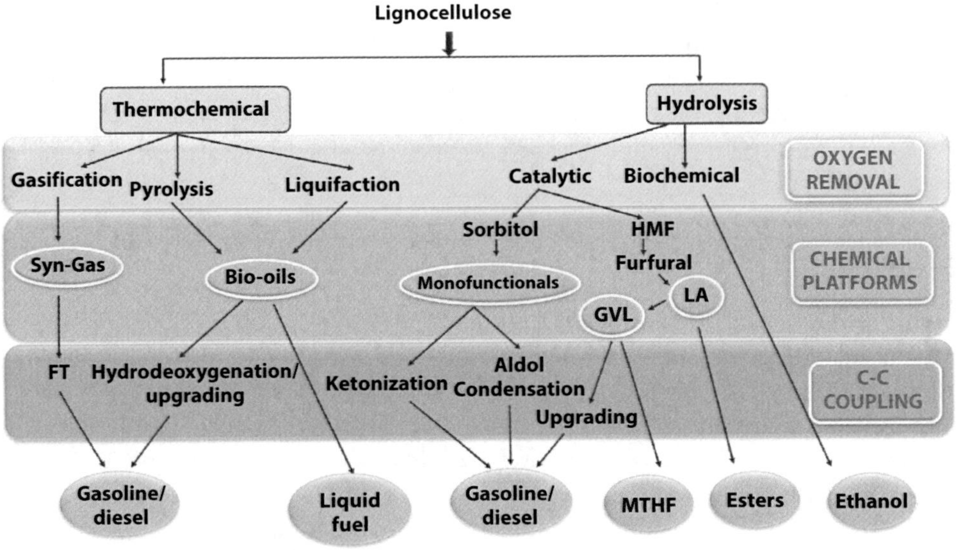

Figure 15.3 Generation of biofuels from lignocellulosic biomass via thermochemical and hydrolysis routes.

hydrodeoxygenation). Another mostly used processing method is fractionation or hydrolysis route that isolates sugar and lignin fractions from lignocellulosic biomass and produces biofuel after a series of chemical treatments or biotreatments.

15.3.1 Gasification

Generation of biofuels via gasification involves partial combustion of biomass feedstock producing a mixture of syngas (mostly H_2 and CO) and producer gas (CO_2, CH_4, and N_2). These can be upgraded to biofuels via Fischer–Tropsch synthesis. Hugo de Lasa and his group[21] and Bridgwater[22] reported that the gasification process is a combination of several subprocesses such as drying, pyrolysis, and partial oxidation. Drying is carried out up to 120°C, and above 500°C, volatile species are generated. In general, gasification is carried out above 1,000 K. In 2006, Soares *et al.*[23] studied low-temperature production of syngas from glycerol utilizing platinum-based catalysts. This endothermic synthesis of syngas coupled with the exothermic Fischer–Tropsch method produces different biofuels. The heat supplied can be intrinsically produced via exothermic combustion reactions or can be supplied from outside sources.[24, 25] Gasification is particularly advantageous, since it is not limited to any specific plant-based feedstock, and thus any type of biomass feedstock can be utilized for biofuel generation. However, the prime disadvantage lies behind the water content of the biomass and production of gaseous feed with impurities that cannot be used in downstream FT processes, since a clean gas feed is required there.

15.3.2 Pyrolysis

Pyrolysis is considered another important inexpensive processing technique for biofuel generation that involves anaerobic thermal degradation of biomass at elevated temperatures.

Depending on various parameters such as temperature, heat rate, and solid residence time, pyrolysis is classified into three categories: slow pyrolysis, fast pyrolysis, and flash pyrolysis.[26] Slow pyrolysis or carbonization process is widely utilized for charcoal production. It requires a long residence time and relatively low temperature (<500°C). Whereas a higher heating rate (10°C/s–200°C/s) and a reduced residence time (<2 s) are needed for fast pyrolysis that produces biofuels in high yield (50%–70%). The yield of biofuels further increases (75%–80%) in the case of flash pyrolysis that requires a concise residence time (<0.5 s) and a higher rate (103°C/s–104°C/s).[27]

Pyrolysis of biomass content proceeds through three vital steps: the initial step involves evaporation of moisture content, the second step involves degradation of primary biomass, and the final step includes secondary reactions such as oil cracking and repolymerization. Maximum degradation of lignocellulosic biomass occurs in the temperature range 200°C–400°C. Pyrolysis of cellulose-based feedstock occurs in the temperature range 325°C–400°C, which produces bio-oil up to 70% yield and levoglucosan as a major by-product. Below 300°C, pyrolysis of cellulose generates low-molecular weight compounds like hydroxyl acetaldehyde, glycoaldehyde, furan, formic acid, water, and CO_2, and at pyrolysis at a higher temperature (<800°C) and a short residence time, biomass turns to produce bio-oil, which on hydrotreating and refining produces biofuel.[28] Figure 15.4 represents the generation of bio-oils via pyrolysis of cellulose-based biomass.

In comparison to cellulose, hemicellulose undergoes easy thermal decomposition (temperature range 250°C–350°C). Below 300°C, thermal decomposition of hemicellulose results in the cleavage of glycosidic bond generating oligosaccharides that on further decomposition produce xylose, CO_2, and liberating water.[29] In addition to this, different low-molecular weight compounds (furan, hydroxyl acetaldehyde, formic acid, water and carbon dioxide, glycoaldehyde, etc.) are produced as obtained during pyrolysis of cellulose. Modification of pyrolysis temperature may produce other light oxygenated compounds such as formic acid, acetic acid, furfural. On further heating (<800°C), biomass produces bio-oil and biochar. Figure 15.5 represents the generation of bio-oils from hemicellulose.[30]

Among cellulose, hemicellulose, and lignin, lignin fraction was found to be thermally most stable that undergoes degradation in the temperature range 300°C–550°C. Pyrolysis of lignin fraction in the different temperature ranges is given in Figure 15.6.

Pyrolysis of lignin majorly produces anisole, cresol, phenol, guaiacol, syringol, etc. In the case of softwood lignin, guaiacol produces in majority during pyrolysis; however, derivatives

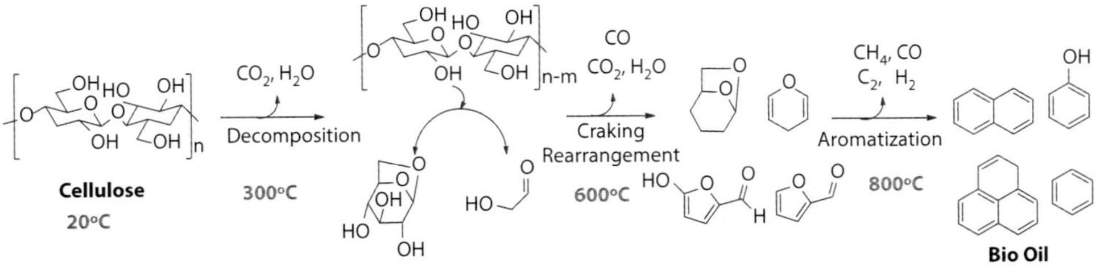

Figure 15.4 Generation of bio-oils via pyrolysis of cellulose-based biomass.

of guaiacol and syringol are produced on pyrolysis of hardwood lignin. Fast pyrolysis of lignin was reported to produce 39% of guaiacol, which is considered a prototype of lignin-derived bio-oil that can be converted to biofuel via catalytic hydrodeoxygenation.[30, 31]

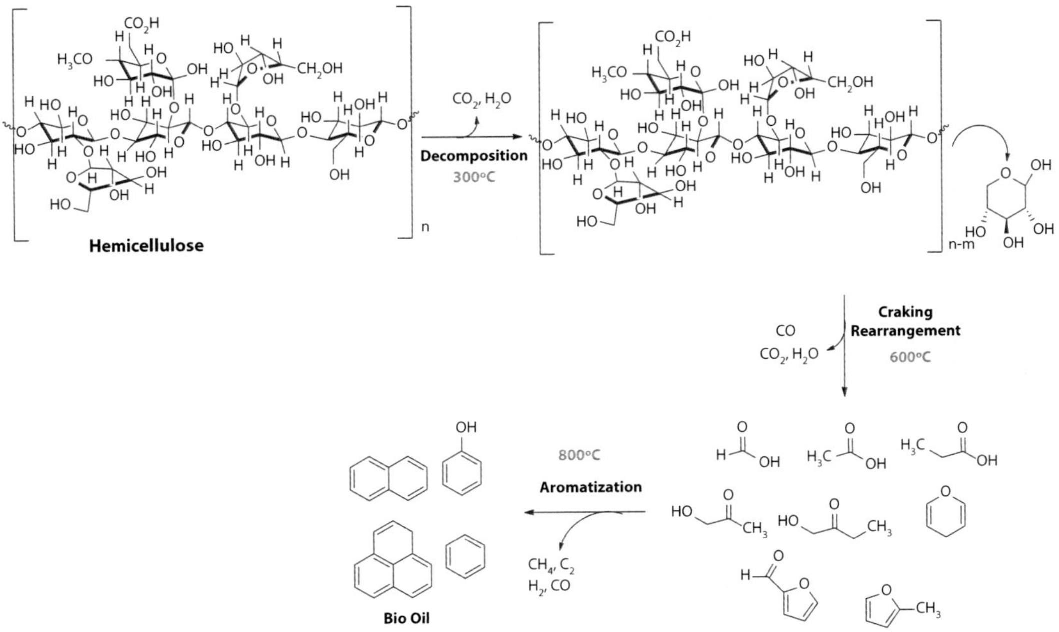

Figure 15.5 Generation of bio-oils via pyrolysis from hemicellulose-based biomass.

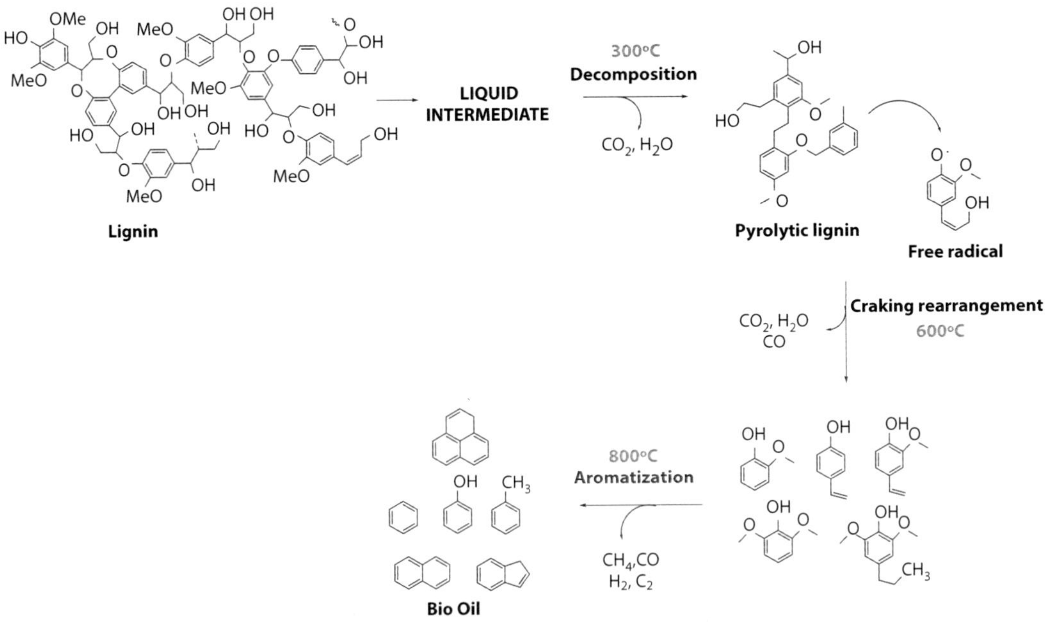

Figure 15.6 Generation of bio-oils via pyrolysis from lignin-based biomass.

Pyrolysis is a cheaper and more efficient way of producing bio-oils and does not require extensive pretreatment steps but requires major treatment before utilization in internal combustion engines, since high acidity and oxygenate content result in corrosive liquid with low energy density that reduces the lifetime and efficiency of existing engines.

15.3.3 Liquefaction

Another effective route for the generation of biofuel is solvent liquefaction, which involves generation of bio-oil via catalytic thermal decomposition of macromolecules producing fragmented species that undergoes polymerization to produce bio-oil. In contrast to pyrolysis, liquefaction requires comparatively lower temperature, higher pressure, and long residence time. Depending on the utilization of the primary solvent employed, solvent liquefaction is specified differently. Water and organic solvents are mostly used solvents for hydrothermal liquefaction.[32] In general, liquefaction techniques are classified into five categories such as hydrothermal liquefaction, organic solvent liquefaction, co-solvent liquefaction, microwave-assisted liquefaction, and plasma electrolytic liquefaction.

Hydrothermal liquefaction is carried out by hydrothermal treatment of a mixture of biomass feedstock, water, and basic catalysts (ex-sodium carbonate) at pressure region 5–20 atm and temperature 525–725 K, producing bio-oil as the major product. To improve the yield percentage of bio-oil, several efforts have been made by different groups of scientists. de Caprariis *et al.*[33] studied that an increase in lignin content in the biomass improves bio-oil yield, whereas Zhu *et al.*[34] reported that an increase in reaction temperature produces good-quality bio-oil with lower oxygen content and high heating values. Different heterogeneous catalysts were used by scientists for biofuel generation. Govindasamy *et al.*[35] reported that among different metal catalysts, bio-oil yield (57.6 wt.%) was maximum for iron-cobalt oxide catalyst.

Apart from water, organic solvents are also widely used for liquefaction of lignocellulosic biomass. Different hydrogen donor solvents (diols, glycerol) and hydrocarbons (tetralin, n-dodecane, naphthalene, 1-methylnaphthalene, polystyrene, decalin) were used by scientists as reaction media. The most widely used solvent among them is sub and supercritical alcohols due to their lower electric constant and low corrosive risk, which make biomass degradation easier and is easily separable from bio-oil. Apart from these, alcohols provide a hydrogen source throughout the reaction that reduces the side reaction and maximizes the yield of bio-oil. To further improve the efficiency of the process, several studies were made by the scientists such as modulation of reaction temperature and utilization of acid catalysts and ionic liquids.[36–38] Acid catalysts promote the cleavage of acetyl and ether bonds between carbohydrates and lignin, facilitating bio-oil generation. Apart from these, different heterogeneous catalysts such as NiMo/Al_2O_3, Pd/Al_2O_3, and Zeolite Y have been used in the liquefaction of lignocellulose in the presence of different hydrogen donor solvents. Among them, the highest yield of bio-oil was obtained for NiMo/Al_2O_3 catalyst.[39]

Sometimes, a mixture of water and organic solvent is used during the liquefaction of lignocellulosic biomass, termed co-solvent liquefaction, which synergistically utilizes the benefits of both organic system and water. Feng *et al.*[40] reported that the ethanol-water mixture behaves as a good solvent for liquefaction of lignocellulosic biomass where water facilitates carbohydrate degradation via hydrolyzing cellulose and hemicellulose, whereas ethanol causes lignin depolymerization by dissolving lignin fragments and renders the recondensation process.[40] Different transition metal-modified catalysts are also used for further improvement of the process.

Often, microwave-assisted liquefaction and plasma electrolytic liquefaction techniques have been utilized for the generation of bio-oil. In the case of microwave-assisted liquefaction, heat generated from microwave energy facilitates the cleavage of glycosidic bonds in biomass resulting in the formation of monomeric units.[41] This process seems advantageous with respect to reaction time, economic feasibility, and energy efficiency.[42] Plasma electrolytic liquefaction techniques have been recently used in the solvent liquefaction of lignocellulosic biomass. Plasma may be defined as a partly ionized gaseous state of a system that shows high chemical reactivity. This process shortened the liquefaction time and makes the process eco-friendlier.[43]

Three different constituents of lignocellulosic biomass, cellulose, hemicellulose, and lignin, decomposed differently during the liquefaction process. Different groups of scientists studied their behavior in the polar and nonpolar media. Xu *et al.*[44] studied the decomposition of cellulose and hemicellulose in an acid-catalyzed methanol liquefaction process. They suggest that the process proceeds through the breakage of glycosidic linkage of both cellulose and hemicellulose producing C_6 sugar derivatives.[44] Ghosh *et al.*[45] suggested that in the polar aprotic solvent, decomposition of cellulose proceeds through depolymerization

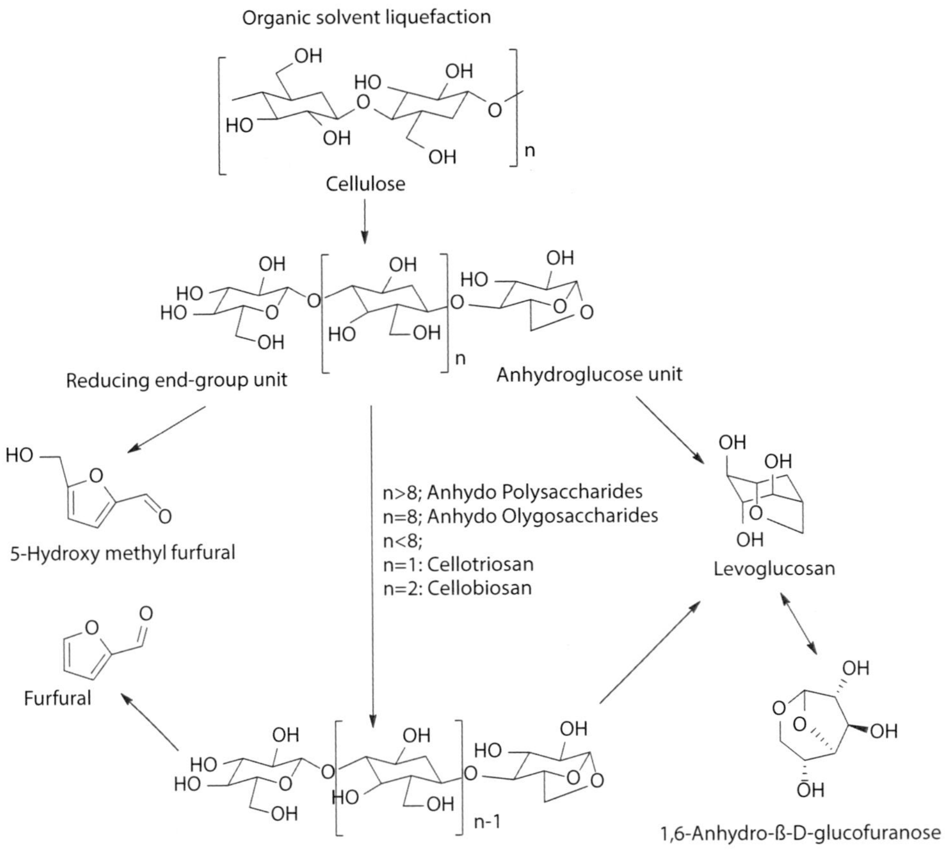

Figure 15.7 Suggested cellulose depolymerization mechanism during organic solvent liquefaction.

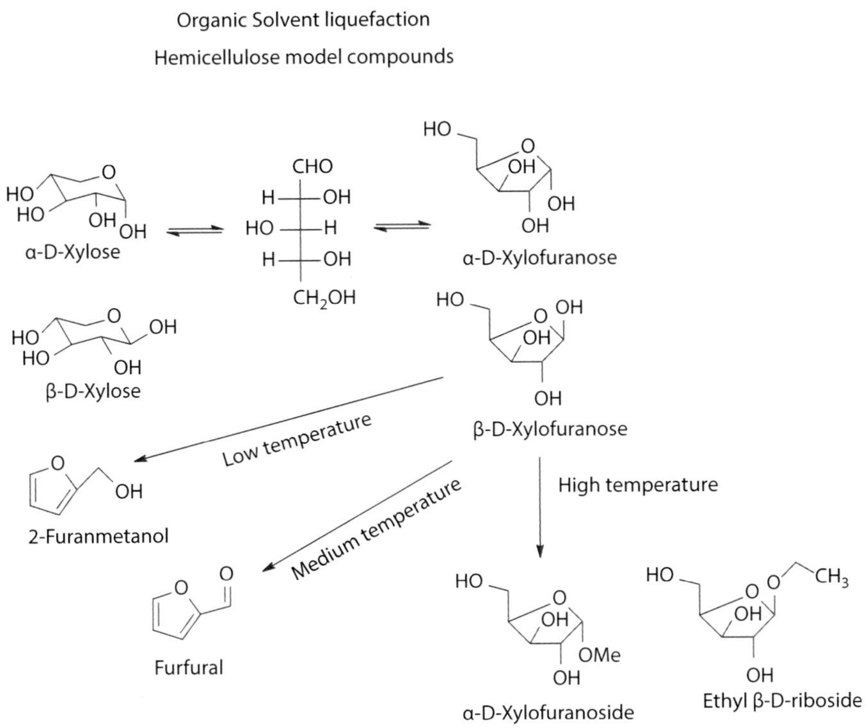

Figure 15.8 Suggested conversion of D-xylose in the presence of supercritical ethanol.

mechanism that produces levoglucosan (LG) as the main sugar monomer along with other anhydro monosaccharides and oligosaccharides. Ring opening, fragmentation, and rearrangement of the reducing ends of anhydro-oligosaccharides produce degradation products 5-HMF and furfural, etc.

In 2015, Brand *et al.*[46] studied metal-catalyzed decomposition of α-cellulose, D-xylose, and lignin in the presence of supercritical alcohols. Figure 15.7 represents the possible products formed and the mechanism involved during depolymerization of lignocellulosic biomass, and Figure 15.8 represents the suggested mechanism for the conversion of D-xylose in the presence of supercritical ethanol. Since a major part (30%) of hemicellulose consists of xylose, therefore, xylose may be considered a representative hemicellulose model.[46, 47] β-D-xylofuranose reacts with ethanol to produce ethyl β-D riboside. The formation of β-D riboside, methyl-a-D-xylofuranoside, and β-methoxy-(S)-2-furanethanol requires a high temperature and a dehydration product such as 2-furanmethanol; furfural produces at low temperature.

Decomposition of lignin is quite complex in reference to cellulose and hemicellulose. Liu *et al.*[48] studied the decomposition of lignin in sub- and supercritical ethanol and showed that decomposition of lignin proceeds through depolymerization involving cleavage of aryl-ether linkages, followed by repolymerization generating carbon-carbon bonds. The group further studied the mechanism of microwave-assisted liquefaction of lignin in isopropanol (Figure 15.9). They suggested that depolymerization of lignin produces oligomers that on hydrogenation produces aromatic monomers. Isopropanol-mediated solvolysis causes cleavage of α-O-4 generating carbonyl groups at the C_α position of the aliphatic side chain.

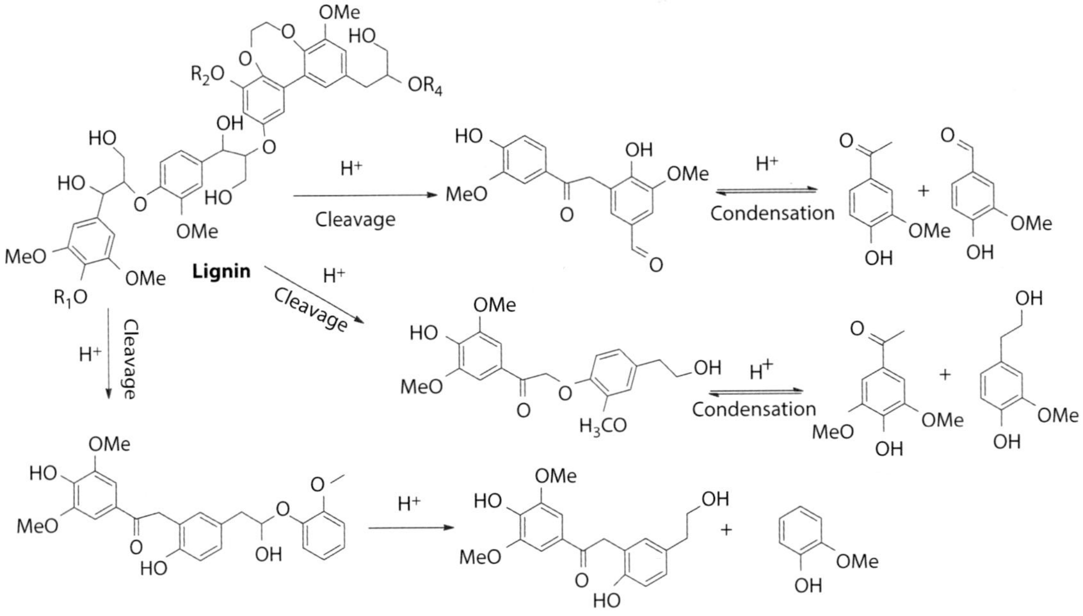

Figure 15.9 Decomposition of lignin proceeds through depolymerization.

The product obtained in case of liquefaction contains lower oxygenate content (12%–14%) than the bio-oil produced by pyrolysis and requires minimal modification for utilization in boilers and engines. The main disadvantage lies behind long-term storing of the bio-oil product. To resolve the stability issues, several techniques have been implemented for bio-oil upgradation such as hydrodeoxygenation and steam reforming.[49]

15.3.4 Hydrolysis

Hydrolysis is another important method for generating targeted biofuels from lignocellulosic biomass. Selective transformation of lignocellulosic biomass requires isolation of cellulose and hemicellulose fraction via microbes or enzyme depolymerization producing sugar constituents. After the separation of sugar monomers, they are processed via different catalytic technologies. Hemicellulose and cellulose undergo easy hydrolysis producing sugar monomers, whereas utilization of lignin remains a challenge. Different catalysts, such as homogeneous (mineral acid, organic acid) and heterogeneous catalysts (metal oxides, H-form zeolites, acidic resin, enzyme) have been utilized for the hydrolysis of lignocellulosic biomass.[50, 51]

One of the most simple and popular methods for the generation of sugar monomers from lignocellulosic biomass is acid hydrolysis. Acidic hydrolysis of cellulose proceeds through the cleavage of β-1-4-glycosidic bond producing shorter chain oligomers that on further degradation generate sugar monomers. The reaction starts with protonation on the β-1-4-glycosidic bond generating corresponding conjugated acid that induces the cleavage of C–O bond and breakdown of the conjugate acid to the cyclic carbonium ion having half-chair conformation. The addition of water produces free sugar. Cellulose hydrolysis produces glucose as a major product that under acidic conditions dehydrated to

5-hydroxymethylfurfural (5-HMF). Compared to cellulose, hemicellulose undergoes easy hydrolysis due to its amorphous branched nature. The hydrolysis proceeds through a similar mechanism as cellulose. After proton diffusion to the lignocellulosic matrix, protonation

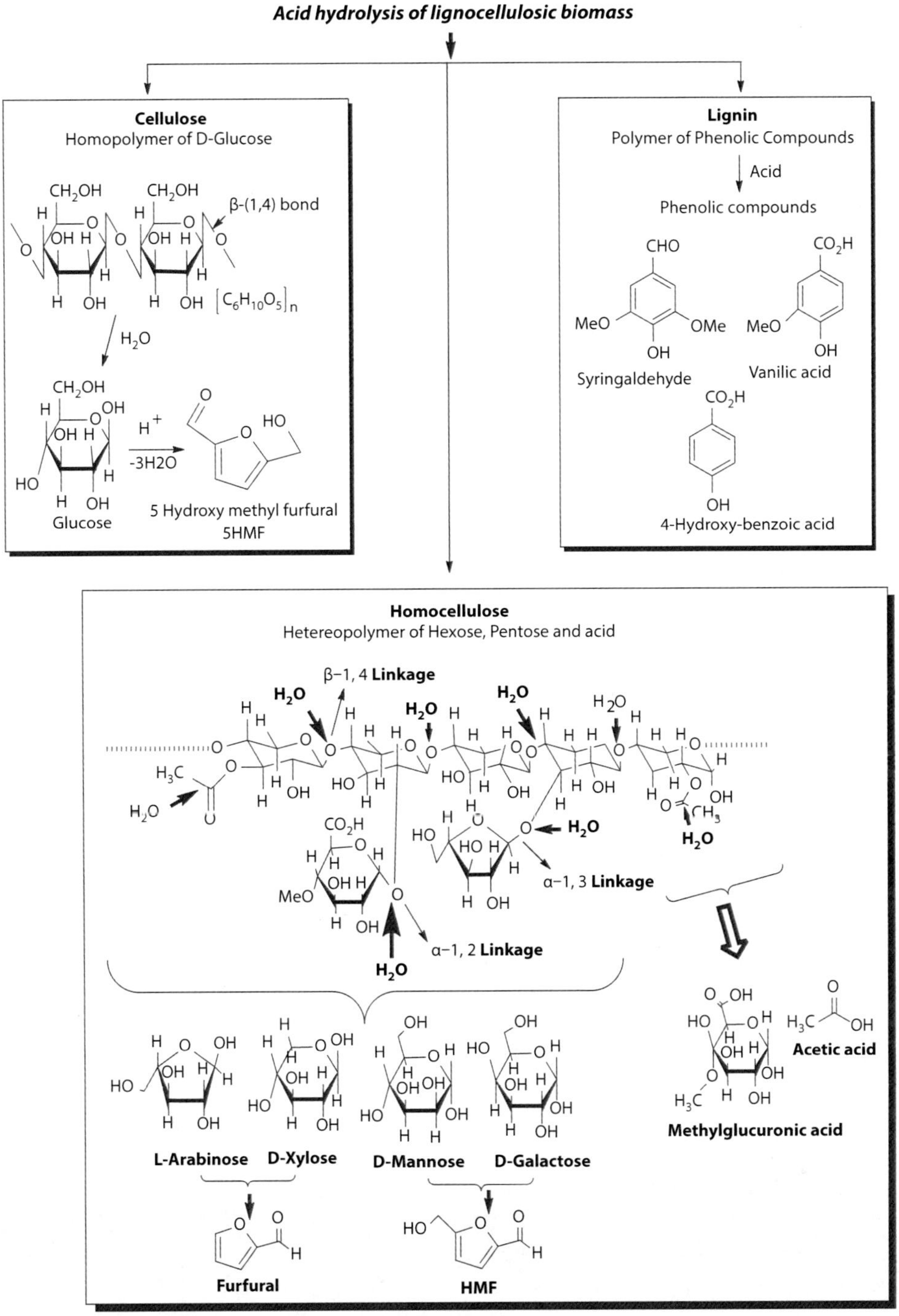

Figure 15.10 Acidic hydrolysis of cellulose, hemicellulose, and lignin.

occurs on the glycosidic bond oxygen that induces cleavage of the glycosidic linkage, that is considered as the rate-limiting step, followed by generation of intermediate carbocation that on water treatment produces mono-, oligo-, or polysaccharide based on the position of glycosidic bonds. Hemicellulose hydrolysis produces xylose as a major product along with other monomers such as arabinose, galactose, and mannose (Figure 15.10).

Acid concentration, temperature, and reaction time were found to largely affect the reaction.[52] It has been observed that mild reaction condition (dilute acid) upgrades xylose yield percentage while suppressing the generation of by-products such as acetic acid, lignin degradation products (LDPs), furfural, and hydroxymethylfurfural (HMF). Rafiqul and Sakinah[53] in 2013 and Amirkhani *et al.*[54] in 2015 utilize dilute sulfuric acid that effectively hydrolyzes hemicellulose into sugar monomers such as xylose, arabinose, and mannose and soluble oligomers. Among different hydrolysis methods, dilute acid hydrolysis is particularly useful because that converts hemicellulose to xylose, which can be transformed into bioxylitol or bioethanol utilizing many microbial strains.[55–57] Phillips and Goss[58] obtained distillate via distillation of lignin (isolated by the fuming hydrochloric acid method) by 12% hydrochloric acid. The distillate gives a precipitate with a mixture of aromatic compounds such as phloroglucinol and barbituric acid and with thiobarbituric acid and formaldehyde.[58]

Another important method for cellulose/hemicellulose degradation is autohydrolysis that is defined as the pretreatment of biomass with hot water or saturated steam (150°C–250°C) inside a pressurized reactor without using any acid or alkaline catalysts. The high temperature and pressure facilitate the autoionization of water, producing hydronium ions, which initially hydrolyzes acetyl residue in xylan producing acetic acid, that further catalyzes the hydrolysis of remaining hemicellulose producing oligosaccharides as the major product.[59, 60] Molecular weight of the oligosaccharide was found to be decreased with increasing reaction temperature in the case of rapeseed straw, and conversion of xylo-oligosaccharides into xylose occurs at lower temperature. Garrote *et al.*[61] reported autohydrolysis of corncob producing xylo-oligosaccharides and other by-products (mainly xylose and furfural). Most of the cellulose content remained in the solid residue, while partial delignification (up to 25% lignin removal) was attained. Lavoie *et al.*[62] reported base-catalyzed depolymerization of lignin producing monomers, dimers, trimers, and residual char. Different pharmaceutically important aromatic compounds are derived from monomers such as catechol, and cycloalkanes are produced from oligomeric fractions.[62]

Enzymatic hydrolysis of cellulose/hemicellulose for the generation of sugar moiety is considered another promising method. Such hydrolysis is carried out in the presence of enzymes xylanases and cellulase under mild reaction conditions. Generally, cellulase consists of three types of enzymes, which synergistically convert cellulose into glucose in a stepwise manner. Initially, endoglucanase cleaved β-1,4-glycosidic bonds of glucan chains producing shorter free chains having reducing and nonreducing ends. These are then degraded by exoglucanase generating cellobiose, which is finally converted to glucose by β-glucosidase. Hydrolysis of hemicellulose is much more complicated. Synergistically, xylanases cleave the β-(1,4)-glycosidic bonds in the xylan molecule producing xylo-oligomers. Release of xylose from hemicellulosic material is facilitated by xylanase blend, consisting of xylosidases, arabinofuranosidases, mannananses, hemicellulolytic esterases, and glucuronidases. Recently, Zhang *et al.*[63] studied enzymatic hydrolysis of lignin fragments facilitated via ligninolytic enzymes. Alkali lignin was used as substrate and ligninolytic enzymes consist of three types of pure enzyme, Lac, LiP, and MnP. These enzymes initiate ring opening of the macromolecular benzene ring structure of lignin and chemical bond cleavage causing generation of various aromatic compounds.[63, 64] Mai *et al.*[65] also studied selective conversion of lignin (EHL) into various aliphatic and aromatic compounds utilizing different catalysts such as $WO_3/\gamma\text{-}Al_2O_3$ catalyst.

In the upcoming section, different aqueous-phase catalytic techniques utilized for generation of hydrocarbon fuels from the processed biomass derivatives containing sugars, polyols, furfurals, levulinic acid, and γ-valerolactone are briefly discussed.

15.3.5 Aqueous-Phase Reforming and Derivative Technologies

Aqueous-phase reforming (APR) involves generation of hydrogen and hydrocarbon fuels from oxygenated biomass feedstock (ethanol, ethylene glycol, polyols, glycerol, sorbitols, and glucose, etc.) in aqueous medium at low temperature and high pressure. Hydrogen (H_2) is a zero-emission fuel. Therefore, generation of renewable H_2 by APR of oxygenated hydrocarbons over supported metal catalysts gains major attention.[66] Selectivity of aqueous phase-reforming process largely depends on the nature of support, promoter, and catalysts. Different platinum-based catalysts such as Pt-black, platinum-supported metal oxides (Al_2O_3, TiO_2, and ZrO_2) have largely been used as catalysts for the generation of hydrogen fuel from ethylene glycol and methanol. Whereas Rh-, Ru-, and Ni-based catalysts facilitate the C-O bond cleavage of polyols rather than the C-C bond, thus producing alkanes as major products and suppressing hydrogen generation. The selectivity for hydrogen generation via APR of polyols is largely affected by the addition of specific promoters. Tin (Sn) is reported as an excellent promoter that accelerates the hydrogen production activity of nickel-supported SiO_2 or Al_2O_3 from polyols.[67] On the other hand, alkane selectivity was found to be improved by lowering the p^H of the reaction system either by the addition of mineral acid or via modulating the solid acid catalyst supports. Increase in acidity of the reaction medium enhances the rate of dehydration and hydrogenation reaction, whereas hydrogenolysis and reforming reactions are suppressed. *Huber et al.*[68] reported generation of hydrogen and light alkanes from biomass-derived oxygenates (glucose, sorbitol, and glycerol) via aqueous phase-reforming reaction. They also studied APR of sorbitol that is obtained via hydrogenation of glucose and modulated the reaction to obtain heavier alkanes (butane, pentane, and hexane).[68] The reforming proceeds through a bifunctional reaction pathway, the initial generation of hydrogen and carbon dioxide on specific metal catalysts, and dehydration on the silica-alumina catalysts. Hydrogen is produced via C-C bond cleavage on specific catalyst followed by water–gas shift (WGS) reaction. Secondary step involves hydrogenation of dehydrated intermediates (isosorbide or enolic intermediates) on specific metal catalysts. Under balanced conditions, hydrogen produced initially is completely utilized in the hydrogenation process converting sorbitol to alkanes, CO_2, and water as shown in Figure 15.11.

Heavier alkanes are found to be produced after a continuous cycle of dehydration and hydrogenation in the presence of hydrogen. However, fast cleavage of C-C bond compared to hydrogenation process was reported to produce lighter alkanes.[68] Metal catalysts such as Ni and Ru also promote the generation of lighter alkane via hydrogenation of CO or CO_2.[69] Xylitol is another important polyol derived from lignocellulosic biomass, involved in the generation of hydrogen via APR. Xylitol shows better selectivity and greater hydrogen yield

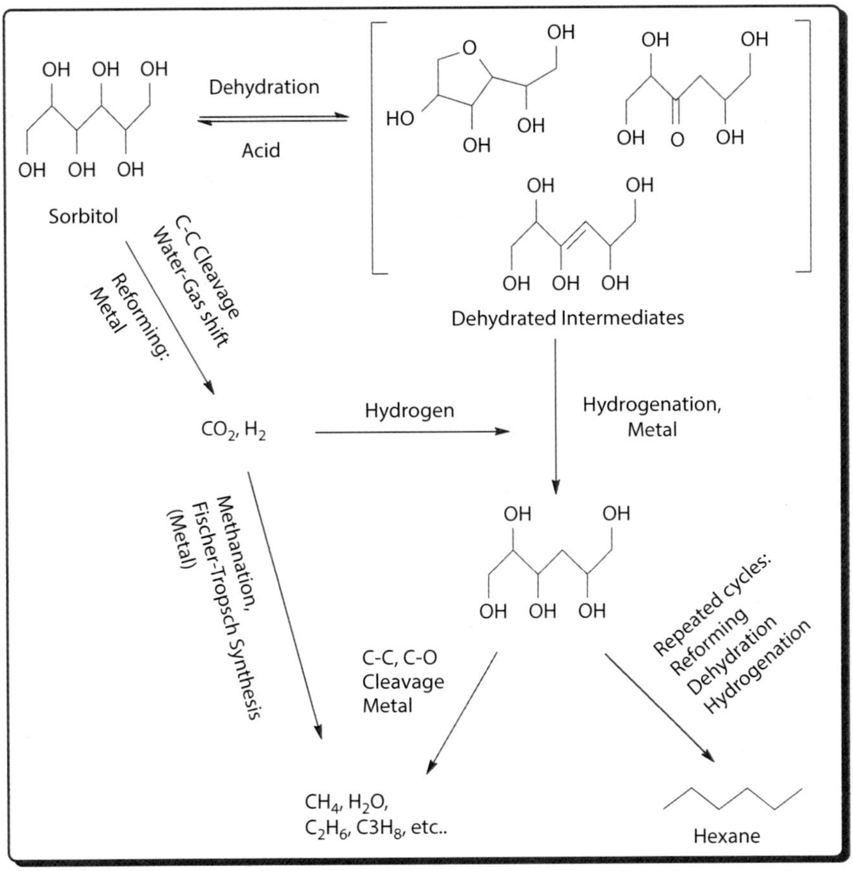

Figure 15.11 Generation of alkanes from sorbitol over metal and acid catalysts.

(83%); however, it undergoes incomplete conversion (ca. 66%). Mechanistic investigation showed that initially, polyols undergo dehydrogenation/dehydration producing corresponding aldehyde, followed by decarbonylation. Selective synthesis of alkanes is reported from xylitol when the reaction is carried out under the flow of hydrogen in the presence of metal catalysts such as Pt, Ni, supported on highly acidic substrates such as mesoporous silica, H-ZSM-5, and MCM-22[70, 71] In the presence of catalyst, 2 wt.% Ni/MCM-22, pentane is obtained from xylitol, whereas pentane along with short-chain alkanes is produced when Pt-based catalysts are used.[72, 73] Alkane production is further facilitated by the addition of Re catalyst, which enhances the polyol adsorption and facilitates selective hydrogenolysis of C-O bonds. Under the condition of high pressure, low temperature, and high oxygenated feed concentration, Pt–Re/C catalyst produces different intermediates (2-ketones, secondary alcohols, heterocycles, and carboxylic acids) via partial reduction of polyols that can be utilized as efficient feedstock for different upgrading strategies.

Being most abundant in nature, glucose also may be used as a convenient raw material for hydrogen generation. Aqueous-phase reforming of glucose produces hydrogen as observed in the following equation:

$$C_6H_{12}O_6 + 6H_2O \rightarrow 12H_2 + 6CO_2$$

The reaction proceeds through the formation of several intermediates and also involves various side reactions. Light alkanes are produced from hydrogenation of CO/CO_2 and hydrogenolysis of polyols, whereas carbonyl compounds (ketones and aldehydes) are produced via dehydration, thereby minimizing hydrogen selectivity. To improve hydrogen selectivity, specific catalysts were used that facilitate the cleavage of C-C, C-H, and O-H bonds and preserve the C-O bonds.[74] In addition, these catalysts must actively participate in water–gas shift reaction, improving hydrogen content and facilitating the conversion of carbon monoxide to carbon dioxide. Nickel and platinum behave as efficient catalysts due to their C-C cleavage properties. Computational studies showed that at 200°C, maximum absorption of C-C bond over Pt catalyst occurs. A computational investigation shows that the C-C bonds of glucose are adsorbed on Pt to a maximum extent at 200°C.[75] However, lower temperatures favor C-O cleavage than C-C bond cleavage.[76] Different scientific groups reported different catalysts for glucose APR. Wen *et al.*[77] studied aqueous-phase reforming of glucose at 533 K in the presence of Ni/Al_2O_3 catalysts. Hydrogen yield was found to be improved by using catalyst obtained via two-step impregnation rather than single impregnation.[77] Apart from this, Pt/C and Pt/Al_2O_3 are also utilized as efficient catalysts for aqueous-phase reforming of glucose. Specifically, metals (Pt, Pd, and Ni) were impregnated over different non-innocent metal oxides, such as γ-Al_2O_3, ZrO_2, CeO_2, and Ce-Al mixed oxides. Isotopic labeling experiment proved that, initially, glucose dehydrates to HMF, followed by conversion to levulinic and formic acid and the formic acid as formed (Figure 15.12) decomposed to hydrogen and carbon dioxide.

Carbon–carbon coupling is one of the most important strategies for the generation of larger hydrocarbon from biomass-derived C_5 and C_6 compounds. C-C coupling can easily be facilitated via aldol condensation of ketones in the presence of different basic catalysts ($MgAlO_x$, Mg/TiO_2, Mg/ZrO_2, Pd-MgO/ZrO_2, La/ZrO_2, Y/ZrO_2, etc.).[78–81] Coupling between secondary alcohols is facilitated in the presence of basic bifunctional metal catalysts and hydrogen. In the presence of metal catalyst, an equilibrium is attained between secondary alcohol and two ketones, the latter of which upgraded to larger methyl ketone via condensation reaction (acid/base catalyzed) with other ketones present in monofunctional products. Hydrogen is utilized to shift the equilibrium toward the condensation product via hydrogenation of the carbon-carbon double bond produced via dehydration of the aldol product. Biomass-derived compounds are also able to undergo C–C coupling via ketonization of carboxylic acid, where in the presence of basic catalyst (Ceria, barium hydroxide), carboxylic acids (C_nOOH) are decomposed to lower ketones (2n-1 carbon atoms), CO_2,

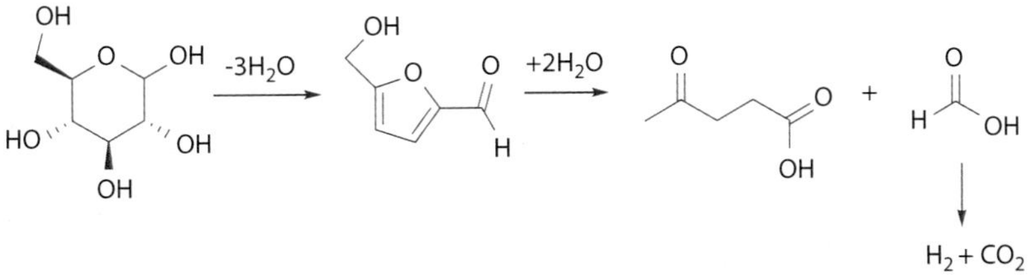

Figure 15.12 Suggested mechanism for hydrogen generation from glucose.

and water.[82–84] James A. Dumesic and his group reported generation of fuels via ketonization of propionic acid obtained from lactic acid in the presence of CeZrOx catalyst. In addition, different groups utilized several catalysts for the upgradation of bio-oil. Cheng *et al.*[85] reported the upgradation of bio-oil in the presence of bifunctional catalysts Fe-Ni/HZSM-5 using supercritical methanol. Catalytically upgraded bio-oil shows modified physicochemical properties in comparison with raw bio-oil.[85] Kunes *et al.*[87] reported generation of C_{6+} gasoline range components via cascade process involving hydrogenation of monofunctional oxygenates followed by conversion directly over H-ZSM-5 at 673 K. A different group of scientists utilizes both the processes, ketonization and aldol condensation, for the generation of C_{7+} oxygenates from glucose-derived monofunctional compounds.[86, 87] James A. Dumesic and his group[88] utilize a dual-bed catalyst system containing CeZrOx and Pd/ZrOx as shown in Figure 15.13.

CeZrOx actively catalyzes the ketonization of carboxylic acids producing CO_2 and water, while Pd/ZrOx catalyzes aldol-condensation/hydrogenation of alcohol and ketones. In the presence of a dual-bed catalyst system, high-molecular weight ketones are produced, which undergo hydrodeoxygenation over Pt/SiO_2–Al_2O_3 catalyst to produce alkanes. Sorbitol (60 wt.%) was found to be converted to liquid alkanes (53%) containing C_{7+} alkanes with minimal branching, preferable for diesel fuel.[88] The generation of biofuels from sugars and polyols in the presence of different catalysts is shown in Figure 15.14.

In 2015, Liu *et al.*[89] reported theoretical and experimental studies regarding catalytic upgradation of biomass via C-C coupling. They investigated the reactions of furan and acetaldehyde in the presence of HZSM-5 zeolite catalyst via periodic density functional theory (DFT) that showed that the reaction proceeds through an alkylation mechanism maximizing C–C coupling and dehydration whereas minimizing catalyst deactivation (e.g., coke formation) via aldehyde condensation.[89] Recently (2021), Yeboah *et al.* reported a versatile and atom-economic strategy for the upgradation of biomass-derived C_3 oxygenates (aldehydes, ketones, and alcohols) to produce jet fuel range aromatics harbors (yield 86.1%, purity 98.2%). The reaction proceeds through the Cu/SiO_2 and TiO_2-catalyzed aldol condensation and ring closure of the propanal, propanol, and acetone producing C_9–C_{12} oxygenates, followed by Ni/HZSM-5-catalyzed hydrogenation–dehydration–aromatization

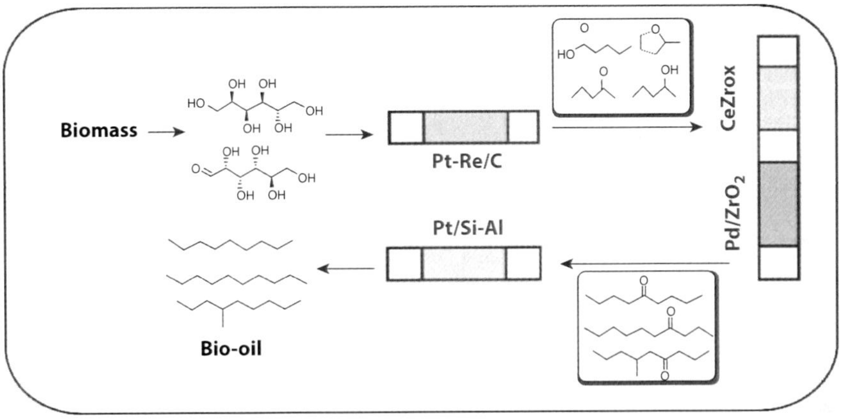

Figure 15.13 C–C coupling reactions of biomass-derived oxygenates over dual-bed catalyst system.

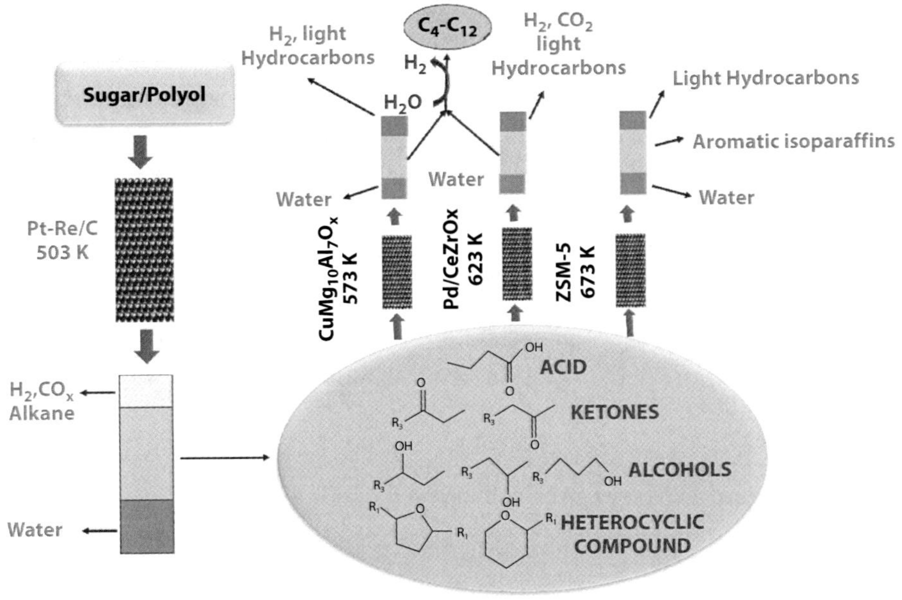

Figure 15.14 Schematic representation of the generation of biofuel from sugars and polyols involving monofunctional intermediates.

of aldol-derived oxygenates.[90] Apart from these, In 2022 Wu *et al.*[91] reported another effective method, electrocatalytic upgradation of biomass for producing biofuels. In 2012, Nilges *et al.*[92] reported electrocatalytic conversion of levulinic acid to octane via a two-step strategy using Pb as cathode and Al as anode, respectively. Two possible mechanistic pathways are suggested for the generation of n-octane as given in Figure 15.15.[92] Recently (2022), Wu *et al.*[93] reported anodic cross-coupling of biomass-derived chemicals in the presence of Ni-based electrocatalysts (Ni_3N) producing jet fuels at room temperature and atmosphere pressure.

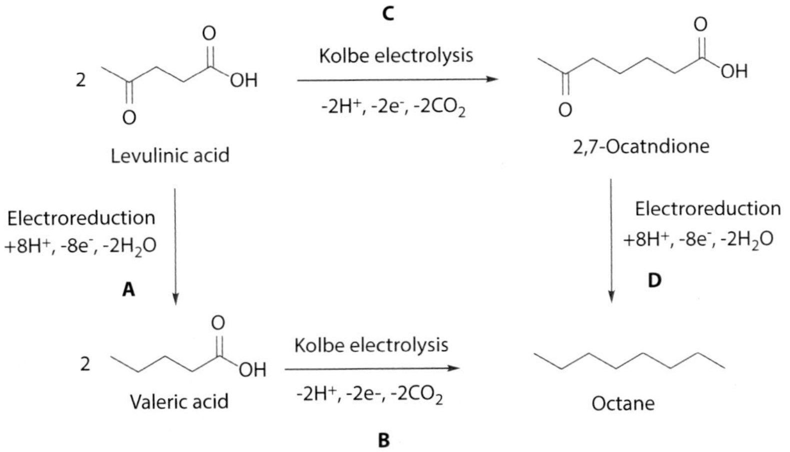

Figure 15.15 Schematic representation of the two electrochemical routes involving generation of n-octane from levulinic acid.

15.4 Selective Transformations of Platform Chemicals

Generation of biofuels from biomass proceeds through the generation of several deoxygenated chemical species. Selective deoxygenated chemical species that can be utilized as substrates for the generation of higher value-added products are termed platform chemicals. As per Department of Energy (DOE), there are 12 biomass-derived building blocks that act as platform chemicals. These include ethanol, 3-hydroxypropionic acid/aldehyde, furfural, HMF (hydroxymethylfurfural), glycerol, 2,5-furandicarboxylic acid, isoprene, succinic acid, lactic acid, levulinic acid, sorbitol, and xylitol. Apart from glycerol and isoprene, other platform chemicals are well-identified. In the upcoming section, the production and processing of a few platform chemicals such as furfural, HMF, levulinic acid (LA), and g-valerolactone (GVL) will be discussed.

15.4.1 Furfural and Hydroxymethylfurfural (HMF)

In lignocellulosic biorefinery, furfural and hydroxymethylfurfural are supreme renewable platform chemicals that are involved in the generation of various value-added chemicals and fuels. Hemicellulose contains large quantities of xylan/xylose, which on hydrolysis and subsequent dehydration produces biobased furfural. Initially, hydrolysis of C_5 polysaccharide produces monomer that on dehydration over an acid catalyst (HCl, 423 K) generated furan compounds. Furfural synthesis is also reported from different biomass feedstock such as corn stalk, sugarcane bagasse, and eucalyptus wood utilizing mineral acids of variable concentration.[94] Hydroxymethylfurfural (HMF) is also obtained from different mono- and polysaccharides utilizing different acid catalysts.[95–97] In biphasic reactors, good selectivity (e.g., 90%) was observed in their generation of furfural from xylose and HMF from fructose.[98] Yield % of HMF from fructose was found to be further improved by the addition of aprotic solvent DMSO.[99, 100] In the presence of excess water, HMF is readily hydrated to produce levulinic acid and formic acid; therefore, it is necessary to minimize water concentration for selective synthesis of HMF. Side reactions can be further suppressed via extracting furfural compounds from aqueous layer utilizing organic solvents (THF, butanol, etc.) or via reducing organic compound solubility in water by the addition of salts.[101, 102] However, in such cases, a downstream separation process will be required that increases the total cost. Román-Leshkov *et al.*[101] utilize phase modifiers to facilitate the generation of HMF from fructose. In a two-phase reactor system, in the aqueous phase, fructose is dehydrated by an acid catalyst and the product HMF is continuously extracted into a modified organic phase (methylisobutylketone) containing 2-butanol as a modifier.[101] Recently (2020), Sayed *et al.*[103] reported an efficient biphasic system containing water/dimethyl carbonate (DMC) for the generation of 5-HMF from fructose, which results in high yield product recovery. Upgradation of HMF to biofuels is given in Figure 15.16. Leshkov *et al.*[104] reported the production of liquid transportation fuel 2,5-dimethylfuran from HMF via hydrogenolysis. In the presence of catalyst Cu–Ru/C, yield was found to be 76%–79%, and over $CuCrO_4$ catalyst, 61% yield was obtained.[104]

In the presence of a basic catalyst such as NaOH, HMF and other furfural products undergo aldol condensation with ketones (acetone) producing long-chain hydrocarbons.[105] C_9 intermediate was formed via condensation of HMF and acetone, which produces C_{15}

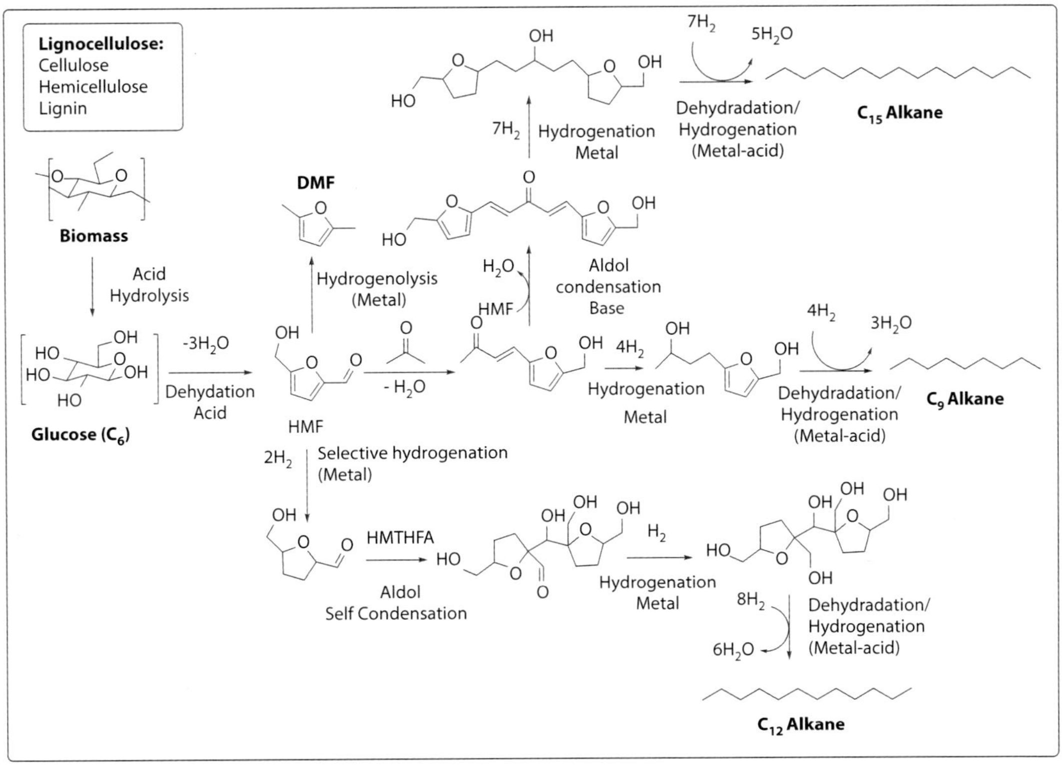

Figure 15.16 Upgradation of HMF to biofuels.

intermediate via reaction with another HMF molecule. Hydrogenation/dehydration over several bifunctional catalysts such as Pd/gAl_2O_3 (373–413 K temperature, 25–52 bar pressure), Pt/$NbPO_5$ (528–568 K temperature, 60 bar pressure) generates long-chain hydrophobic alkanes (C_9, C_{15}) that can be easily separable from water minimizing the purification cost. In the presence of a bifunctional catalyst, Pd/MgO–ZrO_2, carbohydrate-derived compounds such as furfural and 5-hydroxymethylfurfural (HMF) undergo aldol condensation followed by hydrogenation to generate water-soluble intermediates leading to generation of liquid alkanes.[106, 107] HMF and furfural undergo selective hydrogenation to produce 5-hydroxymethyltetrahydrofurfural (HMTHDA) and tetrahydrofurfural (THF2A), which produce C_{12} or C_{10} alkanes after going through self-condensation and hydrogenation/dehydration steps.

15.4.2 Levulinic Acid

Another important platform chemical is levulinic acid (LA), which is obtained from degradation of cellulose, one of the main constituents of biomass. Generation of levulinic acid from the degradation of cellulose is shown in Figure 15.17.[108]

As shown in the above figure, the initial hydrolysis of cellulose produces glucose that is converted to fructose via isomerization. Further dehydration of fructose produces 5-hydroxymethylfurfural (HMF) that on acidic hydrolysis produces levulinic acid. Different high value-added chemicals and biofuels are obtained from levulinic acid such as levulinic

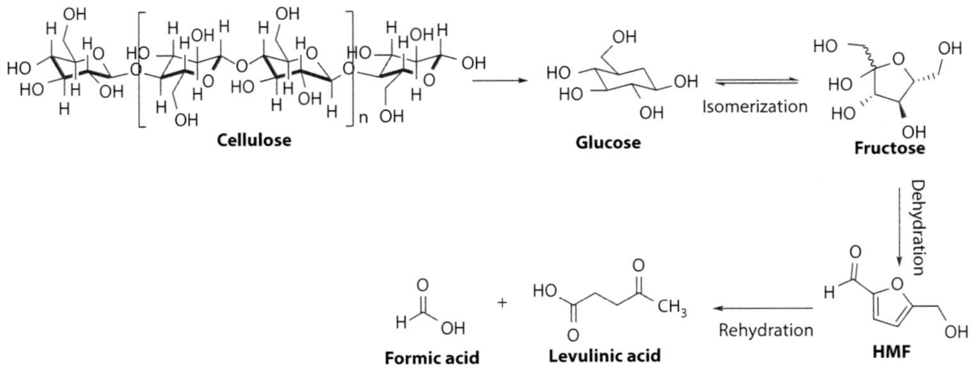

Figure 15.17 Generation of levulinic acid from cellulose.

esters, γ-valerolactone (GVL), valerate esters (VAEs), 2-methyltetrahydrofuran (MTHF), and n-alkanes as shown in Figure 15.18.[109–114] Another effective route for biomass upgradation is base-catalyzed aldol condensation of levulinic acid.[115] Different groups of scientists utilize various catalysts for aldol condensation. MgZr oxide-catalyzed aldol condensation of levulinic acid produces two different condensation products, LA dimer and LA-α-angelica lactone adduct.[116] Base-catalyzed aldol condensation of LA with 5-hydroxymethylfurfural and furfural produces condensation products that on upgradation produce polymers and renewable fuels. GVL produced via hydrogenation of LA also exhibits excellent properties and wide applications. However, higher water solubility and lower octane value limit their activity as fuel additives.[117] GVL may be upgraded to valeric ester through a cascade reaction involving ring opening, hydrogenation, and esterification.[118–120] Conversion of GVL to valeric esters upgrades their properties and widens the applicability in different fields. valerate esters possess higher energy density and acceptable polarity range for conventional fuels, and they behaved as components of gasoline additives or diesel depending on alkyl chain length. Ethyl valerate is considered a second-generation biofuel having valerate esters with shorter alkyl chains.[118] Combustion performance of gasoline was reported to be improved by mixing EV with gasoline in a fixed ratio. VAEs with longer alkyl chain lengths, termed pentyl valerate (PV), behaved as components of diesel based on polarity and ignition properties.

15.4.3 γ-valerolactone

Gamma-valerolactone (GVL) is considered one of the most important platform chemicals derived from lignocellulosic biomass, utilized as fuel additives for the generation of high-density fuels, chemicals, and polymers. GVL can be derived from LA, its esters, and furfural. Conversion of LA to GVL involves a few basic steps such as long-chain hydrogenation, dehydration, and cyclization. The conversion may proceed through the initial hydrogenation of the carbonyl group of LA producing 4-hydroxyvaleric acid, followed by hydroxyl and carboxyl group dehydration and cyclization producing GVL. In a different manner, initially, carbonyl group of LA may undergo enolization that on dehydration by intramolecular esterification produces angelica lactone, and finally, hydrogenation of the lactone produces GVL (Figure 15.19). Depending on the reaction condition, catalysts,

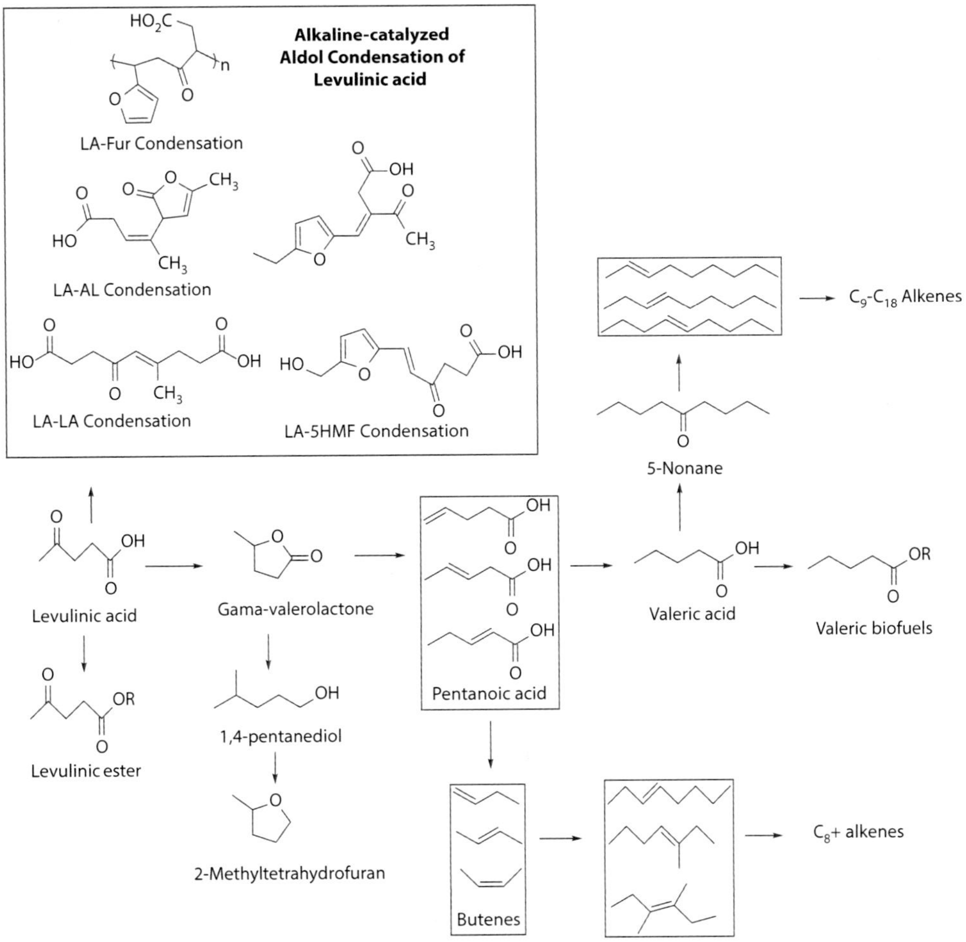

Figure 15.18 Different high value-added chemicals and fuels derived from levulinic acid.

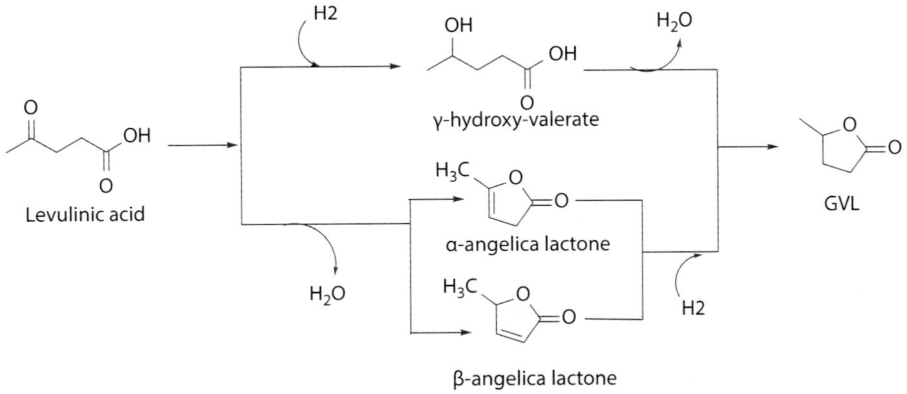

Figure 15.19 Fundamental steps for the synthesis of GVL from levulinic acid.

and solvent used, the reaction may proceed through either the abovementioned pathway. Different supported metal catalysts such as copper/silica, Ni/H-ZSM-5, bimetallic catalysts were used by different groups for conversion of LA to GVL.[121–123] Verma *et al.*[124] utilized bimetallic catalyst AgPd@g-C_3N_4 for visible light-induced conversion of LA to GVL.

Levulinate esters may also be used as precursors for GVL synthesis. GVL is obtained from LA via catalytic transfer hydrogenation utilizing methanol as a hydrogen source, non-noble metal catalysts, and mild reaction conditions.[125–127] Li *et al.*[128] reported low-cost bimetallic catalysts Ni-Fe/AC for hydrogenation of ethyl levulinate (EL) generating γ-valerolactone. Zhou *et al.*[129] utilize highly cross-linked polymer of divinylbenzene-supported Co on TiO_2 to obtain gamma-valerolactone (GVL). In the presence of heterogeneous catalyst N-heterocyclic carbene ruthenium complex grafted on multi-walled carbon nanotube, the conversion efficiency of ethyl levulinate into GVL was found to be increased over 99%.[130]

GVL can also be synthesized from furfural. Furfural is an industrially important compound obtained from acidic hydrolysis of pentosan-rich agricultural wastes. Due to the low cost and high abundance of furfural, researchers paid much interest in different synthetic approaches of GVL starting from furfural. In the generation of GVL from furfural, the initial stage involved metal-catalyzed reduction of furfural into furfuryl alcohol. Furfuryl alcohol thus formed, partly converted to furfuryl ether (FE) in the presence of acid catalyst and via Brønsted acid-catalyzed alcoholysis; these furfuryl alcohol and furfuryl ether produce levulinates (PL). The majority of the levulinate (PL) fraction is converted to 4-hydroxypentanoates (4-HPs) in the presence of an acid catalyst, whereas furanone (MF) is formed in minor amounts, and finally, GVL is formed via lactonization as observed in Figure 15.20. Different groups of scientists utilize Lewis acid catalysts and Brønsted acid catalysts to

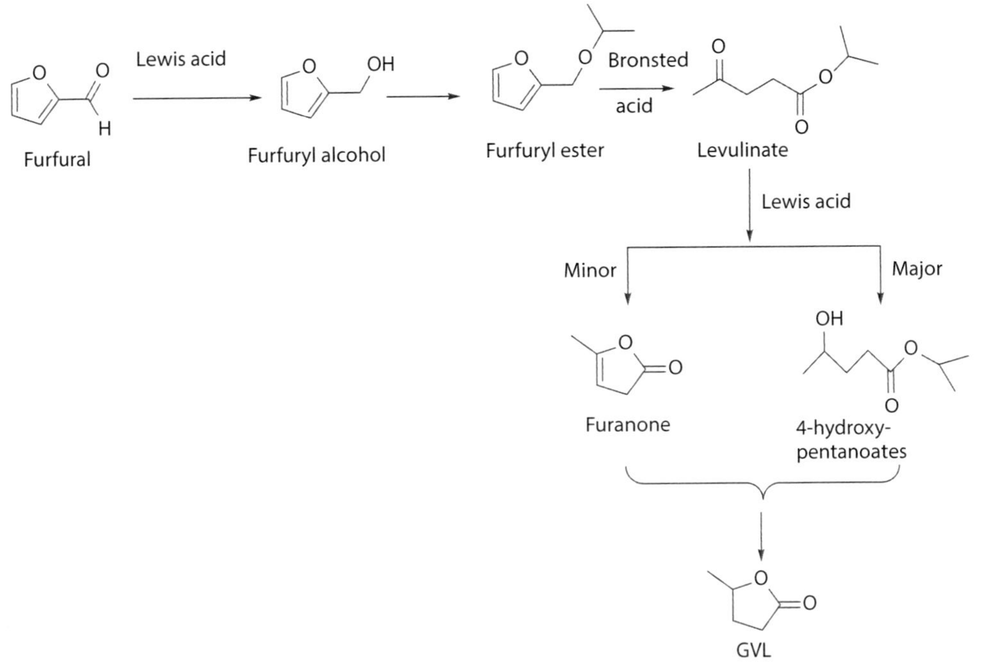

Figure 15.20 Fundamental steps for the synthesis of GVL from furfural.

improve the yield percentage of GVL.[131–133] Zhang *et al.*[134] reported a good yield (79%) of GVL using $FeCl_3·6H_2O$ as a catalyst. Li *et al.*[135] reported $HfCl_4$-mediated acid–base bifunctional catalytic system for the conversion of furfural to GVL. Hydrolysis of $HfCl_4$ generated Lewis acid Hf^{4+}, strong Bronsted acid HCl, and mild base $HfO(OH)_2·xH_2O$, which synergistically contributes to the cascade reaction processes.[135] Zhou *et al.* [136] reported a two-step process for GVL production from furfural. The initial step involved $SnCl_4$-catalyzed conversion of furfural to levulinic acid and formic acid, followed by generation of GVL from this feedstock utilizing Au-Ni/ZrO_2 catalyst.[136]

Although GVL is considered to be a better gasoline additive than ethanol, however, it suffers from several limitations such as high water solubility and lower octane number. Therefore, further upgradation is done by several groups of scientists. Alonso *et al.*[137]

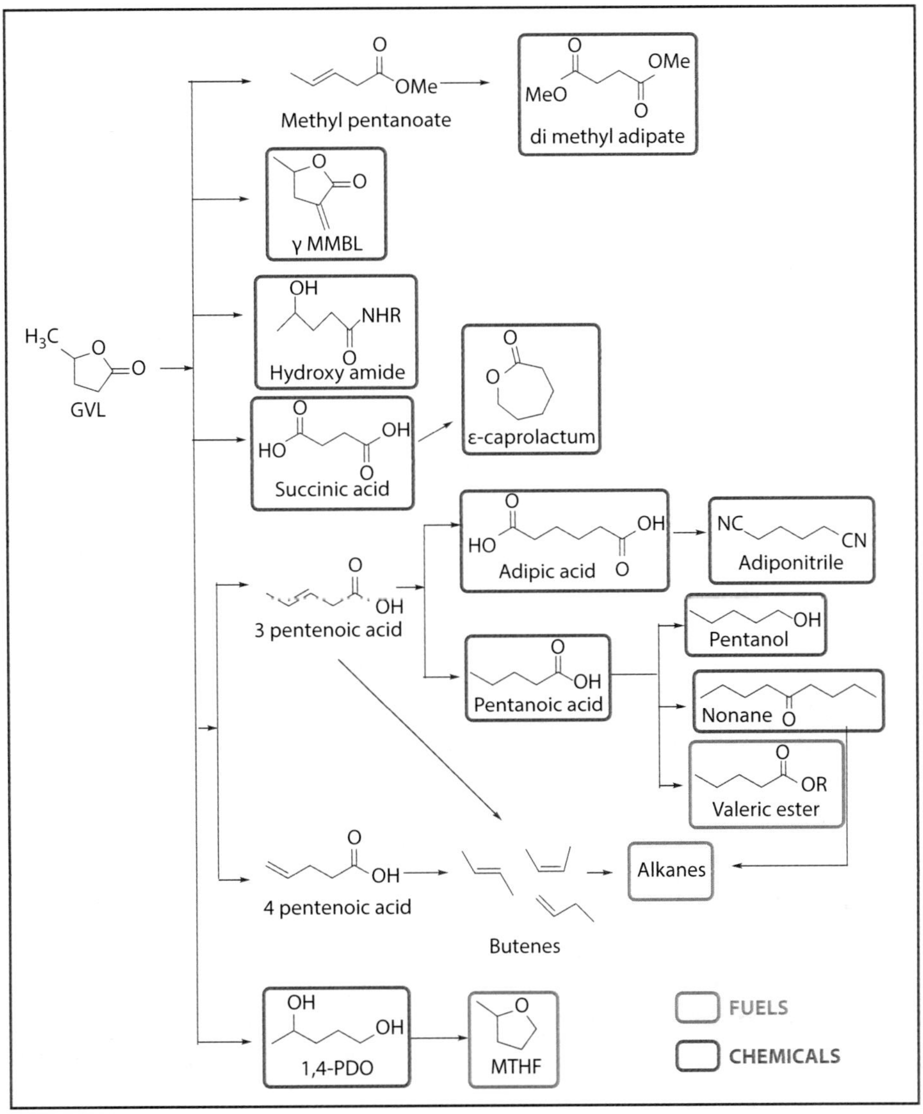

Figure 15.21 Chemicals and fuels obtained from GVL upgradation.

reported the improved conversion of GVL into branched alkanes to make the molecular weight of branched alkanes suitable for liquid transportation fuels. Kang *et al.*[138] reported one-pot conversion of GVL to hydrocarbon oil without using any external hydrogen source. Figure 15.21 shows a few chemicals and fuels that can be obtained via upgradation of GVL.

15.5 Conclusions and Future Perspectives

In this current era of rapid energy consumption, biofuels may be used as sustainable fossil fuel alternatives, alleviating world energy and economic crises. Depending on the nature of biomass feedstock, biofuels are classified into three main categories, first-generation biofuels obtained from edible biomass, second-generation biofuels obtained from nonedible biomass waste, and third-generation biofuels obtained from algae-based feedstock. Among them, second-generation biofuels gain major attention, since they involve conversion of biomass leftovers and trashes to biofuels, thereby reducing waste disposal and environmental issues. The most common type of biomass obtained in such leftovers and trash is lignocellulosic biomass, which acts as abundant feedstock for second-generation biofuel production. The present book chapter mainly focuses on the generation of second-generation biofuels from lignocellulosic feedstock. In comparison to traditional fuels, biofuels are much environment-friendly, nontoxic, less flammable, and associated with high power generation. However, the practical utilization of biofuels suffers from several limitations such as supply and distribution of biofuels. In addition, biofuels are expensive in comparison to fossil fuels and require a special supply chain system. Regular utilization of biofuel requires improvement of lignocellulosic biorefinery techniques via modification of processing strategies producing important platform chemicals and fuels. Extensive catalysis research should be carried out regarding biomass deconstruction, fully integrated biorefinery via coupling of hydrolytic and thermochemical methods, separation and purification techniques, and utilization of the lignin portion of lignocellulosic biomass. Advancement in biological arenas is further necessary. Determination of specific strategy for the conversion of particular biomass is also important. Valued platform chemicals such as levulinic acid and γ-valerolactone are easily prepared via hydrolysis method, but this method is not suitable for biomass containing triglycerides and lignin in larger fraction. For low carbohydrate-containing biomass, other techniques such as hydrotreating, pyrolysis, and gasification may be applied. Finally, proper understanding of chemical reactions occurring during the generation of biofuel from biomass feedstock is necessary, which will help in designing further techniques for the straightforward production of biofuels that added flexibility and integrity to biofuel production. The present book chapter summarizes the generation of biofuels from lignocellulosic biomass, and different chemical reactions that occurred during such conversion are briefly discussed. We hope that the present book chapter provides a good understanding to the reader and will be helpful in modifying existing biomass conversion techniques for large-scale applications.

References

1. Alalwan, H.A., Alminshid, A.H., Aljaafari, H.A.S., Promising evolution of biofuel generations. Subject review, *Renew. Energy Focus.*, 28, 127–139, 2019.
2. Lee, R.A. and Lavoie, J.-M., From first- to third-generation biofuels: Challenges of producing a commodity from a biomass of increasing complexity. *Anim. Front.*, 3, 6–11, 2013.
3. Behera, S., Singh, R., Arora, R., Sharma, N.K., Shukla, M., Kumar, S., Scope of algae as third generation biofuels. *Front. Bioeng. Biotechnol.*, 2, 90, 2015.
4. Alonso, D.M., Bond, J.Q., Dumesic, J.A., Catalytic conversion of biomass to biofuels. *Green Chem.*, 12, 1493–1513, 2010.
5. Rulli, M.C., Bellomi, D., Cazzoli, A., Carolis, G.D., Odorico, P.D., The water-land-food nexus of first-generation biofuels. *Sci. Rep.*, 6, 22521, 2016.
6. Bhuiya, M.M.K., Rasul, M.G., Khan, M.M.K., Ashwath, N., Azad, A.K., Hazrat, M.A., Second Generation Biodiesel: Potential Alternative to-edible Oil-derived Biodiesel. *Energy Proc.*, 61, 1969–1972, 2014.
7. Alaswad, A., Dassisti, M., Prescott, T., Olabi, A.G., Technologies and developments of third generation biofuel production. *Renew. Sustain. Energy Rev.*, 51, 1446–1460, 2015.
8. Battista, F., Frison, N., Pavan, P., Cavinato, C., Gottardo, M., Fatone, F., Eusebi, A.L., Majone, M., Zeppilli, M., Valentino, F., Fino, D., Tommasi, T., Bolzonella, D., Food wastes and sewage sludge as feedstock for an urban biorefinery producing biofuels and added-value bioproducts. *J. Chem. Technol. Biotechnol.*, 95, 328–338, 2020.
9. Bora, A. P., Gupta, D.P., Durbha, K.S., Sewage sludge to bio-fuel: A review on the sustainable approach of transforming sewage waste to alternative fuel. *Fuel*, 259, 116262, 2020.
10. Saravanan, A., Senthil Kumar, P., Jeevanantham, S., Karishma, S., Recent advances and sustainable development of biofuels production from lignocellulosic biomass. *Bioresour. Technol.*, 344, 126203, 2022.
11. Wyman, C.E., Dale, B.E., Elander, R.T., Holtzapple, M., Ladisch, M.R., Lee, Y.Y., Coordinated development of leading biomass pretreatment technologies, *Bioresour. Technol.*, 96, 1959–1966, 2005.
12. Wyman, C.E., Dale, B.E., Elander, R.T., Holtzapple, M., Ladisch, M.R., Lee, Y.Y., Comparative sugar recovery data from laboratory scale application of leading pretreatment technologies to corn stover, *Bioresour. Technol.*, 96, 2026–2032, 2005.
13. Lloyld, T.A. and Wyman, C.E., Combined sugar yields for dilute sulfuric acid pretreatment of corn stover followed by enzymatic hydrolysis of the remaining solids. *Bioresour. Technol.*, 96, 1967–1977, 2005.
14. Rinaldi, R. and Schuth, F., Acid Hydrolysis of Cellulose as the Entry Point into Biorefinery Schemes. *ChemSusChem*, 2, 1096–1107, 2009.
15. Huang, L.Z., Ma, M.G., Ji, X.X., Choi, S.E., Si, C., Recent Developments and Applications of Hemicellulose From Wheat Straw: A Review. *Front. Bioeng. Biotechnol.*, 9, 690773, 2021.
16. Mamman, A.S., Furfural: Hemicellulose/xylosederived biochemical. *Biofuels, Bioprod. Biorefin.*, 2, 438–454, 2008.
17. Saha, B.C., Hemicellulose bioconversion. *J. Ind. Microbiol. Biotechnol.*, 30, 279–291, 2003.
18. Zaldivar, J., Nielsen, J., Olsson, L., Fuel ethanol production from lignocellulose: a challenge for metabolic engineering and process integration. *Appl. Microbiol. Biotechnol.*, 56, 17–34, 2001.
19. Yoo, C.G., Meng, X., Pu, Y., Ragauskas, A.J., The critical role of lignin in lignocellulosic biomass conversion and recent pretreatment strategies: A comprehensive review. *Bioresour. Technol.*, 301, 122784, 2020.
20. Zhang, Y. and Naebe, M., Lignin: A Review on Structure, Properties, and Applications as a Light-Colored UV Absorber. *ACS Sustain. Chem. Eng.*, 9, 4, 1427–1442, 2021.

21. d. Lasa, H., Salaices, E., Mazumder, J., Lucky, R., Catalytic Steam Gasification of Biomass: Catalysts, Thermodynamics and Kinetics, *Chem. Rev.*, 111, 5404–5433, 2011.
22. Bridgwater, A.V., Developments in Thermochemical Biomass Conversion. *Springer Science & Business Media*, vol. 1, Springer, New York City, 2013.
23. Soares, R.R., Simonetti, D.A., Dumesic, J.A., Glycerol as a Source for Fuels and Chemicals by Low-Temperature Catalytic Processing. *Angew. Chem. Int. Ed.* 2006, 45, 3982 –3985.
24. Hayes, D.J., An examination of biorefining processes, catalysts and challenges. *Catal. Today*, 145, 138–151, 2009.
25. Carlson, T.R., Tompsett, G.A., Conner, W.C., Huber, G.W., Aromatic Production from Catalytic Fast Pyrolysis of Biomass-Derived Feedstocks. *Top. Catal.*, 52, 241–252, 2009.
26. Neupane, S., Adhikari, S., Wang, Z., Ragauskas, A.J., Pu, Y., Effect of torrefaction on biomass structure and hydrocarbon production from fast pyrolysis. *Green Chem.*, 17, 2406–2417, 2015.
27. Opatokun, S.A., Kan, T., Shoaibi, A.A., Srinivasakannan, C., Strezov, V., Characterization of Food Waste and Its Digestate as Feedstock for Thermochemical Processing. *Energy Fuels*, 30, 3, 1589–1597, 2016.
28. Wang, X., Liu, Y., Cui, X., Xiao, J., Lin, G., Chen, Y., Yang, H., Chen, H., Production of furfural and levoglucosan from typical agricultural wastes via pyrolysis coupled with hydrothermal conversion: Influence of temperature and raw materials. *Waste Manage.*, 114, 43–52, 2020.
29. Yang, H., Li, S., Liu, B., Chen, Y., Xiao, J., Dong, Z., Gong, M., Chen, H., Hemicellulose pyrolysis mechanism based on functional group evolutions by two-dimensional perturbation correlation infrared spectroscopy. *Fuel*, 2020, 267, 117302.
30. Yogalakshmi, K.N., Sivashanmugam, P., Kavitha, S., Kannah, Y., Varjani, S., AdishKumar, S., Kumar, G., Lignocellulosic biomass-based pyrolysis: A comprehensive review. *Chemosphere*, 286, 131824, 2022.
31. Cui, Y., Wang, W., Chang, J., Study on the product characteristics of pyrolysis lignin with calcium salt additives. *Materials*, 12, 1609, 2019.
32. Hwang, H., Lee, J.H., Choi, I.G., Choi, J.W., Comprehensive characterization of hydrothermal liquefaction products obtained from woody biomass under various alkali catalyst concentrations. *Environ. Technol.*, 40, 1657–1667, 2019.
33. de Caprariis, B., De Filippis, P., Petrullo, A., Scarsella, M., Hydrothermal liquefaction of biomass: Influence of temperature and biomass composition on the bio-oil production. *Fuel*, 208, 618–625, 2017.
34. Zhu, Z., Rosendahl, L., Toor, S.S., Yu, D., Chen, G., Hydrothermal liquefaction of barley straw to bio-crude oil: Effects of reaction temperature and aqueous phase recirculation. *Appl. Energy*, 137, 183–192, 2015.
35. Govindasamy, G., Sharma, R., Subramanian, S., Studies on the effect of heterogeneous catalysts on the hydrothermal liquefaction of sugarcane bagasse to low-oxygen-containing bio-oil. *Biofuels*, 10, 665–675, 2019.
36. Oliveira, A.L.P.C., Almeida, P.S., Campos, M.C.V., Franceschi, E., Dariva, C., Borges, G.R., Thermoliquefaction of palm oil fiber (Elaeis sp.) using supercritical ethanol. *Bioresour. Technol.*, 230, 1–7, 2017.
37. Zhai, Q., Li, F., Wang, F., Xu, J., Jiang, J., Cai, Z., Liquefaction of poplar biomass for value-added platform chemicals. *Cellulose*, 25, 4663–4675, 2018.
38. Liu, D., Li, Q., Zhao, A., Song, L., Wu, P., Yan, Z., Hydro-liquefaction of sawdust and its three components in supercritical ethanol with [BMIM] Cl/NiCl2 catalyst. *Chem. Eng. J.*, 279, 921–928, 2015.
39. Kima, J.-Y., Lee, H.W., Lee, S.M., Jae, J., Park, Y.-K., Overview of the recent advances in lignocellulose liquefaction for producing biofuels, bio-based materials and chemicals. *Bioresour. Technol.*, 279, 373–384, 2019.

40. Feng, S., Wei, R., Leitch, M., Charles, C., Comparative study on lignocellulose liquefaction in water, ethanol, and water/ethanol mixture: Roles of ethanol and water. *Energy*, 155, 234–241, 2018.
41. Feng, J., Jiang, J., Hse, C., Yang, Z., Wang, K., Ye, J., Xu, J., Selective catalytic conversion of waste lignocellulosic biomass for renewable value-added chemicals via directional microwave-assisted liquefaction. *Sustain. Energy Fuels*, 2, 1035–1047, 2018.
42. Xie, J., Hse, C.-Y., Shupe, T.F., Hu, T., Influence of solvent type on microwave-assisted liquefaction of bamboo. *Eur. J. Wood Wood Prod.*, 74, 249–254, 2016.
43. Zhou, R., Zhou, R., Wang, S., Lan, Z., Zhang, X., Yin, Y., Tu, S., Yang, S., Ye, L., Fast liquefaction of bamboo shoot shell with liquid-phase microplasma assisted technology. *Bioresour. Technol.*, 218, 1275–1278, 2016.
44. Xu, J., Xie, X., Wang, J., Jiang, J., Directional liquefaction coupling fractionation of lignocellulosic biomass for platform chemicals. *Green Chem.*, 18, 3124–3138, 2016.
45. Ghosh, A., Brownb, R.C., Bai, X., Production of solubilized carbohydrate from cellulose using non-catalytic, supercritical depolymerization in polar aprotic solvents. *Green Chem.*, 18, 1023–1031, 2016.
46. Brand, S. and Kim, J., Liquefaction of major lignocellulosic biomass constituents in supercritical ethanol. *Energy*, 80, 64–74, 2015.
47. Yang, H., Yan, R., Chen, H., Lee, D.H., Zheng, C., Characteristics of hemicellulose, cellulose and lignin pyrolysis. *Fuel*, 86, 1781–1788, 2007.
48. Liu, H.-M., Li, H.-Y., Li, M.-F., Bio-Oil from Hydro-Liquefaction of Bagasse in Supercritical Ethanol. *J. Energy Inst.*, 90, 734–742, 2017.
49. Elliott, D.C., Historical Developments in Hydroprocessing Bio-oils. *Energy Fuels*, 21, 1792–1815, 2007.
50. Rafiqul, I.S.M., Sakinah, A.M.M., Zularisam, A.W., Waste biomass management-A holistic approach. vol. 2, pp. 73–106, Springer, New York City, 2017.
51. Sasmal, S. and Mohanty, K., Pretreatment of Lignocellulosic Biomass Toward Biofuel Production. *Biorefining of biomass to biofuels: opportunities and perception*, pp. 203–221, Springer, New York City, 2018.
52. Liaw, W.-C., Chen, C.-S., Chang, W.-S., Chen, K.-P., Xylitol Production from Rice Straw Hemicellulose Hydrolyzate by Polyacrylic Hydrogel Thin Films with Immobilized Candida subtropicalis WF79. *J. Biosci. Bioeng.*, 105, 97–105, 2008.
53. Rafiqul, I.S.M. and Sakinah, A.M.M., Processes for the production of xylitol—a review. *Food Rev. Int.*, 29, 127–56, 2013.
54. Amirkhani, H., Yunus, R., Rashid, U., Salleh, S.F., Radhiah, A.B.D., Syam, S., Low-Temperature Dilute Acid Hydrolysis of Oil Palm Frond. *Chem. Eng. Commun.*, 202, 1235–44, 2015.
55. Sun, Y. and Cheng, J.J., Dilute acid pretreatment of rye straw and bermudagrass for ethanol production. *Bioresour. Technol.*, 96, 1599–1606, 2005.
56. Tada, K., Kanno, T., Horiuchi, J-i, Enhanced Production of Bioxylitol from Corn Cobs by Candida magnoliae. *Ind. Eng. Chem. Res.*, 51, 10008–10014, 2012.
57. Rafiqul, I.S.M., Sakinah, A.M.M., Zularisam, A.W., Enzymatic production of bioxylitol from sawdust hydrolysate: screening of process parameters. *Appl. Biochem. Biotechnol.*, 176, 1071–1083, 2015.
58. Phillips, M. and Goss, M.J., The hydrolysis of lignin with 12% hydrochloric acid. *J. Am. Chem. Soc.*, 54, 3374–3377, 1932.
59. Huber, G.W., Chheda, J.N., Barrett, C.J., Dumesic, J.A., Production of liquid alkanes by aqueous-phase processing of biomass-derived carbohydrates. *Science*, 308, 1446–1450, 2005.
60. Huber, G.W., Cortright, R.D., Dumesic, J.A., Renewable alkanes by aqueous-phase reforming of biomass-derived oxygenates. *Angew. Chem. Int. Ed.*, 43, 1549–1551, 2004.

61. Garrote, G., Dominguez, H., Parajó, J.C., Hydrothermal processing of lignocellulosic materials. *Holz als Roh- und Werkstoff.*, 57, 191–202, 1999.
62. Lavoie, J.-M., Baré, W., Bilodeau, M., Depolymerization of steam-treated lignin for the production of green chemicals. *Bioresour. Technol.*, 102, 4917–4920, 2011.
63. Zhang, S., Xiao, J., Wang, G., Chen, G., Enzymatic hydrolysis of lignin by ligninolytic enzymes and analysis of the hydrolyzed lignin products. *Bioresour. Technol.*, 304, 122975, 2020.
64. Li, H., Xue, Y., Wu, J., Wu, H., Qin, G., Li, C., Ding, J., Liu, J., Gan, L., Long, M., Enzymatic hydrolysis of hemicelluloses from Miscanthus to monosaccharides or xylo-oligosaccharides by recombinant hemicellulases. *Ind. Crops Prod.*, 79, 170–179, 2016.
65. Mai, F., Wen, Z., Bai, Y., Ma, Z., Cui, K., Wu, K., Yan, F., Chen, H., Li, Y., Selective Conversion of Enzymatic Hydrolysis Lignin into Alkylphenols in Supercritical Ethanol over a WO3/γ-Al2O3 Catalyst. *Ind. Eng. Chem. Res.*, 58, 10255–10263, 2019.
66. Fasolini, A., Cucciniello, R., Paone, E., Mauriello, F., T. Tabanelli, T., A short overview on the hydrogen production via aqueous phase reforming (APR) of cellulose, C6-C5 sugars and polyols. *Catalysts.*, 9, 917, 2019.
67. Huber, G.W., Shabaker, J.W., Dumesic, J.A., Raney Ni-Sn Catalyst for H2 Production from Biomass-Derived Hydrocarbons. *Science*, 300, 2075–2077, 2003.
68. Huber, G.W., Cortright, R.D., Dumesic, J.A., Renewable alkanes by aqueous-phase reforming of biomass-derived oxygenates. *Angew. Chem. Int. Ed.*, 43, 1549–1551, 2004.
69. Davda, R.R., Shabaker, J.W., Huber, G.W., Cortright, R.D., Dumesic, J.A., Aqueous-phase reforming of ethylene glycol on silica-supported metal catalysts. *Appl. Catal., B*, 43, 13–26, 2003.
70. Jiang, T., Wang, T., Ma, L., Li, Y., Zhang, Q., Zhang, X., Investigation on the xylitol aqueous-phase reforming performance for pentane production over Pt/HZSM-5 and Ni/HZSM-5 catalysts. *Appl. Energy*, 90, 51–57, 2012.
71. Jiang, T., Zhang, Q., Wang, T.J., Zhang, Q., Ma, L.L., High yield of pentane production by aqueous-phase reforming of xylitol over Ni/HZSM-5 and Ni/MCM22 catalysts. *Energy Convers. Manage.*, 59, 58–65, 2012.
72. Godina, L., II, Kirilin, A.V., Tokarev, A.V., Simakova, I.L., Murzin, D.Y., Sibunit-supported mono-and bimetallic catalysts used in aqueous-phase reforming of xylitol. *Ind. Eng. Chem. Res.*, 57, 2050–2067, 2018.
73. Kirilin, A.V., Tokarev, A.V., Manyar, H., Hardacre, C., Salmi, T., Mikkola, J.P., Murzin, D.Y., Aqueous phase reforming of xylitol over Pt-Re bimetallic catalyst: Effect of the Re addition. *Catal. Today*, 223, 97–107, 2014.
74. Kunkes, E.L., Simonetti, D.A., West, R.M., SerranoRuiz, J.C., Gartner, C.A., Dumesic, J.A., Catalytic conversion of biomass to monofunctional hydrocarbons and targeted liquid-fuel classes. *Science*, 322, 417–421, 2008.
75. Soares, R.R., Simonetti, D.A., Dumesic, J.A., Glycerol as a source for fuels and chemicals by low-temperature catalytic processing. *Angew. Chem. Int. Ed.*, 45, 3982–3985, 2006.
76. Patil, P.D. and Deng, S., Optimization of biodiesel production from edible and non-edible vegetable oils. *Fuel*, 88, 1302–1306, 2009.
77. Wen, G., Xu, Y., Xu, Z., Tian, Z., Characterization and Catalytic Properties of the Ni/Al2O3 Catalysts for Aqueous-phase Reforming of Glucose. *Catal. Lett.*, 129, 250–257, 2009.
78. Kumar, P., Barrett, D.M., Delwiche, M.J., Stroeve, P., Methods for pretreatment of lignocellulosic biomass for efficient hydrolysis and biofuel production. *Ind. Eng. Chem. Res.*, 48, 3713–3729, 2009.
79. Simonetti, D.A. and Dumesic, J.A., Catalytic Strategies for Changing the Energy Content and Achieving C-C Coupling in Biomass-Derived Oxygenated Hydrocarbons, Simonetti. *ChemSusChem: Chemistry & Sustainability Energy & Materials*, 1, 8–9, 725–733, 2008.

80. Barrett, C.J., Chheda, J.N., Huber, G.W., Dumesic, J.A., Single-reactor process for sequential aldol-condensation and hydrogenation of biomass-derived compounds in water. *Appl. Catal. B*, 66, 111–118, 2006.
81. Chheda, J.N. and Dumesic, J.A., An overview of dehydration, aldol-condensation and hydrogenation processes for production of liquid alkanes from biomass-derived carbohydrates. *Catal. Today*, 123, 59–70, 2007.
82. Renz, M., Ketonization of carboxylic acids by decarboxylation: mechanism and scope. *Eur. J. Org. Chem.*, 2005, 979–988, 2005.
83. Corma, A., Renz, M., Schaverien, C., Coupling fatty acids by ketonic decarboxylation using solid catalysts for the direct production of diesel, lubricants, and chemicals. *ChemSusChem*, 1, 739–741, 2008.
84. Dooley, K.M., Bhat, A.K., Plaisance, C.P., Roy, A.D., Ketones from acid condensation using supported CeO2 catalysts: Effect of additives. *Appl. Catal. A*, 320, 122–133, 2007.
85. Cheng, S., Wei, L., Julson, J., Muthukumarappan, K., Kharel, P.R., Upgrading pyrolysis bio-oil to hydrocarbon enriched biofuel over bifunctional Fe-Ni/HZSM-5 catalyst in supercritical methanol. *Fuel Process. Technol.*, 167, 117–126, 2017.
86. Blommel, P.G., Keenan, G.R., Rozmiarek, R.T., Cortrigth, R.D., Catalytic conversion of sugar into conventional gasoline, diesel, jet fuel, and other hydrocarbons. *Int. Sugar J.*, 110, 672, 2008.
87. Kunkes, E.L., Gurbuz, E., II, Dumesic, J.A., Vapour-phase C–C coupling reactions of biomass-derived oxygenates over Pd/CeZrOx catalysts. *J. Catal.*, 266, 236–249, 2009.
88. Gürbüz, E., II, Kunkesa, E.L., Dumesic, J.A., Dual-bed catalyst system for C–C coupling of biomass-derived oxygenated hydrocarbons to fuel-grade compounds. *Green Chem.*, 12, 223–227, 2010.
89. Liu, C., Evans, T.J., Cheng, L., Nimlos, M.R., Mukarakate, C., Robichaud, D.J., Assary, R.S., Curtiss, L.A., Catalytic Upgrading of Biomass-Derived Compounds via C–C Coupling Reactions: Computational and Experimental Studies of Acetaldehyde and Furan Reactions in HZSM-5. *J. Phys. Chem. C*, 119, 24025–24035, 2015.
90. Yeboah, I., Feng, X., Rout, K.R., Chen, D., Versatile one-pot tandem conversion of biomass-derived light oxygenates into high-yield jet fuel range aromatics. *Ind. Eng. Chem. Res.*, 60, 15095–15105, 2021.
91. Wu, Y., Guo, Z., Sun, C., Ren, X., Li, Q., High-efficiency electrochemical hydrogenation of biomass-derived benzaldehyde compounds via a durable and versatile dendritic-like Pd/Cu-CF electrocatalyst. *Fuel Process. Technol.*, 237, 107436, 2022.
92. Nilges, P., dos Santos, T.R., Harnischa, F., Schröder, U., Electrochemistry for biofuel generation: Electrochemical conversion of levulinic acid to octane. *Energy Environ. Sci.*, 5, 5231–5235, 2012.
93. Wu, J., Xu, L., Li, Y., Dong, C.-L., Lu, Y., Nga, T.T.T., Kong, Z., Li, S., Zou, Y., Wang, S., Anodic cross-coupling of biomass platform chemicals to sustainable biojet fuel precursors. *J. Am. Chem. Soc*, 144, 23649–23656, 2022.
94. Barbosa, O., Ortiz, C., Murcia, Á. B-, Torres, R., Rodrigues, R.C., Lafuente, R., Glutaraldehyde in bio-catalysts design: A useful crosslinker and a versatile tool in enzyme immobilization. *RSC Adv.*, 4, 1583–1600, 2014.
95. Mulk, S., Sajid, M., Wang, L., Liu, F., Pan, G., Catalytic conversion of sucrose to 5-hydroxymethylfurfural in green aqueous and organic medium. *J. Environ. Chem. Eng.*, 10, 106613, 2022.
96. Zhao, Y., Lu, K., Xu, H., Qu, Y., Zhu, L., Wang, S., Comparative study on the dehydration of biomass-derived disaccharides and polysaccharides to 5-hydroxymethylfurfural. *Energy Fuels*, 33, 9985–9995, 2019.

97. Hu, L., Jiang, Y., Wu, Z., Wang, X., He, A., Xu, J., Xu, J., State-of-the-art advances and perspectives in the separation of biomass-derived 5-hydroxymethylfurfural. *J. Cleaner Prod.*, 276, 124219, 2020.
98. Chheda, J.N. and Dumesic, J.A., An overview of dehydration, aldol-condensation and hydrogenation processes for production of liquid alkanes from biomass-derived carbohydrates. *Catal. Today*, 123, 59–70, 2007.
99. Chheda, J.N., Roman-Leshkov, Y., Dumesic, J.A., Production of 5-hydroxymethylfurfural and furfural by dehydration of biomass-derived mono-and poly-saccharides. *Green Chem.*, 9, 342–350, 2007.
100. Szmant, H.H. and Chundury, D.D., The preparation of 5-hydroxymethylfurfuraldehyde from high fructose corn syrup and other carbohydrates. *J. Chem. Technol. Biotechnol.*, 31, 135–145, 1981.
101. Roman-Leshkov, Y., Chheda, J.N., Dumesic, J.A., Phase modifiers promote efficient production of hydroxymethylfurfural from fructose. *Science*, 312, 1933–1937, 2006.
102. Roman-Leshkov, Y., Barrett, C.J., Liu, Z.Y., Dumesic, J.A., Production of dimethylfuran for liquid fuels from biomass-derived carbohydrates. *Nature*, 447, 982–985, 2007.
103. Sayed, M., Warlin, N., Hulteberg, C., Munslow, I., Lundmark, S., Tunå, P., Zhang, B., Pyo, S.-H., -Kaul, R.H., 5-Hydroxymethylfurfural from fructose: an efficient continuous process in a water-dimethyl carbonate biphasic system with high yield product recovery. *Green Chem.*, 22, 5402–5413, 2020.
104. Leshkov, Y., Barrett, C.J., Liu, Z.Y., Dumesic, J.A., Production of dimethylfuran for liquid fuels from biomass-derived carbohydrates. *Nature*, 447, 982–985, 2007.
105. West, R.M., Liu, Z.Y., Peter, M., Dumesic, J.A., Liquid alkanes with targeted molecular weights from biomass-derived carbohydrates. *ChemSusChem*, 1, 417–424, 2008.
106. Barrett, C.J., Chheda, J.N., Huber, G.W., Dumesic, J.A., Single-reactor process for sequential aldol-condensation and hydrogenation of biomass-derived compounds in water. *Appl. Catal., B*, 66, 111–118, 2006.
107. Dutta, S., Bhat, N.S., Catalytic Transformation of Biomass-Derived Furfurals to Cyclopentanones and Their Derivatives: A Review. *ACS Omega*, 6, 35145-35172, 2021.
108. Muranaka, Y., Suzuki, T., Sawanishi, H., Hasegawa, I., Mae, K., Effective production of levulinic acid from biomass through pretreatment using phosphoric acid, hydrochloric acid, or ionic liquid. *Ind. Eng. Chem. Res.*, 53, 11611–11621, 2014.
109. Nandiwale, K.Y., Sonar, S.K., Niphadkar, P.S., Joshi, P.N., Deshpande, S.S., Patil, V.S., Bokade, V.V., Catalytic upgrading of renewable levulinic acid to ethyl levulinate biodiesel using dodecatungstophosphoric acid supported on desilicated H-ZSM-5 as catalyst. *Appl. Catal. A Gen.*, 460, 90–98, 2013.
110. Nandiwale, K.Y., Niphadkar, P.S., Deshpande, S.S., Bokade, V.V., Esterification of renewable levulinic acid to ethyl levulinate biodiesel catalyzed by highly active and reusable desilicated H-ZSM-5. *J. Chem. Technol. Biotechnol.*, 89, 1507–1515, 2014.
111. Xin, L., Zhang, Z., Qi, J., Chadderdon, D.J., Qiu, Y., Warsko, K.M., Li, W., Electricity Storage in Biofuels: Selective Electrocatalytic Reduction of Levulinic Acid to Valeric Acid or γ-Valerolactone. *ChemSusChem*, 6, 674–686, 2013.
112. Kon, K., Onodera, W., Shimizu, K.-I., Selective hydrogenation of levulinic acid to valeric acid and valeric biofuels by a Pt/HMFI catalyst. *Catal. Sci. Technol.*, 4, 3227–3234, 2014.
113. Obregón, I., Gandarias, I., Miletić, N., Ocio, A., Arias, P.L., One-Pot 2-Methyltetrahydrofuran production from levulinic acid in green solvents using Ni-Cu/Al2O3 Catalysts. *ChemSusChem*, 8, 3483–3488, 2015.

114. Obregón, I., Gandarias, I., Al-Shaal, M.G., Mevissen, C., Arias, P.L., Palkovits, R., The role of the hydrogen source on the selective production of γ-valerolactone and 2-methyltetrahydrofuran from levulinic acid. *ChemSusChem*, 9, 2488–2495, 2016.
115. Amarasekara, A.S., Wiredu, B., Grady, T.L., Obregon, R.G., Margetić, D., Solid acid catalyzed aldol dimerization of levulinic acid for the preparation of C10 renewable fuel and chemical feedstocks. *Catal. Commun.*, 124, 6–11, 2019.
116. Faba, L., Díaz, E., Ordóñez, S., Base-Catalyzed Condensation of Levulinic Acid: A new biorefinery upgrading approach. *ChemCatChem*, 8, 1490–1494, 2016.
117. Alonso, D.M., Wettstein, S.G., Dumesic, J.A., Gamma-valerolactone, a sustainable platform molecule derived from lignocellulosic biomass. *Green Chem.*, 15, 584–595, 2013.
118. Lange, P., Price, R., Ayoub, P.M., Louis, J., Petrus, L., Clarke, L., Gosselink, H., Valeric biofuels: a platform of cellulosic transportation fuels. *Angew. Chem. Int. Ed. Engl.*, 49, 4479–4483, 2010.
119. Yun, G.-N., Takagaki, A., Kikuchi, R., Oyama, S.T., Hydrodeoxygenation of gamma-valerolactone on transition metal phosphide catalysts. *Catal. Sci. Technol.*, 7, 281–292, 2017.
120. Dong, L.-L., He, L., Tao, G.-H., Hu, C., High yield of ethyl valerate from the esterification of renewable valeric acid catalyzed by amino acid ionic liquids, *RSC Adv.*, 3, 4806–4813, 2013.
121. Upare, P.P., Lee, J.-M., Hwang, Y.K., Hwang, D.W., Lee, J.-H., Halligudi, S.B., Hwang, J.-S., Chang, J.-S., Direct Hydrocyclization of Biomass-Derived Levulinic Acid to 2-Methyltetrahydrofuran over Nanocomposite Copper/Silica Catalysts. *ChemSusChem*, 4, 1749–1752, 2011.
122. Mohan, V., Raghavendra, C., Venkata Pramod, C., Rajua, B.D., Rao, K.S.R., Ni/H-ZSM-5 as a promising catalyst for vapour phase hydrogenation of levulinic acid at atmospheric pressure. *RSC Adv.*, 4, 9660–9668, 2014.
123. Okhlopkova, L.B. and Ismagilov, Z.R., Development of catalysts for the hydrogenation of levulinic acid to gamma-valerolactone in the framework of the problem of cellulose conversion into valuable chemicals. *J. Phys.: Conf. Ser.*, 2021, 012008, 1749.
124. Verma, S., Baig, R.B.N., Nadagouda, M.N., Varma, R.S., Sustainable strategy utilizing biomass: visible-light-mediated synthesis of γ-valerolactone. *ChemCatChem.*, 8, 690–693, 2016.
125. Yi, G. and Zhang, Y., One-Pot Selective Conversion of Hemicellulose (Xylan) to Xylitol under Mild Conditions. *ChemSusChem*, 5, 1383–1387, 2012.
126. Wang, X. and Rinaldi, R., Exploiting H-transfer reactions with RANEY® Ni for upgrade of phenolic and aromatic biorefinery feeds under unusual, low-severity conditions. *Energy Environ. Sci.*, 5, 8244–8260, 2012.
127. Shen, Q., Zhang, Y., Zhang, Y., Tan, S., Chen, J., Transformations of biomass-based levulinate ester into γ-valerolactone and pyrrolidones using carbon nanotubes-grafted N-heterocyclic carbene ruthenium complexes. *J. Energy Chem.*, 39, 29–38, 2019.
128. Li, C., Xu, G., Zhai, Y., Liu, X., Ma, Y., Zhang, Y., Hydrogenation of biomass-derived ethyl levulinate into γ-valerolactone by activated carbon supported bimetallic Ni and Fe catalysts. *Fuel*, 203, 23–31, 2017.
129. Zhou, H., Song, J., Kang, X., Hu, J., Yang, Y., Fan, H., Meng, Q., Han, B., One-pot conversion of carbohydrates into gamma-valerolactone catalyzed by highly cross-linked ionic liquid polymer and Co/TiO2. *RSC Adv.*, 5, 15267–15273, 2015.
130. Qiujuan, S., Yi, Z., Yiping, Z., Shaozao, T., Jinzhu, C., Transformations of biomass-based levulinate ester into γ-valerolactone and pyrrolidones using carbon nanotubes-grafted N-heterocyclic carbene ruthenium complexes. *J. Energy Chem.*, 39, 29–38, 2019.
131. Guarinos, J.M., Cirujano, F.G., Rapeyko, A., i Xamena, F.L., Conversion of levulinic acid to γ-valerolactone over Zr-containing metal-organic frameworks: Evidencing the role of Lewis and Brønsted acid sites. *Mol. Catal.*, 515, 111925, 2021.

132. Li, X., Yang, J., Xu, R., Lu, L., Kong, F., Liang, M., Jiang, L., Nie, S., Si, C., Kinetic study of furfural production from Eucalyptus sawdust using H-SAPO3-4 as solid Brønsted acid and Lewis acid catalysts in biomass-derived solvents. *Ind. Crops Prod.*, 135, 196–205, 2019.
133. Bui, L., Luo, H., Gunther, W.R., Román-Leshkov, Y., Domino reaction catalyzed by zeolites with Brønsted and Lewis acid sites for the production of γ-valerolactone from furfural. *Angew. Chem. Int. Ed.*, 52, 8022–8025, 2013.
134. Zhang, L., Yu, H., Wang, P., Li, Y., Production of furfural from xylose, xylan and corncob in gamma-valerolactone using FeCl3· 6H2O as catalyst. *Bioresour. Technol.*, 151, 355–360, 2014.
135. Li, M., Liu, Y., Lin, X., Tan, J., Yang, S., Li, H., One-step upgrading of bio-based furfural to γ-valerolactone via HfCl4-mediated bifunctional catalysis. *RSC Adv.*, 11, 35415–35424, 2021.
136. Zhou, C., Xiao, Y., Xu, S., Li, J., Hu, C., γ-Valerolactone Production from Furfural Residue with Formic Acid as the Sole Hydrogen Resource via an Integrated Strategy on Au-Ni/ZrO2. *Ind. Eng. Chem. Res.*, 59, 17228–17238, 2020.
137. Alonso, D.M., Bond, J.Q., Ruiza, Dumesic, J.A., Production of liquid hydrocarbon transportation fuels by oligomerization of biomass-derived C 9 alkenes. *Green Chem.*, 12, 992–999, 2010.
138. Kang, S., Fu, J., Ye, Y., Liao, W., Xiao, Y., Yang, P., Liu, G., One-pot production of hydrocarbon oils from biomass derived γ-valerolactone. *Fuel*, 216, 747–751, 2018.

16

Detoxification of Industrial Wastewater by Catalytic (Photo/Bio/Nano) Techniques

Mohd Ehtesham[1], Naushad Ansari[2], Gyanendra Kumar[3], Satendra Kumar[4], Panmei Gaijon[5], Sudipta Ghosh[5], M. Ramananda Singh[5] and Arun Kant[5*]

[1]*Jamia Millia Islamia (Central University), Delhi, India*
[2]*MMH College, Ghaziabad, India*
[3]*SSN College, Alipur, University of Delhi, Delhi, India*
[4]*D. S. N. (P. G.) College, Unnao, India*
[5]*Kirori Mal College, University of Delhi, Delhi, India*

Abstract

This chapter aims to detoxify industrial wastewater through techniques like photo/bio and nano-catalysts. Water has a significant role in the sustenance of human life and existence. Human population and its activity are also good factors that incorporate their major factor in water pollution, these major contributions such as the rapid industrialization of forests and unsustainable agricultural practices that are the major global concerns. Rapid industrialization and urbanization result in increased pollution and the buildup of dangerous compounds that damage aquatic and terrestrial ecosystems.

The "International Union of Pure and Applied Chemistry" (IUPAC) has put heavy metals into different groups based on how they are toxic and dangerous. Heavy metals are regarded to represent a severe threat to the environment. There are both organic (like organometallic compounds, organic material from laboratories, dyes, pesticides, hydrocarbon, and medicine from hospitals) and inorganic (like heavy metals from inorganic laboratories, coal-burning power stations, oil refineries, and petrochemical plants) groups of toxic substances that are debauched for the environment and living beings. However, for sustainability, protection of environment and cleaning of contaminated/ waste water are essential. Heavy metals are of two types such as toxic and essential heavy metals. Some trace metals like zinc, copper, and iron are essential for the human body to enhance metabolism. Those heavy metals can persist in the natural aquatic body for a longer period of time because they are nonbiodegradable in nature. Toxic waste material from factories, drainage from villages, municipalities, and household/domestic sectors also must be thrown away, which has become a big problem in modern society.

In the last few years, much research and new ideas have been done in the area of wastewater treatment, and various methods are implemented in this technique such as photocatalytic degradation, bioremediation of toxic material, and nanocatalytic activity for detoxification of wastewater. The goal is to figure out how catalysis works and make it more effective for cleaning the air, soil, and water pollution. These methods are cost-effective for the detoxification of industrial wastewater and easily performed at the industrial level.

**Corresponding author*: arun107941@gmail.com

Mousumi Sen (ed.) Sustainable Green Catalytic Processes, (377–396) © 2024 Scrivener Publishing LLC

The catalytic method is likely effective for treating drinking water and domestic and commercial wastes. Overall, this study intends to project a path of rapid development that provides a broad overview of recent applications of nanotechnologies and biotechnologies as future opportunities that are involved in the field of wastewater treatment.

Keywords: Detoxification of industrial wastewater, catalytic techniques, economic evaluation, nanotechnology

Abbreviations

WHO	World Health Organization
BTEX	xylene, toluene, ethylene, and benzene
PAH	polycyclic hydrocarbons
NPs	nanoparticles
BOD	biological oxygen demand
COD	chemical oxygen demand
CNT	carbon nanotube
ECNMs	electrospun chitosan nanofiber mats
nm	nanometer
RhB	rhodamine B
MO	methyl orange
MB	methylene blue
EBT	Eriochrome Black-T
TFFBR	thin-film fixed-bed reactor
NCMs	nanostructured catalytic membranes
MFCs	microbial fuel cells
MIPs	molecularly imprinted nanospheres
MOF-5	metal–organic framework

16.1 Introduction

Physical-chemical and biological alteration in water or natural resources of water is called water pollution. In general, water is an essential commodity for the survival of human life, plants, and aquatic life as well. Most resource of water is present in nature such as lakes, rivers, groundwater, and seawater. In daily life, water is utilized for drinking, washing, bathing, irrigation, and the amount of water used in the industrial sector. Water has a good property as a solvent to dissolve many kinds of chemicals, molecules, and substances and changes in it to physical, chemical, and biological; therefore, it easily gets polluted.

Water pollution may be caused by two sources such as point sources and nonpoint sources. The source of effluent from industries, coal mines, power plants, and offshore oil wells comes under the point sources of pollution. The nonpoint sources of effluent are not a particular site; it may be rainwater sweeping roads, fields, surface runoff from agricultural fields, drains overflow, etc. Various sources of water pollution such as groundwater and surface water pollution are shown in Figure 16.1.

Various methods are reported for the remediation of hazardous metal, dye, and chemicals from wastewater such as nonexchange, coagulation membrane filtration, electrokinetic coagulation, and bioremediation [1, 2], and all methods mentioned above are very costly and not effective for remediation of dye [3].

Among all methodologies, adsorption is a more appropriate technique for different types of coloring compounds and in the case of some metals also [4–7]. In this chapter, we have discussed about remediation of hazardous material from wastewater; in particular, application of biocomposites and nanocomposite materials as catalyst for detoxification of industrial wastewater through photo-degradation is described.

Nanotechnology and nanoscience improve the various methodologies utilized in the field of wastewater treatment. In comparison to traditional methodology, nanotechnological-based methods are more efficient for the remediation of pollutants from wastewater. Treatment of wastewater by employing nanoparticles is one of the most promising fields for nanotechnology applications [8]. Nanocatalysts such as metal–organic frameworks and other nanocomposites are used for the remediation of metal or other chemicals from wastewater [9]. Nanoparticles and nanocomposites are smaller in size, have high surface area, and have high chemical characteristics [10].

Biosynthesis and green methodology are applied to the synthesis of nanocomposite to improve the properties of its characteristic features [11]. Biogenic nanoparticles and their nanocomposites are eco-friendly, highly efficient, and cost-effective [12]. Biogenic nanoparticles such as biopolymer-coated nanocomposites and metal oxide nanoparticles have significant adsorption properties and catalytic properties and increase the rate of reaction [13, 14]. Biogenic nanocomposites are used for the treatment of wastewater to remediate pollutants [15].

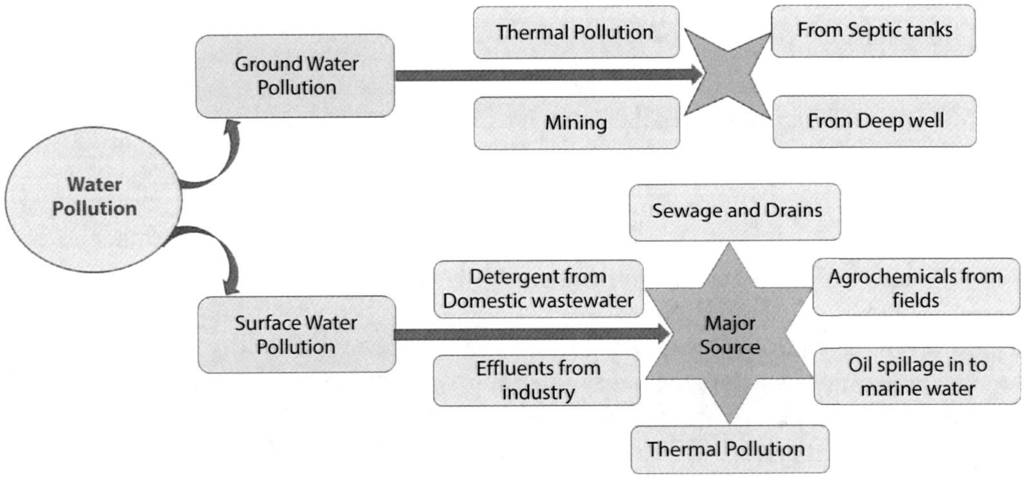

Figure 16.1 Source of water pollution mainly groundwater and surface water pollution.

The photocatalytic decomposition process is also employed to degrade the pollutants, mainly organic chemicals present in the industrial wastewater [16]. In the photocatalytic process, there is no requirement for expansive oxidizing chemicals and no secondary disposal method, which is the advantage of this method [17]. The photocatalytic process can also use to degrade organic material such as trihalomethanes or its precursor that causes unpleasant smell, and test in water bodies [18].

Integration of catalytic processes with renewable energy sources, such as solar or wind power, may offer sustainable and environmentally friendly solutions for wastewater detoxification. The use of renewable energy can help reduce the reliance on fossil fuels and increase the overall efficiency of catalytic reactions. Ongoing research is focused on the development of novel catalysts with enhanced catalytic activity, selectivity, and stability. These catalysts can offer better performance, increased reusability, and reduced costs, making catalytic detoxification more economically viable for large-scale applications. In general, the catalytic detoxification of wastewater offers the potential for efficient and sustainable treatment of various pollutants present in the wastewater for the preservation of natural water resources of the environment.

16.2 Detoxification of Wastewater

Every living organism that exists on Earth depends on water to sustain itself. The most dynamic component of life on Earth is water, which makes up over 71% of the planet's surface [19]. Approximately 844 million people lack contact with hygienic drinking water, with 159 million reliant on surface water, according to "The World Health Organization" (WHO) [20]. A variety of pollutants and pollution sources could contaminate it. Further contamination of water by different hazardous inorganic and organic chemicals, heavy metals, and metalloids is the major issue addressed globally that has adverse effects on natural water resources as well as on aquatic life. In recent years, remediation of these types of pollutants from wastewater using various techniques may be economical or not. Adsorption and degradation are the economical processes for the remediation of pollutants from wastewater. There are several techniques have been described such as a photo/bio/nano-based process to determine the detoxification of wastewater.

16.2.1 Detoxification of Wastewater by Photocatalysis

Hazardous organic contaminants found in wastewater may have adverse effects on aquatic life and human health. Pharmaceuticals and other chemicals, which are persistent organic pollutants like dyes and pesticides, are drawing an increasing amount of attention. The usage of photocatalysis for water purification has recently attracted a lot of interest. According to Ohtani [21], photocatalysis is typically described as the chemical process carried out *via* the absorption of photons through solid material (photocatalysis). The terms "photocatalyst' and catalyst are recurrently used synonymously in the research literature [22–25]. The concepts of catalysis and photocatalysis are separated in terms of thermodynamics. Such as photocatalysis can therefore drive energy-storing reactions ($\partial G > 0$), while catalysis is only capable of driving thermodynamically feasible processes ($\partial G < 0$). According to the review of diverse literature, when related to other types of wastewater treatment technologies,

TiO_2-based photocatalysts are extremely effective at degrading organic and inorganic pollutants from wastewater. A semiconductor's electrical structure is vital for the mechanism of the photocatalyst. This chapter emphasizes a variety of photocatalytic applications, including solid-state photocatalysis for waste plastic mitigation, biological photocatalysis for sewage treatment, and heterogeneous photocatalysis for environmental remediation [26, 27]. Photocatalysis techniques use catalysts to remove organic contaminants from wastewater [28]; a schematic diagram is shown in Figure 16.2. The function of the photocatalytic reaction of TiO_2 [29] is represented in Equation 16.1.

The active electron shifts to the conduction band upon absorption of a photon from the energy source, creating an appositive hole (hVB) in the valence band.

$$TiO_3 + h_v \rightarrow TiO_2^- + h^+vb + (TiO_2) \tag{16.1}$$

The widely known theory can be utilized to describe the fundamental concept of photocatalysis. When photocatalytic material is subjected to light with an energy equal/greater than the photocatalyst's band gap, electron–hole pairs are generated. After creating, electron–hole couples separate into holes (h_+) in the valence band and electrons (e^-) in the conduction band. Molecules that have been adsorbed on the surface of the photocatalytic material are reduced and oxidized as a result of the e^- and h^+. Though electron–hole recombination frequently occurs, it may prevent oxidation and reduction reactions from occurring on the surface of photocatalytic material. Enhanced electron–hole recombination is frequently associated with an increase or decrease in reaction rate, respectively [30].

Due to high catalytic efficiency and stability, as well as nontoxicity, TiO_2 has gathered many interests in the field of photocatalysis. In the crust of the Earth, titanium is the 10th most prevalent element.

Between 1932 and 1934, TiO_2's photoactivity was discovered for the first time. TiO_2 was also utilized as a photo-bleach dye [31]. Due to its exceptional qualities, including a higher

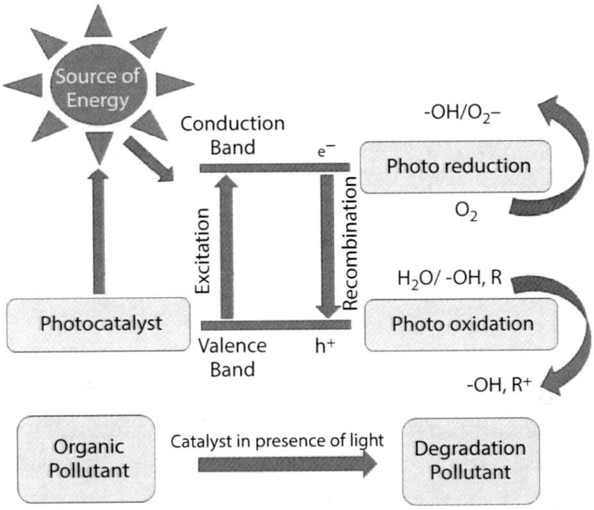

Figure 16.2 Schematic representation of mechanism of photocatalysis.

surface-to-volume ratio, there has been an exponential rise in the number of papers on TiO_2 photocatalysis since 2000, which is found in the current statistical study. Broad titanium dioxide has been widely employed to address current energy and environmental challenges due to its good optical and electrical qualities, such as band gap, band edge, thermal stability, nontoxicity, and strong chemical strength. Additionally, it resists corrosion and is affordable and long-lasting as well as environmentally benign [32].

A World Bank assessment states that 17%–20% of industrial water contamination is caused by textile dyeing. The necessity for ecologically acceptable technology to remove colors from industrial and municipal wastewater has increased along with environmental concerns. Most of the dyes are soluble in water, nonbiodegradable, and dangerous to the environment. The commonly used dye are rhodamine B, methyl orange, and methylene blue, Rose Bengal, Eriochrome Black-T, Victoria blue, indigo red, Red 120, and thymol blue etc. [33]. Mohammadi *et al.* [34] have explored the influence of TiO2/ZnO/rGO nanocomposites on the UV-induced degradation of rhodamine B, methylene blue, and methyl orange and reported the mechanism for their degradation. Their analysis reveals that a heterojunction of TiO_2/ZnO/rGO materials can perform better than a single material in terms of practical application, stability, and recyclability for photocatalysis. Ozone, UV-vis light, and a photocatalyst are all essential ingredients for photocatalytic ozonation to occur. Ozone molecules have also been adsorbed on the surface of the photocatalyst, in addition to the photocatalytic procedure that is brought on by the photoexcitation of the catalyst's surface. As a result, active oxygen radicals are created. Water molecules combine with active oxygen radicals to produce hydroxyl radicals. Moreover, ozone absorbs a wavelength that is shorter than 300 nm, and active oxidizing species are created [35]; see Equation 16.2.

$$O_3 + h_v \rightarrow O + O_2 \tag{16.2}$$

Nanomaterial-based photocatalysis is widely used nowadays for water detoxification. Nanoparticles comprise a size ranging within the 100-nm range. They possess characteristics such as huge surface area, elevated surface-to-volume ratio, specific affinity, and active surface sites. These nanoparticles have larger surfaces that are comparable to the bulk material; hence, heavy metals and various hazardous effluents are being removed using nanomaterial-based catalysts.

In wastewater treatment, nanomaterials are used in the photocatalysis and adsorption technique. TiO_2 is the ubiquitous and superlative material among the enormous photocatalytic substances reported for photocatalysis. TiO_2 comprises high photocatalytic activity in hydrogen production, is less toxic, and is cost-effective, but it has a band gap of a higher range nearly 3.2 eV, which corresponds to the ultraviolet-visible spectrum (comprising only 4% of the solar spectrum) region of the spectrum. Hence, it is vital to bring into being a photocatalyst that is responsive to the visible range of the light spectrum (comprising 43% of the solar spectrum). In the present need, it is essential to bring in the photocatalysts that cover the visible range of the spectrum. Various metal oxides that comprise a wide band gap such as tungsten oxides, bismuth oxides, strontium oxides, titanium oxides, and zinc oxides promote the photocatalytic activity in an ultraviolet range of light spectrum [36]. The use of photocatalytic methods is the most common treatment of wastewater in recent years.

The discharges of heavy metals into aquatic environments have a negative effect on the environment, human health, and other creatures. Heavy metal concentrations in industrial effluents are substantial and ought to be treated using various technologies. Heavy metals like bismuth vanadate are some of the best photocatalysts known owing to their ideal optical and electrical properties and comprising a band gap of 2.4 eV. Bismuth vanadate comprises enormous properties required for a photocatalyst such as higher stability, low toxicity, cost-effectiveness, chemically inert, biologically stable, and active in the visible and ultraviolet range of spectrum [37]. Pure bismuth vanadate has a limited range of applications in the field of photocatalysis but due to its highly encouraging photocatalytic activity in the oxidation of water, monoclinic bismuth vanadate has recently attracted the attention of researchers. According to the general theory, the genuine valence band gap of this material is located at 2.4 eV when compared to a reversible hydrogen cathode, with a close to the optimal vitality band gap of 2.3–2.4 eV. Numerous initiatives and methods, including doping, new nanoparticles, surface titration with electrocatalysts, and morphological management, have been developed to improve photocatalytic exhibits. The electrical characteristics of materials are typically the reason why ion doping changes greatly. For instance, doping a semiconducting material with metal ions will increase its conductivity. When a strong metallic particle fuse occurs in the host lattice cross-section, the doping of bismuth vanadate through advanced metal particles activates core vitality levels inside the band gap [38]. There are various kinds of detoxification of water from wastewater treatment plants that are widely utilized nowadays; some of given below:

(i) **Photocatalytic Solar Reactor**
The main expense during the operation of photocatalytic treatment of wastewater facilities is the photocatalytic solar reactor, which artificially produces the photons required for the detoxification of dirty water. This shows that using the sun as a low-cost environmentally friendly source of illumination is possible [39]. The vast majority of these photons are utilized in photocatalytic reactors to purge water from contaminants. Over the past 15 years, several solar photocatalytic water treatment reactors have been developed and tested.

(ii) **Parabolic Trough Reactor**
A plug flow reactor called a parabolic trough reactor (PTR) uses 30–50 parallel (direct) beams from the sun's active ultraviolet zone. The invention of solar detoxifying loops was initiated in Albuquerque, New Mexico.

(iii) **Thin-Film Fixed-Bed Reactor**
The most common type of reactor that can be used without concentrating equipment is the thin-film fixed-bed reactor (TFFBR). Thus, diffuse and direct UV-A rays from the sun are both used in TFFBR photocatalysis. A key element in the TFFBR process is a slanted plate with a photocatalyst like Degussa P25 and a very thin layer (100 m) rinsed with unclean water. Testing the TFFBR was essential to make sure it functioned properly. It has been demonstrated that TFFBR is more efficient than PTR in removing the same pollutants that are dissolved in pure water and real wastewater that is obtained from various industrial units.

(iv) **Design of Photocatalytic Reactors**
Photocatalysts, light irradiations, auxiliaries, additives, and other components are all included in the setup technique for photocatalytic reactors. When it comes to semiconductor photocatalysts, morphology and composition are the two most crucial factors to take into account. The composition has an impact on the charge carriers' mobility during photocatalytic activity.

16.2.2 Detoxification of Wastewater by Biogenic Nanoparticles

Nanoparticles (biogenic NPs) are employed to increase the exclusion effectiveness of noxious and hazardous pollutants from the treated effluent. Numerous studies also looked into the encapsulation method on silica, polyacrylamide, polyurethane, and zeolites used for reactor setups and the stabilization of nanoparticles on the bacterial cell wall. Ordinarily, activated sludge, which is used to treat wastewater, has limited relevance for the remediation of hazardous and nonbiodegradable industrial contaminants. However, it produces effective results for treating municipal wastewater. Chlorinated organic pollutants are removed using zero-valent iron (ZVI) nanoparticles, and degradation of organic pollutants is a new development in wastewater treatment. A schematic representation of the remediation of pollutants by adsorption method using bio-nanocomposite is shown in Figure 16.3.

Zero-valent iron nanoparticles have a dual action on eliminating contaminants when utilized in a bioreactor system to treat industrial wastewater with immobilized aerobic microorganisms. By acting as electron donors, nanobimetallic structures cause aerobic bacteria to partially digest organic molecules, reducing them to nontoxic chemicals. Granular iron breaks down chlorinated solvents and petroleum hydrocarbons into chlorinated ethenes and short hydrocarbon chains through dichlorination and oxidation. It is being used to treat azo dye and nitroaromatic compounds using the synergistic impact of ZVI-NPs and bioactive components. The removal efficiency was 96.5% for BOD and 86.6% for COD. According to the study's findings, the pretreatment using a ZVI reactor and biological treatment afterward represent the most workable solution for the purification of industrial wastewater [40].

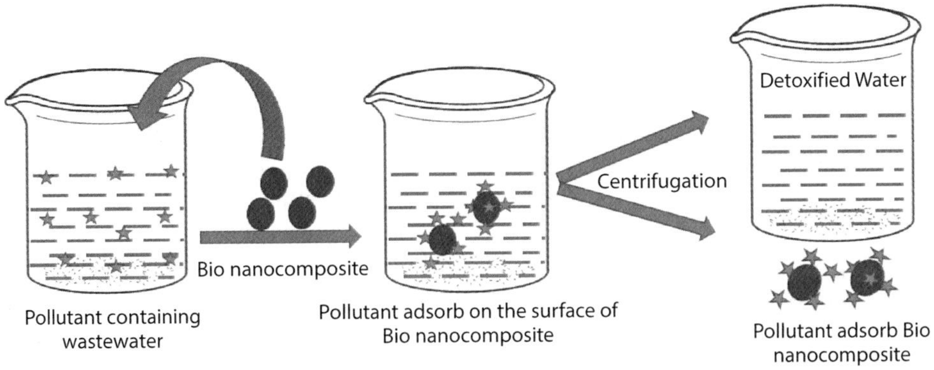

Figure 16.3 Schematic representation of remediation of pollutants using bio-nanocomposite material.

A wide variety of microorganisms are used in microbial fuel cells (MFCs) in biocatalysts to treat wastewater and then produce energy from organic and inorganic contaminants. Acetate, propionate, and butyrate are some examples of complicated compounds that the bacterial cell aids in decomposing into less complex forms, such as H_2O and CO_2. Therefore, by utilizing the energy created by MFCs, the demand for unadventurous technology for power production is decreased. Although MFC efficiency is subpar compared to other fuel cells and component costs are substantially greater, these factors restrict its commercial viability. An important problem is the optimization of microbial fuel cell commotion in the substance utilized for substrate reaction and digestion. The best alternative material for entrapping and digesting pollutants is called nanoparticles (NPs), and nanoparticles with a larger surface area interact with biological systems more effectively. Many nanocomposites and nanostructured carbon electrodes have shown improved MFC performance. The commercially available cathode catalyst for Pt in microbial fuel cells can be replaced with CNT/Pt due to its distinct properties. In a different study, CNT nanocomposites with polymers like polyaniline and polypyrrole were utilized to improve microbial adherence and lessen the toxicity of NPs. The presence of positively charged polycationic polymer results in electrostatically bonded MFC anodes coated with CNT nanocomposites. Owing to the PPy-CNT nanocomposite electrodes' high surface area for bacterial connection and biofilm growth for transfer of electrons to the anode surface, the active regions for microbial adhesion and electrochemical reactions are tailored. As a result, nanocomposite MFCs offer a superior chance for the production of electricity and the handling of wastewater [41].

The bioremediation of wastewater utilizing nanotechnological methods leads to effective treatment processes and contamination removal, whether by utilizing nanoparticles or joint handling with microbial cells. The biomachinery, which consists of bioactive small molecules, nanosized catalytic membranes, nanotubes, nanopowder, nanosorbents, and nanocatalysts, efficiently traps and removes contaminants from wastewater, cleans the system, and enhances the water eminence. Bioreactors based on algal membranes with nanoparticles were utilized as integrated sustainable remediation and energy production technology in green chemistry for wastewater treatment. Contaminated wastewater bodies offer nutrient-rich surroundings for the growth of algae, including salts of PO_4^{-3} as well as NO_3 with NH^{4+}, Na, Ca, K, and micronutrients such as vitamins such as thiamine and cyanocobalamin.

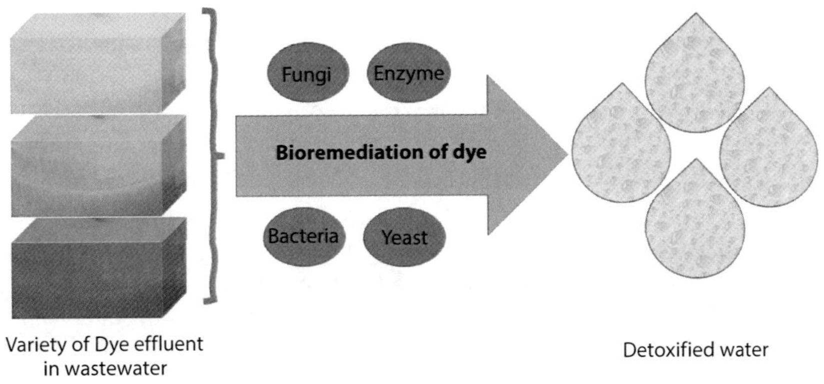

Figure 16.4 Schematic representation of detoxification of wastewater by microbial nanotechnology.

As a result, the growth of algal cells in sewage aids in the removal of contaminants by adsorbing [42]. There are various methods for isolating bacteria, yeast, fungi, and algal biomass, although they are more expensive for large-scale treatment, shown in Figure 16.4.

Physical methods include centrifugation, sedimentation, and air flotation or chemical methods such as flocculation. Membrane fouling, which lowers performance as a result of particles' addition or solute in the membrane surface or the membrane pores, is a key downside of algal membrane-based bioreactors [43].

16.2.3 Detoxification of Wastewater by Nanocatalysis

Nanomaterials are materials with sizes between one and hundreds of nanometers, and because of their special characteristics, such as high surface area and high catalytic performance, they are extremely helpful in all fields. Water contamination is a significant global issue that is currently being challenged [44]. The utilization of nanomaterials to purge various contaminants from wastewater and many different types of nanomaterials have the potential to efficiently cure contaminated water (such as metal toxins and various organic and inorganic contaminants) because of their special qualities, such as having a larger surface area and the ability to function at low concentrations. While the technologies based on nanostructured catalytic membranes, nanosorbents, and nano-photocatalysts to remove contaminants from wastewater are efficient and environmentally benign, they require more energy and money to clean the wastewater. The treatment of wastewater is fraught with difficulties and problems. Currently, the WHO ("World Health Organization") reported that 4 billion cases of various health conditions were reported annually owing to waterborne infections and that over 1.7 million people died as a result of water pollution [45]. Recently, more advanced developments in nanomaterials that included nanocomposites, nanosorbents, nanostructured catalytic membranes (NCMs), and bioactive nanoparticles are efficiently usable for the treatment process of polluted water.

In 2021, Kumar and Masram [46] have fruitfully synthesized MOF-5@GO nanocomposite employed for the exclusion of toxic dye. The MOF-5@GO nanocomposite has an admirable adsorption capacity for application in the remediation of RhB dye. The physical and chemical properties of nanocomposite were explained with diverse structural and morphological characterization techniques. Furthermore, the morphology of RhB adsorbed MOF-5@GO nanocomposite reveals the growth of irregular nanoparticles (MOF-5) with sizes varying between 50 and 200 nm. The extent of adsorption capacity was found to be decrease in the presence of ion such as Ni^{2+} and Cu^{2+}. Adsorption capacity of nanocomposite was found 99.58% at pH 2. It can be concluded that MOF-5@GO nanocomposite is best used as an adsorbent material due to its cost-effective highly efficient adsorbent material with excellent reusability, eco-friendly, one-step synthesis and potential applications in wastewater treatment. The appropriate management of consumption costs is crucial in the context of water treatment using nanoparticles, given the increased technological demands and the need to align with prevailing market competition [47, 48]. Nanoclay composite, zeolite, and carbon nanotube are also used as adsorbent materials for different varieties of effluent such as dye, and metal ion adsorption as well as intercalation takes place [49–53]. A schematic representation of clay nanocomposite material used as an adsorbent is shown in Figure 16.5.

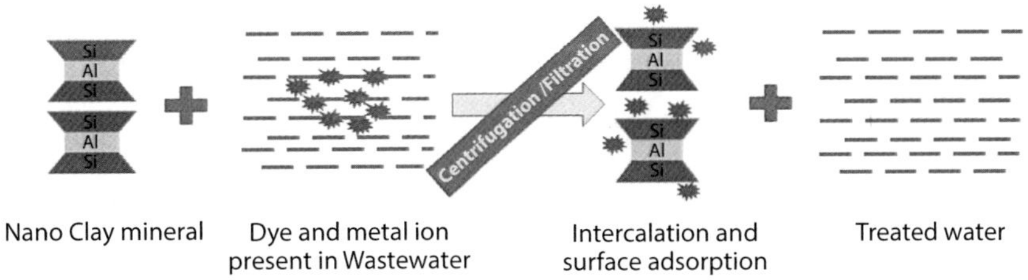

Figure 16.5 A schematic representation of clay nanocomposite used as an adsorbent.

Water contamination has significant implications not just for the environment and human health but also for commercial. There are numerous commercially and non-commercially employed strategies to combat the escalating problem that is perpetually expanding as a result of technological advancements. Nanotechnology has emerged as a highly effective and advanced method for wastewater treatment. The success of nanotechnology can be attributed to a multitude of factors, and ongoing efforts by scientists continue to focus on enhancing its applications. Nanoparticles possess significant absorption, interaction, and reactivity capabilities as a result of their small size and large concentration of atoms on their surface. It can be combined with aqueous solutions, hence exhibiting colloidal behavior. Nanoparticles can enhance energy savings due to their small size, thereby potentially leading to economic advantages through cost reduction. Nanoparticles offer a notable benefit in the remediation of water in geographically isolated and inaccessible areas, a potentiality that is often disregarded by conventional approaches.

16.3 Miscellaneous Types of Adsorbent

Miscellaneous types of adsorbent material and membranes also used as remediation of pollutants from wastewater are as follows:

(a) Nanosorbents

Nanosorbents exhibit a notable sorption capacity that is both high and selective, rendering them very applicable in various processes such as water purification and in the remediation of pollutants. The availability of commercialized nanosorbents is currently limited, and extensive research efforts are underway to develop a wide range of nanosorbents that specifically target particular contaminants found in water. Several advances and applications of nanosorbents are provided below.

The utilization of magnetic nanosorbents is a valuable approach in the treatment of wastewater, which is particularly for the elimination of organic pollutants [54, 55]. Filtration aids are commonly employed to absorb nonmagnetic impurities, which are subsequently subjected to magnetic separation. The preparation of nanosorbents for magnetic separation involves the application of particular ligands onto magnetic nanoparticles [56]. Nanosorbents are consistently recognized for their cost-effectiveness and are increasingly advocated for commercialization purposes.

Nanocatalysts are a class of catalysts that possess nanoscale dimensions. Nanocatalysts are extensively employed in the field of water treatment due to their ability to enhance catalytic activity on the surface. This is attributed to their unique qualities, which include a larger surface zone and shape-dependent properties. Which facilitates the enhancement of both the reactivity and degradation processes of pollutants. Semiconductor materials have zero-valence metals, and bimetallic nanoparticles are frequently employed as catalysts for the deprivation of various environmental contaminants, including but not limited to polychlorinated biphenyls, halogenated aliphatic compounds, azo dyes, organochlorine pesticides, halogenated herbicides, and the nitro functional group-containing aromatic compound [57, 58].

Various catalysts, including silver nanocatalysts, ZrO_2 and N-doped TiO_2, nanoparticle catalysts, were synthesized and have demonstrated a high level of efficiency in the destruction of microbiological pollutants present in water [59]. Nanocatalysts composed of silver and amidoxime fibers, synthesized through coordination contacts, exhibit high efficiency in the degradation of organic dyes [55, 60]. The using of photocatalytic ZnO nanoparticles with Pd as well as nano-WO_3 prepared by sol-gel method has been effective for the remediation of E. coli from water [58, 61]. The efficiency of Palladium nanoparticles as a catalyst has been reported to be effective in remediation of chromium ion from wastewater [62].

Nanosorbent material has efficient activity as nanocatalyst, enabling the simultaneous sorption and degradation of pollutants, and has demonstrated high efficacy in the field of water treatment.

(b) Nanostructured Catalytic Membranes

Nanostructured catalytic membrane (NCM) is widely employed in the field of water contaminant treatment. This method has many significances such as a high level of consistency among catalytic sites, the ability to optimize and act in a short period of time, and the ability to make sequential reactions easier. Nanostructured TiO_2 films and membranes have been found to exhibit many activities when exposed to UV and visible light irradiation. These functions include the degradation of organic pollutants and inactivation of microorganisms [63].

The N-doped ZnO nanostructured substantial that exhibits a “nut-like” morphology has been found to possess excellent efficacy in the removal of water pollutants and enhancement of photodegradation activity of pollutants under visible light irradiation and has anti-bacterial properties [64]. Several research studies have been conducted to investigate the immobilization of metallic nanoparticles in different types of membranes, including cellulose acetate, polyvinylidene fluoride, and chitosan. The utilization of metallic nanoparticles in membranes offers numerous benefits, such as increased reactivity, improved organic partitioning, prevention of nanoparticle loss, and prevention of agglomeration [65].

Nanocomposite films were fabricated using polyetherimide and palladium acetate, and further investigations focused on examining the hydrogen-specific interactions with Pd-based nanoparticles, thereby demonstrating their efficacy in water treatment applications. The metal nanoparticles were formed within the matrix through the process of annealing the precursor film under various circumstances, employing both *in situ* and *ex situ* methods [66].

The field of nanotechnology has witnessed significant progress, leading to the development of numerous innovative nanostructured catalytic membranes that exhibit enhanced permeability and selectivity [67].

(c) Bioactive Nanoparticles

Water contamination has been attributed to the proliferation of numerous pathogenic agents, resulting in a significant increase in the incidence of infectious diseases. Numerous microorganisms that function as pathogens have antibiotic resistance, hence posing significant challenges in their eradication from water sources [68]. The silver nanoparticles is reported to have significant antibacterial properties against the bacterium Bacillus cereus [69]. The combination of MgO nanoparticles besides cellulose acetate fibers containing Ag nanoparticles exhibits significant biocidal activity against Gram-positive and Gram-negative bacteria and bacterial spores [70].

The present and developing nanotechnology methodologies for the identification of microbial pathogens will contribute to the enhancement of microbial and pathogen detection, as well as diagnostic capabilities.

(d) Molecularly Imprinted Polymers

Recently, molecularly imprinted polymers (MIPs) have gained recognition as a highly effective method for a wide range of applications in the fields of biology, pharmaceuticals, and environmental science. The polymers exhibit a high level of selectivity as a result of their synthesis approach, which involves the attachment of a template molecule to appropriate monomers containing functional groups through covalent, semi-covalent, or non-covalent bonds. Molecular imprinting polymers exhibit a high degree of selectivity and possess significant capacity as remediation of water contaminants, even at extremely low concentrations, and are utilized as water purification [71, 72]. Molecularly imprinted nanospheres (nano-MIPs) have been synthesized by the utilization of the mini-emulsion polymerization process (particles size 50–500 nm) that also have the efficiency in the adsorption of micropollutants present in hospital wastewater [73].

(e) Biomimetic Membrane for Water Treatment

Biomimetic membranes have been established through a collaborative effort between Sandia National Laboratories, located in Albuquerque, and the University of New Mexico. The aforementioned technology signifies a novel and sophisticated approach to water filtration [74].

The invention employs a combination of self-assembly and atomic layer deposition techniques to create nanopores that are specifically optimized for high-flux desalination. The membranes facilitate the removal of contaminants, such as salt and other substances, from water through the application of pressure generated by electrical energy.

Significant advancements have been observed in the field of water purification in terms of cost efficiency owing to the integration of nanofabrication techniques and the emulation of protein channels found in biological membranes. Robust synthetic porous membranes were developed by incorporating certain molecular design concepts observed in natural porous systems.

A biomimetic membrane can be created through the process of vesicle fusion over a dense water-permeable substrate, such as a nanofiltration membrane. The procedure employs the notion of electrostatic stitching, which aims to optimize the attraction and minimize the repulsion between head groups. This membrane possesses a combination of high permeability and selectivity, along with chemical stability [75]. Chemical stability can be achieved through the utilization of several synthetic biomimetic counterparts such as aquaporins and carbon nanotubes.

16.4 Adsorption Isotherm and Its Kinetics

An absorption isotherm for dye, drug, and metal ion absorption is a graphical representation of the relationship between the concentration of dye or metal ion in a solution and the amount of dye adsorbed onto a solid surface at a constant temperature. It helps to understand the interaction between the dye molecules and the surface and can provide information on factors such as adsorption capacity, equilibrium adsorption, and surface coverage. Freundlich adsorption isotherm and Langmuir adsorption isotherm are the basic models to explain the characteristics of monolayer or multilayer adsorption taking place [76–78].

Kinetics for dye adsorption refers to the study of the rate at which dye, metal ions, and drug molecules are adsorbed onto a solid surface that is adsorbent. It examines how quickly the dye molecules attach themselves to the surface and reach equilibrium. Kinetic studies help in understanding the mechanisms and factors that affect the adsorption process, such as temperature, concentration, surface area, and agitation.

There are various models used to describe the kinetics of dye adsorption, such as pseudo-first-order and pseudo-second-order kinetics. The pseudo-first-order model assumes that the rate of adsorption is directly proportional to the difference in concentration between the adsorbate on the surface and the bulk solution. The pseudo-second-order model suggests that the rate of adsorption is dependent on the square of the concentration, leading to a better fit for many adsorption processes [79–81]. By studying the kinetics of dye adsorption, researchers can optimize the adsorption process to achieve efficient and effective removal of dyes and metal ions from the solutions [82].

16.5 Significance of Adsorption Technique for Remediation of Hazardous Effluents

Adsorption processes are highly significant for dyes and for metal ions of effluent from wastewater in several ways. Here are some important key points:

- Adsorption is commonly used as a method to remove dyes from wastewater and other solutions. Adsorbent materials, such as activated carbon or clay minerals, can effectively adsorb and remove dyes from aqueous solutions, helping to treat and purify wastewater.
- Dye pollution is a significant environmental concern, as it can adversely impact water bodies and ecosystems. Adsorption processes provide a way to mitigate this issue by removing dyes from wastewater before it is discharged into the environment.
- Adsorption of dyes onto fabrics or other substrates is an essential step in coloring and dyeing processes. In these applications, adsorption allows dyes to

attach to the surface of the material, leading to coloration and achieving the desired dyeing effects.

- Understanding the adsorption processes for dyes is crucial for optimizing adsorption efficiency. Factors such as the choice of adsorbent material, contact time, pH, temperature, and concentration of dyes can all impact the adsorption efficiency. By studying and optimizing these processes, researchers can develop more effective adsorbents and methods for dye removal.

Adsorption processes play a crucial role in the removal of dyes from wastewater, as well as in the dyeing and coloring industry. They contribute to environmental protection, water treatment, and the production of colored textiles and materials.

16.6 Future Prospects of Detoxification of Wastewater Through Catalysis

The detoxification of wastewater through catalysis (photo/bio/nano) has promising future prospects due to its potential advantages and advancements in catalytic technologies.

- Advanced oxidation processes, that is, catalytic processes, such as heterogeneous catalysis and photocatalysis, are being actively researched and developed for wastewater detoxification. Advanced oxidation processes involve the use of catalysts to generate highly reactive species, such as hydroxyl radicals, which can effectively degrade and detoxify various organic pollutants in wastewater.
- Catalytic processes can significantly enhance the degradation efficiency of toxic compounds compared to traditional treatment methods. Catalysts provide active sites that facilitate the oxidation or reduction of pollutants, leading to their complete breakdown into harmless by-products.
- Catalytic processes have the potential to treat diverse types of pollutants found in wastewater, including organic dyes, pharmaceuticals, pesticides, and industrial chemicals. Different catalysts can be tailored to target specific pollutants, providing a versatile approach to detoxification.

The catalytic detoxification of wastewater is one of the important technique for efficient and sustainable treatment of various pollutants, contributing to the preservation of water resources and the protection of the environment.

16.7 Conclusion

In conclusion, the detoxification of water through catalysis offers promising prospects for the future. Through advanced oxidation processes, catalytic reactions can efficiently degrade and detoxify various pollutants in wastewater. This technology has a wide range of applications and can be tailored to target specific pollutants, making it a versatile approach to water treatment. Integration with renewable energy sources further enhances the sustainability of catalytic detoxification. Ongoing research and development are focused on improving catalyst performance and reducing costs, making this method economically viable for large-scale applications. Overall, catalytic detoxification has the potential to provide efficient and sustainable solutions for water pollution, safeguarding water resources and protecting the environment.

References

1. Haque, N., Morrison, G., Jorge, I.C.A., Gardea-Torresday, L., Iron Modified Light Expanded Clay Aggre-gates for the Removal of Arsenic (V) from Ground Water. *Microchem. J.*, 88, 1, 7–13, 2008.
2. Maiti, A., Gupta, S.D., Basu, J.K., De, S., Adsorption of Arsenite Using Natural Laterite as Adsorbent. *Sep. Purif. Technol.*, 55, 3, 350–359, 2007.
3. Banat, M., Nigam, P., Singh, D., Marchant, R., Microbial decolorization of textile-dye containing effluents: A review. *Bioresour. Technol.*, 58, 217–227, 1996.
4. Jain, A.K., Gupta, V.K., Bhatnagar, A., Suhas, Utilization of industrial waste products as adsorbents for the removal of dyes. *J. Hazard. Mater.*, 101, 1, 31–42, 2003.
5. Ho, Y.S. and McKay, G., Sorption of dyes and copper ions onto biosorbents. *Process Biochem.*, 38, 1047–1061, 2003.
6. Deliyanni, E.A., Peleka, E.N., Matis, K.A., Effect of Cationic Surfactant on Adsorption of Arsenates onto Aka-ganeite Nanocrystals. *Sep. Sci. Technol.*, 42, 5, 993–1012, 2007.
7. Khalaf, H., Bouras, O., Perrichon, V., Synthesis and Characterization of Al-Pillared and Cationic surfactant Modified Al-Pillared Algerian Bentonite, microporous. *Materials*, 8, 3-4, 141–150, 1997.
8. Suleiman, M., Mousa, M., Hussein, A., II, Wastewater Disinfection by Synthesized Copper Oxide Nanoparticles Stabilized with Surfactant. *J. Mater. Environ. Sci.*, 6, 7, 1924–1937, 2015.
9. Sikiru, S., Abiodun, O.A., Sanusi, Y.K., Sikiru, Y.A., Soleimani, H., Yekeen, N., Haslija, A.A., A comprehensive review on nanotechnology application in wastewater treatment a case study of metal-based using green synthesis. *J. Environ. Chem. Eng.*, 10, 108065, 2022.
10. Yadav, V.K., Gupta, N., Kumar, P., Dashti, M.G., Tirth, V., Khan, S.H., Yadav, K.K., Islam, S., Choudhary, N., Algahtani, A., Recent Advances in Synthesis and Degradation of Lignin and Lignin Nanoparticles and Their Emerging Applications in Nanotechnology. *Materials*, 15, 953, 2022.
11. Bhat, S.A., Sher, F., Hameed, M., Bashir, O., Kumar, R., Vo, D.-V.N., Ahmad, P., Lima, E.C., Sustainable nanotechnology based wastewater treatment strategies: Achievements, challenges and future perspectives. *Chemosphere*, 288, 132606, 2022.
12. Razack, S.A. and Duraiarasan, S., A way to create a sustainable environment: Green nanotechnology– with an emphasis on noble metals, in: *The ELSI Handbook of Nanotechnology: Risk, Safety, ELSI and Commercialization*, C.M. Hussain, (Ed.), pp. 359–425, Wiley-Scrivener, Austin, TX, USA, 2020.
13. Khan, S., *Green Materials for Wastewater Treatment*, Springer: Berlin/Heidelberg, Germany, 2020.
14. Nahak, B.K., Preetam, S., Sharma, D., Shukla, S., Syvajarvi, M., Toncu, D.-C., Tiwari, A., Advancements in net-zero pertinency of lignocellulosic biomass for climate neutral energy production. *Renew. Sustain. Energy Rev.*, 161, 112393, 2022.
15. Kaminski, A., Edwards, C., Chrapusta-Srebrny, E., Lawton, L.A., Anatoxin-a degradation by using titanium dioxide. *Sci. Total Environ.*, 756, 143590, 2021.

16. Yatmaz, H.C., Dizge, N., Kurt, M.S., Combination of photocatalytic and membrane distillation hybrid processes for reactive dyes treatment. *Environ. Technol.*, 38, 21, 2743–2751, 2017.
17. Abhilasha And D, J., Vaya, Photocatalytic Activity Of TiO2 Nanomaterial. *J. Chil. Chem. Soc.*, 62, 4, 3683–3690, 2017.
18. Zhang, Y., Crittenden, J.C., Hand, D.W., The solar photocatalytic decontamination of water. *Chem. Indust.*, 18, 714–717, 1994.
19. Panchal, H., Patel, K., Elkanawy, M., Bastawissi, H., A use of various phase change materials on performance of solar still: A Review. *Int. J. Ambient Energy*, 42, 13, 1575–1580, 2019, https://doi.org/10.1080/01430750.2019.1594376
20. Al-Gamal, A.Q., Falath, W.S., Saleh, T.A., Enhanced efficiency of polyamide membranes by incorporating TiO2 graphene oxide for water purification. *J. Mol. Liq.*, 323, 114922, 2021.
21. Ohtani, B., Chapter10- Photocatalysis by Inorganic Solid Materials: Revisiting Its Definition, Concepts and Experimental Procedures. *Adv. Inorg. Chem.*, 63, 395–430, 2011, https://doi.org/10.1016/B978-0-12-385904-4.00001-9
22. Rajczykowski, K. and Loska., K., Stimulation of Heavy Metal Adsorption Process by Using a Strong Magnetic Field. *Water Air Soil Pollut.*, 229, 20, 2018.
23. Kumar, G., Dutta, A., Goswami, M., Meena, B., Parasuboyina, S., Nongkhlaw, R., Masram, D.T., Synthesis of Zr-MOF/rGO-nanocomposite used for spirooxindole scaffolds. *J. Mol. Struct.*, 135653, 2023.
24. Mogha, N.K., Subodh, Mogha, K., Chaudhary, G., Masram, D.T., Fur-Imine-Functionalized Graphene Oxide-Immobilized Copper Oxide Nanoparticle Catalyst for the Synthesis of Xanthene Derivatives. *ACS Omega*, 16377–1638, 2018.
25. Kumar, G., Mogha, N.K., Kumar, M., Subodh, Masram, D.T., NiO nanocomposites/rGO as heterogeneous catalysis for imidazole scaffolds with their applications in inhibiting DNA binding activity. *Dalton Trans.*, 49, 1963–1974, 2020.
26. Castilla-Caballero, D., Sadak, O., Martínez-Díaz, J., Martínez-Castro, V., Colina-Márquez, J., Machuca-Martínez, F., Gunasekaran, S., Solid-state photocatalysis for plastics abatement: a review. *Mater. Sci. Semicond. Process*, 149, 106890, 2022.
27. Wang, H., Li, X., Zhao, X., Li, C., Song, X., Zhang, P., Huo, P., A review on heterogeneous photocatalysis for environmental remediation: from semiconductors to modification strategies. *Chin. J. Catal.*, 43, 2, 178–214, 2022.
28. Lee, C.M., Palaniandy, P., Dahlan, I., Pharmaceutical residues in aquatic environment and water remediation by TiO2 heterogeneous photocatalysis: a review. *Environ. Earth Sci.*, 76, 17, 611, 2017.
29. Sujatha, G., Shanthakumar, S., Chiampo, F., UV light-irradiated photocatalytic degradation of Coffee processing wastewater using TiO_2 as a catalyst. *Environments,* 7, 47, 1–13, 2020; doi:10.3390/environments7060047
30. Ohtani, B., Titania photocatalysis beyond recombination: a critical review. *Catalysts*, 4, 942e953, 2013.
31. Hashimoto, K., Irie, H., Fujishima, A., TiO2 photocatalysis: a histopical overview and future prospects. *Jpn. J. Appl. Phys.*, 44, 12R, 8269–8285, 2005.
32. Othman, S., Salam, N.R.A., Zainal, N., Basha, R.K., Talib, R., Antimicrobial activity of TiO2 nanoparticle coated film for potential food packaging applications. *Int. J. Photoenergy*, 1–6, 2014.
33. Carey, J.H., Lawrence, J., Tosine, H.M., Photodechlorination of PCB's in the presence of titanium dioxide in aqueous suspensions. *Bull. Environ. Contam. Toxicol.*, 16, 697–701, 1976.
34. Mohammadi, M., Roknabadi, M.R., Behdani, M., Kompany, A., Enhancement of visible and UV light photocatalytic activity of rGO-TiO2 nanocomposites: The effect of TiO2/Graphene oxide weight ratio. *Ceram. Int.*, 45, 10, 12625–12634, 2019.

35. Mehrjouei, M., Müller, S., Meoller, D., A review on photocatalytic ozonation used for the treatment of water and wastewater. *Chem. Eng. J.*, 209e219, 2015.
36. Darwent, J.R. and Mills, A., Photooxidation of water sensitized by tungsten trioxide powder. *J. Chem. Soc., Faraday Trans.*, 2, 78, 359–367, 1982.
37. Ibhadon, A.O. and Fitzpatrick, P., Heterogeneous photocatalysis: Recent advances and applications. *Catalysts*, 3, 1, 189–218, 2013.
38. Tahir, M.B., Kiran, H., Iqbal, T., The detoxification of heavy metals from aqueous environment using nano-photocatalysis approach: a review. *Environ. Sci. Pollut. Res.*, 26, 11, 10515–10528, 2019.
39. Enesco, A., The influence of photocatalytic reactors design and operating parameters on the wastewater organic pollutants removal: a mini-review. *Catalysts*, 11, 556, 2021.
40. Malik, S., Dhasmana, A., Preetam, S. *et al.*, Exploring Microbial-Based Green Nanobiotechnology for Wastewater Remediation: *A Sustainable Strategy. Nanomaterials*, 12, 4187, 1–26, 2022.
41. Sivaraj, R., Salam, H.A., Rajiv, P., Rajendran, V., Green Nanotechnology: The Solution to Sustainable Development of Environment, in: *Environmental Sustainability*, pp. 311–324, Springer, India, 2015.
42. Preetam, S., Dash, L., Sarangi, S.S., Sahoo, M.M., Pradhan, A.K., Application of Nanobiosensor in Health Care Sector, in: *Bio-Nano Interface*, pp. 251–270, Springer, Singapore, 2022, doi: 10.1007/978-981-16-2516-9_14.
43. Medina-Cruz, D., Mostafavi, E., Vernet-Crua, A., Cheng, J., Shah, V., Luis Cholula-Diaz, J., Guisbiers, G., Tao, J., Miguel García-Martín, J., Webster, J.T., Green nanotechnology-based drug delivery systems for osteogenic disorders. *Expert Opin. Drug Deliv.*, 17, 3, 341–356, 2020.
44. Rizwan, M., Ali, S., Qayyum, M.F., Ok, Y.S., Zia-ur-Rehman, M., Abbas, Z., Hannan, F., Use of Maize (Zea mays L.) for phytomanagement of Cd-contaminated soils: a critical review. *Environ. Geochem. Health*, 39, 2, 259–277, 2017.
45. Briggs, A.M., Cross, M.J., Hoy, D.G., Blyth, F.H., Woolf, A.D., March, L., Musculoskeletal Health Conditions Represent a Global Threat to Healthy Aging: A Report for the 2015 World Health OrganizationWorld Report on Ageing and Health. *Gerontologist*, 56, 243–255, 2016.
46. Kumar, G. and Masram, D.T., Sustainable Synthesis of MOF-5@GO Nanocomposites for Efficient Removal of Rhodamine B from Water. *ACS Omega*, 6, 14, 9587–9599, 2021.
47. Crane, R.A. and Scott, T.B., Nanoscale zero-valent iron: Future prospects for an emerging water treatment technology. *J. Hazard Mater.*, 211–212, 2012.
48. Kumar, G., Ehtesham, M., Kumar, S., Meena, B., Rai, G.J., Masram, D.T., Metal-Organic Framework Sponges for Water Remediation, in: *Nanosponges for Environmental Remediation*, S. Gulati, (Ed.), pp. 247–261, Springer, Cham, Springer, One New York Plaza, Suite 4600, New York, USA, 2023. DOI:10.1007/978-3-031-41077-2_11.
49. Gaijon, P., Singh, M.R., Ghosh, S., Kant, A., Synthesis, haracterization of the valuable Hydroxysodaliteand its application for remediation of toxic dye: *Rhodamine B. Eur. Chem. Bull.*, 12, 4, 680–706, 2023.
50. Yadav, A., Bagotia, N., Yadav, S., Sharma, A.K., Adsorptive studies on the removal of dye s from single and binary systems using saccharummunja plant-based novel functionalized CNT composites. *Environ. Technol. Innovation*, 24, 102015, 2021.
51. Kaur, M. and Datta, M., Adsorption Characteristics of Acid Orange 10 from Aqueous Solutions onto Montmorillonite Clay. *Adsorpt. Sci. Technol.*, 29, 3, 2011.
52. Kaur, M. and Datta, M., Adsorption Equilibrium and Kinetics of Toxic Dye- Erythrosine B Adsorption onto Montmorillonite. *Sep. Sci. Technol.*, 48, 1370–1381, 2013.
53. Nadeem, U. and Datta, M., Adsorption Studies of Zinc(II) Ions On Biopolymer Composite Beads Of Alginate-Fly Ash. *Eur. Chem. Bull.*, 3, 7, 682–691, 2014.

54. Campos, A.F.C., Aquino, R., Cotta, T.A.P.G., Tourinho, F.A., Depeyrot, J., Using speciation diagrams to improve synthesis of magnetic nanosorbents for environmental applications. *Bull. Mater. Sci.*, 34, 1357–1361, 2011.
55. Wua, Z.C., Zhanga, Y., Taoa, T.X., Zhangb, L., Fongb, H., Silver nanoparticles on amidoxime fibers for photo-catalytic degradation of organic dyes in waste water. *Appl. Surf. Sci.*, 257, 3, 1092–1097, 2010.
56. Apblett, A.W., Al-Fadul, S.M., Chehbouni, M., Trad, T., Removal of Petrochemicals from Water Using Magnetic Filtration, *Proceedings of the 8th International Environmental Petroleum Consortium 2001,* Office of Science Advisor, Policy & Engagement U.S. Environmental Protection Agency 1200 Pennsylvania Avenue NW, Washington, D.C., 2021.
57. Zhaoa, X., Lva, L., Pana, B., Zhanga, W., Zhanga, S., Zhanga, Q., Polymer-supported nanocomposites for environmental application: A review. *Chem. Eng. J.*, 170, 2–3, 381–394, 2011.
58. Khalil, A., Gondal, M.A., Dastageer, M.A., Synthesis of nano-WO3 and its catalytic activity for enhanced antimicrobial process for water purification using laser induced photo-catalysis. *Catal. Commun.*, 11, 3, 214–219, 2009.
59. Chaturvedi, A.S., Pragnesh, N., Dave, A., Shah, N.K., Applications of nano-catalyst in new era. *J. Saudi Chem. Soc.*, 16, 3, 307–325, 2012.
60. Hildebrand, H., Mackenzie, K., Kopinke, F.-D., Novel nano-catalysts for wastewater treatment. *Glob. NEST J.*, 10, 1, 47–53, 2008.
61. Khalil, A., Gondal, M.A., Dastageer, M.A., Augmented photocatalytic activity of palladium incorporated ZnO nanoparticles in the disinfection of Escherichia coli microorganism from water. *Appl. Catal., A*, 402, 1–2, 162–167, 2011.
62. Omole, A.M., Owino, K., II, Sadik, A.O., Nanostructured Materials for Improving Water Quality: Potentials and Risks, in: *Nanotechnology Applications for Clean Water*, pp. 233–247, 2009.
63. Choi, H., Souhail, R., Abed, A., Dionysiou, D.D., Nanostructured Titanium Oxide Film and Membrane-Based Photocatalysis for Water Treatment, in: *Nanotechnology Applications for Clean Water*, pp. 39–46, 2009.
64. Bai, H., liu, Z., Sun, D.D., Hierarchical ZnO nanostructured membrane for multifunctional environmental applications. *Colloids Surf., A*, 410, 20, 11–17, 2012.
65. Xu, J., Bachas, L., Bhattacharyya, D., Synthesis of Nanostructured Bimetallic Particles in Poly ligand Functionalized Membranes for Remediation Applications, in: *Nanotechnology Applications for Clean Water*, pp. 311–335, 2009.
66. Clemenson, S., Espuche, E., David, L., Leonard, L., Nanocomposite membranes of polyetherimide nanostructured with palladium particles: Processing route, morphology and functional properties. *J. Membr. Sci.*, 361, 1–2, 167–175, 2010.
67. Tarabara, V.V., Multifunctional Nanomaterial-Enabled Membranes for Water Treatment, in: *Nanotechnology Applications for Clean Water*, pp. 59–75, 2009.
68. Yadav, K.K., Singh, J.K., Gupta, N., Kumar, V., A Review of Nano bioremediation Technologies for Environmental Cleanup: A Novel Biological Approach. *J. Mater. Environ. Sci.*, 8, 2, 740–757, 2017.
69. Prakash, A., Sharma, S., Ahmad, N., Ghosh, A., Sinha, P., Synthesis of Agnps by Bacillus Cereus Bacteria and Their Antimicrobial Potential. *J. Biomater. Nanobiotechnol.*, 2, 2, 155–161, 2011.
70. Savage, N. and Diallo, M.S., Nanomaterials and water purification: Opportunities and challenges. *J. Nanopart. Res.*, 7, 331–342, 2005.
71. Caro, E., Marce, R.M., Borrull, F., Cormack, P.A.G., Sherrington, D.C., Application of molecularly imprinted polymers to solid-phase extraction of compounds from environmental and biological samples. *TrAC, Trends Anal. Chem.*, 25, 2, 143–154, 2006.

72. Kiparissides, C., Nanotechnology Meets Water Treatment. *Dissemination Workshop of the Nano4water Cluster*, 2010.
73. Schreibera, T., Webera, A., Niedergalla, K., Rieglera, J., Brynioka, D., Hirtha, T., Guenter Tovara, E.M., Water treatment by molecularly imprinted polymer nanoparticles. MRS Spring Meeting. *Cambridge J. Online*, 11, 69, 2009.
74. Fuwad, A., Ryu, H., Malmstadt, N., Kim, S.M., Jeon, T.-J., Biomimetic membranes as potential tools for water purification: *Preceding and future avenues. Desalination*, 15, 97–115, 2019.
75. Kaufman, Y. and Freger, V., Supported Biomimetic Membranes for Pressure- Driven Water Purification, in: *On Biomimetics*, L. Pramatarova (Ed.), pp. 271–5, InTech, 2011, ISBN:978-953-307-Available from: http://www.intechopen.com/books/on-biomimetics/supportedbiomimetic-membranes-for-pressure-driven water-purification, InTech China, Shanghai, China. And InTech Europe University Campus STeP Ri Slavka Krautzeka 83/A 51000 Rijeka, Croatia.
76. Kant, A. and Datta, M., Adsorption Characteristics Of Victoria Blue On Low Cost Natural Sand And Its Removal From Aqueous Media. *Eur. Chem. Bull.*, 3, 8, 752–759, 2014.
77. Vermeulan, T.H., Vermeulan, K.R., Hall, L.C., Fundamental. *Indian Eng.Chem.*, 5, 212–223, 1966.
78. Easton, J.R. and Cooper, P., On Site Application of Pseudomonas Aeruginosa ETL-1942 and Bacillus Cereus ETL-1949 in Decolorization and Degradation of Remazol Black-B, in: *Society of dyers and colourists*, pp. 9–21, 1995.
79. Kant, A., Gaijon, P., Nadeem, U., Adsorption Equilibrium and Kinetics of Crystal Violet Dye from Aqueous Media onto Waste Material. *Chem. Sci. Rev. Lett.*, 3, 11S, 1–13, 2014.
80. Lagergren Zur, S., theories der sogenannten adsorption gelosterstoffe. *Handlingar*, 24, 4, 1–39, 1898.
81. Ho, Y.S. and McKay, G., Pseudo-second order model for sorption processes. *Process Biochem.*, 34, 451–465, 1999.
82. Kant, A., Kumar, G., Ehtesham, M., Ghosha, S., Singha, M.R., Gaijona, P., *Application of Carbonbased Nano-composite materials for the Wastewater treatment, Book Chapter 14*, p. 335, IGI Global publisher, USA, May, 2022, doi: DOI: 10.4018/978-1-6684-4553-2, ISBN: 9781668445532.

17

New Trends in Asymmetric Catalysis: Chiral Hypervalent Iodine Compounds as Green and Sustainable Catalysts

Vikas Yadav, Rohit Kumar, Amrit Gond, Ashvani Yadav, Mitushree Ghosh, Ram Singh Kuri and Virendra Prasad*

Department of Chemistry, Institute of Science, Banaras Hindu University (BHU), Varanasi, India

Abstract

Catalysis is widely recognized as a fundamental aspect of synthesis owing to the significant advancements made in the development of numerous catalysts, catalytic protocols, and methodologies over the past few decades. Among the diverse forms of catalysis, asymmetric catalysis has experienced significant growth in contemporary times. In the literature, many significant catalysts have been established for asymmetric synthesis. Organocatalysis is being employed as a preferred alternative to asymmetric enzymatic and metallic catalytic processes. Significantly, with regard to asymmetric redox reactions, organocatalysis has yielded remarkably exceptional outcomes. However, it is regrettable that asymmetric organocatalytic oxidations have proven to be exceedingly arduous. In order to address the challenges associated with asymmetric oxidations, chiral hypervalent iodine compounds have emerged as superior alternatives. Currently, numerous chiral hypervalent iodine compounds have been established and widely recognized for asymmetric synthesis. Chiral hypervalent iodine (III/V)-catalyzed asymmetric oxidations represent an efficient approach for enantioenriched compounds and intermediates of significant importance. Such reactions hold great promise for the development of novel therapeutic agents. The construction of chiral hypervalent iodines (III/V) through asymmetric organocatalytic oxidations has established a promising area in asymmetric chemistry. Several noteworthy asymmetric reactions include the oxidative dearomatization of phenols, ketone oxygenation, and alkene dioxygenation. This chapter presents a concise overview of several catalytic properties exhibited by hypervalent iodines (III/V) and their use for stereoselective oxidative rearrangement of disubstituted unactivated olefins. A catalytic method for the multi-O-C coupling of alkyne in a simple one-step process has been developed. This method is utilized for the synthesis of valuable α-ketoesters and their chiral analogs. Furthermore, an expanding yet insufficiently researched group of stereogenic substances known as axially chiral C-N compounds possesses the capacity to serve as chiral ligands and biologically active frameworks that can be developed. Moreover, a noteworthy progression in enantioselective catalysis pertains to the asymmetric oxidative dearomatization of naphthol carboxylic acids, which leads to the formation of spirolactones. This process is facilitated by a recently developed conformationally rigid oxo-bridged hypervalent iodine (III) species. Hypervalent iodine (III/V) reagents, characterized by their ease of handling, lack of toxicity, mildness, environmental friendliness (as green oxidants), and high stability, have

**Corresponding author*: vprasadbhu@gmail.com

Mousumi Sen (ed.) Sustainable Green Catalytic Processes, (397–446) © 2024 Scrivener Publishing LLC

been employed in numerous reactions and have facilitated the exploration of previously unexplored enantioselective transformations. This chapter aims to collate the reactions documented in the field of asymmetric hypervalent iodine chemistry, with a particular emphasis on the synthesis of chiral iodine (III and V) reagents or chiral iodoarenes.

Keywords: Asymmetric catalysis, chiral hypervalent iodine, enantioselective, asymmetric oxidations, stereoselective, asymmetric dearomatization, chiral reagents

17.1 Introduction

Catalysis plays a crucial role in chemistry and has led to the growth of numerous catalysts and catalytic methodologies in recent decades. Asymmetric synthesis has made significant advances through catalysis,[1, 2] including asymmetric metal catalysis, enzymatic catalysis, and organocatalysis.[3–12] Organocatalysis, which uses small organic molecules as catalysts, has emerged as a well-established approach for a wide spectrum of enantioselective transformations.[13] While organocatalysis has shown excellent results in asymmetric reductions,[14–17] asymmetric organocatalytic oxidations have posed greater challenges. Controlling the stereoselectivity in such processes is often more difficult than in metal-catalyzed[18–20] or enzymatic oxidations.[21, 22] Therefore, the development of more efficient asymmetric organocatalytic oxidations remains a major challenge.

In light of this, the investigation of hypervalent iodine reagents has proved prolific and interesting, especially in the area of oxidation. Hypervalent iodine compounds possess distinctive features like simple handling, mild reaction conditions, environmental friendliness, and high stability. These reagents offer advantages over metal-catalyzed oxidations.[23–35] Since the first synthesis of hypervalent iodine(III) molecules by Willgerodt[36] in 1885, a wide range of hypervalent iodine species have been developed (Figure 17.1). Hypervalent iodine compounds can be classified into two groups based on the oxidation state of the iodine atom:

(i) Compounds with iodine in the +3 oxidation state (λ3-iodanes) (Figure 17.1).
(ii) Compounds with iodine in the +5 oxidation state (λ5-iodanes) (Figure 17.2).

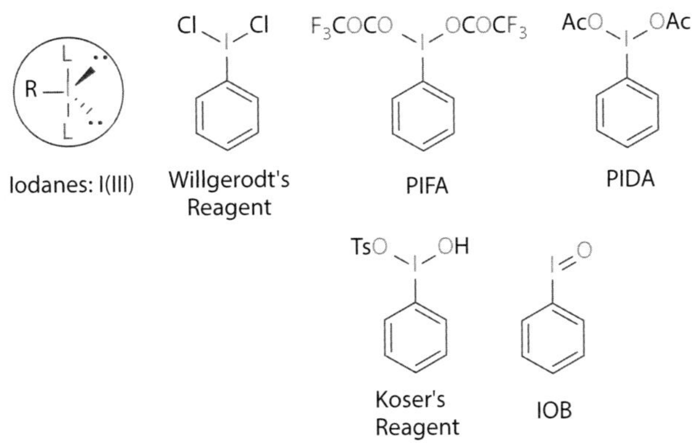

Figure 17.1 Iodanes with iodine in +3 oxidation state.

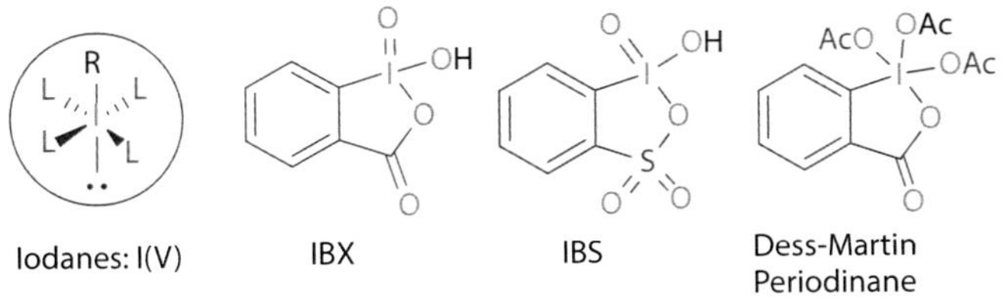

Figure 17.2 Iodanes with iodine in +5 oxidation state.

The λ^3-iodanes, such as PIDA[37] and PIFA,[38] have a pseudo-trigonal bipyramidal geometry, while λ5-iodanes such as IBS,[39] IBX,[40] and Dess-Martin periodinane,[41] have a square-pyramidal geometry.[13] These hypervalent iodine reagents have become powerful and efficient oxidants that find extensive application in the domain of organic synthesis. Although the inclusion of organic hypervalent iodine reagents into asymmetric synthesis took some time to establish, this may be due to some unsuccessful results in early attempts, which will be described later. In recent years, there has been significant growth in the development and application of chiral hypervalent iodines(III/V) associated with ligands or chiral structures. This thriving field of research has led to the development of various chiral catalysts of hypervalent and substrate that can enable complex enantioselective synthetic transformations. Extensive reviews have been published on this topic, highlighting the reactivity, synthetic results, synthesis, and proposed mechanisms of these chiral iodine compounds.[42–50] Overall, the use of chiral hypervalent iodine compounds holds great promise for future transformations and represents an exciting and important research area.

17.2 Role of Hypervalent Iodines in Asymmetric Synthetic Approach

17.2.1 Historical Aspects

To understand the initial connection between chirality and an iodine(III) molecule, we need to delve into the early few years of the previous century. In the year 1907, the very first application and characteristics of substances formed by a combination of iodosobenzene and L-tartaric acid were reported.[51] This discovery marked an important milestone in the elucidation of the relationship between chirality and the specific properties of iodine(III) compounds. Although no structures were known at this point, it was presumed that the cyclic structures depicted in Figure 17.3 were possibly identified with the help of this initial iodine(III) compound. Extensive research was conducted on optical rotation to demonstrate the chirality of the novel structure.

The chemistry pertaining to this topic did not gain acceptance among chemists for nearly six decades following the aforementioned report. It was not until Merkushev *et al.*[52] published their findings about the first amino acid containing chiral hypervalent iodine(III) reagents that the field saw significant progress (Figure 17.4). Merkushev *et al.*[52] conducted experiments involving the reaction of PIDA [phenyliodine(III) diacetate] with various chiral N-protected amino acids. They successfully isolated and characterized several products resulting from these reactions. Notably, they obtained derivatives of L-alanyl and L-leucyl

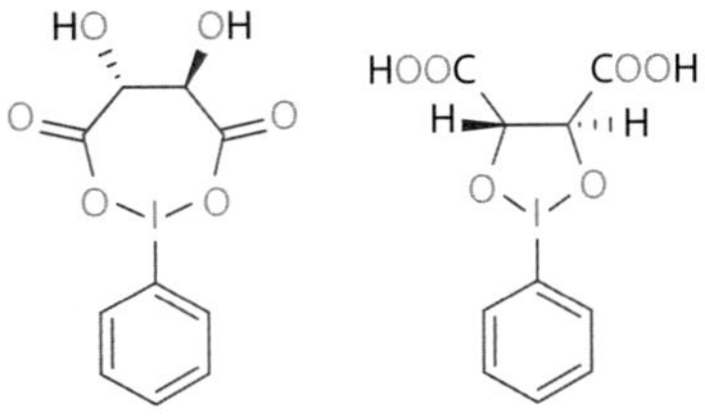

Figure 17.3 First reported chiral hypervalent iodine(III) compounds.

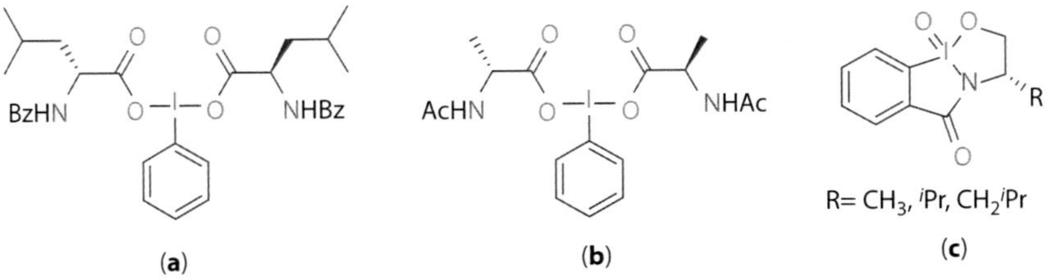

Figure 17.4 (a, b) First reported hypervalent iodine(III) derivatives of amino acid. (c) First reagents of chiral hypervalent iodine(V).

in addition to others (Figures 17.4a, 17.4b). The characterization of these compounds was achieved by infrared spectroscopy. Despite making a significant addition to the scientific work, these substrates remained untapped in asymmetric transformations until the 1990s. However, it is worth noting from a historical perspective the pioneering achievement of the first compound of hypervalent iodine possessing chirality with +5-oxidation number (as depicted in Figure 17.4c). As far as we know, this milestone was not reached until much later, specifically in the year 2000, by Zhdankin *et al.*[53] They successfully synthesized multiple reagents of chiral hypervalent iodine (V), utilizing the chiral nature of amino acids. This approach aimed to enhance stereocontrol and reactivity. The preparation and applications of these reagents will be elaborated upon in the relevant section.

17.2.2 Strategies for Stereoselectivity Control

The high reactivity of hypervalent iodine compounds is primarily due to the remarkable electrophilicity of the iodine atom and the extraordinary efficacy of the leaving groups, which is approximately 10^6 times higher as compared to triflate.[54] The reactivity and stereoinduction of these compounds are determined by the properties and arrangement of the ligands surrounding the iodine. Therefore, understanding the reactivity and structure relationships of such compounds is crucial to comprehend the diverse strategies employed to develop the preferred catalysts and reagents in the field of asymmetric synthesis. Contextually, when considering hypervalent iodine (III), there are two major categories of compounds that can be prominently displayed (Scheme 17.1).[55]

(1) Category I: Hypervalent iodine reagents feature a substituent of carbon compound and two ligands of a heteroatom. These ligands of a heteroatom are consistently located at the peak place within the geometry of pseudotrigonal bipyramidal.

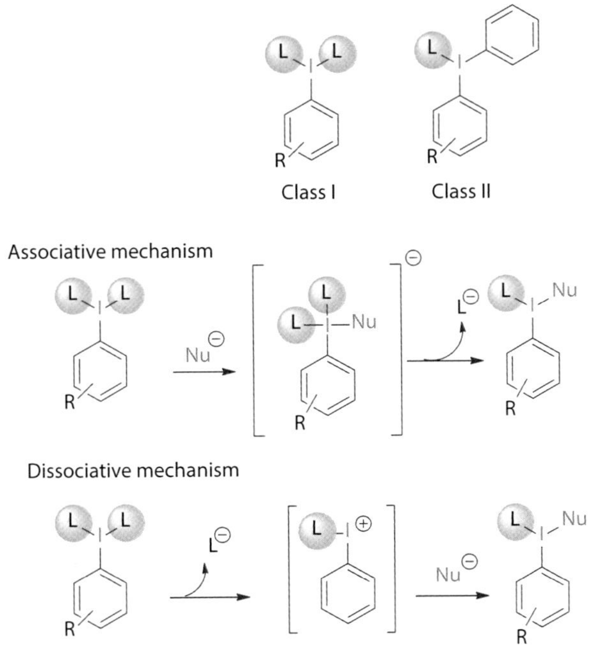

Scheme 17.1 Common reaction mechanistic pattern.

(2) Category II: Hypervalent iodine molecules are produced by the association of two different substituents of carbon compounds and a ligand of a heteroatom.

Category I hypervalent iodine reagents are highly effective for conducting the reactions of oxidation because of the presence of ligands containing a heteroatom that facilitate reactivity through ligand exchange. On the other hand, class II reagents are not as potent as oxidants, but they find utility in transferring one substituent of carbon. With respect to the stereocontrol of these chiral compounds, the chiral surrounding of iodine, created by the ligands containing a heteroatom, has a controlled impact on inducing stereochemistry. Instead, it actively participates in the oxidation process, following either a dissociative or an associative pathway[56] (Scheme 17.1). Indeed, the initial attempts to incorporate chirality into hypervalent iodine reagents and catalysts yielded limited success in terms of enantioselectivity. Consequently, it is reasonable to presume that the introduction of chirality components into the fragments of carbon of these catalysts would lead to solid and proficient outcomes. To grasp the advancement and design of catalysts known to facilitate the enantioselective transformations with the help of iodine (III/V), it becomes essential to comprehend the diverse structural chirality introduced into these molecules, particularly focusing on those where chirality arises from carbon fragments. Upon meticulous analysis of reported chiral hypervalent iodine catalysts and reagents, approaches can be categorized into four distinct groups based on the chirality type involved (Figure 17.5).

Category I encompasses chiral hypervalent iodine catalysts or reagents that are derived from chiral iodoarenes by *in situ* oxidation. These catalysts utilize the chirality arising from one or more stereogenic elements in the carbon portion, enabling stereocontrol during asymmetric transformation. In pursuit of this goal, a variety of chiral features have been

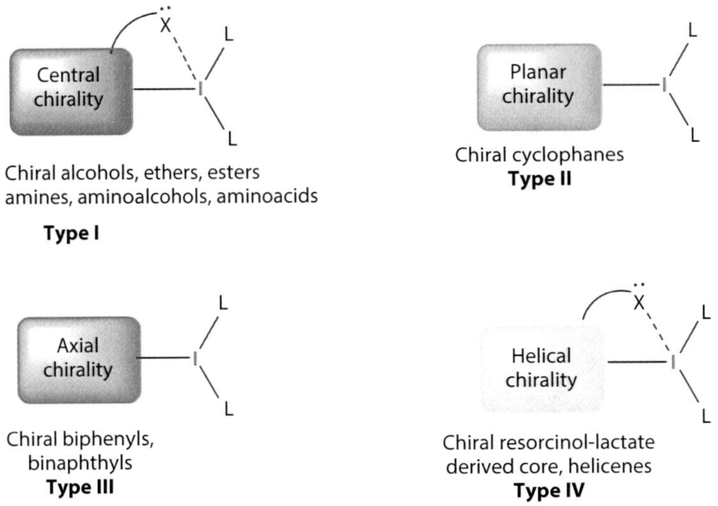

Figure 17.5 Approaches to stereoselectivity.

incorporated into these catalysts, including chiral amines, alcohols, amino acids, and amino alcohols, along with other functional groups. These chiral elements play a crucial role in imparting stereochemistry and achieving enantioselectivity in the desired asymmetric transformations. Furthermore, these incorporated functional groups not only contribute to stereocontrol but also possess coordinative capacity. This coordinative ability enables them to stabilize the intramolecular hypervalent iodine intermediate through interactions within the catalyst structure. Despite the preparation of numerous chiral reagents using this approach, the achieved enantioinduction has been limited, with only a few transformations yielding high enantiomeric excesses. This indicates that further improvements are necessary to enhance the enantioselectivity of these reactions.

Despite the compounds of chiral hypervalent iodine with central chirality (Category I), there are other types of structures that are important and involve the inclusion of planar and axial chiralities, known as category II and III (Figure 17.5). In these instances, chiral cyclophanes, chiral binaphthyls, or chiral biphenyls were applied as chiral inducers to achieve better stereoselectivity. Even though these approaches have achieved a reasonable level of success in enantiocontrol, there remains considerable scope for improvement.

However, the most successful and promising strategy for conducting asymmetric hypervalent iodine-promoted oxidations involves the assimilation of helical chirality[57] (Category IV) around the atoms of hypervalent iodine. This innovative approach was first introduced by Fujita *et al.*[58] and Uyanik *et al.*[59] The structures that exhibit the fascinating helical chirality feature with C_2 symmetry are obtained due to resorcinol core. This core bears chiral lactamide I/lactate H arms that are conformationally flexible (Figure 17.6). This incorporation of helical chirality has demonstrated remarkable success and holds great potential for advancing asymmetric hypervalent iodine-promoted oxidations.

In the scenario of hypervalent iodine compounds, it is possible for intramolecular n-σ* interactions to take place between lone pair electrons of carbonyl groups of amides and esters and electron-deficient hypervalent iodane species (via the σ* orbital of the C-I bond)

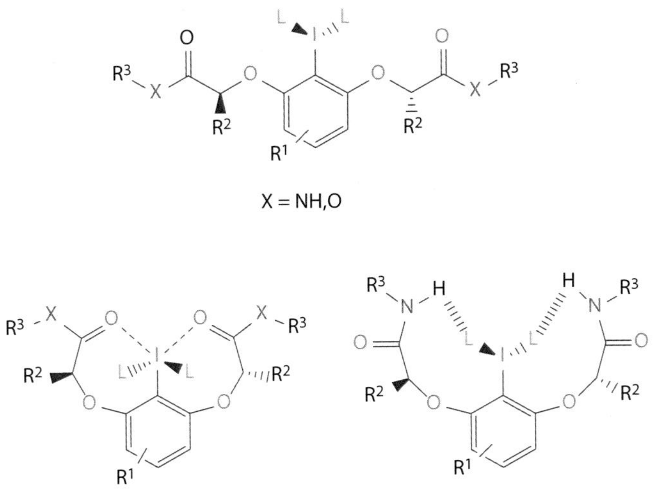

Figure 17.6 Hypervalent iodine molecules with helical chirality.

(Figure 17.6).[13] Furthermore, it is possible to establish hydrogen bonding interactions between the basic atoms of ligands and the N-H group (Figure 17.6). These noncovalent interactions, including π-π stacking, can contribute to an assembly of three dimensions, resulting in a distinctive helical conformation. By integrating helical conformational chiral arms, a distinctive topology is created that exerts a significant influence on the stereodetermining step of numerous asymmetric conversions. This exceptional attribute has brought a revolutionary impact in the field, empowering improved stereocontrol and unveiling fresh opportunities for the advancement of highly enantioselective transformations. Significant contributions are made using DFT calculations,[60] confirming the helical nature of several structures mentioned. Specifically, extensive studies have been conducted on the stereoelectronic and confirmational features of structure L shown in Figure 17.8, which is formed *in situ* from the reaction of mCPBA as oxidizing agent and respective chiral iodine in the presence of CF_3COOH (Figure 17.7).[60]

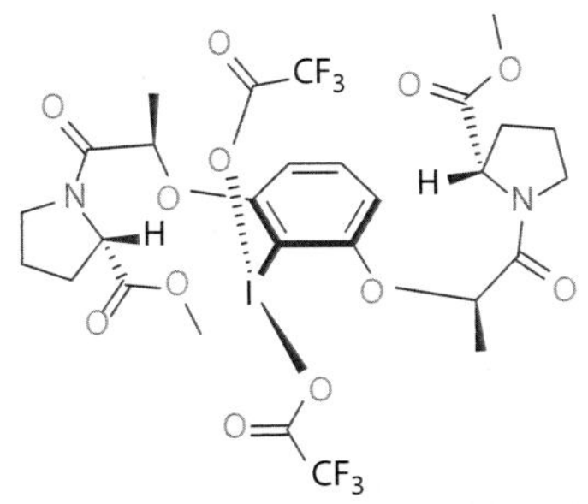

Figure 17.7 Structure of hypervalent iodine presenting helical chirality.

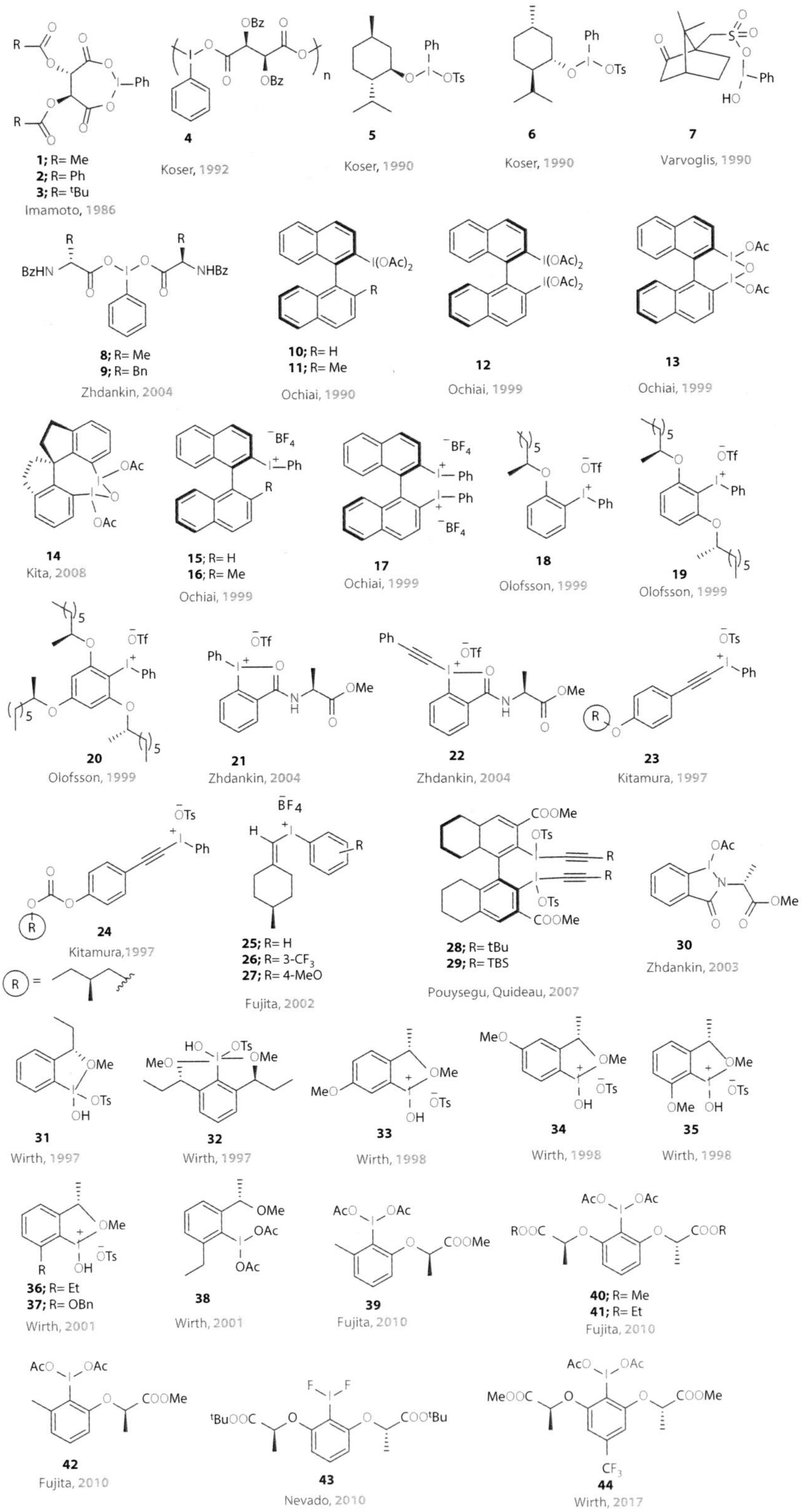

Figure 17.8 Various reagents of chiral hypervalent iodine(III).

17.2.3 Review of Catalysts

Presently, the area of research focusing on hypervalent iodine in asymmetric catalysis has become a well-established and flourishing area, resulting in a broad range of notable reactions. Organic chemists have exhibited fascinating creativity in incorporating hypervalent iodine(III/V) within a chiral environment, employing various approaches and strategies. Figure 17.8 illustrates the progression of these strategies, starting with the utilization of chiral alcohols or acids to impart chirality in the skeleton and advancing toward the integration of chiral ionic salts, axial chirality, chiral coordinative groups, and chiral amino acids. Among these approaches, the combination based on resorcinol/lactate has proven to be highly successful, exhibiting remarkable enantioselectivity in numerous hypervalent iodine(III)-catalyzed asymmetric reactions. Throughout this review, these compounds are referred to as chiral hypervalent iodines. One potential drawback of these methodologies, which has been subsequently addressed, was their initial usage in stoichiometric quantities. However, through later advancements, this limitation has been overcome. Figures 17.8 and 17.9 present a comprehensive overview of the structures of chiral hypervalent iodines classified by families, accompanied by the corresponding research teams with years responsible for their development. These outlines serve as a valuable resource, providing a concise summary of the structures that have been synthesized and employed to date. As a result, they can be instrumental in guiding the design and discovery of novel and potential catalysts in the future.

In a similar manner, a diverse chiral hypervalent iodine(V) molecules have been created, incorporating various elements of chirality derived from privileged catalyst frameworks like bornyl or fenchyl, proline, resorcinol/lactate, oxazoline, trans-1,2-cyclohexanediamine, biphenyls, and so on (Figure 17.9). The compounds of chiral hypervalent with iodine in the +5-oxidation state have been comparatively less explored than the lower oxidation state of +3. Nevertheless, this comprehensive overview will serve as inspiration for researchers to explore and devise novel chiral hypervalent iodines in the coming time. It is important to note that all chiral hypervalent iodines are employed stoichiometrically.

Over the years, a primary drawback associated with the use of these reagents in both non-asymmetric and asymmetric reactions has been the necessity of stoichiometric or even superstoichiometric amounts, which leads to the generation of considerable waste. Consequently, this drawback has spurred the quest for catalytic alternatives. In 2005,

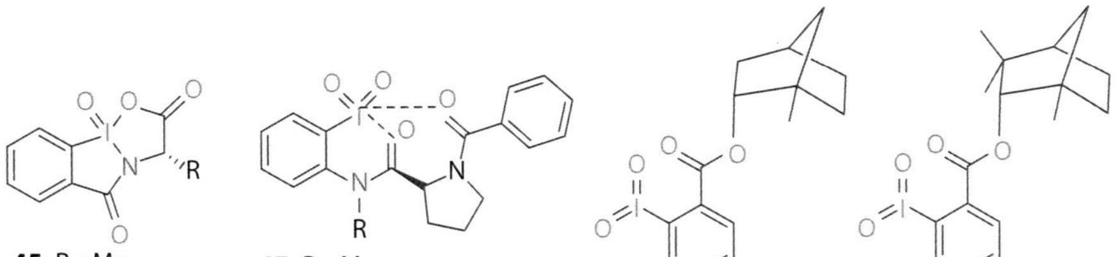

45; R= Me
46; R= $CH_2{}^iPr$
Zhadankin, 2000

47; R= Me
48; R= 4-Br-benzyl
Zhadankin, 2006

49
Wirth, 2010

50
Wirth, 2010

Figure 17.9 Various reagents of chiral hypervalent iodine(V).

Ochiai *et al.*[61] and Dohi *et al.*[62] independently reported the very first use of iodoarene quantities as a catalyst with other terminal oxidants to produce the catalysts of hypervalent iodine *in situ*. The concept pertains to the utilization of catalytic quantities of the corresponding iodoarenes in conjunction with stoichiometric quantities of sacrificial oxidants for the purpose of conducting reactions (Scheme 17.2). Various sacrificial oxidants have been employed, with Selectfluor, mCPBA, and peracetic acid demonstrating the highest degree of efficacy. Due to the significant benefits offered by this approach, it expeditiously attained eminence within the realm of chiral hypervalent iodine chemistry. Consequently, a diverse range of chiral iodoarenes were synthesized, devised, and used as catalysts in catalytic conversion in conjunction with terminal oxidants. The adoption of the catalytic approach has facilitated the investigation of numerous supplementary catalysts and diverse methodologies for their synthesis in recent times.

Figure 17.10 illustrates the chiral iodoarenes that have been utilized as catalysts in the context of asymmetric hypervalent iodine oxidations. Similarly, a wide range of chirality components and privileged structures have been incorporated to identify the most promising candidates with respect to chemical and optical yields. The classification of these candidates into distinct families, as depicted in Figure 17.10, provides a comprehensive perspective on the modern cutting edge and serves as a valuable reference point for future

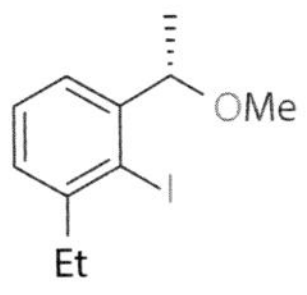

51
Wirth, 2007

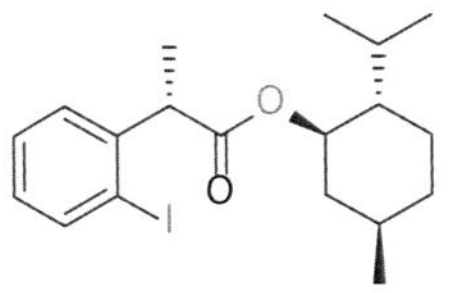

52
Wirth, 2006

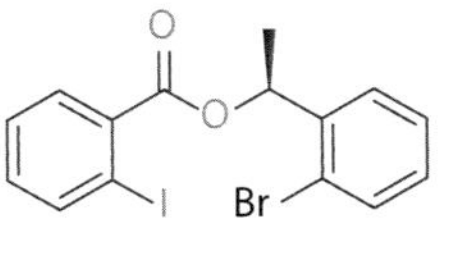

53
Wirth, 2010

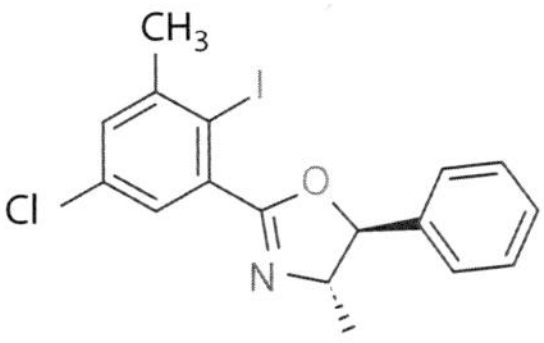

54
Legault, 2012

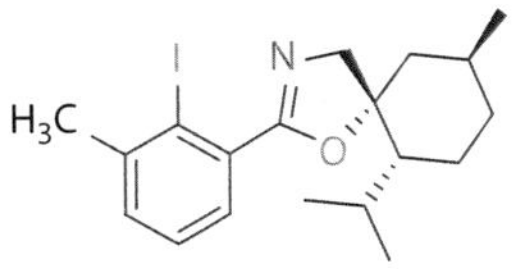

55
Legault, 2013

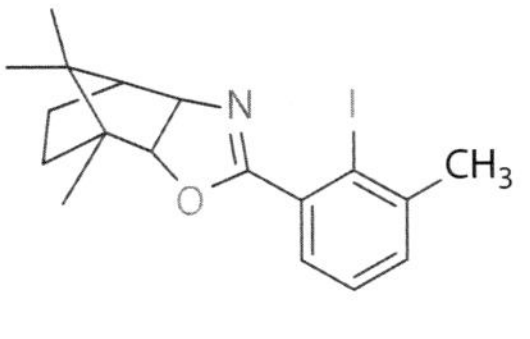

56
Legault, 2013

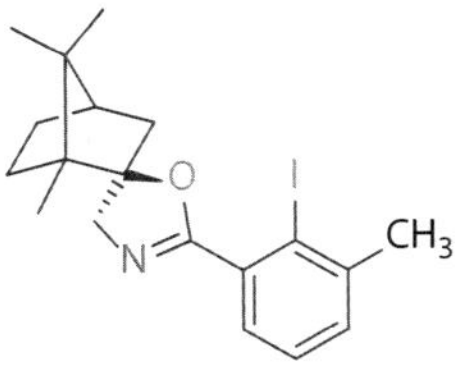

57
Legault, 2013

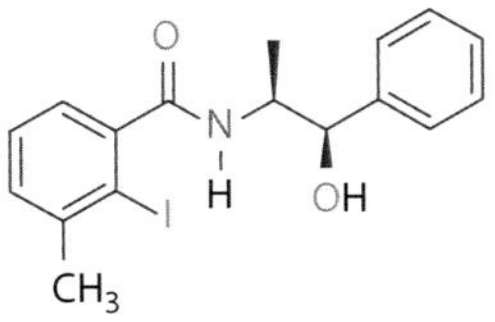

58
Moran, 2012

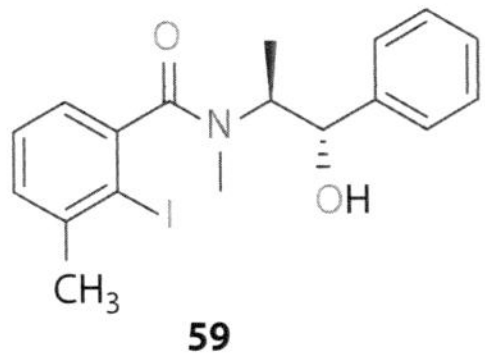

59
Moran, 2012

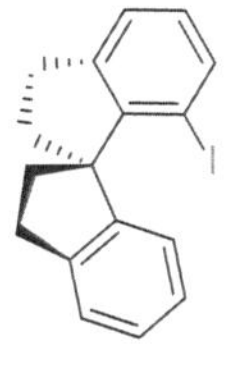

60
Zhang, 2011

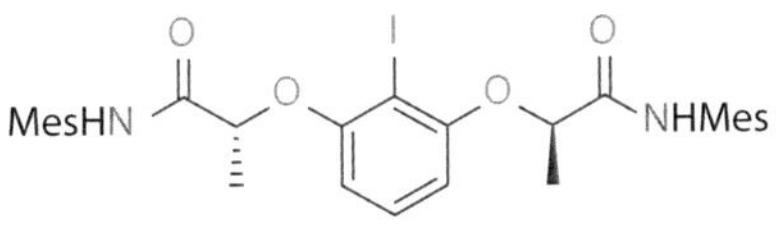

61
Ishihara, 2010

Figure 17.10 Chiral iodoarene precursors.

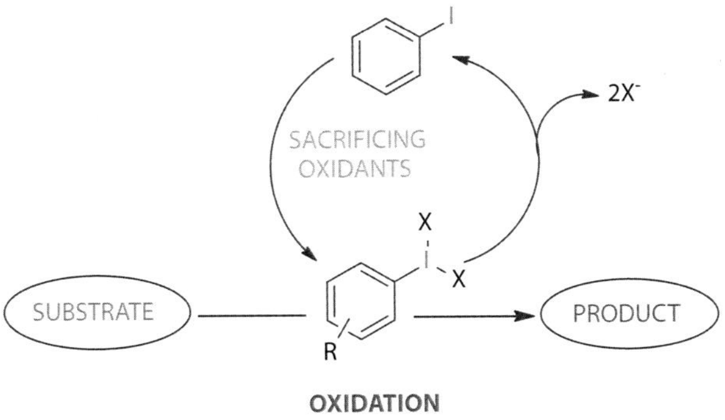

Scheme 17.2 A general presentation of the catalytic form of iodoarene in oxidations.

investigations into the pursuit of novel catalysts. The above catalysts are classified as chiral iodoarenes in order to differentiate them from chiral hypervalent iodines. The distinction between chiral hypervalent iodines and chiral iodoarenes aims to streamline the discourse on noteworthy advancements in this area.

17.3 Synthesis and Reactivity

17.3.1 Oxidations of Sulfides via Asymmetric Catalysis

Enantiopure sulfoxides are highly valued as chiral auxiliaries and find extensive applications in various reactions.[63] Asymmetric synthesis using chiral hypervalent iodines began in 1986 with the pioneering work of Imamoto *et al.*[64] They prepared iodoarenes (**1-3**) by reacting iodosobenzene with various O-diacylated L-tartaric anhydrides in suitable solvents and proposed a cyclic structure with a seven-membered ring for these moieties (Scheme 17.3). These chiral λ^3-iodanes were used as such for the synthesis of prochiral sulfides (**63a-b**) by asymmetric oxidation with excellent product yields and better enantioselectivities. In particular, the oxidation reaction using acetyl-L-lactic acid-derived λ^3-iodane and iodosobenzene resulted in a nearly racemic sulfoxide, suggesting that the C_2 symmetry present in **1-3** plays a major role in regulating enantioselectivity. Although the induction of asymmetry in sulfoxides (**53**) was modest, this method acted as a stimulus for the design of various chiral λ^3-hypervalent iodines for use in asymmetric synthesis.

Ray and Koser[65] conducted a detailed investigation of the structural conversion developed by Imamoto *et al.*[64] Koser[66] focused on iodinanes and bis(acyloxy)iodinanes, which exhibited a T-shaped geometry, and suggested that cyclic structures of **1-3** may be associated with high-energy geometry (Scheme 17.4). They subsequently said that an acyclic structure of **4** appears to be a more sensible candidate for involvement in such conversions.

Ray and Koser[67] also achieved the preparation of λ^3-iodanes having ligands with homochiral alkoxy for the first time using the methoxy group in **66** (Scheme 17.5). The resulting yields of **5-6** were nearly moderate to good. These chiral hypervalent iodanes were subsequently utilized in the synthesis of chiral sulfoxides (R) or (S)-**64**, with substantial yields and fair enantiomeric excesses achieved through base-catalyzed hydrolysis of enantioexcess (menthyloxy)sulfonium tosylates (-)- or (+)-**67**.

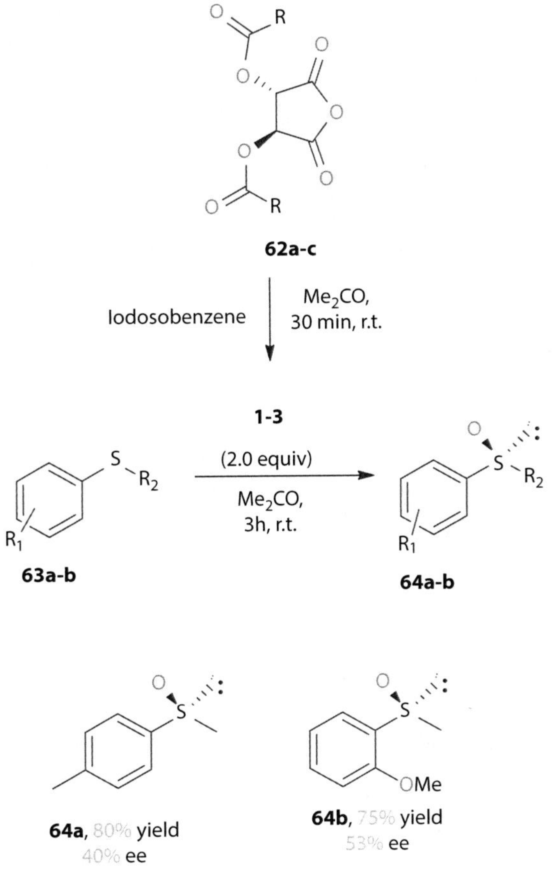

Scheme 17.3 Asymmetric sulfide oxidation.

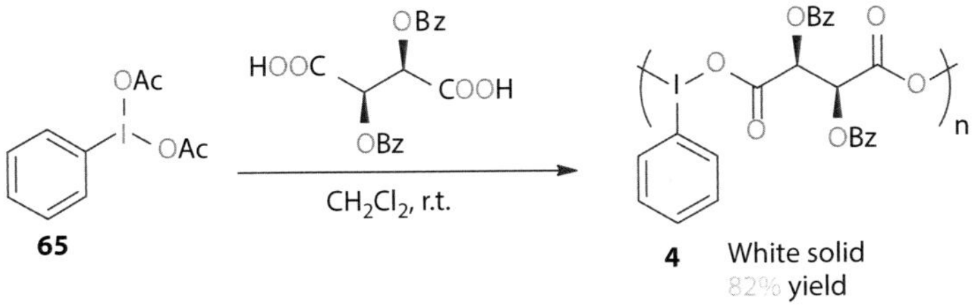

Scheme 17.4 Extraction of dibenzoyl tartrates of iodine(III).

Taking an opposite approach to Koser's work, Hatzigrigoriou *et al.*[68] incorporated a sulfonic acid ligand bearing a chiral scaffold at iodine(III). The reaction of (+)-10-camphorsulfonic acid **68** with (diacetoxyiodo)benzene **5** resulted in the formation of an iodoarene with chirality **7**, with a yield of 80% (Scheme 17.6). Xia and Chen[69] later employed compound **7** for the purpose of asymmetric oxidation of sulfoxide or sulfide, resulting in high yields with limited enantioselectivity. Notably, none of these research groups reported any instances of overoxidation to the sulfone.

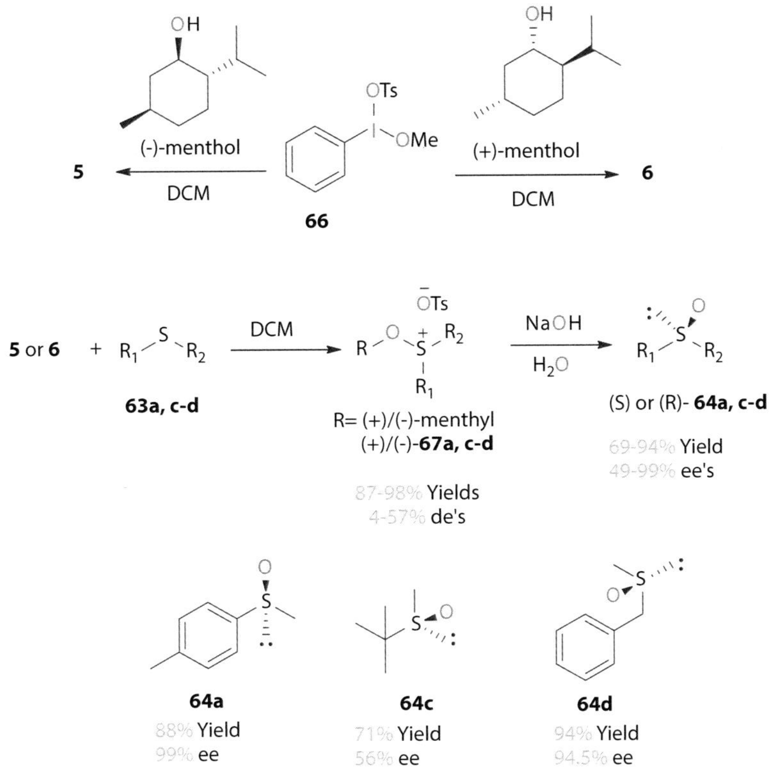

Scheme 17.5 Ray and Koser[67] λ³-iodanes with homochiral alkoxy ligands.

Tohma *et al.*[70] used iodosobenzene (PhI=O)-catalyzed controlled oxidation for the preparation of sulfoxides (**64a, i, j**) by catalysis using a system of reverse micelle. They found that the solubility of the iodoarene reagents was regulated by a phase transfer catalyst like cetyltrimethylammonium bromide (CTAB) along with an activated PhI=O. Furthermore, they applied this protocol to asymmetrically oxidize the sulfides (**63a, i** and **j**) in the context of di(2-methoxy)benzoyl-L-tartaric acid (S, S)-**69** presence (Scheme 17.7).[71] Particularly after investigating numerous oxidants, they discovered that iodoxybenzene ($PhIO_2$), an iodine(V) compound, yielded better results compared to its homologous counterpart, iodosobenzene like (PhI=O). Hence, enantioexcess sulfoxides (**64a, i, j**) were prepared in excellent yields with modest to fair enantiomeric excesses by this new and interesting protocol. This was reported to be the first instance of iodine(V) catalyzed non-symmetric oxidation of sulfur to achieve reasonable optical results in the absence of species of transition metal. Subsequently, the same group conducted further experiments and found that similar results were obtained when iodoxybenzene ($PhIO_2$) and catalytic amounts of $MgBr_2$ were used in this conversion.[72]

In 2000, Zhdankin *et al.*[53] synthesized a range of more stable catalysts of chiral hypervalent iodine(V) with better solubility of **56** and **57** based on the structures of IBX and Dess-Martin periodinane (Scheme 17.8). Asymmetric oxidation of sulfide **63k** using these molecules as catalysts was carried out that led to the formation of the product in excellent yields but with limited enantioselectivities.

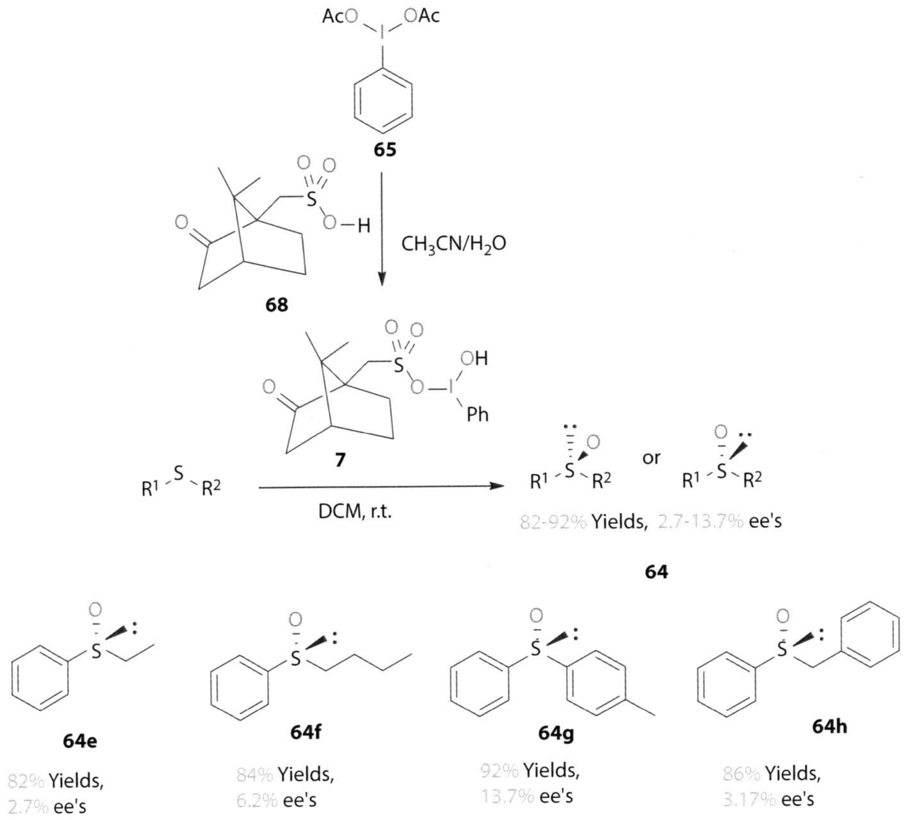

Scheme 17.6 Xia and Chen[69] λ^3-iodanes with homochiral sulfonate ligands.

In subsequent years, Ladziata *et al.*[73] came up with a new design for the synthesis of chiral λ^5-iodanes (**58** and **59)** using a catalyst based on the chiral skeleton of amino acid (S)-proline (Scheme 17.9). They aimed to improve the enantioselectivity achieved with the previous catalysts **56** and **57**. The proposed structure of **58** and **59** involved two chiral pseudo-benziodoxazine derivatives, wherein the upper side center of the hypervalent iodine was jammed, allowing the attack to take place favorably from the lower side (Scheme 17.9). This structure was maintained with the assistance of oxygen of the amide group of carbonyl of proline, forming a ring with seven sides. A similar asymmetric oxidation of sulfide **63a** using **59** resulted in higher enantioselectivity. Additionally, meso-diol **74** was subjected to kinetic resolution using **58** and resulted with a 41% ee.

Alterman *et al.*[74] developed a protocol for the construction of novel catalysts of hypervalent iodine(V), **60** and **61** (Scheme 17.10). Synthesis of these catalysts was carried out by oxidizing the respective iodines **76** or **77** with dimethyldioxirane. The yields for these catalysts ranged from good to very good. However, when these catalysts were used in oxidation of sulfide-sulfoxide, an enantiomeric excess or a racemic mixture was obtained that could not be accurately calculated.

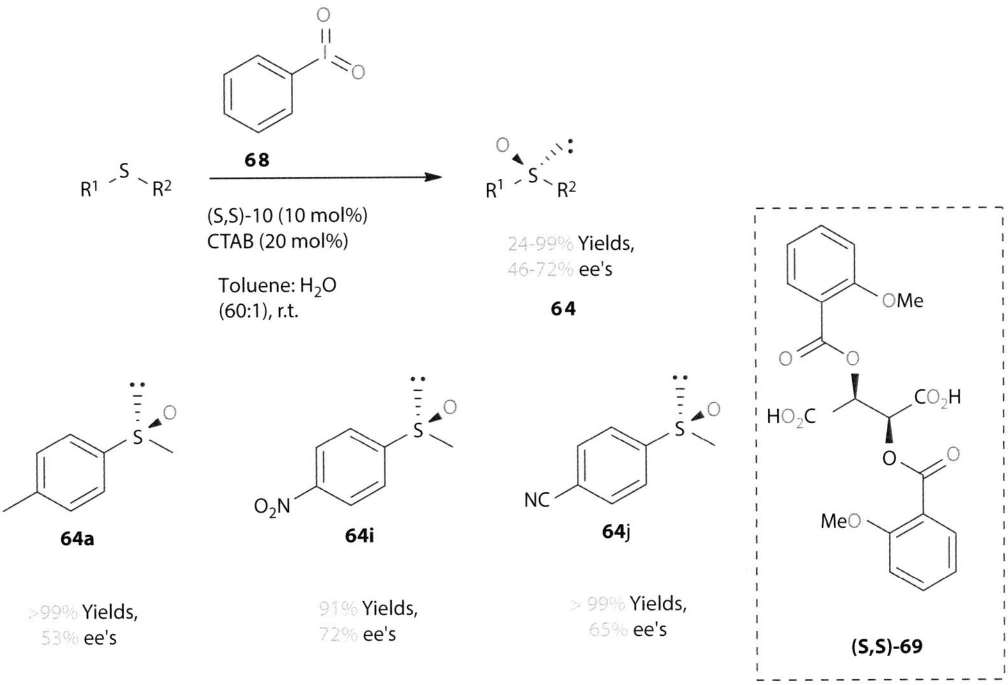

Scheme 17.7 Oxidation of sulfides in asymmetric manner.

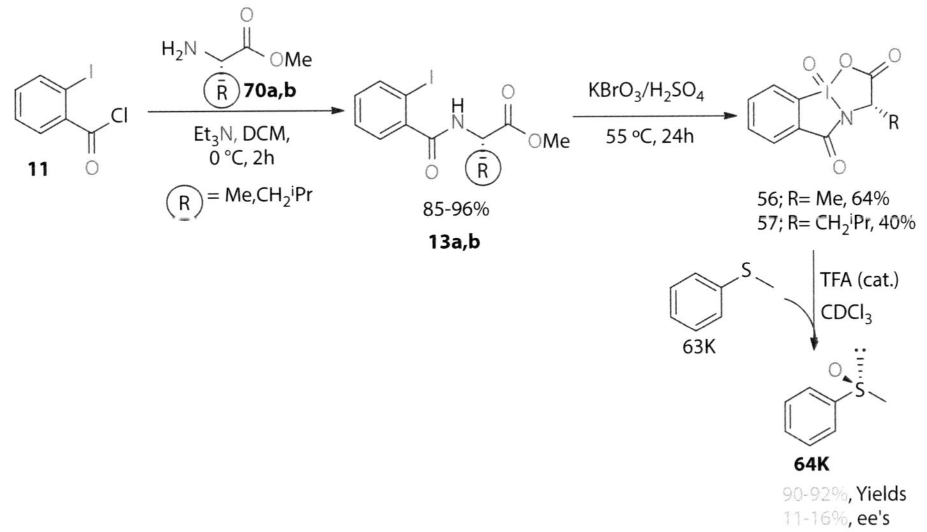

Scheme 17.8 Synthetic protocol for chiral hypervalent iodines(V).

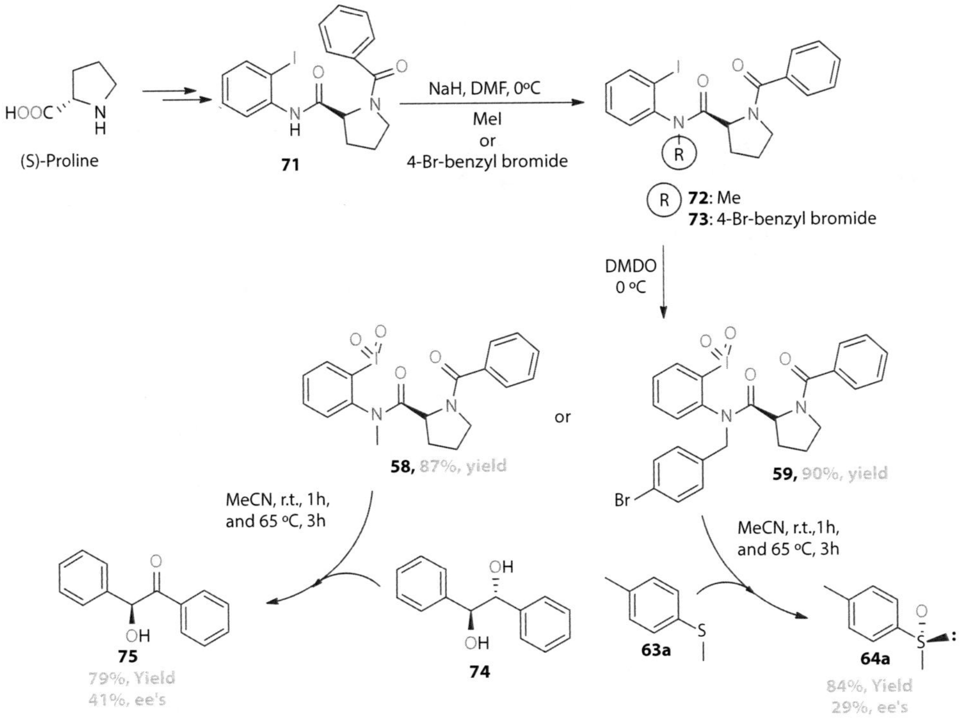

Scheme 17.9 Asymmetric oxidations of chiral pseudo-benziodoxazines.

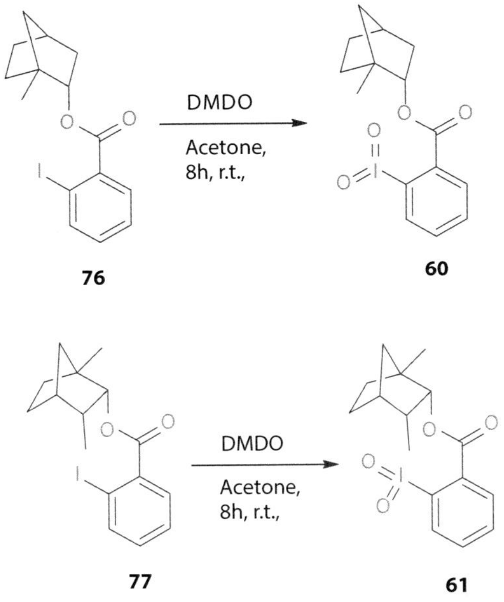

Scheme 17.10 Catalysts based on new chiral hypervalent iodine(V).

17.3.2 Reactions of Arylation and Alkynylation

17.3.2.1 Reactions of Arylation

In 1990, Ochiai *et al.*[75] conducted innovative studies on chiral hypervalent iodine(III) reagents with axial chirality (Scheme 17.11). They prepared chiral λ^3-iodane **10** through the oxidation of compound **78**, yielding excellent yields. More intriguing is the synthesis of chiral hyperdiiodine **12** with C_2 symmetry, which was obtained in excellent yield from compound **79** using sodium perborate. However, compound **12** was observed to exhibit instability, characterized by rapid decomposition into the cyclic μ-oxo-diiodine **13** possessing chirality. The stability of the configuration of iodine(III)-based hypervalent molecules was evaluated by means of NMR analysis. An isomerization of degenerative nature was observed in acetoxy moieties of **10** and **12** through a unimolecular mechanism, which suggests the occurrence of pseudorotation around the iodine atom in the structures **80** and **81** (as depicted in Scheme 17.11 at the bottom).

In 1999, Ochiai *et al.*[76] developed an innovative method for arylation of enolates of β-keto ester asymmetrically using these chiral catalysts of hypervalent iodine(III) as precursors (Scheme 17.12). Compounds **10**, **11**, and **12** were reacted with Ph_4Sn to synthesize the respective tetrafluoroborates of chiral iodonium(III) (**18-20**) in high yields. These salts exhibited high reactivity caused by nucleofugacity, which is likely due to the presence of aryliodonio substituent. This study reports on the arylation of enolate generated from **82a** using chiral reagents of hypervalent iodine(III). The resulting α-phenylated compound exhibited fair enantioselectivities and yields while demonstrating complete oxy-versus carbo-elective arylation of the enolate. Although the enantioselectivity obtained was lower, it represented a significant milestone in the field.

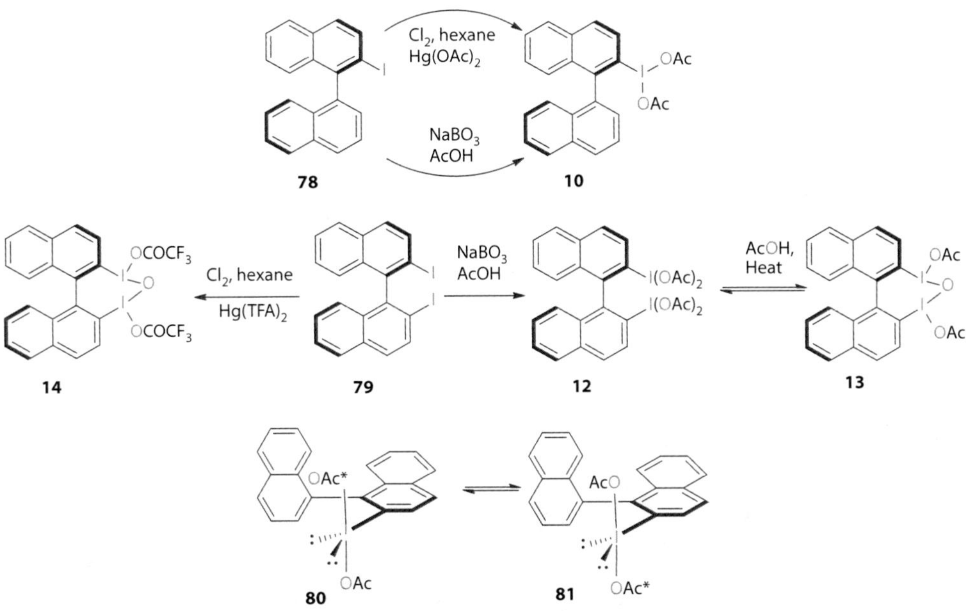

Scheme 17.11 Synthesis of novel chiral hypervalent iodine(III) catalysts featuring binaphthyl moieties.

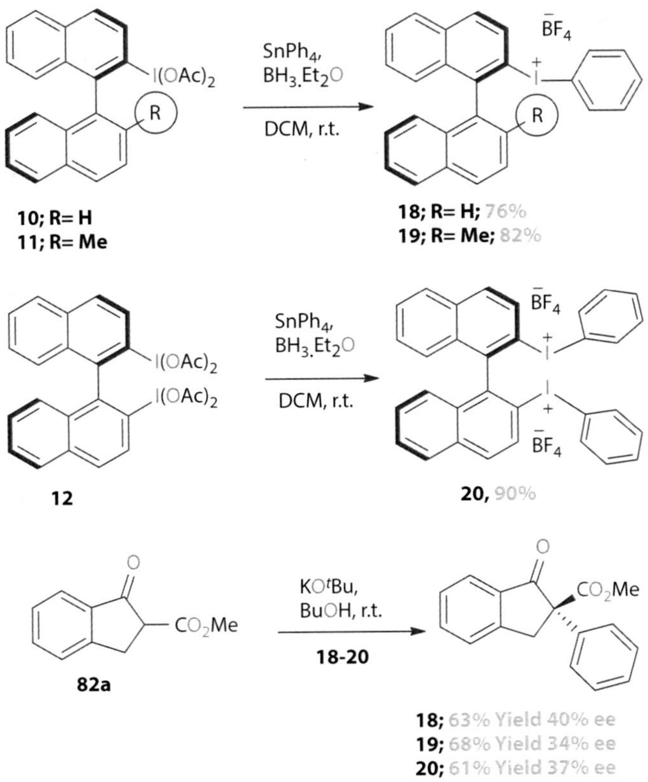

Scheme 17.12 Asymmetric arylation.

In a recent study, Jalalian and Olofsson[77] described the preparation of a range of salts of chiral iodonium (**21-23**) as alternative reagents for α-phenylation of carbonyl compounds in asymmetrical manner with better yield (Scheme 17.13). The authors employed various techniques to sequentially introduce hindered substitution around the iodine atom.

Zhdankin *et al.*[78] conducted an investigation into the preparation of novel hypervalent iodane(III) with the aim of broadening the scope of chiral reagents of hypervalent iodine for future use (Scheme 17.14). Compound **33** is expeditiously synthesized through 2-iodobenzamide (**91**) oxidation, which is prepared using respective chiral amino acid. Subsequently, **33** underwent a transformation into phenyliodonium salt **24** via a reaction with tributylphenyltin and TMSOTf, resulting in fair yields. Same substrate was used in the synthesis of alkenyliodonium salt 25. The X-ray crystallographic analysis of compound **24** demonstrated the presence of an intramolecular interaction between the iodine atom and the oxygen atom of the amido group.

17.3.2.2 Reactions of Alkynylation

In 1997, Kitamura *et al.*[79] made a groundbreaking contribution by introducing chiral iodonium salts for alkynylation reactions, specifically focusing on the synthesis of diynes containing chiral centers within their structures. This pioneering work opened new avenues for asymmetric alkynylation reactions, enabling the creation of complex molecules with

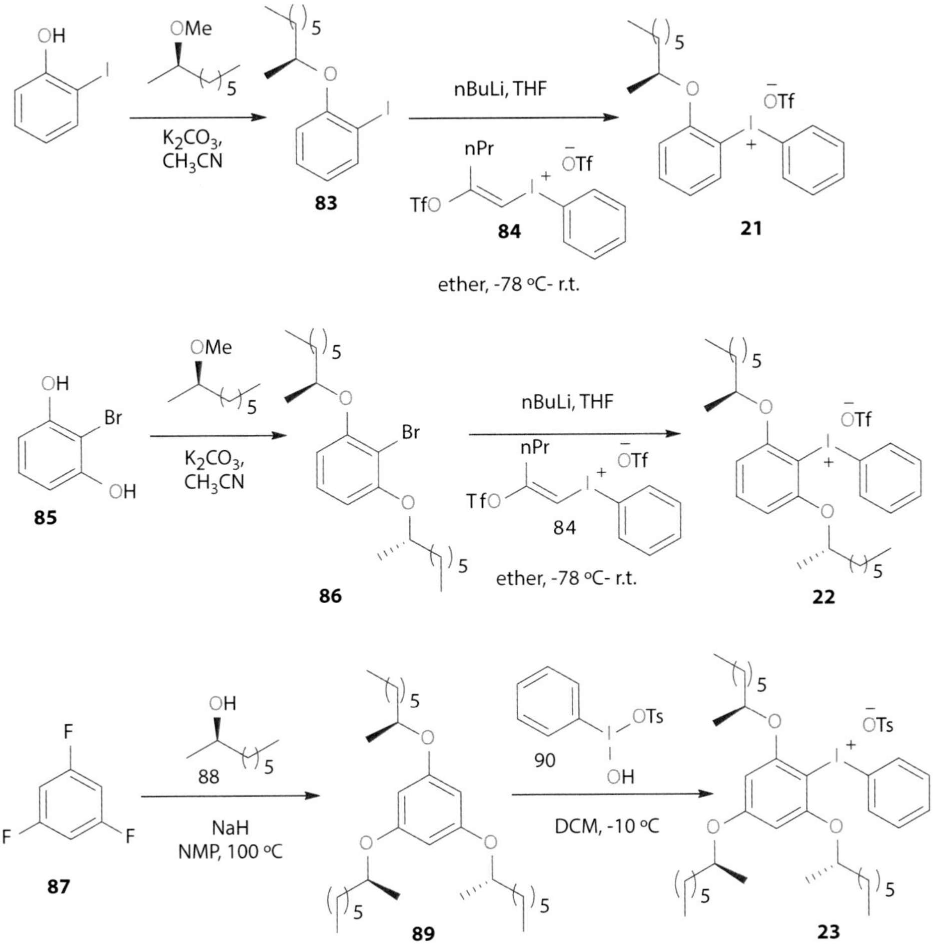

Scheme 17.13 Iodonium salts of chiral hypervalent.

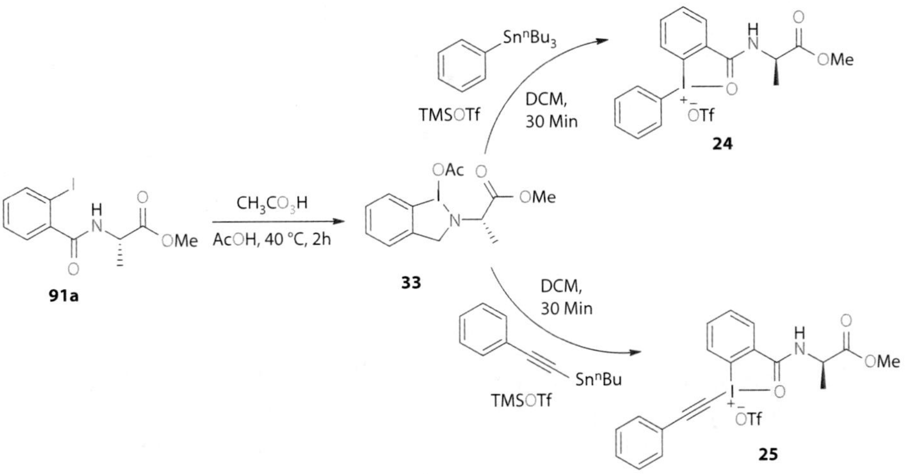

Scheme 17.14 Iodonium salts of chiral hypervalent developed by Zhdankin *et al.*[78]

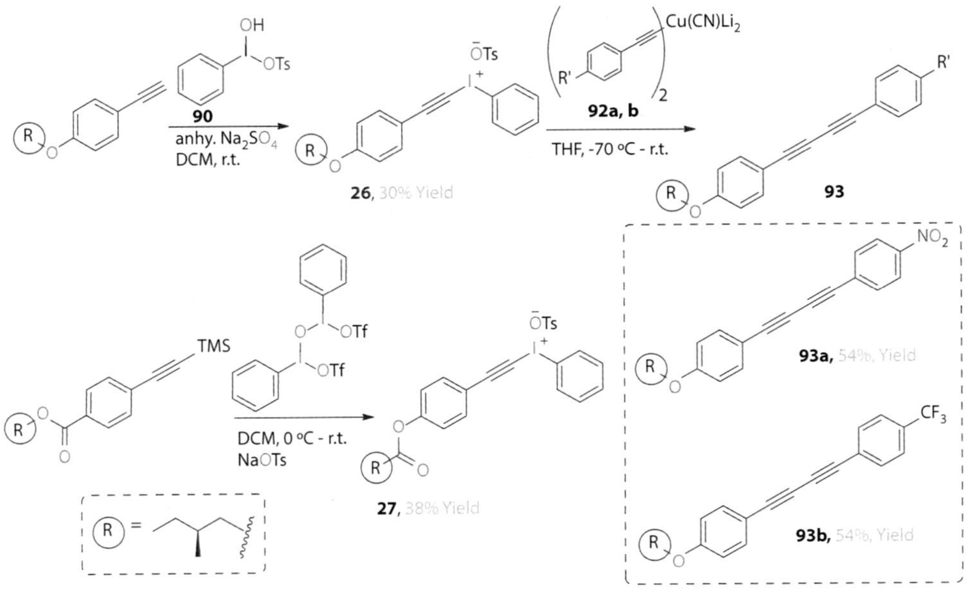

Scheme 17.15 Iodonium salts of chiral hypervalent by Kitamura *et al.*[79]

chirality in the alkyne segment (Scheme 17.15). Compounds **26** and **27** were prepared based on protocols established by Rebrovic and Koser[80] and Bachi *et al.*,[81] yielding gentle yields. These iodanes were found to be stable to moisture and air. These were then employed as the reagents of alkynylation in reactions with higher-order cyanocuprates **92a** and **92b**, resulting in the formation of the corresponding products **93a** and **93b** in substantial yields. These molecules exhibited interesting characteristics within the realm of chiral liquid crystals, a field with a multitude of material applications. However, these iodonium salts with chirality were not utilized in any transformations that exhibit enantioselectivity.

In 2017, Companys *et al.*[82] published a report on the alkynylation of β-ketoesters enantioselectively using hypervalent iodine reagents **31** and **32** with chirality (Scheme 17.16). This work built upon the alkynylation studies conducted by Fernandez *et al.*[83] and Wu *et al.*[84] and others, which employed achiral hypervalent iodine(III) reagents under chiral phase-transfer catalysis. In a novel and contrasting method, scientists combined alkynyl transfer groups with chiral λ^3-iodanes. They successfully synthesized biphenylic bis(alkynyl-λ^3-iodanes) with axial chirality labeled as **31** and **32** using diiodine (S)-**94**, which is enantiomerically pure, resulting in remarkable yields. These reagents exhibited significant reactivity, leading to the formation of α-alkynylated products **97a-c** with highly favorable yields, although the enantioselectivities were moderate (Scheme 17.16).

17.3.2.3 Preparation of Chiral Alkenyl Hypervalent Iodine

Fujita *et al.*[85] reported on the synthesis of compounds 28-30, which are chiral compounds of the 4-methyl cyclohexylidenemethyliodonium salts (Scheme 17.17). These salts represent a unique example of chiral alkenyl hypervalent iodine compounds. Optically active vinyl iodonium salts **28** to **30** are prepared with substantial yields with enantiomeric excesses by reacting vinylsilane-(R) **99** with the respective hypervalent reagents while retaining

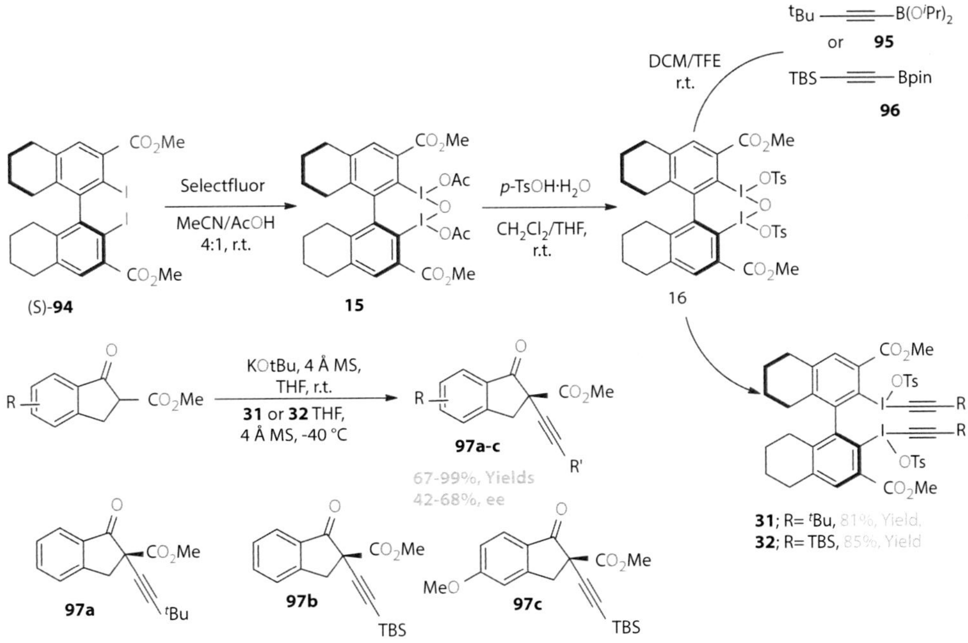

Scheme 17.16 Reaction of alkynylation.

the configuration. The synthesis of these chiral salts aimed to investigate the presence and potential rearrangements of primary vinyl cation intermediates. The chiral salts **28** to **30** on solvolysis in a mixture of methanol and water give a combination of unrearranged enol derivatives (S)- or (R)-**101** and rearranged 4-methylcycloheptanone compound (R)-**100** (Scheme 17.17). Interestingly, the 4-methylcycloheptanone product (R)-**100** was consistently obtained while preserving the optical purity of initial chiral hypervalent iodine. In contrast, a combination of enantiomers was obtained for the unrearranged compounds (R)- or (S)-**101**. Fujita *et al.*[85] proposed that the rearranged compound (R)-**100** is generated via a rearrangement process triggered by σ-bond involvement during the loss of the nucleofuge, while retaining its optical purity. They concluded that the involvement of a vinyl carbocation in this mechanism is improbable.

In a separate study, the same group investigated the reaction of chiral cyclohexylidenemethyliodonium compounds toward sulfonates, including tosylates mesylates, and triflates.[86] The findings indicate that a cycloheptyne intermediate may be implicated in accounting for the optical outcomes observed, despite their similarity in nature. Alkynes have numerous uses in synthetic chemistry, biology, and the study of materials.

17.3.3 Carbonyl Compound Reactions of α-Functionalization

Hypervalent iodine reagents display a diverse spectrum of reactivity, with α-functionalization of carbonyl molecules standing out as the first reaction to be explored in this field as well as utmost extensively studied. This reaction involves attaching various functional groups like hydroxy, alkoxy,[56] sulfonyloxy, phosphoryloxy, acetoxy, and others to the α-position of carbonyl compounds.[87–90] These transformations have been successfully achieved across a

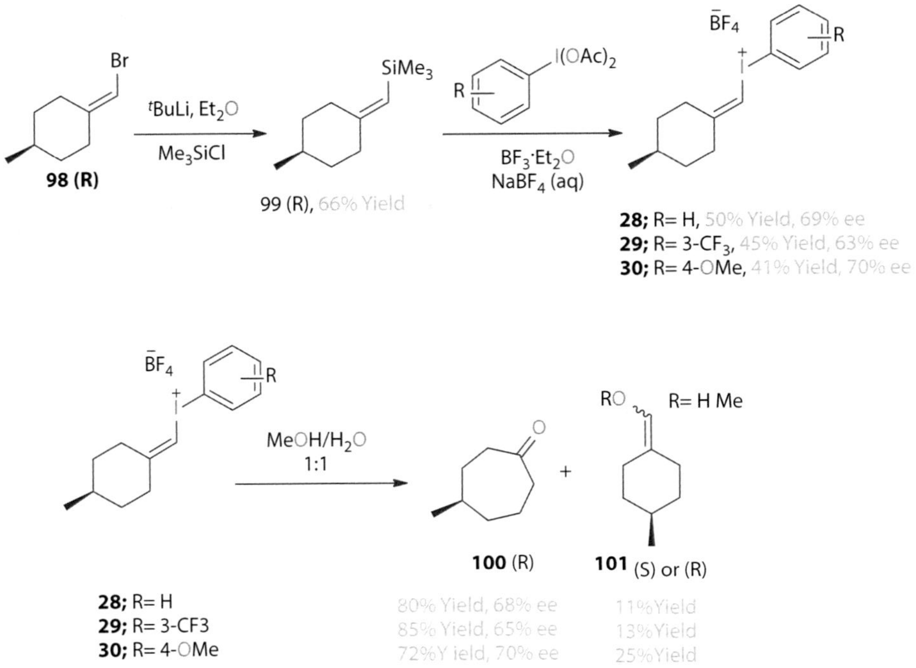

Scheme 17.17 Chiral salts of 4-methylcyclohexylidenemethyliodonium.

series of carbonyl compounds. Notably, Hatzigrigoriou *et al.*[68] were pioneers in attempting an α-sulfonyloxylation of β-diketone in asymmetric manner using chiral hypervalent iodine molecules. This work utilized the chiral catalyst 7, synthesized with remarkable efficiency following the procedure by Koser and Wettach.[91] The preparation of 7 yielded very positive results, as evidenced by its excellent yield in the reaction (Scheme 17.18). Subsequently, compound **7** was employed in the α-sulfonyloxylation of compound **102**, resulting in a highly favorable yield. However, the diastereoselective ratio achieved was initially moderate at 75:25, which diminished to 50:50 following purification. Despite this somewhat less pronounced outcome, it holds importance as a noteworthy advancement within this field.

In 1997, Wirth and Hirt[92] introduced the subsequent development in α-functionalization of ketones. They synthesized two distinct catalysts, the C2 symmetry-based **35** and **34**, with well-achieved yields through traditional reactions (as illustrated in Scheme 17.19). Notably, Rabah and Koser[93] also synthesized similar structures, yet their reactivity was not subjected to evaluation. Upon analyzing the X-ray structure of **34**, the researchers made an intriguing observation. Aside from the commonly recognized T-shaped configuration around the iodine atom, the presence of a methoxy group caused displacement of the tosylate group, resulting in an elongated distance between oxygen and iodine in the tosylate moiety. Once isolated, these catalysts were tested with ketone **108a**. Regrettably, the catalyzed reaction exhibited minimal asymmetric induction, with enantiomeric excesses reaching only up to 15% for both catalysts.

Following this, Hirt *et al.*[94] continued their innovation by designing additional catalysts based on the same conceptual foundation (Scheme 17.20). This effort led to the successful

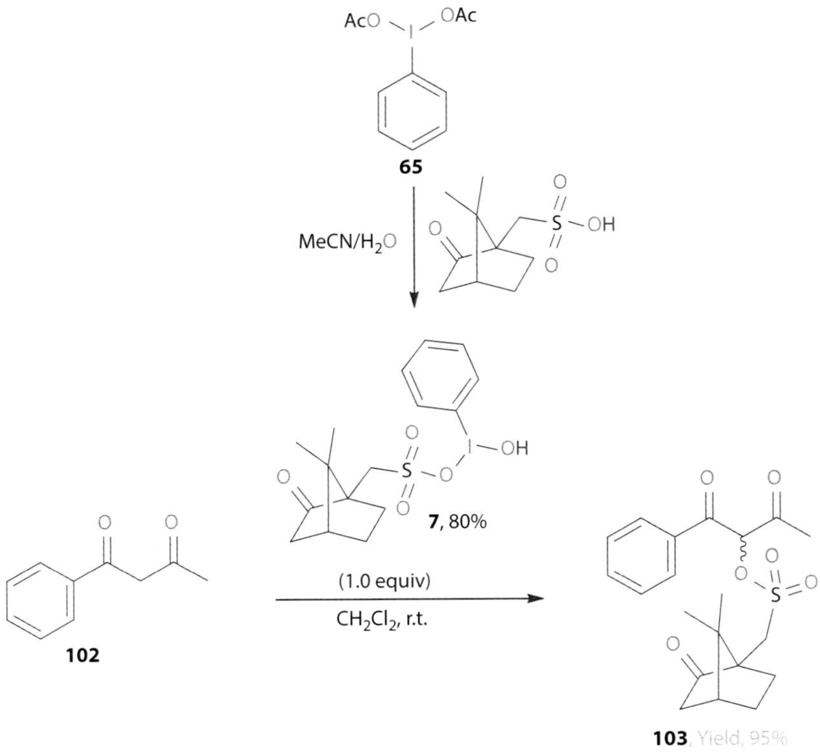

Scheme 17.18 Hatzigrigoriou *et al.*[68]: asymmetric α-sulfonyloxylation.

synthesis of **36** to **38** using a comparable synthetic route, yielding consistently positive outcomes across the board. Nevertheless, when these catalysts were applied to the task of asymmetric α-sulfonyloxylation involving propiophenone **108a**, the resulting enantiomeric excesses persisted at lower levels, albeit marginally higher than those obtained with **34** and **35** (Scheme 17.19).[97] The researchers postulated that this enhancement in enantioselectivity might find its roots in an electronic influence.

Several years later, Hirt *et al.*[95] continued to explore the potential of this catalyst family, this time incorporating *ortho*-substituted iodoarenes **118** and **119** into their approach (Scheme 17.21). While they synthesized other catalysts, we will focus on those that yielded the highest enantioselectivity. Notably, by substituting the methoxy moiety with benzoxy and ethyl moieties in **40** and **39**, respectively, they managed to enhance the optical outcome of α-sulfonyloxylation reaction involving propiophenone **108a**. This improvement suggested a collaborative interplay between chiral moiety and ortho substituent. Intriguingly, attempts to utilize bulkier catalysts did not yield superior enantioselective outcomes.

In 2007, Richardson *et al.*[96] pioneered catalytic chiral iodoarenes for α-sulfonyloxylation of ketones asymmetrically (Scheme 17.22). This marked the first use of catalytic chiral hypervalent iodines. Compound **51** emerged as the optimal catalyst after thorough screening of chiral iodoarenes. Employing the chiral catalyst along with mCPBA as an oxidizing agent, they achieved high yields in asymmetric catalytic ketone α-sulfonyloxylation **108b-e**. However, enantioselectivity remained limited, particularly for non-propiophenone ketones. Two possible pathways were proposed: in **path A**, chiral hypervalent iodine **122** formation

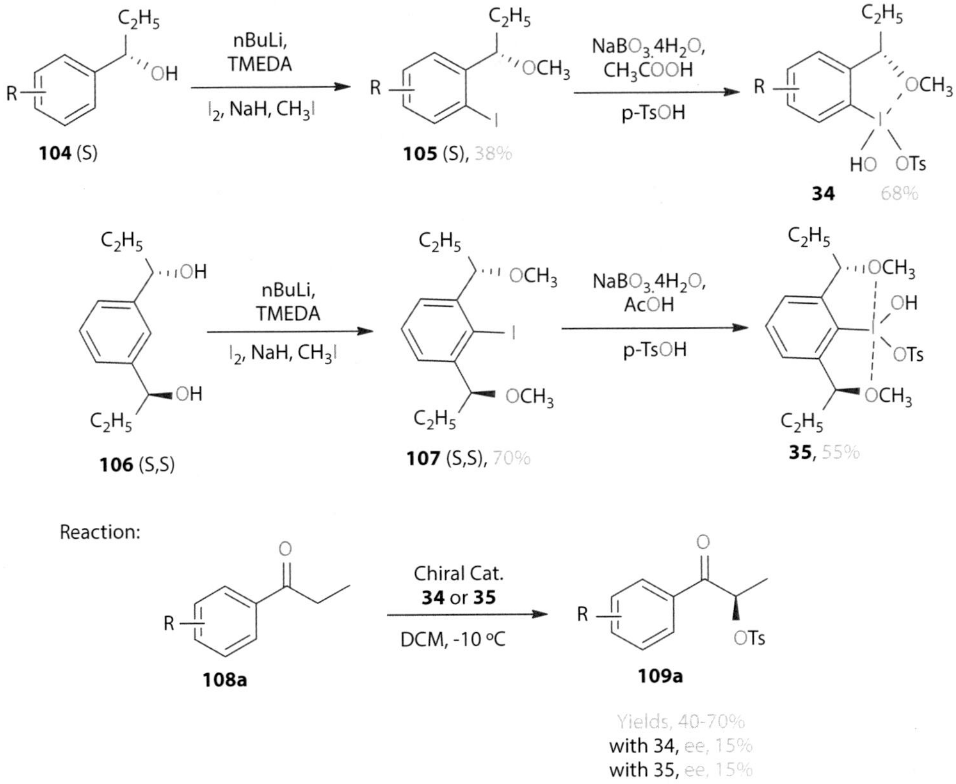

Scheme 17.19 Wirth *et al.*[92]: asymmetric α-sulfonyloxylation.

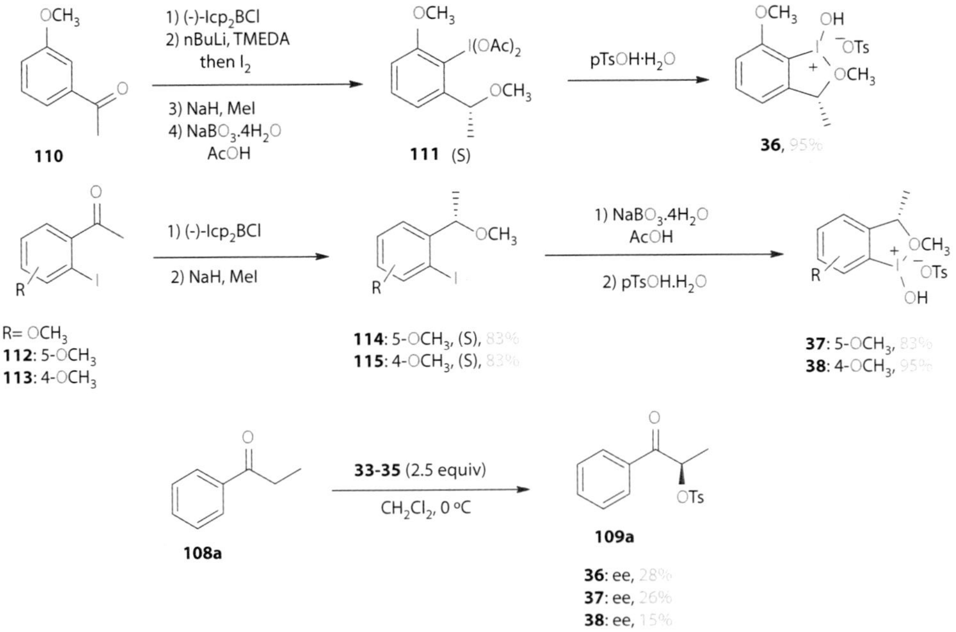

Scheme 17.20 Hirt *et al.*[94]: asymmetric α-sulfonyloxylation.

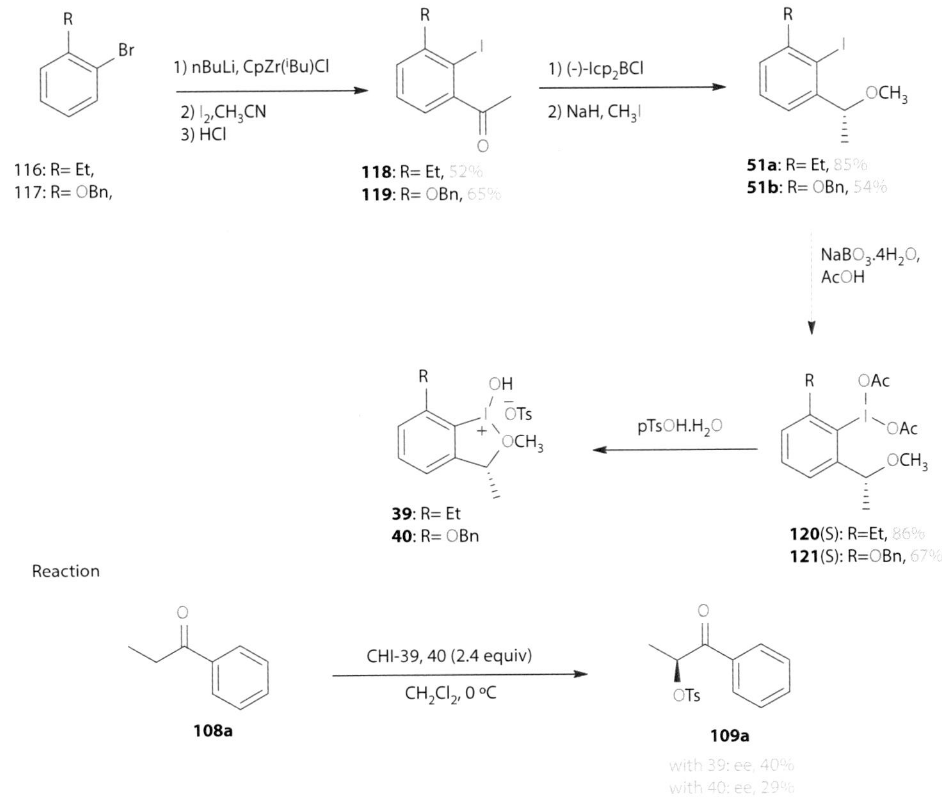

Scheme 17.21 Wirth *et al.*[95]: asymmetric α-sulfonyloxylation.

led to intermediate **123** and eventual product **125** via S_N2' with tosylate anion; **path B** suggested alkyl iodane intermediate **124** reacting with tosylate anion via a hyper-leaving group, yielding the desired product.[97] Pathway A seemed less likely due to spatial constraints between prochiral center and chiral surrounding in intermediate **123** (Scheme 17.22).

Following this, the Altermann *et al.*[98] extended their investigation to other chiral iodoarenes within the established protocol. Enhanced enantioselectivity, reaching up to 40%, is attained in propiophenone (**108a**) α-sulfonyloxylation using a novel catalyst **52** (Scheme 17.23).[98] This catalyst was derived using enantiomerically pure (S)-2-(2-iodophenyl)propanoic acid (S)-**126** and (-)-menthol. Despite conducting tests on a broad array of chiral iodoarenes, the attained enantioselectivities were comparatively lower across all instances. Additionally, Farooq *et al.*[99] synthesized further chiral iodoarenes, including **53** featuring an iodoarene ring with a chiral ortho-ester moiety (Scheme 17.24).

Despite its efficient preparation through classical coupling reactions, this catalyst yielded no noteworthy enhancement in the enantioselective α-sulfonyloxylation of propiophenone **108a** and enantioselectivity of the catalytic compared to the previously reported outcomes. Farooq *et al.*[99] also investigated the asymmetric variant of the oxylactonization process of ketocarboxylic acids into ketolactones, which had been formerly elucidated by Uyanik *et al.*[100] Despite an acceptable result achieved for compound **130**, the reaction advanced with no enantioselectivity. In 2011, Yu *et al.*[101] presented an alternative approach to

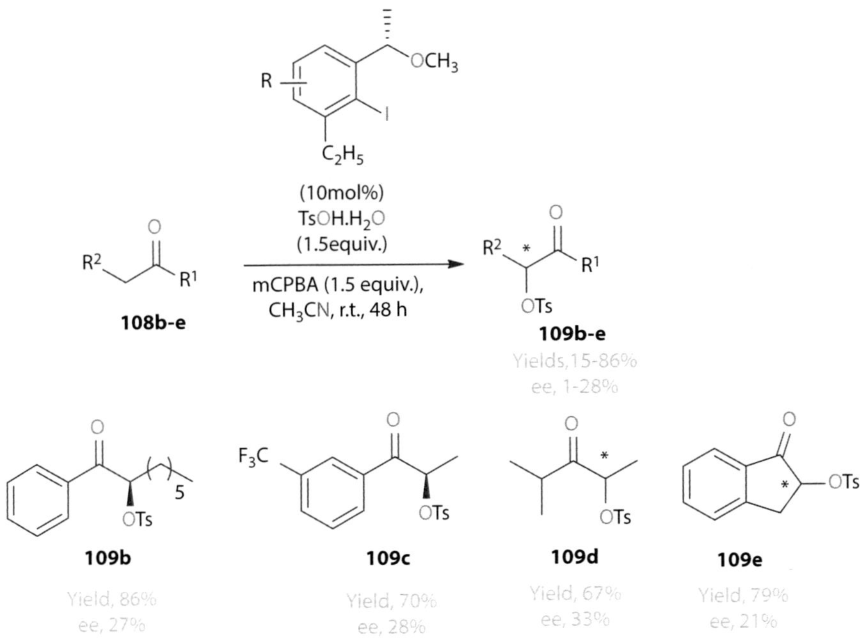

Scheme 17.22 Richardson *et al.*[96]: asymmetric α-sulfonyloxylation.

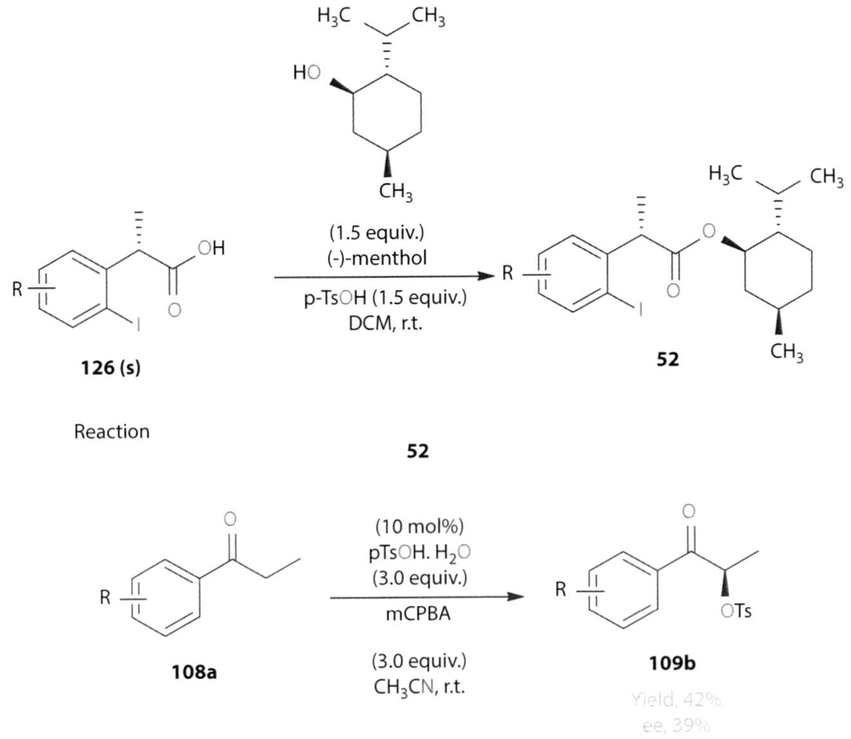

Scheme 17.23 Altermann *et al.*[98]: asymmetric α-sulfonyloxylation.

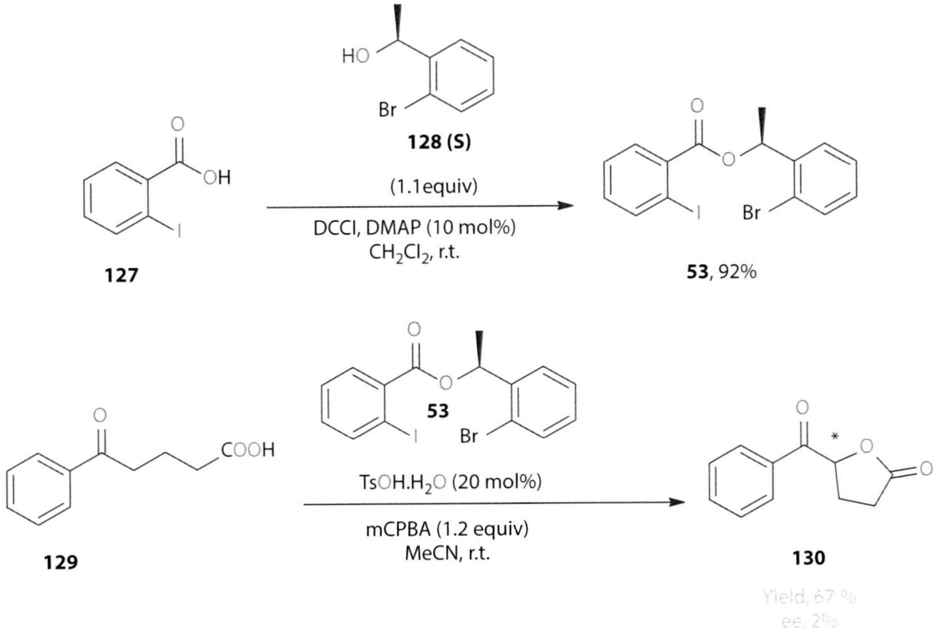

Scheme 17.24 Farooq *et al.*[99]: asymmetric α-sulfonyloxylation.

α-functionalization of ketones. The α-sulfonyloxylation of ketones **108a, f–h** using the catalyst **54** was reported by these researchers (Scheme 17.25).[101] Designed around a spiroindane backbone, this chiral catalyst aimed to evaluate its effectiveness for ketone α-oxytosylation. The preparation of **54** involved several steps, originating from the earlier prepared chiral diamine (S)-**131**,[102] with the utmost difficult step being the monodeamination of (S)-**132**. Employing 10 mol% of the catalyst in tandem with mCPBA as oxidizing agent yielded α-tosylated propiophenone **108a** with a noteworthy 58% enantiomeric excess (ee), marking the highest enantioselectivity achieved at that point utilizing a hypervalent iodine possessing chirality for this conversion. Under the optimized parameters, the investigators explored various ketones, albeit generally achieving minor yield and ee's. Chen *et al.*[102] introduced a stereochemical-based model rooted in steric effect to rationalize the attained enantioselectivity. Interestingly, the investigators proved that the inclusion of *p*-TsOH led to a reduction of optical activity in enantioenriched final product **109a**, potentially explaining the modest enantioselectivity observed in this process.

Simultaneously, Guilbault and Legault[103] introduced a novel class of catalysts (Scheme 17.26). Their approach revolved around utilizing a chiral oxazoline structure, akin to PHOX chiral ligands (phosphinooxazolines) that have proven effective in diverse asymmetric transformations.[104] Among the various chiral iodoarenes crafted and evaluated for α-oxytosylation of propiophenone **108a**, **55** emerged as highly effective with respect to enantioselectivity. The catalyst was derived from a chiral amino alcohol (R,R)-**135** and benzoic acid **191**, achieving a satisfactory result (Scheme 17.26). Various ketones underwent assessment against this catalyst, yielding both fair yields and better enantiomeric excess (ee) values compared to prior reports. The researchers[103] confirmed that the introduction of an alkyl group, like methyl moiety adjacent to iodine atom on the aryl ring, yielded

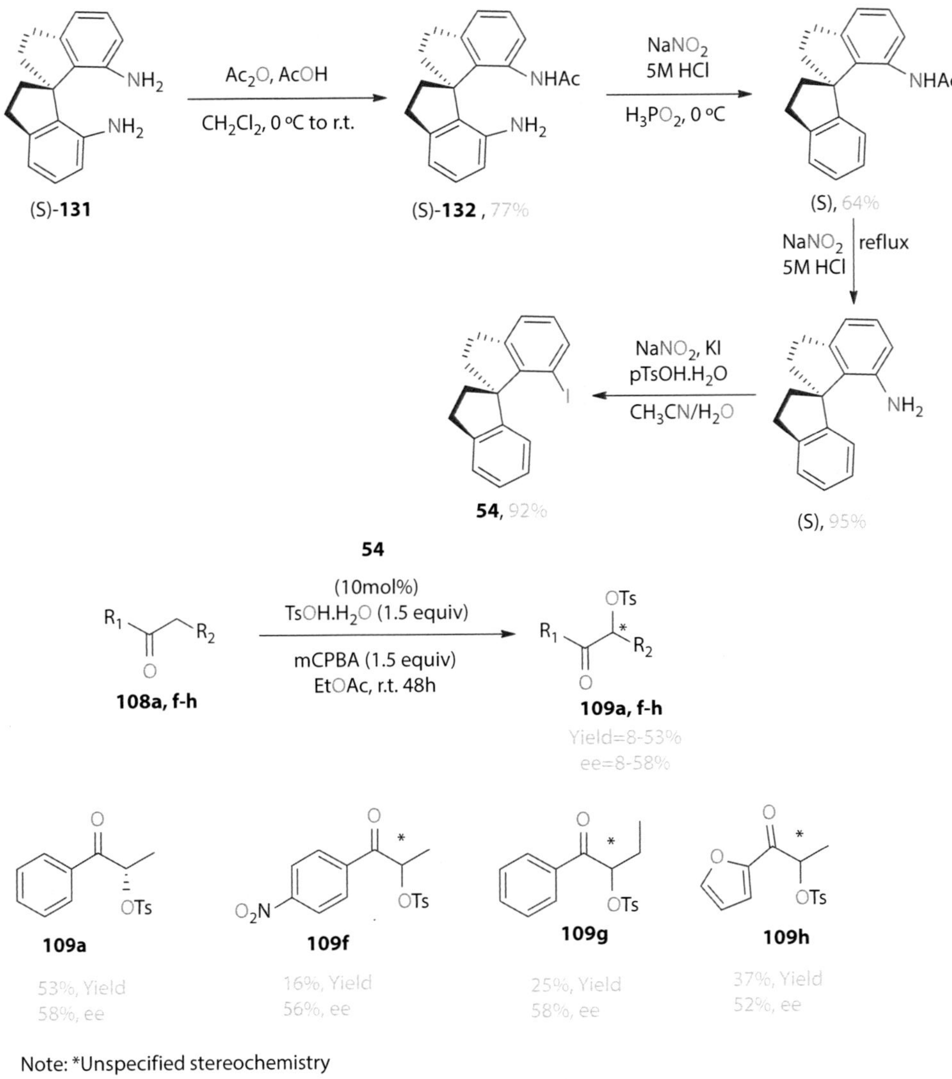

Scheme 17.25 Yu *et al.*[101]: asymmetric α-sulfonyloxylation.

optimal stereoinduction and catalytic activity. The influence of this ortho methyl group was examined using DFT calculations.[103] The discussion revolves around iodoarene **55** and its hypervalent derivative, designated as species **137**. The prevailing notion was that, due to the excess of TsOH, these entities would experience protonation at the nitrogen atom within the oxazoline moiety. In the specific instance of **55**, the process was found to be energetically favorable, resulting in the formation of the protonated variant denoted as **136**. Subsequent treatment with mCPBA and oxidation would yield the hypervalent iodane **137**, where protonation might ensue. However, this protonation step becomes energetically unfavorable, representing an endergonic process (Scheme 17.26). Upon nitrogen protonation, the oxazoline ring must undergo rotation to align the coordinated oxygen atom with the hypervalent atom. This conformational adjustment maximizes Coulombic interactions between the two entities. Consequently, the protonation leads to a heightened reactivity of the resulting intermediate **138** when compared to the initial **137**. This is attributed to the fact that the N–I interaction now prevails over the weaker I–O interaction due to the protonation. However, the introduction of a methyl moiety at ortho position serves to facilitate the protonation process. This substitution accelerates the reaction more than nonsubstituted counterpart, thereby favoring the shift of balance toward the more reactive substrate **138**.

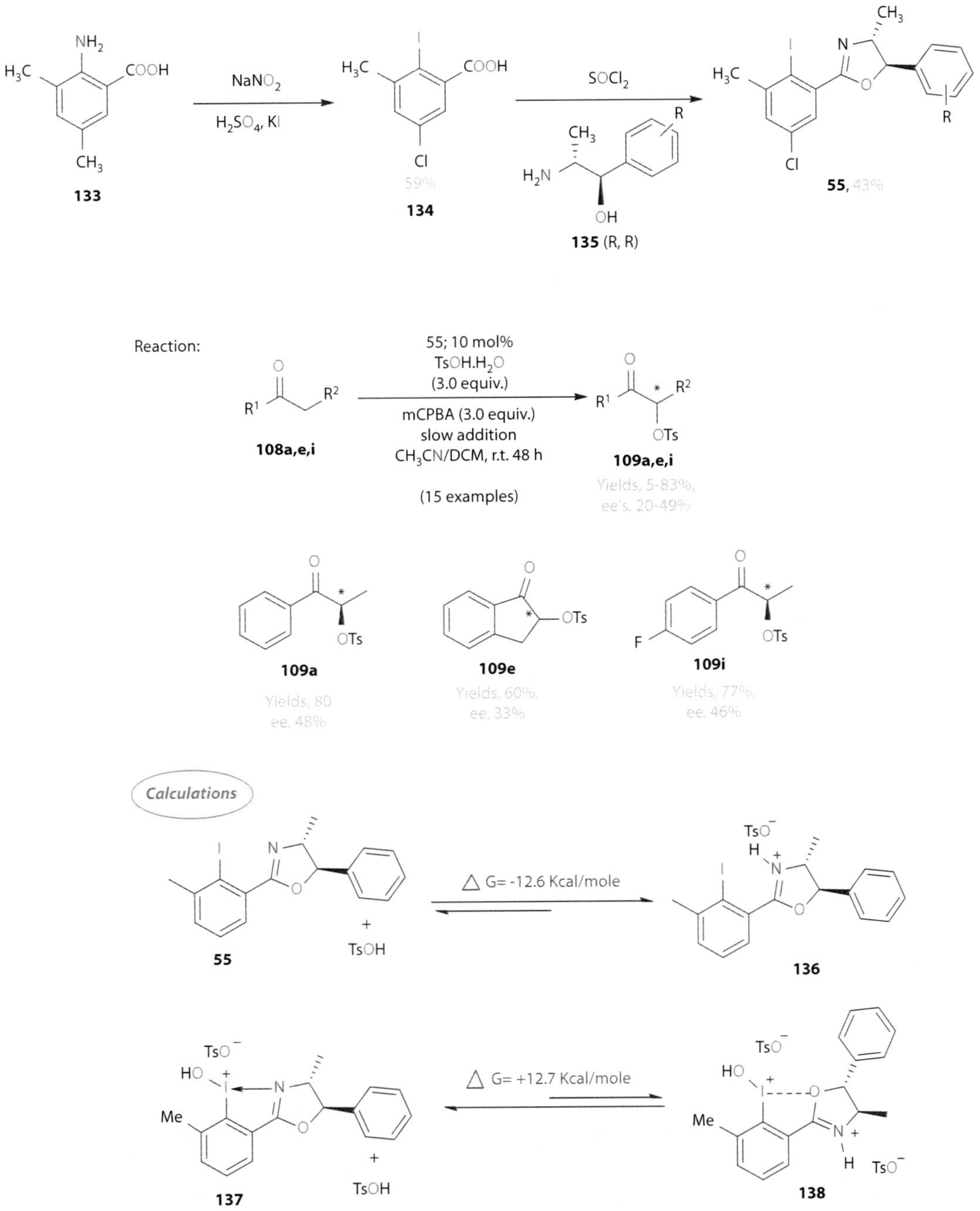

Scheme 17.26 Guilbault *et al.*[103]: asymmetric α-oxidation of ketones.

Furthermore, the presence of a chiral surrounding close to oxygen atom within the oxazoline ring assumes significance in this context. This chiral influence is manifest in the enantioselectivity observed, further underscoring the intricate interplay of structural features in these reactions.

Following this reasoning, Therien *et al.*[105] embarked on designing and synthesizing alternative promising catalysts, as outlined in Scheme 17.27. They explored diverse chiral amino alcohols, designated as **140**, which, upon reaction with aldehyde **139**, yielded chiral iodo-oxazolines **56-58**. Notably, these newly formed compounds exhibited significant variations in their structural environments.

However, despite their efforts, the introduction of bulkier frameworks in proximity to the oxazoline-oxygen did not lead to improved enantioexcesses compared to the outcome achieved with **55** in the context of propiophenone (**108a**) α-oxytosylation (Scheme 17.26). In a series of reactions detailed in Schemes 22–24, as previously reported by Rodríguez and Moran,[106] the synthesis of **59** and **60** was disclosed.[96, 98, 99] These compounds were synthesized through a straightforward amination process involving iodoarene **141** and either (1R,2S)-norephedrine **135** or (1S,2S)-pseudonorephedrine hydrochloride **142**, respectively (Scheme 17.28). Upon *in situ* generation of chiral hypervalent molecules from iodoarene **59**, propiophenone α-oxytosylation was revisited as a representative reaction. However, the resulting compound displayed a modest enantiomeric excess of only 18%. In contrast, when employing **60** in the oxidative cyclization of compound **129**, a more favorable outcome was achieved. This led to an enantioselectivity reaching up to 51%, which represented the highest level of enantioselectivity reported for this reaction up to that point in time.

17.3.4 Reaction of Dearomatization

In recent decades, dearomatization reactions have gained significant importance as valuable synthetic strategies, particularly in total synthesis.[107] These reactions have sparked special interest in developing intriguing disconnections for complex molecule construction.[108] A prominent category of dearomatization reactions involves phenols, which are extensively used as substrates, particularly in asymmetric versions.[109] One highly compatible and

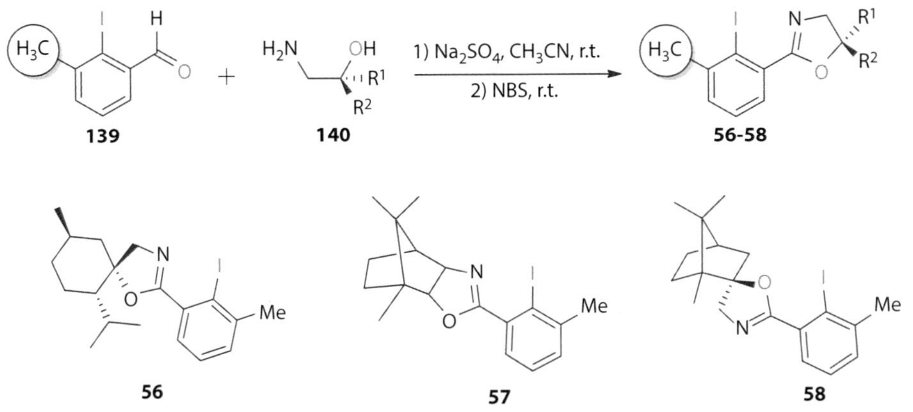

Scheme 17.27 Therien *et al.*[105]: asymmetric α-sulfonyloxylation of propiophenone.

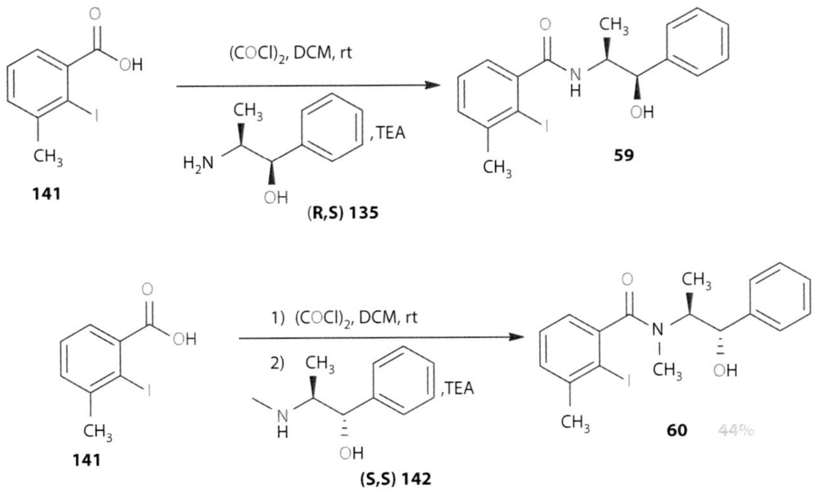

Scheme 17.28 Rodríguez *et al.*[106]: asymmetric α-oxidation of ketones.

successful approach for performing dearomatization reactions utilizes hypervalent iodine reagents.[110] These reagents have demonstrated remarkable efficacy, especially due to the use of milder reaction conditions compared to conventional oxidants. This success has made them popular choices as promoters in dearomatization reactions, opening new avenues for creative and efficient synthetic pathways in organic chemistry. Kürti *et al.*[111] made the initial effort to accomplish an enantioselective dearomatization reaction of phenols using catalysts of chiral hypervalent iodine (Scheme 17.29).

They employed λ^3-iodane ent-**2** and **2**, which were synthesized using PIDA and the respective D- or L-dibenzoyltartaric acid. These reagents were used to oxidize various phenols, but unfortunately, all of the reactions resulted in racemic products, indicating a failure in case of enantioselectivity. To explain the results, the authors proposed two possible reaction pathways (Scheme 17.29). In the first pathway (associative path A), the oxygen of phenol displaces an alkoxy ligand bound to the hypervalent iodine center, followed by a nucleophilic attack by an alcohol. The second pathway (dissociative path B) involves the formation of an intermediate phenoxenium ion **146**, which undergoes a reaction with alcohol to form the desired product. The absence of enantioselectivity suggests the possible involvement of intermediate **146**, leading to a loss of chiral environment in the substrate. The authors also considered the possibility that dibenzoyltartaric acid could rapidly displace from **2** or ent-**2** by methanol, resulting in the formation of a reagent of achiral hypervalent $PhI(OMe)_2$. However, this was ruled out based on the obtained results when utilizing a solvent like chiral (S)-2-methylbutan-1-ol, as product **144c** was obtained as a 50% mixture of diastereoisomers, indicating lack of enantioselectivity. The initial disappointing results with chiral hypervalent iodine catalysts may have indeed differed the progress of asymmetric phenolic oxidation. A decade elapsed before Dohi *et al.*[112] made a significant advancement in the field, as they described the dearomatization of phenols in seminal enantioselective manner through the utilization of a novel chiral organoiodine(III) catalyst, **17** (Scheme 17.30).

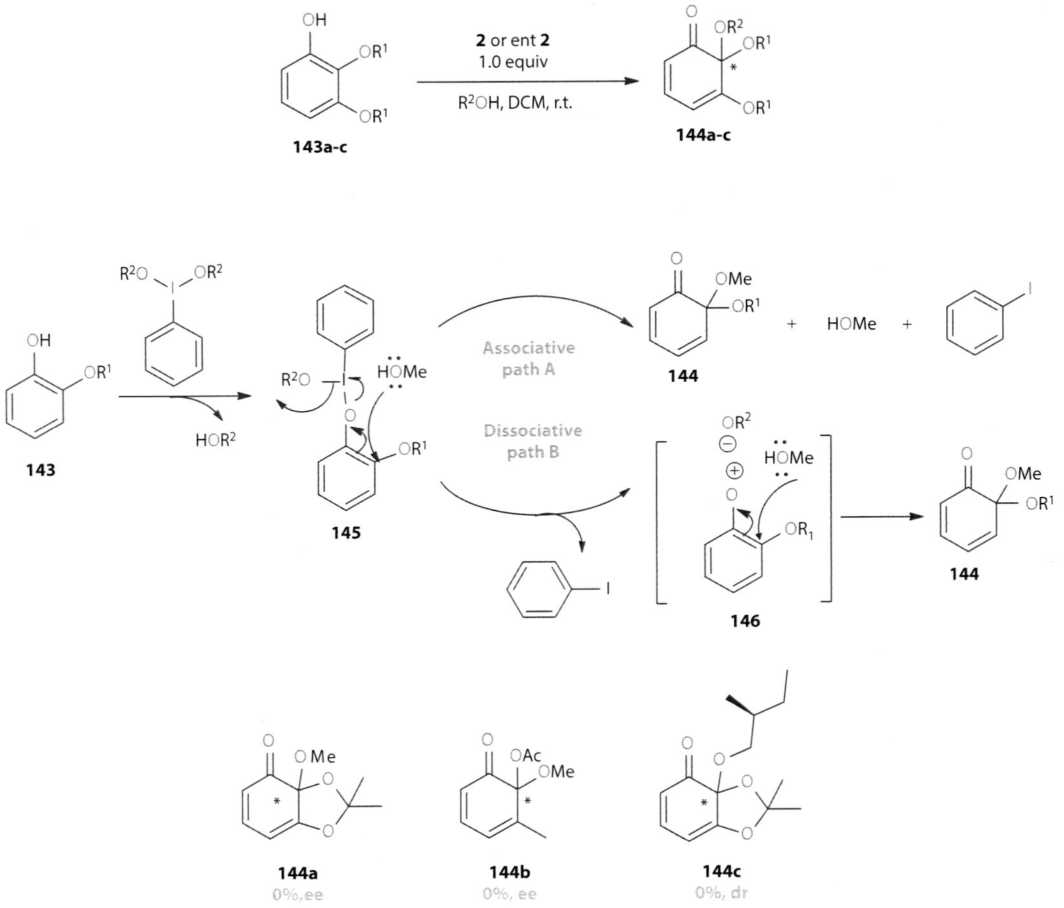

Scheme 17.29 Kürti *et al.*[111]: asymmetric phenol oxidation.

Compound **17** featured a conformationally inflexible structure and was easily synthesized using enantiopure ditriflate (R)-**147** through a series of iodation and amination steps, resulting in a satisfactory result. Selectfluor served as the ideal oxidizing agent, producing the μ-oxo-bridged λ^3-iodane **17** in a highly respectable 90% yield. When compound **17** was employed stoichiometrically, it promoted o-spirolactonization of α-naphthols **148a-d**, yielding products with fair to substantial yields with optimal enantiomeric excess (ee). The choice of solvent was found to be critical, as higher polarity solvents provided lower enantioselectivity, suggesting that a dissociative mechanism was at play, and phenoxenium ions were unlikely to be involved (path **B** in Scheme 17.29). A notable achievement was that Kita *et al.*[110] successfully performed this transformation using only catalytic amounts of mCPBA and **17** as a stoichiometric oxidant, yielding similar results as before. This landmark study by Kita *et al.*[110] laid the foundation for further advancements in asymmetrical phenol oxidation using a reagent of chiral hypervalent iodine, opening exciting possibilities in asymmetric synthesis.

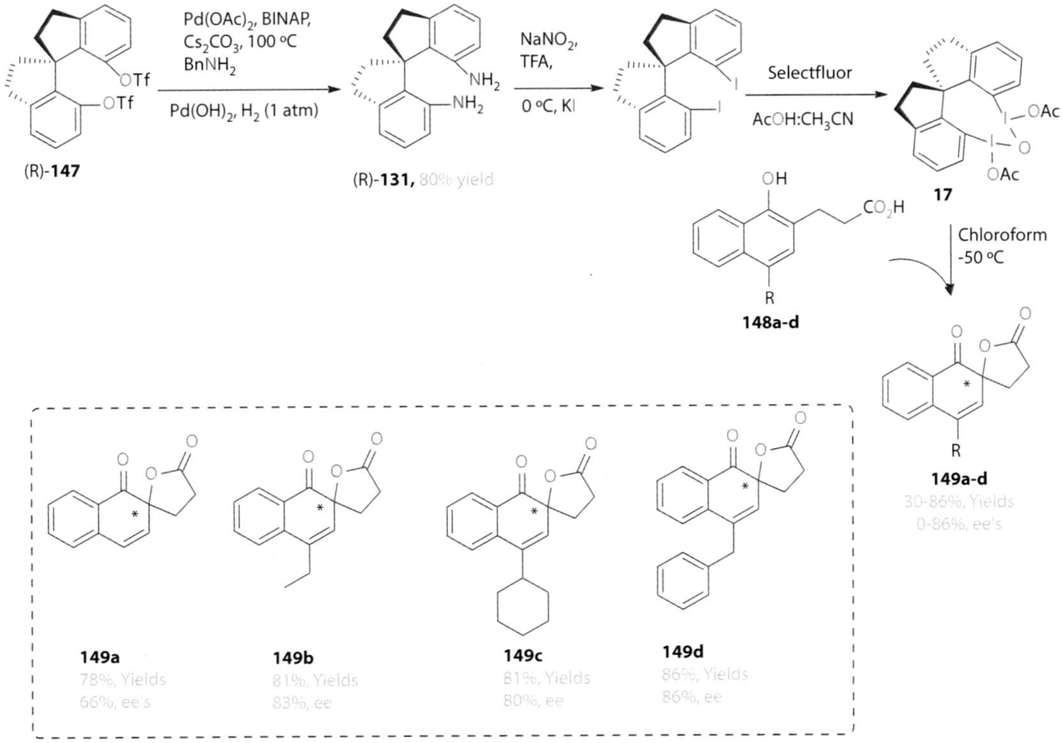

Scheme 17.30 Dearomatization of phenols by Dohi *et al.*[112]

17.3.5 Functionalization of Alkene

Alkenes hold a significant position among the most crucial functional groups in organic chemistry owing to their ease of conversion into various functional groups. Consequently, achieving the functionalization of alkene asymmetrically has become a highly sought-after and challenging objective. Among the methods used for the expeditious synthesis of chiral molecules with polyfunctional, asymmetrical alkene oxidation stands out as a particularly noteworthy approach. To provide a comprehensive understanding, this section will be divided into two parts. The first part will focus on intermolecular alkene functionalization and delve into its details. This approach involves the introduction of functional groups onto the alkene by using an external reagent or catalyst in a selective and asymmetric manner. In the subsequent part, the intramolecular variant of this transformation is explored, wherein the alkene and the functionalization reagent are part of the same molecule. This intramolecular approach presents its own set of challenges and opportunities for creating complex chiral molecules efficiently.[13]

17.3.5.1 Intermolecular Functionalization of Alkenes

In the area of chiral hypervalent iodine catalysis, Wirth and Hirt[92] made a significant breakthrough by attempting the enantioselective difunctionalization of alkenes from an intermolecular perspective. Their pioneering work was reported in conjunction with the previously mentioned α-functionalization of propiophenone (Scheme 17.19).

The study involved the evaluation of hypervalent iodines (III) having chirality, namely, **34** and **35**, in the transformation of styrene **150a**. Through their catalysis, styrene **150a** was converted into two different products: the ditosylated compound **151** and the monotosylated compound **150** (Scheme 17.31). However, despite these remarkable achievements, the exact mechanism of the formation of the products and the origin of the obtained opposite configuration in compound **152** remained somewhat unclear. This research represents a significant step forward in the progress of enantioselective difunctionalization reactions of alkenes utilizing chiral catalysis-based hypervalent iodine. Further investigations and understanding of the underlying mechanisms will undoubtedly contribute to the advancement of this exciting and promising area of organic chemistry. The initial attempts reported by Wirth *et al.*[92] in conducting difunctionalization of alkenes enantioselectively using a catalyst of chiral hypervalent iodine resulted in relatively low yields and enantioselectivities. However, this groundbreaking research laid the foundation for future advancements in the field, offering promising prospects for asymmetric alkene difunctionalization.

Subsequently, in 1998 and 2001, Hirt *et al.*[94, 95] further explored the catalytic potential of chiral iodines(III) **36** and **39**. These catalysts were not only used for α-functionalization of propiophenone but also used for asymmetrical ditosylation of styrene **150a**. Notably, these new catalysts demonstrated a significant improvement in enantioselectivity compared to the earlier **34** catalyst (Schemes 17.20 and 17.21). With the use of **36** and **39**, enantioselectivities of up to 65% were achieved, marking a notable enhancement in the selectivity of the reactions. This progress in achieving higher enantioselectivities underlines the potential and importance of chiral hypervalent iodine catalysis in facilitating asymmetric difunctionalization reactions. The continuous exploration of such catalysts and further refinement of reaction conditions hold great promise for advancing this area of research in organic chemistry. Koposov *et al.*[113] successfully synthesized a different reagent category of chiral hypervalent iodine known as **8** and **9**. These reagents were synthesized by reacting (diacetoxyiodo) benzene **65** with naturally occurring L-N-protected amino acid (S)-**153**. The synthesis is accomplished with satisfactory isolated yields (Scheme 17.32).[113] The newly developed **8** and **9** catalysts exhibited remarkable catalytic activity in the oxidative conversion of cyclic alkenes. This transformation resulted in the formation of anti-iodine-amino acid esters **155** and **157**. The compounds were prepared with high yields, and importantly, they showed complete anti-stereoselectivity. However, despite the excellent stereoselectivity, it should be

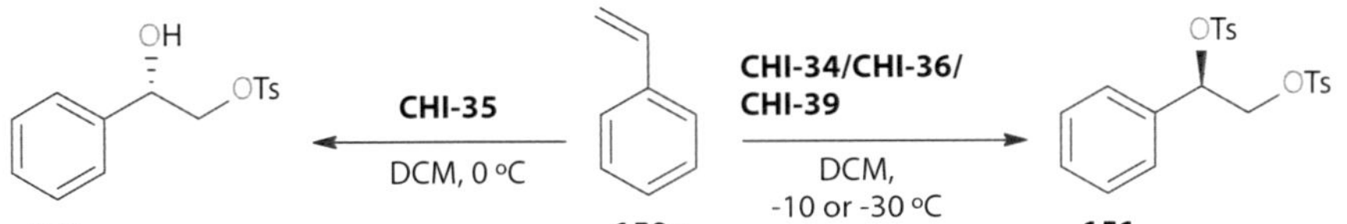

152 **150a** **151**

17% ee

34, 21% ee
36, 75% yield, 53% ee
39, 65% ee

Scheme 17.31 Difunctionalization of enantioselective alkenes.

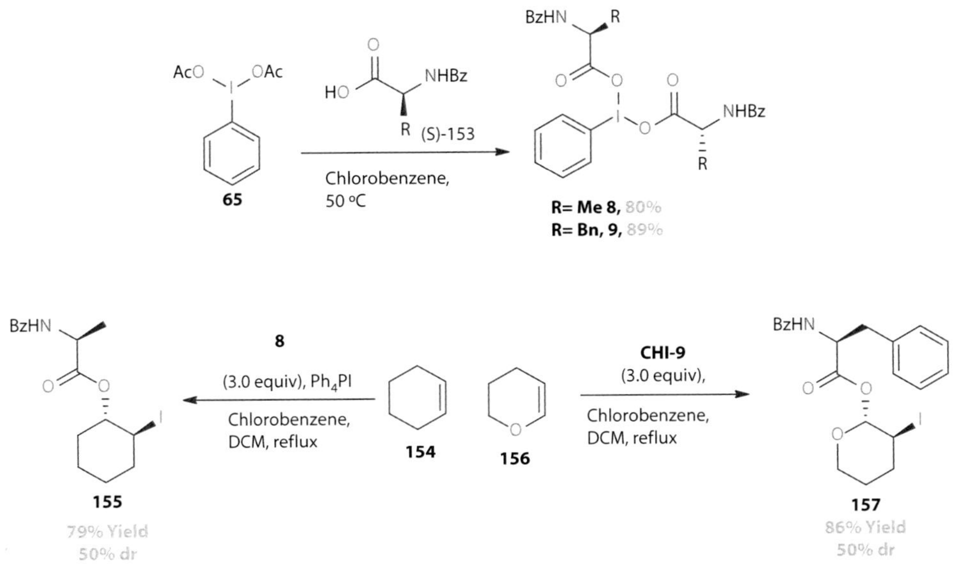

Scheme 17.32 Koposov *et al.*[113]: β-iodocarboxylation of alkenes.

noted that the products were obtained as 50% mixtures of diastereoisomers. This diastereomeric mixture is a significant outcome to consider, as it may impact the application of these products in various synthetic and medicinal chemistry contexts.

Overall, the successful synthesis of compound **8** and **9** reagents and their efficient catalytic performance in achieving anti-stereoselectivity represent a significant advancement in the area of chiral hypervalent iodine catalysis. Further studies may be required to address the diastereomeric mixture issue and fully exploit the potential of these catalysts in various synthetic applications.

17.3.5.2 Intramolecular Functionalization of Alkenes

In 2003, Boye *et al.*[114] reported the initial instance of functionalization of alkenes using an intramolecular method. They demonstrated the chiral iodine(III) **41**-catalyzed lactonization of 4-phenyl-4-pentenoic acid **158** (Scheme 17.33). The resulting lactone **159** was obtained in an average yield but showed practical racemization, with only 4% enantiomeric excess (ee). An interesting observation in the final product was the rearrangement of the phenyl ring. Boye *et al.*[114] attributed this rearrangement to the hypernucleofuge capability of hypervalent iodine possessing chirality, which leads to the formation of compound **160**, and the possible involvement of phenonium ion **161**-based intermediates.

In 2007, Fujita *et al.*[115, 116] introduced a noteworthy family of chiral iodine(III) catalysts. Among them, chiral iodine(III) **42** stood out as an especially useful catalyst and was synthesized using enantiopure (S)-methyl lactate and 2-iodophenol **250** through a Mitsunobu reaction followed by oxidation, yielding a very good yield of the desired catalyst (Scheme 17.34). Inspired by this new group of chiral catalysts, Ishihara's precatalyst **61** was successfully developed as the first chiral generation.[58] To assess the catalyzing potential of chiral iodine(III) **42**, the asymmetrical tetrahydrofuranylation of alkenes **164a-c** was selected as a

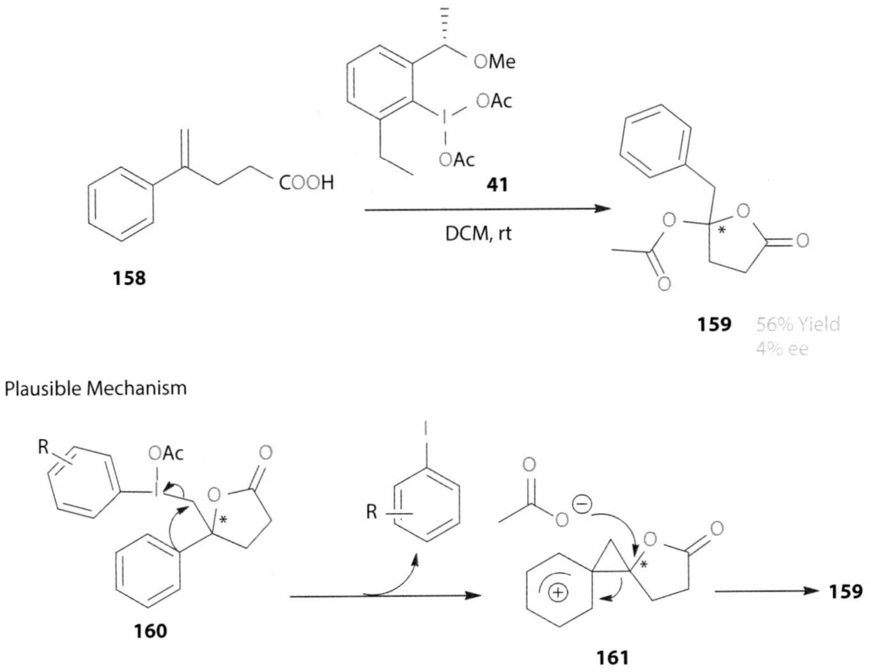

Scheme 17.33 Alkene lactonization.

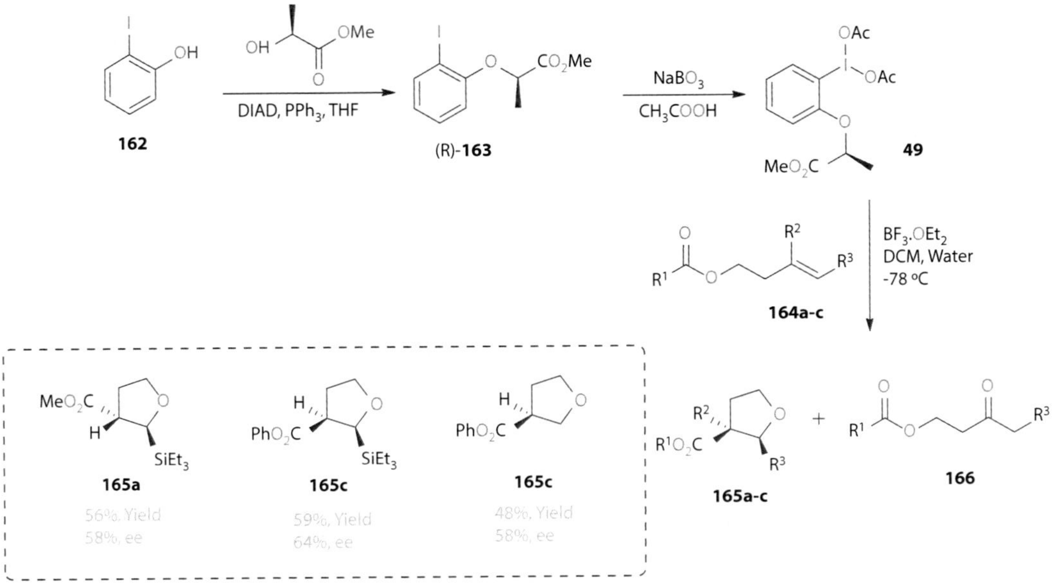

Scheme 17.34 Asymmetric tetrahydrofuranylation.

representative reaction. The results showed the promising potential of compound **42** as an effective and selective catalyst for this transformation.

These discoveries represent significant milestones in the area of chiral hypervalent iodine catalysis, paving the way for future developments in asymmetric intramolecular functionalization of alkenes. Further research and optimization of these catalysts hold great promise for advancing synthetic methodologies in organic chemistry. In order to achieve a high conversion in the reaction, the addition of $BF_3{\cdot}OEt_2$ was found to be essential as an additive. However, in certain instances, an unwanted by-product in the form of ketone **166** was observed. To gain a deeper understanding of the reaction pathway and the formation of this by-product. In 2019, Parra conducted further experiments using $H_2{}^{18}O$ and alkenylsilanes as substrates. Additionally, they employed density functional theory (DFT) calculations to propose a plausible reaction mechanism.[13]

In 2013, Kong *et al.*[117] introduced an innovative approach for the enantioselective aminofluorination of alkenes (Scheme 17.35). This notable conversion was facilitated by a new

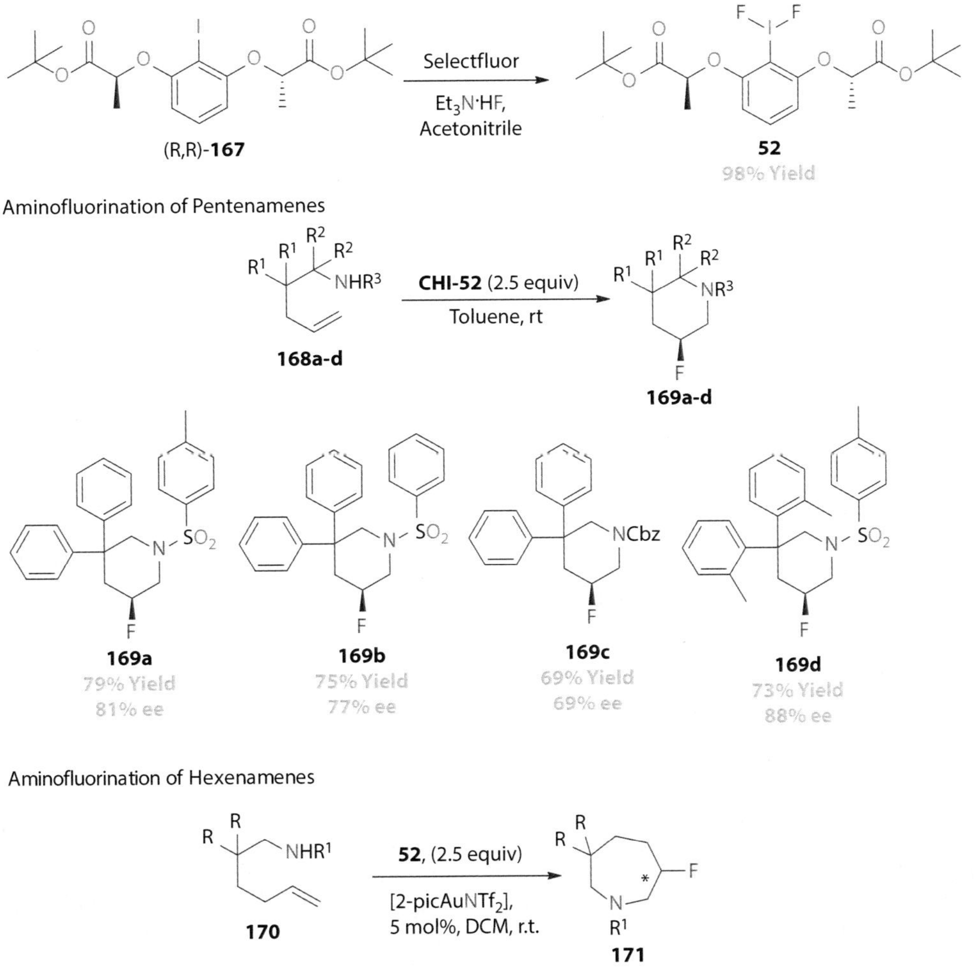

Scheme 17.35 Asymmetric aminofluorination.

chiral reagent, namely, difluoroiodonium salt **52**, which they specifically designed for this purpose. The **52** reagent was synthesized in a highly efficient manner, with a very high yield by the reaction of $Et_3N{\cdot}HF$/Selectfluor with Ishihara's precatalyst, an iodoarene-based analogue of a chiral first-generation catalyst.[58]

Initially, pentenamines **168a-d** were employed as substrates in the reaction, resulting in the formation of enantioenriched 3-fluoropiperidines **169a-d** in good yields. These products showed moderate enantioselectivity, which could be enhanced to over 90% ee through a straightforward recrystallization step. Of particular interest was the total 6-endo-selectivity observed in the reaction, leading to the formation of the desired products. Next, the team extended the application of the chiral hypervalent reagent **52** to the aminofluorination of hexenamines, affording azepane rings **171a–d** via a 7-endo-trig cyclization. Although this reaction was less favorable, the addition of 5 mol% of [2-picAuNTf$_2$] served as a necessary catalyst to promote the transformation. Various experimental results supported a possible mechanistic pathway involving the preparation of an aziridinium compound, which gets subsequently opened through a fluoride ion attack, leading to the observed anti-selectivity in the products.[13]

This enantioselective aminofluorination of alkenes using the novel chiral hypervalent iodine reagent **52** represents a significant advancement in synthetic methodology. The results obtained in this study open new possibilities for the construction of fluorinated nitrogen-based heterocycles with potential applications in medicinal chemistry and other fields.

17.3.6 Rearrangements

Investigations into rearrangements facilitated by reagents of hypervalent iodine have garnered significant attention in the past decade. This research has unveiled noteworthy and unconventional transformations, including typical rearrangements,[118–121] expansion of cyclic structures,[122] contraction of rings,[123] and migration of aryl groups.[124] The introduction of asymmetry into these oxidation-driven rearrangements was pioneered by Farid *et al.*[125] Building upon the precedents set by prior transformations,[126] this research team achieved the enantioselectivity in ketone's **(172a-c)** rearrangement, yielding α-arylated compounds **173a-c** (Scheme 17.36). Chiral iodine(III) **45** served as the catalyst for this migration, leading to moderate to good yields and, generally, exceptional enantiomeric excess. Notably, the inclusion of TMSOTf was crucial for the activation of hypervalent iodine, resulting in improved conversion compared to the previously employed Brønsted acid in similar rearrangements. Various aryl rings, encompassing both electron-withdrawing and electron-donating moiety, underwent testing under this process. While the previous group exhibited fair selectivities, the yields were relatively lower possibly due to reduced reactivity.

In a subsequent development, in the year 2016, Brown *et al.*[127] presented another intriguing example of an asymmetric process catalyzed by chiral iodine(III) species in the realm of rearrangements. Here, diarylalkenes **174a–c** engaged with chiral iodine(III) **47**, yielding α-arylated ketones **175a–c** in a highly efficient manner. A diverse array of substrates underwent evaluation, yielding impressive outcomes with respect to both enantiomeric excesses and chemical yields (Scheme 17.37).

Notably, the involvement of hydrated TsOH was essential for activating the chiral iodine(III) molecules presumably via a ligand exchange process. Without its presence, no

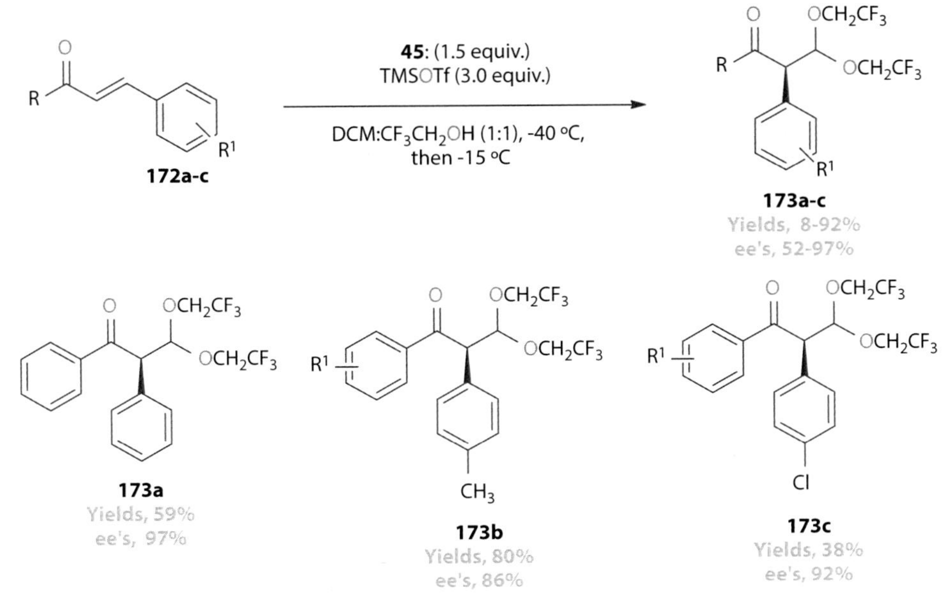

Scheme 17.36 Farid *et al.*[125]: asymmetric rearrangement.

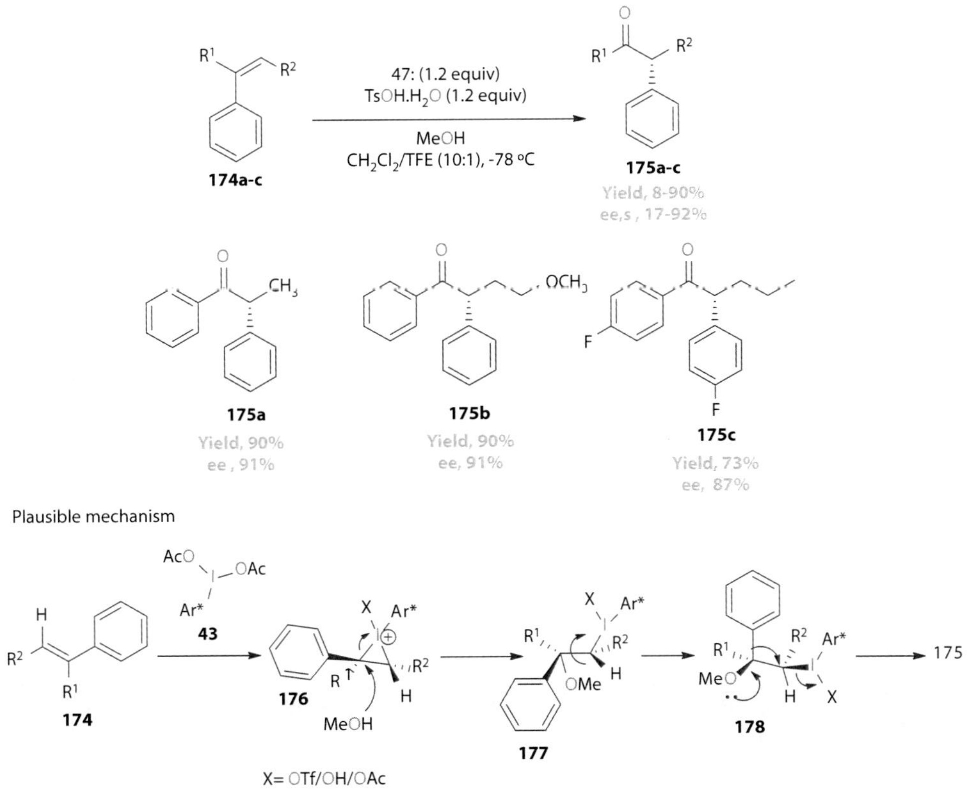

Scheme 17.37 Asymmetric rearrangement by Brown *et al.*[127]

transformation takes place. The proposed pathway commences with the formation of iodonium species **176**, which can be opened by a methanol molecule, yielding intermediate **177**. Subsequent rotation of a simple bond sets the stage for a 1,2-aryl migration, leading to the final product **175** with inversion of configuration. In order to shed light on the arrangement of iodine atom and aryl group during the rearrangement process, density functional theory (DFT) calculations were undertaken. It is important to note that Brown *et al.*[126] do not dismiss the possibility of non classical carbenium ion generation in this transformative process. Driven by the pursuit of potent catalysts of chiral iodine(III), Qurban *et al.*[128] introduced Fujita's **47** as a novel chiral variant with electron deficiency in the year 2017 (Scheme 17.38). The strategic incorporation of CF_3 at position-4 of the iodine(III) moiety induced a pronounced effect on the electrophilic potential of the hypervalent iodine species, thereby enhancing its reactivity. Consequently, compound **54** was efficiently synthesized using compound **179** first by simple iodination followed by the Mitsunobu reaction, affording a remarkable overall result. To gauge the performance in terms of reaction

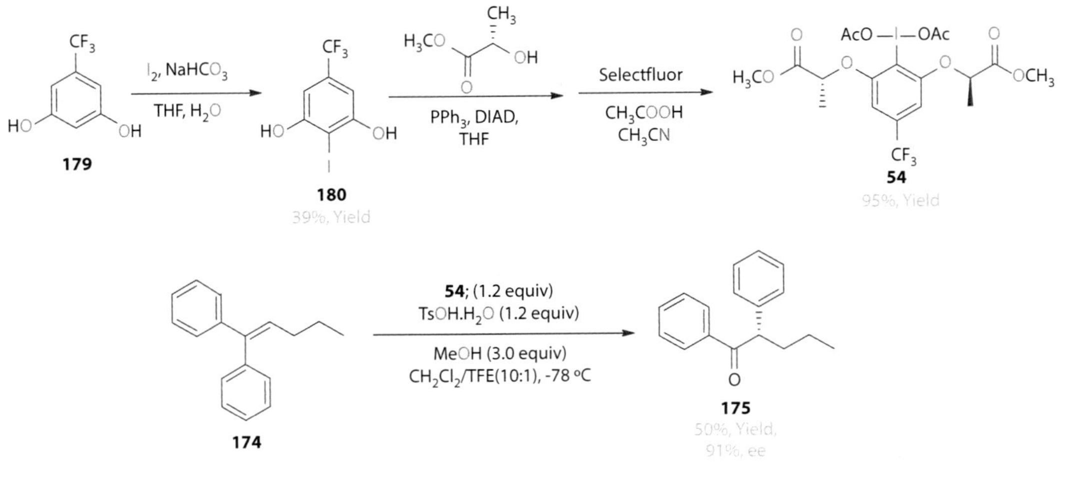

Scheme 17.38 Qurban *et al.*[128]: asymmetric rearrangement.

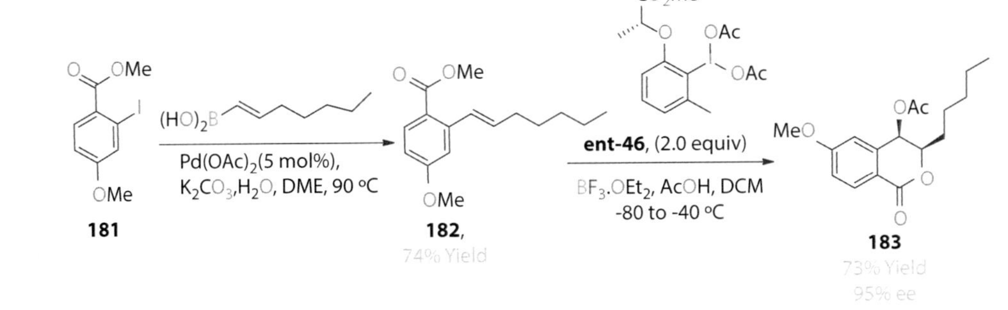

Scheme 17.39 Synthesis of a naturally occurring compound.

rate and enantioselectivity of this novel catalyst, the previously described rearrangement outlined in Scheme 17.37 was chosen as a test case. As the numerical results suggest, this iodine(III) variant exhibited analogous behavior to **47** in the given context, presenting no immediate additional advantages. Nonetheless, its potential for forthcoming applications remains promising.

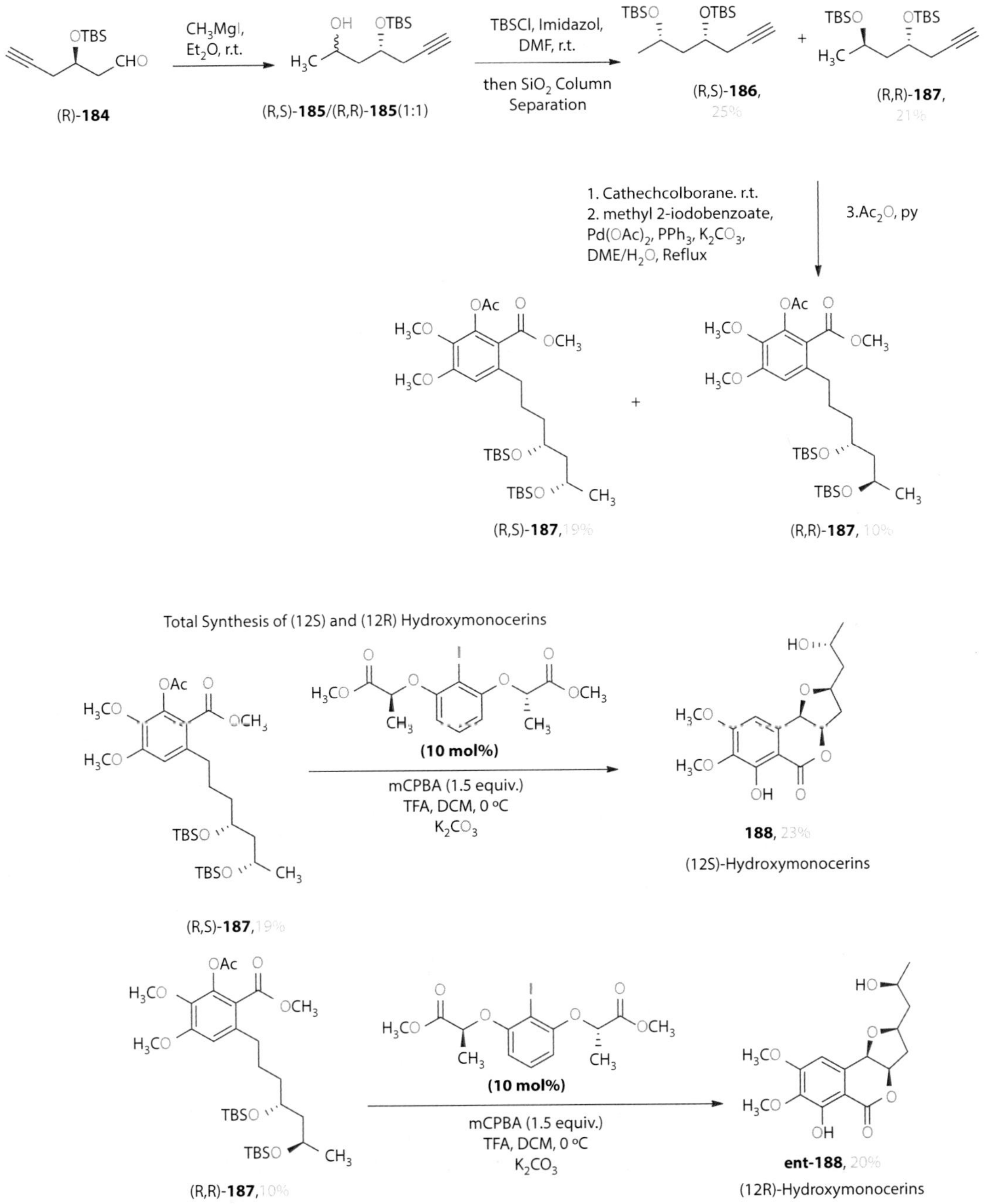

Scheme 17.40 Synthesis of hydroxymonocerins: natural products.

17.3.7 Total Synthesis Using Chiral Hypervalent Iodines

Chiral hypervalent iodines have found captivating utility in the realm of total synthesis, both for naturally occurring compounds and their enantioselective counterparts. Hypervalent iodine chemistry has proven particularly effective in the complete construction of numerous compounds, particularly within dearomatization reactions. Pouysegu *et al.*[129] have extensively documented these achievements in their comprehensive review. The prominent approach in this context revolves around the methodology pioneered by Fujita and collaborators,[58] specifically tailored for effecting the oxylactonization of styrenes. This inventive strategy, as exemplified in the pioneering work by Fujita and colleagues[58] in 2010, marks the initial instance wherein chiral hypervalent iodine reagents were employed for the dedicated preparation of natural products of biological importance (Scheme 17.39).[57]

Endringer *et al.*[130] successfully synthesized compound **183**, a natural product derived from *Xris pterygoblephara*, which exhibits selective inhibition of aromatase activity. The styrene precursor, **182**, was prepared *via* a cross-coupling reaction of Suzuki–Miyaura involving iodoarene **181** and (E)-hept-1-enylboronic acid, with a yield of 73%. This styrene substrate was then subjected to a reaction with **ent-46** and acetic acid, resulting in the formation of compound **183** with an impressive isolated yield of 73% and a remarkably high enantiomeric excess of 95%. The preferred stereochemical configuration of the naturally occurring compound required the use of the antipodal spatial arrangement of compound **46**. This approach was effectively extended for the total synthesis of both (12R)- and (12S)-hydroxymonocerins **188**, natural compounds sourced from *Exserohilum rostratum* and endophytic fungus *Microdochium bolleyi*, recognized for their potent antifungal, antibacterial, and antialgal properties (Scheme 17.40).[131, 132]

17.4 Conclusion

Currently, the application of hypervalent iodine(III/V) as a catalyst for asymmetric oxidations is widely recognized as a basic technique for obtaining significant compounds possessing high enantioselectivity, as well as synthetic precursors of considerable importance in medicinal chemistry and pharmaceuticals. The growth of a wide variety of reagents and catalysts of iodine(III/V) has established a new and fertile field in asymmetric synthesis. This has led to the performance of many oxidation reactions catalyzed by iodine(III/V) with highly favorable outcomes. The development of novel chiral hypervalent iodides and chiral iodide (III/V) precursors using various structures and chirality has led to numerous efficacious strategies. This chapter presents the overview of asymmetric hypervalent iodine chemistry. We anticipate that new catalysts will emerge in the coming period that will allow previously unknown transformations to be performed.

References

1. Beller, M. and Bolm, C. (Eds.), *Transition Metals for Organic Synthesis*, Wiley-VCH, Weinheim, 2004.

2. Blaser, H.-U. and Schmidt, E. (Eds.), *Asymmetric Catalysis on Industrial Scale*, Wiley-VCH, Weinheim, 2004.
3. Berkessel, A. and Goger, H., *Asymmetric Organocatalysis*, Wiley-VCH, Weinheim, Germany, 2005.
4. Dalko, P., II, *Enantioselective Organocatalysis*, Wiley-VCH, Weinheim, 2007.
5. Erkkila, A., Majander, I., Pihko, P.M., Iminium Catalysis. *Chem. Rev.*, 107, 5416–5470, 2007.
6. MacMillan, D.W.C., The Advent and Development of Organocatalysis. *Nature*, 455, 304–308, 2008.
7. Barbas, C.F., III Organocatalysis Lost: Modern Chemistry, Ancient Chemistry, and an Unsee Biosynthetic Apparatus. *Angew. Chem., Int. Ed.*, 47, 42–47, 2008.
8. Melchiorre, P., Marigo, M., Carlone, A., Bartoli, G., Asymmetric Aminocatalysis-Gold Rush in Organic Chemistry. *Angew. Chem., Int. Ed.*, 47, 6138–6171, 2008.
9. Dondoni, A. and Massi, A., Asymmetric Organocatalysis: From Infancy to Adolescence. *Angew. Chem., Int. Ed.*, 47, 4638–4660, 2008.
10. Aleman, J. and Cabrera, S., Applications of Asymmetric Organo-catalysis in Medicinal Chemistry. *Chem. Soc Rev.*, 42, 774–793, 2013.
11. Halskov, K.S., Donslund, B.S., Paz, B.M., Jørgensen, K.A., Computational Approach to Diarylprolinol-Silyl Ethers in Amino-catalysis. *Acc. Chem. Res.*, 49, 974–986, 2016.
12. Reyes-Rodríguez, G.J., Rezayee, N.M., Vidal-Albalat, A., Jørgensen, K.A., Prevalence of Diarylprolinol Silyl Ethers as Catalysts in Total Synthesis and Patents. *Chem. Rev.*, 119, 4221–4260, 2019, and references cited therein.
13. Parra, A., Chiral Hypervalent Iodines: Active Players in Asymmetric Synthesis. *Chem. Rev.*, *119*, 24, 12033–12088, 2019.
14. Ouellet, S.G., Walji, A.M., MacMillan, D.W., Enantioselective Organocatalytic Transfer Hydrogenation Reactions using Hantzsch Esters. *Acc. Chem. Res.*, 40, 1327–1339, 2007.
15. Wang, Z. and Jiang, Z., Recent Advances in Enantioselective Organocatalytic Reduction of C-N Bonds with Hantzsch Esters as the Hydride Source. *Asian J. Chem.*, 22, 4141–4149, 2010.
16. de Vries, J.G. and Mrsic, N., Organocatalytic Asymmetric Transfer Hydrogenation of Imines. *Catal. Sci. Technol.*, 1, 727–735, 2011.
17. Rueping, M., Dufour, J., Schoepke, F.R., Advances in Catalytic Metal-free Reductions: From Bio-inspired Concepts to Applications in the Organocatalytic Synthesis of Pharmaceuticals and Natural Products. *Green Chem.*, 13, 1084–1105, 2011, and references cited therein.
18. Beller, M. and Bolm, C. (Eds.), *Transition Metals for Organic Synthesis*, Wiley-VCH, Weinheim, Germany, 2004.
19. Punniyamurthy, T., Velusamy, S., Iqbal, J., Recent Advances in Transition Metal Catalyzed Oxidation of Organic Substrates with Molecular Oxygen. *Chem. Rev.*, 105, 2329–2364, 2005.
20. Bryliakov, K.P. and Talsi, E.P., Transition Metal Catalyzed Asymmetric Oxidation of Sulfides: From Discovery to Recent Trends. *Curr. Org. Chem.*, 16, 1215–1242, 2012.
21. Hollmann, F., Arends, I.W.C.E., Buehler, K., Schallmey, A., Buehler, B., Enzyme-mediated Oxidations for the Chemist. *Green Chem.*, 13, 226–265, 2011.
22. De Gonzalo, G., Orden, A.A., Bisogno, F.R., New trends in Organic Synthesis with Oxidative Enzymes. *Curr. Org. Chem.*, 16, 2598–2612, 2012.
23. Varvoglis, A., Commentary: Organic Hypervalent Iodine Chemistry Spanning Three Centuriesa Historical Overview. *Tetrahedron*, 66, 5739–5744, 2010.
24. Zhdankin, V.V. and Stang, P.J., Recent Developments in the Chemistry of Polyvalent Iodine Compounds. *Chem. Rev.*, 102, 2523–2584, 2002.
25. Wirth, T., Hypervalent Iodine Chemistry in Synthesis: Scope and New Directions. *Angew. Chem., Int. Ed.*, 44, 3656–3665, 2005.

26. Richardson, R.D., Desaize, M., Wirth, T., Hypervalent IodineMediated Aziridination of Alkenes: Mechanistic Insights and Requirements for Catalysis. *Chem. Eur. J.*, 13, 6745–6754, 2007.
27. Ladziata, U., Zhdankin, V. V. N-(2-iodylphenyl)acylamides-New Pentavalent Iodine Oxidizing Reagents. *Synlett*, 2007, 0527–0537, 2007.
28. Zhdankin, V.V. and Stang, P.J., Chemistry of Polyvalent Iodine. *Chem. Rev.*, 108, 5299–5358, 2008.
29. Uyanik, M. and Ishihara, K., Hypervalent Iodine-mediated Oxidation of Alcohols. *Chem. Commun.*, 16, 2086–2099, 2009.
30. Zhdankin, V.V., Organoiodine(V) Reagents in Organic Synthesis. *J. Org. Chem.*, 76, 1185–1197, 2011.
31. Singh, F.V. and Wirth, T., Oxidative Rearrangements with Hypervalent Iodine Reagents. *Synthesis*, 45, 2499–2511, 2013.
32. Zhdankin, V.V., *Hypervalent Iodine Chemistry: Preparation, Structure, and Synthetic Applications of Polyvalent Iodine Compounds*, Wiley: Chichester, U.K, 2014.
33. Zheng, Z., Zhang-Negrerie, D., Du, Y.F., Zhao, K., The Applications of Hypervalent Iodine(III) Reagents in the Construction of Heterocyclic Compounds Through Oxidative Coupling Reactions. *Sci. China: Chem.*, 57, 189–214, 2014.
34. Yoshimura, A. and Zhdankin, V.V., Advances in Synthetic Applications of Hypervalent Iodine Compounds. *Chem. Rev.*, 116, 3328–3435, 2016.
35. Singh, F.V. and Wirth, T., *In Patai's Chemistry of Functional Groups, Hypervalent Halogen Compounds*, I. Marek, B. Olofsson, Z. Rappoport, (Eds.), John Wiley & Sons, Ltd, Chichester, 2018.
36. Willgerodt, C., Ueber einige aromatische Jodidchloride. *J. Prakt. Chem.*, 33, 154–160, 1885.
37. Prakash, O. and Singh, S.P., Iodobenzene Diacetate and Related Hypervalent Iodine Reagents in the Synthesis of the Heterocyclic compounds. *Aldrichimica Acta*, 27, 15–23, 1994.
38. Tellitu, I. and Dominguez, E., The Application of [Bis-(trifluoroacetoxy)iodo]benzene (PIFA) in the Synthesis of NitrogenContaining Heterocycles. *Synlett*, 23, 2165–2175, 2012.
39. Uyanik, M. and Ishihara, K., 2-Iodoxybenzenesulfonic acid(IBS) Catalyzed Oxidation of Alcohols. *Aldrichimica Acta*, 43, 83–91, 2010.
40. Nicolaou, K.C., Montagnon, T., Baran, P.S., *In Electronic Encyclopedia of Reagents for Organic Synthesis (e-EROS)*, L.A. Paquette, (Ed.), WileyVCH, Weinheim, Germany, 2012, and references cited therein.
41. Dess, D.B. and Martin, J.C., Readily Accessible 12-I-5 Oxidant for the Conversion of Primary and Secondary Alcohols to Aldehydes and Ketones. *J. Org. Chem.*, 48, 4155–4156, 1983.
42. Ngatimin, M. and Lupton, D.W., The Discovery of Catalytic Enantioselective Polyvalent Iodine Mediated Reactions. *Aust. J. Chem.*, 63, 653–658, 2010.
43. Liang, H. and Ciufolini, M.A., Chiral Hypervalent Iodine Reagents in Asymmetric Reactions. *Angew. Chem., Int. Ed.*, 50, 11849–11851, 2011.
44. Parra, A. and Reboredo, S., Chiral Hypervalent Iodine Reagents: Synthesis and Reactivity. *Chem. Eur. J.*, 19, 17244–17260, 2013.
45. Singh, F.V. and Wirth, T., Hypervalent Iodine-Catalyzed Oxidative Functionalizations Including Stereoselective Reactions. *Chem. Asian J.*, 9, 950–971, 2014.
46. Romero, R.M. and Woste, T.H., Vicinal Difunctionalization of Alkenes with Iodine(III) Reagents and Catalysts. *Chem. Asian J.*, 9, 972–983, 2014.
47. Rawling, M.J. and Tomkinson, N.C., Metal-Free syn-Dioxygenation of Alkenes. *Org. Biomol. Chem.*, 11, 1434–1440, 2013.
48. Ghosh, S., Pradhan, S., Chatterjee, I.A., Survey of Chiral Hypervalent Iodine Reagents in Asymmetric Synthesis. *Beilstein J. Org. Chem.*, 14, 1244–1262, 2018.

49. Flores, A., Cots, E., Berges, J., Muñiz, K., Enantioselective Iodine (I/III) Catalysis in Organic Synthesis. *Adv. Synth. Catal.*, 361, 2–25, 2019.
50. Cai, Q. and Ma, H., Recent Advances of Chiral Hypervalent Iodine Reagents. *Huaxue Xuebao*, 77, 213–230, 2019.
51. Pribram, R., Ueber das optische Drehungsvermogen des Jodoniumtartrates. *Justus Liebigs Ann. Chem.*, 351, 481–485, 1907.
52. Merkushev, E.B., Novikov, A.N., Makarchenko, S.S., Moskal'chuk, A.S., Glushkova, V.V., Kogai, T., II, Polyakova, L.G., Organic Compounds of Polyvalent Iodine. Simple Synthesis of (Diacyloxyiodo)benzenes. *J. Org. Chem. USSR (Engl. Trans.)*, 11, 1246–1249, 1975.
53. Zhdankin, V.V., Smart, J.T., Zhao, P., Kiprof, P., Synthesis and Reactions of Amino Acid-derived Benziodazole Oxides: New Chiral Oxidizing Reagents. *Tetrahedron Lett.*, 41, 5299–5302, 2000.
54. Okuyama, T., Takino, T., Sueda, T., Ochiai, M., Solvolysis of Cyclohexenyliodonium Salt, a New Precursor for the Vinyl Cation: Remarkable Nucleofugality of the Phenyliodonio Group and Evidence for Internal Return from an Intimate Ion–Molecule Pair. *J. Am. Chem. Soc.*, 117, 3360–3367, 1995.
55. Wirth, T. (Ed.),Hypervalent Iodine Chemistry, in: *Modern Developments in Organic Synthesis; Topics in Current Chemistry*, vol. 224, Springer, Berlin, 2003.
56. Moriarty, R.M. and Prakash, O., Hypervalent Iodine in Organic Synthesis. *Acc. Chem. Res.*, 19, 244–250, 1986.
57. Narcis, M.J. and Takenaka, N., Helical-Chiral Small Molecules in Asymmetric Catalysis. *Eur. J. Org. Chem.*, 2014, 21–34, 2014.
58. Fujita, M., Yoshida, Y., Miyata, K., Wakisaka, A., Sugimura, T., Enantiodifferentiating endo-Selective Oxylactonization of ortho-Alk-1- enylbenzoate with a Lactate-Derived Aryl-λ3 - Iodane. *Angew. Chem., Int. Ed.*, 49, 7068–7071, 2010.
59. Uyanik, M., Yasui, T., Ishihara, K., Enantioselective Kita Oxidative Spirolactonization Catalyzed by In Situ Generated Chiral Hypervalent Iodine(III) Species. *Angew. Chem., Int. Ed.*, 49, 2175–2177, 2010.
60. Sreenithya, A., Patel, C., Hadad, C.M., Sunoj, R.B., Hypercoordinate Iodine Catalysts in Enantioselective Transformation: The Role of Catalyst Folding in Stereoselectivity. *ACS Catal.*, 7, 4189–4196, 2017.
61. Ochiai, M., Takeuchi, Y., Katayama, T., Sueda, T., Miyamoto, K., Iodobenzene-Catalyzed α-Acetoxylation of Ketones. In Situ Generation of Hypervalent (Diacyloxyiodo)benzenes Using mChloroperbenzoic. *Acid. J. Am. Chem. Soc.*, 127, 12244–12245, 2005.
62. Dohi, T., Maruyama, A., Yoshimura, M., Morimoto, K., Tohma, H., Kita, Y., Versatile Hypervalent-Iodine(III)-Catalyzed Oxidations with m-Chloroperbenzoic Acid as a Cooxidant. *Angew. Chem., Int. Ed.*, 44, 6193–6196, 2005.
63. O'Mahony, G.E., Ford, A., Maguire, A.R., Asymmetric Oxidation of Sulfides. *J. Sulfur Chem.*, 34, 301–341, 2013, and references cited therein.
64. Imamoto, T. and Koto, H., Asymmetric Oxidation of Sulfides to Sulfoxides with Trivalent Iodine Reagents. *Chem. Lett.*, 15, 967–968, 1986.
65. Ray III, D.G. and Koser, G.F., Iodinanes with Chiral Ligands. Synthesis and Structure of Iodine (III) Dibenzoyl Tartrates. *J. Org. Chem.*, 57, 1607–1610, 1992.
66. Koser, G.F., Chapter 18, in: *Hypervalent Halogen Compounds. In The Chemistry of Functional Groups. Suplement D*, S. Patai, and Z. Rappoport, (Eds.), pp. 729–740, Wiley, Chichester, UK, 1983.
67. Ray III, D.G. and Koser, G.F., Iodinanes with Iodine(III)-Bound Homochiral Alkoxy Ligands: Preparation and Utility for the Synthesis of Alkoxysulfonium Salts and Chiral Sulfoxides. *J. Am. Chem. Soc*, 112, 5672–5673, 1990.

68. Hatzigrigoriou, E., Varvoglis, A., Bakola-Christianopoulou, M., Preparation of [Hydroxy(+)-10-camphorsulfonyl)oxy)iodo]benzene and Its Reactivity towards Carbonyl Compounds. *J. Org. Chem.*, 55, 315–318, 1990.
69. Xia, M. and Chen, Z.-C., Hypervalent Iodine in Synthesis XXIII: Oxidation with [Hydroxy(tosyloxy) iodo]benzene: Selective Oxidation of Sulfide to Sulfoxides. *Synth. Commun.*, 27, 1315–1320, 1997.
70. Tohma, H., Takizawa, S., Watanabe, H., Kita, Y., Hypervalent Iodine(III) Oxidation Catalyzed by Quaternary Ammonium Salt in Micellar Systems. *Tetrahedron Lett.*, 39, 4547–4550, 1998.
71. Tohma, H., Takizawa, S., Watanabe, H., Fukuoka, Y., Maegawa, T., Kita, Y., Hypervalent Iodine(V)-Induced Asymmetric Oxidation of Sulfides to Sulfoxides Mediated by Reversed Micelles: Novel Nonmetallic Catalytic System. *J. Org. Chem.*, 64, 3519–3523, 1999.
72. Tohma, H., Takizawa, S., Morioka, H., Maegawa, T., Kita, Y., Novel Catalytic Asymmetric Sulfoxidation in Water Using the Hypervalent Iodine Reagent Iodoxybenzene. *Chem. Pharm. Bull.*, 48, 445–446, 2000.
73. Ladziata, U., Carlson, J., Zhdankin, V.V., Synthesis and Oxidative Reactivity of New Chiral Hypervalent Iodine(V) Reagents Based on (S)-Proline. *Tetrahedron Lett.*, 47, 6301–6304, 2006.
74. Altermann, S.M., Schafer, S., Wirth, T., New Chiral Hypervalent Iodine(V) Compounds as Stoichiometric Oxidants. *Tetrahedron*, 66, 5902–5907, 2010.
75. Ochiai, M., Takaoka, Y., Masaki, Y., Nagao, Y., Shiro, M., Synthesis of Chiral Hypervalent Organoiodinanes, Iodo(III)-binaphthyls, and Evidence for Pseudorotation on Iodine. *J. Am. Chem. Soc*, 112, 5677–5678, 1990.
76. Ochiai, M., Kitagawa, Y., Takayama, N., Takaoka, Y., Shiro, M., Synthesis of Chiral Diaryliodonium Salts, 1,1′-Binaphthyl-2-yl-(phenyl)iodonium Tetrafluoroborates: Asymmetric α-Phenylation of β-Keto Ester Enolates. *J. Am. Chem. Soc.*, 121, 9233–9234, 1999.
77. Jalalian, N. and Olofsson, B., Design and Asymmetric Synthesis of Chiral Diaryliodonium Salts. *Tetrahedron*, 66, 5793–5800, 2010.
78. Zhdankin, V.V., Koposov, A.Y., Su, L., Boyarskikh, V.V., Netzel, B.C., Young, V.G., Jr., Synthesis and Structure of Amino AcidDerived Benziodazoles: New Hypervalent Iodine Heterocycles. *Org. Lett.*, 5, 1583–1586, 2003.
79. Kitamura, T., Lee, C.H., Taniguchi, Y., Fujiwara, Y., Matsumoto, M., Sano, Y.A., New Route to Chiral Diaryldiacetylenic Liquid Crystals. Preparation of Iodonium Salts Bearing Chiral Alkynyl Ligands and Utility for Chiral Alkynyl Transfer Agents. *J. Am. Chem. Soc.*, 119, 619–620, 1997.
80. Rebrovic, L. and Koser, G.F.A., Alkynylaryliodonium Tosylates and Aryl[β-(tosyloxy)vinyl] iodonium Tosylates from Reactions of Terminal Alkynes with [Hydroxy(tosyloxy)iodo]benzene. *J. Org. Chem.*, 49, 4700–4702, 1984.
81. Bachi, M.D., Bar-Ner, N., Crittell, C.M., Stang, P.J., Williamson, B.L., Synthesis of Alkynyl(phenyl) iodonium Triflates and Their Reaction with Diethyl 2-aminomalonate. *J. Org. Chem.*, 56, 3912–3915, 1991.
82. Companys, S., Peixoto, P.A., Bosset, C., Chassaing, S., Miqueu, K., Sotiropoulos, J.-M., Pouysegu, L., Quideau, S., Asymmetric Alkynylation of β-Ketoesters and Naphthols Promoted by New Chiral Biphenylic Iodanes. *Chem. Eur. J.*, 23, 13309–13313, 2017.
83. Fernandez González, D.F., Brand, J.P., Mondiére, R., Waser, J., Ethynylbenziodoxolones (EBX) as Reagents for the Ethynylation of Stabilized Enolates. *Adv. Synth. Catal.*, 355, 1631–1639, 2013.
84. Wu, X., Shirakawa, S., Maruoka, K., Efficient Asymmetric Synthesis of Spiro-2(3H)-furanones via Phase-transfer-catalyzed Alkynylation. *Org. Biomol. Chem.*, 12, 5388–5392, 2014.

85. Fujita, M., Sakanishi, Y., Ishii, M., Yamataka, H., Okuyama, T., Solvolysis of 4-Methylcyclohexylidenemethyliodonium Salt: Chirality Probe Approach to a Primary Vinyl Cation Intermediate. *J. Org. Chem.*, 67, 8130–8137, 2002.
86. Fujita, M., Sakanishi, Y., Nishii, M., Okuyama, T., Cycloheptyne Intermediate in the Reaction of Chiral Cyclohexylidenemethyliodonium Salt with Sulfonates. *J. Org. Chem.*, 67, 8138–8146, 2002.
87. Mizukami, E., Ando, M., Tanaka, T., Imamura, J., The Acetoxylation of p-Substituted Acetophenones and β-Diketones with (Diacetoxyiodo)benzene. *Bull. Chem. Soc. Jpn.*, 51, 335–336, 1978.
88. Koser, G.F., Lodaya, J.S., Ray, D.G., III; Kokil, P. B. Direct.alpha.-Phosphoryloxylation of Ketones and (Phosphoryloxy)-lactonization of Pentenoic Acids with Hydroxy[(bis(phenyloxy)- phosphoryl)oxy]iodo]benzene. *J. Am. Chem. Soc.*, 110, 2987–2988, 1988.
89. Lodaya, J.S. and Koser, G.F., Direct.alpha.-mesyloxylation of ketones and.beta.-dicarbonyl compounds with [hydroxy(mesyloxy)-iodo]benzene. *J. Org. Chem.*, 53, 210–212, 1988.
90. Dong, D.-Q., Hao, S.-H., Wang, Z.-L., Chen, C., Hypervalent Iodine: a Powerful Electrophile for Asymmetric α-Functionalization of Carbonyl Compounds. *Org. Biomol. Chem.*, 12, 4278–4289, 2014.
91. Koser, G.F. and Wettach, R.H., Hypervalent Organoiodine. Reactions of Silver Arylsulfonates with Iodosobenzene Dichloride. *J. Org. Chem.*, 42, 1476–1478, 1977.
92. Wirth, T. and Hirt, U.H., Chiral Hypervalent Iodine Compounds. *Tetrahedron: Asymmetry*, 8, 23–26, 1997.
93. Rabah, G.A. and Koser, G.F., Facile Synthetic Entry into the 1,3-Dihydro-3-methyl-3-phenyl-l,2- benziodoxole Family of λ3-iodanes. *Tetrahedron Lett.*, 37, 6453–6456, 1996.
94. Hirt, U.H., Spingler, B., Wirth, T., New Chiral Hypervalent Iodine Compounds in Asymmetric Synthesis. *J. Org. Chem.*, 63, 7674–7679, 1998.
95. Hirt, U.H., Schuster, M.F.H., French, A.N., Wiest, O.G., Wirth, T., Chiral Hypervalent Organo-Iodine(III) Compounds. *Eur. J. Org. Chem.*, 2001, 1569–1579, 2001.
96. Richardson, R.D., Page, T.K., Altermann, S., Paradine, S.M., French, A.N., Wirth, T., Enantioselective α-Oxytosylation of Ketones Catalysed by Iodoarenes. *Synlett*, 2007, 0538–0542, 2007.
97. Moriarty, R., Prakash, O., Duncan, M.P., Carbon-Carbon Bond Formation Using Hypervalent Iodine Under Lewis Acid Conditions: Scope of the Method for the Synthesis of Butane-1.4-diones. *J. Chem. Soc., Perkin Trans.*, 11, 559–561, 1987.
98. Altermann, S.M., Richardson, R.D., Page, T.K., Schmidt, R.K., Holland, E., Mohammed, U., Paradine, S.M., French, A.N., Richter, C., Bahar, A.M., Witulski, B., Wirth, T., Catalytic Enantioselective α-Oxysulfonylation of Ketones Mediated by Iodoarenes. *Eur. J. Org. Chem.*, 2008, 5315–5328, 2008.
99. Farooq, U., Schafer, S., Shah, A.-U.-H. A., Freudendahl, D.M., Wirth, T., Synthesis of New Enantiomerically Pure Organoiodine Catalysts and Their Application in the α-Functionalization of Ketones. *Synthesis*, 2010, 1023–1029, 2010.
100. Uyanik, M., Yasui, T., Ishihara, K., Hypervalent Iodine-catalyzed Oxylactonization of Ketocarboxylic Acids to Ketolactones. *Bioorg. Med. Chem. Lett.*, 19, 3848–3851, 2009.
101. Yu, J., Cui, J., Hou, X.-S., Liu, S.-S., Gao, W.-C., Jiang, S., Tian, J., Zhang, C., Enantioselective α-Tosyloxylation of Ketones Catalyzed by Spirobiindane Scaffold-based Chiral Iodoarenes. *Tetrahedron: Asymmetry*, 22, 2039–2055, 2011.
102. Chen, C., Zhu, S.-F., Wu, X.-Y., Zhou, Q.-L., Preparation and Application of Chiral Spiro Nitrogen-containing Ligands for Cobaltcatalyzed Asymmetric Michael Addition. *Tetrahedron: Asymmetry*, 17, 2761–2767, 2006.

103. Guilbault, A.-A. and Legault, C.Y., Drastic Enhancement of Activity in Iodane-Based α-Tosyloxyl- ation of Ketones: Iodine(III) Does the Hypervalent Twist. *ACS Catal.*, 2, 219–222, 2012.
104. Helmchen, G. and Pfaltz, A., Phosphinooxazolines-A New Class of Versatile, Modular P,N-Ligands for Asymmetric Catalysis. *Acc. Chem. Res.*, 33, 336–345, 2000.
105. Therien, M.-E., Guilbault, A.-A., Legault, C.Y., New Chiral Iodooxazoline Catalysts for the I(III)-Mediated α-Tosyloxylation of Ketones: Refining the Stereoinduction Model. *Tetrahedron: Asymmetry*, 24, 1193–1197, 2013.
106. Rodríguez, A. and Moran, W.J., Chiral Aryl Iodide-Catalyzed Enantioselective α-Oxidation of Ketones. *Synthesis*, 44, 1178–1182, 2012.
107. Lopez-Ortiz, F., Iglesias, M.J., Fernández, I., Andújar-Sánchez, C.M., Ruiz Gomez, G., Nucleophilic Dearomatizing (D NAr) Reactions of Aromatic C,H-Systems. A Mature Paradigm in Organic Synthesis. *Chem. Rev.*, 107, 1580–1691, 2007.
108. For an excellent review concerning dearomatization reaction in total synthesis of natural products, see: Roche, S. P.; Porco, J. A., Jr. Dearomatization Strategies in the Synthesis of Complex Natural Products. *Angew. Chem., Int. Ed.*, 50, 4068–4093, 2011.
109. Zhuo, C.-X., Zhang, W., You, S.-L., Catalytic Asymmetric Dearomatization Reactions. *Angew. Chem., Int. Ed.*, 51, 12662–12686, 2012, and references cited therein.
110. Kita, Y. and Dohi, T., Pioneering Metal-Free Oxidative Coupling Strategy of Aromatic Compounds Using Hypervalent Iodine Reagents. *Chem. Rec.*, 15, 886–906, 2015.
111. Kurti, L., Herczegh, P., Visy, J., Simonyi, M., Antus, S., Pelter, A., New Insights Into the Mechanism of Phenolic Oxidation with Phenyliodonium(III) Reagents. *J. Chem. Soc., Perkin Trans.*, 1, 1, 379–380, 1999.
112. Dohi, T., Maruyama, A., Takenaga, N., Senami, K., Minamitsuji, Y., Fujioka, H., Caemmerer, S.B., Kita, Y.A., Chiral Hypervalent Iodine(III) Reagent for Enantioselective Dearomatization of Phenols. *Angew. Chem., Int. Ed.*, 47, 3787–3790, 2008.
113. Koposov, A.Y., Boyarskikh, V.V., Zhdankin, V.V., Amino Acid Derived Iodobenzene Dicarboxylates: Reagents for Oxidative Conversion of Alkenes to Amino Acid Esters. *Org. Lett.*, 6, 3613–3615, 2004.
114. Boye, A.C., Meyer, D., Ingison, C.K., French, A.N., Wirth, T., Novel Lactonization with Phenonium Ion Participation Induced by Hypervalent Iodine Reagents. *Org. Lett.*, 5, 2157–2159, 2003.
115. Fujita, M., Okuno, S., Lee, H.J., Sugimura, T., Okuyama, T., Enantiodifferentiating Tetrahydrofuranylation of But-3-enyl Carboxylates Using Optically Active Hypervalent Iodine(III) Reagents via a 1,3-Dioxan-2-yl Cation Intermediate. *Tetrahedron Lett.*, 48, 8691–8694, 2007.
116. Fujita, M., Ookubo, Y., Sugimura, T., Asymmetric Cycloetherification Based on a Chiral Auxiliary for 4-Acyloxy-1-butene Substrates During Oxidation with Iodosylbenzene via a 1,3-Dioxan-2-yl cation. *Tetrahedron Lett.*, 50, 1298–1300, 2009.
117. Kong, W., Feige, P., de Haro, T., Nevado, C., Regio- and Enantioselective Aminofluorination of Alkenes. *Angew. Chem., Int. Ed.*, 52, 2469–2473, 2013.
118. Vatele, J.M., Lewis Acid Promoted Oxidative Rearrangement of Tertiary Allylic Alcohols with the PhIO/TEMPO System. *Synlett*, 2008, 1785–1788, 2008.
119. Zagulyaeva, A.A., Banek, C.T., Yusubov, M.S., Zhdankin, V.V., Hofmann Rearrangement of Carboxamides Mediated by Hypervalent Iodine Species Generated in Situ from Iodobenzene and Oxone: Reaction Scope and Limitations. *Org. Lett.*, 12, 4644–4647, 2010.
120. Beaulieu, M.-A., Guerard, K.C., Maertens, G., Sabot, C., Canesi, S., Oxidative Prins-Pinacol Tandem Process Mediated by a Hypervalent Iodine Reagent: Scope, Limitations, and Applications. *J. Org. Chem.*, 76, 9460–9471, 2011.

121. Jacquemot, G. and Canesi, S., Oxidative ipso-Rearrangement Performed by a Hypervalent Iodine Reagent and Its Application. *J. Org. Chem.*, 77, 7588–7594, 2012.
122. Silva, L.F., Vasconcelos, R.S., Nogueira, M.A., Iodine(III)-Promoted Ring Expansion of 1-Vinylcycloalkanol Derivatives: A Metal-Free Approach toward Seven-Membered Rings. *Org. Lett.*, 10, 1017–1020, 2008.
123. Silva, L.F., Siqueira, F.A., Pedrozo, E.C., Vieira, F.Y.M., Doriguetto, A.C., Iodine(III)- Promoted Ring Contraction of 1,2-Dihydronaphthalenes: A Diastereoselective Total Synthesis of (±)-Indatraline. *Org. Lett.*, 9, 1433–1436, 2007.
124. Miki, Y., Fujita, R., Matsushita, K., II, Oxidative Rearrangement of Pentaalkoxychalcones with Phenyliodine(III) Bis(trifluoroacetate) (PIFA): Synthesis of (±)-10-Bromopterocarpin and (±)-Pterocarpin. *J. Chem. Soc., Perkin Trans.*, 1, 2533–2536, 1998.
125. Farid, U., Malmedy, F., Claveau, R., Albers, L., Wirth, T., Stereoselective Rearrangements with Chiral Hypervalent Iodine Reagents. *Angew. Chem., Int. Ed.*, 52, 7018–7022, 2013.
126. Singh, F.V., Rehbein, J., Wirth, T., For a seminal work concerning this transformation in achiral version see: Facile Oxidative Rearrangements Using Hypervalent Iodine Reagents. *Chem. Open*, 1, 245–250, 2012.
127. Brown, M., Kumar, R., Rehbein, J., Wirth, T., Enantioselective Oxidative Rearrangements with Chiral Hypervalent Iodine Reagents. *Chem. Eur. J.*, 22, 4030–4035, 2016.
128. Qurban, J., Elsherbini, M., Wirth, T., Electron-Deficient Chiral Lactic Acid-Based Hypervalent Iodine Reagents. *J. Org. Chem.*, 82, 11872–11876, 2017.
129. Pouysegu, L., Deffieux, D., Quideau, S., Hypervalent iodine- mediated phenol dearomatization in natural product synthesis. *Tetrahedron*, 66, 2235–2264, 2010.
130. Endringer, D.C., Guimaraes, K.G., Kondratyuk, T.P., Pezzuto, J.M., Braga, F.C., Selective Inhibition of Aromatase by a Dihydroisocoumarin from Xyris pterygoblephara. *J. Nat. Prod.*, 71, 1082–1084, 2008.
131. Fujita, M., Mori, K., Shimogaki, M., Sugimura, T., Total Synthesis of (12R)- and (12S)-12 Hydroxymonocerins: Stereoselective Oxylactonization using a Chiral Hypervalent Iodine(III) Species. *RSC Adv.*, 3, 17717–17725, 2013.
132. Zhang, W., Krohn, K., Draeger, S., Schulz, B., Bioactive Isocoumarins Isolated from the Endophytic Fungus Microdochium bolleyi. *J. Nat. Prod.*, 71, 1078–1081, 2008.

18

High-Turnover Palladium Catalysts: Accelerating C-H Activation for Sustainable Green Catalysis

Biswajit Panda

Department of Chemistry, City College, Kolkata, India

Abstract

Efficient and sustainable catalytic processes are pivotal in green chemistry, striving to minimize environmental impact and promote lasting chemical transformations. High-turnover catalysts exhibit the ability to efficiently participate in cycles, enabling multiple reactions without significant degradation. They accelerate reactions while maintaining activity over an extended period, aligning with the objective of maximizing efficiency and minimizing waste in sustainable catalysis. These catalysts enhance overall sustainability by enabling more reactions with less catalyst, reducing environmental impact and costs. Achieving high turnover involves optimizing factors like activity, selectivity, and stability, with catalyst design and engineering crucial for enhancing turnover numbers through features such as well-defined active sites and tailored ligands. The field of C-H activation is particularly noteworthy for its potential to transform unreactive carbon-hydrogen bonds directly, garnering significant attention in the pursuit of revolutionizing synthetic chemistry.

Palladium catalysts have shown exceptional versatility and efficiency in a broad spectrum of C-H activation processes, such as arylation, alkenylation, alkynylation, and cross-dehydrogenative coupling reactions. The diverse applications of high-turnover palladium catalysts, showcasing their ability to facilitate the synthesis of various complex molecules more efficiently and atom-economically, have been explored. The significance of C-H activation lies in its ability to eliminate the need for prefunctionalization steps, thereby streamlining synthetic routes and reducing waste generation. By directly targeting C-H bonds, high-turnover palladium catalysts enable more sustainable and environmentally friendly catalytic methodologies. The importance of high-turnover palladium catalysts in accelerating C-H activation, emphasizing their role in advancing sustainable green catalysis, is also highlighted here.

This chapter highlights the crucial role of high-turnover palladium catalysts in expediting C-H activation for sustainable green catalysis. Efficient catalysts, such as palladium, not only facilitate complex molecule synthesis but also mitigate the environmental impact of chemical processes. Ongoing research and innovation are imperative for advancing greener synthetic methodologies, with broad benefits for industries like pharmaceuticals, materials, and agrochemicals. The chapter, offering a comprehensive overview of recent advancements, challenges, and future prospects, serves as a vital resource for researchers in catalysis and green chemistry. It underscores the potential of high-turnover palladium catalysts to propel the development

Email: biswajitchem@gmail.com

Mousumi Sen (ed.) Sustainable Green Catalytic Processes, (447–472) © 2024 Scrivener Publishing LLC

of sustainable and environmentally friendly catalytic methodologies, contributing to a greener future in chemical synthesis.

Keywords: C–H activation, catalysis, efficiency, turnover, Palladium, green catalysis

18.1 Introduction

The activation of C–H bonds is a really attractive and practical technique for the functionalization of simple starting materials to complex molecules [1]. Among the various C-H activation processes, metal-catalyzed selective C-H activation has appeared as one of the utmost promising areas of synthetic organic chemistry due to its ability to provide novel functionalized molecules cost-effectively and thereby ensuring a smooth strategy for organic synthesis [2]. Though research has primarily centered on the functionalization of C-H bonds directly, it has now expanded to include other types of C-H activation. Although C-H activation is a booming method for organic transformations, 10% is a common catalyst loading in this area, thus the turnover number (TON) of these catalytic processes is generally near 10. Commonly, the C-H activation processes are catalyzed by palladium catalysts, and generally, cyclopentadienyl, N-heterocyclic carbenes, or bulky phosphenes are used as ligands. The necessity of higher percentage of expensive catalyst loadings poses a significant barrier to the extensive adoption of C-H activation reaction for the application in synthetic chemistry and industry. The expense of these compounds restricts their accessibility and limits the potential for large-scale implementation. Although some methods have been reported to achieve turnover numbers of 100 with only 1 mol% of suitable catalyst, instances of even higher TONs remain rare. Nevertheless, there have been exceptional cases where catalysts exhibit tens of thousands of turnovers.

This chapter aims to offer a comprehensive overview of transformations that involve remarkably high catalyst turnovers, shedding light on the underlying patterns that could aid in future catalyst development. Despite the progress made in understanding catalyst behavior, several factors influencing the lifetime of the catalyst in C-H activation still elude us. Among the critical parameters affecting TONs, stability of the catalyst, the void of catalyst poisons, and in general high catalytic activity stand out as advantageous attributes for achieving impressive TONs. However, the rational design of systems that embody all of these characteristics is still in its early stages. Further research and innovation are needed to fully unlock the potential of catalysts used in C-H activation reactions and make them highly feasible for broader application. As we delve deeper into the mechanisms behind high catalyst turnovers, the identification of novel catalyst materials, the optimization of reaction conditions, and the understanding of catalyst–substrate interactions hold great promise for enhancing catalytic efficiency. By bridging the gap between theoretical insights and experimental findings, researchers can expedite the development of cost-effective and sustainable catalyst systems.

Raising awareness within the scientific community about the critical and often underestimated aspect of C-H activation holds the potential to ignite future research endeavors focused on developing efficacious catalysts and encouraging the widespread use of C-H activation methodologies. An improved understanding of this aspect will undoubtedly foster significant advancements in the field.

One important metric used to assess the efficiency of catalysts is the turnover number, which is determined by dividing the reaction yield by the amount of catalyst used. However, there are instances where the catalyst's initial form is a dimer, but during the catalytic process, it converts to mononuclear species. This transformation is frequently observed with a catalyst like [$Pd_2(dba)_3$]. In such cases, loading of the catalyst is often twice the quantity of the dimer, resulting in a doubling of the TON.

To exemplify the concept of high-turnover catalysts for C-H activation, we have carefully selected typical cases that exhibit exceptionally high TONs. These examples showcase specific circumstances that lead to the highest recorded TON in published studies. It is essential to note that these instances typically apply to a singular case within the respective study.

The significance of these exemplary cases lies not only in their impressive TONs but also in the valuable insights they provide into the factors influencing catalyst efficiency. By understanding the conditions and mechanisms that facilitate high TONs, researchers can develop strategies to enhance catalytic performance and expand the application of C-H activation reactions. We would like to note that under alternative circumstances, greater yields than those indicated were achieved in the majority of the given studies. The reactions generally proceed through the generation of organometallic compound intermediates having carbon metal bond through the metal-mediated activation of a C-H bond. We will not go into finer mechanistic aspects, since the actual processes are often unclear, questioned, or reliant on certain conditions [3].

The progress of efficient catalytic methods for C-H activation has revolutionized organic synthesis by enabling functionalization of C-H bonds, traditionally considered inert. Among the transition metals, palladium has emerged as a prominent catalyst because of its ability to activate C-H bonds under mild reaction conditions and promote large numbers of transformations. One key aspect in the design of effective palladium catalysts for C-H activation reactions is achieving high turnover numbers and turnover frequencies (TOFs). High-TON catalysts exhibit prolonged catalytic activity by undergoing multiple catalytic cycles, while high-TOF catalysts demonstrate rapid turnover rates, ensuring efficient substrate conversion within a given time frame. The use of high-turnover palladium catalysts in C-H activation reactions has garnered significant attention, as it provides enhanced reaction efficiency, improved productivity, and reduced catalyst loadings, ultimately contributing to more sustainable and cost-effective synthetic processes. This chapter focuses on the recent advances and strategies employed to develop high-TON and high-TOF palladium catalysts for C-H activation reactions. We discuss the key factors influencing catalyst efficiency, such as ligand design, substrate scope, reaction conditions, and mechanistic insights. Furthermore, we highlight the applications and synthetic utility of these high-turnover catalysts in various C-H activation transformations, including cross-coupling, functionalization of heterocycles, and selective C-H bond activation in complex molecules. The utilization of high-turnover palladium catalysts in C-H activation opens up new avenues for the synthesis of valuable organic compounds and holds promise for the development of highly sustainable and effectual synthetic methodologies.

Moreover, the use of high-turnover palladium catalysts in C-H activation reactions addresses the challenges associated with limited substrate availability and reactivity. Traditional approaches often require the use of prefunctionalized substrates, which involve multiple synthetic steps, increased waste generation, and potential issues with functional group compatibility. In contrast, high-TON and high-TOF palladium catalysts enable the activation of C-H bonds directly, expanding the scope of accessible starting materials and streamlining synthetic paths. The development of high-turnover catalysts has been facilitated by advances in ligand design and optimization. Tailoring the ligand structure allows for fine-tuning of catalyst properties, such as steric and electronic effects, which significantly impact catalytic activity and selectivity. Furthermore, ligand design plays a crucial role in stabilizing reactive intermediates, controlling catalyst speciation, and minimizing undesired side reactions.

In addition to ligand design, reaction conditions and solvent choice play essential roles in achieving high catalyst efficiency. Optimization of various reaction parameters, including temperature, pressure, and additive selection, can further enhance catalyst performance and increase TON and TOF values. Solvent effects, including polarity, coordinating ability, and hydrogen-bonding interactions, also influence reaction rates, selectivity, and catalyst stability. Understanding and leveraging these solvent effects can lead to the development of efficacious and sustainable C-H activation reactions.

The application of high-turnover palladium catalysts in C-H activation reactions extends beyond academic research. Industries such as pharmaceuticals, agrochemicals, and materials science benefit from the ability to access structurally diverse molecules in a more efficient and cost-effective manner. High-TON and high-TOF catalysts enable the synthesis of complex scaffolds, facilitate late-stage functionalization, and enable the rapid generation of compound libraries for drug discovery and development.

In brief, the use of high-turnover palladium catalysts in C-H activation reactions has emerged as an elegant strategy in organic synthesis. It offers enhanced reaction efficiency, improved productivity, and increased substrate scope. The continuous development of innovative ligands, optimization of reaction conditions, and mechanistic insights into catalytic processes contribute to the advancement of high-TON and high-TOF palladium catalysts. As this field continues to evolve, the utilization of high-turnover catalysts in C-H activation reactions will pave the way for more sustainable, efficient, and versatile synthetic methodologies with broad applications in diverse areas of chemical synthesis.

Turnover number and turnover frequency are significant factors used to characterize the efficiency and performance of a catalyst in a chemical reaction.

Turnover number (TON): The turnover number represents the number of times a catalyst molecule participates in a chemical reaction before becoming deactivated or consumed. It is defined as the ratio of the total moles of desired product formed to the moles of catalyst used. TON provides an indication of the catalyst's longevity and efficiency in facilitating the desired transformation. A higher TON indicates that the catalyst is more efficient and can undergo multiple catalytic cycles before losing activity.

$$\text{TON} = (\text{moles of product formed})/(\text{moles of catalyst used})$$

Turnover frequency (TOF): The turnover frequency is an evaluation of the catalytic activity of a catalyst. It represents the number of catalytic cycles that a single active site on the catalyst undergoes per unit of time. TOF is calculated by dividing the TON by the reaction time and is often expressed in units of moles of product formed per mole of catalyst per unit of time.

$$\text{TOF} = \text{TON}/(\text{reaction time})$$

TOF provides a measure of the catalyst's efficiency in converting reactants into products within agiven time frame. Higher TOF values indicate faster catalytic turnover and increased productivity. Both TON and TOF are crucial parameters for evaluating and comparing different catalysts' performance in a reaction. They help in assessing the catalytic efficiency, activity, and stability, providing valuable insights for catalyst design and optimization. Currently, one of the highly explored research areas has been Pd-catalyzed direct arylation reaction through C-H activation process. A palladium (II) species usually activates C-H bonds in the catalytic cycle, followed by transmetalation, reductive elimination, and the creation of C-C bonds to produce the desired product (Figure 18.1).

Several factors can influence the catalytic efficiency of a catalyst in a chemical reaction. These factors can be broadly categorized into intrinsic factors related to the catalyst itself and extrinsic factors related to the reaction conditions and the nature of the reactants. Here are some key factors that influence catalytic efficiency:

- (i) Catalyst Design: The design and structure of the catalyst play a significant role in its efficiency. Factors such as the choice of metal, ligands, and supporting materials can influence the catalyst's activity, selectivity, and stability. Catalyst design involves tailoring the electronic, steric, and geometric properties of the catalyst to enhance its interaction with the reactants and promote the desired reactions.
- (ii) Substrate Interaction: The nature and strength of the interaction between the catalyst and the reactants or substrates are crucial for catalytic efficiency. Catalysts with optimal binding affinity and selectivity toward the target substrates can facilitate faster and more selective reactions. The ability of

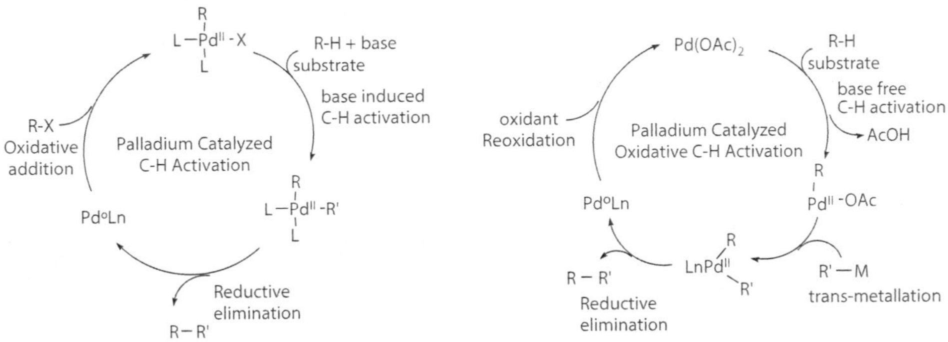

Figure 18.1 General mechanism of palladium-catalyzed C-H activation process.

the catalyst to activate the bonds of the substrate and stabilize reactive intermediates is essential for efficient catalysis.

(iii) Reaction Conditions: Reaction conditions such as temperature, pressure, solvent choice, and pH can significantly impact catalytic efficiency. Optimizing the reaction conditions can enhance the activity, selectivity, and stability of the catalyst. The choice of suitable reaction conditions depends on the specific requirements of the reaction and the catalyst's sensitivity to external factors.

(iv) Catalyst Loading: The amount of catalyst used in a reaction can affect its efficiency. Too little catalyst may result in slow reaction rates, while excessive catalyst loading may lead to waste and increased costs. Balancing the catalyst loading to provide the optimum catalytic activity and selectivity is essential for maximizing efficiency.

(v) Reactant Concentration and Reactivity: The concentration and reactivity of the reactants can influence catalytic efficiency. Higher concentrations of reactant can enhance the rate of reaction, although excessively high concentrations can result in undesirable side reactions. Reactant reactivity, including bond strengths and steric hindrance, affects the ease of bond activation and subsequent transformations.

(vi) Catalyst Deactivation: Catalyst deactivation can significantly impact catalytic efficiency over time. Deactivation can occur due to factors such as catalyst poisoning, leaching, aggregation, or structural changes. Understanding and mitigating catalyst deactivation mechanisms are crucial for maintaining long-term catalytic efficiency.

(vii) Mass Transport: Efficient mass transport of reactants and products to and from the catalyst surface is vital for catalytic efficiency. Diffusion limitations can affect reaction rates and selectivity. Proper reactor and catalyst design, as well as appropriate stirring or flow conditions, can help overcome mass transport limitations and enhance catalytic efficiency.

18.2 High-TON Pd Catalysis for C-H Arylation of Arenes

The palladium-catalyzed C-H arylation of arenes has garnered significant importance in organic synthesis due to its transformative impact on the construction of arylated compounds. This synthetic methodology allows for the functionalization of a particular C-H bonds directly in aromatic molecules, by passing the need for prefunctionalized substrates. By enabling the direct installation of functional aryl or heteroaryl groups onto specific C-H bonds, this process provides unparalleled regioselectivity and site selectivity, allowing for precise control over the desired chemical transformations. The importance of palladium-catalyzed C-H arylation is underscored by its synthetic efficiency and atom economy. It minimizes waste generation by utilizing the C-H bonds as the starting point for functionalization, resulting in excellent atom economy and reducing the number of synthetic steps required. Furthermore, this process shows a wide substrate scope, accommodating various types of arenes containing different functional groups, making it a useful tool for the synthesis of various arylated products. The functional group tolerance of palladium catalysts in C-H arylation reactions facilitates the incorporation of multiple functionalities in a single step, streamlining synthetic routes and reducing the reliance on protecting group strategies. Overall, the importance of palladium-catalyzed C-H activation reaction for the arylation of arenes lies in its synthetic versatility, regioselectivity, atom economy, and the direct access it provides to complex arylated compounds, making it an invaluable tool for modern organic synthesis.

Biaryl motifs are ubiquitous in various molecules, ranging from essential medicines to cutting-edge functional materials, playing crucial roles in advancing pharmaceutical and materials sciences. Palladium-catalyzed cross-coupling reactions of (hetero)arene have become highly reliable and specific in directing chemical reactions, leading to a significant impact on the synthesis of various hetero-biaryl or biaryl compounds and advancing the field of synthetic chemistry as a whole [4]. This efficient process has been driven by extensive research on catalyst efficiency, resulting in the advancement of remarkably effective catalysts capable of undergoing millions of turnovers [5]. A drawback of this approach is the generation of excess metal salts as by-products, necessitating additional synthetic steps and causing stoichiometric imbalances. To overcome these limitations and enhance the overall sustainability of cross-coupling methodologies, scientists have dedicated substantial efforts to explore alternative strategies that involve utilizing C-H bond activation reactions [6]. By harnessing the power of C-H activation, researchers aim to introduce one or both coupling partners directly, thereby minimizing waste and improving the overall efficiency of the synthetic process. Among these innovative approaches, particularly noteworthy are the highly effective methods for the direct arylation reaction of arenes and (hetero)arenes using corresponding aryl and (hetero)aryl halides, which have emerged as key contenders in the pursuit of greener and more streamlined chemical transformations. The continuous advancements in this area are driving the field of synthetic chemistry toward more sustainable and efficient practices, opening up exciting possibilities for the future synthesis of valuable biaryl compounds and beyond. In 2005, Daugulis *et al.* [7] reported a novel method for double arylation of anilide through C-H activation process with high efficiency (Scheme 18.1). Although the scope of this conversion was shown using 5 mol% catalyst, a reaction using 0.2 mol% that obtained a TON of 950 was also mentioned in this report.

Compounds with Lewis-basic functional groups perform an important role in enhancing C-H activation processes by coordinating with the catalyst. This coordination effect leads to an increased local concentration of the catalyst at the substrate, promoting the efficiency of the reaction and stabilizing important organometallic intermediates in the process [8]. In 2017, Kim and Hong [9] introduced a noteworthy protocol for the direct arylation of basic arenes, demonstrating remarkable results at relatively low catalyst loadings. By employing a mere 0.1 mol% of the catalyst and a diimine ligand, they achieved an impressive turnover number of 290, indicating the extraordinary catalytic efficiency of their

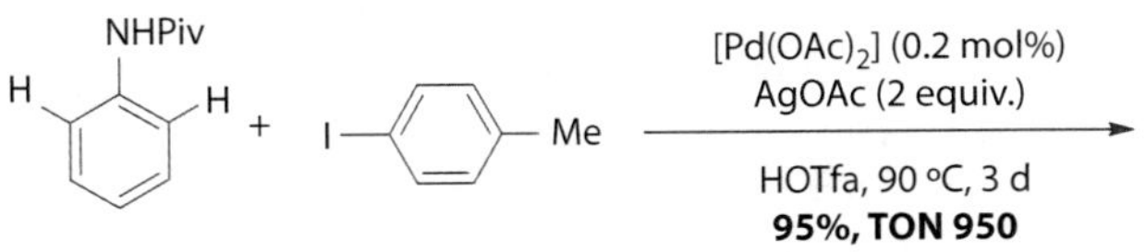

Scheme 18.1 NHPiv-group-directed double C-H arylation of arenes.

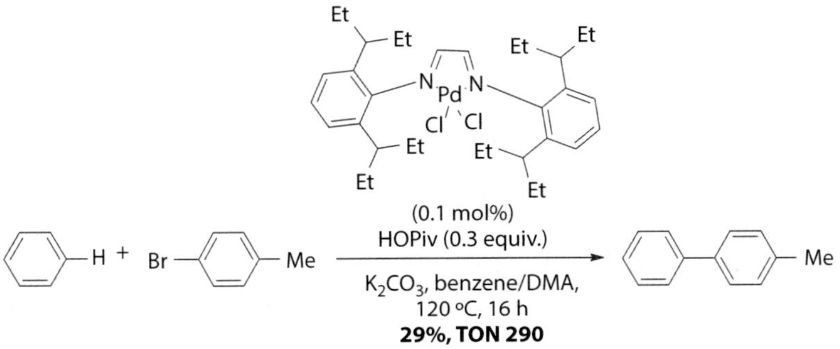

Scheme 18.2 The C-H arylation of simple arenes using an efficient Pd catalyst.

method (Scheme 18.2) [9]. This innovative approach paves the way for more sustainable and economically viable synthetic strategies in the field of organic chemistry.

The inert nature of simple arenes has long been recognized as a challenge in direct arylation processes, making it challenging to efficiently introduce new aryl groups directly onto these molecules. On the other hand, heteroarenes, containing atoms such as nitrogen, oxygen, or sulfur in their aromatic ring structure, exhibit higher electron density, rendering them more reactive and amenable to direct arylation. Notably, the Doucet group [10] has made remarkable contributions by developing a series of highly efficient strategies for direct arylation of diverse heteroarenes, achieving high turnover numbers of up to 85,000 (Scheme 18.3). Despite these significant advancements, the precise mechanism underlying these reactions remains an ongoing subject of research. The exact nature of the catalyst–substrate interactions and the intermediates involved might differ based on the specific heteroarene substrate molecule and the particular catalyst employed [11].

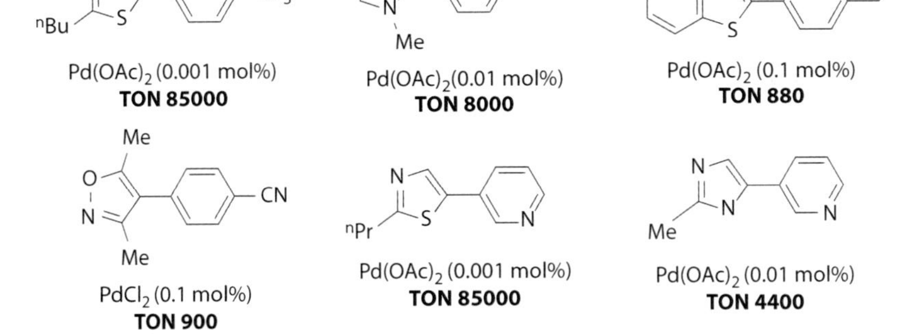

Scheme 18.3 Highly efficient C-H arylation reaction of heteroarenes.

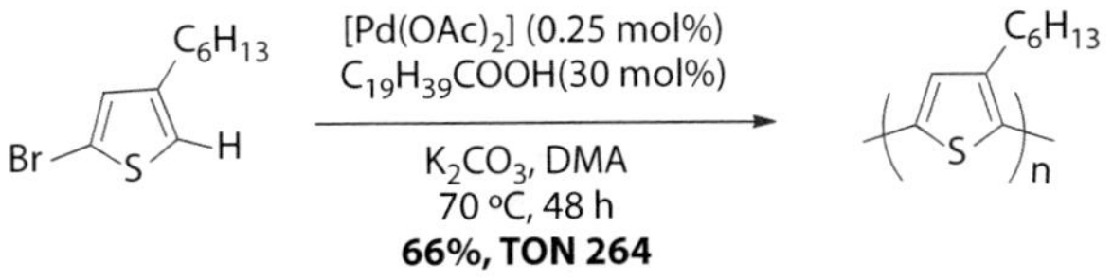

Scheme 18.4 Direct arylation polymerization of bromothiophene.

Polythiophenes hold immense potential as organic semiconductor materials, finding applications in various electronic devices [12]. The conventional methods for synthesizing polythiophenes involve multistep procedures, which can be laborious and often lead to low yields. However, the recent advancement of direct arylation polymerization (DArP) has provided an innovative alternative, streamlining the synthesis process and offering the prospect of higher yields and enhanced control over the polymer's properties [13]. In 2014, Rudenko *et al.* [14] reported an effective method for the DArP reactions of functionalized 2-bromothiophenes to synthesize polythiophenes using $Pd(OAc)_2$ as catalyst with low catalyst loading (Scheme 18.4). The catalyst facilitates the direct coupling of 2-bromothiophene derivatives, enabling the controlled growth of the polymer chain with tailored molecular structures. The simplicity and effectiveness of the palladium-catalyzed DArP approach have growing interest among researchers, leading to continuous exploration and optimization of the reaction parameters for the synthesis of novel polythiophenes with diverse functionalities.

Intramolecular direct arylation is a powerful chemical transformation that leads to the formation of polycyclic compounds within molecules carrying both an aryl halide and another properly positioned arene moiety. This process involves the oxidative addition of palladium to the aryl halide, facilitating C-H activation intramolecularly in a manner similar to the application of a directing group. Cyclic ethers are an important class of compounds used in the synthesis of diverse natural products and pharmaceutical compounds [15]. Notably, Campeau *et al.* [16] have exemplified the exceptional efficacy of palladium catalysts in synthesizing benzochromenes and other complex polycyclic heterocycles, achieving impressive turnover numbers of up to 960 (Scheme 18.5).

Ferrocene has become a widely versatile organometallic scaffold, finding applications in catalysis, materials science, and medicinal chemistry [17]. Notably, ferrocenes with planar chirality seemed extensively used as ligands (e.g., Josiphos and PPFA) or catalysts; some of them have been used industrially in pharmaceutical and agrochemical production [18].

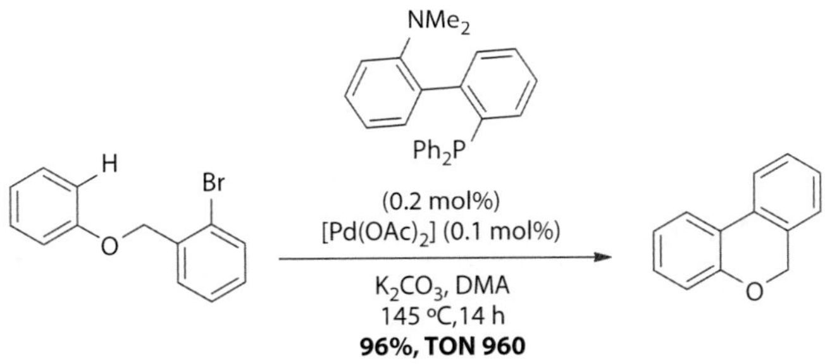

Scheme 18.5 Palladium-catalyzed intramolecular C-H arylation cyclization.

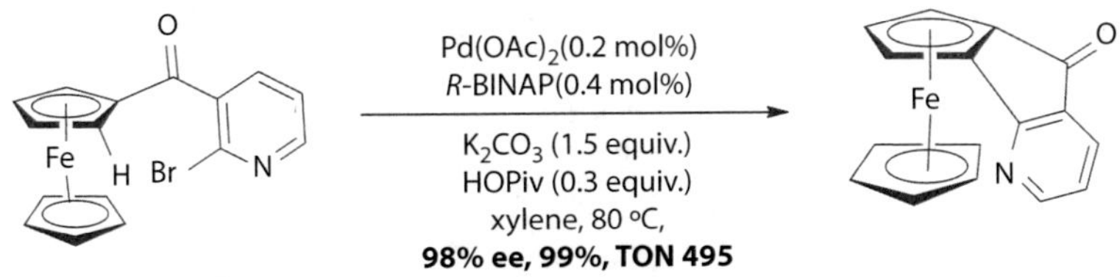

Scheme 18.6 Pd-catalyzed intramolecular enantioselective direct C-H arylation.

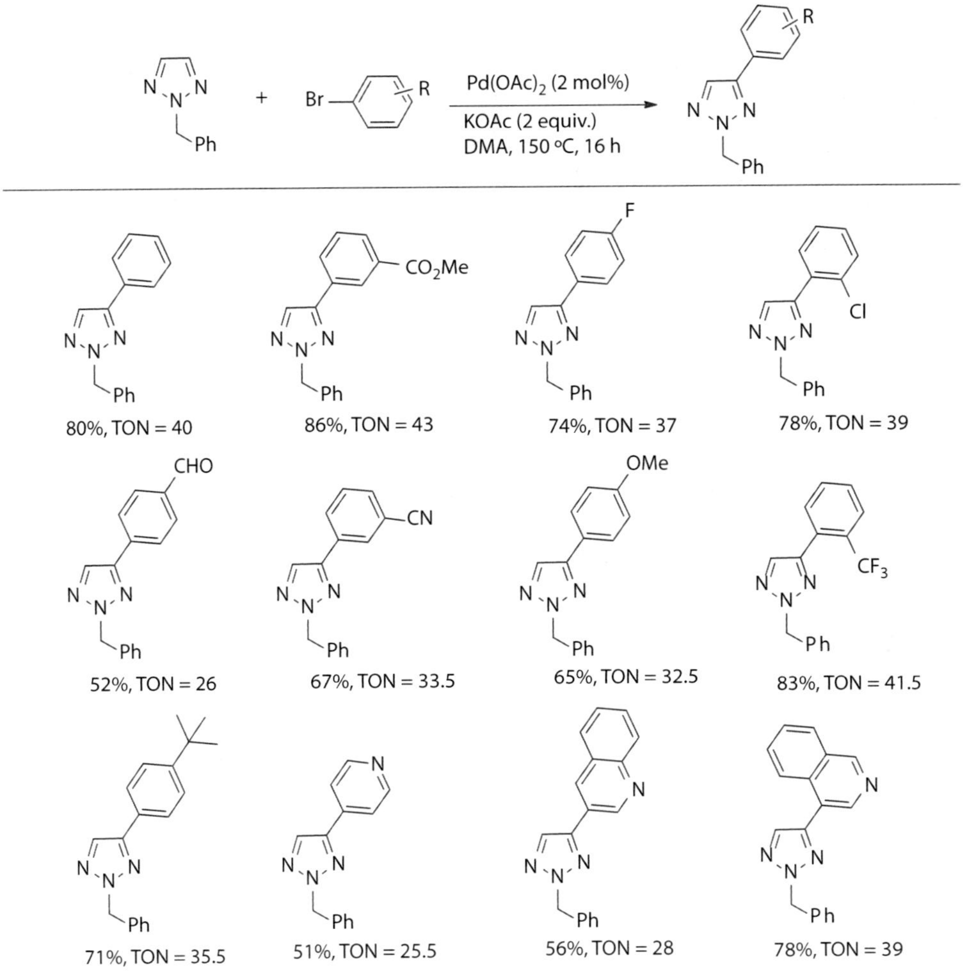

Scheme 18.7 Palladium-catalyzed C-H arylation of 2-benzyl, 1, 2,3-triazoles.

Over the years, significant efforts have been dedicated to synthesizing planar chiral ferrocenes [19]. Several methods have been employed, such as diastereoselective directed ortho-metalation (DoM) [20], asymmetric DoM [21], followed by resolution [22]. While these strategies have been successful and extensively used, they often require installation of chiral auxiliaries prior to the reaction and also need to use external chiral bases in stoichiometric amount. In 2014, Gao *et al.* [23] made a notable advancement by developing a highly efficient method for asymmetric synthesis of planar chiral ferrocenes through direct C–H arylation reaction using readily available starting materials using Pd(0) catalyst (Scheme 18.6). Utilizing 0.2 mol% $Pd(OAc)_2$ as precatalyst in the presence of a chiral bisphosphine ligand, a TON of 495 was achieved at complete conversion in a gram-scale experiment. This innovative method achieved yields of up to 99% and enantiomeric excess of 99%. The synthesized chiral planar ferrocene ligand was successfully utilized in Pd-catalyzed asymmetric allylic alkylation and amination reactions.

Shi *et al.* [24] demonstrated a dignified approach for the efficient palladium-catalyzed direct C4-arylation of 2-benzyl-1,2,3-triazoles (Scheme 18.7). In this method, 2% $Pd(OAc)_2$ was employed as effective catalyst and KOAc as suitable base in N,N-dimethyl acetamide solvent at 150°C. Substituted aryl and heteroaryl halides with varied functional groups such as –Cl, -F, -OMe, -t-Bu, -CHO, -CO_2Me, -CF_3, and -CN are highly tolerable under this experimental condition.

18.3 Palladium-Catalyzed Activation of Csp^3-H Bonds

Janody *et al.* [25] discovered an interesting conformational effect that enabled an effective activation reaction of a Csp^{3}H bond (Scheme 18.8). The efficient C-H activation at a ^{t}Bu group using only 0.5 mol% of a $Pd(PPh_3)_4$ catalyst precursor resulted in a notable turnover number of 196. This demonstrates the remarkable catalytic proficiency of the system. Even minor changes, such as employing a substrate with an iPr group, resulted in lower catalytic efficiency.

18.4 Palladium-Catalyzed Cross-Dehydrogenative Coupling

In recent years, palladium-catalyzed cross-dehydrogenative coupling (CDC) has garnered significant interest as a potent approach for synthesizing both carbon-carbon and carbon-heteroatom bonds directly from two distinct C-H bonds [26]. This transformative strategy offers a versatile and efficient means to construct complex molecular structures. This synthetic methodology offers a direct and efficient way for the synthesis of complex molecules, eliminating the need for prefunctionalized substrates. By harnessing the reactivity of two C-H bonds, CDC enables the selective formation of new bonds, allowing for the rapid assembly of molecular frameworks. Palladium catalysts play a key role in facilitating this transformation by activating the C-H bonds and mediating the cross-coupling process. The versatility of CDC is demonstrated by its broad substrate scope, accommodating a range of C-H bonds and coupling partners, including various heteroatoms and functional groups. The development of practical and selective catalyst systems, as well as the studying

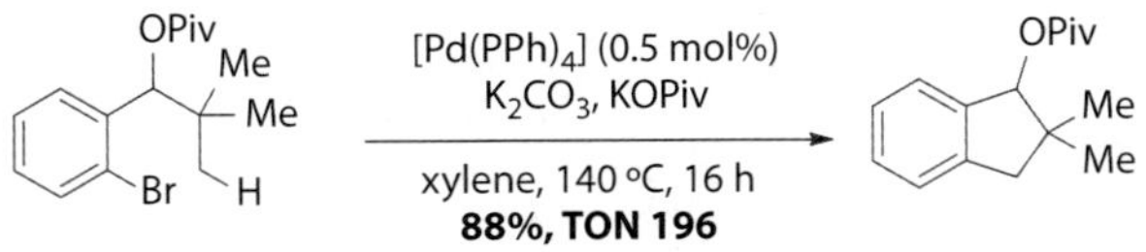

Scheme 18.8 Pd(0)-catalyzed highly efficient intramolecular Csp^3-H activation.

of the reaction pathway, has paved the way for the function of CDC in the synthesis of pharmaceuticals, agrochemicals, and other valuable organic compounds. Palladium-catalyzed CDC represents an innovative approach to bond formation, offering a streamlined and atom-economical process for the construction of complex molecular architectures.

Van Helden and Verberg [27] reported the first example of palladium-mediated CDC in the 1960s, in which they used stoichiometric $PdCl_2$ to generate isomeric mixtures of biphenyls. Since then, significant development has been produced in catalytic CDC [28]. The efficiency of catalyst in these chemical transformations faces a crucial limitation due to catalyst leaching and deactivation through the agglomeration and deposition process of Pd(0) generated in the reductive elimination step. Consequently, the oxidant employed to recycle Pd(II) is critical for the catalyst's life span. While silver- or copper-based oxidants are frequently used, considerable advancement has been achieved in the utilization of oxygen or air as green oxidant [29].

On the basis of mechanistic analysis, the Stahl group [30] designed a better very active catalytic system that allows the highly regulating CDC of o-xylene and the relatively short feasible route to Upilex. As weakly coordinating ligands, triflates and 2-fluoropyridine are utilized to increase the electrophilicity of the metal center. The introduction of Fe(OTf)3 as source of triflate counter ions boosts the TON over 150 (Scheme 18.9) [31].

Heterocycles play a vital role as essential structures in numerous natural products, biologically active molecules, and also find utility as valuable compounds in organic and material chemistry [32]. In 2012, Huet *et al.* [33] demonstrated an efficient route for the total synthesis of carbazole alkaloid murrayastine. In this synthesis, they have used a palladium-catalyzed intramolecular CDC coupling process with high TONs (up to 230) for the construction of carbazole moiety (Scheme 18.10). In this step, air was used as green oxidant [33].

Palladium catalyzed the oxidative olefination of arenes commonly known as the Fujiwara Moritani or oxidative Heck reaction. Jia *et al.* [34] made a remarkable advancement in this field when they employed tBuOOH as the oxidant, resulting in a substantial increase in catalyst efficiency, achieving impressive turnover numbers of up to 280 (Scheme 18.11). Building upon this progress, the Jacobs group [51] further enhanced the catalyst efficacy by

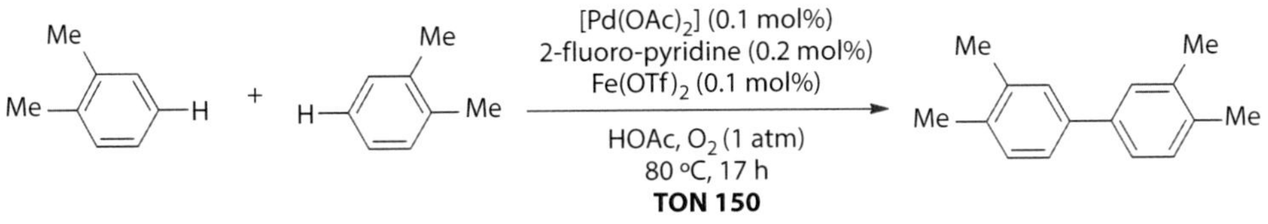

Scheme 18.9 Palladium-catalyzed oxidative dimerization of o-xylene.

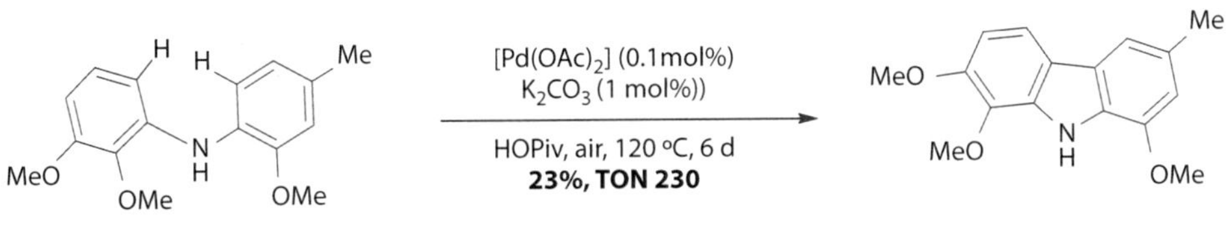

Scheme 18.10 Catalyzed oxidative cyclization for carbazole synthesis.

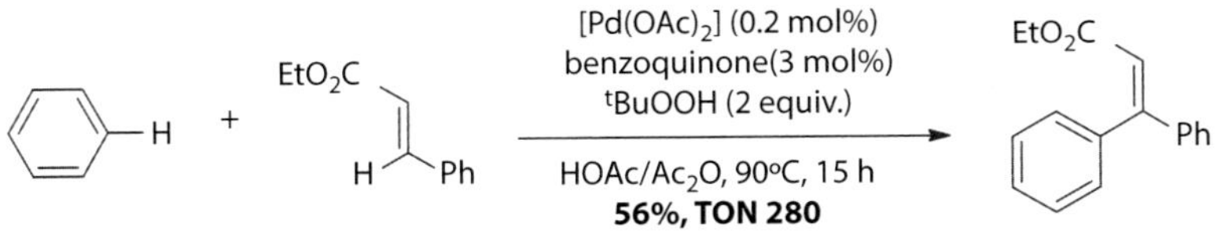

Scheme 18.11 Pd-catalyzed oxidative olefination of arenes.

incorporating benzoic acid, employing an excess amount of the arene, and carrying out the reaction under high pressure of oxygen gas. These modifications led to even higher yields and improved performance of the oxidative olefination reaction, showcasing the potential for continued advancements in C-H functionalization methodologies.

The inclusion of a monoprotected amino acid as suitable ligand substantially improved the TON as well as reaction scope of the reaction (Scheme 18.12), avoiding the formation of regioisomericproducts in C-H activation of arenes [35]. In the process of olefination, a substituted phenylacetic acid was subjected to a noteworthy transformation, achieving a high turnover number of 415. This highly efficient reaction was made possible through the use of $Pd(OAc)_2$ and a mere 0.4 mol% of N-acetyl isoleucine as a ligand. The success of this methodology was further broadened to a large-scale reaction, demonstrating its scalability and practical applicability. The key to this enhanced reactivity and the widespread applicability of the ligand was uncovered through meticulous kinetic measurements. By introducing appropriate ligands, the undesired formation of catalytically less active palladium species, resulting from the coordination of carboxylates generated from the starting materials or the product, was effectively inhibited. This strategic approach opened up new possibilities for achieving improved catalytic performance and broadening the scope of C-H functionalization reactions.

In 1999, Jia *et al.* [36] demonstrated a nice method for palladium-catalyzed C-H activation of arenes with olefins to produce arylated olefins in good yields (Scheme 18.13). In their method, catalytic amounts of palladium acetate having very low catalyst loading and catalytic amount of benzoquinone (BQ) were used, and tert-butyl hydroperoxide was used as effective oxidant. Broad substrate scopes and high turnover number of the catalyst are merits of this protocol. Additionally, several heterocycles including furan, indoles, etc., are coupled with olefin to produce desired heteroarylated olefin in good yields. Interestingly, this highly efficient method provides trans-olefins predominantly with high regio- and stereoselectivity.

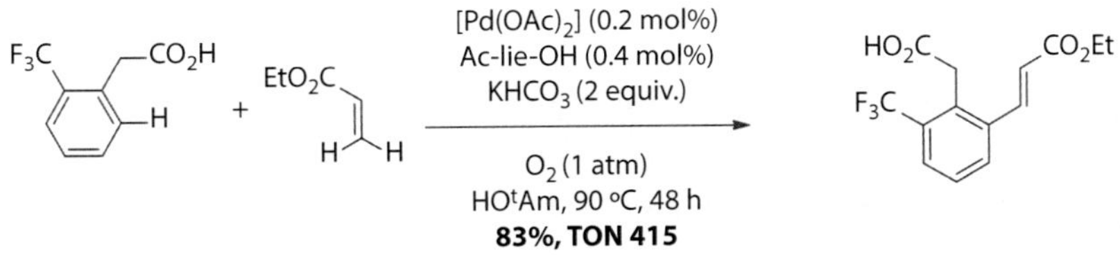

Scheme 18.12 Directed oxidative olefination of arenes catalyzed by palladium.

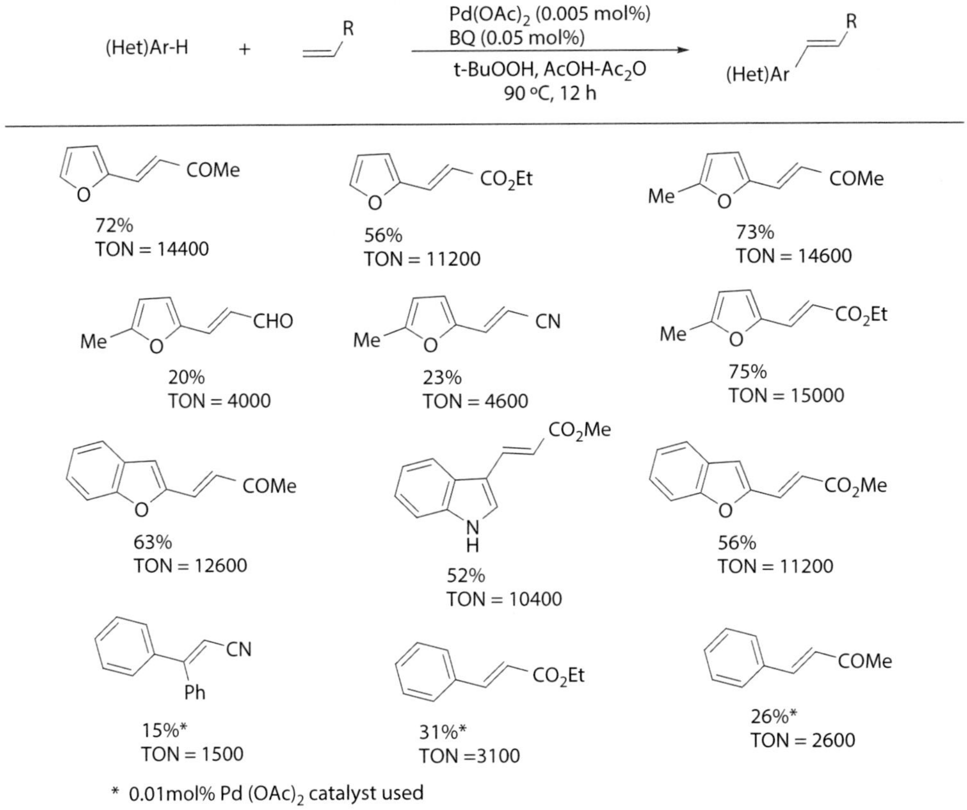

Scheme 18.13 Palladium-catalyzed oxidative olefination of arenes.

18.5 Oxidative Alkynylation Reactions

Alkynes exhibit remarkable versatility as substrate, easily undergoing diverse transformations due to the broad reactivity of triple bond toward various electrophiles and nucleophiles. Utilizing alkynes as suitable substrates, numerous natural products [37–39], biologically important molecules [40], and carbo- [41] and heterocycles [42] have been synthesized. Consequently, developing an effective and sustainable technique for alkyne synthesis has become a pivotal focus in organic chemistry [43]. The palladium-catalyzed oxidative direct C-H alkynylation of arenes and heteroarenes holds immense importance in modern synthetic chemistry due to its ability to efficiently construct carbon-carbon triple bonds precisely from C-H bonds [44, 45]. This powerful transformation provides a direct and atom-economical strategy for the synthesis of alkynylated compounds, which are valuable building blocks in the preparation of numerous biologically active molecules, pharmaceuticals, and materials. By bypassing the need for prefunctionalized substrates, the direct C-H alkynylation simplifies synthetic routes, reduces the number of steps, and minimizes production of waste materials. Palladium catalysts enable activation of C-H bond selectivity and facilitate the introduction of alkynyl groups, offering control over regioselectivity and site selectivity.

In 2013, Jie *et al.* [46] developed a valuable procedure for alkynylation reaction of various heterocyclic compounds using various terminal alkynes under oxidative conditions

(Scheme 18.14). This diverse range of functionalized thiophenes, pyrroles, and furans had successfully undergone oxidative alkynylation in this method. In this reaction, $Pd_2(dba)_3$ serves as the catalyst, while silver oxide acts as the activator in the reaction. The buffer system was used in this protocol that includes Cs_2CO_3, pivalic acid, and Et_3N. When employing different highly functionalized aryl and silyl acetylenes, the reaction yields alkynylated products in satisfactory yields.

A key feature of this protocol is the use of a low catalyst loading, specifically 0.2 mol% of the Pd catalyst. This strategic choice effectively minimizes the production of unlikely homo-coupling by-products from the terminal alkynes, enhancing the selectivity of the reaction. Moreover, the method exhibits remarkable compatibility with an extensive variety of functional groups, such as iodo, bromo, chloro, aldehyde, keto, ester, trifluoroacetyl, and alkenyl, leading to the formation of alkynylated products in good yields. The protocol's versatility is further demonstrated by its successful alkynylation of various heterocyclic substrates, including thiazoles, oxazoles, N-methylpyrazole, and N-fused imidazo[1,2-a]pyridine, yielding satisfactory yields in these cases as well. Such wide-ranging utility in organic synthesis is attributable to the protocol's straightforward experimental conditions, high tolerance toward multifarious functional groups, and the use of a low catalyst loading. These combined attributes make this methodology a valuable and efficient tool for the synthesis of an extensive array of functionalized compounds.

In a study conducted by Shibahara *et al.*, [47] a promising method for the oxidative alkynylation reaction of a variety of heteroarenes, for example, oxazole, imidazole, thiazole,

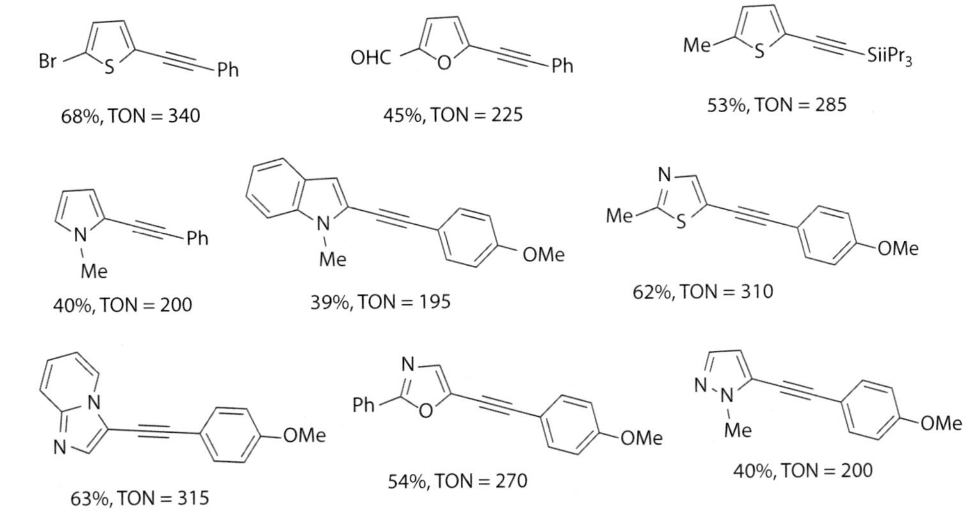

Scheme 18.14 Alkynylation of various heterocycles.

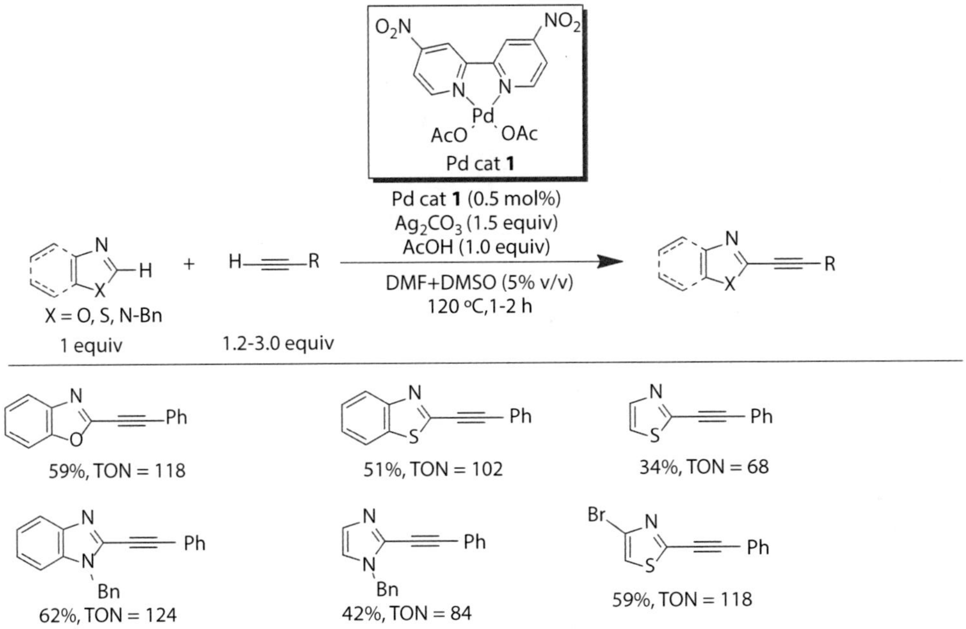

Scheme 18.15 Palladium-catalyzed alkynylation reaction of heteroarenes.

benzimidazole, benzoxazole, and benzothiazole, was demonstrated in the presence of palladium catalyst and silver carbonate as oxidant (Scheme 18.15). The researchers observed that the bromine atom present in the substrate molecule remained inert under the given reaction conditions. Furthermore, it was found that classical Sonogashira coupling product was not isolated from the crude reaction mixture, which attests to the compatibility of the CDC protocol. However, it was noted that 5-bromothiazole displayed inert behavior in this protocol, allowing for the recovery of the starting compound in quantitative yield.

In a noteworthy investigation, Shibahara *et al.* [47] explored the oxidative alkynylation reaction of imido[1,5-a]pyridine, a compound featuring two potentially reactive C-H bonds. They discovered that the reaction selectively occurred at the C3 position, leading to the formation of alkynylated products with satisfactory yields (Scheme 18.16) [47]. Interestingly, it has been found that bromo group is highly tolerable under this protocol, opening the possibility for further functionalization through various cross-coupling reactions.

In 2015, Parsharamulu *et al.* [48] introduced a novel approach to directly alkynylate heteroarenes using terminal alkynes. The method involved the utilization of an anionic/amidocarbocyclic carbene Pd(II) complex as a suitable precatalyst (Scheme 18.17) [48]. Through their investigation, the researchers achieved successful alkynylation of various azoles in heteroarenes, leading to the formation of the desired alkynylated products in good yields. To facilitate the reaction, they employed Ag_2O, which played a crucial role in oxidizing Pd(0) to Pd(II) through the reductive elimination step. This innovative protocol holds promise for the efficient synthesis of functionalized heteroarenes, potentially opening up new avenues in the field of organic synthesis.

It was observed that benzoxazoles, containing a halogen atom in the benzene ring, also participated in this double C-H functionalization reaction, providing the expected

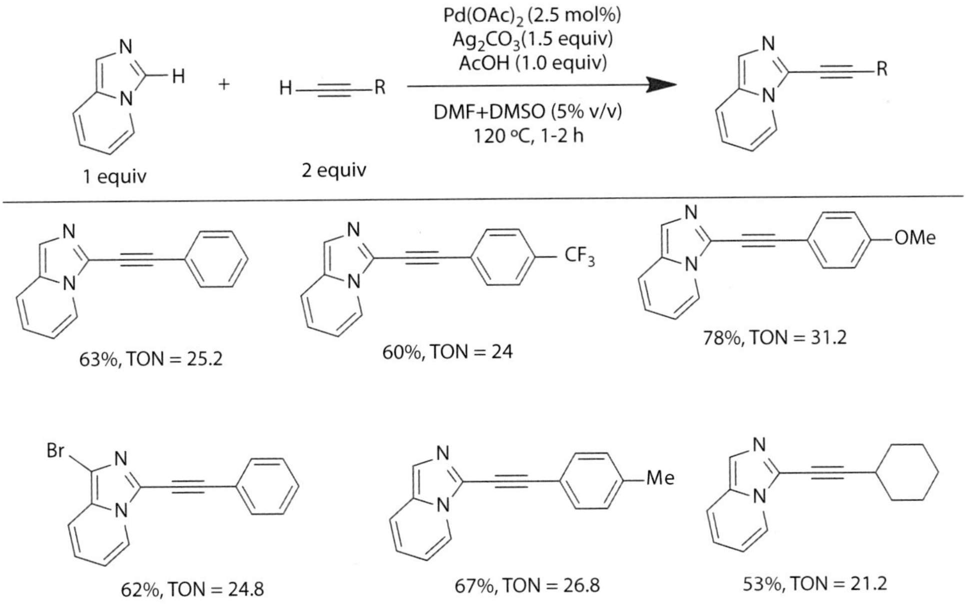

Scheme 18.16 Palladium-catalyzed alkynylation reaction of imido[1,5-a]pyridines.

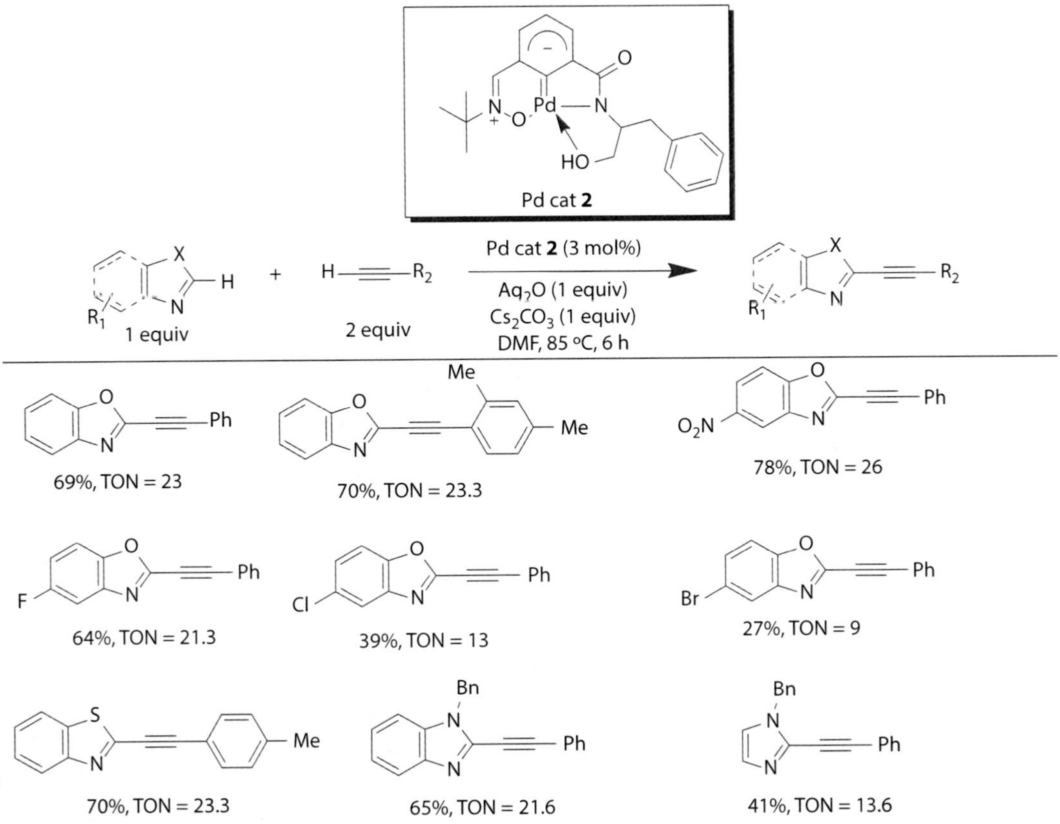

Scheme 18.17 Alkynylation of heteroarenes.

alkynylated products. Notably, their reactivity followed the trend F > Cl > Br, with fluorine exhibiting the highest reactivity. A fascinating observation emerged from the study, as electron-rich terminal alkynes displayed significantly higher yields when reacting with azoles, in contrast to the usage of electron-deficient alkynes. However, the reaction between benzoxazoles and aliphatic terminal alkynes did not lead to the formation of the intended products, suggesting limitations in this specific combination. On the other hand, benzothiazole, N-benzyl benzamidazole, and N-benzyl imidazole exhibited excellent reactivity with alkynes, proving highly effective in producing the desired alkynylated products. These findings shed light on the influence of electronic factors in the alkynylation process and highlight the specific substrate combinations that lead to successful outcomes, providing valuable insights for future reaction design and optimization.

Recently, Ha *et al.* [49] introduced a graceful method for the palladium-catalyzed cross-dehydrogenative alkynylation of C-H bond of nitro-pyrazoles with alkynes, resulting in the formation of alkynylated products with good yields (Scheme 18.18). In this protocol, 2 mol% of $Pd(OAc)_2$ was utilized as the catalyst, PPh_3 as the effective ligand, in the presence of additive isobutyric acid. Additionally, Ag_2CO_3 was used as the activator and Cs_2CO_3 as the base. An interesting observation was made when excess (three equivalents) amounts of terminal alkynes were employed. Under these conditions, a regio- and stereoselective hydroarylation reaction of nitropyrazoles occurred as the major product. Thus, the proper equilibrium of the ratio of the catalysts, alkynes, and oxidants was essential for this unusual

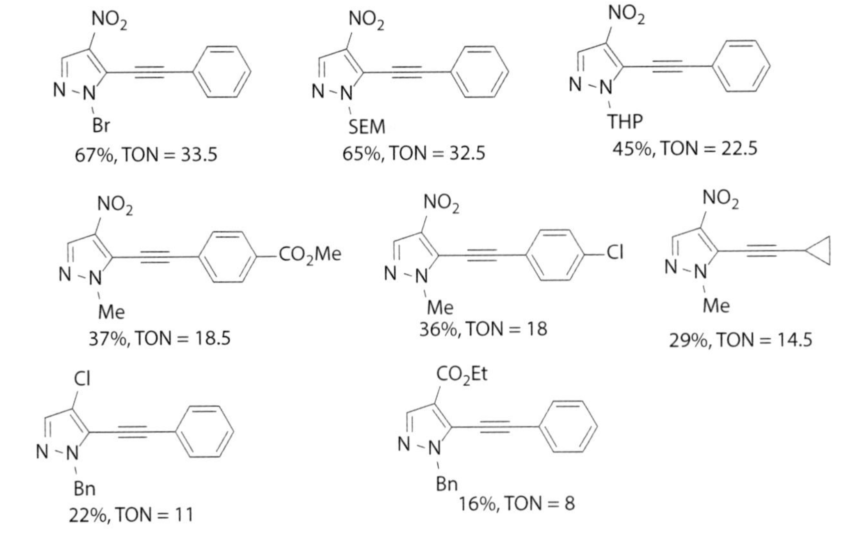

Scheme 18.18 Alkynylation of pyrazoles.

sequencing process, which included the homo-dimerization reaction of terminal alkynes and hydroarylation.

18.6 Tandem C–H and N–H Activation

In 2016, Arumugam *et al.* [50] described a graceful method for the synthesis of functionalized carbazoles via a one-pot tandem process involving C–H and N–H activation reaction of acetanilides with aryl boronic acids (Scheme 18.19). This transformation takes place under aerobic conditions, utilizing an NNO pincer-type palladium(II) complex as a catalyst. Remarkably, this reaction can be carried out in neat water with an exceptionally low catalyst loading of just 0.01 mol%. Moreover, the catalyst exhibits impressive recyclability, allowing for up to six consecutive runs without any loss in efficiency. Furthermore, the title heterocycles can be isolated in excellent yields, and the process eliminates the need for column chromatography.

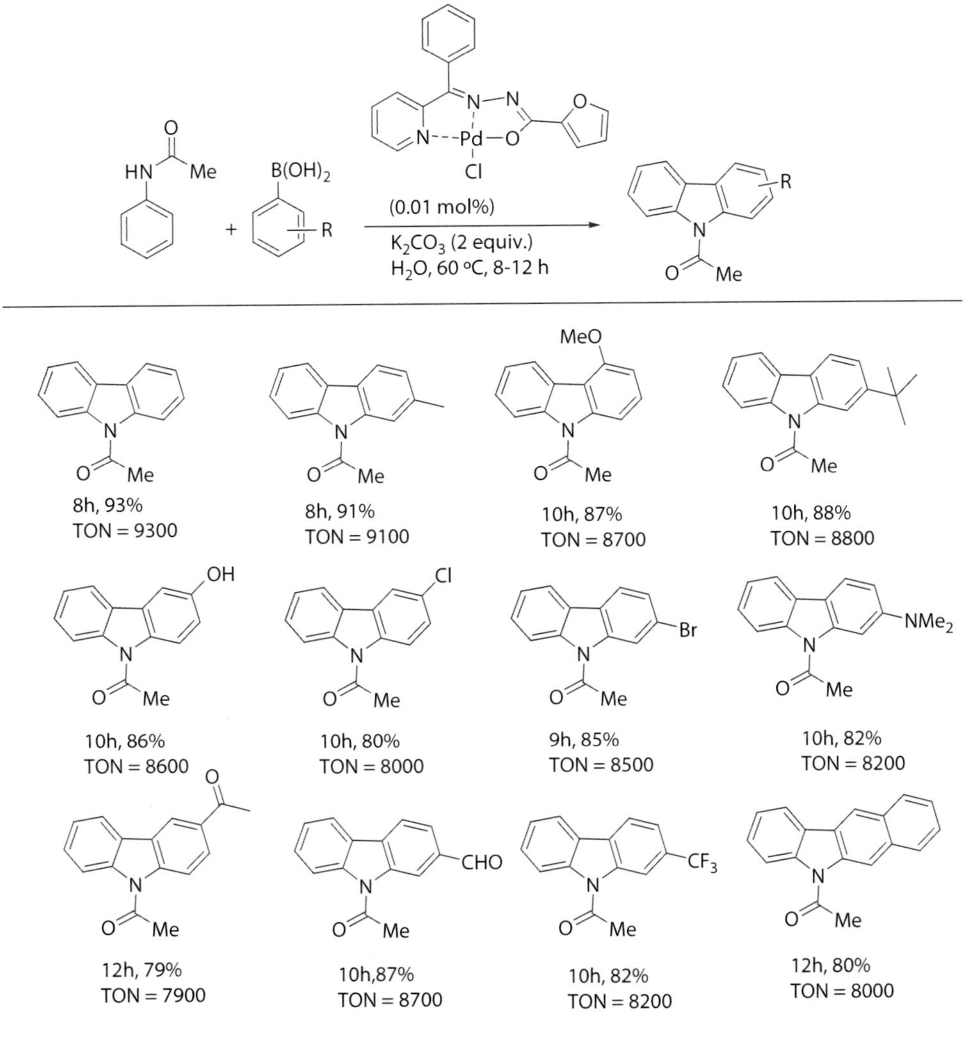

Scheme 18.19 Palladium-catalyzed synthesis of functionalized carbazoles.

In summary, we can consider several strategies to enhance catalyst efficiency in C-H activation reactions such as the following:

1. Catalyst Design and Ligand Optimization: The design and selection of appropriate catalysts and ligands play a vital role in improving efficiency. Rational design andoptimization of catalyst structures allow for enhanced reactivity and selectivity. Ligand modifications can fine-tune the electronic and also steric properties of the metal catalyst, influencing its ability for the C-H activation and control regioselectivity. Developing ligands with improved chelating ability and directing effects can enhance catalyst efficiency.
2. Catalyst Immobilization and Supported Systems: Immobilization of catalysts onto solid supports or incorporating them into heterogeneous systems can improve their efficiency. Supported catalysts offer several advantages, including easier catalyst separation and recycling, increased stability, and reduced catalyst leaching. Immobilization strategies can enhance catalyst performance by providing a controlled environment, improving mass transfer, and minimizing deactivation.
3. Cooperative Catalysis and Bimetallic Systems: Cooperative catalysis involving the synergistic interaction between two or more catalysts can significantly enhance efficiency. Combining different metal catalysts or ligands can enable new reactivity, enhance catalyst stability, and control selectivity. Bimetallic systems exploit the unique properties of each metal to achieve higher catalytic efficiency and selectivity.
4. Reaction Conditions and Solvent Optimization: Modifying reaction conditions, such as temperature, pressure, and solvent choice, can influence catalyst efficiency. Optimization of these parameters can accelerate reaction rates, enhance selectivity, and improve catalyst performance. The choice of an appropriate solvent can influence substrate solubility, catalyst activity, and stability, thereby impacting efficiency.
5. Catalyst Recycling and Sustainability: Developing catalyst systems that can be easily recovered and reused is essential for sustainable catalysis. Implementing recycling strategies minimizes catalyst waste and reduces costs. Efficient catalyst recovery methods, such as immobilization techniques or solid-phase extraction, contribute to overall process efficiency.
6. Computational Approaches: Computational methods, for example, density functional theory (DFT), have become valuable tools for understanding reaction mechanisms and catalyst–substrate interactions. Computational modeling allows for the prediction and optimization of catalyst efficiency by providing insights into transition states, reaction energetics, and selectivity-determining factors.
7. Directed Evolution and Catalyst Discovery: Directed evolution and high-throughput screening methods have been employed to develop new catalysts with improved efficiency. These techniques enable the screening of large catalyst libraries and the identification of catalyst variants with enhanced reactivity and selectivity.

Thus, efforts to increase catalyst efficiency in C-H activation process are driven by the desire to advance synthetic methodologies, improve sustainability, and enable the synthesis of complex molecules efficiently. Continued research in catalyst design, ligand development, reaction optimization, and computational modeling will contribute to further enhancing the efficiency of C-H activation reactions.

18.7 Conclusions

Highly efficient C-H activation or functionalization reactions have revolutionized the field of organic synthesis by providing a powerful tool for selective and direct functionalization of inert carbon-hydrogen bonds. These reactions offer several advantages, including streamlined synthetic routes, atom economy, and reduced waste generation. The development of efficient catalysts, such as transition metals like palladium, has enabled the activation of a widespread range of C-H bonds at gentle reaction conditions. The high catalytic efficiency of these reactions is evident in their ability to deliver the desired functionalization with high yields and regioselectivity, often requiring only low catalyst loadings. Additionally, the compatibility of C-H activation with a diversity of functional groups further enhances its utility in complex molecule synthesis. Highly efficient C-H activation reactions have been fruitfully applied in the synthesis of pharmaceuticals, agrochemicals, and materials, among other areas. Ongoing research efforts are focused on expanding the substrate scope, improving selectivity, and developing sustainable catalytic systems to further enhance the efficiency and applicability of C-H activation reactions. It is believed that this chapter will serve as a motivating factor for scientists to explore catalyst efficiency, thereby facilitating the increased adoption of C-H activations in the industry.

References

1. (a) Dalton, T., Teresa Faber, T., Glorius, F., C–H Activation: Toward Sustainability and Applications, *ACS Cent. Sci.*, 7, 2, 245–261, 2021. (b) Rogge, T., Kaplaneris, N., Chatani, N., C–H activation. *Nat. Rev. Methods Primers.* 1, 43, 2021. (c) Cernak, T., Dykstra, K.D., Tyagarajan, S., Vachal, P., Krska, S.W., The medicinal chemist's toolbox for late stage functionalization of drug-like molecules, *Chem. Soc. Rev.*, 45, 546–576, 2016. (d) Davies, H. M. L., Morton, D., Recent Advances in C–H Functionalization, *J. Org. Chem.*, 81, 343–350, 2016.
2. (a) Wencel-Delord, J.; Glorius, F., C–H bond activation enables the rapid construction and late-stage diversification of functional molecules, *Nat. Chem.*, 5, 369–375, 2013. (b) Schönherr, H., Cernak, T., Profound Methyl Effects in Drug Discovery and a Call for New C-H Methylation Reactions, *Angew. Chem. Int. Ed.*, 52, 47, 12256–12267, 2013.
3. (a) Carr, K.J.T., Macgregor, S.A., McMullin, C.L., Computational Studies on Heteroatom-Assisted C–H Activation and Functionalisation at Group 8 and 9 Metal Centres, *Top. Organomet. Chem.*, 48, 53–76, 2015. (b) Davies, D.L., Macgregor, S.A., McMullin, C.L., Computational Studies of Carboxylate-Assisted C–H Activation and Functionalization at Group 8–10 Transition Metal Centers, *Chem. Rev.*, 117, 8649–8709, 2017.
4. (a) Negishi, E., Palladium- or nickel-catalyzed cross coupling. A new selective method for carbon-carbon bond formation, *Acc. Chem. Res.*, 15, 340–348, 1982. (b) Miyaura, N., Suzuki, A., Palladium-Catalyzed Cross-Coupling Reactions of Organoboron Compounds, *Chem. Rev.*,

95, 2457–2483, 1995.;(c) Hassan, J., Sévignon, M., Gozzi, C., Schulz, E., Lemaire, M., Aryl–Aryl Bond Formation One Century after the Discovery of the Ullmann Reaction, *Chem. Rev.*, 102, 1359–1470, 2002.

5. Farina, V., High-Turnover Palladium Catalysts in Cross-Coupling and Heck Chemistry: A Critical Overview, *Adv. Synth. Catal.*, 346, 1553–1582, 2004.
6. (a) Wu, Y., Wang, J., Mao, F., Kwong, F.Y., Palladium-Catalyzed Cross-Dehydrogenative Functionalization of C(sp2)-H Bonds, Chem. Asian J., 9, 26–47, 2014; (b) Li, C.-J. (Ed.), From C-H to C-C Bonds: Cross-Dehydrogenative Coupling, Royal Society Of Chemistry, Cambridge, 2014; (c) Liu, C., Yuan, J., Gao, M., Tang, S., Li, W., Shi, R., Lei, A., Oxidative Coupling between Two Hydrocarbons: An Update of Recent C–H Functionalizations, *Chem. Rev.*, 115, 12138–12204, 2015.
7. Daugulis, O., Zaitsev, V.G., Anilide ortho-Arylation by Using C-H Activation Methodology, *Angew. Chem. Int. Ed.*, 44, 4046–4048, 2005.
8. Chen, Z., Wang, B., Zhang, J., Yu, W., Liu, Z., Zhang, Y., Transition metal-catalyzed C–H bond functionalizations by the use of diverse directing groups, *Org. Chem. Front.*, 2, 1107–1295, 2015.
9. Kim, J., Hong, S.H., Ligand-Promoted Direct C–H Arylation of Simple Arenes: Evidence for a Cooperative Bimetallic Mechanism, *ACS Catal.*, 7, 3336–3343, 2017.
10. (a) Roger, J., Pozgan, F., Doucet, H., Ligand-Free Palladium-Catalyzed Direct Arylation of Thiazoles at Low Catalyst Loadings, *J. Org. Chem.*, 74, 1179–1186, 2009. (b) Roger, J., Požgan, F., Doucet, H., Ligand-less palladium-catalyzed direct 5-arylation of thiophenes at low catalyst loadings, Green Chem., 11, 425-432, 2009. (c) Fall, Y., Reynaud, C., Doucet, H., Santelli, M., Ligand-Free-Palladium-Catalyzed Direct 4-Arylation of Isoxazoles Using Aryl Bromides, *Eur. J. Org. Chem.*, 4041–4050, 2009. (d) Roger, J., Doucet, H., Regioselective C-2 or C-5 Direct Arylation of Pyrroles with Aryl Bromides using a Ligand-Free Palladium Catalyst, *Adv. Synth. Catal.*, 351, 1977–1990, 2009. (e) Roger, J., Doucet, H., Phosphine-free palladium-catalysed direct 5-arylation of imidazole derivatives at low catalyst loading, *Tetrahedron*, 65, 9772–9781, 2009. (f) Zhao, L., Bruneau, C., Doucet, H., Phosphine-free palladium-catalysed direct C2-arylation of benzothiophenes with aryl bromides, *Tetrahedron*, 69, 7082–7089, 2013.
11. Bheeter, C.B., Chen, L., Soulé, J.-F., Doucet, H., Regioselectivity in palladium-catalysed direct arylation of 5-membered ring heteroaromatics, *Catal. Sci. Technol.*, 6, 2005–2049, 2016.
12. Kaloni, T.P., Giesbrecht, P.K., Schreckenbach, G., Freund, M. S., Polythiophene: From Fundamental Perspectives to Applications, *Chem. Mater.*, 29, 24, 10248–10283, 2017.
13. Pouliot, J.-R., Grenier, F., Blaskovits, J.T., Beaupré, S., Leclerc, M., Direct (Hetero)arylation Polymerization: Simplicity for Conjugated Polymer Synthesis, *Chem. Rev.*, 116, 14225–14274, 2016.
14. Rudenko, A.E., Khlyabich, P.P., Thompson, B.C., Random Poly(3-hexylthiophene-co-3-cyanothiophene) Copolymers via Direct Arylation Polymerization (DArP) for Organic Solar Cells with High Open-Circuit Voltage, *ACS Macro Lett.*, 3, 387–392, 2014.
15. Panda, B., Sarkar, T.K., A one-pot tandem oxidation–reduction protocol for the synthesis of cyclic ethers from their diols, *Tetrahedron Lett.*, 49, 47, 6701–6703, 2008.
16. Campeau, L.C., Parisien, M., Leblanc, M., Fagnou, K., Biaryl Synthesis via Direct Arylation: Establishment of an Efficient Catalyst for Intramolecular Processes, *J. Am. Chem. Soc.*, 126, 9186–9187, 2004.
17. Dai, L.-X. and Hou, X.-L. (Eds.), *Chiral Ferrocenes in Asymmetric Catalysis*, Wiley, New York, 2010.
18. (a) Fu, G.C., Asymmetric Catalysis with “Planar-Chiral” Derivatives of 4-(Dimethylamino)pyridine, *Acc. Chem. Res.*, 37, 542–547, 2004. (b) Arrayas, R.G., Adrio, J., Carretero, J. C., Recent Applications of Chiral Ferrocene Ligands in Asymmetric Catalysis *Angew. Chem. Int. Ed.*, 45,

46, 7674–7715, 2006. (c) Fu, G.C., Applications of Planar-Chiral Heterocycles as Ligands in Asymmetric Catalysis, *Acc. Chem. Res.*, 39, 11, 853–860, 2006. (d) Arae, S., Ogasawara, M., Catalytic Asymmetric Synthesis of Planar-Chiral Transition-Metal Complexes, *J. Synth. Org. Chem.*, 70, 593-605, 2012.

19. (a) Dendele, N., Bisaro, F., Gaumont, A.-C., Perrio, S., Richards, C.J., Synthesis of a [2.2]-paracyclophane based planar chiral palladacycle by a highly selective kinetic resolution/C–H activation reaction, *Chem. Commun.*, 48, 1991, 2012. (b) Takebayashi, S., Shizuno, T., Otani, T., Shibata, T., Rh(III)-catalyzed directed C–H bond amidation of ferrocenes with isocyanates, *Beilstein J. Org. Chem.*, 8, 1844–1848, 2012. (c) Singh, K.S., Dixneuf, P.H., Ruthenium(II)-Catalyzed Alkenylation of Ferrocenyl Ketones via C–H Bond Activation, *Organometallics*, 31, 7320–7323, 2012. (d) Piotrowicz, M., Zakrzewski, J., erobic Dehydrogenative Heck Reaction of Ferrocene with a $Pd(OAc)_2$/4,5-Diazafluoren-9-one Catalyst, *Organometallics*, 32, 20, 5709–5712, 2013.
20. Bolm, C., Kesselgruber, M., Muñiz, K., Raabe, G., Diastereoselective Synthesis of Ferrocenyl Sulfoximines with Planar and Central Chirality, Organometallics, 19, 9, 1648–1651, 2000. (b) Bolm, C., Kesselgruber, M., Raabe, G., The First Planar-Chiral Stable Carbene and Its Metal Complexes, *Organometallics*, 21,4, 707–710, 2002.
21. (a) Laufer, R.S., Veith, U., Taylor, N.J., Snieckus, V., C2-Symmetric Planar Chiral Ferrocene Diamides by (–)-Sparteine-Mediated Directed ortho-Lithiation. Synthesis and Catalytic Activity, *Org. Lett.*, 2, 629–631, 2000. (b) Genet, C., Canipa, S.J., O'Brein, P., Taylor, S., Catalytic Asymmetric Synthesis of Ferrocenes and P-Stereogenic Bisphosphines, *J. Am. Chem. Soc.*, 128, 9336, 2006.
22. (a) Alba, A. -N. R.; Rios, R., Kinetic Resolution: A Powerful Tool for the Synthesis of Planar-Chiral Ferrocenes, *Molecules*, 14, 4747, 2009. (b) Bueno, A., Rosol, M., García, J., Moyano, A., Asymmetric Dihydroxylation of 2-Substituted 1-Vinylferrocenes: The First Non-Enzymatic Kinetic Resolution of Planar-Chiral Ferrocenes, *Adv. Synth. Catal.*, 348, 2590–2596, 2006. (c) Alba, A.-N., Gomez- Sal, P., Rios, R., Moyano, A., Organocatalytic kinetic resolution of a planar-chiral ferrocenecarbaldehyde, *Tetrahedron: Asymmetry*, 20, 1314–1318, 2009.
23. (a) Gao, D.W., Yin, Q., Gu, Q., You, S.L., Enantioselective Synthesis of Planar Chiral Ferrocenes via Pd(0)-Catalyzed Intramolecular Direct C–H Bond Arylation, *J. Am. Chem. Soc.*, 136, 4841–4844, 2014. (b) Gao, D.-W., Zheng, C., Gu, Q., You, S.-L., Pd-Catalyzed Highly Enantioselective Synthesis of Planar Chiral Ferrocenylpyridine Derivatives, *Organometallics*, 34, 4618–4625, 2015.
24. Shi, X., Zhang, J., Roisnel, T., Soulé, J.-F., Doucet, H., Palladium-Catalyzed Direct Diarylation of 2-Benzyl-1,2,3-triazole: a Simple Access to 4-Aryl- or 4,5-Diaryl-2-benzyl-1,2,3-triazoles and Phenanthro[9,10-d][1,2,3]triazoles, *Eur. J. Org. Chem.*, 17, 2375–2382, 2021.
25. Janody, S., Jazzar, R., Comte, A., Holstein, P.M., Vors, J.P., Ford, M.J., Baudoin, O., Synthesis of 1-Indanols and 1-Indanamines by Intramolecular Palladium(0)-Catalyzed $C(sp^3)$-H Arylation: Impact of Conformational Effects, *Chem. Eur. J.*, 20, 11084–11090, 2014.
26. (a) Li, C.-J., Cross-Dehydrogenative Coupling (CDC): Exploring C–C Bond Formations beyond Functional Group Transformations, *Acc. Chem. Res.*, 42, 2, 335–344, 2009. (b) Tian, T., Li, Z., Li, C.-J., Cross-dehydrogenative coupling: a sustainable reaction for C–C bond formations, *Green Chem.*, 23, 6789–6862, 2021. (c) Peng, K.; Dong, Z.-B., Recent Advances in Cross-Dehydrogenative Couplings (CDC) of C–H Bond in Aqueous Media, *Adv. Synth. Catal.*, 363, 5, 1185–1201, 2021.
27. van Helden, R., Verberg, G., The oxidative coupling of aromatic compounds with palladium salts, *Recl. Trav. Chim. Pays-Bas*, 84, 1263–1273, 1965.
28. (a) Yeung, C.S., Dong, V.M., Catalytic Dehydrogenative Cross-Coupling: Forming Carbon–Carbon Bonds by Oxidizing Two Carbon–Hydrogen Bonds, *Chem. Rev.*, 111, 1215–1292, 2011.

(b) Bugaut, X., Glorius, F., Palladium-Catalyzed Selective Dehydrogenative Cross-Couplings of Heteroarenes, *Angew. Chem. Int. Ed.*, 50, 7479–7481, 2011. (c) Cho, S.H., Kim, J.Y., Kwak, J., Chang, S., Recent advances in the transition metal-catalyzed twofold oxidative C–H bond activation strategy for C–C and C–N bond formation, *Chem. Soc. Rev.*, 40, 5068–5083, 2011. (d) Stuart, D.R.; Fagnou, K., The Discovery and Development of a Palladium(II)-Catalyzed Oxidative Cross-Coupling of Two Unactivated Arenes, *Top. Organomet. Chem.*, 44, 91–119, 2013. (e) Yang, Y., Lan, J., You, J., Oxidative C–H/C–H Coupling Reactions between Two (Hetero)arenes, *Chem. Rev.*, 117, 8787–8863, 2017.

29. (a) Baslé, O., From C-H to C-C Bonds: Cross-Dehydrogenative Coupling, C. -J. Li (Ed.), pp. 197–218, Royal Society Of Chemistry, Cambridge, 2015. (b) Gulzar, N., Schweitzer-Chaput, B., Klussmann, M., Oxidative coupling reactions for the functionalisation of C–H bonds using oxygen, *Catal. Sci. Technol.*, 4, 2778–2796, 2014; c) Wang, D., Weinstein, A.B., White, P.B., Stahl, S.S., Ligand-Promoted Palladium-Catalyzed Aerobic Oxidation Reactions, *Chem. Rev.*, 118, 5, 2636–2679, 2018.
30. (a) Izawa, Y.; Stahl, S.S., Aerobic Oxidative Coupling of o-Xylene: Discovery of 2-Fluoropyridine as a Ligand to Support Selective Pd-Catalyzed C-H Functionalization, *Adv. Synth. Catal.*, 352, 3223–3229, 2010. (b) Wang, D., Izawa, Y., Stahl, S.S., Pd-Catalyzed Aerobic Oxidative Coupling of Arenes: Evidence for Transmetalation between Two Pd(II)-Aryl Intermediates, *J. Am. Chem. Soc.*, 136, 9914–9917, 2014.
31. Wang, D., Stahl, S.S., Pd-Catalyzed Aerobic Oxidative Biaryl Coupling: Non-Redox Cocatalysis by Cu(OTf)2 and Discovery of Fe(OTf)3 as a Highly Effective Cocatalyst, *J. Am. Chem. Soc.*, 139,16, 5704–5707, 2017.
32. (a) Panda, B., Basak, S., Hazra, A., Sarkar, T.K., A domino Michael-Dieckmann-Peterson approach to the synthesis of substituted hydroxyquinolines and hydroxyisoquinolines, *J. Chem. Res.*, 34, 109–113, 2010. (b) Panda, B.; Gianluigi, A., DMF as CO Surrogate in Carbonylation Reactions: Principles and Application to the Synthesis of Heterocycles, *Catalysts*, 11, 12, 1531, 2021. (c) Panda, B., Gianluigi, A., Synthetic Methods for the Preparation of Conformationally Restricted Analogues of Nicotine, *Molecules*, 26, 24, 7544, 2021. (d) Peng, J. J; Panda, B.; Satyanarayana, K., Yang, H-R., Huang, S. L., Huang, M., Chen, C. H., Lai, G., Lai, Y. Y., Luh, T. Y., Stereospecific Synthesis of Poly(methylene-E-vinylene) by Ring Opening Metathesis Polymerization (ROMP) of Substituted Cyclopropene Using Grubbs *Catalysts, Macromolecules*, 52, 7749–7755, 2019.
33. Huet, L., Forke, R., Jäger, A., Knölker, H.-J., Transition Metals in Organic Synthesis, Part 100: Highly Efficient Palladium(II)-Catalyzed Oxidative Cyclization to the 1,7,8-Trioxygenated Carbazole Alkaloid Murrayastine, *Synlett*, 23, 1230–1234, 2012.
34. Jia, C., Lu, W., Kitamura, T., Fujiwara, Y., Highly Efficient Pd-Catalyzed Coupling of Arenes with Olefins in the Presence of tert-Butyl Hydroperoxide as Oxidant, *Org. Lett.*, 1, 2097–2100, 1999.
35. (a) Wang, D.-H., Engle, K.M., Shi, B.-F., Yu, J.-Q., Ligand-Enabled Reactivity and Selectivity in a Synthetically Versatile Aryl C–H Olefination, *Science*, 327, 315–319, 2010. (b) Engle, K.M.,Wang, D.H., Yu, J.-Q., Ligand-Accelerated C–H Activation Reactions: Evidence for a Switch of Mechanism, *J. Am. Chem. Soc.*, 132, 14137–14151, 2010. (c) Baxter, R.D., Sale, D., Engle, K.M., Yu, J.-Q., Blackmond, D.G., Mechanistic Rationalization of Unusual Kinetics in Pd-Catalyzed C–H Olefination, *J. Am. Chem. Soc.*, 134, 4600–4606, 2012.
36. Jia, C., Lu, W., Kitamura, T., Fujiwara, Y., Highly Efficient Pd-Catalyzed Coupling of Arenes with Olefins in the Presence of tert-Butyl Hydroperoxide as Oxidant, *Org. Lett.*, 1, 13, 2097–2100, 1999.

37. Panda, B.; and Sarkar, T.K., Gold Catalysis: Regio- and Stereoselective Total Synthesis of Xyloketals D and G and the Related Natural Product Alboatrin, *J. Org. Chem.*, 78, 2413–2421, 2013.
38. Panda, B., Total Synthesis of Xyloketals and Related Natural Product Alboatrin: Strategies and Tactics, *ChemistrySelect*, 4, 9143–9164, 2019.
39. Panda, B., Gooyee, A.K., Bioactivity of Marine Natural Product Xyloketals, *Lett. Org. Chem.*, 18, 507–512, 2021.
40. Panda, B., Total synthesis of Bruguierols: Strategies and tactics, *Arkivoc, i*, 293–303, 2019.
41. Panda, B., Small-Ring Metathesis: Delights and Difficulties, *Asian J. Org. Chem.*, 7, 2386–2396, 2018.
42. (a) Panda, B., Bhadra, J., Sarkar, T.K., An Approach to Highly Functionalized Quinolines and Isoquinolines via a Gold-Catalyzed Benzannulation, *Synlett*, 5, 689–693, 2011. (b) Panda, B., Sarkar, T.K., Gold-Catalyzed Benzannulation of Electronically Rich/Rich and Deficient/Deficient Oxoalkynes with Alkynes, *Synthesis*, 45, 1227–1234, 2013. (c) Panda, B., Use of Gold Nanoparticles in the Synthesis of Heterocyclic Compounds, *Lett. Org. Chem.*, 20, 18–27, 2023.
43. (a) Panda, B., Sarkar, T.K., On the catalytic duo $PdCl_2(PPh_3)_2/AuCl(PPh_3)$ that cannot effect a Sonogashira-type reaction: a correction, *Tetrahedron Lett.*, 51, 301–305, 2010. (b) Panda, B., Sarkar, T.K., Gold and Palladium Combined for the Sonogashira Coupling of Aryl and Heteroaryl Halides, *Synthesis*, 45, 817–829, 2013. (c) Panda, B., Sarkar, T.K., Gold and palladium combined for the Sonogashira-type cross-coupling of arenediazonium salts, *Chem. Commun.*, 46, 3131–3133, 2010. (d) Panda, B., Sonogashira coupling of arenediazonium salts: discovery and developments, *Arkivoc, ix*, 177–199, 2021. (e) Panda, B., Catalytic alkynylations through the activation of C-N bonds *Arkivoc, ii*, 1–36, 2023. https://doi.org/10.24820/ark.5550190.p012.011
44. Panda, B., Joy and Challenges of Alkynylation of Arenes and Heteroarenes through Double C–H Functionalizations, *Asian J. Org. Chem.*, 9, 492–507, 2020.
45. Panda, B., Progress in Catalytic Decarboxylative Oxidative C-H Alkynylation, *Curr. Catal.*, 11, 2–15, 2022.
46. Jie, X., Shang, Y., Hu, P., Su, W., Palladium-Catalyzed Oxidative Cross-Coupling between Heterocycles and Terminal Alkynes with Low Catalyst Loading, *Angew. Chem. Int. Ed.*, 52, 3630–3633, 2013.
47. Shibahara, F., Dohke, Y., Murai, T., Palladium-Catalyzed C–H Bond Direct Alkynylation of 5-Membered Heteroarenes: A Well-Defined Synthetic Route to Azole Derivatives Containing Two Different Alkynyl Groups, *J. Org. Chem.*, 77, 5381–5388, 2012.
48. Parsharamulu, T., Reddy, P.V., Likhar, P.R., Lakshmi Kantam, M., Dehydrogenative and decarboxylative C–H alkynylation of heteroarenes catalyzed by Pd(II)–carbene complex, *Tetrahedron*, 71, 1975–1981, 2015.
49. Ha, H., Shin, C., Bae, S., Joo, J.M., Divergent Palladium-Catalyzed Cross-Coupling of Nitropyrazoles with Terminal Alkynes, *Eur. J. Org. Chem.*, 2018, 2645–2650, 2018.
50. Arumugam, V., Kaminsky, W., Nallasamy, D., Pd(ii) pincer type complex catalyzed tandem C–H and N–H activation of acetanilide in aqueous media: a concise access to functionalized carbazoles in a single step, *Green Chem.*, 18, 3295–3301, 2016.
51. Dams, M., De Vos, D.E., Celen, S., Jacobs, P.A., Toward Waste-Free Production of Heck Products with a Catalytic Palladium System under Oxygen, *Angew. Chem. Int. Ed.*, 42, 3512–3515, 2003.

19

Thin-Film Fabrication Techniques in Dye-Sensitized Solar Cells for Energy Harvesting

Aman Kumar[1], Anamika Chaudhari[1], Sudhanshu Kumar[1], Suman Kushwaha[1*] and Sudip Mandal[2†]

[1]*Department of Chemistry, Institute of Science, Banaras Hindu University, Varanasi, Uttar Pradesh, India*

[2]*Department of Chemistry, School of Applied Sciences and Humanities, Vignan's Foundation for Science, Technology and Research (Deemed to be University), Vadlamudi, Guntur, Andhra Pradesh, India*

Abstract

From the last few decades, dye-sensitized solar cells (DSSCs) have emerged out as economical and reliable substitutes for energy source owing to their simple preparation, cheap cost, low toxicity, and competitive efficiency. The key components of DSSCs include the semiconducting thin film, sensitizer, redox couple, and counter electrode. The photoanode, which is made of a semiconducting thin film with a dye adsorbed on it, plays a crucial role in DSSCs, and its morphology depends on the deposition techniques used. Various established physical and chemical coating/deposition methods are promising for attaining high-quality thin films for DSSC fabrication. The morphology and particle size of the thin films strongly depend on the deposition techniques employed. Surface and interfacial effects significantly impact the whole performance of these films, leading to improved properties and characteristics compared to bulk materials. Highly functionalized thin films have their benefits when used as photoanodes in DSSCs, as the boosted functional properties of materials can be exploited at the nanoscale level. There are several techniques for depositing thin films, including physical, chemical, and colloidal routes. Some widely used methods include spray pyrolysis, sol-gel, plasma spray, electrophoretic deposition, screen printing, and dip coating. Vacuum deposition techniques such as magnetron sputtering and pulsed laser deposition are successfully established in the thin-film coating industry due to the high quality and reproducibility of the fabricated thin films. While these techniques often require expensive equipment and complex experimental setups, they offer the advantage of producing thin and dense films. Colloidal, doctor blade, and dip coating techniques are advantageous because of their low cost and simplified experimental setup and equipment. These techniques produce thin films with controlled structures and particle size distributions. However, they may be prone to issues such as cracks, pores, or layer separation, which can be overcome by repeating the coating process several times intermittently. Sol-gel and slurry methods are extensively used to fabricate nanocrystalline thin films in DSSCs due to their lower cost than gas-phase methods. However, the meager stability of the precursors against hydrolysis remains a concern. The SILAR (successive ionic layer adsorption and reaction) method is another solution-based thin-film

**Corresponding author*: sumankush@bhu.ac.in

†*Corresponding author*: drs_sh@vignan.ac.in

Mousumi Sen (ed.) Sustainable Green Catalytic Processes, (473–520) © 2024 Scrivener Publishing LLC

deposition technique, a modified version of chemical bath deposition. The resulting thin films exhibit improved grain structure orientation. In short, various deposition techniques are employed to fabricate thin films for DSSCs, each with its advantages and challenges. Researchers continue to explore and optimize these techniques/methods to improve the overall performance of DSSCs.

Keywords: Thin films, colloidal process, gas-phase processes, liquid-phase processes, sputtering, PLD, SILAR

19.1 Introduction

19.1.1 Energy Crisis and the World Scenario

Energy is indeed a fundamental requirement for human survival and development, but the current supply is inadequate to meet the increasing demands. Conventional energy resources like fossil fuels, hydroelectric power, and nuclear power have traditionally met the world's energy needs. However, these resources are finite and depleting quickly and have significant environmental concerns. Fossil fuels, including coal, natural gas, and oil, have been the dominant sources of power for several decades. Their excessive consumption produces greenhouse gases and air pollution and contributes to global warming and ozone layer depletion [1]. While offering a high energy density, nuclear power poses concerns about releasing radioactive material and managing nuclear waste. In light of the issues associated with conventional energy sources, the focus has shifted toward renewable energy sources that are nonpolluting and economically viable. Renewable energy sources have several advantages, including their perennial nature, local availability, decentralized applications, and suitability for use in remote areas. They offer a viable and environmentally friendly option to conventional energy sources. Sun energy, in particular, has immense potential, total solar radiation reaching the Earth's surface is several thousand times more than world's annual energy need [2]. According to the 2017 survey, renewables contributed 19.3% to global energy consumption, with various sources like traditional biomass, modern biomass, solar heat and geothermal, hydroelectricity, and wind, solar, and geothermal electricity playing a role. Figure 19.1 shows the world's electricity generation by different sources [3]. Renewable energy systems have been rapidly improving in terms of efficiency and cost-effectiveness. The renewable energy industry has also created a significant number of jobs, with solar photovoltaics being a major job provider [4–6]. The share of renewable

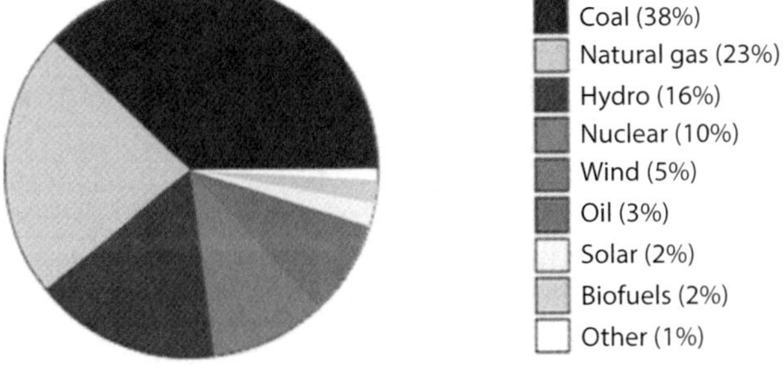

Figure 19.1 World electricity generation by different sources in 2018 [3].

energy in total energy consumption is continuously increasing, and by 2050, more than two-thirds of newly installed electricity capacity worldwide will come from renewable sources. The consumption of coal and oil is expected to decline in the future due to the increased adoption of renewables and natural gas. This transition to renewable energy sources is crucial for mitigating the environmental impact of energy production and ensuring a sustainable energy future [7–9].

Solar energy is indeed a universally available and convenient source of renewable energy. The sun emits an enormous amount of solar radiation, and a significant portion of it reaches the Earth's surface. Solar radiation received on a surface is measured in solar irradiance (W/m^2). Solar irradiance covers a wide spectral range, from 300 to 2,100 nm. However, not all of the solar radiation that spreads on the upper atmosphere of the Earth's surface is utilized effectively. Approximately 30% is reflected back into space, while the remaining portion is captivated by various elements, including clouds, oceans, land, and living organisms [10, 11]. To quantify the total amount of solar energy received over a specific period, solar irradiance can be integrated or summed up. Integrated solar irradiance is referred to as solar irradiation, solar exposure, solar insolation, or simply insolation. It is typically reported in units of J/m^2 for a given time interval. By harnessing solar energy through technologies like photovoltaic panels or solar thermal systems, we can convert sunlight into usable forms of energy, such as electricity or heat, without emitting greenhouse gases or depleting finite resources. This makes solar energy a clean and renewable alternative to fossil fuels for meeting our energy needs [12]. Figure 19.2 shows the solar radiation spectrum on the Earth and above the atmosphere [13].

Approximately half of the incident solar energy reaches the Earth apparently and can be converted into usable forms of energy. Solar thermal energy and solar power, which involve converting sunlight into electricity through technologies like photovoltaics (PVs) or concentrated solar power (CSP), are widely used solar technologies [14–16]. Photovoltaic systems utilize the photoelectric effect to convert incident light into direct current (DC) electricity. Including all available renewable energy technologies, photovoltaic technology is particularly attractive due to its versatility and widespread applicability. While extremely efficient silicon-solar cells are commonly used in terrestrial photovoltaic applications, their

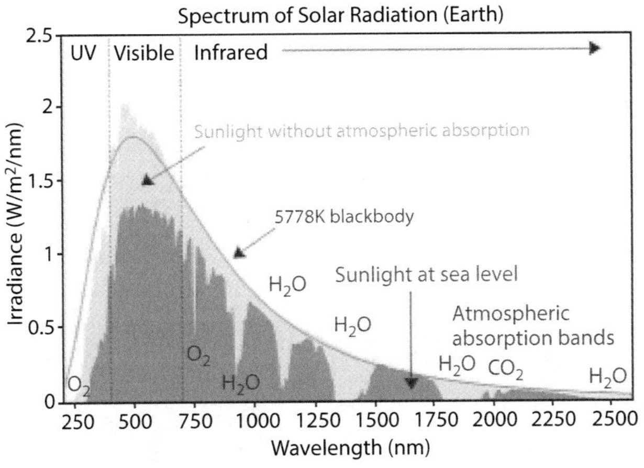

Figure 19.2 Solar radiations on the Earth's surface [13].

production and environmental costs can be high. As an alternative, dye-sensitized solar cells (DSSCs) offer a low-cost option due to their inexpensive materials and simple fabrication process. This review article would likely discuss the structure and working principle of DSSCs, as well as various methods employed in fabricating semiconducting thin films for DSSCs. Some commonly used techniques for growing these films include chemical bath deposition (CBD), electrodeposition, dip coating, radio frequency (RF) sputtering, spray pyrolysis, hydrothermal atomic layer deposition, sol-gel method, successive ionic layer adsorption and reaction (SILAR), and spin coating. These techniques allow for the deposition of semiconducting thin films with specific characteristics required for efficient DSSC operation.

19.2 Structure and Operation Principle of DSSCs

Typical DSSCs consist of several key components:

i. Conducting Substrate: The DSSC requires a conducting substrate, typically transparent conducting oxide (TCO) such as fluorine-doped tin oxide (FTO). This substrate allows light to penetrate through and provides electrical conductivity.
ii. Photoanode: The photoanode is the layer where the light-harnessing dye molecules are adsorbed. It usually comprises a porous film of n-type semiconductor nanoparticles like titanium dioxide (TiO_2). The dye molecules absorb light and initiate the electron transfer process.
iii. Redox Electrolyte: The redox couple plays a crucial role in the DSSC by facilitating the movement of electrons between the photoanode and the counter electrode (generally platinum). It is typically a liquid or gel electrolyte containing redox couples that can undergo reversible oxidation and reduction reactions.
iv. Counter Electrode: The counter electrode, often referred to as the cathode, is usually a platinized electrode that facilitates the reduction of the redox species from the electrolyte. It acts as a catalyst for the regeneration of the dye molecules.

The working principle of a DSSC involves the absorption of light by the dye molecules, which generates excited electrons. These electrons are injected into the semiconductor photoanode's conduction band, creating a current flow. The redox electrolyte then carries the electrons from the photoanode to the counter electrode, which is used to reduce the redox species. The completed circuit allows the flow of electrons, resulting in the generation of electricity. It is important to note that DSSCs offer a cheap alternative to silicon-based solar cells, as they can be fabricated using inexpensive materials and simpler manufacturing processes. However, their overall efficiency is currently lower compared to silicon solar cells. Ongoing research aims to improve the efficiency and stability of DSSCs to make them more competitive in the renewable energy market. The working principle of DSSC is represented in Figure 19.3.

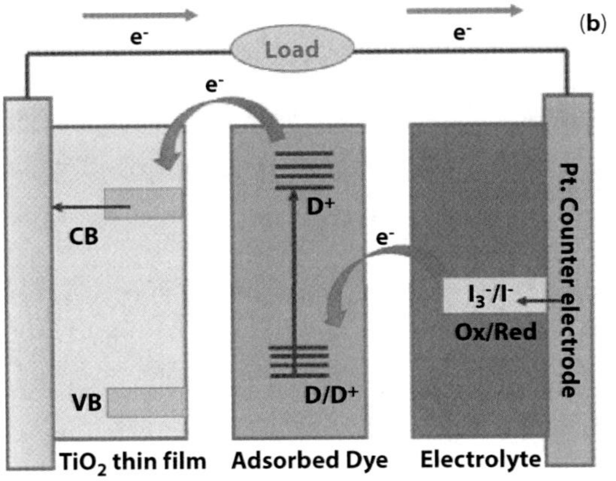

Figure 19.3 The diagram presenting working principles of a DSSC.

19.3 Various Methods for Fabricating Thin Films for DSSCs

The reviewed section focuses on fabricating thin films as the photoanode in DSSCs. The choice of deposition method significantly affects the morphology and properties of the thin films, ultimately influencing the efficiency of the DSSC. Thin films with well-defined structures and functionalities are crucial for enhancing the performance of DSSCs at the nanoscale level. Various physical and chemical methods are employed for depositing these films. Some commonly used techniques include doctor blading, spraying, and screen printing for depositing the semiconducting layer on conducting substrate. These deposition techniques, among others, enable the growth of semiconducting layers with specific properties required for efficient DSSC operation. They offer versatility in film thickness control, scalability, and compatibility with different substrates. It is worth noting that the choice of deposition method depends on several influences, including the specific material used, the desired film thickness, the substrate type, and the overall device design. Researchers continue to explore and develop novel deposition techniques to improve the performance and scalability of DSSCs.

19.3.1 Physical and Chemical Deposition Techniques

Physical deposition techniques like magnetron sputtering, plasma spray, and laser ablation involve bringing atoms from a solid or molten target into gas phase through physical processes like evaporation, sputtering, or laser ablation. These techniques are commonly used for depositing thin films onto substrates. The resulting deposited films from these physical deposition techniques are typically polycrystalline, meaning they consist of many small crystalline domains or grains. These films often exhibit a columnar structure, with the grains aligned perpendicularly to the substrate surface. The size of these grains are controlled by regulating the deposition conditions, such as the deposition rate, temperature, or pressure. By optimizing these parameters, researchers can tailor the properties of the deposited films to meet specific requirements.

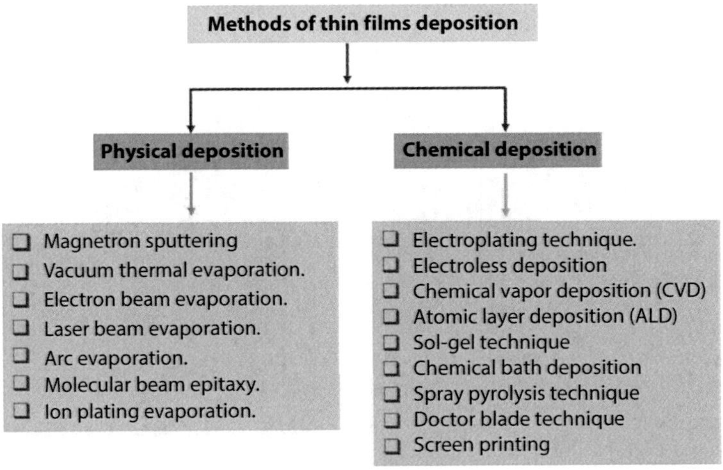

Figure 19.4 Brief classification of thin-film deposition techniques used in DSSCs.

On the other hand, chemical deposition methods offer advantages in controlling the chemical composition and achieving a dense film structure. These methods involve chemical reactions between precursor materials in the gas or liquid phase, forming a solid film on the substrate surface. Chemical deposition techniques are known for their ability to provide accurate control over the composition of deposited material, permitting the fabrication of complex materials or multilayered structures. Chemical deposition methods are also suitable for mass production, as they can be easily scaled up and automated, enabling efficient and consistent film production. Chemical vapor deposition (CVD), atomic layer deposition (ALD), electroplating, and spin coating are some of the chemical deposition techniques. These methods find wide applications in various industries, including semiconductor manufacturing, display technology, optoelectronics, and energy storage. The flowchart of thin-film deposition technique classification is briefly shown in Figure 19.4.

19.3.1.1 Magnetron Sputtering

Magnetron sputtering is a well-established deposition technique for fabricating coatings of various materials. Thin films produced through magnetron sputtering often exhibit superior quality compared to films fabricated by other methods. One advantage of magnetron sputtering is the ability to achieve low-defect-density films [17, 18]. The sputtering process allows for precise control over the coating parameters, such as the sputtering power, gas pressure, and substrate temperature, which can influence film quality. By carefully optimizing these parameters, it is possible to minimize defects, such as voids, impurities, and grain boundaries, resulting in films with high structural integrity. Another advantage is the capability to grow thin films of materials of high melting point on substrates that cannot be heated. In magnetron sputtering technique, the growth of the deposited film and phase formation are primarily governed by kinetic processes rather than thermodynamics. This means that without heating the substrate, the energetic sputtered atoms have sufficient mobility to rearrange themselves and form a desired crystal structure. This is particularly advantageous for high-melting point materials, as it avoids the need for elevated temperatures that could lead to substrate damage or phase transformations. The versatility of magnetron sputtering makes it widely used in many applications, including semiconductor manufacturing, optical coatings, thin-film solar cells, magnetic storage devices, and decorative coatings, among others. However, it is important to consider that the quality of deposition technique should be centered on the specific requirements of desired coating and the properties of the target material [18, 19]. Figure 19.5 represents the schematic illustration of the apparatus, principle, and deposition of the magnetron sputtering technique.

In magnetron sputtering, the target material's surface is bombarded by Ar^+ ions, which are generated through collisions among Ar gas atoms and trapped e- in a magnetic field atmosphere. The Ar^+ ions are attracted to negatively charged target (cathode), causing target particle ejection. These particles are deposited as a thin layer on substrate that is kept on the anode. This process allows the controlled deposition of thin layer with chosen properties. Based on the power source used, magnetron sputtering technique is mainly classified into two main types: (a) direct current magnetron sputtering and (b) radio frequency magnetron sputtering. In DC magnetron sputtering, a DC power supply creates the magnetic field and provides the necessary energy for ionization and sputtering. On the other hand, RF magnetron sputtering employs an RF power source to generate the magnetic field and sustain the plasma for ionization and sputtering. In electron-blocking layers in DSSCs [20, 21], magnetron sputtering is considered a favorite choice due to its simplicity and repeatability. Figure 19.6 shows the schematic illustration of DSSC with compact blocking layer to avoid recombination.

Materials such as SnO_2 [22], ZnO [23], Nb_2O_5 [24], and TiO_2 can be used as electron-blocking layers in DSSCs. Among them, TiO_2 is often recommended as the best electron-blocking layer due to its low cost, chemical stability in electrolytes, availability, good n-type conductivity, and appropriate electronic band structure that matches the sensitizer. Magnetron sputtering provides a reliable and versatile method for depositing thin

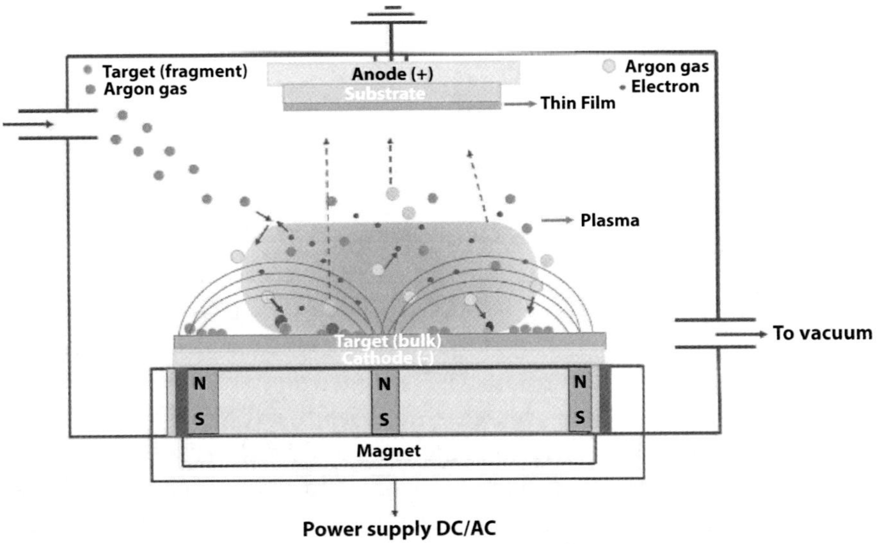

Figure 19.5 Schematic illustration of the apparatus, principle, and deposition of the magnetron sputtering technique.

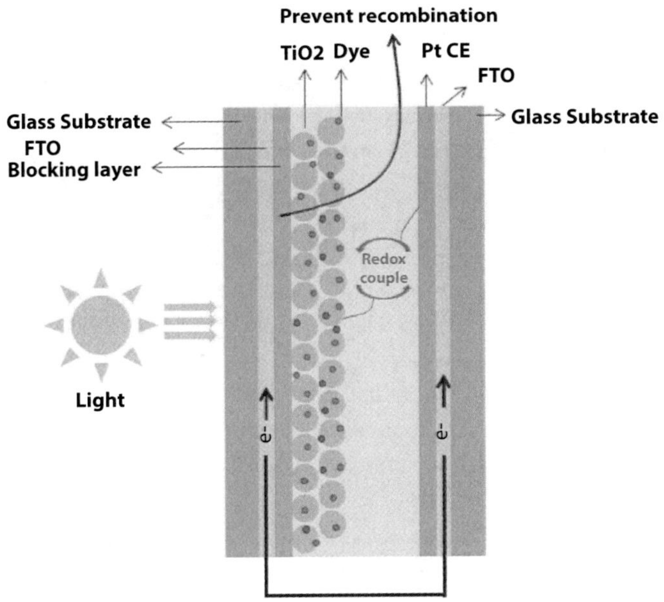

Figure 19.6 Schematic illustration of DSSC with compact blocking layer to avoid recombination.

films, including electron-blocking layers in DSSCs, offering control over film thickness, composition, and quality. Table 19.1 summarizes the references for thin-film synthesis with magnetron sputtering and RF magnetron sputtering for DSSCs. Meng *et al.* [25a] prepared TiO_2 nanorod arrays using DC reactive magnetron sputtering on the ITO substrate. The average diameter of the nanorod arrays was modified by adjusting the target-substrate distances. The highest power conversion efficiency of 4.78% was achieved using the prepared TiO_2 nanorod arrays at an 80-mm target-substrate distance. Sutthana *et al.* [25b] explored DC magnetron sputtering to fabricate a multilayer film of Ag deposited on Al-doped ZnO (AZO) and coated with AZO and labeled as AZO/Ag/AZO multilayer films. They achieved (IPCE) incident photon-to-current conversion efficiency of 0.61% using AZO/Ag/AZO multilayer films. The poor performance of the DSSC is attributed to the higher sheet resistance (8 nm Ag on FTO substrate) and lower transmittance value.

Li *et al.* [26] reported an enhanced efficiency of the NiO-TiO_2 electrodes coated using magnetron sputtering for DSSC. NiO sputtered on the TiO_2 electrode reduced the trap states on TiO_2, which improved the electrical performance from 3.21% to 4.16%, exhibiting the positive role of NiO coating on TiO_2. Using RF magnetron sputtering, Kang *et al.* [27] prepared columnar-structured rutile TiO_2 on a rugged FTO substrate. The enhanced power conversion efficiency was attained for rutile TiO_2 (2.4%) compared to the anatase TiO_2 (1.78%). Wang *et al.* [28] used reactive DC magnetron sputtering technique to assemble TiO_2 electrodes with insulating MgO layer. The sputtered MgO coating improved the electrical performance (power conversion efficiency of 7.57%) by reducing the trap states, enhancing the dye adsorption, and suppressing the interfacial recombination. Wu *et al.* [29] explored a surface modification by ZnO coating on the TiO_2 electrode using RF magnetron sputtering. The surface modification enhanced the efficiency of DSSC from 4.76%

Table 19.1 List of references for thin-film synthesis with magnetron sputtering and RF magnetron sputtering for DSSCs.

S. no.	Photoanode substrate	Efficiency (%)	Reference
1	TiO_2 nanorod	4.78	[25]
2	AZO/Ag/AZO multilayer film	0.61	[25]
3	TiO_2/NiO	4.16	[26]
4	Rutile TiO_2	2.4	[27]
5	TiO_2/MgO	7.57	[28]
6	TiO_2/ZnO	6.55	[29]
7	TiO_2/ZnO	5.1	[29]
8	TiO2/Al_2O_3	5.91	[30]
9	Nanofibrous ZnO	1.66	[31]
10	Anatase (Core)/Rutile (Shell) Nanostructured TiO_2	-	[32]
11	ZnO:Eu^{3+}	1.33	[33]
12	Anatase TiO_2	-	[34]
13	Titanium-doped indium oxide	5.79	[35]
14	Core-shell TiO_2/ZnO	0.72	[36]
15	Nb-doped TiO_2	6	[37]

to 6.55% by suppressing interfacial recombination. Kouhestanian *et al.* [30] explored ZnO as an efficient blocking layer on the TiO_2 electrode using DC magnetron sputtering, which enhanced the electron transfer process. The optimum thickness of the ZnO blocking layer enhanced the efficiency by 5.1%. Wu *et al.* [1] reported that reactive DC magnetron sputtering of Al_2O_3-modified TiO_2 electrodes improved the efficiency from 3.93% to 5.91%. The enhanced performance was achieved due to the increased dye adsorption, reduced trap states on TiO_2, and suppression of the interfacial recombination. Lupan *et al.* [32] fabricated nanofibrous-like ZnO layers using magnetron sputtering. The device made of ZnO photoelectrode yielded a maximum efficiency of ~1.7%. Li *et al.* [33] synthesized anatase (core)/rutile (shell) nanostructured TiO_2 thin films using magnetron sputtering for DSSC applications. Using RF magnetron sputtering, Zou *et al.* [34] fabricated a thin Eu-doped ZnO (ZnO:Eu^{3+}) film. High optical transparency was achieved with ZnO:Eu^{3+} thin film and yielded an efficiency of 1.33% compared to the pristine ZnO without Eu (1.19%). Senthilkumar *et al.* [35] prepared anatase TiO_2 thin films by DC magnetron sputtering using Ti target source material for DSSC application. Using RF magnetron sputtering, Heo *et al.* [36] fabricated titanium-doped indium oxide (ITiO) thin films on a glass substrate. Higher transmittance and lower resistivity of the ITiO thin film showed a conversion efficiency of 5.79%. Zhang *et al.* [37] synthesized core-shell TiO_2/ZnO photoanode using a

combination of chemical growth and DC magnetron sputtering. The power conversion efficiency of 0.72% was achieved by optimizing the ZnO magnetron sputtering and chemical bath deposition times. Horie *et al.* [38] prepared a column-like Nb-doped TiO_2 (NTO) dense thin layer by DC magnetron sputtering at a low-temperature process. The maximum increase efficiency of ~6% was attained for the columnar NTO compact layer compared with the granular NTO layer fabricated by the sol-gel route. Figure 19.6 shows the schematic illustration of DSSC with compact blocking layer to avoid recombination.

19.3.1.2 Laser Ablation or Pulsed Laser Deposition

PLD is a versatile technique that has several applications such as optoelectronics, nanotechnology, and materials science, allowing for the deposition of high-quality thin films with tailored properties. It involves using a laser with a high power density and a narrow frequency bandwidth to vaporize the desired deposition material. In PLD, beam of pulsed laser, typically in the UV wavelengths range, is focused on desired thin-film composition target material. The laser ablation vaporizes the target material, creating a cloud of vaporized materials. These vaporized species get condensed and deposited on substrate held contrary to the target. PLD can synthesize various materials, including nanotubes, nanopowders, and quantum dots. In PLD, the thickness of the deposited layer is typically between 10 and 500 nm, depending on the specific requirements and application. One common application of PLD is the deposition of TiO_2 films. However, it is worth noting that using PLD for TiO_2 blocking layers or buffer layers may be less common compared to other deposition methods, depending on researchers' or manufacturers' specific needs and preferences. PLD offers advantages such as precise control over film composition, stoichiometry, and crystal structure. It enables the deposition of complex materials and heterostructures with good film quality and interface properties [39–42]. Combinatorial PLD and matrix-assisted pulsed laser evaporation are variations or extensions of the PLD technique, incorporating additional features to enable high-throughput screening, combinatorial synthesis, or specific material preparation approaches. Table 19.2 summarizes the references using laser ablation or pulsed laser deposition technique for thin-film fabrication for DSSC application. Kim *et al.* [43] fabricated laser-processing nanocrystalline TiO_2 films for DSSC application. A pulsed UV laser of 355 nm wavelength was used to forward transfer a suspension of TiO_2 nanopowder (P25) onto the FTO substrate. The DSSCs fabricated with PLD-deposited TiO_2 thin film showed the efficiency of 4.3% at 10 mW/cm^2 illumination. Gondal *et al.* [44] used pulsed layer ablation technique and synthesized SiC-TiO_2 nanoparticles and revealed application in photocatalytic water purification and DSSCs. They used SiC-TiO_2 nanocomposite as a photoanode in DSSCs and as a photocatalyst in the photodegradation of methyl orange (MO) dye in water. The photoconversion efficiency of DSSCs was increased to 1.65% from 0.6%, and the percentage removal of methyl orange dye in water was improved to 77% from 22% at 24 minutes under solar spectrum. Sima *et al.* [45] fabricated ZnO photoanode by laser ablation from a Zn pure target on conducting glass substrate in an O_2 atmosphere at room temperature. The power conversion efficiency of ZnO film deposited by laser ablation was measured as 0.22%, which was less as compared to the photoelectrode made using the conventional doctor blade method (efficiency 0.43%). Ghosh *et al.* [46] reported Ta-doped TiO_2 photoelectrode prepared by laser ablation, which increased the photocurrent in DSSCs. The efficiency of the DSSCs with N719 dye adsorbed on hierarchical nanostructures of Ta-doped TiO_2 was improved up to 6.7% owing to the large concentration of the charge carriers and reduced recombination rate, which enhanced the photocurrent of the device. Gondal *et al.* [47] revealed that pulsed laser ablation-deposited ZnO/TiO_2 nanocomposite enhanced the photovoltaic efficiency and photocatalytic performance. Here, 6.7% efficiency was acquired using ZnO/TiO_2 nanocomposite in DSSCs, which was improved substantially by 63.73% in comparison to the pure ZnO, and the photodegradation rate was boosted by more than 2-fold. The performance improvement in the ZnO/TiO_2 nanocomposite is mainly due to the exceptional charge carrier separation and fast interfacial electron flow. Ghosh *et al.* [48] reported nano forest Nb_2O_5 photoanode deposited by pulsed laser ablation technique for DSSCs. An efficiency of 2.41% was obtained for an optimal structure using DSSCs made with N3 dye. This work revealed that the pressure and gas composition are important parameters during the deposition. Sahu *et al.* [49] have successfully grown vertically aligned core-shell Au-TiO_2 nano-architecture using PLD techniques and showed its application in DSSCs. Due to the presence of Au nanowires inside the highly crystalline TiO_2 nanoshell, the electron transport was facile between the TiO_2 and the conducting substrate, which provided a direct conduction path. An efficiency of 6.7% was obtained from DSSCs made with N535 dye and triiodide/iodide redox electrolyte. Albu *et al.* [50] prepared DSSC structures using a pulsed laser-deposited TiO_2 thin film in O_2, N_2, He, or Ar gas. The nature of the gas plays a key parameter in the DSSC architecture and the performance of the deposited thin films. The best photovoltaic performance of 2.32% was obtained using TiO_2 thin film grown in He owing to the film's roughness (of 44 ± 2.9 mm RMS), resulting in deeper dye penetration inside the nanostructured TiO_2 coatings. The electro-optical performances of the TiO_2 coating in O_2, N_2, and Ar were observed as 1.18%, 1.02%, and 1.42%, respectively. Sacco *et al.* [52] revealed an enhanced photovoltaic performance in DSSCs using a pulsed layer-deposited Nb_2O_5 as blocking layer on the conducting substrate. With optimal thickness of the Nb_2O_5 blocking layer, they achieved 6.15% efficiency. The open-circuit voltage decay and electrochemical impedance spectroscopy studies proved that electron recombination was suppressed at the substrate/electrolyte interface by depositing Nb_2O_5 blocking layer. Garvey *et al.* [53] fabricated porous nano-carpets of In_2O_3:Sn (ITO) films by pulsed laser-deposited that serve as charge collectors in core shell (ITO/TiO_2) architecture for DSSCs. The deposition of a conformal TiO_2 shell layer by atomic layer deposition (ALD) achieved a photovoltaic efficiency of 0.432%. TiO_2 shell acted as a blocking layer, decreasing recombination, and overall mosh-like discrete nanostructures (core-shell ITO/TiO_2) boosted the electron transport. Ilyas *et al.* [54] reported a facile synthesis of TiO_2/CdS nanocomposite using pulsed laser ablation in liquid and showed its application in photocatalytic degradation of methyl orange in water and DSSCs. The concentration of CdS was varied, and photocatalytic and photovoltaic performance was studied. The nanocomposite made with 10 wt. % CdS in anatase phased TiO_2 showed the improved photovoltaic performance in comparison to the pure TiO_2, and the overall efficiency of the DSSCs was improved to 4.3% from 0.6% while the percentage of the photocatalytic degradation of dye increased from 58% to 82% after 36-minute irradiation using UV-vis solar light. The improved performance was achieved owing to the efficient conduction of charge carriers, retarded electron-hole recombination, higher lifetime, enhanced UV-vis light activity, decreased photocorrosion but enhanced pore size. Hara *et al.* [55] reported Ta-doped TiO_2 nanobrush photoanode using the PLD technique for DSSCs. A patterned ITO substrate was fabricated using nanoimprint lithography and PLD techniques. Extra deposition of Ta:TiO_2

Table 19.2 References using laser ablation or pulsed laser deposition technique for thin-film fabrication for DSSC application.

S. no.	Photoanode substrate	Efficiency (%)	Reference
1.	Nanocrystalline TiO_2	4.3	[43]
2.	SiC/TiO_2	1.65	[44]
3.	ZnO	0.22	[45]
4.	Ta-doped TiO_2	6.7	[46]
5.	ZnO/TiO_2 nanocomposite	6.7	[47]
6.	Nanoforest Nb_2O_5	2.41	[48]
7.	Core-shell Au–TiO_2	6.7	[49]
8.	$TiO_{2\_}O_2$	1.18	[50]
	$TiO_{2\_}N_2$	1.02	
	$TiO_{2\_}Ar$	1.42	
	$TiO_{2\_}He$	2.32	
9.	NiO/ZnO	1.26	[51]
10.	Nb_2O_5	6.15	[52]
11.	ITO/TiO_2 (core/shell)	0.432	[53]
12.	TiO_2/CdS	4.3	[54]
13.	Ta:TiO_2	3.2	[55]

by PLD onto these patterned ITO leads to growing of discrete brushlike arrangements. Here, 3.2% efficiency was achieved using these patterned ITO substrates.

19.3.1.3 Chemical Bath Deposition

Chemical bath deposition is a deposition method that involves the use of precursor solutions containing the species to be deposited. The method typically needs solution vessels and substrate-holding apparatus with heating and stirring capabilities. In its simplest form, the substrate is soaked in the precursor solution, and the solution is then hydrolyzed under heating and stirring conditions. This promotes particle nucleation and growth on the coating substrate surface. Controlling factors such as temperature and deposition interval can manipulate the deposited film's size, shape, morphology, and thickness. This allows for precise control over the desired film properties. Chemical bath deposition produces stable and uniform thin film with good adherence reproducibility. The process enables the formation of thin films on various substrates, including metals, semiconductors, and insulators. One advantage of CBD is its versatility, as it favors depositing a wide range of materials, metal oxides, sulfides, chalcogenides, etc. The technique is suitable for applications in optoelectronics, photovoltaics, sensors, and energy storage. Therefore, chemical bath deposition is a widely used film preparation technique that offers control over film properties and provides stable and uniform film deposition with good reproducibility [56, 57]. Ku and Wu [58] reported chemical bath-deposited ZnO nanowire-layer basic zinc acetate (LBZA) nanocomposite electrodes for DSSCs. The composite film, ZnO NW-LBZA/ZnO NP, retains a considerably enlarged surface area by NP growth compared to the ZnO NW array. The electron transport was not compromised due to an enlarged surface area; it also acted as a promising photoanode in DSSCs. The photoconversion efficiency of 2.77% was obtained using ZnO NW-LBZA/ZnO NP composite film. Ariyanto *et al.* [59] fabricated ZnO-based DSSCs using the chemical bath deposition method. ZnO film was prepared by translating a flower-structured morphology of zinc carbonate hydroxide using the above technique. Flowered morphology facilitated the charge transport in the photoanode of DSSCs. I-V characteristics showed that the deposition time and deposition temperature of CBD influence the overall efficiency. The finest efficiency of 2.45% was obtained from the DSSCs fabricated using the N719 dye-adsorbed ZnO film due to the small electron lifetime and due to higher electron diffusion coefficient revealed by electrochemical impedance spectroscopy. In another work, replace with Zhu *et al.* [60] reported the growth of rutile TiO_2 nanosheet array thin films on an FTO substrate using chemical bath deposition in a short time. DSSCs fabricated using this film exhibited an efficiency of 1.82%. Thambidurai and Muthukumarasamy [61] reported a synthesis of ZnO nanorods for DSSCs using a CBD technique. The DSSCs fabricated using ZnO nanorods sensitized with natural dye extracted from Lantana camara showed 0.71% efficiency. Wang *et al.* [62] reported the preparation of cauliflower-structured ZnO films on the FTO glass by CBD. DSSCs fabricated with cauliflower-like ZnO films/N719 dye exhibited a Jsc of 6.08 mA/cm^2, Voc of 0.66 V, an FF of 0.55, and an η of 2.18%. Using chemical bath deposition, Chu *et al.* [63] prepared flower-like columnar nanostructured ZnO thin films in basic aqueous ammonia solutions. The DSSCs fabricated using flower-like columnar nanostructured ZnO film exhibited an efficiency of 0.84%.

Table 19.3 List of references used in thin-film fabrication by CBD technique for DSSC application.

S. no.	Photoanode substrate	Efficiency (%)	Reference
1	ZnO nanowires	2.77	[58]
2	ZnO	2.45	[59]
3	TiO_2 nanosheet	1.82	[60]
4	ZnO nanorod	0.71	[61]
5	ZnO	2.18	[62]
6	ZnO	0.84	[63]
7	WS_2 QDs on TiO_2 in DSSCs	5.8	[64]
8	Zinc stannate (Zn_2SnO_4) film	3.45	[65]

Powar *et al.* [64] prepared CBD-deposited WS_2 quantum dots coated on TiO_2 for DSSCs. WS_2 quantum dots exhibited absorption between wavelength 350 and 800 nm. A DSSC fabricated by integrating 2D-layered WS_2 quantum dots and N719 dye exhibited an efficiency of 5.8%. Chen *et al.* [65] reported efficient DSSCs fabricated with Zn_2SnO_4 film using the chemical bath deposition method. An efficiency of 3.45% was obtained due to the decreased charge recombination rate and prolonged electron lifetime. Moreover, the slightly decreased conduction band edge of Zn_2SnO_4 provided the extra driving force for the electron injection, enhancing short-circuit current density and hence the efficiency. Table 19.3 summarizes the list of references used in thin-film fabrication by the CBD technique for DSSC application.

19.3.1.4 Chemical Vapor Deposition

Chemical vapor deposition is a versatile technique used for coating thin films of various materials onto heated substrates. It involves the deposition of solids through chemical process in the vapor phase. Chemical vapor deposition is commonly employed to fabricate thin films for microelectronics, wear-resistant coatings, etc. [65, 67]. A typical chemical vapor deposition setup selects a mixture of chemical precursors that can vaporize stably. A carrier gas (inert in nature) like argon (Ar) or nitrogen (N_2) transports the vaporized precursors into reaction chamber or evacuated furnace. Then, additional reactive gas or gases were mixed with these precursors. For example, for the deposition of metal oxide thin films, O_2 gas is used, while a reducing atmosphere with hydrogen (H_2) gas is created for metallic thin films such as copper (Cu). Figure 19.7 shows the pictorial representation of the CBD experimental setup. Unlike techniques such as magnetron sputtering and pulsed laser deposition (PLD), CVD does not require ultrahigh-vacuum systems. It is a kinetically focused process that minimizes cluster and offers faster deposition rates compared to atomic layer deposition (ALD) [68–70]. However, there are some drawbacks to the CVD technique. It can be challenging to deposit multicomponent thin films because different precursors have different evaporation rates. Additionally, chemical and safety hazards may be associated with corrosive, toxic, and flammable precursors or exhaust gases. Another drawback is the potential for poor adhesion strength of the deposited thin films. To deposit thin films, CVD can utilize several chemical reactions such as reduction, oxidation, pyrolysis, hydrolysis, carburization, disproportionation, and nitridation. The reactions in CVD are controlled by factors like thermodynamics, kinetics of the chemical reactions, pressure, and temperature [71–75]. CVD techniques are classified into different categories based on their methods of providing the required activation energy (A.E) to start the reaction. Some common types of CVD techniques include metal-organic CVD (MOCVD), molecular beam epitaxy (MBE), low-pressure CVD (LPCVD), and atmospheric pressure CVD (APCVD). MOCVD and MBE are widely used for depositing semiconductor thin films, while LPCVD and APCVD are also utilized. MOCVD is particularly attractive due to its relative simplicity compared to other growth deposit methods. It offers exceptional control over thin-film composition by precisely regulating the amounts of introduced species. MOCVD is highly effective in producing heterostructures, multi-quantum wells, superlattices, and alloys with tailored doping profiles. However, one drawback of MOCVD is the use of harmful gases and the production of toxic gases, which require careful handling and safety measures. Cheng *et al.* [76] reported thermal CVD-grown quasi-aligned one-dimensional ZnO nanostructures for DSSCs.

They have established that a heavily packed ZnO can be made on the conducting glass substrate by a double-source double-tube chemical vapor deposition method. The η of 0.6% was achieved from the device. Zhang and Li [77] fabricated ZnO/TiO_2 core-shell nanostructured photoanode for DSSCs by the mist CVD method. The coating time influenced the morphological, optical, structural, and photoluminescence properties of the ZnO/TiO_2 core-shell. Kim *et al.* [78] reported thermal CVD-deposited nanocrystalline ZnO-covered TiO_2 film electrodes (ZnO/TiO_2) for DSSCs. These electrodes enhanced the photocurrent density and open-circuit voltage of DSSCs, and the efficiency of 4.3% was achieved. Shakir *et al.* [79] fabricated un-doped and Mg-doped TiO_2 thin films by aerosol-assisted chemical vapor deposition method for DSSCs. Mg doping replaced some of the Ti metals from its lattice and a further enhancement of the concentration of the dopant leading to red-shifted absorption band of the films toward the visible region and thus lowering the band gap of the TiO_2. Mg doping also increased porosity, enhancing dye loading. These positive effects of Mg doping suppressed the charge recombination, improved the charge carrier generation, and improved the photoabsorption. There was an overall rise in open-circuit voltage, photocurrent density, and efficiency of the DSSCs made up using Mg-doped TiO_2 photoanode, among which 2 mol% Mg doping exhibited maximum efficiency of 6.10%. Ameen *et al.* [80] reported vertical growth of aligned ZnO nanorods on graphene oxide thin-layered substrate by hot filament CVD method for DSSC application and obtained 2.5% efficiency. Murakami *et al.* [81] reported fabrication of mesoporous TiO_2 films at low temperature for efficient photoelectrode in DSSCs using CVD combined with UV light irradiation and obtained an efficiency of 3.8%. Lee and Kim [82] reported the deposition of ZnO photoelectrodes using plasma-enhanced CVD for its application in solar device as photoanode. The maximum power conversion efficiency of 2.63% was achieved with a moderately thick (8.06 μm) ZnO. Lee and Kim [83] reported the deposition of the TiO_2 passivation layer by plasma-enhanced CVD method between the FTO and TiO_2 electrode. The optimized thickness of the TiO_2 passivation layer enhanced the connection and reduced the recombination on the surface of the FTO glass. The DSSCs fabricated with a 40-nm-thick TiO_2 passivation layer exhibited a maximum efficiency of 6.93% owing to the enhanced connectivity between the mesoporous TiO_2 and conducting substrate FTO and the retardation of the electron recombination from the FTO substrate to the electrolyte. Table 19.4 summarizes the list of references for thin-film fabrication by chemical vapor deposition technique for DSSC application.

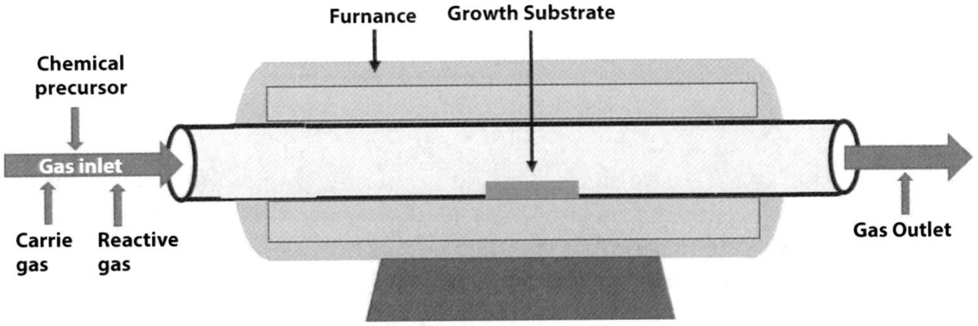

Figure 19.7 Schematic representation of chemical vapor deposition experimental setup.

Table 19.4 List of references for thin-film fabrication by chemical vapor deposition technique for DSSC application.

S. no.	Photoanode substrate	Efficiency (%)	Reference
1.	ZnO nanostructure	0.6	[76]
2.	ZnO/TiO_2 core-shell nanostructure		[77]
3.	ZnO/TiO_2	4.3	[78]
4.	Mg-doped TiO_2	5.88	[79]
5.	ZnO nanorods	2.5	[80]
6.	TiO_2	3.8	[81]
7.	ZnO	2.63	[82]
8.	TiO_2 passivation layer	6.93	[83]

19.3.1.5 Molecular Beam Epitaxy

Molecular beam epitaxy is a method used to grow epitaxial films with high precision and control. In MBE, elemental sources are evaporated at controlled rates through heating, and the evaporated atoms or molecules are then condensed onto substrate surface. The process takes place under ultrahigh-vacuum (UHV) conditions to ensure film purity. There are two main advantages of operating under UHV conditions in MBE. Firstly, the atoms and molecules that reach the growth surface are clean, since they are not contaminated by impurities from the surrounding environment. This allows for the growth of high-quality films with excellent crystalline properties. Secondly, the UHV environment enables *in situ* diagnostic techniques to monitor the growth process in real time. This means that crystal growth can be observed and controlled at the atomic layer level, providing precise control over layer thickness and doping profiles. MBE has become an established and highly regarded crystal growth technology due to its ability to produce complex and diverse structures. It is particularly well-suited for the fabrication of semiconductor devices and heterostructures. The precise control over layer thickness and doping profile offered by MBE allows for the creation of tailored structures with specific electronic properties. This level of control has enabled the development of advanced electronic and optoelectronic devices, such as quantum wells, superlattices, and high-performance transistors. In short, MBE is a powerful technique for the growth of epitaxial layer offering exceptional control and precision in fabricating complex structures with well-defined properties. Chu *et al.* [84] reported flexible DSSCs centered on vertical ZnO nanowire arrays synthesized by MBE for DSSC application and obtained an η of 0.44%. Mathew *et al.* [85] studied photoluminescence characteristics of europium-doped ZnO and Zinc-MgO thin films. These films were grown on sapphire (a-orientated) by O_2 plasma-MBE. Naseem *et al.* [86] reported epitaxial growth of cobalt-doped titanium oxide thin films on $LaAlO_3$ (100) substrate by MBE. The 8% cobalt-doped thin film exhibited 91% and 88% degradation with methylene blue and Azo dye, respectively. Jang *et al.* [87] reported the effects of Al doping on the different

Table 19.5 List of references for thin-film fabrication by molecular beam epitaxy technique for DSSC application.

S. no.	Photoanode substrate	Efficiency (%)	Reference
1	ZnO nanowires	0.44	[84]
2	Eu:ZnO and Eu:ZnMgO	-	[85]
3	Co:TiO_2 on $LaAlO_3$	-	[86]
4	Al:ZnCdO	-	[87]

properties of rock-salt ZnCdO thin films grown by MBE. The expansion of optical band gap and increasing of the electrical conductivity rock-salt ZnCdO were achieved through Al doping. Table 19.5 summarizes the list of references for fabrication of thin films by MBE technique for DSSC application.

19.3.1.6 Atomic Layer Deposition

ALD is a thin-film deposition method that shares similarities with chemical vapor deposition but operates with a different approach. In atomic layer deposition, precursor gases are sequentially presented subsequently one another, unlike CVD, where all gases may be present simultaneously in the chamber. ALD involves well-ordered sinusoidal pulses of gas-phase precursors that react with the surface of the depositing substrate. The reaction between the precursor and surface of substrate occurs step by step, with each precursor individually reacting with the substrate surface. Individual gas-surface reactions occurring between precursor and substrate are half-reactions and contribute to forming only a portion of the desired thin layer. In each half-reaction, under vacuum conditions, a precursor is inserted into a chamber for a particular duration to let it react with the substrate surface. After each half-reaction, inert carrier gas is purged in the chamber to remove unreacted precursor and by-products. Purge step ensures that the chamber is ready for the subsequent half-reaction. The process continues by introducing the next precursor for its respective half-reaction, followed by another purge step. The above depositing steps are repeated until the film of desired thickness is obtained. The advantage of this sequential deposition process is the precise control over layer thickness and composition. By carefully controlling the exposure time and pulsing sequence of different precursors, achieving highly uniform and conformal thin films with atomic-level precision is possible. ALD has found applications in various fields, *viz.*, energy storage, catalysis, optoelectronics, photoanode for DSSCs, etc. [88–95]. ALD was employed to deposit ZnO and TiO_2 on a silicon dioxide aerogel network [96, 97] to enhance the photoanode performance in the DSSC. Overall, ALD offers excellent film thickness and composition control, making it a valuable technique for precise thin-film deposition in various applications. Prasittichai and Hupp [98] reported ALD-deposited ultrathin layers of Al_2O_3 on SnO_2 photoelectrodes in DSSCs to improve photovoltage. The highest efficiency of 3.7% was achieved using this photoelectrode, and the photovoltage was increased from 230 mV to 510 mV. Chandiran *et al.* [99] reported a subnanometer gallium oxide tunneling layer by ALD to obtain Voc of 1.1 V and η of 4.6%

in DSSCs. Hamann *et al.* [100] used the ALD technique to deposit TiO_2-blocking layer to enhance the photovoltaic performance. Liu *et al.* [101] reported TiO_2 inverse-opal electrode *via* ALS and achieved 2.22% efficiency. Ganapathy *et al.* [102] reported ALD-deposited TiO_2/alumina core-shell electrode to enhance the performance of DSSCs. The efficiency was enhanced from 6.2% to 8.4% owing to 2-nm alumina deposition by ALD. Chandiran *et al.* [103] reported low-temperature crystalline TiO_2 on SiO_2 film by ALD technique for DSSCs. In this work, over 8-μm-thick SiO_2 film, a 15-nm TiO_2 layer was deposited that led to a higher efficiency of 7.1%. Park *et al.* [104] reported a fabrication of TiO_2 layer on ZnO thin film by ALD process and showed its application in DSSCs. An enhanced efficiency of 6.3% was achieved in this layer owing to increasing surface areas of the photoelectrode resulting in a large quantity of dye accumulation, which led to increased light harvesting. Hamann *et al.* [96] reported a new photoelectrode made of ALD-deposited TiO_2 on aerogel templates and obtained an efficiency of 4.3%. Shanmugam *et al.* [105] reported the effect of ALD-deposited HfO_2 and Al_2O_3 interfacial layers on the DSSC performance. The enhancement of 7.1% and 5% efficiencies was observed for ALD-deposited HfO_2 and Al_2O_3 electrodes, respectively, as compared with the reference DSSCs. Bills *et al.* [106] also reported ALD-deposited HfO_2 compact layer and obtained an efficiency of 6% in DSSCs.

Table 19.6 List of references for photoanode preparation by ALD technique for DSSC application.

S. no.	Photoanode substrate	Efficiency (%)	Reference
1	Al_2O_3/SnO_2	3.7	[98]
2	Ga_2O_3 tunneling layer	4.6	[99]
3	TiO_2 blocking layer	-	[100]
4	TiO_2	2.22	[101]
5	TiO_2/Al_2O_3 core-shell	8.4	[102]
6	ALD TiO_2 on SiO_2 film	7.1	[103]
7	ALD TiO_2 on ZnO film	6.3	[104]
8	TiO_2	4.3	[96]
9	TiO_2/HfO_2	7.1	[105]
	TiO_2/Al_2O_3	5	
10	ALD HfO_2	6	[106]
11	ALD ZrO_2/TiO_2	1.08	[107]
12	Al_2O_3	6.5	[108]
13	ALD Al_2O_3 on TiO_2	6.5	[109]
14	Ta:TiO_2	1.56	[110]
15	ZnO nanobelts/TiO_2	4.68	[111]

Li *et al.* [107] used ALD of ZrO_2 for the surface passivation of TiO_2 for solid-state DSSCs and achieved an efficiency of 1.08%. Lin *et al.* [108] stated an efficiency of 6.5% of DSSCs by an Al_2O_3 charge-recombination barrier *via* a low-temperature ALD technique. Tien *et al.* [109] reported an ALD layer of Al_2O_3 on nanoporous TiO_2 electrodes and achieved an efficiency of 6.5% in DSSCs. Choi *et al.* [110] reported ALD-deposited Ta-doped TiO_2 electrodes for DSSCs, and the obtained efficiency was 1.56%. Lu *et al.* [111] employed interface engineering through the ALD of TiO_2 film surrounding the ultrathin ZnO nanobelts and TiO_2 nanoparticles toward the highly improved performance of DSSCs. The photoelectrodes exhibited the highest efficiency of 6.88% owing to barrier-free access of electrons to conductive electrodes. Table 19.6 summarizes the list of references for photoanode preparation by ALD technique for DSSC application.

19.3.1.7 *Thermal Oxidation*

The thermal oxidation technique is usually used to grow nanotubes and nanorods on the substrate surface, particularly for materials like zinc oxide (ZnO) and titanium dioxide (TiO_2). This process involves the oxidation of substrate surfaces to promote the formation of desired nanostructures. In the thermal oxidation process, the substrate is heated to an elevated temperature, typically in the presence of specific additives or gases that help control the growth mechanism of the nanostructures. The additives or gases can influence parameters such as the resulting nanostructures' growth rate, morphology, and crystalline structure. By carefully controlling the temperature and the environment during thermal oxidation, it is possible to induce the formation of nanotubes and nanorods on the substrate surface. The oxidation process causes the surface atoms of the substrate material to react with oxygen, leading to the growth of oxide nanostructures. The resulting nanotubes and nanorods have unique properties such as high aspect, excellent structural integrity ratio, and large surface area. These characteristics make them attractive for various applications, including nanostructured photoanodes, where they can enhance light absorption and electron transport properties. Overall, the thermal oxidation process is a valuable technique for the growth of nanotubes and nanorods, enabling the development of nanostructured materials with tailored properties. By controlling the oxidation temperature and employing suitable additives, researchers can achieve precise control over the growth mechanism and

Table 19.7 List of references for thin-film fabrication by thermal oxidation technique for DSSC application.

S. no.	Photoanode substrate	Efficiency (%)	Reference
1	MgO	2.49	[112]
2	Cu_2O@CuO core-shell NPs	0.0165	[113]
3	ZnO	6.67	[114]
4	TiO_2	1.7	[115]
5	ZnO-CuO core-shell	0.108	[116]

morphology of the nanostructures, opening up possibilities for various nanoscale applications. Nilphai *et al.* [112] synthesized MgO nanoparticles by microwave-supported thermal oxidation process and showed its application in DSSCs. The DSSCs fabricated with MgO layer exhibited higher photocurrent, photo voltage, and 2.49% efficiency. Jiang *et al.* [113] synthesized Cu_2O@CuO core-shell NPs *via* the precipitation-thermal oxidation method and showed its application in p-type DSSCs. The DSSC based on Cu_2O@CuO NPs exhibited 0.0165% efficiency. Table 19.7 shows a list of references for thin-film fabrication by thermal oxidation technique for DSSC application.

19.3.1.8 Successive Ionic Layer Adsorption and Reaction Method

The SILAR method was first reported in 1985 by Pathan and Lokhande [117]. SILAR is a technique used for the deposition of uniform, compact, and crystalline thin films. The process involves immersing the substrate alternately in anionic and cationic precursor solutions, with subsequent rinsing steps. In the SILAR method, the substrate is first immersed in a cationic precursor solution. During immersion, cations from the solution adsorb onto the substrate's surface. After the immersion step, the substrate is removed, allowing any unabsorbed or precipitated material to separate from the surface. Similarly, the substrate is then dipped in an anionic precursor solution, and here, the anions react with the reabsorbed cations on the substrate surface [68, 118, 119]. The residual unreacted material is removed through rinsing. This cycle of soaking the substrate in the cationic and anionic precursor solutions, followed by rinsing, is considered a complete SILAR process. A multilayer film with the desired thickness will be formed by repeating the deposition process several times. Figure 19.8 represents the SILAR growth method schematically. The SILAR method offers several advantages over other deposition techniques. Firstly, it does not require specialized quality target materials or substrates or a vacuum at any stage of the process. Secondly, thin-film thickness and the deposition rate can be easily controlled. Thirdly, it provides a convenient way to dope films, allowing for the introduction of desired dopants. Fourthly, there are no limitations on the substrate material type, surface profile that make it versatile for various applications. Finally, this method is suitable for large-area deposition, making it convenient for industrial-scale production. SILAR has been successfully used to deposit thin films of various chalcogenide materials, including binary and ternary chalcogenides from I-VI, II-VI, III-VI, V-VI, and VIII-VI groups [68]. It has also been employed for the deposition of composite film preparation. The growth progression in SILAR can be understood in four main steps: adsorption, rinsing (I), reaction, and rinsing (II). These steps ensure the controlled deposition and formation of crystalline thin layers.

Therefore, SILAR is a versatile and relatively simple method for depositing thin films, offering control over film properties and the ability to dope films easily. Its wide applicability and compatibility with various substrates make it a valuable technique in materials science and thin-film technology. Kale *et al.* [68] prepared amorphous TiO_2 thin films on a glass substrate using TiO_2 precursor solution *via* the method. Park *et al.* [119] fabricated thin film using TiO_2 NPs and oppositely charged polyelectrolytes or titanium (IV) bis (ammonium lactato) dihydroxide (TALH) and characterized for different properties by the SILAR method. Gaikwad *et al.* [120] prepared nanostructured ZnO photoelectrodes *via* the green route M-SILAR method for DSSCs. DSSC fabricated with ZnO nanorods exhibited the highest efficiency of 0.59%. Gaikwad *et al.* [121] prepared the ZnO photoanode by M-SILAR

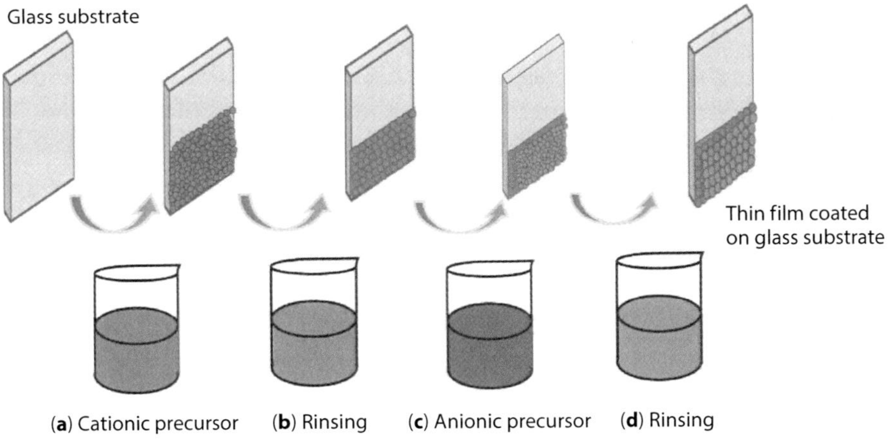

Figure 19.8 Basis of the SILAR method and its different steps.

method and studied the influence of Zn concentration and dye adsorption time on the photovoltaic performance. The DSSC fabricated using photoanode of 0.1 M Zn concentration showed the highest efficiency of 0.70%. Guo *et al.* [122] prepared photoelectrodes comprising Ag_2S quantum dots as an interfacial layer for solid-state bifacial DSSCs and obtained an efficiency of 6.5% from both side illuminations. Patil *et al.* [123] synthesized photosensitive nanograined TiO_2 thin films by the method. The annealed film exhibited an efficiency of 0.047%. Chung *et al.* [124] synthesized ZnS shells on ZnO nanowires using the SILAR method and measured the photovoltaic performance using the nanowire-based DSSCs. The DSSC made with ZnS shell on ZnO nanowires exhibited an efficiency of 0.01%. Isah *et al.* [125] incorporated Ag nanoparticles in (synthesized by SILAR method) TiO_2 film electrodes and studied the effect on the photovoltaic performance of DSSCs. Incorporation of Ag nanoparticles into the pores of mesoporous TiO_2 electrodes by SILAR method produced plasmonic enhanced DSSC exhibiting a power conversion efficiency of 0.27%. Table 19.8 summarizes the list of references for thin-film fabrication by SILAR technique for DSSC application.

Table 19.8 List of references for thin-film fabrication by SILAR technique for DSSC application.

S. no.	Photoanode substrate	Efficiency (%)	Reference
1.	ZnO	0.59	[120]
2.	ZnO	0.7	[121]
3.	Ag_2S QDs	6.5	[122]
4.	TiO_2	-	[123]
5.	ZnS shell on ZnO	0.01	[124]
6.	Ag NPs	0.27	[125]

19.3.2 Colloidal and Ceramic Powder Techniques

These techniques are cost-effective, need less complicated process, and are widely used methods for manufacturing nanocrystalline thin films. These techniques include various coating methods such as spin coating, dip coating, screen printing, and slurry coating. In these methods, ceramic powders or nanoparticles are dispersed in a liquid medium to form a colloidal particle suspension. Dispersants and surfactants are added to stabilize the suspension and prevent particle agglomeration. Using dispersants and surfactants helps achieve a suitable and stable suspension with a uniform dispersion of the particles. The colloidal suspension is applied onto the substrate to deposit the thin film using the chosen coating method. Each method has its advantages and is selected based on the application's specific requirements. For example, spin coating involves spinning the substrate at high speeds to spread the suspension evenly, while dip coating immerses the substrate in the suspension and then withdraws it at a controlled rate. Screen printing employs a stencil or mask to apply the suspension onto the substrate in a desired pattern. After the deposition, watchful drying and heat treatment processes are crucial for removing the organic additives and toxins and achieve a crack-free film. The drying process should be performed gradually and with control to prevent rapid evaporation, which can lead to cracks and defects. Heat treatment, often carried out in a furnace, is employed to promote the sintering and bonding of the particles, resulting in the thickness of film. Depending on the desired thickness and film quality, the coating procedure is repeated multiple times to build a compact and pore-free layer of the desired thickness. This repetition allows for the gradual forming of a uniform thin film with improved integrity and properties. Overall, colloidal and ceramic powder techniques are highly valuable for manufacturing nanocrystalline thin films due to their simplicity, cost-effectiveness, and versatility. By these methods, along with careful drying and heat treatment steps, compact and high-quality thin films can be fabricated for vast applications. González Granados *et al.* [126] prepared a crack-free and porous TiO_2 thin film from colloidal suspension by colloidal method for DSSC application. A photovoltaic efficiency of 5.7% was achieved with 12.8-μm thickness coating with a dye loading of 4.52×10^{-10} mol/mm^2. Zama *et al.* [127] prepared TiO_2 based semitransparent photoanode for DSSCs from the paste of organic colloidal suspension. The prepared DSSCs showed an efficiency of 5.22% and 2.39% for illuminated areas 2 cm^2 and 90 cm^2, respectively.

19.3.2.1 Immersion/Dip Coating

It is a widely used and durable technique for thin-film fabrication including the preparation of one or more layers [128]. It is a simple, cheap, and versatile method for depositing uniform thin films with controlled thickness and roughness at the nanometer scale. In dip coating, a solution is prepared using NPs of the precursor material mixed in a volatile solvent. The solvent is of low viscosity, which allows the coating to run back into the bath as the substrate emerges, ensuring a smooth and even coating. The substrate, which has been cleaned and dried, is soaked at a steady speed into the solution and detained for a specific duration to allow the nanoparticles to adhere and form a layer on the substrate surface. After immersion, the substrate is slowly withdrawn from the solution, allowing excess solution to drain. The coated substrate is then dried to evaporate the solvent, leaving behind a thin layer of nanoparticles adhered to the substrate surface. The drying conditions must be

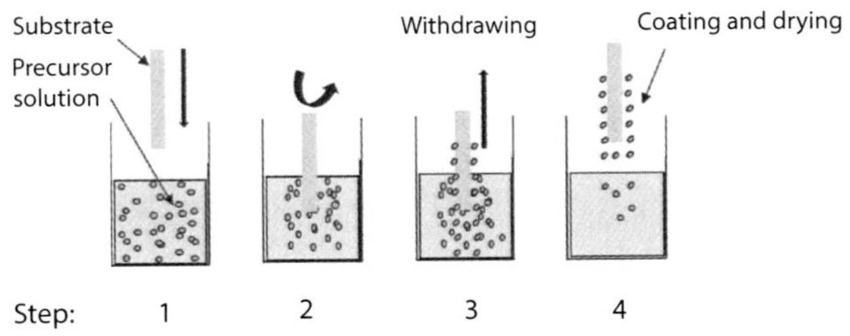

Figure 19.9 Basis of dip coating method and its different steps.

controlled to ensure a uniform and crack-free film formation. During the dip coating process, several forces come into play. Inertia, viscous drag, gravitational force, and surface tension are the major forces utilized to control the coating process. The speed and pulling out rate of the substrate, the properties of the material, and the substrate surface can be adjusted to control the thickness and morphology of the deposited layer. Figure 19.9 represents the basis of the immersing coating method and its different steps. Dip coating offers several advantages, such as simplicity, low cost, and the ability to coat large areas. It is a versatile technique used with various materials and substrates. However, it is important to note that dip coating may not be suitable for all types of materials or for achieving highly precise film thickness control. Other coating methods, such as spin coating or spray coating, may be preferred in certain applications because of these drawbacks.

Lim *et al.* [129] prepared TiO_2 thin film by a simple dip coating method and fabricated DSSCs. The tandem cell exhibited Voc of 0.79 V and 7.1% efficiency. Chou *et al.* [130] used dip coating to prepare ZnO-coated TiO_2 electrodes for DSSC application and achieved 6.62% efficiency. Toe *et al.* [131] studied the consequences of dip coating cycles in ZnO thin film-based DSSC, which exhibited a maximum efficiency of 0.68%. Table 19.9 summarizes the list of references for dip coating (immersion) technique-based thin-film preparation for DSSC application.

Table 19.9 List of references for dip coating (immersion) technique-based thin-film preparation for DSSC application.

S. no.	Photoanode substrate	Efficiency (%)	Reference
1	PAN/PDC-TiO_2	7.1	[129]
2	ZnO-coated TiO_2 electrode	6.62	[130]
3	ZnO thin film	0.68	[131]
4	Nb_2O_5 nanolayer coating on ZnO	5.19	[132]
5	TiO_2 blocking layer	5.49	[133]
6	ZnO coated TiO_2	6.7	[134]
7	ZnO nanowire/TiO_2	0.79	[135]

19.3.2.2 Spin Coating Technique

It is a laboratory method used to produce thin film membranes using centrifugal forces. The process involves depositing a precursor solution or suspension of nanoparticles onto a rotating substrate. During spin coating, a washed and dried substrate is placed on the chuck of the spin coater, and a vacuum is applied beneath the substrate to fix it in position. The substrate is then rotated at approximately 3,000 rpm, creating centrifugal forces. The solution or suspension of NPs is carefully dropped onto the rotating substrate. Figure 19.10 shows the procedure of the spin coating of the thin film with different stages. The layer thickness of the resulting thin film can be controlled by adjusting three experimental factors: (a) Solvent Evaporation Rate: The degree at which the solvent in the solution evaporates affects the drying time and, ultimately, the thickness of the film. The evaporation rate is influenced by humidity, temperature, and the volatility of the solvent used. (b) Spin Rate: Speed at which the substrate rotates, known as the spin rate, also affects the film thickness. Higher spin rates tend to result in thinner films, while lower spin rates can lead to thicker films. (c) The Coating Solution's Viscosity: The solution or suspension's viscosity affects the flow behavior and spreading of the material on the substrate. Higher viscosities can lead to thicker films, while lower viscosities may result in thinner films. This process can be divided into four distinct stages: (i) Dispense Stage: The liquid suspension is distributed drop-wise onto the rotating substrate. (ii) Substrate Acceleration Stage: The substrate begins to accelerate, spreading the solution or suspension over the surface. (iii) Constant-Rate Spinning Stage With Fluid Viscous Forces Dominating: At this stage, the substrate continues to spin constantly and the fluid viscous drag governs the thinness of the coating. (iv) Constant-Rate Spinning Stage With Solvent Evaporation Dominating: As the solvent evaporates, the coating starts to thin primarily due to solvent evaporation, forming a thin film of nanoparticles. By controlling the experimental parameters and the spin coating process stages, one can achieve precise control over the thickness and morphology of the deposited thin film. This coating method is widely applied in various research and manufacturing applications successfully for its simplicity, versatility, and ability to produce uniform thin films. In this coating method, the thickness of the film obtained can be measured by the equation represented as follows [135]:

$$h = \frac{h_0}{\left(1 + \frac{4\,\omega^2 h_0^{\,2} t}{3\eta}\right)^{1/2}}$$

Here, h_0 is the initial film thickness, ω is the spin rate, and η is the viscosity.

Sadikin *et al.* [136] studied the effect of the spin-coating cycle on the properties of TiO_2 thin film and the performance of DSSCs. The DSSC made with seven times spin coating number demonstrated the best efficiency of 0.7%. Merazga *et al.* [137] reported the effect of sol-gel MgO spin coating on photovoltaic performance. An efficiency of 3.9% was obtained from the DSSC made with TiO_2/MgO electrodes sensitized with N3 dye. Al-Juaid *et al.* [138] reported the effect of sol-gel ZnO spin coating on the performance of TiO_2-based DSSCs and recorded an efficiency of 3%. Yildiz *et al.* [139] fabricated TiO_2 thin-film

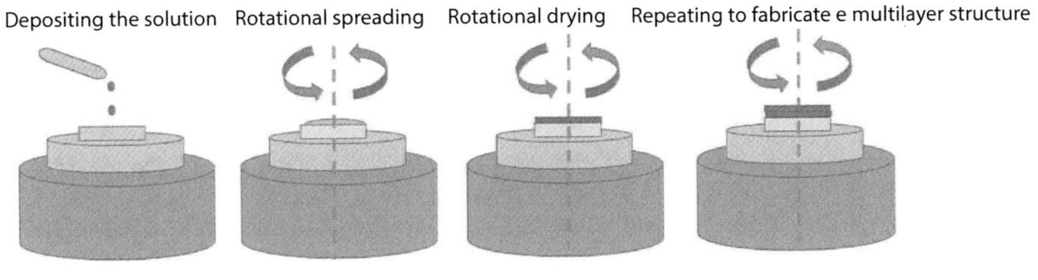

Figure 19.10 Schematic representation of spin coating of the thin film with different stages.

photoanodes by spin coating and recorded the best power conversion efficiency of 1.3%. Biswas and Chatterjee [140] carried out the surface modification *via* sol-gel spin coating ZnO nanoparticles, and the performance of modified WO_3 photoanode-based DSSCs was recorded as 1.21%. Lee *et al.* [141] used the spin coating method to prepare TiO_2 blocking layer to improve the photovoltaic performance of DSSCs. The maximum efficiency of 5.62% was obtained from the DSSC fabricated, including TiO_2 blocking layer. Singh *et al.* [142] reported the performance of spin-coated Ag-doped ZnO photoanode-based DSSCs and achieved an efficiency of 0.44%. Table 19.10 summarizes the list of references for thin-film fabrication by spin coating technique for the DSSC application.

19.3.2.3 Doctor Blade Technique

The doctor blade technique, also known as blade coating or knife coating, is a commonly used coating method for preparing lab-scale DSSCs (dye-sensitized solar cells). It is an adaptation of the technology used to fabricate ceramic capacitors. In the doctor blade technique, a substrate is coated by displacing a blade over its surface at a constant velocity and with a specific contact angle and height. The blade serves as a metering device to control the thickness of the coating. The coating material, often in the form of a paste or suspension, is applied onto the substrate and uniformly distributed as the blade moves across it. The key advantage of the doctor blade technique is its simplicity and ease of operation, making it a preferred choice for lab-scale DSSC fabrication. It involves fewer moving parts and is a cost-effective method. The layer thickness achieved through the doctor blade technique typically ranges from 10 to 500 μm. This range allows for the deposition of relatively thick films, which can be beneficial in certain applications where a higher amount of material is desired. The doctor blade technique is widely used in various fields beyond DSSCs, including coating applications in the production of batteries, fuel cells, thin-film solar cells, and

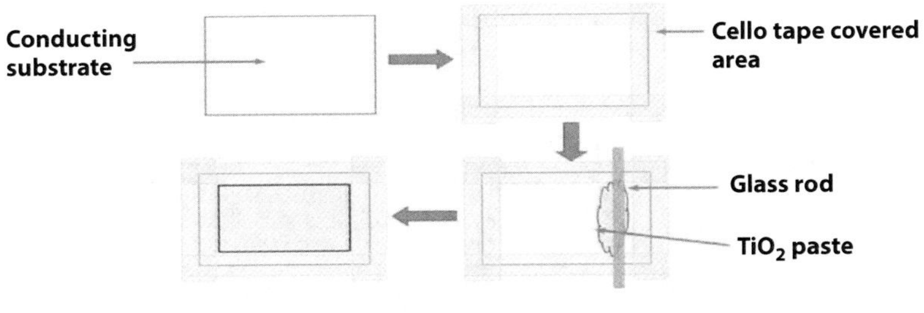

Figure 19.11 Schematic representation of the doctor blade technique process for the fabrication of thin films.

Table 19.10 List of references for thin-film fabrication by spin coating technique for DSSC application.

S. no.	Photoanode substrate	Efficiency (%)	Reference
1	TiO_2	0.7	[136]
2	MgO	3.9	[137]
3	ZnO	3	[138]
4	TiO_2	1.3	[139]
6	ZnO	1.21	[140]
7	TiO_2	5.62	[141]
8	Ag-doped ZnO	0.44	[142]

other electronic devices. It offers a practical and efficient method for producing consistent and controlled coatings on substrates. The doctor blade technique process is schematically shown in Figure 19.11.

Agrawal *et al.* [143] reported TIO_2-based DSSCs prepared by the doctor blade method and recorded an enhanced efficiency of 4.86%. During the photovoltaic measurement, the photovoltaic performance was further increased to 6.13% by keeping a simple white paper on the back side of DSSCs. Kontos *et al.* [144] prepared TiO_2 films for DSSCs by combining doctor blade and sol-gel techniques and showed a maximum efficiency of 1.9%. Azizi *et al.* [145] used the doctor blade method to prepare the TiO_2 layers for DSSCs, and the performance of 0.2% was achieved from the cell. Kumari and Sanjeevadharshini Ceylon [146] reported the effect of TiO_2 photoanode film thickness on the photovoltaic performance of DSSCs. The DSSC made with 12.73-μm TiO_2 layer thickness exhibited the best efficiency of 6.07%. Table 19.11 summarizes the list of references for thin-film fabrication by doctor blade technique for DSSC application.

19.3.2.4 Solvothermal/Hydrothermal Method

Solvothermal and hydrothermal methods are effective techniques used in materials synthesis, particularly for controlling the resulting materials' size, distribution, surface morphology,

Table 19.11 List of references for thin-film fabrication by doctor blade technique for DSSC application.

S. no.	Photoanode substrate	Efficiency (%)	Reference
1	TiO_2	4.86	[143]
2	TiO_2	1.9	[144]
3	TiO_2	0.2	[145]
4	TiO_2	6.07	[146]

and shape. The solvothermal process involves carrying out a synthesis reaction in a solvent other than water. This is particularly important when synthesizing non-oxide materials, as the use of water as a solvent can lead to the oxidation of certain species. Using a nonaqueous solvent, the solvothermal method allows for better control over the reaction conditions and prevents unwanted oxidation reactions. Solvents commonly used in solvothermal synthesis include organic solvents such as alcohols, ethers, or hydrocarbons. On the other hand, hydrothermal synthesis refers to conducting reactions in aqueous media at temperatures above 100°C and under elevated pressure (typically approximately 1 bar). In this method, the reaction occurs in a closed vessel, such as a high-pressure autoclave, which allows for maintaining high temperature and pressure conditions. Hydrothermal synthesis is widely used to fabricate fine oxide powders, enabling the controlled growth of crystalline materials with desired properties. The hydrothermal method offers several advantages in the synthesis of pure oxide powders. It allows for the precise control of reaction parameters, such as temperature, pressure, and reaction time, which influence the formation and crystallinity of the resulting oxide materials. Additionally, using water as a solvent promotes the dissolution and diffusion of reactants, leading to the growth of high-quality crystalline structures. Both solvothermal and hydrothermal methods are versatile and have been widely employed for the synthesis of a range of materials, including nanoparticles, nanowires, and thin films. These methods allow researchers to tailor the properties of the synthesized materials by adjusting the reaction conditions, leading to materials with desired characteristics for various applications in fields such as catalysis, energy storage, and electronics. Cheng *et al.* [147] synthesized branched ZnO nanowires from the solvothermal method for DSSCs. An efficiency of 1.51% was achieved due to the enlargement of the surface area, which allowed to obtain higher dye loading. Aiswarya *et al.* [148] synthesized nickel cobalt sulfide nanoparticles by varying the amount of thiourea by solvothermal method for DSSCs. The best power conversion efficiency of 1.87% was obtained among the three samples. Ramakrishnan and Pitchaiya [149] compared the performance of TiO_2 nanoparticles synthesized by microwave and solvothermal methods as photoanode in DSSCs. The DSSCs constructed using TiO_2 nanoparticles synthesized by microwave and solvothermal methods exhibited efficiencies of 7.44% and 5.98%, respectively. Zhao *et al.* [150] prepared TiO_2 nanorods by the solvothermal method and studied the effect of thermal treatment for DSSCs. Efficiency of 3.21% was obtained using TiO_2 nanorod-based DSSC owing to higher electron transport. Lee *et al.* [151] compare the photovoltaic performance of DSSC made of TiO_2 thin film fabricated by solvothermal and sol-gel methods. Here, 8.51% efficiency was recorded using the solvothermal-prepared TiO_2-based DSSC. Shen *et al.* [152] prepared anatase TiO_2 thin film by microwave-assisted solvothermal method for DSSCs and obtained an efficiency of 7.8%. Rui *et al.* [153] used a solvothermal method to synthesize size-tunable TiO_2 nanorod microspheres as a scattering layer for DSSCs and achieved an efficiency of 8.22%. Rashad *et al.* [154] employed a solvothermal method to transform TiO_2 nanoparticles into nanotubes for DSSC application. The DSSC fabricated using the prepared TiO_2 nanotubes showed an efficiency of 7.2% due to enhanced surface area resulting in higher dye adsorption. Ramakrishnan and Muthukumarasamy [155] performed capping agent-assisted hydrothermal method and obtained different kinds of ZnO nanostructured materials for DSSCs. Sheet-sphere ZnO exhibited higher crystallinity surface area and uniform film morphology; an overall efficiency of 2.61% was recorded. Akhtar *et al.* [156] reported the synthesis of TiO_2 hollow sphere using a one-pot hydrothermal synthesis for DSSCs.

Table 19.12 List of references for thin-film fabrication by solvothermal technique for DSSC application.

S. no.	Photoanode substrate	Efficiency (%)	Reference
1	ZnO	1.51	[147]
2	$NiCo_2S_4$	1.87	[148]
3	TiO_2	7.44	[149]
4	TiO_2	3.21	[150]
5	TiO_2	8.51	[151]
6	TiO_2	7.8	[152]
7	TiO_2	8.22	[153]
8	SnO_2	3.2	[154]
9	TiO_2	7.2	[155]
10	ZnO	2.61	[156]
11	TiO_2	9.1	[157]
12	TiO_2	1.87	[158]
13	TiO_2	1.86	[159]
14	ZnO	0.231	[160]
15	SnO_2	0.475	[161]

A DSSC fabricated using a bilayer structured photoelectrode comprising shell-in-shell TiO_2 film as a scattering layer on the top of TiO_2 thin film exhibited an efficiency of 9.10% (due to higher light harvesting) under AM 1.5G one sun condition in comparison to the 7.65% efficiency for the pristine P25 photoanode. Table 19.12 summarizes the list of references for thin-film fabrication by solvothermal technique for the DSSC application.

19.3.2.5 Spray Pyrolysis Method

Spray pyrolysis is a promising method for preparing thin films and has gained significant attention in various research areas. This technique involves depositing a thin layer by spraying a solution onto a preheated substrate. This process is relatively simple and cost-effective, making it attractive for large-scale production. In spray pyrolysis, the precursor solution is atomized into small droplets, which are then sprayed on preheated substrate. The substrate's heat causes the solvent in the droplets to evaporate rapidly, leaving behind solid particles or molecules that undergo thermal decomposition. This decomposition leads to the growth of thin film on the substrate surface. The thickness of film can be controlled by adjusting parameters like solution's concentration, spraying time, and the temperature gradient between the substrate and the solution. The technique offers several advantages.

First, it does not require sophisticated instruments or complex setups, making it accessible for laboratories with limited resources. Second, it allows the deposition of layer on different kinds of substrates such as glass, silicon, and various other materials. Third, spray pyrolysis allows the deposition of uniform films with good adherence over large areas. In spray pyrolysis, the deposition parameters can be optimized to achieve desired film properties. The choice of atomization technique, such as ultrasonic or pneumatic atomization, affects the droplet size and spray pattern, influencing the film characteristics. The temperature gradient between the substrate and the sprayed solution determines the solvent evaporation rate and affects the film's microstructure. Solution concentration and heating time also play a crucial role in controlling the film thickness and morphology. Due to its simplicity, scalability, and versatility, spray pyrolysis has applications in various fields, including thin-film solar cells, gas sensors, catalysts, and semiconductor devices. Researchers continue to explore and optimize this technique to enhance film quality and tailor the properties of the deposited films for specific applications. Okuya *et al.* [162] used spray pyrolysis deposition (SPD) technique to fabricate TiO_2 thin films to act as photoanode for DSSCs and achieved an efficiency of 3.2%, attributing the enhanced dye loading owing to the increased surface area by the addition of aluminum acetylacetonate in the source solution. Zhang *et al.* [163] synthesized porous TiO_2 thin films by SPD technique and recorded 5.1% efficiency. Dhamodharan *et al.* [164] fabricated thin film of high optical transmittance in the visible region and low resistivity of indium-doped ZnO films using SPD method for DSSC photoanode and achieved an efficiency of 1.09%. Shaban *et al.* [165] prepared ZnO blocking layer by the spray pyrolysis method to improve the efficiency of DSSCs. The film annealed at 350°C exhibited lower resistance and high crystallinity and showed higher efficiency of 7.03%. Noh *et al.* [166] fabricated F-doped SnO_2 substrate by the SPD method, which exhibited a photovoltaic efficiency of 5.5%. Widiyandari *et al.* [167] prepared ZnO hollow sphere by SPD technique serving as an antireflection layer that enhances the efficiency of DSSCs. The DSSC using ZnO hollow sphere annealed at 900°C and coated onto TiO_2 film showed an efficiency of 2.61%. Jiang *et al.* [168] deposited TiO_2 blocking layer by the SPD

Table 19.13 List of references for thin-film fabrication by spray pyrolysis deposition technique for DSSC application.

S. no.	Photoanode substrate	Efficiency (%)	Reference
1.	TiO_2	2.44	[169]
2.	TiO_2	3.2	[162]
3.	TiO_2	5.1	[163]
4.	In:ZnO	1.09	[164]
5.	ZnO	7.03	[165]
6.	F-doped SnO_2	5.5	[166]
7.	ZnO	2.61	[167]
8.	TiO_2	3.77	[168]

method for solid-state DSSCs and recorded an efficiency of 3.77%. Table 19.13 summarizes the list of references for film fabrication by SPD technique and its application in DSSCs.

19.3.2.6 Sol-Gel Technique

It is a renounced wet chemical technique that is applied in the dye-sensitized solar cells to form TiO_2 photoanodes. The sol-gel method offers several merits, including the ability to control the composition and morphology of the synthesized materials. In the sol-gel process, metal oxides such as TiO_2 are synthesized by the hydrolysis and condensation reactions of metal alkoxide precursors in a solution. Common precursors used for TiO_2 synthesis in the sol-gel process include titanium isopropoxide, titanium tert-butoxide, and titanium tetrachloride. These precursors undergo hydrolysis, leading to the formation of metal-hydroxo or metal-oxo complexes. These complexes can then undergo further condensation reactions to form a gel network. During the gelation process, the metal-hydroxo or metal-oxo complexes undergo cross-linking, forming a three-dimensional gel structure. This gel can be directly coated onto a substrate, such as a conductive glass or flexible substrate, and subsequently dried to produce a thin and uniform coating. The sol-gel technique is particularly advantageous for synthesizing doped nanoparticles and composites in DSSCs. By incorporating dopants or other materials into the sol-gel precursor solution, it is possible to obtain thin films of desired properties. Doping with elements such as nitrogen or carbon can modify the electronic properties of TiO_2, enhancing its photoactivity and improving the performance of DSSCs. Researchers have combined other coating techniques, *viz.*, dip coating, spin coating, or electrospinning with sol-gel method to further control the film morphology and thickness. These techniques can be used in conjunction with the sol-gel process to achieve specific film architectures or to create multilayer structures with different functionalities. Overall, the sol-gel technique provides a versatile and effective approach for synthesizing TiO_2 nanoparticles and films, allowing for the fabrication of tailored photoanodes in DSSCs and offering opportunities for developing improved

Table 19.14 List of references for thin-film fabrication by sol-gel method for its application in DSSCs.

S. no.	Photoanode substrate	Efficiency (%)	Reference
1	TiO_2	1.9	[144]
2	TiO_2	5.93	[151]
3	TiO_2	3.94	[169]
4	TiO_2	5.3	[170]
5	ZnO	1.11	[171]
6	TiO_2	1.57	[172]
7	TiO_2	3.7	[173]
8	$SrTiO_3$	0.58	[174]

solar cell devices. Kontos *et al.* [144] fabricated TiO_2 thin films by combining doctor blade with sol-gel techniques for DSSC application. The modified films with high roughness and controlled porosity obtained 1.9% efficiency. Lee *et al.* [151] compared the efficiencies of DSSCs fabricated using solvothermal and sol-gel methods based TiO_2 photoanode. The efficiencies of 8.51% and 5.93% were recorded for solvothermal- and sol-gel-prepared TiO_2 electrodes sensitized with N719 dye. The higher performance for the DSSC was achieved due to the rapid electron transfer to the surface of the TiO_2 thin film prepared with solvothermal method compared with TiO_2 thin film modified with sol-gel. Hočevar *et al.* [170] developed a TiO_2 paste modified with Pechini sol-gel method and recorded an improved efficiency of 5.3%. Laranjo *et al.* [173] prepared TiO_2 and TiO_2/SiO_2 NPs by the sol-gel technique for DSSCs and achieved 3.7% efficiency. Table 19.14 shows the list of references for thin-film fabrication by the sol-gel technique for its use in DSSCs.

19.3.2.7 Screen Printing

Screen printing is widely used for depositing thin films or patterns onto substrates in various industries, including electronics, textiles, and graphics. The screen is typically made of a mesh material, commonly polyester, etc., which is stretched strongly on a rigid frame prepared of metallic or wooden material. The procedure starts by placing a desired amount of semiconducting paste or ink onto the screen. The screen is then positioned over the ITO/FTO substrate. A squeegee applies pressure and moves the paste across the screen, forcing it through the pattern openings onto the substrate below. This transfers the desired pattern onto the substrate while the excess paste is retained on the screen. With the help of screen printing technique, thin film of thickness 2–4 μm can be obtained. The thickness of the film depends on properties of the screen, such as thread diameter and thread counts, distance between threads, and squeezer's pressure and speed. This control over thickness helps achieve uniformity and desired film properties. Screen printing technique is schematically shown in Figure 19.12. Screen printing method has some limitations. Screen printing requires a relatively larger amount of paste or ink than other deposition techniques. This is because the paste needs to fill the screen pattern and transfer it onto the substrate, and due to this, the initial wet thickness of the deposited layer can be relatively high. However, subsequent processes such as drying or curing can reduce the thickness to the desired level. The viscosity of the deposited material used in this method is an important factor. A higher viscosity is often preferred to prevent the paste from flowing too easily through the screen and to achieve better control during printing. However, this requirement can limit the range of materials that can be effectively used in screen printing. Screen printing is generally suitable for materials with low volatility, as volatile components may evaporate during the printing process, leading to changes in the composition of the deposited film. Materials with high volatility may not be suitable for screen printing.

Despite these limitations, screen printing remains a popular and versatile technique for depositing thin films and patterns due to its advantages, including thickness uniformity, ease of operation, and the ability to handle a wide range of substrates and inks. Researchers and engineers continue to explore advancements and improvements in screen printing processes to overcome these limitations and enhance their capabilities. Agrawal *et al.* [143] reported the performance analysis of TiO_2-based DSSCs prepared by doctor blade and screen printing methods. The overall 6.72% efficiency was achieved with screen-printed

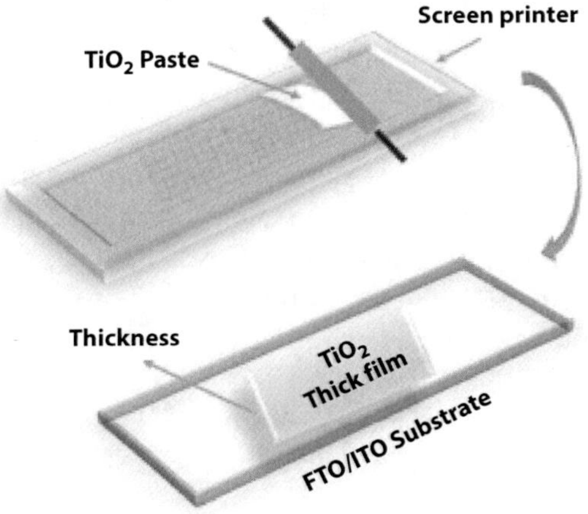

Figure 19.12 A schematic diagram of the screen printing technique for thin-film fabrication.

Table 19.15 List of references for thin-film fabrication by screen printing method for its application in DSSCs.

S. no.	Photoanode substrate	Efficiency (%)	Reference
1	TiO_2	6.72	[143]
2	TiO_2	0.25	[176]
3	TiO_2	5.13	[175]
4	ZnO	0.94	[177]

TiO_2-based DSSC. Khalifa *et al.* [175] prepared single-layered TiO_2-deposited film using the screen printing method for DSSC. The results revealed that the dye adsorption was increased by increasing the film thickness, and highest efficiency of 5.13% was achieved with the 12-μm-thick TiO_2 film. Table 19.15 listed references for thin-film fabrication by screen printing method for its application in DSSCs.

19.3.2.8 Electrophoretic Deposition

Electrophoretic deposition (EPD) is a technique that has gained significant attention due to its low cost, short fabrication time, and ability to control the semiconductor layer thickness effectively. Here is a further explanation of the process: EPD involves the movement followed by deposition of charged particles suspended in a liquid medium onto a substrate under the influence of an applied DC voltage. The process is commonly used to fabricate high-performance DSSCs by optimizing the characteristics of thin film. The EPD process offers several merits, including controlled thickness and surface morphology of the thin film, simplicity of the apparatus, short preparation time, homogeneity of the deposition, and low cost. An EPD cell typically consists of four main components: the working electrode (substrate), a counter electrode, a colloidal suspension containing the charged particles, and a power source (AC/DC). EPD is usually performed in a two-electrode cell system. The EPD process generally follows a two-step mechanism. In the first step, the suspension's solid particles become charged under an electric field's influence. These charged particles then migrate toward the working electrode, which possesses an opposite charge. This migration of charged particles is known as electrophoresis. Second step includes attraction of charged particles to the anode and coagulate to form a deposited layer. This process is referred to as particle coagulation. The solid deposition and growth occur on the working electrode as the charged particles come together and adhere to the substrate. By utilizing electrophoresis, EPD allows for the controlled movement and deposition of charged particles dispersed in a liquid suspension, forming a solid layer on the working electrode. EPD offers advantages such as flexibility in controlling the thickness and morphology of the deposited layer, as well as its suitability for various materials and low-cost fabrication [178]. These characteristics make EPD a promising technique in applications such as solar cells, coatings, and material synthesis. Ongoing research continues to explore and optimize the EPD process for various materials and applications. Shikoh *et al.* [179] used the EPD process to deposited TiO_2 layer on conducting substrate and obtained 4.6 % efficiency substrates varying voltage range between 2.5 V DC and 7.5 V DC using an and obtained 4.6 % efficiency. Jarernboon *et al.* [180] prepared the TiO_2 film by EPD technique for DSSC application. The thickness of the TiO_2 film was optimized w.r.t. time and voltage, and the optimum thickness of 14 μm showed a maximum efficiency of 2.8%. Grinis *et al.* [181] reported an EPD TiO_2 thin film for DSSCs. Alcoholic suspension was used to prepare thick, uniform, adherent TiO_2 films in two steps: (i) alcoholic suspension of NPs was treated with a small amount of acetylacetone; (ii) the suspension thus obtained was added charging solution (additives of I_2, acetone, and water). The fabrication of multilayer TiO_2 electrodes with a thickness of 25 μm exhibited an efficiency of 8.5%. Kim *et al.* [182] reported the EPD of titanate nanotubes from commercial titania nanoparticles for DSSCs and achieved 6.71% efficiency. Luan *et al.* [183] prepared thin film of reduced graphene oxide (RGO) nanosheets on TiO_2 nanotube by EPD for DSSCs. The DSSC made with RGO deposited on bamboo-type TiO_2 nanotube arrays (16.5 μm long) exhibited an efficiency of 6.01% due to enhanced surface area, resulting in increased dye loading.

19.3.2.9 Electrochemical Deposition Technique

Electrochemical deposition (ECD) is found to be one of the most cost-effective techniques for depositing thin films. It involves oxidation and reduction reactions of solution of starting materials in various types of electrolytes, *viz.*, aqueous solutions, organic solutions, or molten salts. Low fabrication temperature, purity of deposited films, controlled thickness, controlling rate of deposition are some of the advantages of the ECD method. The electrochemical technique can be classified into two types: electrolytic plating and electroless plating (autocatalytic). Electrolytic plating process requires an external power source for deposition to occur. The power applied causes to dissolve the metal from the anode, and the reduced metal ions are coated onto the cathode, which is substrate where the film is formed. Whereas electroless plating does not need an external power but it uses catalysts to trigger the surface. The reduction and deposition of the metal occur through autocatalytic reactions initiated by the catalytic surface. Electroplating, electrolytic anodization, and electrophoretic deposition are ECD techniques that can be used to synthesize various materials.

Table 19.16 List of references for thin-film fabrication by ECD method for its application in DSSCs.

S. no.	Photoanode substrate	Efficiency (%)	Reference
1	ZnO	0.32	[184]
2	NiO	0.166	[185]
3	ZnO	2.91	[186]
4	ZnO	1.66	[187]

In electrophoretic deposition, the dissociated colloidal cations and anions disperse onto a conductive substrate. When an electric field is applied, the charged particles move toward the conducting substrate, get discharged, and form a thin film. The electrochemical deposition technique has been extensively studied and modified according to the application, such as fabricating photoanodes for DSSCs. By controlling the deposition parameters and the composition of the electrolyte, researchers can achieve desired film properties for particular applications. Ongoing research continues to explore and refine the electrochemical deposition technique to enhance its capabilities further and expand its applications. Zi *et al.* [184] prepared ZnO photoanodes with different morphologies by ECD for DSSCs and recorded 0.32% efficiency. Bonomo *et al.* [185] prepared electrochemically deposited NiO films as a blocking layer for its application in DSSCs and achieved efficiency of 0.166%. Yu *et al.* [186] reported ECD of ZnO porous nanoplate network of 27-μm thicknesses for DSSC application and achieved 2.91% efficiency. Meng *et al.* [187] synthesized ZnO nanoparticle-modified ZnO nanorods for surface attachment by electrodeposition. The DSSC fabricated with the modified ZnO film achieved 1.66% efficiency due to more dye adsorption. Table 19.16 summarizes the list of references for thin-film fabrication by the ECD method for its application in DSSCs.

19.3.3 Flame-Assisted Colloidal Process

19.3.3.1 Electrospray Deposition

In electrospray deposition (ESD) technique, charged particles generated by atomizing a liquid through an electrical force are deposited uniformly on the substrate. The electrical force used for atomization enables the production of tiny particles with a narrow size distribution, which can be evenly dispersed over the coating substrate. ESD performed at high temperatures, thin film will be formed due to the thermal decomposition of the precursor solution during deposition. This variant of ESD is referred to as electrospray pyrolysis (ESP). Typically, the ESD setup consists of a pump for pressurizing the solution, a nozzle for droplet atomization, and power supply for charging the atomized droplets. The charged droplets are adjusted in such a way that they point toward the grounded substrate and deposited onto its surface. By adjusting various parameters like potential applied voltage, gap between the atomizing nozzle and the substrate, rate of flow, concentration of used precursor solution, thin film of desired properties can be prepared. Figure 19.13 shows a schematic diagram of electrospray deposition techniques for thin-film fabrication. ESD is an efficient technique that requires fewer nanoparticles, resulting in less waste (approximately 5%–8%). ESD offers advantages such as achieving high-quality uniform coatings, precise control over film properties, and flexibility in adjusting deposition variables. It has applications in various fields, including materials science, nanotechnology, and thin-film fabrication. Hence, ESD is a promising technique for the deposition of charged particles providing control over deposition parameters to achieve desired film properties. Ongoing research continues exploring and optimizing ESD for specific applications and further enhancing its capabilities. Lee *et al.* [188] fabricated hierarchically structured TiO_2(HS-TiO_2) electrodes by ESD technique for flexible DSSCs. After compression treatment, the chemical sintering treatment of HS-TiO_2 cell with titanium n-tetra butoxide showed 5.57% efficiency due to faster electron diffusion. Tang and Gomez [189] synthesized mesoporous controlled structure of TiO_2 by ESD for DSSCs, which yielded a maximum efficiency of 2.087%. Liu *et al.* [190] prepared TiO_2 photoanode with stacked graphene *via* ESD and showed its application in DSSCs. Pure graphene (three layers) in TiO_2 photoanode yielded a dramatic improvement of the efficiency of 8.9%. This enhanced efficiency is due to fast electron transport

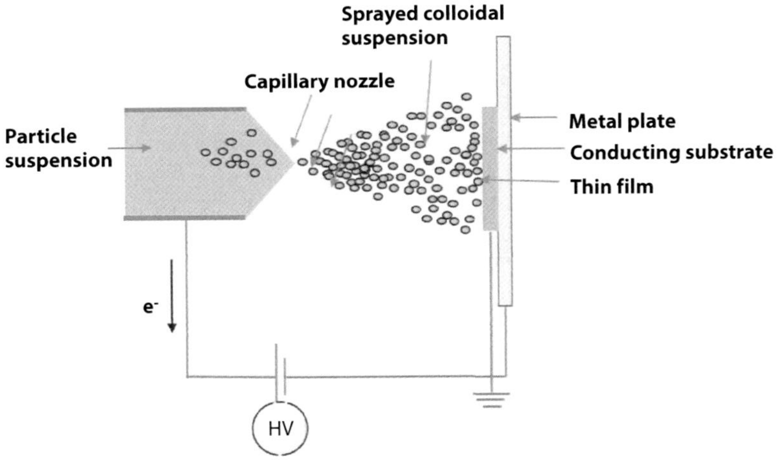

Figure 19.13 Schematic diagram of electrospray deposition techniques for the fabrication of thin film.

Table 19.17 List of references for thin-film fabrication by electrospray deposition method for its application in DSSCs.

S. no.	Photoanode substrate	Efficiency (%)	Reference
1	TiO_2	5.57	[185]
2	TiO_2	2.087	[189]
3	Graphene-TiO_2	8.9	[190]
4	TiO_2	6.81	[191]
5	TiO_2	10.57	[192]

and reduced charge recombination at the electrode/dye and electrode/electrolyte interfaces. Zhu *et al.* [191] prepared dense electrospray suspension of TiO_2 nanoparticles for DSSCs. DSSC fabricated using this photoanode exhibited 6.81% efficiency. Hwang *et al.* [192] prepared ESD of HS-TiO_2 spheres for DSSCs. Because of faster electron diffusion coefficient and longer lifetime of HS-TiO_2 than those in nanostructured TiO_2, DSSC fabricated using HS-TiO_2 photoelectrodes showed 10.57% efficiency. Table 19.17 listed the references for thin-film fabrication by electrospray deposition method for its application in DSSCs.

19.4 Concluding Remarks

This work highlighted various thin-film fabrication techniques used in the dye-sensitized solar cell (DSSC) fabrication and discussed their advantages and challenges. The spray pyrolysis technique involves spraying a precursor solution onto a heated substrate. It is a versatile and cost-effective method, but it may lead to less film thickness and uniformity control. The plasma spray method uses a plasma jet to melt and deposit materials onto the substrate. It is commonly used for depositing thick films or coatings and can provide good film adhesion. However, it may require high temperatures and specialized equipment. The electrophoretic deposition technique utilizes the migration of charged particles in a suspension under an applied electric field to deposit the particles onto a substrate. With the help of this technique, film with controlled thickness can be obtained and also this method is cheap and simple to implement. In sol-gel method, there is the conversion of a liquid precursor into a solid film through a series of chemical reactions. It is a low-cost technique that enables the fabrication of nanocrystalline thin films. However, the stability of chemical precursors against hydrolysis can be a concern. Screen printing and dip coating techniques are widely applied in industries because of their low cost and simplicity. With the help of these techniques, thin films with controlled structures and particle size distribution can be fabricated. However, they may be prone to issues such as cracks, pores, or layer separation, which can be overcome by repeating the coating process. Vacuum deposition techniques (e.g., magnetron sputtering and pulsed laser deposition) are known for producing high-quality, thin, and dense films. They require more advanced equipment and setup, but they offer unique film properties and control advantages. SILAR is a solution-based method, and one can obtain thin films with improved grain structure and orientation. It offers better control over film formation compared to gas-phase methods. Each technique has its strengths and limitations, and its suitability for DSSC fabrication depends on the specific requirements and desired film properties. Researchers continue to explore and optimize these techniques to enhance the performance of DSSCs and other thin-film applications.

References

1. Lian, H., Hou, Z., Shang, M., Geng, D., Zhang, Y., Lin, J., Rare earth ions doped phosphors for improving efficiencies of solar cells. *Energy*, 57, 270–283, 2013.
2. Gerlak, A.K., Weston, J., McMahan, B., Murray, R.L., Mills-Novoa, M., Climate risk management and the electricity sector. *Clim. Risk Manage.*, 19, 12–22, 2018.

3. Sawin, J.L., Sverrisson, F., Seyboth, K., Adib, R., Murdock, H.E., Lins, C., Brown, A. *et al.*, Renewables 2016 Global Status Report. Key findings. A Record Breaking Year for Renewable Energy: New Installations, Policy Targets, Investment and Jobs. Mainstreaming renewables: guidance for policy makers, 2016.
4. Mastoi, M.S., Munir, H.M., Zhuang, S., Hassan, M., Usman, M., Alahmadi, A., Alamri, B., A critical analysis of the impact of pandemic on China's electricity usage patterns and the global development of renewable energy. *Int. J. Environ. Res. Public Health*, 19, 8, 4608, 2022.
5. Khan, K., Kumar, R., Richa, R., Present status and future opportunity in renewable energy-A review. *SKUAST J. Res.*, 22, 1, 19–33, 2020.
6. Khan, S. and Majeed, M.T., Drivers of decoupling economic growth from carbon emission: Empirical analysis of ASEAN countries using decoupling and decomposition model. *Pak. J. Commer. Soc. Sci. (PJCSS)*, 14, 2, 450–483, 2020.
7. Ravichandran, N., Ravichandran, N., Panneerselvam, B., Comparative assessment of offshore floating photovoltaic systems using thin film modules for Maldives islands. *Sustainable Energy Technol. Assess.*, 53, 102490, 2022.
8. C,.P.E., Cioccolanti, L., François, B., Jurasz, J., Zhang, Y., Varini, M., Stridh, B., Yan, J., Li-ion batteries for peak shaving, price arbitrage, and photovoltaic self-consumption in commercial buildings: A Monte Carlo Analysis. *Energy Convers. Manage.*, 234, 113889, 2021.
9. Faruque, R.B., Scudder, M.F., Ula, S., Real-world implementation of advanced pv curtailment and reactive power control using non-smart pv inverter: A case study, in: *2020 IEEE Power & Energy Society General Meeting (PESGM)*, IEEE, pp. 1–5, 2020.
10. Hassan, Q., Algburi, S., Sameen, A.Z., Tariq, J., Al-Jiboory, A.K., Salman, H.M., Ali, B.M., Jaszczur, M., A comprehensive review of international renewable energy growth. *Energy Built Environ.*, 2024.
11. Al-Shetwi, A.Q., Hannan, M.A., Jern, K.P., Mansur, M., Mahlia, T.M.I., Grid-connected renewable energy sources: Review of the recent integration requirements and control methods. *J. Clean. Prod.*, 253, 119831, 2020.
12. Novotny, J., Ganju, R.K., Smiley, S.T., Hussey, R.E., Luther, M.A., Recny, M.A., Siliciano, R.F., Reinherz, E.L., A soluble, single-chain T-cell receptor fragment endowed with antigen-combining properties. *Proc. Natl. Acad. Sci.*, 88, 19, 8646–8650, 1991.
13. Change, on Climate, *Intergovernmental panel on climate change*, vol. 52, pp. 1–43, World Meteorological Organization, Nairobi, 2007.
14. Karuppu, K. and Venk Sitaraman, N.V.I.C.O., *Solar assessment guidance: A guide for solar trainee, trainer & assessor examination*, Notion Press, India, 2019.
15. Li, H., *Pavement Materials for heat Island Mitigation*, pp. 79–96, Science Direct, Butterworth-Heinemann, 2016.
16. Nižetić, S., Djilali, N., Papadopoulos, A., Rodrigues, J.J.P.C., Smart technologies for promotion of energy efficiency, utilization of sustainable resources and waste management. *J. Clean. Prod.*, 231, 565–591, 2019.
17. Kelly, P.J. and Derek Arnell, R., Magnetron sputtering: a review of recent developments and applications. *Vacuum*, 56, 3, 159–172, 2000.
18. Swann, S., Magnetron sputtering. *Phys. Technol.*, 19, 2, 67, 1988.
19. Bräuer, G., Szyszka, B., Vergöhl, M., Bandorf, R., Magnetron sputtering–Milestones of 30 years. *Vacuum*, 84, 12, 1354–1359, 2010.
20. Meng, L. and Li, C., Blocking layer effect on dye-sensitized solar cells assembled with TiO2 nanorods prepared by DC reactive magnetron sputtering. *Nanosci. Nanotechnol. Lett.*, 3, 2, 181–185, 2011.

21. Kment, S., Krysova, H., Hubicka, Z., Kmentova, H., Kavan, L., Zboril, R., Very thin thermally stable TiO2 blocking layers with enhanced electron transfer for solar cells. *Appl. Mater. Today*, 9, 122–129, 2017.
22. Duong, T.-T., Choi, H.-J., He, Q.-J., Le, A.-T., Yoon, S.-G., Enhancing the efficiency of dye sensitized solar cells with an SnO_2 blocking layer grown by nanocluster deposition. *J. Alloys Compd.*, 561, 206–210, 2013.
23. Yeoh, M.-E. and Chan, K.-Y., Efficiency enhancement in dye-sensitized solar cells with ZnO and TiO_2 blocking layers. *J. Electron. Mater.*, 48, 4342–4350, 2019.
24. Chun, J.H. and Kim, J.S., Comparison of different structures of niobium oxide blocking layer for dye-sensitized solar cells. *J. Nanosci. Nanotechnol.*, 14, 8, 6226–6230, 2014.
25. (a) Meng, L., Ren, T., Li, C., The control of the diameter of the nanorods prepared by dc reactive magnetron sputtering and the applications for DSSC. *Appl. Surf. Sci.*, 256, 11, 3676–3682, 2010. (b) Sutthana, S., Hongsith N., Choopun S.. AZO/Ag/AZO multilayer films prepared by DC magnetron sputtering for dye-sensitized solar cell application. *Curr. Appl. Phys.*, 10, 813–816, 2010.
26. Li, L., Chen, R., Jing, G., Zhang, G., Wu, F., Chen, S., Improved performance of TiO_2 electrodes coated with NiO by magnetron sputtering for dye-sensitized solar cells. *Appl. Surf. Sci.*, 256, 14, 4533–4537, 2010.
27. Kang, S.H., Kang, M.-S., Kim, H.-S., Kim, J.-Y., Chung, Y.-H., Smyrl, W.H., Sung, Y.-E., Columnar rutile TiO_2 based dye-sensitized solar cells by radio-frequency magnetron sputtering. *J. Power Sources*, 184, 1, 331–335, 2008.
28. Wang, Y., Xu, H., Zhang, J., Li, G., Electrochemical sensors for clinic analysis. *Sensors*, 8, 4, 2043–2081, 2008.
29. Wu, S., Han, H., Tai, Q., Zhang, J., Chen, B.L., Xu, S., Zhou, C., Yang, Y., Hu, H., Zhao, X.-Z., Improvement in dye-sensitized solar cells with a ZnO-coated TiO_2 electrode by rf magnetron sputtering. *Appl. Phys. Lett.*, 92, 122106, 2008.
30. Kouhestanian, E., Mozaffari, S.A., Ranjbar, M., Amoli, H.S., Enhancing the electron transfer process of TiO_2-based DSSC using DC magnetron sputtered ZnO as an efficient alternative for blocking layer. *Org. Electron.*, 86, 105915, 2020.
31. Wu, S., Han, H., Tai, Q., Zhang, J., Xu, S., Zhou, C., Yang, Y., Hu, H., Chen, BL., Zhao, X.-., Improvement in dye-sensitized solar cells employing TiO_2 electrodes coated with Al_2O_3 by reactive direct current magnetron sputtering. *J. Power Sources*, 182, 1, 119–123, 2008.
32. Lupan, O., Guerin, V.M., Ghimpu, L., Tiginyanu, I.M., Pauporté, T., Nanofibrous-like ZnO layers deposited by magnetron sputtering and their integration in dye-sensitized solar cells. *Chem. Phys. Lett.*, 550, 125–129, 2012.
33. Li, J., Luo, S., Wang, C., Tang, Q., Wang, Y., Han, X., Ran, H. *et al.*, Low Li ion diffusion barrier on low-crystalline FeOOH nanosheets and high performance of energy storage. *Nano Res.*, 13, 759–767, 2020.
34. Zou, K.H., Warfield, S.K., Bharatha, A., Tempany, C.M.C., Kaus, M.R., Haker, S.J., Wells III, W.M., Jolesz, F.A., Kikinis, R., Statistical validation of image segmentation quality based on a spatial overlap index1: scientific reports. *Acad. Radiol.*, 11, 2, 178–189, 2004.
35. Senthilkumar, V., Jayachandran, M., Sanjeeviraja, C., Preparation of anatase TiO_2 thin films for dye-sensitized solar cells by DC reactive magnetron sputtering technique. *Thin Solid Films*, 519, 3, 991–994, 2010.
36. Heo, J.-H., Jung, K.-Y., Kwak, D.-J., Lee, D.-K., Sung, Y.-M., Fabrication of titanium-doped indium oxide films for dye-sensitized solar cell application using reactive RF magnetron sputter method. *IEEE Trans. Plasma Sci.*, 37, 8, 1586–1592, 2009.
37. Zhang, Z., Hu, Y., Qin, F., Ding, Y., DC sputtering assisted nano-branched core–shell TiO_2/ZnO electrodes for application in dye-sensitized solar cells. *Appl. Surf. Sci.*, 376, 10–15, 2016.

38. Horie, Y., Daizaka, K., Mukae, H., Guo, S., Nomiyama, T., Enhancement of photocurrent by columnar Nb-doped TiO_2 compact layer in dye sensitized solar cells with low temperature process of dc sputtering. *Electrochim. Acta*, 187, 348–357, 2016.
39. Lowndes, D.H., Geohegan, D.B., Puretzky, A.A., Norton, D.P., Rouleau, C.M., Synthesis of novel thin-film materials by pulsed laser deposition. *Science*, 273, 5277, 898–903, 1996.
40. Smith, Howard, M. and Turner, A.F., Vacuum deposited thin films using a ruby laser. *Appl. Opt.*, 4, 1, 147–148, 1965.
41. Singh, R.K., Holland, O.W., Narayan, J., Theoretical model for deposition of superconducting thin films using pulsed laser evaporation technique. *J. Appl. Phys.*, 68, 1, 233–247, 1990.
42. Willmott, P.R. and Huber, J.R., Pulsed laser vaporization and deposition. *Rev. Mod. Phys.*, 72, 1, 315, 2000.
43. Kim, H., Kushto, G.P., Arnold, C.B., Kafafi, Z.H., Piqué, A., Laser processing of nanocrystalline TiO_2 films for dye-sensitized solar cells. *Appl. Phys. Lett.*, 85, 3, 464–466, 2004.
44. Gondal, M.A., Ilyas, A.M., Baig, U., Facile synthesis of silicon carbide-titanium dioxide semiconducting nanocomposite using pulsed laser ablation technique and its performance in photovoltaic dye sensitized solar cell and photocatalytic water purification. *Appl. Surf. Sci.*, 378, 8–14, 2016.
45. Sima, C., Grigoriu, C., Toma, O., Antohe, S., Study of dye sensitized solar cells based on ZnO photoelectrodes deposited by laser ablation and doctor blade methods. *Thin Solid Films*, 597, 206–211, 2015.
46. Ghosh, R., Hara, Y., Alibabaei, L., Hanson, K., Rangan, S., Bartynski, R., Meyer, T.J., Lopez, R., Increasing photocurrents in dye sensitized solar cells with tantalum-doped titanium oxide photoanodes obtained by laser ablation. *ACS Appl. Mater. Interfaces*, 4, 9, 4566–4570, 2012.
47. Gondal, M.A., Ilyas, A.M., Baig, U., Pulsed laser ablation in liquid synthesis of ZnO/TiO_2 nanocomposite catalyst with enhanced photovoltaic and photocatalytic performance. *Ceram. Int.*, 42, 11, 13151–13160, 2016.
48. Ghosh, R., Kyle Brennaman, M., Uher, T., Ok, M.-R., Samulski, E.T., McNeil, L.E., Meyer, T.J., Lopez, R., Nanoforest Nb_2O_5 photoanodes for dye-sensitized solar cells by pulsed laser deposition. *ACS Appl. Mater. Interfaces*, 3, 10, 3929–3935, 2011.
49. Sahu, G., Wang, K., Gordon, S.W., Zhou, W., Tarr, M.A., Core-shell Au–TiO_2 nanoarchitectures formed by pulsed laser deposition for enhanced efficiency in dye sensitized solar cells. *RSC Adv.*, 2, 9, 3791–3800, 2012.
50. Albu, D.F., Lungu, J., Popescu-Pelin, G., Mihăilescu, C.N., Socol, G., Georgescu, A., Socol, M., Bănică, A., Ciupina, V., Mihailescu, I.N., Thin film fabrication by pulsed laser deposition from TiO_2 Targets in O_2, N_2, He, or Ar for dye-sensitized solar cells. *Coatings*, 12, 3, 293, 2022.
51. Mahajan, A., Deshpande, P., Butee, S., Synthesis and characterization of NiO/ZnO composite prepared by solid-state reaction method. *Mater. Today: Proc.*, 50, 1912–1917, 2022.
52. Sacco, A., Di Bella, M.S., Gerosa, M., Chiodoni, A., Bianco, S., Mosca, M., Macaluso, R., Calì, C., Pirri, C.F., Enhancement of photoconversion efficiency in dye-sensitized solar cells exploiting pulsed laser deposited niobium pentoxide blocking layers. *Thin Solid Films*, 574, 38–42, 2015.
53. Garvey, T.R., Farnum, B.H., Lopez, R., Pulsed laser deposited porous nano-carpets of indium tin oxide and their use as charge collectors in core–shell structures for dye sensitized solar cells. *Nanoscale*, 7, 6, 2400–2408, 2015.
54. Ilyas, A.-M., Gondal, M.A., Yamani, Z.H., Baig, U., Facile synthesis of titanium dioxide-cadmium sulfide nanocomposite using pulsed laser ablation in liquid and its performance in photovoltaic and photocatalytic applications. *Int. J. Energy Res.*, 41, 10, 1422–1435, 2017.

55. Hara, Y., Garvey, T., Alibabaei, L., Ghosh, R., Lopez, R., Controlled seeding of laser deposited Ta: TiO_2 nanobrushes and their performance as photoanode for dye sensitized solar cells. *ACS Appl. Mater. Interfaces*, 5, 24, 13140–13145, 2013.
56. Lee, Y.-M., Hsu, C.-H., Chen, H.-W., Structural, optical, and electrical properties of p-type NiO films and composite TiO_2/NiO electrodes for solid-state dye-sensitized solar cells. *Appl. Surf. Sci.*, 255, 8, 4658–4663, 2009.
57. Zhang, Y., Shen, Y., Gu, F., Wu, M., Xie, Y., Zhang, J., Influence of Fe ions in characteristics and optical properties of mesoporous titanium oxide thin films. *Appl. Surf. Sci.*, 256, 1, 85–89, 2009.
58. Ku, C.-H. and Wu, J.-J., Chemical bath deposition of ZnO nanowire–nanoparticle composite electrodes for use in dye-sensitized solar cells. *Nanotechnology*, 18, 50, 505706, 2007.
59. Ariyanto, N.P., Abdullah, H., Syarif, J., Yuliarto, B., Shaari, S., Fabrication of zinc oxide-based dye-sensitized solar cell by chemical bath deposition. *Funct. Mater. Lett.*, 3, 04, 303–307, 2010.
60. Zhu, H., Yang, J., Feng, S., Liu, M., Zhang, J., Li, G., Growth of TiO_2 nanosheet-array thin films by quick chemical bath deposition for dye-sensitized solar cells. *Appl. Phys. A*, 105, 769–774, 2011.
61. Thambidurai, M., Muthukumarasamy, N., Velauthapillai, D., Lee, C., Chemical bath deposition of ZnO nanorods for dye sensitized solar cell applications. *J. Mater. Sci.: Mater. Electron.*, 24, 1921–1926, 2013.
62. Wang, Y., Cui, X., Zhang, Y., Gao, X., Sun, Y., Preparation of cauliflower-like ZnO films by chemical bath deposition: photovoltaic performance and equivalent circuit of dye-sensitized solar cells. *J. Mater. Sci. Technol.*, 29, 2, 123–127, 2013.
63. Chu, J.B., Huang, S.M., Zhang, D.W., Bian, Z.Q., Li, X.D., Sun, Z., Yin, X.J., Nanostructured ZnO thin films by chemical bath deposition inábasic aqueous ammonia solutions for photovoltaic applications. *Appl. Phys. A*, 95, 849–855, 2009.
64. Powar, N.S., Gopakumar, G., Ramanathan, K.V., Nair, S.V., Shanmugam, M., Chemical bath deposited WS_2 quantum dots on TiO 2 for dye sensitized solar cell applications. *Opt. Quantum Electron.*, 52, 1–8, 2020.
65. Chen, C., Li, Y., Sun, X., Xie, F., Wei, M., Efficiency enhanced dye-sensitized Zn_2SnO_4 solar cells using a facile chemical-bath deposition method. *New J. Chem.*, 38, 9, 4465–4470, 2014.
66. Huerta-Flores, A.M., García-Gómez, N.A., de la Parra, S.M., Sánchez, E.M., Comparative study of Sb_2S_3, Bi_2S_3 and In_2S_3 thin film deposition on TiO_2 by successive ionic layer adsorption and reaction (SILAR) method. *Mater. Sci. Semicond. Process.*, 37, 235–240, 2015.
67. Ristov, M., Sinadinovski, G.J., Grozdanov, I., Chemical deposition of Cu2O thin films. *Thin Solid Films*, 123, 1, 63–67, 1985.
68. Kale, S.S., Mane, R.S., Chung, H., Yoon, M.-Y., Lokhande, C.D., Han, S.-H., Use of successive ionic layer adsorption and reaction (SILAR) method for amorphous titanium dioxide thin films growth. *Appl. Surf. Sci.*, 253, 2, 421–424, 2006.
69. Jilani, A., Abdel-Wahab, M.S., Hammad, A.H., Advance deposition techniques for thin film and coating, in: *Modern technologies for creating the thin-film systems and coatings*, vol. 2, no. 3, pp. 137–149, 2017.
70. Lasia, A., Electrochemical impedance spectroscopy and its applications, in: *Modern aspects of electrochemistry*, pp. 143–248, Springer US, Boston, MA, 2002.
71. Macdonald, J.R., Impedance spectroscopy. *Ann. Biomed. Eng.*, 20, 289–305, 1992.
72. Macdonald, R.W., Harner, T., Fyfe, J., Recent climate change in the Arctic and its impact on contaminant pathways and interpretation of temporal trend data. *Sci. Total Environ.*, 342, 1-3, 5–86, 2005.
73. Fabregat-Santiago, F., Bisquert, J., Palomares, E., Otero, L., Kuang, D., Zakeeruddin, S.M., Grätzel, M., Correlation between photovoltaic performance and impedance spectroscopy of dye-sensitized solar cells based on ionic liquids. *J. Phys. Chem. C*, 111, 17, 6550–6560, 2007.

74. Muñoz-García, A.B., Benesperi, I., Boschloo, G., Concepcion, J.J., Delcamp, J.H., Gibson, E.A., Meyer, G.J. *et al.*, Dye-sensitized solar cells strike back. *Chem. Soc. Rev.*, 50, 22, 12450–12550, 2021.
75. Sarker, S., Seo, H.W., Kim, D.M., Calculating current density–voltage curves of dye-sensitized solar cells: a straight-forward approach. *J. Power Sources*, 248, 739–744, 2014.
76. Cheng, A.-J., Tzeng, Y., Zhou, Y., Park, M., Wu, T.-., Shannon, C., Wang, D., Lee, W., Thermal chemical vapor deposition growth of zinc oxide nanostructures for dye-sensitized solar cell fabrication. *Appl. Phys. Lett.*, 92, 092113, 2008.
77. Zhang, Q. and Li, C., TiO_2 coated ZnO nanorods by mist chemical vapor deposition for application as photoanodes for dye-sensitized solar cells. *Nanomaterials*, 9, 9, 1339, 2019.
78. Kim, K.E., Jang, S.-R., Park, J., Vittal, R., Kim, K.-J., Enhancement in the performance of dye-sensitized solar cells containing ZnO-covered TiO_2 electrodes prepared by thermal chemical vapor deposition. *Sol. Energy Mater. Sol. Cells*, 91, 4, 366–370, 2007.
79. Shakir, S., Abd-ur-Rehman, H.M., Yunus, K., Iwamoto, M., Periasamy, V., Fabrication of un-doped and magnesium doped TiO_2 films by aerosol assisted chemical vapor deposition for dye sensitized solar cells. *J. Alloys Compd.*, 737, 740–747, 2018.
80. Ameen, S., Shaheer Akhtar, M., Song, M., Shin, H.S., Vertically aligned ZnO nanorods on hot filament chemical vapor deposition grown graphene oxide thin film substrate: solar energy conversion. *ACS Appl. Mater. Interfaces*, 4, 8, 4405–4412, 2012.
81. Murakami, T.N., Kijitori, Y., Kawashima, N., Miyasaka, T., Low temperature preparation of mesoporous TiO_2 films for efficient dye-sensitized photoelectrode by chemical vapor deposition combined with UV light irradiation. *J. Photochem. Photobiol., A*, 164, 1-3, 187–191, 2004.
82. Lee, S.Y. and Kim, S.H., Deposition of zinc oxide photoelectrode using plasma enhanced chemical vapor deposition for dye-sensitized solar cells. *J. Nanosci. Nanotechnol.*, 14, 12, 9272–9278, 2014.
83. Lee, S.Y. and Kim, S.H., Deposition of TiO2 passivation layer by plasma enhanced chemical vapor deposition between the transparent conducting oxide and mesoporous TiO_2 electrode in dye sensitized solar cells. *Jpn. J. Appl. Phys.*, 51, 10S, 10NE19, 2012.
84. Chu, S., Li, D., Chang, P.-C., Lu, J.G., Flexible dye-sensitized solar cell based on vertical ZnO nanowire arrays. *Nanoscale Res. Lett.*, 6, 1–4, 2011.
85. Mathew, J.A., Tsiumra, V., Sajkowski, J.M., Wierzbicka, A., Jakiela, R., Zhydachevskyy, Y., Przezdziecka, E., Stachowicz, M., Kozanecki, A., Photoluminescence of Europium in ZnO and ZnMgO thin films grown by Molecular Beam Epitaxy. *J. Lumin.*, 251, 119167, 2022.
86. Naseem, S., Pinchuk, I.V., Luo, Y.K., Kawakami, R.K., Khan, S., Husain, S., Khan, W., Epitaxial growth of cobalt doped TiO_2 thin films on LaAlO3 (100) substrate by molecular beam epitaxy and their opto-magnetic based applications. *Appl. Surf. Sci.*, 493, 691–702, 2019.
87. Jang, HC., Saito, K., Guo, Q., Yu, K.M., Tanaka, T., Effects of Al doping on the structural, electrical, and optical properties of rock-salt ZnCdO thin films grown by molecular beam epitaxy. *J. Phys. Chem. Solids*, 163, 110571, 2022.
88. Suntola, T., Atomic layer epitaxy. *Mater. Sci. Rep.*, 4, 5, 261–312, 1989.
89. Puurunen, R.L., Surface chemistry of atomic layer deposition: A case study for the trimethylaluminum/water process. *J. Appl. Phys.*, 97, 121301, 2005.
90. Kim, H., Atomic layer deposition of metal and nitride thin films: Current research efforts and applications for semiconductor device processing. *J. Vac. Sci. Technol. B: Microelectron. Nanometer Struct. Process. Meas. Phenom.*, 21, 6, 2231–2261, 2003.
91. Leskelä, M. and Ritala, M., Atomic layer deposition chemistry: recent developments and future challenges. *Angew. Chem. Int. Ed.*, 42, 45, 5548–5554, 2003.
92. Leskelä, M. and Ritala, M., Atomic layer deposition (ALD): from precursors to thin film structures. *Thin Solid Films*, 409, 1, 138–146, 2002.

93. George, S.M., Ott, A.W., Klaus, J.W., Surface chemistry for atomic layer growth. *J. Phys. Chem.*, 100, 31, 13121–13131, 1996.
94. Bakke, J.R., Pickrahn, K.L., Brennan, T.P., Bent, S.F., Nanoengineering and interfacial engineering of photovoltaics by atomic layer deposition. *Nanoscale*, 3, 9, 3482–3508, 2011.
95. Hamann, T.W., Martinson, A.BF., Elam, J.W., Pellin, M.J., Hupp, J.T., Aerogel templated ZnO dye-sensitized solar cells. *Adv. Mater.*, 20, 8, 1560–1564, 2008.
96. Hamann, T.W., Martinson, A.BF., Elam, J.W., Pellin, M.J., Hupp, J.T., Atomic layer deposition of TiO_2 on aerogel templates: new photoanodes for dye-sensitized solar cells. *J. Phys. Chem. C*, 112, 27, 10303–10307, 2008.
97. Han, T.H., Moon, H.-S., Hwang, J.O., Seok, S.I., Im, S.H., Kim, S.O., Peptide-templating dye-sensitized solar cells. *Nanotechnology*, 21, 18, 185601, 2010.
98. Prasittichai, C. and Hupp, J.T., Surface modification of SnO_2 photoelectrodes in dye-sensitized solar cells: significant improvements in photovoltage via Al_2O_3 atomic layer deposition. *J. Phys. Chem. Lett.*, 1, 10, 1611–1615, 2010.
99. Chandiran, A.K., Tetreault, N., Humphry-Baker, R., Kessler, F., Baranoff, E., Yi, C., Nazeeruddin, M.K., Grätzel, M., Subnanometer Ga_2O_3 tunnelling layer by atomic layer deposition to achieve 1.1 V open-circuit potential in dye-sensitized solar cells. *Nano Lett.*, 12, 8, 3941–3947, 2012.
100. Hamann, T.W., Farha, O.K., Hupp, J.T., Outer-sphere redox couples as shuttles in dye-sensitized solar cells. Performance enhancement based on photoelectrode modification via atomic layer deposition. *J. Phys. Chem. C*, 112, 49, 19756–19764, 2008.
101. Liu, L., Karuturi, S.K., Su, L.T., Tok, A.I.Y., TiO_2 inverse-opal electrode fabricated by atomic layer deposition for dye-sensitized solar cell applications. *Energy Environ. Sci.*, 4, 1, 209–215, 2011.
102. Ganapathy, V., Karunagaran, B., Rhee, S.-W., Improved performance of dye-sensitized solar cells with TiO_2/alumina core–shell formation using atomic layer deposition. *J. Power Sources*, 195, 15, 5138–5143, 2010.
103. Chandiran, A.K., Yella, A., Stefik, M., Heiniger, L.-P., Comte, P., Nazeeruddin, M.K., Grätzel, M., Low-temperature crystalline titanium dioxide by atomic layer deposition for dye-sensitized solar cells. *ACS Appl. Mater. Interfaces*, 5, 8, 3487–3493, 2013.
104. Park, K., Zhang, Q., Garcia, B.B., Zhou, X., Jeong, Y.-H., Cao, G., Effect of an ultrathin TiO_2 layer coated on submicrometer-sized ZnO nanocrystallite aggregates by atomic layer deposition on the performance of dye-sensitized solar cells. *Adv. Mater.*, 22, 21, 2329, 2010.
105. Shanmugam, M., Baroughi, M.F., Galipeau, D., Effect of atomic layer deposited ultra thin HfO_2 and Al_2O_3 interfacial layers on the performance of dye sensitized solar cells. *Thin Solid Films*, 518, 10, 2678–2682, 2010.
106. Bills, B., Shanmugam, M., Baroughi, M.F., Effects of atomic layer deposited HfO_2 compact layer on the performance of dye-sensitized solar cells. *Thin Solid Films*, 519, 22, 7803–7808, 2011.
107. Li, T.C., Góes, M.S., Fabregat-Santiago, F., Bisquert, J., Bueno, P.R., Prasittichai, C., Hupp, J.T., Marks, T.J., Surface passivation of nanoporous TiO_2 via atomic layer deposition of ZrO_2 for solid-state dye-sensitized solar cell applications. *J. Phys. Chem. C*, 113, 42, 18385–18390, 2009.
108. Lin, C., Tsai, F.-Y., Lee, M.-H., Lee, C.-H., Tien, T.-C., Wang, L.-P., Tsai, S.-Y., Enhanced performance of dye-sensitized solar cells by an Al_2O_3 charge-recombination barrier formed by low-temperature atomic layer deposition. *J. Mater. Chem.*, 19, 19, 2999–3003, 2009.
109. Tien, T.-C., Pan, F.-M., Wang, L.-P., Lee, C.-H., Tung, Y.-L., Tsai, S.-Y., Lin, C., Tsai, F.-Y., Chen, S.-J., Interfacial energy levels and related properties of atomic-layer-deposited Al_2O_3 films on nanoporous TiO_2 electrodes of dye-sensitized solar cells. *Nanotechnology*, 20, 30, 305201, 2009.
110. Choi, J.-H., Kwon, S.-H., Jeong, Y.-K., Kim, I., Kim, K.-H., Atomic layer deposition of Ta-doped TiO_2 electrodes for dye-sensitized solar cells. *J. Electrochem. Soc.*, 158, 6, B749, 2011.

111. Lu, H., Tian, W., Guo, J., Li, L., Interface engineering through atomic layer deposition towards highly improved performance of dye-sensitized solar cells. *Sci. Rep.*, 5, 1, 12765, 2015.
112. Nilphai, S., Thepnurat, M., Hongsith, N., Ruankham, P., Phadungdhitidhada, S., Gardchareon, A., Wongratanaphisan, D., Choopun, S., Synthesis and characterization of MgO by microwave-assisted thermal oxidation for dye-sensitized solar cells. *Key Eng. Mater.*, 675, 158–162, 2016.
113. Jiang, T., Bujoli-Doeuff, M., Gautron, E., Farré, Y., Cario, L., Pellegrin, Y., Boujtita, M., Odobel, F., Jobic, S., Cu2O@ CuO core-shell nanoparticles as photocathode for p-type dye sensitized solar cell. *J. Alloys Compd.*, 769, 605–610, 2018.
114. Sacco, A., Lamberti, A., Gazia, R., Bianco, S., Manfredi, D., Shahzad, N., Cappelluti, F., Ma, S., Tresso, E., High efficiency dye-sensitized solar cells exploiting sponge-like ZnO nanostructures. *PCCP*, 14, 47, 16203–16208, 2012.
115. Zhu, K., Neale, N.R., Halverson, A.F., Kim, J.Y., Frank, A.J., Effects of annealing temperature on the charge-collection and light-harvesting properties of TiO_2 nanotube-based dye-sensitized solar cells. *J. Phys. Chem. C*, 114, 32, 13433–13441, 2010.
116. Jung, K., Lim, T., Li, Y., Martinez-Morales, A.A., ZnO-CuO core-shell heterostructure for improving the efficiency of ZnO-based dye-sensitized solar cells. *MRS Adv.*, 2, 15, 857–862, 2017.
117. Pathan, H.M. and Lokhande, C.D., Deposition of metal chalcogenide thin films by successive ionic layer adsorption and reaction (SILAR) method. *Bull. Mater. Sci.*, 27, 85–111, 2004.
118. Lee, H.J., Wang, M., Chen, P., Gamelin, D.R., Zakeeruddin, S.M., Gratzel, M., Nazeeruddin, M.K., Efficient CdSe quantum dot-sensitized solar cells prepared by an improved successive ionic layer adsorption and reaction process. *Nano Lett.*, 9, 12, 4221–4227, 2009.
119. Park, S., DiMasi, E., Kim, Y.-I., Han, W., Woodward, P.M., Vogt. The preparation, T., and characterization of photocatalytically active TiO_2 thin films and nanoparticles using Successive-Ionic-Layer-Adsorption-and-Reaction. *Thin Solid Films*, 515, 4, 1250–1254, 2006.
120. Gaikwad, M.A., Suryawanshi, M.P., Maldar, P.S., Dongale, T.D., Moholkar, A.V., Nanostructured zinc oxide photoelectrodes by green routes M-SILAR and electrodeposition for dye sensitized solar cell. *Opt. Mater.*, 78, 325–334, 2018.
121. Gaikwad, M.A., Suryawanshi, M.P., Nikam, S.S., Bhosale, C.H., Kim, J.H., Moholkar, A.V., Influence of Zn concentration and dye adsorption time on the photovoltaic performance of M-SILAR deposited ZnO-based dye sensitized solar cells. *J. Photochem. Photobiol., A, 329*, 246–254, 2016.
122. Guo, X., Gao, J., Zhang, Z., Xiao, S., Pan, D., Zhou, C., Shen, J., Hong, J., Yang, Y., Highly efficient interfacial layer using SILAR-derived Ag_2S quantum dots for solid-state bifacial dye-sensitized solar cells. *Mater. Today Energy*, 5, 320–330, 2017.
123. Patil, U.M., Gurav, K.V., Joo, O.-S., Lokhande, C.D., Synthesis of photosensitive nanograined TiO_2 thin films by SILAR method. *J. Alloys Compd.*, 478, 1-2, 711–715, 2009.
124. Chung, J., Myoung, J., Oh, J., Lim, S., Synthesis of a ZnS shell on the ZnO nanowire and its effect on the nanowire-based dye-sensitized solar cells. *J. Phys. Chem. C*, 114, 49, 21360–21365, 2010.
125. Isah, K.U., Jolayemi, B.J., Ahmadu, U., Kimpa, M.I., Alu, N., Plasmonic effect of silver nanoparticles intercalated into mesoporous betalain-sensitized-TiO_2 film electrodes on photovoltaic performance of dye-sensitized solar cells. *Mater. Renew. Sustain. Energy*, 5, 1–9, 2016.
126. González, Z., Yus, J., Sanchez-Herencia, A.J., Dewalque, J., Manceriu, L., Henrist, C., Ferrari, B., A colloidal approach to prepare binder and crack-free TiO_2 multilayer coatings from particulate suspensions: Application in DSSCs. *J. Eur. Ceram. Soc.*, 39, 2-3, 366–375, 2019.
127. Zama, I., Martelli, C., Gorni, G., Preparation of TiO_2 paste starting from organic colloidal suspension for semi-transparent DSSC photo-anode application. *Mater. Sci. Semicond. Process.*, 61, 137–144, 2017.

128. Padinger, F., Brabec, C.J., Fromherz, T., Hummelen, J.C., Serdar Sariciftci, N., Fabrication of large area photovoltaic devices containing various blends of polymer and fullerene derivatives by using the doctor blade technique. *Opto-Electron. Rev.*, 8, 4, 280–283, 2000.
129. Lim, J., Lee, M., Balasingam, S.K., Kim, J., Kim, D., Jun, Y., Fabrication of panchromatic dye-sensitized solar cells using pre-dye coated TiO_2 nanoparticles by a simple dip coating technique. *RSC Adv.*, 3, 14, 4801–4805, 2013.
130. Chou, C.-S., Chou, F.-C., Kang, J.-Y., Preparation of ZnO-coated TiO_2 electrodes using dip coating and their applications in dye-sensitized solar cells. *Powder Technol.*, 215, 38–45, 2012.
131. Toe, M.Z., Pung, S.-Y., Yaacob, K.A.B., Han, S.S., Effect of dip-coating cycles on the structural and performance of ZnO thin film-based DSSC. *Arabian J. Sci. Eng.*, 46, 7, 6741–6751, 2021.
132. Ueno, S. and Fujihara, S., Effect of an Nb2O5 nanolayer coating on ZnO electrodes in dye-sensitized solar cells. *Electrochim. Acta*, 56, 7, 2906–2913, 2011.
133. Hart, J.N., Menzies, D., Cheng, Y.-B., Simon, G.P., Spiccia, L., TiO2 sol–gel blocking layers for dye-sensitized solar cells. *C.R. Chim.*, 9, 5-6, 622–626, 2006.
134. Chou, C.-S., Chou, F.-C., Ding, Y.-G., Wu, P., The effect of ZnO-coating on the performance of a dye-sensitized solar cell. *Sol. Energy*, 86, 5, 1435–1442, 2012.
135. Gan, X., Li, X., Gao, X., Zhuge, F., Yu, W., ZnO nanowire/TiO_2 nanoparticle photoanodes prepared by the ultrasonic irradiation assisted dip-coating method. *Thin Solid Films*, 518, 17, 4809–4812, 2010.
136. Sadikin, S.N., Rahman, M.Y.A., Umar, A.A., Salleh, M.M., Effect of spin-coating cycle on the properties of TiO_2 thin film and performance of DSSC. *Int. J. Electrochem. Sci.*, 12, 6, 5529–5538, 2017.
137. Merazga, A., Al-Subai, F., Albaradi, A.M., Badawi, A., Jaber, A.Y., Alghamdi, A.A.B., Effect of sol–gel MgO spin-coating on the performance of TiO_2-based dye-sensitized solar cells. *Mater. Sci. Semicond. Process.*, 41, 114–120, 2016.
138. Al-Juaid, F., Merazga, A., Al-Baradi, A., Abdel-Wahab, F., Effect of sol–gel ZnO spin-coating on the performance of TiO_2-based dye-sensitized solar cell. *Solid State Electron.*, 87, 98–103, 2013.
139. Yildiz, Z.K., Atilgan, A. B. D. U. L. L. A. H., Atli, A., Özel, K., Altinkaya, C., Yildiz, A., Enhancement of efficiency of natural and organic dye sensitized solar cells using thin film TiO_2 photoanodes fabricated by spin-coating. *J. Photochem. Photobiol., A*, 368, 23–29, 2019.
140. Biswas, R. and Chatterjee, S., Effect of surface modification via sol-gel spin coating of ZnO nanoparticles on the performance of WO_3 photoanode based dye sensitized solar cells. *Optik*, 212, 164142, 2020.
141. Lee, J.G., Cheon, J.H., Yang, H.S., Lee, D.K., Kim, J.H., Enhancement of photovoltaic performance in dye-sensitized solar cells with the spin-coated TiO_2 blocking layer. *J. Nanosci. Nanotechnol.*, 12, 7, 6026–6030, 2012.
142. Singh, A., Mohan, D., Ahlawat, S.D., Performances of spin coated silver doped ZnO photoanode based dye sensitized solar cell. *Process. Appl. Ceram.*, 11, 3, 213–219, 2017.
143. Agrawal, A., Siddiqui, S.A., Soni, A., Khandelwal, K., Sharma, G.D., Performance analysis of TiO_2 based dye sensitized solar cell prepared by screen printing and doctor blade deposition techniques. *Sol. Energy*, 226, 9–19, 2021.
144. Kontos, A.I., Kontos, A.G., Tsoukleris, D.S., Bernard, M.-C., Spyrellis, N., Falaras, P., Nanostructured TiO_2 films for DSSCS prepared by combining doctor-blade and sol–gel techniques. *J. Mater. Process. Technol.*, 196, 1-3, 243–248, 2008.
145. Azizi, T., Touihri, A.E., Ben Karoui, M., Gharbi, R., Comparative study between dye-synthesized solar cells prepared by electrophoretic and doctor blade techniques. *Optik*, 127, 10, 4400–4404, 2016.

146. Kumari, J.M.K.W., Sanjeevadharshini, N., Dissanayake, M.A.K.L., Senadeera, G.K.R., Thotawatthage, C.A., The effect of TiO_2 photo anode film thickness on photovoltaic properties of dye-sensitized solar cells. *Ceylon J. Sci.*, 45, 1, 33–41, 2016.
147. Cheng, H.-M., Chiu, W.-H., Lee, C.-H., Tsai, S.-Y., Hsieh, W.-F., Formation of branched ZnO nanowires from solvothermal method and dye-sensitized solar cells applications. *J. Phys. Chem. C*, 112, 42, 16359–16364, 2008.
148. Aiswarya, K.M., Raguram, T., Rajni, K.S., Synthesis and characterisation of nickel cobalt sulfide nanoparticles by the solvothermal method for dye-sensitized solar cell applications. *Polyhedron*, 176, 114267, 2020.
149. Ramakrishnan, V.M., Pitchaiya, S., Muthukumarasamy, N., Kvamme, K., Rajesh, G., Agilan, S., Pugazhendhi, A., Velauthapillai, D., Performance of TiO2 nanoparticles synthesized by microwave and solvothermal methods as photoanode in dye-sensitized solar cells (DSSC). *Int. J. Hydrogen Energy*, 45, 51, 27036–27046, 2020.
150. Zhao, J., Yao, J., Zhang, Y., Guli, M., Xiao, L., Effect of thermal treatment on TiO_2 nanorod electrodes prepared by the solvothermal method for dye-sensitized solar cells: Surface reconfiguration and improved electron transport. *J. Power Sources*, 255, 16–23, 2014.
151. Lee, Y., Chae, J., Kang, M., Comparison of the photovoltaic efficiency on DSSC for nanometer sized TiO_2 using a conventional sol–gel and solvothermal methods. *J. Ind. Eng. Chem.*, 16, 4, 609–614, 2010.
152. Shen, P.-S., Tai, Y.-C., Chen, P., Wu, Y.-C., Clean and time-effective synthesis of anatase TiO_2 nanocrystalline by microwave-assisted solvothermal method for dye-sensitized solar cells. *J. Power Sources*, 247, 444–451, 2014.
153. Rui, Y., Li, Y., Zhang, Q., Wang, H., Size-tunable TiO_2 nanorod microspheres synthesised via a one-pot solvothermal method and used as the scattering layer for dye-sensitized solar cells. *Nanoscale*, 5, 24, 12574–12581, 2013.
154. Rashad, M.M., Ibrahim, I.A., Osama, I., Shalan, A.E., Distinction between SnO_2 nanoparticles synthesized using co-precipitation and solvothermal methods for the photovoltaic efficiency of dye-sensitized solar cells. *Bull. Mater. Sci.*, 37, 903–909, 2014.
155. Ramakrishnan, V.M., Muthukumarasamy, N., Balraju, P., Pitchaiya, S., Velauthapillai, D., Pugazhendhi, A., Transformation of TiO_2 nanoparticles to nanotubes by simple solvothermal route and its performance as dye-sensitized solar cell (DSSC) photoanode. *Int. J. Hydrogen Energy*, 45, 31, 15441–15452, 2020.
156. Akhtar, M.S., Alam Khan, M., Jeon, M.S., Yang, O.-B., Controlled synthesis of various ZnO nanostructured materials by capping agents-assisted hydrothermal method for dye-sensitized solar cells. *Electrochim. Acta*, 53, 27, 7869–7874, 2008.
157. Wu, X., Lu, G.Q.M., Wang, L., Shell-in-shell TiO_2 hollow spheres synthesized by one-pot hydrothermal method for dye-sensitized solar cell application. *Energy Environ. Sci.*, 4, 9, 3565–3572, 2011.
158. Zhao, Y., Gu, X., Qiang, Y., Influence of growth time and annealing on rutile TiO_2 single-crystal nanorod arrays synthesized by hydrothermal method in dye-sensitized solar cells. *Thin Solid Films*, 520, 7, 2814–2818, 2012.
159. Iraj, M., Nayeri, F.D., Asl-Soleimani, E., Narimani, K., Controlled growth of vertically aligned TiO_2 nanorod arrays using the improved hydrothermal method and their application to dye-sensitized solar cells. *J. Alloys Compd.*, 659, 44–50, 2016.
160. Kuo, S.-Y., Yang, J.-F., Lai, F.-I., Improved dye-sensitized solar cell with a ZnO nanotree photoanode by hydrothermal method. *Nanoscale Res. Lett.*, 9, 1–6, 2014.
161. Dadkhah, M. and Salavati-Niasari, M., Dye-sensitized solar cells based on tin dioxide nanoparticles prepared by a facile hydrothermal method. *Mater. Sci. Semicond. Process.*, 20, 41–48, 2014.

162. Okuya, M., Nakade, K., Kaneko, S., Porous TiO_2 thin films synthesized by a spray pyrolysis deposition (SPD) technique and their application to dye-sensitized solar cells. *Sol. Energy Mater. Sol. Cells*, 70, 4, 425–435, 2002.
163. Zhang, X., Zhang, J.-J., Xia, Y.-Y., Molecular design of coumarin dyes with high efficiency in dye-sensitized solar cells. *J. Photochem. Photobiol., A*, 194, 2-3, 167–172, 2008.
164. Dhamodharan, P., Chen, J., Manoharan, C., Fabrication of In doped ZnO thin films by spray pyrolysis as photoanode in DSSCs. *Surf. Interfaces*, 23, 100965, 2021.
165. Shaban, Z., Majles Ara, M.H., Falahatdoost, S., Ghazyani, N., Optimization of ZnO thin film through spray pyrolysis technique and its application as a blocking layer to improving dye sensitized solar cell efficiency. *Curr. Appl Phys.*, 16, 2, 131–134, 2016.
166. Noh, S.I., Ahn, H.-J., Riu, D.-H., Photovoltaic property dependence of dye-sensitized solar cells on sheet resistance of FTO substrate deposited via spray pyrolysis. *Ceram. Int.*, 38, 5, 3735–3739, 2012.
167. Widiyandari, H., Wijayanti, S., Prasetio, A., Purwanto, A., ZnO hollow sphere prepared by flame spray pyrolysis serves as an anti-reflection layer that improves the performance of dye-sensitized solar cells. *Opt. Mater.*, 107, 110077, 2020.
168. Jiang, C., Koh, W.L., Leung, M.Y., Hong, W., Li, Y., Zhang, J., Influences of alcoholic solvents on spray pyrolysis deposition of TiO_2 blocking layer films for solid-state dye-sensitized solar cells. *J. Solid State Chem.*, 198, 197–202, 2013.
169. Aboulouard, A., Gultekin, B., Can, M., Erol, M., Jouaiti, A., Elhadadi, B., Zafer, C., Demic, S., Dye sensitized solar cells based on titanium dioxide nanoparticles synthesized by flame spray pyrolysis and hydrothermal sol-gel methods: a comparative study on photovoltaic performances. *J. Mater. Res. Technol.*, 9, 2, 1569–1577, 2020.
170. Hočevar, M., Krašovec, U.O., Berginc, M., Dražič, G., Hauptman, N., Topič, M., Development of TiO_2 pastes modified with Pechini sol–gel method for high efficiency dye-sensitized solar cell. *J. Sol-Gel Sci. Technol.*, 48, 156–162, 2008.
171. Rani, S., Suri, P., Shishodia, P.K., Mehra, R.M., Synthesis of nanocrystalline ZnO powder via sol–gel route for dye-sensitized solar cells. *Sol. Energy Mater. Sol. Cells*, 92, 12, 1639–1645, 2008.
172. Dubey, R.S., Krishnamurthy, K.V., Singh, S., Experimental studies of TiO_2 nanoparticles synthesized by sol-gel and solvothermal routes for DSSCs application. *Results Phys.*, 14, 102390, 2019.
173. Laranjo, M.T., Ricardi, N.C., Arenas, L.T., Benvenutti, E.V., de Oliveira, M.C., Santos, M.J.L., Costa, T.M.H., TiO_2 and TiO_2/SiO_2 nanoparticles obtained by sol–gel method and applied on dye sensitized solar cells. *J. Sol-Gel Sci. Technol.*, 72, 273–281, 2014.
174. Gholamrezaei, S., Niasari, M.S., Dadkhah, M., Sarkhosh, B., New modified sol–gel method for preparation $SrTiO_3$ nanostructures and their application in dye-sensitized solar cells. *J. Mater. Sci.: Mater. Electron.*, 27, 118–125, 2016.
175. Khalifa, A., Shafie, S., Hasan, W.Z.W., Lim, H.N., Rusop, M., Pandey, S.S., Vats, A.K., Al Sultan, H.A., Samaila, B., Comprehensive performance analysis of dye-sensitized solar cells using single layer TiO_2 photoanode deposited using screen printing technique. *Optik*, 223, 165595, 2020.
176. Fan, K., Liu, M., Peng, T., Ma, L., Dai, K., Effects of paste components on the properties of screen-printed porous TiO_2 film for dye-sensitized solar cells. *Renewable Energy*, 35, 2, 555–561, 2010.
177. Sarkar, K., Braden, E.V., Bonke, S.A., Bach, U., Müller-Buschbaum, P., Screen-Printing of ZnO Nanostructures from Sol–Gel Solutions for Their Application in Dye-Sensitized Solar Cells. *ChemSusChem*, 8, 16, 2696–2704, 2015.
178. Nyongesa, F. and Aduda, B., Electrophoretic deposition of titanium dioxide thin films for photocatalytic water purification systems. *Adv. Mater.*, 6, 4, 31–37, 2017.

179. Shikoh, A.S., Ahmad, Z., Touati, F., Shakoor, R.A., Al-Muhtaseb, S.A., Optimization of ITO glass/TiO_2 based DSSC photo-anodes through electrophoretic deposition and sintering techniques. *Ceram. Int.*, 43, 13, 10540–10545, 2017.
180. Jarernboon, W., Pimanpang, S., Maensiri, S., Swatsitang, E., Amornkitbamrung, V., Optimization of titanium dioxide film prepared by electrophoretic deposition for dye-sensitized solar cell application. *Thin Solid Films*, 517, 16, 4663–4667, 2009.
181. Grinis, L., Dor, S., Ofir, A., Zaban, A., Electrophoretic deposition and compression of titania nanoparticle films for dye-sensitized solar cells. *J. Photochem. Photobiol., A*, 198, 1, 52–59, 2008.
182. Kim, G.-S., Seo, H.-K., Godble, V.P., Kim, Y.-S., Yang, O.-B., Shin, H.-S., Electrophoretic deposition of titanate nanotubes from commercial titania nanoparticles: application to dye-sensitized solar cells. *Electrochem. Commun.*, 8, 6, 961–966, 2006.
183. Luan, X., Chen, L., Zhang, J., Qu, G., Flake, J.C., Wang, Y., Electrophoretic deposition of reduced graphene oxide nanosheets on TiO_2 nanotube arrays for dye-sensitized solar cells. *Electrochim. Acta*, 111, 216–222, 2013.
184. Zi, M., Zhu, M., Chen, L., Wei, H., Yang, X., Cao, B., ZnO photoanodes with different morphologies grown by electrochemical deposition and their dye-sensitized solar cell properties. *Ceram. Int.*, 40, 6, 7965–7970, 2014.
185. Bonomo, M., Di Girolamo, D., Piccinni, M., Dowling, D.P., Dini, D., Electrochemically deposited NiO films as a blocking layer in p-type dye-sensitized solar cells with an impressive 45% fill factor. *Nanomaterials*, 10, 1, 167, 2020.
186. Yu, W.-C., Sabastian, N., Chang, W.-C., Tsia, C.-Y., Lin, C.-M., Electrochemical deposition of ZnO porous nanoplate network for Dye-Sensitized solar cells. *J. Nanosci. Nanotechnol.*, 18, 1, 56–61, 2018.
187. Meng, Y., Lin, Y., Lin, Y., Electrodeposition for the synthesis of ZnO nanorods modified by surface attachment with ZnO nanoparticles and their dye-sensitized solar cell applications. *Ceram. Int.*, 40, 1, 1693–1698, 2014.
188. Lee, H., Hwang, D., Jo, S.M., Kim, D., Seo, Y., Kim, D.Y., Low-temperature fabrication of TiO_2 electrodes for flexible dye-sensitized solar cells using an electrospray process. *ACS Appl. Mater. Interfaces*, 4, 6, 3308–3315, 2012.
189. Tang, J. and Gomez, A., Control of the mesoporous structure of dye-sensitized solar cells with electrospray deposition. *J. Mater. Chem. A*, 3, 15, 7830–7839, 2015.
190. Liu, J., Fu, X., Cao, D.-P., Mao, L., Wang, J., Mu, D.-H., Mi, B.-X., Zhao, B.-M., Gao, Z.-Q., Stacked graphene–TiO_2 photoanode via electrospray deposition for highly efficient dye sensitized solar cells. *Org. Electron.*, 23, 158–163, 2015.
191. Zhu, T., Li, C., Yang, W., Zhao, X., Wang, X., Tang, C., Mi, B., Gao, Z., Huang, W., Deng, W., Electrospray dense suspensions of TiO_2 nanoparticles for dye sensitized solar cells. *Aerosol Sci. Technol.*, 47, 12, 1302–1309, 2013.
192. Hwang, D., Lee, H., Jang, S.-Y., Jo, S.M., Kim, D., Seo, Y., Kim, D.Y., Electrospray preparation of hierarchically-structured mesoporous TiO_2 spheres for use in highly efficient dye-sensitized solar cells. *ACS Appl. Mater. Interfaces*, 3, 7, 2719–2725, 2011.

Index

Printed in the USA/Agawam, MA
September 21, 2026
910661.177